AF333344

COTTON

AGRONOMY

A Series of Monographs

The American Society of Agronomy and Academic Press published the first six books in this series. The General Editor of Monographs 1 to 6 was A. G. Norman. They are available through Academic Press, Inc., 111 Fifth Avenue, New York, NY 10003.

1. C. EDMUND MARSHALL: The Colloid Chemical of the Silicate Minerals. 1949
2. BYRON T. SHAW, *Editor*: Soil Physical Conditions and Plant Growth. 1952
3. K. D. JACOB, *Editor*: Fertilizer Technology and Resources in the United States. 1953
4. W. H. PIERRE and A. G. NORMAN, *Editors*: Soil and Fertilizer Phosphate in Crop Nutrition. 1953
5. GEORGE F. SPRAGUE, *Editor*: Corn and Corn Improvement. 1955
6. J. LEVITT: The Hardiness of Plants. 1956

The Monographs published since 1957 are available from the American Society of Agronomy, 677 S. Segoe Road, Madison, WI 53711.

7. JAMES N. LUTHIN, *Editor*: Drainage of Agricultural Lands. 1957
8. FRANKLIN A. COFFMAN, *Editor*: Oats and Oat Improvement. 1961
9. A. KLUTE, *Editor*: Methods of Soil Analysis. 1986
   Part 1—Physical and Mineralogical Methods. Second Edition.
   A. L. PAGE, R. H. MILLER, and D. R. KEENEY, *Editor*: Methods of Soil Analysis. 1982
   Part 2—Chemical and Microbiological Properties. Second Edition.
10. W. V. BARTHOLOMEW and F. E. CLARK, *Editors*: Soil Nitrogen. 1965
    (Out of print; replaced by no. 22)
11. R. M. HAGAN, H. R. HAISE, and T. W. EDMINSTER, *Editors*: Irrigation of Agricultural Lands. 1967
12. FRED ADAMS, *Editor*: Soil Acidity and Liming. Second Edition. 1984
13. E. G. HEYNE, *Editor*: Wheat and Wheat Improvement. Second Edition. 1987
14. A. A. HANSON and F. V. JUSKA, *Editors*: Turfgrass Science. 1969
15. CLARENCE H. HANSON, *Editor*: Alfalfa Science and Technology. 1972
16. J. R. WILCOX, *Editor*: Soybeans: Improvement, Production, and Uses. Second Edition. 1987
17. JAN VAN SCHILFGAARDE, *Editor*: Drainage for Agriculture. 1974
18. G. F. SPRAGUE and J. W. DUDLEY, *Editors*: Corn and Corn Improvement, Third Edition. 1988
19. JACK F. CARTER, *Editor*: Sunflower Science and Technology. 1978
20. ROBERT C. BUCKNER and L. P. BUSH, *Editors*: Tall Fescue. 1979
21. M. T. BEATTY, G. W. PETERSEN, and L. D. SWINDALE, *Editors*: Planning the Uses and Management of Land. 1979
22. F. J. STEVENSON, *Editor*: Nitrogen in Agricultural Soils. 1982
23. H. E. DREGNE and W. O. WILLIS, *Editors*: Dryland Agriculture. 1983
24. R. J. KOHEL and C. F. LEWIS, *Editors*: Cotton. 1984
25. N. L. TAYLOR, *Editor*: Clover Science and Technology. 1985
26. D. C. RASMUSSON, *Editor*: Barley. 1985
27. M. A. TABATABAI, *Editor*: Sulfur in Agriculture. 1986
28. R. A. OLSON and K. J. FREY, *Editors*: Nutritional Quality of Cereal Grains: Genetic and Agronomic Improvement. 1987
29. A. A. HANSON, D. K. BARNES, and R. R. HILL, JR., *Editors*: Alfalfa and Alfalfa Improvement. 1988
30. B. A. STEWART and D. R. NIELSEN, *Editors*: Irrigation of Agricultural Crops. 1990
31. JOHN HANKS and J. T. RITCHIE, *Editors*: Modeling Plant and Soil Systems, 1991
32. D. V. WADDINGTON, R. N. CARROW, and R. C. SHEARMAN, *Editors*: Turfgrass, 1992
33. H. G. MARSHALL and M. E. SORRELLS, *Editors*: Oat Science and Technology, 1992

COTTON

Editors

R. J. Kohel and C. F. Lewis

Managing Editor
DOMENIC A. FUCCILLO

Editor-in-Chief ASA Publications
DWAYNE R. BUXTON

Number 24 in the series
AGRONOMY

American Society of Agronomy, Inc., Crop Science Society of America, Inc., Soil Science Society of America, Inc., Publishers
Madison, Wisconsin, USA
1984

American Society of Agronomy, Inc.
Crop Science Society of America, Inc.
Soil Science Society of America, Inc.
677 South Segoe Road, Madison, Wisconsin 53711 USA

Second Printing 1993

Library of Congress Cataloging in Publication Data

Cotton

    (Agronomy; no. 24)
    Bibliography
    Includes index.
    Contents: Cotton

IV. Series.
83-073245
ISBN 0-89118-077-X

Printed in the United States of America

Cotton in full bloom (top) and mechanical picking of cotton (bottom).

CONTENTS

5 Quantitative Genetics

WILLIAM R. MEREDITH, JR.

6 Physiology

C. R. BENEDICT

7 Breeding

G. A. NILES AND C. V. FEASTER

8 Crop Growing Practices

B. A. WADDLE

12 Fiber

HENRY H. PERKINS, JR., DON E. ETHRIDGE, AND
CHARLES K. BRAGG

13 Seed

JOHN P. CHERRY AND HARRY R. LEFFLER

14 Marketing and Economics

ARLIE L. BOWLING

FOREWORD

Cotton is a unique crop species that has been a participant in many epics of history. It is one of only a few species that were domesticated in both the Old and New Worlds. Cotton was central to the success of the industrial revolution in Northern Europe, and it was a major thread that wove through the development of colonial empires. Cotton played a major role in the U.S. Civil War and in the economic recovery of the South after the war.

Cotton also has participated in the development of several technologies. The saw gin, which separates lint from seeds, permitted cotton to become a fiber for the masses. Competition from man-made synthetic fibers resulted in the development of processing technology that made cotton into a "wash-and-wear" fabric. Also, the myriad of insect pests and diseases that attack cotton spawned "Integrated Pest Management," a technology for the combined use of biological, mechanical, and chemical agents of control.

In plant research, cotton has shared in the establishment of many principles. Cytogenetic studies on cotton have prompted similar research on other species and provided the knowledge needed for the introgression of genes from ancestral and sister species. The elucidation of the nature of insect tolerance in cotton has led to new techniques in resistance breeding of crop species.

Yes, cotton is unique—it has survived history, been central to development of technologies, and played key roles in plant research. In the monograph, all aspects of cotton are discussed. The taxonomy, morphology, and physiology of this species are major topics of discussion; its genetics and the breeding of superior cultivars are treated thoroughly; growing and harvesting the crop are subjects examined; and the techniques of processing cotton and the crop's impact on world trade are presented.

The American Society of Agronomy, the Crop Science Society of America, and the Soil Science Society of America are pleased to be the sponsors and publishers of the Cotton monograph. This compendium of the literature on cotton will provide a knowledge base for scientists who work in agronomic, crop, and soils research. It is a scholarly treatise from which new insights in plant research will occur, and a knowledge platform from which new cotton research and technologies will grow.

Kenneth J. Frey, *President*
American Society of Agronomy

Wayne F. Keim, *President*
Crop Science Society of America

Donald R. Nielsen, *President*
Soil Science Society of America

PREFACE

Cotton is at once a fiber, food, and feed crop. The cotton plant is a warm-season woody perennial shrub that is grown as an annual field crop. Worldwide, over 30 million ha of cotton are grown between 47° N and 32° S Lat with over 50% of the production above 30° N Lat. From these crop plants comes cotton lint, an industrial raw material. Thus a renewable agricultural resource enters into competition with synthetic fibers in the textile industry. These crop plants produce not only lint but also the world's second most important oilseed.

Cotton has played a great role in the economy and politics of the world. Today the increased trade in grains and competition from synthetic fibers has reduced cotton's relative importance, but its world consumption continues to grow. Cotton will continue to be a significant commodity in future world trade.

The American Society of Agronomy, in recognition of the importance of cotton, established an editorial committee for a monograph consisting of C. R. Benedict, J. N. Jenkins, I. W. Kirk, R. J. Kohel, C. F. Lewis, P. A. Miller, and T. C. Tucker.

The editorial committee met and enthusiastically developed a comprehensive subject matter outline and limited the monograph to a single volume treating the major subject matter areas.

The monograph starts with the history of cotton. Succeeding chapters treat the biology of the plant and the production of the crop. The remaining chapters are concerned with the industrial utilization of fiber and seed and the economics of cotton. The intent was to produce a timely monograph that would serve to introduce the world of cotton to those unfamiliar with the complexity of this agricultural-industrial commodity and to provide a technical reference for cotton specialists.

The Editors wish to thank the editorial committee for their assistance and contributions and to especially thank the authors for their dedicated efforts that produced the substance of this endeavor.

R. J. Kohel
C. F. Lewis
Editors

CONTRIBUTORS

Roy V. Baker
Research Leader, South Plains Ginning Research Laboratory, Agricultural Research Service, U.S. Department of Agriculture, Lubbock, Texas

Alois A. Bell
Research Leader, Agricultural Research Service, U.S. Department of Agriculture, Texas A&M University, College Station, Texas

C. R. Benedict
Professor, Department of Plant Sciences, Texas A&M University, College Station, Texas

Arlie L. Bowling
Director of Cotton Foundations and Senior Economist, National Cotton Council of America, Memphis, Tennessee

Charles K. Bragg
Research Textile Technologist, Cotton Quality Research Station, Agricultural Research Service, U.S. Department of Agriculture, Clemson, South Carolina

J. M. Chandler
Associate Professor, Department of Soil and Crop Sciences, Texas A&M University, and Texas Agricultural Experiment Station, College Station, Texas

John P. Cherry
Associate Director, Eastern Regional Research Center, Agricultural Research Service, U.S. Department of Agriculture, Philadelphia, Pennsylvania

Rex F. Colwick
Agricultural Engineer, Agricultural Research Service, U.S. Department of Agriculture, and Mississippi State University, Mississippi State, Mississippi (now retired)

J. E. Endrizzi
Professor, Plant Science Department, University of Arizona, Tucson, Arizona

Don E. Ethridge
Agricultural Economist, Department of Agricultural Economics, Texas Tech University, Lubbock, Texas

C. V. Feaster
Research Agronomist, Cotton Research Center, Agricultural Research Service, U.S. Department of Agriculture, Phoenix, Arizona

Paul A. Fryxell
Research Botanist, Agricultural Research Service, U.S. Department of Agriculture, and Texas A&M University, College Station, Texas

A. Clyde Griffin, Jr.
Laboratory Director, U.S. Cotton Ginning Laboratory, Agricultural Research Service, U.S. Department of Agriculture, Stoneville, Mississippi

R. J. Kohel
Research Geneticist, Agricultural Research Service, U.S. Department of Agriculture, and Texas A&M University, College Station, Texas

William F. Lalor	Director of Processing Research, Cotton Incorporated, Raleigh, North Carolina
Joshua A. Lee	Geneticist, Agricultural Research Service, U.S. Department of Agriculture, and Department of Crop Science, North Carolina State University, Raleigh, North Carolina
Harry R. Leffler	Plant Physiologist, Cotton Physiology and Genetics Research, Delta States Research Center, Agricultural Research Service, U.S. Department of Agriculture, Stoneville, Mississippi
Jack R. Mauney	Plant Physiologist, Western Cotton Research Laboratory, Agricultural Research Service, U.S. Department of Agriculture, Phoenix, Arizona
William R. Meredith, Jr.	Research Geneticist, Cotton Physiology and Genetics, Delta States Research Center, Agricultural Research Service, U.S. Department of Agriculture, Stoneville, Mississippi
G. A. Niles	Professor, Department of Soil and Crop Sciences, Texas A&M University, College Station, Texas
Henry H. Perkins, Jr.	Research Chemist, Cotton Quality Research Station, Agricultural Research Service, U.S. Department of Agriculture, Clemson, South Carolina
R. L. Ridgway	Laboratory Chief, Beltsville Agricultural Research Center, Agricultural Research Service, U.S. Department of Agriculture, Beltsville, Maryland
E. L. Turcotte	Research Geneticist, Cotton Research Center, Agricultural Research Service, U.S. Department of Agriculture, Phoenix, Arizona
Joseph A. Veech	Research Plant Physiologist/Hematologist, Agricultural Research Service, U.S. Department of Agriculture, and Texas A&M University, College Station, Texas
B. A. Waddle	Altheimer Professor, Department of Agronomy, University of Arkansas, Fayetteville, Arkansas
Lambert H. Wilkes	Professor, Department of Agricultural Engineering, Texas A&M University, College Station, Texas

1   Cotton as a World Crop

Joshua A. Lee

ARS-USDA and North Carolina State University
Raleigh, North Carolina

Cottons (from the arabic *quotn*) are useful in that some species produce spinnable fibers (lint) on the seed coats. Such fibers begin as elongated cells growing outward from the surface of the ovule. As the ovule enlarges, successive layers of cellulose are laid down in a helical pattern by the protoplast. As the fiber matures, the protoplast dies, and the cell wall, which is virtually pure cellulose, collapses inward to form a convoluted ribbon. The flattening and convolution of the dried cell wall promotes adhesion when the fibers are twisted together in yarn bundles during spinning.

There are four domesticated species of cotton. *Gossypium arboreum* L. and *G. herbaceum* L. (Fig. 1–1 and 1–2), both diploids, are native to the Old World. *Gossypium arboreum* remains an important crop in India, whereas *G. herbaceum,* important in earlier times, is today grown mostly for local use in the drier areas of Africa and Asia. *Gossypium barbadense* L. and *G. hirsutum* L. (Fig. 1–3 and 1–4) evolved in the New World. Both are allotetraploids. *Gossypium barbadense,* commonly known as extra-long-staple, Egyptian, and Pima cotton, and other names, supplies about 8% of the current world production of fiber. The fiber is used mostly for the production of luxury fabrics and sewing thread. *Gossypium hirsutum,* known most widely as Upland cotton, contributes about 90% of the current world production of 65 million bales of fiber weighing about 218 kg/bale. Upland cotton fibers are used in the manufacture of a variety of textile products, cordage, and other non-woven products. Linters, the short fibers removed from seeds before crushing, are an important source of industrial cellulose.

Although cotton is grown mostly for fiber, the seeds are also important. Cottonseed oil is used for culinary purposes, and the oilcake residue is a protein-rich feed for ruminant livestock. Cotton is grown virtually around the world in tropical latitudes, and as far north as 43°N lat. in the USSR, and 45°N in the People's Republic of China (PRC).

Contribution from ARS-USDA in cooperation with the North Carolina Agricultural Research Service. Paper No. 8035 of the Journal Series of the North Carolina Agricultural Experiment Station. Published in *Cotton,* Agronomy Monograph No. 24, © ASA-CSSA-SSSA, 677 South Segoe Road, Madison, WI 53711, USA.

1-1 THE ORIGINS OF COTTON

1-1.1 Theoretical Considerations and Archaeological Evidence

Seemingly, the presence of spinnable seed fibers should have provided sufficient reason for using wild cottons and the eventual domestication of the four cultivated species. However, Chowdhury and Burth (1971) pointed out that a diploid species—possibly *G. arboreum*—might have been domesticated in Nubia for the seeds as a nutritious feed for domesticated animals. Hutchinson et al. (1947) argued that the presence of seed fibers was the chief impetus for the domestication of *G. herbaceum* in the Old World but hypothesized that spinnable fiber represented a mutant, presumably rare, phenotype in a wild, or perhaps early domesticated, form of the cotton. These workers concluded further that the New World tetraploid cottons evolved under domestication after the introduction of a cultivated Old World species, *G. arboreum* being the most likely candidate. The cultivated diploid then crossed with a New World diploid taxon, one similar to, or identical with, the extant species, *G. raimondii* Ulb., thereby spawning the prototype from which the tetraploid species of *Gossypium* diverged. Presumably the event occurred in South America. Modern views support the hypothesis of primitively wild linted cottons in both the Old and the New Worlds and that the presence of lint provided the chief impetus for domesticating the cottons, perhaps the only reason for domestication in the New World (Stephens, 1967; Fryxell, 1979).

Evidence for the beginnings of agriculture in the Near East date with certainty from about 7000 B.C. (Harlan, 1975). No remains of cotton have been found in the Near Eastern Center. Indeed, there is no evidence that cotton ever grew wild in the region. The chief interest of the Near Eastern Center in relation to the domestication of cotton in the Old World is that the agriculture evolving in the area—a grain, pulse, livestock, and linen economy—supported the societies into which cotton was later introduced as a wild, or perhaps newly domesticated, plant. Thus Hutchinson et al. (1947) reasoned that the growing and spinning of cotton evolved in economies that already had a technology for the spinning and weaving of flax (*Linum usitatissimum* L.) and wool. This conclusion was based largely upon the fact that the only known wild form of a domesticated diploid cotton, *G. herbaceum* var. *africanum* (Watt) Hutchinson and Ghose, has fiber length in the range of 15 mm or less, well below the staple length of modern *G. herbaceum*. Presumably cotton was grown initially for uses other than textile yarn, for wound dressings, wadding, and the like. Longer staples had to be selected before cotton fiber could become competitive with either flax or wool.

The findings of cotton remains in Nubia (Chowdhury and Burth, 1971) and at Mohenjo-Daro (Fig. 1-5) dating from about 2700 B.C. (Gulati and Turner, 1928) give credence to the idea that cottons were domesticated at sites that already knew a productive agriculture and the arts of spinning and weaving. The Nubia finds might represent cotton at the interface of domestication; the Mohenjo-Daro finds certainly do not. The cottons of

Fig. 1–1. *Gossypium herbaceum* L. This figure and Fig. 1–2 to Fig. 1–4 from Parlatore (1866).

Fig. 1–2. *Gossypium arboreum* L.

Fig. 1–3. *Gossypium barbadense* L.

Fig. 1–4. *Gossypium hirsutum* L.

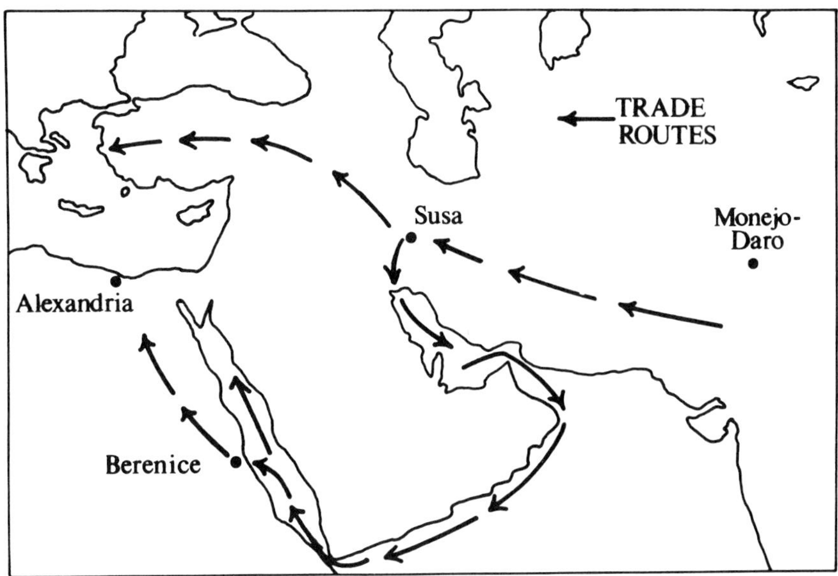

Fig. 1–5. The Near East in Alexandrian and post-Alexandrian times with points of archaeological interest, trade routes, and commercial centers.

Mohenjo-Daro, which seems to have been the hub city of an ancient Indus civilization, were advanced diploids, very likely *G. arboreum.* There is no clue as to whether the cottons had advanced to an annual habit of production, as is standard for modern cottons of all species, but the latitude where discovered—about 27°N lat.—suggests that the cottons might have evolved toward an annual habit.

Mohenjo-Daro was a flourishing metropolis of commerce and industry. Irrigation was practiced and the agriculture obviously sustaining. Spinning and weaving were practiced arts. Cotton lint was separated from seeds on the *churka,* a kind of hand-cranked roller gin, and the fibers processed into fabrics with drop spindles and primitive looms. The quality of the textiles produced appears to have been excellent.

Gossypium herbaceum var. *africanum* grows as a wild plant in the bushveld of southern Africa and possibly in the Sahel of north Africa (Hutchinson et al., 1947) and, according to these workers and Fryxell (1979), is the wild form from which the domesticated species most likely arose. The cotton is linted, but the seed fibers are sparse, short, and dingy-grey in color. Nevertheless, the fiber of *africanum* is spinnable.

South Africa is distant from the Near East and an unlikely place for *G. herbaceum* to have been domesticated. Fryxell (1979) points out that the cotton could have been introduced to the Near East in ancient times where the plant was subsequently domesticated. Trade between Arabia and southern Africa is very old (Hutchinson et al., 1947) but just how ancient is not known. Alternatively, the domestication of *G. herbaceum* might have begun in north Africa, or wild *G. herbaceum* might once have been distributed to Arabia, Baluchistan, and perhaps the Sind.

No forms of *G. arboreum* are extant in the wild. Hutchinson et al. (1947) proposed that the species evolved from *G. herbaceum* after the latter was domesticated. Other than through ploidy manipulation, the derivation of one species from another, or others, under domestication is unprecedented. Hybrids between *G. arboreum* and *G. herbaceum* have apparently regular chromosomal pairing, except for the presence of one translocation (Gerstel, 1953), and an uncomplicated meiosis. There is evidence of developmental incompatibility ("genetic breakdown") in ensuing generations. Where the two species overlap in distribution as cultivars, as in India, the identity of each is maintained, presumably through the combined effects of genetic and temporal barriers (Hutchinson et al., 1947). Fryxell (1979) reasoned that the time allowable for such a degree of divergence between the two taxa seemed insufficient to indicate speciation in the domesticated state.

Today *G. barbadense* grows as a wild plant on the coasts of Peru and Ecuador and perhaps the Galapagos Islands. Commensal and marginally cultivated forms of the species occur through much of northern South America, the West Indies and parts of Middle America (Fig. 1–6). The latter site, along with the warmer parts of the Old World, were populated with *G. barbadense* in post-Columbian times (Hutchinson et al., 1947). The center of diversity for the species is northwestern South America. *Gossypium darwinii* Watt grows wild in the Galapagos Islands. This taxon is a near-relative of *G. barbadense* (Fryxell, 1979) and is regarded as being conspecific with the cultivated species by some authorities, Hutchinson et al. (1947) for example.

Stephens and Moseley (1974) examined fruits, foliar parts, and fiber of cottons taken from archaeological contexts, dating from about 2500 to 1750 B.C., in the Ancon-Chillon district of Peru. Based upon fiber characteristics and colors, fruit shape and size, and the absence of trichomes from the floral nectaries, the authors concluded that the remains were from primitive cultivated forms of *G. barbadense,* some barely beyond the threshold of domestication. The only other cultivated species recovered was the hard gourd, *Lagenaria* spp. The impetus for the utilization of cotton in the area seems to have been the need for cordage for fishing lines and nets.

Ancon-Chillon thus appears to be a site where cotton was domesticated. Of special interest was the fact that the cotton was *G. barbadense* and not an Old World diploid species. This finding does much to negate the possibility that one of the ancestors of New World tetraploid cottons was introduced by agricultural people. Moreover, the primitive state of the Ancon-Chillon cottons suggest that a linted form of *G. barbadense* grew wild in the area. The Ancon-Chillon people, even if they showed no evidence of farming for food, were, nonetheless, a settled folk who practiced a kind of agriculture suited to their needs, namely, the gathering of seafoods. *Gossypium barbadense* thus seems to have evolved in South America as a wild, and later as a cultivated, species.

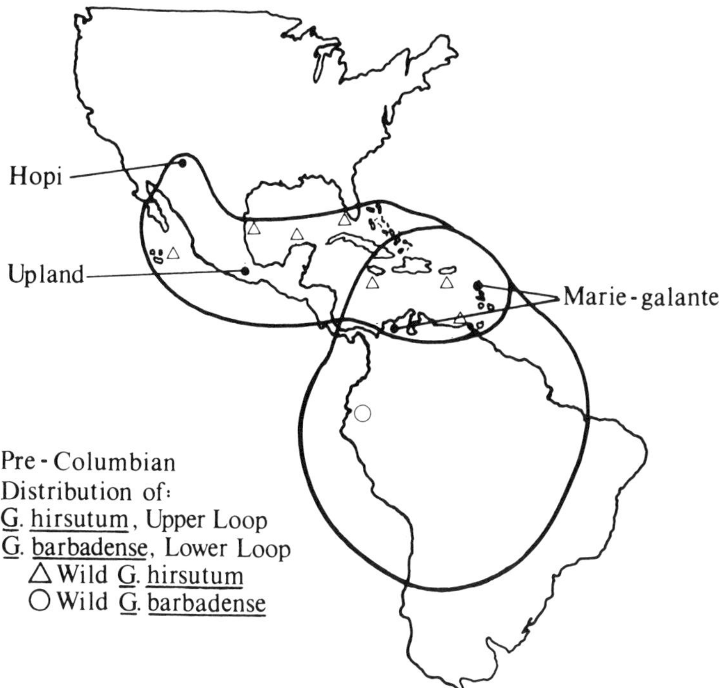

Fig. 1-6. Approximate distribution of *G. barbadense* and *G. hirsutum* in the New World in pre-Colombian times with notations on the distribution of wild populations, and sites of domestication.

Gossypium hirsutum in its wild and commensal forms is distributed in the drier areas of Middle America, northern South America, the West Indies, the southern tip of Florida, Polynesia, and through introduction, north Africa and southern Asia. The wild populations are rare and widely dispersed. All grow on beach strands, or else are confined to small islands. Archaeological remains of *G. hirsutum* have been found mostly in Mexico, the oldest from the Tehaucan Valley contexts (Smith and Stephens, 1971). These were dated from the Abejas phase, about 3500 to 2300 B.C., and appeared to have been from domesticated stocks. The Tehaucan cottons were believed to have been introductions, inasmuch as there is no evidence that *G. hirsutum* ever grew wild in the valley.

Evidently *G. hirsutum* was domesticated in at least two, perhaps three, separate areas. Modern Upland cottons stem from a center of diversity near the border of Mexico and Guatamala (Hutchinson et al., 1947). The domesticated forms of *G. hirsutum* var. *marie-galante* (Watt) Hutch. might have been derived from a center of diversity for the race in northern Colombia (Stephens, 1974a) or from introgression between West Indian wild forms of *G. hirsutum* and introduced *G. barbadense* (Stephens, 1967). Wild and commensal forms of *marie-galante* along the Pacific coast of Middle America overlap the distribution of other races of *G. hirsutum* intruding from the north only slightly. That, plus the fact that domesticated stocks of *marie-galante* are distributed in northern South America and the West Indies, argues for a separate domestication for the race and the Upland cottons.

The Hopi cottons of the American Southwest are annual forms placed in *G. hirsutum* var. *punctatum* (Schumacher) Hutch. by Hutchinson et al. (1947). The *punctatum* cottons are the most widely distributed of the wild and commensal forms of *G. hirsutum.* The distinctiveness of the Hopi cottons, when compared with Upland cultivars, suggests that the Hopi forms had an origin independent of modern Uplands and *marie-galante.* No form of *G. hirsutum* as a perennial plant has a history of growing north of the Mexican border in what is now the United States. Therefore, it is likely that the Hopi cottons were part of a northward migration of what was basically a corn, bean and squash agriculture into the American Southwest, about 100 to 500 A.D.

1-1.2 Domestication and Spread of Cotton

Most wild cottons inhabit regions where rainfall is strictly seasonal, occurring in most such places in summer. Characteristically such cottons are perennial shrubs or small trees. Mauney and Phillips (1963) showed that a series of wild cottons required 15 or more nodes along the main plant axis before reaching a transition from the vegetative to a fruitful phase of growth. Therefore, a seedling emerging during the rains would rarely have time to develop the plant stature required for flowering at the end of the wet season but could flower seasonally thereafter.

Typically a wild cotton, such as the Jackson Bay population of *G. hirsutum* of the Caribbean strand of Jamaica, lapses into dormancy during the winter dry season and becomes vegetative with the onset of rain in early summer. At the end of the rainy season, vegetative recruitment ceases, and the plants enter the reproductive phase. There is a short flush of flowering, followed by fruit maturation, and re-entry into the dormant phase.

At least three environmental factors interact to trigger an orderly flowering process in wild cottons. First, flowering does not proceed as long as there is sufficient soil moisture to sustain vegetative increase (Hutchinson et al., 1947; Mauney and Phillips, 1963). Second, a differential between day and night temperatures is needed. During the wet season in the tropics the temperature differential between day and night is minimized by the presence of high humidity and cloud cover. At the onset of the dry season radiation on clear nights increases the differential between daily maximum and nightly minimum temperatures. According to Mauney and Phillips (1963) a differential of at least 12°C is needed to maintain orderly bud set and subsequent flowering in many wild cottons. The third requirement is for a decreasing photoperiod. However, it is difficult to argue for a photoperiod requirement for a taxon such as *G. darwinii* which grows within 1° of the equator. Nevertheless, when the cotton was subjected to variable photoperiods by Mauney and Phillips (1963), *G. darwinii* was found to require short days for flowering and setting fruit.

Based upon the present distribution of the wild forms in the New World, *G. hirsutum* must have diversified several degrees north of the equator, at latitudes where photoperiod varied seasonally. Thus photoperiodic response has evolved as an important criterion in timing the onset of reproduction in many of these cottons (Stephens, 1967).

Most modern cottons behave as annuals; that is to say, the crop is harvested within the same year planted. However, "annual" cottons are not true annuals in the sense that death of the plant is a natural consequence of seed ripening. Modern cultivars of cotton can be maintained indefinitely under glasshouse culture. As one might expect with taxa that have been transformed from long-lived perennials with a slow rate of plant development to cultivars that can be cropped annually from seeding, a broad spectrum of fruiting habits, relative to timing after seedling emergence, can be observed. Thus Hutchinson et al. (1947) spoke of early annuals, late annuals, and perennials. Stephens (1974b) listed a group of cottons as annuals apparently because the cultivars shared the property of fruiting at low enough node for the crop to be harvested within the same year planted. Some of the cottons listed by Stephens are not "day neutral" as far as photoperiodic requirement for fruiting is concerned. For example, the *G. barbadense* cultivar 'Lengupa' does not set fruit when grown in the field at College Station, Texas (Lewis and Richmond, 1960), but sets fruits at low nodes under the short-day conditions of northern South America. Therefore Stephens (1974b) seemed to be defining annual habit as the trait of setting fruit precociously under at least some environmental conditions. In a later paper (1975) Stephens chose to reserve the term, annual habit, for some cultivars of *G. hirsutum* that cropped precociously in the tropics and then failed to survive the dry season. Under the revised definition cottons such as the Sea Island and Egyptian forms of *G. barbadense* would be regarded as perennials which are grown as annuals in subtropical latitudes.

From the standpoint of understanding how cottons fitted for extra-tropical production might have evolved, the definition implicit in Stephen's (1974b) reference seems most meaningful. Therein annual habit is interpreted as the consequence of a downward shift in the number of nodes required along the main plant axis for the onset of fruiting. Such a change must precede, or at least accompany, loss of short-day response if the cotton is to adapt to latitudes that have killing frosts.

The loss of the short-day requirement for flowering and fruiting might have accompanied the shift to annual habit in some cultivated cottons, but most certainly not in others. Lengupa is one such example, and some accessions of *G. hirsutum* from the Mexican center are annual in fruiting habit but retain the short-day requirement for flowering (Kohel and Richmond, 1962). The evidence shows that the shift to "day neutrality" in *G. hirsutum* involved multigenic changes (Kohel and Richmond, 1962), whereas the one case where contrasts were compared in *G. barbadense* showed evidence for a monogenic change (Lewis and Richmond, 1960). Obviously, both changes to annual habit and "day neutrality" were required before cottons could be grown in extra-tropical environments with cold winters.

The cottons of the Old World might have begun the route to domestication as wild plants introduced into what were sophisticated agricultural economies. In the New World the indigenous peoples doubtless took from the wild cotton plants the materials needed for trivial domestic uses, whether fiber or seeds. From such utilization the cultivation of cotton might have started as an adaptation to "dump heap" agriculture as described by Anderson (1954). The dropping of seeds on such middens, likely with no initial intent to cultivate, could have evolved into a handy way to grow a few plants away from the cotton's native ground. The wild forms of the domesticated New World cottons seem opportunistic toward disturbed land. Thus I once noted that Jackson Bay wild cotton, which typically grows at the interface between saline marsh and xeric scrub in Jamaica, had spread inland a short distance along roads cut by charcoal gatherers.

Cottons commensal with human settlement, arising perhaps from vicarious plantings of wild material, evolved into forms with somewhat more fibers on seeds, and frequently, but not always, larger fruits (Hutchinson et al., 1947; Fryxell, 1979). Human selection is indicated inasmuch as there seems to be no conceivable way that increased fibers could be advantageous to cottons naturally propagating, but would be of value to the incipient, or actual, cultivator. Such plants remained perennial in growth and reproductive habit, and they fitted in well with small-plot subsistence agriculture. A few plants supplied a family's needs, and perenniality meant a definite savings in management.

The citizens of Mohenjo-Daro were clothed in cotton raiment, as were those of southern Mexico in the days of Toltec, Aztec, and Maya. The Inca of the Andes also used cotton lint extensively. Annual habit in cotton was selected in conjunction with intensive cultivation of the crop by these peoples. Annual production allowed for the transition of cotton-growing from a marginal, home-based enterprise to production on a commercial scale (Hutchinson et al., 1947).

The loss of the short-day sensitivity requirement for fruiting in the various cotton species remains, as to time and place, a mystery, but apparently occurred in tropical, or near-tropical, latitudes. Such an inference seems sound because planting cottons with stringent photoperiodic requirements in latitudes such as those of the American Cotton Belt could not result in production. However, it is possible, inasmuch as the photoperiodic trait of *G. hirsutum* is multigenic and expressed in an additive manner (Kohel and Richmond, 1962), that some of the earlier introductions of Upland cottons to the American Cotton Belt were only partly converted to "day neutrality", that is to say, were perhaps capable of producing a crop as daylength decreased in late summer. Such cottons could then have been selected for earlier performance in the new environment. The introduction of the Sea Island form of *G. barbadense* into the south Atlantic coastal region of the American colonies is documented (Stephens, 1976). With these cottons annual habit coupled with "day neutrality" was definitely a matter of preadaptation.

Stephens (1975) proposed a model whereby the Sea Island cottons might have lost their short-day response and simultaneously acquired the annual habit. A cross between a West Indian commensal form of *G. barbadense* and wild accession of *G. hirsutum,* WH—24, produced some segregates in later generations that set fruits under the long-day summer environment of North Carolina. Neither parent was "day neutral" in flowering habit. Earlier Kohel et al. (1974) recovered a few summer-flowering plants from the F_2 of a cross between an accession of *G. hirsutum* var. *marie-galante* and the *G. barbadense* cultivar, Lengupa.

Some of the Stephens segregates resembled Sea Island in both plant form and fiber properties. Thus Stephens reasoned that Sea Island cotton might have acquired both fitness for production in South Carolina and Georgia and the cotton's distinctive fiber properties from introgression of germplasm from a short-day-flowering form of *G. hirsutum* into a coarse-fibered form of *G. barbadense.* If such introgression occurred, the event must have taken place in the West Indies before Sea Island cotton was introduced into a region where it gained such renown as a crop.

There is no evidence that *G. hirsutum* in the Mexican center of diversity was ever exposed to interspecific introgression. The potential for interracial mixing is evident. Hutchinson et al. (1947) hypothesized that as the need for cotton grew in Middle America, the crop was gradually introduced into altitudinal zones not well suited for annual cottons adapted for short-day flowering and fruiting during the dry season. The result was selection for a tendency to set fruit during a part of the vegetative cycle of growth, a trait of modern Upland cultivars. The short-day response diminished under the new regime of growth. Table 1-1 compares wild and modern cultivated *G. hirsutum.*

Table 1–1. Comparison of wild and modern cultivated forms of *G. hirsutum.*

Characteristic	Wild	Domesticated
Distribution	Mostly tropical; populations rare; insular and strand habitats.	Tropical to temperate; widespread as a crop; cultivated land.
Germination response	Impervious seeds and sporadic germination.	Seeds imbibe water and germinate rapidly.
Phenology	Sharply seasonal pattern of vegetative increase followed by fruit production and dormancy.	Short vegetative phase followed by overlap in vegetative and fruit increase and eventual "cut-out" and fruit ripening; plants usually destroyed after producing.
Fruiting habit	Perennial fruiting; 15, or more, nodes to first fruiting branch.	Annual fruiting; 5 to 7 nodes to first fruiting branch.
Fruiting response	Requires short days, cool nights, and increasing drought stress for fruiting.	"Day-neutral"; prefers cool nights and ample soil moisture for best production.
Fruit size	Small, 2 g of seed and lint.	Large; 6 to 8 g of seed and lint.
Lint quantity	Sparse; 15 to 20% gin out-turn.	Copious; 35 to 40% gin out-turn.
Fiber properties	Lint short and fine; dingy brown in color.	Lint length variable up to 28 mm; coarse or fine; most often white.

Gossypium arboreum spread to China, and *G. herbaceum* to China, Turkestan, and Persia prior to the 13th century (Fryxell, 1979), where some very precocious annual forms of both species were selected. *Gossypium hirsutum* became a crop of local importance in the Southeastern U.S. by the mid-18th century. The early stocks, known as "Georgia Green Seed", were probably of Mexican origin (Hutchinson et al., 1947). Similar stocks were taken to southern Asia at about the same time. These cottons became known as Cambodias, and, unlike Georgia Green Seed, were short-day-sensitive in flowering response. The Cambodia cottons are virtually obsolete today. Later introduction of Mexican stocks to the American Cotton Belt were "white-seeded" stocks with larger fruits and longer lint than the "green-seeded" cottons. Hybridization between the stocks produced the explosion of Upland cottons that are the foundation of the world's cotton industry today.

The first "day neutral", annual forms of *G. barbadense* were the Sea Island cottons of South Carolina and Georgia. In the early 19th century, Sea Island hybridized with Jumel's tree cotton, apparently a perennial form of Peruvian *G. barbadense,* to spawn the distinctive annuals known today as Egyptian cottons (Balls, 1912). Egyptian cottons later gave rise through further hybridization to the Pima cottons of the Southwestern U.S. (Bryan, 1955) and to the limbless cultivars of *G. barbadense* grown today in the USSR.

1-2   THE ADVANCE OF COTTON IN WORLD TRADE
AND MANUFACTORIES

1-2.1   From Alexander to the Industrial Revolution

Much in the following sections was extracted from J. A. B. Scherer's (1916) *Cotton as a World Power.* Scherer points out that Alexander of Macedonia introduced cotton and cotton goods to the then-known Western World. Evidence for the existence of a cotton industry in the Meroitic empire of the Sudan dates from about 400 B.C. (Hutchinson et al., 1947). Nevertheless, cotton was apparently unknown in ancient Egypt. Nearchus, a general in Alexander's army during the march to the Indus, described cotton apparel thusly: ". . .a shirt, or a tunic, reaching the middle of the leg, a sheet folded over the shoulders, and a turban wound around the head."

Theophrastus (373 to 288 B.C.) described cotton plants from material brought to Greece by Alexander. Cotton had spread to Persia in pre-Hellenic times, cotton textiles being mentioned in the Book of Esther. That cotton was grown and utilized by the Assyrians in the time of King Sennacherib (706 to 681 B.C.) is attested to by a cylinder now in the British Museum. The cylinder reads: "The trees that bore wool they clipped, and they carded it into garments". Neither Persia nor Assyria seemed to have traded with the civilizations to the west. As mentioned before, the spinning of cotton was an ancient art in India by the time of Alexander. Some of the fabrics produced on the looms of the Indus people have been described as "webs of woven wind."

Alexander established trade routes between East and West. Susa, one of the capital cities of Persia, was selected to serve as the hub of a trading empire with a highway overland through Asia Minor and water routes through the Euphrates and Tigris to the Persian Gulf and thence to the Eastern Mediterranean (Fig. 1–5).

After Alexander, the Ptolemaic pharoahs made Alexandria the chief city of the eastern Mediterranean and the leading center for dispersing goods from the East. Ptolemy Philadelphus envisioned a Suez canal, but built instead the port of Berenice on the Red Sea coast. Goods from India were landed at Berenice and transported overland by caravan to the Nile whence they were floated down river to Alexandria. Cotton goods thus became commonplace, although luxurious, items for the ruling classes of Greco-Roman times.

The growing of cotton in the lands about the eastern Mediterranean followed the introduction of cotton goods. The crop became established in lower Egypt and in the Levant. Cotton (most likely *G. herbaceum*) was introduced into Spain by the Moors in 712 A.D. where it remained an important crop until the final expulsion of Islam from Iberia in 1492.

Cotton fabrics developed to the point of promoting style in clothing during Renaissance times in Italy. The state of Venice forthwith became the leading commercial center of the Mediterranean, depending largely upon cotton goods from India as a staple of commerce. From Venice cotton cloth found markets as far away as northern Europe.

In 1492 Christopher Columbus sailed under the flag of Isabella the Catholic to the westward, seeking India, but discovered the New World instead. Among the first products of the New World examined by the Admiral were cotton fabrics woven by the indigenous people, his "Indians." In 1497 Vasco DaGama sailed around Africa to India, thereby establishing the sovereignty of Portugal over the sea routes to the East. Lisbon replaced Venice as purveyor of cotton goods and spices to Europe, and the Alexandrian trade routes collapsed into disuse.

England eventually became predominant in cotton trade and manufacturies, and one can justly state that cotton rode to prominence on the back of wool. As a result of royal guarantees concerning private property, English yeomen of post-medieval times had no fear that the crown, or the nobility, would confiscate such readily purloinable property as sheep, apparently a genuine hazard for the husbandmen of continental Europe at the time. Thus did England come to enjoy a virtual monopoly on wool production in Europe. However, England had no commercial spinning and weaving industry until Edward III (1327 to 1377) invited Flemish weavers, who worked virtually as serfs for the continental guilds, to settle in the Midlands. Shortly thereafter England became not only the chief producer of European wool but a leading manufacturer of woolens as well.

The incursion of cotton into the domain of wool was, at first, neither welcomed nor tolerated. However, the utility of cotton apparel for summer wear overwhelmed the opposition so that by the turn of the 18th century the cottage spinning and weaving industry of Lancashire was heavily involved in the processing of the "tree wool," and the West Indies had become a prominent source of supply.

1-2.2   Cotton and the Industrial Revolution

Up to the late 18th century the cottage textile manufactures in England comprised a labor-intensive industry employing hand cards, spinning wheels, and manually operated looms. Two kinds of spinning wheels have evolved through the ages (Anderson, 1976), the Old, or Jersey, wheel and the Saxony wheel. The Jersey wheel uses a system of spinning hand-drawn roving (slightly twisted strands) followed by winding of the spun yarn onto the base of an open spindle. This kind of spinning is an intermittent process and represents a kind of mechanization of the ancient drop-spindle technique of spinning yarn. The Saxony wheel spins and winds continuously, the roving being twisted by a U-shaped flyer while being wound onto a bobbin inside the flyer. Both the intermittent and continuous spinning principles live on in modern textile machinery.

In the latter half of the 18th century problems of supply in the cottage weaving industry spurred inventions which, in turn, begot further supply problems. The concatenation of events that followed revolutionized the production of textiles, and, indeed, spawned the factory system and the Industrial Revolution. Warp for the loom was most commonly produced on the Saxony wheel, whereas shuttle weft, requiring fewer twists and lower counts, was produced largely on the Jersey wheel. Because the Saxony wheel, with its continuous process of spinning was more productive than the Jersey wheel, there was frequently a shortage of weft. The shortage was made acute for both warp and weft after John Kay's invention of the fly shuttle in 1738, a system whereby the shuttle was routed rapidly through the warp by a system of springs (Anderson, 1976).

As if in answer to the shortage of yarn, James Hargreaves patented the spinning jenny in 1770 (Aspin and Chapman, 1964). The jenny (Fig. 1-7) employed the same intermittent spinning system as the Jersey wheel, the difference being that the jenny had a bank of 12 or more spindles, and spinning was accomplished by twisting against the draft; i.e., a bank of rovings, each attached to a spindle, was attenuated by drawing the clamped rovings away from the spindles with a sliding clasp, or clove. Once enough twists had been applied, the spindles were reversed to "back off" the twists on the spindles knives, and a "faller" wire was dropped to depress the spun yarn onto the body of the spindles. The yarn was then wound up by pushing the clove toward the spindles as the spindles were cranked slowly by hand.

In 1738, Lewis Paul built a machine that multiplied yarn through a bank of flyers and bobbins, the principle of the Saxony wheel. The machine was apparently not successful. However, Paul was granted a patent for roller drafting, a system of attenuating roving by passing it through sets of rollers travelling at progressively higher speeds (Anderson, 1976). In 1769, Richard Arkwright led a group that built a spinning frame employing roller drafting and the flyer and bobbin principle. The machine was successful and came to be called the "water frame" after it was adapted to water powering. The coarse yarns spun by the frame were useful as warp, whereas the jenny spun fine-count weft at an accelerated rate.

The spinning mule became obsolete in recent times. This machine was a kind of hybrid between the water frame and the jenny; that is to say, rovings were drafted by rollers and spinning was of the open-knife, intermittent kind as in the jenny (Fig. 1-8). Samuel Crompton, who invented the mule, was operating the prototype by 1779 (Catling, 1970). The mule differed from the jenny in that the spindles travelled on a carriage that ran outward on the spinning and drafting cycle, and inward on the wind.

The jenny, used as late as 1926 in the woolen industry of Yorkshire, was always hand-cranked. The spinning frame and the mule were originally hand-cranked machines. Obviously, the amount of yarn one worker could spin on such machines was limited. The perfection of the steam engine by James Watt around 1790 changed that. By the beginning of the 19th century powered mules and spinning frames, as well as powered looms, perfected by Edmund Cartwright around 1792, were the tools of the burdgeoning textile industry in England, and increasingly, abroad. But where were the mills to get the cotton needed to keep the machines operating?

Fig. 1-7. The Hargreaves spinning jenny. (Courtesy of Smithsonian Institution.)

1–2.3   Eli Whitney and the First White Revolution

The grievances that spawned the American Revolution were many, prominent among which were attempts by Britain to exclude manufactures from the colonies, to levy import taxes on certain goods bound for the colonies, and to restrict coinage. Britain envisioned the colonies as producers of raw materials and consumers of finished goods and little more. Be that as it may, the American Revolution was fought and won, and by 1793, the year Eli Whitney invented the saw gin, the newly united states were well on their way to self-sufficiency in both agriculture and manufactures.

Eli Whitney was an educated man, seemingly destined for a career as a schoolmaster. While visiting Mrs. Nathaniel Greene near Savannah, Georgia, he became intrigued with the problem of separating the lint of Upland cotton from the seed. The naked seed of Sea Island cotton could be separated with a churka, but that device did not work well with the fuzzy seed of Upland. Sea Island cotton, the basis of a modest industry along the coasts of Georgia and South Carolina at the time, was poorly adapted for growing on inland sites; by contrast, the potential for growing Upland cotton seemed unlimited.

Whitney's gin employed a system of rollers. One roller was equipped with teeth for plucking lint from seeds, the lint then being doffed by a counter-rotating roller equipped with brushes. That Whitney envisioned and developed the basis principles of the saw gin is a matter of history. Less well known is the fact that a mechanic in Augusta, Georgia, Hodgen Holmes, developed a roller equipped with saw teeth, and the Holmes saws proved to be more serviceable than the Whitney pluckers. Litigation over patent rights to the various particulars of the saw gin's development seem

Fig. 1–8. Crompton's mule. (Courtesy of Harold Catling.)

endless. Suffice it to say that both Whitney and Holmes were well rewarded for their efforts.

In 1793 the Southern states produced about 10,000 bales of cotton of some 227 kg each. By 1810, 177,824 bales were produced, and Southern cotton had become important in world trade. The result of the invention of the saw gin was to reverse the trend toward emancipation of slavery in the South and to turn the region from a growing system of manufactures to a nearly total agricultural economy based upon the production of cotton and slaves. The South changed from a region strongly supporting the concept of the Federal Union to one espousing the narrowest conception of States Rights. The North, and particularly the New England States, became centers of manufactures and began to demand protective tariffs. The South, as producer and supplier of cotton both to the North and Britain, found it good economics to load vessels on the return voyage from Britain with British goods. As the North gained in population and political strength, tariffs became a reality, and the profit margin from the cotton trade began to shrink. The doctrine of nullification of Federal statutes, if the law in question did not serve a particular state's interests, was advocated by John C. Calhoun of South Carolina. That, growing sectionalism, and the question of the status of slavery in Western lands, exacerbated tensions between the North and the South and did much to precipitate the War Between the States.

That the Civil War was ruinous to the South can scarcely be exaggerated. The Confederacy had hoped that a shortage of cotton would compel the industrial nations of Europe to recognize the soveriegnty of the Confederate States and lend military assistance. That did not happen. Apparently the working classes of Britain came to despise slavery more than they craved bread. More subtle opinions held that Britain preferred a united and strong English speaking nation in the western hemisphere to prevent incursions by other European nations into what the British had come to accept as their sphere of influence. Be that as it may, cotton, if it had caused the rise and fall of the Confederacy, most certainly rescued the economy of the Southern states after reunion. After a short hiatus of cotton delivery and consumption during and after the war, demand rebounded for the fiber for which there was no substitute. Cotton production continued to climb, both in the South and abroad.

1-3   COTTON INTO MODERN TIMES—THE SECOND WHITE REVOLUTION

1-3.1   The Development of Modern Spinning Technology

William Kelly of Lanark first applied power to the spinning mule around 1790 (Catling, 1970). However, power was used only with the drawing and spinning cycle. Human labor was still required to "put up" the mule, i.e., to back yarn twists off the spindles and then wind the spun yarn onto spindle cops, or packages. The invention of the quadrant arm, and the attendant nosing and strapping actions, by Richard Roberts between 1830 and 1836 provided for automatic cop winding and the "self-acting" mule was born (Catling, 1970). Mules reached the zenith of their development in Britain near the end of the 19th century when units driving 1000 or more spindles were commonplace. The mule was used widely in Britain up to modern times.

With the advent of steam power, the speed possible on spinning frames increased dramatically, thereby creating the need for more durable equipment. The wooden flyer was replaced by one of metal, a device that warbled as it ran at high speeds. Mill workers likened the sound to a babel of song thrushes, the "throstle" of Shakespeare, and forthwith christened the new machines throstles.

The throstle was to provide the prototype from which sprang a series of important inventions in America. In 1828 Charles Danforth and John Thorp independently patented a cap-spinning throstle. The cap sat over the bobbin, replacing the flyer of the spinning frame. A guide at the base of the cap fed spun yarn to the bobbin which traveled up and down on a lifter. Air resistance caused the yarn to "balloon" free of the cap.

Although the cap throstle increased spinning speed, virtually doubling per-spindle output in comparison with the mule, the cap had to be removed for doffing the filled spools. Thus the major importance of the cap throstle principle was that it provided the basic impetus for the development of modern ring spinning (Anderson, 1976).

In 1828 John Thorp, one of the inventors of cap spinning, patented the basic elements of ring spinning. Thorp's mechanism featured a stationary ring fitted with a rotating wire band carrying a hook to guide the ballooning yarn onto the bobbin. The wire band rotated within the ring. The major advantage of the ring over the cap was that the bobbin was easily doffed. Indeed, automatic doffing became commonplace in later years.

The advantage of the ring frame over the mule was that the continuous spinning processes of the frames were about twice as productive as the mule. A disadvantage was that the yarn quality was generally inferior to that produced on the mule. More recent developments have erased much of the difference so that ring spinning is the method most widely used in America today. Moreover, ring spinning has largely replaced mule spinning in Britain, the place of the mule's birth (Anderson, 1976). Newer methods promise to replace the ring frame.

In 1807 a patent was granted to a British inventor, Samuel Williams, for a spinning system that foreshadowed modern open-end, or break, spinning (Anderson, 1976). Apparently Williams's proposal met with little reception, hence no success. By 1975 the concept of twisting without rotating the take-up package—bobbin or cop—was being pursued seriously in many textile laboratories. An open-end "rotor" technique has been the result.

In the open-end technique fibers are plucked from a continuous sliver by a wire-covered cylinder. The fibers are then lifted from the cylinder and transported inside a rotor by an air stream. Within the rotor the fibers are continuously reassembled in a groove in the rim of the rotor and twisted onto the end of a yarn skien exiting at the center of the rotor.

A much higher draft ratio (attenuation of roving) is possible with the open-end process than with the mule or with the ring frame. Hence finer yarns can be spun at a more rapid rate. Currently open-end yarns comprise about 30% of the world's supply of yarns from all methods of spinning. These yarns are spun mostly in Europe and Japan.

Other yarn-making processes are in the offing. The Repco self-twist process, developed by CSIRO of Australia, is useful for making two-fold yarns for certain purposes. Some processes, called zero-twist, require adhesives to hold yarn together during weaving. The adhesives are washed away after the cloth is woven. Open-end techniques using fluids as carriers instead of air have been developed. Some of these techniques are better adapted for use with synthetics, or cotton-synthetic blends, than with pure cotton.

1-3.2 Development of Fiber Evaluation

Over the millennia cotton farmers have made impressive strides in fitting the various taxa of cotton for production under domesticated conditions. Indeed, selection at the hands of people innocent of any kind of formal training transformed cottons from wild plants closely adapted to seasonally arid conditions to productive crop plants fitted for enriched soils and abundant soil moisture. Moreover, some of the best cotton fibers even known—those of the Sea Island growths—were selected largely by the senses of sight and touch. Nonetheless, processors of cotton over the centuries have suspected that measures of cotton quality needed to depend upon more than "feel" and "pull".

W. L. Balls (1928) recognized the problems involved in attempting to spin variable lots of cotton and developed the micro-spinning test for the purpose of evaluating small lots of fiber. Aside from its usefulness in mills, the micro-spinning test inaugurated the breeding of cottons, particularly Egyptian cotton, fitted for specific end uses.

Research and development in cotton fiber evaluation began in the United States about 50 years ago (Lewis and Richmond, 1968). Robert W. Webb organized a fiber and spinning laboratory in the USDA Bureau of Agricultural Economics. H. W. Barre, Head of the Cotton Division of the Bureau of Plant Industry at the time, saw the potential usefulness of fiber technology in cotton breeding and entered into close cooperation with Webb. To these workers must go the lion's share of the credit for the subsequent establishment of fiber technology laboratories by cotton merchandising firms, plant breeding establishments, and universities around the United States. Notable among the achievements in fiber technology in modern times are those of Hertel (summarized by Duckett, 1974). Among the instruments developed to modern design in Hertel's laboratory were the digital fibergraph, the Stelometer breaker, and the Micronaire

(Arealometer) instrument. In Britain the Shirley Institute, organized primarily to serve the needs of the British textile industry, has made great contributions to cotton textile technology.

Modern cotton merchants routinely sample and extract a full round of information on the fiber properties of each bale of cotton that enters the trade. Such meticulous operations indicate the strides that have been made in relating the properties of raw cotton to spinnability and yarn appearance.

Using fiber evaluation as a guide, plant breeders can now set goals for fiber quality in their cultivars with a reasonable expectation of achieving them. That is because basic genetical studies have shown that fiber properties are highly heritable and are thus readily selectable (Miller and Rawlings, 1967). Unfortunately, desirable fiber properties may be associated with low fiber yield. However, some of these correlations have been broken. Modern fiber technology made possible the breeding for substantial increase in fiber yield in Pima cottons while maintaining acceptable fiber quality (Feaster and Turcotte, 1968; Feaster et al., 1979), and the combining of acceptable lint yield with increased fiber strength in Upland cotton (Culp et al., 1979). Thus can cotton breeders, cotton merchants, and processors and spinners of cotton appreciate differences in the quality of cotton from the breeding nursery through to the finished textile product.

1–3.3 Modern Cotton Production

Cotton production recovered at a remarkable pace after 1865, but production techniques changed little over the following 60 years. Planting and cultivation continued to be performed with mule-drawn gear, and the hoe remained the nemesis of the weed. The use of fertilizers was sporadic, and the crop was picked by hand. Production in the United States hovered around 224 kg/ha, or slightly less, from around 1870 to about 1935 (Lewis and Richmond, 1968).

Beginning in the mid-1930's, yields began to rise. That was because of incentives offered by the newly enacted price support programs coupled with restrictions in acreage. Beyond that, the use of fertilizers increased. There were, in at least some regions, attempts at insect control, and production rose in the irrigated West where production practices could be controlled more precisely than in the rainfed areas to the eastward.

The period just prior to World War II saw the first tentative steps toward mechanization of the crop. The post-war era saw a move toward total mechanization, not only in the United States, but in many of the other cotton producing areas of the world. Multi-row planting and cultivating equipment became standard. Weeds are controlled today largely through the use of herbicides, and insect control through the use of various insecticides gives a degree of plant protection earlier cotton growers could not

Table 1-2. Average production costs and yields per hectare of cotton lint for various regions in the USA for 1978 (From McArthur and Cooke, 1980).

Region	Cost/kg	Lint
	$ USA	kg/ha
Southeast	1.73	520
Delta	1.65	522
Southern Plains	1.42	317
Far West	1.70	878

Table 1-3. Cotton production by continents for 1980 in bales of 218 kg (From USDA, 1981).

Asia, including U.S.S.R.	39 498 000
North America	13 842 000
South America	4 762 000
Europe	863 000
Australia	385 000

have dreamed of. By 1966 yields of lint in the United States had reached 561 kg/ha, a two-fold increase in yield in 30 years. Since that time the yield has plateaued. The reasons for the leveling off in yield has been a vexing problem and one that is now under attack by research workers. Some reasons frequently cited are that insect control has become more difficult, especially with the *Heliothis* (bollworm) complex which has adapted to resist pesticides, and that larger areas of traditionally low productivity per acre—the High Plains of Texas for example—have come increasingly into production. Perhaps another factor is that cotton farmers have become more conscious of production costs and now seek to optimize lint yield rather than maximize it. Table 1-2 compares cotton production in various cotton-growing areas of the United States. Cotton growing in some regions is, today, scarcely a profitable venture.

On a world basis, the future of cotton growing and utilization seems bright. The crop continues to increase in volume as the world's population grows. Table 1-3 gives world production by continents in 1980. The United States and the U.S.S.R lead in production on a per-country basis with an annual crop that fluctuates between 11 and 14 million bales. The PRC contributes 10 million bales, or slightly more, to the total, and has a growing interest in expanding and mechanizing the crop.

The future of cotton in the United States seems less bright than in the remainder of the world. Only about six million bales of the annual production disappear into the domestic textile manufactures. Another 6 to 7 million bales are exported. Thus the current health of the American cotton growing industry is dependent upon exports. Ironcially, the United States has been, and will likely remain, the chief source of germplasm for the improvement of the world's cotton crop. Thus the well-being of the world's cotton crop seems tied to the health of the American cotton growing and marketing industry.

REFERENCES

Anderson, Edgar. 1954. Plants, life, and man. Andrew Melrose, London. p. 120–133.

Anderson, G. C. 1976. The evolution of spinning. The Edward R. Schwartz Lecture. Conf. Amer. Soc. Mech. Eng.

Aspin, Christopher, and S. D. Chapman. 1964. James Hargreaves and the spinning jenny. The Guardian Press, Preston, Great Britain.

Balls, W. L. 1912. The cotton plant in Egypt. McMillan and Co., Ltd., London.

————. 1928. Studies of quality in cotton. McMillan and Co., Ltd., London.

Bryan, W. E. 1955. Prospects for Pima S-1 cotton. Prog. Agric. Ariz. 7:6.

Catling, Harold. 1970. The spinning mule. David L. Charles Ltd., Newton Abbott, Britain.

Chowdhury, K. A., and G. M. Burth. 1971. Cotton seeds from the Neolithic of Egyptian Nubia and the origin of Old World cotton. Linn. Soc. London. Biol. J. 3:303–312.

Culp, T. W., D. C. Harrell, and T. Kerr. 1979. Some genetic implications in the transfer of high fiber strength genes in Upland cotton. Crop Sci. 19:481–484.

Duckett, K. E. 1974. Cotton fiber instrumentation research in Tennessee. Univ. of Tennessee Agric. Exp. Stn. Bull. 536.

Feaster, C. V., and E. L. Turcotte. 1968. Association of color, yield, and quality of lint in Pima cotton. Crop Sci. 8:197–199.

————, J. E. Ross, E. L. Turcotte, and C. K. Bragg. 1979. Effects of fiber length uniformity on processing Pima cotton. USDA Prod. Res. Rep. No. 176.

Fryxell, P. A. 1979. The natural history of the cotton tribe. Texas A&M Univ. Press, College Station.

Gulatti, A. M., and A. J. Turner. 1928. A note on the early history of cotton. Indian Central Cotton Committee, Tech. Lab. Bull. No. 17.

Gerstel, D. U. 1953. Chromosomal translocations in interspecific hybrids of the genus Gossypium. Evolution 7:234–244.

Harlan, J. R. 1975. Crops and man. Crop Science Society of America, Madison, Wis.

Hutchinson, J. B., R. A. Silow, and S. G. Stephens. 1947. The evolution of *Gossypium* and the differentiation of the cultivated cottons. Oxford University Press, London.

Kohel, R. J., and T. R. Richmond. 1962. The genetics of flowering response in cotton. IV. Quantitative analysis of photoperiodism of Texas 86, *Gossypium hirsutum* race *latifolium*, in a cross with an inbred line of cultivated American Upland cotton. Genetics 47: 1535–1542.

————, ————, and C. F. Lewis. 1974. Genetics of flowering response in cotton. VI. Flowering behavior of *Gossypium hirsutum* L. and *Gossypium barbadense* L. hybrids. Crop Sci. 14: 696–699.

Lewis, C. F., and T. R. Richmond. 1960. The genetics of flowering response in cotton. II. Inheritance of flowering response in a *Gossypium barbadense* cross. Genetics. 45:79–85.

————, and ————. 1968. Cotton as a crop. p. 1–21. *In* F. C. Elliot, M. Hoover, and W. K. Porter, Jr. (eds.) Advances in the production and utilization of quality cotton. Iowa State University Press, Ames.

Mauney, J. R., and L. L. Phillips. 1963. Influence of daylength and night temperature on flowering in *Gossypium.* Bot. Gaz. 124:278–283.

McArthur, W. C., and F. T. Cooke, Jr. 1980. Cotton production practices and costs in the United States. Univ. Georgia Res. Rep. No. 365.

Miller, P. A., and J. O. Rawlings. 1967. Selection for increased lint yield and correlated responses in Upland cotton, *Gossypium hirsutum* L. Crop Sci. 7:637–640.

Parlatore, F. 1866. Le specie dei cotoni. Stamperia Reale, Firenze.

Scherer, J. A. B. 1916. Cotton as a world power. F. A. Stokes Co., New York.

Smith, C. E., Jr., and S. G. Stephens. 1971. Critical identification of Mexican archaeological cotton remains. Econ. Bot. 25:160–168.

Stephens, S. G. 1967. Evolution under domestication of the New World cottons (*Gossypium* spp.). Cienc. Cult. (Sao Paulo) 19:118–134.

----. 1974a. Geographical distribution of cultivated cottons relative to probable centers of domestication in the New World. p. 239–254. *In* A. M. Srb (ed.) Genes, enzymes, and population. Plenum Publishing Corp., New York.

----. 1974b. The use of two polymorphic systems, nectary fringe hairs and corky alleles, as indicators of phylogenetic relationships in New World cottons. Biotropica 6:194–201.

----. 1975. Some observations on photoperiodism and the development of annual forms of domesticated cottons. Econ. Bot. 30:409–418.

----. 1976. The origin of Sea Island cotton. Agric. Hist. 50:391–399.

----, and M. E. Moseley. 1974. Early domesticated cottons from achaeological sites in central coastal Peru. Am. Antiquity 39:109–122.

U.S. Dep. of Agriculture. 1981. World cotton situation. USDA Rep. FAS, FC-7-81.

2 Taxonomy and Germplasm Resources

Paul A. Fryxell

ARS-USDA and Texas A&M University
College Station, Texas

This chapter deals with the biological diversity that exists in the genus *Gossypium,* the genus that includes the cultivated cottons and numerous wild relatives. This diversity is examined first from a taxonomic viewpoint, considering the range of variation in the genus and the relationships among the species, and providing botanical descriptions of all of the species. Second, the diversity is considered from the viewpoint of germplasm resources, their accumulation, management, and utilization.

2-1 TAXONOMY OF *GOSSYPIUM*

2-1.1 History

The genus *Gossypium* is moderately large (39 species) and very diverse. Coupled with the great economic importance of the cottons, this diversity has ensured that the genus should receive considerable scientific study. Taxonomy is basic to the scientific study of so diverse a group of plants. As Edgar Anderson (1969, p. 69–70) wrote: "The cotton breeders of the world, for practical reasons if for no other, have had to take an intense interest in the classification of all cottons, whether they seemed to be directly concerned with cultivated varieties or not. As a result, the classification of wild and cultivated cotton is more truly critical and on a more global basis than for any other world crop." *Gossypium* has been the subject of taxonomic study since the mid-18th century, when Linnaeus described the genus, and it has been the subject of relatively intensive study since the mid-19th century, continuing to the present day as new species are discovered and new techniques provide additional data for evaluating relationships.

Contribution from ARS-UDSA. Published in *Cotton,* Agronomy Monograph 24, © ASA-CSSA-SSSA, 677 South Segoe Road, Madison, WI 53711, USA.

Serious taxonomic study of *Gossypium* began with the work of two 19th century Italian botanists, Parlatore and Todaro. Parlatore's monograph, *Le Specie dei Cotoni* (1866), was relatively brief and focused attention almost solely on the species that bear fibers, i.e. the cottons of agricultural or potential agricultural value. His study was primarily historical, and he surveyed the literature of the time thoroughly. Todaro's work was based on his study of living plants that he had assembled from all parts of the world growing in the Royal Botanic Garden at Palermo. He published a series of publications on *Gossypium* that culminated in his principal work, *Relazione sulla cultura dei Cotoni in Italia sequita da una Monografia del genere Gossypium* (1877). The first part of this work dealt with cotton cultivation in Italy, and the second part was a complete taxonomic monograph of the genus *Gossypium*. It was the first such monograph to be published and still stands as a significant reference.

In 1907 the English botanist, Sir George Watt, published *The Wild and Cultivated Cotton Plants of the World*. This monumental work was based on his familiarity with cotton as an agricultural crop during his years of residence in India, followed by subsequent herbarium and library research after his return to England. It is a remarkable storehouse of information on the history, taxonomy, and agricultural use of cotton. Although Watt's taxonomic conceptions are now outdated and have been superseded by new understandings based on new knowledge about these plants, his book continues to be an important source because of his thoroughgoing scholarship.

The study of *Gossypium* taxonomy may be divided into an early and a modern period. The "watershed" dividing these periods and ushering in our modern conceptions of *Gossypium* taxonomy was a 1928 paper entitled, "A Contribution to the Classification of the Genus *Gossypium* L.," by the Soviet agricultural botanist G. S. Zaitzev. This seminal paper laid the groundwork for our current understanding of the genus, and was based on insights derived from studies of cotton cytology. Confining his attention to the fiber-bearing cottons, Zaitzev noted that the cottons of the Old World are diploid ($2n = 26$) and those of the New World are tetraploid ($2n = 52$). He further perceived that each of these could be subdivided into two groups, the Old World cottons into an African and an Indo-Chinese group, and the New World cottons into a Central American and a South American group. Thus, he took the manifold variation of the cultivated cottons and reduced it to a view recognizing four major groups, each with morphological and geographic integrity, which subsequent studies have shown to be interpretable as four species, with minor modifications based on subsequent knowledge. It is easy to regard Zaitzev's simplifying insight as obvious (most strokes of genius are obvious by hindsight) and therefore not of major significance, but I think that view widely misses the mark. Zaitzev's predecessors "could not see the forest for the trees," whereas he was able to grasp the pattern underlying the variability; he paved the way for subsequent work and enabled us to advance in our taxonomic understanding of *Gossypium*.

Zaitzev's insight was built upon by Sir Joseph Hutchinson, a British agricultural botanist, who presented a taxonomic treatment of *Gossypium* (Hutchinson et al., 1947) that has been widely accepted, and by F. M. Mauer, a student of Zaitzev, whose 1954 monograph in Russian differed only slightly from Hutchinson's treatment. My own taxonomic interpretation of *Gossypium* incorporates information on newly described species (since the works of Hutchinson and Mauer) and newly understood relationships (Fryxell, 1979b).

2-1.2 Aims of Systematics

Systematic biology or taxonomy is concerned in a broad sense with biological existence on the earth and the past and continuing evolution of the earth's inhabitants. More specifically, taxonomic research seeks to discover diversity, to find patterns within this diversity, and to study the processes that have produced these patterns. The study brings together all available kinds of data concerning the organisms so that degrees of relationship can be evaluated in the least biased manner. Since new kinds of data are continually becoming available, the methodology of systematics is aptly described as "the unending synthesis" (Constance, 1964). The purposes of a good classification are to permit making generalizations and to facilitate the storage and retrieval of information about the organisms.

Thus, in *Gossypium* taxonomy, when cytological techniques first provided data on chromosome numbers and chromosome pairing behavior, these data advanced our understanding of relationships within the genus. More recently this understanding has been further refined, for example, by the development of phytochemical techniques for studying flavonoids, terpenes, and seed proteins. In addition, field studies have revealed phytogeographical and ecological information that can be added to the synthesis, and new species have been discovered that amplify our understanding of the patterns of variability within the genus. Thus, the "unending synthesis" of taxonomic research continues to broaden and deepen our knowledge of *Gossypium*.

2-1.3 Geography of the Genus

The various species of *Gossypium* occur in widely separated parts of the world. Typically they occur in relatively arid parts of the tropics and subtropics. The wild diploid species may be subdivided into three geographical groups, as follows (Fryxell, 1979b).

a. The Australian group includes 11 species, some from the dry central deserts, and some from the northwestern coasts with seasonally dry monsoon climates.

b. The American group includes 12 species, 10 of which occur in western Mexico (one of which extends into Arizona), the other two from Peru and the Galapagos Archipelago. Some of these species occur in extremely arid habitats (e.g. Baja California), others in areas with alternating rainy and dry seasons.

c. The Afro-Arabian groups includes eight species that are widely scattered around Africa and the Arabian peninsula. One species occurs in the Cape Verde Islands, and another extends from East Africa (including part of the Arabian peninsula) to Pakistan.

The geography of the cultivated species of *Gossypium* is considerably obscured by the spread of these species in domestication and by their introduction into parts of the world distant from their original centers. The details of these distributions and migrations are too complex to analyze here, beyond noting that the Old World cultivated diploids, *G. arboreum* and *G. herbaceum,* have their centers principally in India and southeast Asia (*G. arboreum*) and in the Levant countries and East Africa (*G. herbaceum*). The tetraploids of the New World include both wild and cultivated species, as follows:

G. tomentosum—endemic to Hawaii,

G. mustelinum—northeastern Brazil,

G. darwinii—the Galapagos Archipelago,

G. lanceolatum—western Mexico, in houseyard cultivation,

G. barbadense—with a South American center, but now occurring elsewhere in cultivation, and

G. hirsutum—from Central America, but cultivated in virtually all cotton-growing areas of the world.

2–1.4 Subdivisions of the Genus

To best achieve the aims of systematics, species relationships may be indicated by subdividing a group in a hierarchical way. Thus a genus can be subdivided into subgenera, and these in turn into sections and subsections, if such a treatment proves to serve a better understanding of the relationships among the species. Such is the case in *Gossypium*, where the subgenera generally conform to the geographic groups previously noted, with three subgenera being basically Australian, American, and African in their geography, with a fourth subgenus comprising the tetraploid species. This is not to say that the subgenera were defined in geographical terms. They were defined and delimited on the basis of data from comparative morphology, cytology, genetics, and other fields, and their phytogeography was found generally to support the resulting classification. This classification can be presented in summary form as follows:

Subgenus *Sturtia* (R. Brown) Todaro

Section *Sturtia* (*G. sturtianum, G. sturtianum* var. *nandewarense, G. robinsonii*)

Section *Grandicalyx* Fryxell (*G. costulatum, G. pulchellum, G. populifolium, G. pilosum, G. cunninghamii*)

Section *Hibiscoidea* Todaro (*G. australe, G. nelsonii, G. bickii, G. triphyllum*)
Subgenus *Houzingenia* Fryxell
Section *Houzingenia*
Subsection *Houzingenia* (*G. trilobum, G. thurberi*)
Subsection *Integrifolia* (Todaro) Todaro (*G. klotzschianum, G. davidsonii*)
Subsection *Caducibracteolata* Mauer (*G. harknessii, G. armourianum, G. turneri*)
Section *Erioxlum* (Rose & Standley) Prokhanov
Subsection *Erioxylum* (Rose & Standley) Fryxell (*G. aridum, G. lobatum, G. laxum*)
Subsection *Selera* (Ulbrich) Fryxell (*G. gossypioides*)
Subsection *Austroamericana* Fryxell (*G. raimondii*)
Subgenus *Gossypium*
Section *Gossypium*
Subsection *Gossypium* (*G. arboreum, G. herbaceum*)
Subsection *Anomala* Todaro (*G. anomalum, G. capitis-viridis*)
Section *Pseudopambak* Prokhanov
Subsection *Pseudopambak* (*G. stocksii, G. somalense, G. incanum, G. areysianum*)
Subsection *Longiloba* Fryxell (*G. longicalyx*)
Subgenus *Karpas* Rafinesque (*G. tomentosum, G. lanceolatum, G. mustelinum, G. darwinii, G. barbadense, G. hirsutum*)

2–1.5   An Enumeration of the Species

The following enumeration lists the species alphabetically, giving their correct names, botanical descriptions, and brief statements concerning their geographical distribution. Relationships among the species may be evaluated by reference to Section 2–1.4. Synonyms for each species, a key for the identification of species, and lists of illustrations may be found in Fryxell (1979b). Nomenclatural details and information about type specimens may be found in Fryxell (1976b).

Gossypium anomalum Wawra & Peyritsch

Shrub 1 to 2 m tall, the *stems* stellate-pubescent, gland-dotted; *leaves* petiolate, pedately seven-nerved, deeply five-lobed, the lobes ovate, basally constricted (the sinuses closed), entire, acuminate, stellate-pubescent above and beneath, with one to three inconspicuous *foliar nectaries* beneath on principal veins; *stipules* subulate, 4 to 5 mm long; flowers and fruits solitary in the leaf axils or borne on sympodial inflorescences, the *pedicel* 5 to 15 mm long, with pubescence like that of stem, surmounted by three *involucellar nectaries; bracts of the involucel* three, inserted above the nectaries, ligulate, more or less reflexed, 11 to 22 mm long, 2 to 5 mm wide, entire or apically trifid; *calyx* 7 to 10 mm long, five-lobed, the lobes acuminate, stellate-pubescent, gland-dotted; *petals* 25 to 40 mm long, pale yellow with a dark-red spot at base, sparsely and minutely gland-dotted; *staminal column* pallid, glabrous, non-glandular, ca. 10 mm long, the filaments 1 mm long; *style* exceeding the androecium, sparsely gland-dotted, the stigmatic lobes decurrent; *capsules* ovoid-beaked, 15 to 20 mm long, three-celled, glabrous, prominently gland-dotted, the glands few, raised, up to 1 mm diam; *seeds* ca. 6 mm long, densely pubescent, the fibers brownish, appressed. From Africa: disjunct, in sub-Saharan Africa and in southwestern Africa.

Gossypium arboreum Linnaeus

Shrubs or subshrubs up to 2 m tall, highly variable in morphology and pubescence, the *stems* usually puberulent to pubescent, gland-dotted; *leaves* petiolate, pedately seven to nine-nerved, moderately to deeply palmately three to seven-lobed, sometimes with small accessory lobes in the sinuses, the lobes broadly ovate to narrowly lanceolate and more or less constricted basally, entire, acute or acuminate, the *foliar nectary* inconspicuous or absent; *stipules* 5 to 15 mm long, narrowly or broadly falcate, caducous; flowers and fruits borne on sympodial inflorescences, the *pedicels* 10 to 60 mm long, the *involucellar nectaries* commonly absent; *bracts of the involucel* three, broadly ovate, more or less connate basally forming a cup, pubescent externally, glabrous internally, entire or three to seven-laciniate, 20 to 35 mm long, almost as broad; a trio of *nectaries* sometimes present within the involucel at the juncture of involucel and calyx, alternate with the bracts; *calyx* 7 to 12 mm long, glabrous, prominently gland-dotted, truncate to distinctly 5-lobed; *petals* 25 to 45 mm long, usually yellow with a purplish basal spot, sometimes white or flushed purplish or with basal spot absent, minutely gland-dotted, ciliate on claw, pubescent externally where exposed in bud, glabrous internally; *staminal column* glabrous, pallid, usually gland-dotted, half length of petals or less, the filaments 2 to 4 mm long, *style* only slightly exceeding androecium; *capsules* usually three-celled, ovoid to elongate, usually beaked, glabrous, gland-dotted, the glands sunken, flaring widely at maturity; *seeds* bearing white or tan fibers.

Gossypium areysianum (Deflers) Deflers

Shrub 0.5 to 1.5 m tall, the *stems* minutely puberulent and obscurely gland-dotted; *leaves* petiolate, cordate, broadly rotund-ovate and weakly trilobed, entire, rounded-acute or acuminate, usually broader than long, pedately seven-nerved, sparsely pubescent to glabrate, with an obscure *foliar nectary* near base of midrib beneath; *stipules* linear-falcate, 5 to 15 mm long, caducous; flowers and fruits borne on short (usually one- or two-flowered) sympodial inflorescences, the *pedicels* slender, 5 to 10 mm long, surmounted by three *involucellar nectaries; bracts of the involucel* three, inserted above the nectaries, ligulate, 10 to 15 mm long, 1 to 2 mm wide, entire or irregularly ligulate, 10 to 15 mm long, 1 to 2 mm wide, entire or irregularly trifid apically, gland-dotted; *calyx* 6 to 8 mm long, gland-dotted, shallowly five-lobed, the sinuses rounded; *petals* 20 to 25 mm long; yellow with a dark-red spot at base, nonglandular, pubescent externally especially where exposed in bud; *staminal column* 10 to 15 mm long, pallid, glabrous, the filaments subsessile or up to 0.5 mm long, the anthers reddish; *style* exceeding the androecium, gland-dotted, the stigmatic lobes decurrent; *capsules* ovoid, three-celled; *seeds* 6 to 7 mm long, densely pubescent, the fibers brownish-gray, appressed. From Arabia.

G. aridum (Rose & Standley) Skovsted

Trees 4 to 10 m tall (occasionally shrubs), black-glanded more or less throughout, the *stems* minutely puberulent, eventually glabrate; *leaves* petiolate, broadly to narrowly ovate (rarely obscurely trilobulate), truncate or weakly cordate, entire, acuminate, more or less minutely stellate-pubescent; *flowers* solitary or paired in the leaf axils, the pedicels very short, yellowish-puberulent, surmounted by three prominent bordered *nectaries*; *involucel* of three bractlets, each inserted above a nectary, 2 to 4 mm long, triangular, appressed; *calyx* 6 to 9 mm long, yellowish-farinose, 5-dentate; *corolla* funnel-form, 35 to 50 mm long, rose or lavender with a dark red or purplish throat; *androecium* included, the column pallid, the filaments and anthers purplish, the pollen yellow-orange; *style* single, exceeding the androecium, gland-dotted, the stigmatic lobes decurrent; *capsule* 20 to 25 mm long, 10 to 15 mm diameter, glabrous, very prominently black-gland-dotted, three- or four-celled, fully dehiscent but not flaring widely; *seeds* numerous, 4 to 6 mm long, with dense brown fibers tightly appressed to the seed. From the Pacific coast of Mexico, from Sinaloa to the Isthmus of Tehuantepec and also in Puebla and in lowland Veracruz, the only diploid cotton to occur in Mexico on the Atlantic watershed.

Gossypium armourianum Kearney

Compact branched shrub ca. 1 m tall, the *stems* (and most plant parts) obscurely puberulent to glabrate, gland-dotted; *leaves* petiolate, thick-textured, the glands on the petioles raised, small (15 to 30 mm long, as broad or broader), ovate, more or less cordate, unlobed, entire, acute or somewhat acuminate, with a *foliar nectary* near base of midrib; *stipules* 1.5 to 2.5 mm long, subulate, caducous; *peduncles* solitary in the axils, up to 65 mm long (exceeding the subtending leaf)/articulate near the middle, with a foliar bract at articulation, prominently gland-dotted, the glands raised, surmounted by three prominent *involucellar nectaries*; *involucellar bracts* ligulate (ca. 8 mm long, 2 mm wide), prominently gland-dotted, caducous well before anthesis leaving scars; *calyx* 5 to 7 mm long, subtruncate or five-toothed, evenly gland-dotted; *petals* 25 to 45 mm long, minutely gland-dotted, yellow with red spot at base, externally yellow with reddish margins where exposed in bud, ciliate on margin of claw, otherwise glabrous; *staminal column* glabrous, pallid, sparsely gland-dotted, the filaments numerous, 2 to 4 mm long, densely inserted in upper half of column; *style* single, somewhat shorter than petals, exceeding androecium, densely gland-dotted, the stigmatic lobes decurrent; *capsules* ovoid, apiculate, three- or four-celled, prominently glandular, the glands sunken, glabrous except the inner suture margin with long (1 mm or more) whitish hairs; *seeds* one to three per locule, ca. 8 mm long, brownish with tightly appressed fibers. From the Gulf of California on San Marcos Island, and from Baja California near San Francisquito Bay.

Gossypium australe F. von Mueller

Shrubs 1 to 2 m tall, the *stems* softly stellate-tomentose, the hairs mostly concealing the underlying black glands; *leaves* petiolate, narrowly to broadly ovate-elliptic, entire, acute or mucronate, densely stellate-tomentose above and beneath, with a prominent elongated often reddish *foliar nectary* 3 to 10 mm long beneath; *stipules* subulate, 5 to 9 mm long, caducous; flowers and fruits borne on a sympodial inflorescence, the flowers sometimes cleistogamic, the *pedicels* 5 to 15 mm long, stellate-tomentose, surmounted by three prominent reddish *involucellar nectaries* 1 to 3 mm long; *bracts of the involucel* three, inserted above the nectaries, linear, 8 to 15 mm long, equaling to shorter than the calyx; *calyx* campanulate, basally constricted, 9 to 15 mm long, deeply five-lobed, the lobes acuminate, the sinuses rounded, gland-dotted, the glands few, often distally placed, sometimes lacking; petals 35 to 40 mm long, minutely gland-dotted, pink with dark-red spot at base, ciliate on claw, pubescent externally where exposed in bud, otherwise glabrous; *staminal column* pallid, glabrous, 10-nerved, almost without glands, the filaments pallid, 2 mm long, the anthers pinkish; *style* exceeding androecium, densely gland-dotted, the stigmatic lobes decurrent; *capsules* three- or four-celled, 15 to 20 mm long, gland-dotted but the glands concealed by short stellate tomentum, ovoid to obovoid, more or less apiculate, the inner suture pubescent, the hairs 2 to 4 mm long; seeds 4 to 5 mm long, densely pubescent, the fibers brownish, straight and spreading. From northern and central Australia.

Gossypium barbadense Linnaeus

Shrubs 1 to 3 m tall, sometimes arborescent, the *stems* sparsely stellate-pubescent to glabrate, prominently gland-dotted; *leaves* petiolate, cordate, three to seven-lobed, palmately seven to nine-nerved, glabrate, the lobes ovate, entire, acuminate, with one to five *foliar nectaries* beneath; *stipules* subulate to falcate, 10 to 50 mm long, often prominent; flowers solitary or in sympodial inflorescences, the *pedicels* 10 to 40 mm long, gland-dotted, usually glabrate, surmounted by three *involucellar nectaries*; *bracts of the involucel* three, inserted above the nectaries, ovate, up to 60 mm long, 45 mm broad, seven to 19-laciniate; *calyx* 6 to 10 mm long, undulate or truncate, prominently gland-dotted, ciliate on margin, otherwise glabrous, a trio of *nectaries* often present at juncture of calyx and involucel, alternate with bracts; *petals* up to 80 mm long, usually yellow with dark-red spot at base, minutely gland-dotted; *staminal* column ca. 25 mm long, pallid, glabrous, gland-dotted, the filaments 2 to 4 mm long; *styles* exceeding the androecium, gland-dotted; *capsules* three-celled, glabrous, prominently pitted, usually narrowly elongate (35 to 60 mm long) and beaked; *seeds* several per cell, free or fused together, lanate, the fibers usually white. From South America and parts of Central America and the Antilles, now cosmopolitan in cultivation.

Gossypium bickii Prokhanov

Spreading shrub ca. 0.5 m tall, the *stems* gland-dotted, stellate-pubescent; *leaves* petiolate, ovate-elliptic, sometimes deeply three (rarely five)-lobed, truncate, entire, acute or mucronate, with scattered stellate pubescence above and beneath, with inconspicuous *foliar nectary* 0.5 to 2.0 mm long on midrib beneath; *stipules* subulate, 8 to 12 mm long, caducous; flowers and fruits borne on sympodial inflorescences, the flowers sometimes cleistogamic, the *pedicels* 10 to 30 mm long, stellate-pubescent, sparsely gland-dotted, surmounted by three prominent reddish *involucellar nectaries* 1 to 3 mm long; *bracts of the involucel* three, inserted above the nectaries, linear, 7 to 13 mm long, usually shorter than the calyx; *calyx* broadly campanulate, basally rounded, 18 to 24 mm long, deeply five-lobed, the lobes acuminate-caudate, the sinuses acute to rounded, prominently gland-dotted, the glands evenly distributed; *petals* 30 to 40 mm long, pink (sometimes white) with dark-red spot at base, minutely gland-dotted, ciliate on claw, pubescent externally where exposed in bud, otherwise glabrous; *staminal column* pallid, glabrous, non-glandular, the filaments pallid, 1 to 2 mm long, the anthers pallid; *style* exceeding the androecium, gland-dotted, the stigmatic lobes decurrent; *capsules* three to five-celled, 15 to 20 mm long (or smaller if resulting from cleistogamic flowers), glabrous and prominently glanded externally, the glands raised and sometimes coalesced, pubescent on inner suture margin, the hairs 1 to 2 mm long, subglobose (if small) to ovoid and long-beaked; *seeds* ca. 5 mm long, densely pubescent with tightly appressed fibers. From central Australia.

Gossypium capitis-viridis Mauer

Shrubs 1.0 to 2.5 m tall, the *stems* stellate-pubescent (densely canescent when young), gland-dotted; *leaves* petiolate, pedately seven- to nine-nerved, deeply five to seven-lobed, the lobes ovate, basally constricted (the sinuses closed), the central lobe itself sometimes irregularly lobed, entire, acuminate, stellate-pubescent above and beneath, with one to three inconspicuous *foliar nectaries* beneath on principal veins; *stipules* subulate, 4 to 7 mm long; flowers and fruits solitary in the axils or borne on sympodial inflorescences, the *pedicels* 5 to 12 mm long, with pubescence like that of stem, surmounted by three *involucellar nectaries*; *bracts of the involucel* three, inserted above the nectaries, deeply three-parted, 15 to 18 mm long; *calyx* 7 to 10 mm long, five-lobed, the lobes acute to acuminate, stellate-pubescent, gland-dotted, the calyx abscises circumscissely at base and is more or less deciduous in fruit; *petals* 25 to 30 mm long, yellow with a dark-red spot at base, sparsely and minutely gland-dotted; *staminal column* pallid, glabrous, non-glandular, ca. 10 mm long, the filaments 1 mm long; *style* exceeding the androecium, sparsely gland-dotted, the stigmatic lobes decurrent; *capsules* ovoid, beaked, 15 to 20 mm long, three to five-celled, glabrous, prominently gland-dotted, the glands few, raised, ca. 0.5 mm diam; *seeds* ca. 6 mm long, densely pubescent, the fibers brownish, appressed. From the Cape Verde Islands.

Gossypium barbasanum was also described from the Cape Verde Islands and was distinguished from *G. capitis-viridis* primarily on the basis of having no petal spot and five-locular capsules. However, a photograph of the type specimen (*Stewart 63,* LE) of the latter species shows a single fruit with four locules and no flowers. Moreover, a photograph of a specimen from the type locality of *G. barbosanum* (*Barbosa 6964*) shows a plant with three-, four- and five-loculed fruits. Consequently, it is concluded that only a single species is present in the Cape Verde Islands and that its correct (older) name is *G. capitis-viridis.*

Gossypium costulatum Todaro

Arching to decumbent shrub, the *stems* stellate-tomentose, gland-dotted; *leaves* petiolate, ovate, truncate, entire, rounded-acute and apiculate, palmately five-nerved, subglabrous, with a small obscure *foliar nectary* at base of blade beneath; *pedicels* 40 to 50 mm long, fluted, articulated below the middle, bracteate at articulation; *bracts of the involucel* three, subulate, shorter than the calyx, erect or somewhat reflexed; *calyx* ca. 25 mm long, tomentose, costulate, deeply five-lobed, the lobes triangular; *petals* 40 to 50 mm long, white with a dark-red spot at base internally, pink externally on margin where exposed in bud, tomentose externally; *capsules* glabrous, three-celled, subglobose, obtuse; *seeds* subglabrous, three per locule, arillate, 4 to 6 mm long, black with raised venation. From northwestern Australia.

Gossypium cunninghamii Todaro

Shrub ca. 0.5 m tall, the *stems* glabrate, gland-dotted; *leaves* short-petiolate (the petiole 2 to 4 mm), elliptic, up to 90 mm long, 50 mm broad, basally cuneate, entire, rounded-acute and minutely apiculate, penninerved, glabrous and gland-dotted above and beneath, with a *foliar nectary* at the base of the blade; *stipules* 6 to 7 mm long, narrowly lanceolate to subulate, caducous; *peduncles* solitary in the leaf axils, prominently gland-dotted, usually 15 to 30 mm long, articulated 10 to 13 mm below the flower; *bracts of the involucel* three (rarely four) whorled at base of calyx or irregularly inserted on pedicel, ligulate, 8 to 17 mm long, 1.5 to 2.5 mm broad, sometimes more or less reflexed; calyx 20 to 25 mm long, campanulate, deeply five-lobed (ca. 3/4-lobed), gland-dotted, glabrous, the lobes triangular, acute; *petals* 45 to 50 mm long, apparently pinkish with dark-red spot at base, nearly without glands, externally pubescent where exposed in bud, ciliate on claw; *capsules* ovoid, three-celled, glabrous, gland-dotted, 20 mm long, 15 mm diam; *seeds* black, minutely pubescent, 8 mm long. From the Cobourg Peninsula, northernmost Australia.

Gossypium darwinii Watt

Shrubs up to 2 m tall, the *stems* slender, slightly angular, gland-dotted, stellate-pubescent to glabrate; *leaves* petiolate, cordate, three to five-lobed (rarely simple), palmately seven-nerved, gland-dotted, the lobes ovate, entire, acuminate, sparsely pubescent with an inconspicuous *foliar nectary* on midrib beneath; *stipules* subulate, 6 to 10 mm long, caducous; flowers and fruits borne on sympodial inflorescences, the *pedicel* 20 to 50 mm long, slender, gland-dotted, somewhat pubescent to glabrate; *bracts of the involucel* three, ovate, seven to 15-laciniate, 25 to 30 mm long, somewhat longer than wide; a trio of *nectaries* often present within the involucel at the juncture of involucel and calyx, alternate with the bracts; *calyx* ca. 5 mm long, undulate or subtruncate, the margin ciliate, otherwise glabrous, prominently gland-dotted; *petals* 40 to 50 mm long, yellow with a small maroon spot at base, ciliate on claw and pubescent externally where exposed in bud, glabrous internally, minutely gland-dotted; *staminal column* 20 mm long, glabrous, pallid, gland-dotted, the filaments 1.5 to 2.5 mm long; *style* exceeding the androecium, somewhat shorter than the petals, sparsely gland-dotted, the stigmatic lobes decurrent; *capsules* ovoid, beaked, less than 30 mm long, gland-dotted, the glands sunken; *seeds* pubescent, the fibers brownish, sparse, but the pubescence pattern highly variable. From the Galapagos Archipelago.

Gossypium davidsonii Kellogg

Shrub 1 to 2 m tall, sparingly branched; *stems* minutely soft-puberulent at least when young, gland-dotted; *leaves* petiolate, ovate, cordate, simple or sometimes three-lobed, entire, acute or acuminate, pubescent above and beneath, with a small *foliar nectary* beneath; flowers usually borne on sympodial inflorescences, the *pedicels* with pubescence like that of stem, surmounted by three involucellar nectaries; *bracts of the involucel* pubescent, cordate-ovate, 7- to 11-laciniate, 15 to 25 mm long, about as broad; *calyx* 4 mm long, truncate or undulate, gland-dotted and sparsely pubescent; *petals* 25 to 30 mm long, yellow with small red spot at base (or spot sometimes suppressed), minutely gland-dotted, ciliate on margin, minutely pubescent externally, glabrous internally; *staminal column* pallid, glabrous or sparsely pubescent near base, 7 to 11 mm long, the filaments 1 to 2 mm long; *style* exceeding androecium, the stigmatic lobes decurrent; *capsule* commonly four-celled, ovoid, apiculate, ca. 15 mm long, gland-dotted, glabrous except ciliate on inner suture margin, the hairs 1 to 2 mm long; *seeds* 6 mm long, with sparse appressed fibers. From southern Baja California, around the Cape region, and from near Guaymas, Sonora.

Gossypium gossypioides (Ulbrich) Standley

Large shrubs or small trees 3 to 6 m tall; *stems* minutely puberulent at least when young; leaves petiolate, prominently cordate, pedately seven to nine-nerved, more or less deeply three-lobed, the lobes entire, acuminate, subglabrous except for nerves, gland-dotted; flowers generally borne on sympodial inflorescences, the *pedicels* 20 to 30 mm long, minutely puberulent and obscurely gland-dotted; *involucellar nectaries* lacking; *bracts of the involucel* three, 25 to 45 mm long, almost as broad, cordate-ovate, entire, acuminate, many-veined, gland-dotted, enclosing bud and fruit; *calyx* 4 to 5 mm long, glabrous, prominently gland-dotted, truncate or undulate; *petals* 45 to 55 mm long, rose-lavender with dark red-purple spot on lower half within, minutely gland-dotted, ciliate on margin especially at base; *staminal column* glabrous, prominently gland-dotted, pallid, the filaments often purplish, ca. 2 mm long, the yellowish anthers forming a columnar mass; *style* exceeding the androecium, prominently glandular, the stigmatic lobes decurrent; *capsule* three-celled, ca. 15 mm long, glabrous, prominently gland-dotted (the glands sometimes as much as 1 mm diam), flaring widely at dehiscence; *seeds* 7 mm long with grayish appressed fibers. From central Oaxaca, Mexico.

Gossypium harknessii Brandegee

Spreading shrubs up to 2.5 m tall and 3 to 4 m broad, the *stems* minutely puberulent becoming glabrate, obscurely gland-dotted; *leaves* thick-textured, petiolate, 20 to 60 mm long, somewhat broader than long, deeply cordate, more or less three-lobed, entire, the lobes obtuse to acute, gland-dotted, with a small *foliar nectary* near base of midrib beneath; *stipules* subulate, 2 to 3 mm long, caducous; *peduncles* solitary in the leaf axils, equaling or shorter than subtending petiole (up to

26 mm long), articulated near the middle, obscurely gland-dotted; *involucellar bracts* ovate, 9 to 18 mm long, 4 to 10 mm broad, entire, gland-dotted, caducous at anthesis or shortly after, leaving scars; *calyx* 6 to 8 mm long, truncate or obscurely five-toothed, evenly gland-dotted; *petals* 25 to 40 mm long, minutely gland-dotted, yellow with red spot at base, ciliate on margin; *staminal column* glabrous, pallid, sparsely gland-dotted, the filaments 2 to 3 mm long, densely inserted in upper two-thirds of column; *style* single, somewhat shorter than the petals, exceeding the androecium, densely gland-dotted, the stigmatic lobes decurrent; *capsule* three to four-celled, ovoid, apiculate, prominently glandular, the glands sunken, glabrous except the inner suture margin with long (1 to 2 mm) whitish hairs; *seeds* 8 to 10 mm long, grayish with tightly appressed fibers. From Baja California in the Cape region and on the adjacent islands of Isla Coronado and Isla Carmen.

Gossypium herbaceum Linnaeus

Shrubs and subshrubs up to 1.5 m tall, the *stems* usually hirsute, more or less gland-dotted; *leaves* petiolate, pedately seven to nine-nerved, moderately palmately three to five-lobed, the lobes broadly ovate, usually somewhat constricted basally, entire, acuminate, the *foliar nectary* inconspicuous; *stipules* subulate, 8 to 13 mm long, caducous; flowers and fruits borne on sympodial inflorescence, the *pedicels* 10 to 30 mm long, the *involucellar nectaries* commonly absent; *bracts of the involucel* three, broadly ovate, cordate, frequently flaring, five to 13-dentate, pubescent externally, glabrous internally except on teeth, 20 to 35 mm long, almost as broad, a trio of *nectaries* sometimes present within the involucel at the juncture of involucel and calyx, alternate with the bracts; *calyx* 5 to 7 mm long, glabrous, prominently gland-dotted, truncate to undulately five-lobed; *petals* 30 to 40 mm long, yellow with a purplish basal spot, minutely gland-dotted to virtually without glands, ciliate on claw, pubescent externally where exposed in bud, glabrous internally; *staminal column* glabrous, pallid, densely gland-dotted to non-glandular, half length of petals or less, the filaments 2 to 3 mm long; *style* barely exceeding androecium; *capsules* three or four-celled, more or less globose, glabrous, obscurely gland-dotted, not opening widely at maturity; *seeds* bearing usually white fibers.

Gossypium hirsutum Linnaeus

Shrubs 1 to 2 m (or more) tall, usually widely branching, more or less stellate-pubescent, gland-dotted throughout; *leaves* long-petiolate, cordate, weakly three to five-lobed, the lobes broadly triangular to ovate, acute to acuminate; *stipules* subulate, 5 to 15 (rarely to 20) mm long; flowers usually in sympodial inflorescences, the pedicels 20 to 40 mm long, surmounted by three *involucellar nectaries; bracts of the involucel* inserted above each nectary, foliaceous (enclosing the bud), ovate, three to 19-laciniate; *calyx* truncate or five-toothed, 5 to 6 mm long (excluding teeth); *petals* up to 50 mm long, cream-colored or pale yellow, with or without a dark spot at base; *androecium* included; *style* single with decurrent stigmatic lobes, more or less enclosed by androecium or somewhat exceeding androedium; *capsule* three to five-celled, glabrous, smooth, broadly ovoid or subglobose; *seeds* several per locule, lanate, the seed fibers white, tan, or red-brown. Indigenous to Middle America and the Antilles and in certain Pacific Islands (Socorro, the Marquesas, Samoa, etc.); now virtually cosmopolitan in cultivation.

Gossypium incanum (Schwartz) Hillcoat

Sparsely branched shrub 1.0 to 1.5 m tall, the *stems* stellate-tomentose; *leaves* petiolate, oblong-ovate to shallowly two or three-lobed, truncate or subcordate, palmately five to seven-nerved, entire, abruptly acuminate, softly tomentose above and beneath, the *foliar nectary* obscure; *stipules* subulate, 6 to 8 mm long, 1 mm broad, caducous; flowers and fruits borne on short (usually one or two-flowered) sympodial inflorescences, the *pedicels* slender, 6 to 14 mm long, bracteate at articulation, surmounted by three *involucellar nectaries; bracts of the involucel* three, inserted above the nectaries, ovate, narrowed to a short claw, 20 to 30 mm long, 10 to 20 mm wide, distally five to 11-toothed, tomentose, obscurely gland-dotted; *calyx* ca. 10 mm long, pubescent, obscurely gland-dotted, five-lobed, the lobes ovate-acuminate; *petals* 25 mm long, yellow with dark-red spot at base, minutely gland-dotted distally, pubescent externally, especially where exposed in bud, ciliate on claw, glabrous internally; *staminal column* glabrous, pallid, non-glandular, ca. 10 mm long, the filaments subsessile or up to 1 mm long; *style* exceeding the androecium, gland-dotted, the stigmatic lobes decurrent; *capsules* ovoid, beaked, 10 mm long, three-celled, glabrous, gland-dotted; *seeds* ca. 4 mm long, densely pubescent, the fibers tan, appressed. From Arabia.

Gossypium klotzschianum **Andersson**

Shrubs up to 4 m tall, the *stems* minutely stellate-tomentose to glabrate, obscurely gland-dotted; *leaves* petiolate, ovate, more or less cordate, entire, acuminate, stellate-pubescent above and beneath with a small *foliar nectary* on midrib beneath; *stipules* subulate, 15 mm long, 1.5 mm wide; *peduncles* solitary in the leaf axils, articulated ca. 10 mm below the flower, with a reduced leaf at the articulation; *bracts of the involucel* three, pubescent, cordate-ovate, 10 to 15-laciniate, 20 to 30 mm long, about as broad; *calyx* 4 to 5 mm long, truncate to undulate, gland-dotted; *petals* 35 to 50 mm long, yellow (or sometimes reddish at base), more or less plicate, minutely gland-dotted; *staminal column* 10 to 12 mm long, pallid, glabrous, non-glandular, the filaments 1 to 2 mm long; *style* exceeding androecium, gland-dotted, the stigmatic lobes decurrent; *capsule* ovoid-fusiform, to 25 mm long, commonly four-celled, ciliate on inner suture margins; *seeds* 5 to 6 mm long, with sparse inconspicuous fibers. From the Galapagos Islands.

Gossypium lanceolatum **Todaro**

Shrubs 1 to 2 m tall, the stems glabrate, black-gland-dotted; *leaves* petiolate, deeply digitately five-parted (or reduced above to three-parted or even a narrowly lanceolate simple leaf), the lobes narrowly lanceolate, entire, acuminate, the margins ciliate, otherwise sparsely pubescent especially on nerves, gland-dotted, with a *foliar nectary* on the midrib basally; *stipules* falcate, 8 to 10 mm long, acuminate, caducous; *flowers* borne on sympodia, the *pedicels* 0.5 to 1.0 cm long, surmounted by ovate, three to seven-laciniate, 15 to 20 mm long, glabrate, sparsely gland-dotted; *calyx* 5 to 6 mm long, undulately five-toothed (subtruncate), prominently gland-dotted, glabrous; *petals* 25 to 30 mm long, yellowish with indistinct reddish spot at base, with minute black glands scattered evenly; *staminal column* ca. 6 mm long, glabrous, pallid, gland-dotted; anthers and pollen yellow-orange; *style* slender, exceeding androecium by 10 mm, the stigmatic lobes decurrent; *capsule* three-loculate, subglobose, apiculate, glabrous, pitted; *seeds* several per locule, lanate, the fibers white. In houseyard cultivation principally in western Mexico (Nayarit to Oaxaca).

Gossypium laxum **Phillips**

Small tree up to 7 m tall, the young twigs stellate-pubescent and obscurely gland-dotted, the older branches with brownish bark and paler lenticels; *leaves* petiolate, deeply cordate, pedately seven-nerved, manifestly three (or sometime five)-lobed, the lobes entire, acuminate, stellate-pubescent beneath, subglabrous above, the *foliar nectary* weakly developed or absent, the principal veins densely gland-dotted, up to 25 cm long, as wide as or wider than long; *stipules* subulate, ca. 5 mm long, caducous; *pedicel* solitary in the leaf axils, 4 to 7 mm long, stout, glabrate, surmounted by three prominent *involucellar nectaries,* the nectaries bordered horizontal slits; *bracts of the involucel* three, inserted above the nectaries, triangular, 1.5 to 3.0 mm long, 3 to 4 mm broad; *calyx* 8 to 10 mm long, subtruncate with five apiculate teeth, glabrous, gland-dotted; *petals* 50 to 80 mm long, pink with dark-red spot on lower half within, minutely gland-dotted, densely pubescent externally where exposed in bud, mostly glabrous within; *staminal column* glabrous, pallid, ca. 25 mm long, the filaments numerous, reddish, 4 to 8 mm long, *style* slightly exceeding androecium, the stigmatic lobes decurrent; *capsules* three to five-celled, 30 to 40 mm long, narrowly to broadly ovoid and acuminoid, glabrous, gland-dotted (the glands not notably raised), dehiscing to the base but not flaring widely; *seeds* several per locule, 6 to 8 mm long, densely pubescent, the fibers tan, 6 to 8 mm long. From Cañon de Zopilote in central Guerrero, Mexico, where it is relatively abundant.

Gossypium lobatum Gentry

Trees 3 to 7 m tall, the branches lax, minutely stellate-pubescent, becoming glabrate in age; *leaves* petiolate, distichously arranged, usually broader than long, deeply cordate (the sinus open or closed), divaricately three or five-lobed, pedately nerved, obscurely gland-dotted, minutely stellate-pubescent; flowers and fruits in fascicles of one to five in the leaf axils, the *pedicels* 3 to 8 mm long, stout, glabrate, gland-dotted, surmounted by three prominent sunken *nectaries; bracts of the involucel* inserted above the nectaries, broadly triangular, 4 to 5 mm long, acute to obtuse, gland-dotted, nearly glabrous; *calyx* yellowish, densely stellate-pubescent, 12 to 15 mm long, manifestly five-lobed, the lobes one-nerved, acuminate, the sinuses rounded; *petals* 40 to 50 mm long, lavender with dark purplish spot covering lower half of petal (internally); *staminal column* ca. 20 mm long, glabrous, the filaments 3 to 5 mm long, the anthers purplish, the pollen yellow-orange; *style* exceeding androecium, slender, the stigmatic lobes decurrent; *capsules* 20 to 30 mm long, 15 mm diameter, three-celled, minutely pubescent becoming glabrate, prominently gland-dotted, the glands raised, 0.5 to 1.0 mm diam; *seeds* several per locule, densely pubescent, the fibers whitish to tan, appressed. From central Michoacan, Mexico.

Gossypium longicalyx Hutchinson & Lee

Scandent shrub usually supported by other vegetation, the *stems* sparsely pubescent to glabrate, prominently gland-dotted; *leaves* petiolate, ovate, truncate to cordate, entire, palmately seven-nerved, acuminate, gland-dotted, with an inconspicuous *foliar nectary* on midrib beneath; *stipules* 4 to 8 mm long, 1 to 4 mm wide, acuminate, caducous; flowers and fruits borne on sympodial inflorescences, the *pedicels* 8 to 20 (rarely 30) mm long, sparsely pubescent and gland-dotted; *involucellar nectaries* lacking; *bracts of the involucel* three, ovate-cordate, entire, acute to acuminate, gland-dotted, 15 to 20 (rarely 30) mm long, almost as broad; *calyx* 8 to 12 mm long, prominently gland-dotted, deeply divided, the lobes narrowly triangular; *petals* 15 to 20 mm long, yellow, minutely gland-dotted, ciliate on claw, pubescent externally where exposed in bud, glabrous internally; *staminal column* pallid, glabrous, non-glandular, 8 to 9 mm long, the filaments 1 to 2 mm long; *style* exceeding androecium, the stigmatic lobes decurrent; *capsule* ovoid, 10 to 15 mm long, three-loculed, gland-dotted, somewhat pubescent; *seeds* two to three per locule, 5 to 6 mm long, densely pubescent, the fibers grayish, appressed. From East Africa.

Gossypium mustelinum Watt

Shrub or small tree up to 5 m tall, the *stems* stellate-tomentose; *leaves* petiolate, cordate, three or five-lobed, the lobes ovate, entire, acuminate, densely soft-tomentose, obscurely gland-dotted, with one to three inconspicuous *foliar nectaries* beneath; *stipules* ca. 10 to 12 mm long, subulate or falcate, caducous; flowers and fruits borne on sympodial inflorescences, the *pedicels* densely tomentose, gland-dotted, surmounted by three *involucellar nectaries; bracts of the involucel* three, inserted above the nectaries, ovate, ca. 25 mm long, 16 mm wide, nine to 17-laciniate, externally tomentose, internally glabrate, with three additional *nectaries* at the juncture of involucel and calyx, alternate with the bracts; *calyx* 5 to 6 mm long, gland-dotted, glabrous except ciliate on margin, undulate or subtruncate; *petals* 40 to 45 mm long, yellow with dark-red spot at base, minutely gland-dotted, externally pubescent, internally glabrous; *staminal column* 18 mm long, gland-dotted, pallid, glabrous; *style* exceeding androecium, gland-dotted; *capsules* three or four-celled, glabrous, narrowly ovoid, beaked, gland-dotted, the glands raised; *seeds* bearing sparse, brownish fibers. From northeastern Brazil.

Gossypium nelsonii Fryxell

Erect subshrub to 2 m tall, the *stems* densely stellate-pubescent (the hairs ca. 0.5 mm long); *leaves* petiolate, elliptic to ovate (rarely weakly lobed), truncate or slightly cordate, entire, acute to obtuse but often apiculate, up to 90 mm long, half as wide to as wide as long, with a single reddish *foliar nectary* on the midrib beneath; *stipules* linear, up to 17 mm long, caducous; flowers solitary or in sympodial inflorescences; *pedicels* 2 to 5 mm long, tomentose, surmounted by three reddish *involucral nectaries; involucellar bracts* inserted above the nectaries, filiform, 2 to 8 mm long, subequal to calyx or shorter; *calyx* campanulate, 6 to 12 mm long, about half-divided, manifestly gland-dotted, the glands sparse and evenly distributed; flowers commonly cleistogamous, the *petals* closed, 4 to 5 mm long, whitish (rarely chasmogamous, the *petals* 25 to 30 mm long, lavender with a purplish base), *staminal column* glabrous, the filaments purplish, 1 to 2 mm long; *capsules* three to five-celled, 10 to 15 mm long, apically depressed and short-apiculate, the beak less than 1 mm long, nearly glabrous externally with hairs 2 to 4 mm long on suture within after dehiscence, prominently glandular, the glands black, 0.3 mm diameter, so dense as to be frequently coalesced; *seeds* 5 mm long, densely invested with straight fibers ca. 5 mm long. From central Australia, in the southern part of the Northern Territory.

It has been suggested (Valíček, 1979) that *G. nelsonii* is simply a hybrid between *G. australe* and *G. bickii* because it is intermediate in several characters, neglecting the fact that a sterility barrier evidently prevents the hybridization of these two species. Others (unpublished data) consider *G. nelsonii* and *G. australe* to represent a single species because of their superficial resemblance, failing to note the series of at least eight characters that distinguish them.

Gossypium pilosum Fryxell

Erect or arching shrub, the *stems* gland-dotted, prominently pilose (the hairs simple, up to 2 mm long), with an understory of minute stellate hairs, sometimes glabrate; *leaves* petiolate, ovate, cordate or rarely truncate, entire or subcrenulate, acute, up to 70 mm (or more) long, about as broad, somewhat pilose to glabrate, usually ciliate on the margin, with a single *foliar nectary* near the base of the midrib beneath; *stipules* falcate, 10 to 15 mm long, 1 to 2 mm broad, nine-nerved, sparsely ciliate to glabrate; *peduncles* axillary, one-flowered, articulated near the base, bracteate at articulation; *pedicel* 10 to 90 mm long, punctate, with pubescence like that of stem; *involucellar nectaries* absent; *involucellar bracts* three, 10 to 20 mm long, 2 to 3 mm wide, equaling to shorter than the calyx, linear-lanceolate, entire, more or less reflexed, usually pilose without, glabrous within; *calyx* 15 to 22 mm long, gland-dotted, densely pilose externally (hairs to 2 mm), glabrous internally, deeply five-lobed, the lobes 10 to 16 mm long; *petals* 3 to 4 cm long, pinkish with a maroon base; *staminal column* 10 to 15 mm long, pallid, glabrous, the filaments 1 mm long; *style* exceeding the androecium, prominently gland-dotted, the stigmatic lobes decurrent; *capsule* glabrous, prominently gland-dotted, three-celled, 15 mm long, ovoid to globose; *seeds* blackish, arillate, sparsely and minutely pubescent. From Western Australia, on the northern edge of the Mitchell Plateau.

Gossypium populifolium F. von Mueller

Prostrate perennial subshrub, the *stems* trailing, glabrous to densely pilose, the hairs 0.5 to 1.0 mm long, obscurely gland-dotted; *leaves* petiolate, deeply cordate, rotund to broadly ovate, entire, acute to obtuse-apiculate, commonly broader than long, glabrous except ciliate on margin, obscurely gland-dotted, palmately seven to nine-nerved, with an obscure *foliar nectary* on midrib beneath; *stipules* subulate, 2 to 8 mm long; flowers and fruits on one or two-flowered sympodial branches, the *pedicels* 5 to 15 mm long with pubescence like that of stem, the *involucellar nectaries* lacking; *bracts of the involucel* three, ligulate, commonly reflexed, 6 to 11 mm long, 1.5 to 2.5 mm wide; *calyx* 15 to 20 mm long, campanulate, deeply five-lobed (ca. 3/4-lobed), gland-dotted, glabrate, the lobes triangular to linear-lanceolate, acute; *petals* 40 to 50 mm long, pink (sometimes white) with dark-red spot at base, nonglandular, ciliate on claw, pubescent externally where exposed in bud, otherwise glabrous; *staminal* column glabrous, reddish (?), nearly without glands, 15 mm long, the filaments 2 to 3 mm long, the anthers pinkish; *style* exceeding the androecium, the stigmatic lobes pubescent, decurrent; *capsules* three or four-celled, subglobose, ca. 10 mm long, glabrous, prominently gland-dotted, the glands sunken, dehiscent to the base and flaring widely; *seeds* 5 to 6 mm long, black, minutely pubescent, with a small aril attached to hilum. From the Kimberley region of northwestern Australia.

Gossypium pulchellum (C. A. Gardner) Fryxell

Shrub up to 1 m tall, the *stems* minutely tomentose, obscurely gland-dotted; *leaves* petiolate, broadly ovate, truncate or subcordate, entire, short-acuminate, palmately five-nerved, about as broad as long, with a small *foliar nectary* near base of midrib; flowers in the axils of the leaves, the *pedicels* 10 to 15 mm long, the *involucellar nectaries* absent; *bracts of the involucel* three, subulate, 4 to 7 mm long, more or less reflexed; *calyx* 9 to 14 mm long, minutely tomentose, with large black glands (these obscured by tomentum), deeply five-lobed, the lobes acuminate, 1-nerved; *petals* 25 to 30 (rarely 40) mm long, white (fading pink), prominently gland-dotted, externally pubescent where exposed in bud; *staminal column* ca. 10 mm long, glabrous, gland-dotted, more or less purplish toward the base, the filaments ca. 1 mm long, the pollen yellow; *style* exceeding the androecium, gland-dotted, the stigmatic lobes decurrent; *capsule* ovoid to globose, beaked, 10 to 13 mm long, three (rarely four)-celled, glabrous, gland-dotted; *seeds* arillate, 5 to 7 mm long, black with raised venation, subglabrous. From northwestern Australia.

Gossypium raimondii Ulbrich

Large shrub 2 to 3 m tall, the *stems* softly stellate-tomentose, gland-dotted (though glands partially concealed by tomentum), weakly pentangular; *leaves* petiolate, broadly ovate, cordate, entire, acuminate, softly tomentose above and beneath, palmately 9 to 11-nerved, the nerves prominently and the lamina obscurely gland-dotted, with one to three *foliar nectaries* on principal nerves beneath near base; *stipules* subulate to falcate, 7 to 10 mm long, 1 to 3 mm wide, caducous; flowers and fruits borne on one or two-flowered sympodia, the *pedicels* 5 to 20 mm long, stout and fluted, stellate-tomentose; *bracts of the involucel* three, broadly ovate, gland-dotted, pubescent, deeply fimbriately divided, the divisions 15 to 20, caudate-acuminate; *calyx* 5 mm long, truncate, gland-dotted, pubescent; *petals* 40 to 45 mm long, lavender with dark-red spot at base, minutely gland-dotted, ciliate on margin of claw and pubescent externally where exposed in bud, otherwise glabrous; *staminal column* glabrous, sparsely gland-dotted, more or less purplish (especially the filaments); *style* exceeding the androecium, non-glandular, the stigmatic lobes decurrent; *capsules* four-celled, narrowly ovoid, 20 to 25 mm long, glabrous, densely gland-dotted; *seeds* 8 mm long, densely pubescent, the fibers tan, more or less appressed. From Peru.

Gossypium robinsonii F. von Mueller

Erect shrub 2 m tall, the *stems* glabrous, gland-dotted, the glands slightly raised; *leaves* petiolate, truncate or subcordate, deeply five-lobed, entire, palmately five to seven-nerved, glabrous, gland-dotted, the lobes narrowly lanceolate, acuminate, with a prominent often reddish *foliar nectary* on principal nerve of each lobe medially to distally placed; *stipules* falcate, acuminate, 15 to 18 mm long, ca. 1.5 mm wide, caducous; inflorescences sympodial, the pedicels glabrous, gland-dotted, 10 to 20 mm long, surmounted by three *involucellar nectaries; bracts of the involucel* three, lanceolate, entire, acuminate, 15 to 20 mm long, 3 to 10 mm wide, glabrous, gland-dotted; *calyx* 8 to 12 mm long, glabrous, gland-dotted, about half-divided into five caudate lobes; *petals* 35 to 50 mm long, mauve with small dark-red spot at base, gland-dotted on margin where exposed in bud, otherwise non-glandular; *staminal column* ca. 10 mm, more or less purplish, glabrous, non-glandular, the filaments 2 mm long, the anther mass columnar, the pollen orange; *style* exceeding the androecium, pallid, non-glandular; *capsule* four-celled, ovoid, 20 mm long, glabrous, prominently gland-dotted; *seeds* several per locule, 6 to 7 mm long, densely pubescent, the fibers appressed, grayish. From Western Australia.

Gossypium somalense (Gürke) Hutchinson

Shrub to 1.5 m tall, the *stems* stellate-tomentose; *leaves* petiolate, more or less rotund, cordate, shallowly three-lobed, entire, pedately five to seven-lobed, the lobes acute to obtuse (sometimes mucronate), stellate-tomentose above and beneath, with an obscure *foliar nectary* on midrib beneath; *stipules* 6 to 7 mm long, subulate, caducous; flowers and fruits on short (one or two-flowered) sympodial inflorescences, the *pedicels* 6 to 12 mm long, bracteate at articulation, surmounted by three *involucellar nectaries; bracts of the involucel* three, inserted above the nectaries, broadly ovate and deeply cordate, narrowed to a very short claw, 20 to 30 mm long, 30 to 35 mm broad, coarsely five to 13-dentate, tomentose, obscurely gland-dotted; *calyx* 5 to 7 mm long, undulately five-lobed, somewhat tomentose, with scattered black glands, the lobes obtuse; *petals* ca. 20 mm long (subequal to involucel), yellow with prominent dark-red spot at base, non-glandular or nearly so, pubescent externally especially where exposed in bud, glabrous internally; *staminal column* pallid, glabrous, non-glandular, the filaments subsessile to 1 mm long; *style* exceeding the androecium, gland-dotted, the stigmatic lobes decurrent; *capsules* ovoid, beaked, 10 mm long, three- or four-celled, more or less appressed-hirsute, gland-dotted, pubescent on inner suture line, the hairs 2 mm long, included in persisting involucel; *seeds* solitary, ca. 7 mm long, densely pubescent, the fibers brownish, appressed. From Somalia.

Gossypium stocksii Masters

Decumbent or trailing shrub, the *stems* softly tomentose, gland-dotted; *leaves* petiolate, cordate, broadly rotund in outline, pedately seven to nine-lobed, moderately five-lobed, the lobes entire, rounded-apiculate, the sinuses mostly closed, obscurely pubescent to glabrate, gland-dotted (obscurely so above), with a single inconspicuous *foliar nectary* beneath; *stipules* subulate, 5 to 8 mm long, caducous; flowers and fruits borne on short (one or two-flowered) sympodial inflorescences, the *pedicels* slender, 5 to 10 mm long, surmounted by three *involucellar nectaries; bracts of the involucel* three, inserted above the nectaries, cuneate-elliptic, seven- to nine-laciniate, 15 to 20 mm long, 7 to 9 mm wide, pubescent, gland-dotted; *calyx* 6 to 7 mm long, pubescent, gland-dotted, five-lobed, the lobes acuminate, the sinuses rounded; *petals* 25 mm long, yellow with red spot at base, non-glandular or nearly so, pubescent externally especially where exposed in bud, glabrous internally;

staminal column glabrous, pallid, non-glandular or sparsely gland-dotted, the filaments up to 1 mm long, the pollen yellow; *style* exceeding the staminal column, gland-dotted, the stigmatic lobes decurrent; *capsules* subglobose, beaked, three-celled, gland-dotted; *seeds* 6 mm long, densely pubescent, the fibers brownish, appressed. From Pakistan, Arabia, and East Africa.

Gossypium sturtianum J. H. Willis

Shrub, widely branching, 1 to 2 m tall, the *stems* glabrous, prominently glanded, the glands raised; *leaves* petiolate, basally truncate or subcordate, simple (rarely lobed), ovate-elliptic to subrotund, longer than broad, entire, palmately five (or seven)-nerved, glabrous and gland-dotted above and beneath, obtuse to acute and apiculate, with a *foliar nectary* 1 to 2 mm long on midrib beneath; *stipules* falcate, 5 to 9 mm long, 1.0 to 1.5 mm wide, caducous; flowers and fruits borne on sympodial inflorescences, the *pedicels* 10 to 20 mm long, glabrous, with raised glands, surmounted by three reddish *involucellar nectaries; bracts of the involucel* three, inserted above the nectaries, lanceolate-ovate, 18 to 22 mm long, 9 to 12 mm broad, entire, acute to acuminate, glabrous, gland-dotted; *calyx* 8 to 9 mm long, prominently gland-dotted, glabrous, shallowly five-toothed, the lobes acuminate; *petals* 35 to 45 mm long, mauve with dark-maroon spot at base within, ciliate on claw, pubescent externally where exposed in bud, otherwise glabrous, non-glandular or with glands along margin exposed in bud; *staminal column* pallid, glabrous, non-glandular, ca. 20 mm long, the filaments uniformly 1.0 to 1.5 mm long, the anthers pinkish, the anther mass columnar, the pollen yellow-orange; *style* exceeding the androecium, sparsely gland-dotted, the stigmatic lobes decurrent, pubescent; *capsules* ovoid, 20 mm long, five-celled, glabrous, prominently glanded; *seeds* 4 to 5 mm long, densely pubescent, the fibers brownish, appressed. From central Australia.

Gossypium sturtianum var. *nandewarense* (Derera) Fryxell

Shrub, widely branching, 1 to 2 m tall, the *stems* glabrous, prominently glanded, the glands raised; *leaves* petiolate, basally truncate to subcordate with two auriculate appendages, simple or three-lobed, broader than long, entire, palmately five (or seven)-nerved, glabrous and gland-dotted above and beneath, the lobes rounded-apiculate, with an elongate *foliar nectary* 2.0 to 2.5 mm long on midrib beneath; *stipules* falcate, 6 to 12 mm long, 1 to 2 mm wide, caducous; flowers and fruits borne on sympodial inflorescences, the *pedicels* 10 to 20 mm long, glabrous, with raised glands, surmounted by three prominent reddish *involucellar nectaries; bracts of the involucel* three, inserted above the nectaries, broadly cordate-ovate, 18 to 20 mm broad, entire or obscurely toothed apically, glabrous, gland-dotted; *calyx* 7 to 8 mm long, prominently gland-dotted, glabrous, shallowly five-lobed, the lobes apiculate; *petals* 45 to 55 mm long, mauve with dark-maroon spot at base within, densely ciliate on claw, pubescent externally where exposed in bud, otherwise glabrous, gland-dotted on margin where exposed in bud, the glands sparse to absent elsewhere; *staminal column* pallid, glabrous, non-glandular, ca. 15 mm long, the filaments uniformly 1 mm long, the anthers pinkish, the anther mass columnar, the pollen yellow-orange; *style* exceeding the androecium, non-glandular, the stigmatic lobes pubescent, decurrent; *capsules* ovoid, 20 mm long, five-celled, glabrous, with prominent raised glands; *seeds* 4 to 5 mm long, densely pubescent, the fibers grayish, appressed. From southeastern Australia.

Gossypium thurberi Todaro

Shrubs ca. 2 m tall, the *stems* minutely gland dotted, pentangular when young, green, glabrate; *leaves* with two-ridged petioles, subcordate, deeply three to five-lobed (nearly to the base), glabrous, gland-dotted, the central lobe basally constricted, narrowly lanceolate (four to six times as long as broad), long-accuminate, with a single foliar *nectary* on midrib near base; *stipules* subulate, 5 mm long, caducous; flowers and fruits borne on sympodial inflorescences, the *pedicels* 10 to 30 mm long, erect, obscurely three-angled above, glabrous, surmounted by three prominent triangular *nectaries; bracts of the involucel* three, inserted above the nectaries, ligulate, entire or sometimes apically trifid, acuminate, 8 to 12 mm long, 2 to 4 mm broad; *calyx* 3 mm long, truncate, gland-dotted, glabrous; *petals* 15 to 25 mm long, cream colored or pale yellow, with a vestigial red spot at base, or spot lacking, minutely gland-dotted; *androecium* included, pallid, the column glabrous, gland-dotted, the filaments 3 to 4 mm long; *style* slightly exceeding the androecium, the stigmatic lobes decurrent; *capsule* glabrous, three-loculed, 10 to 15 mm long, 8 to 12 mm broad, subglobose to oblong, densely hairy on inner suture, the hairs ca. 2 mm long; *seeds* 6 to 8 per locule, 3 to 4 mm long, blackish, subglabrous. From Arizona and Sonora.

Gossypium tomentosum Nuttall ex Seemann

Shrubs 1 to 2 m tall, the *stems* minutely tomentose, obscurely gland-dotted; *leaves* petiolate, deeply cordate, moderately three- or five-lobed, pedately seven to nine-nerved, densely and minutely tomentose above and beneath, the lobes ovate, entire, acute or acuminate, the *foliar nectary* lacking; *stipules* subulate, tomentose, 7 to 8 mm long, caducous; flowers and fruits borne on sympodial inflorescences, the *pedicels* 15 to 30 mm long, tomentose, obscurely gland-dotted, the *involucellar nectaries* lacking; *bracts of the involucel* three, ovate, apically three to nine-dentate, tomentose, 12 to 18 mm long, 9 to 12 mm wide; *calyx* 5 to 7 mm long, prominently gland-dotted, subtruncate to five-undulate; petals yellow and shiny (internally) but often drying brown and dull, minutely gland-dotted, pubescent externally especially where exposed in bud; *staminal column* 9 to 12 mm long, glabrous, pallid, nonglandular, the filaments 6 to 9 mm long; *style* exceeding the androecium, subequal to petals, slender, gland-dotted or glands lacking; *capsules* globose or ovoid, beaked, three (or four)-celled, 15 (rarely to 20) mm long, glabrous, prominently gland-dotted; *seeds* ca. 7 to 8 mm long, copiously pubescent, the fibers reddish-brown. From the Hawaiian Archipelago.

Gossypium trilobum (Mociño & Sessé ex DeCandolle) Skovsted

Shrubs or small trees 3 to 4 m tall, the *stems* minutely gland-dotted, pentangular or five-ridged when young, green, minutely stellate-pubescent becoming glabrate; *leaves* with quadrangular petioles, cordate, five-lobed below to three-lobed in the inflorescence, with a single foliar nectary, ca. 4/5-divided, the lobes ovate, acuminate, up to 35 mm broad, ciliate-margined; flowers and fruits borne on sympodial branches, the *pedicels* 10 to 15 mm long, erect, ridged, glabrate, surmounted by three *nectaries; bracts of the involucel* three, inserted above the nectaries, cordate-ovate, entire, acuminate, 15 to 20 mm long, 10 to 12 mm broad; *calyx* prominently gland-dotted, 5 to 6 mm long, with an irregular number (5 to 10) of aristate teeth 1 to 4 mm long; *petals* pale yellow with a small red basal spot, 22 to 35 mm long, gland-dotted; *androecium* included, pallid, the column gland-dotted, the filaments 2 to 3 mm long; *style* slender, exceeding the androecium, the stigmatic lobes decurrent; *capsule* glabrous, usually three (rarely two)-loculed, 15 to 18 mm long, 10 to 12 mm broad, oblong, beaked, densely hairy on inner suture of capsule, the hairs 2 mm long; *seeds* 8 to 10 per locule, 3 to 4 mm long, angularly turbinate, blackish with minute tan pubescence. From western Mexico, Sinaloa to Morelos.

Gossypium triphyllum (Harvey ex Harvey & Sonder) Hochreutiner

Slender-branched shrub, the *stems* stellate-tomentose, obscurely gland-dotted; *leaves* short-petiolate, palmately trifoliolate, the leaflets narrowly elliptic, entire, basally and apically acute, with a single principal nerve, obscurely gland-dotted especially along margins, tomentose, the *foliar nectaries* one to three, obscure;

stipules subulate, 2 to 4 mm long, tomentose, caducous; *peduncles* axillary, one-flowered, tomentose, obscurely gland-dotted, articulated 5 to 20 mm below the flower, bracteate at articulation, the *pedicel* surmounted by three *involucellar nectaries; bracts of the involucel* three, inserted above the nectaries, triangular to ligulate, sometimes apically tridentate, tomentose, more or less reflexed, 7 to 9 mm long; *calyx* 6 to 8 mm long, tomentose, gland-dotted, five-lobed, the lobes acuminate, the sinuses rounded; *petals* 30 mm long, pale yellow suffused with lavender, with dark-purple spot on lower half, gland-dotted, externally pubescent where exposed in bud, internally glabrous; *staminal column* glabrous, pallid or somewhat purplish, sparsely gland-dotted at base, the filaments ca. 1 mm long; *style* slender, exceeding the androecium, non-glandular or nearly so; *capsule* ovoid, beaked, three-celled, densely pubescent; seeds 6 to 7 mm long, densely pubescent, the fibers creamy-tan, appressed. From southwestern Africa.

Gossypium turneri Fryxell

Spreading shrub ca. 1 m tall; twigs stellate-tomentulose, gland-dotted, the glands somewhat raised, the *stems* glabrate with red-brown bark and prominent lenticels; *leaves* thick-textured, petiolate, cordate, shallowly trilobed, ca. 20 mm long, about as broad, entire, acute to obtuse, subglabrous, gland-dotted, with a single *foliar nectary* near base of blade; *stipules* linear, 2 to 4 mm long, caducous; *peduncles* solitary in the leaf axils, ca. 20 mm long (equaling or exceeding the subtending petiole), articulate at or above middle with stipuliform bract at articulation, surmounted by three prominent *nectaries; bracts of involucel* inserted above the nectaries, ovate, usually laciniate, 9 to 26 mm long, 7 to 9 mm wide, caducous at anthesis; *calyx* subtruncate, 5 to 8 mm long, prominently gland-dotted, subglabrous; *petals* 40 to 45 mm long, bright yellow with small red spot at base; *staminal column* glabrous, pallid, 26 to 27 mm long; filaments 5 to 6 mm long, anthers purplish, pollen yellow-orange; *style* exceeding androecium by 5 to 15 mm, prominently gland-dotted, the stigmatic lobes decurrent; *capsule* three to five-celled, 20 to 25 mm long, globose to ovoid, prominently glandular, glabrous exept for long (2 mm) white hairs along suture margin after dehiscence; *seeds* 7 to 8 mm long, brownish with tightly appressed fibers. From the coast of Sonora, Mexico, near San Carlos Bay.

2–1.6   Range of Variation Available

The more than 30 wild species of *Gossypium* present an impressive range of variation in many characters, as do the numerous variants within the cultivated species, all of which is potentially available for exploitation in cotton improvement programs. A review of the salient features of this variation may provide some guideposts for such exploitation.

Growth habits in *Gossypium* are highly variable and include weak sub-shrubs (*G. triphyllum, G. bickii*), procumbent perennials (*G. populifolium, G. stocksii*), erect robust shrubs (perhaps a majority of the species), scandent shrubs (*G. longicalyx*), fully arborescent species (*G. aridum, G. laxum, G. lobatum*), and a variety of intermediate types.

Various levels of pubescence on vegetative parts may be found from essentially glabrous to densely hirsute or finely tomentose. The leaves may have a waxy epidermis (*G. sturtianum, G. robinsonii*), a double palisade layer (*G. harknessii, G. armourianum, G. turneri*), or the foliar nectaries absent (*G. gossypioides, G. tomentosum*) or bright red (e.g. *G. australe, G. robinsonii*). Leaf form varies from broadly ovate and more or less shallowly lobed in a number of species to deeply dissected in such species as *G. thurberi, G. lanceolatum,* and *G. robinsonii* to fully trifoliolate in *G. triphyllum.*

Gossypol glands are found in all species more or less distributed throughout the plant, but interesting variations occur. (It will be noted in following paragraphs that glandless types have been created through selective breeding and transgressive segregation.) The glands are often especially large and prominent on the calyx but sometimes the calyx glands are few (e.g., *G. australe*) or small (e.g. *G. aridum*) or both (e.g. *G. lobatum*). The gossypol glands are often found in the petals but in certain species the petals lack glands (e.g. *G. areysianum, G. populifolium*) or have very few glands (e.g. *G. incanum, G. cunninghamii*). Similarly, gossypol glands may be present or absent in the staminal column or in the style in different species. The capsules in most species have gossypol glands embedded in the carpel walls, but in some species these glands are raised (e.g. *G. lobatum, G. sturtianum, G. capitis-viridis, G. anomalum*) whereas in others they are sunken (e.g. *G. barbadense, G. populifolium, G. lanceolatum, G. harknessii*). Perhaps of greatest applied interest are those species having the gossypol glands reduced or absent from the embryos (*G. bickii, G. australe, G. sturtianum*) while generally present in other plant parts.

The form of the involucel is highly variable in *Gossypium.* It is always present, always trimerous, and usually associated with involucellar nectaries. (The nectaries are absent in a few species: *G. tomentosum, G. gossypioides, G. longicalyx, G. pilosum, G. populifolium, G. pulchellum.*) The bracts may be inconspicuous triangular scales (*G. aridum, G. laxum, G. lobatum*), they may be narrowly ligulate (*G. thurberi, G. areysianum, G. anomalum*) or even filiform (*G. australe, G. bickii*), or they may be broadly foliaceous, generally enclosing the bud and sometimes the fruit (*G. somalense, G. trilobum, G. gossypioides, G. davidsonii, G. raimondii, G. hirsutum, G. barbadense,* etc.). In a few species the bracts of the involucel are reflexed (e.g. *G. populifolium, G. pilosum*), in a few deciduous (*G. armourianum, G. harknessii, G. turneri*).

The calyx, as previously stated, is commonly gland-dotted somewhat more prominently than are other parts of the plant. In many species (e.g. *G. barbadense, G. thurberi, G. raimondii*) the calyx is truncate or nearly so, showing little evidence of its pentamerous nature. Other species range from obscurely five-toothed to deeply five-lobed, reaching the extreme of expression in *G. lobatum, G. longicalyx, G. cunninghamii, G. costulatum, G. populifolium,* and *G. pilosum.*

Petal colors include various intensities of yellow from very pale (*G. thurberi*) to very bright (*G. turneri, G. longicalyx, G. tomentosum*), lavender or rose (*G. aridum, G. gossypioides*), mauve (*G. sturtianum*), and sometimes white (*G. pulchellum* and certain strains of *G. bickii, G. arboreum, G. populifolium,* etc.). The petals have a dark spot (dark-red maroon, or purplish) at the base of the petal in many species or no spot in some (*G. longicalyx, G. tomentosum, G. klotzschianum*). The size of the spot may vary from vestigial or very small (e.g. *G. thurberi*) to covering more than half the petal (*G. triphyllum, G. laxum, G. lobatum, G. gos-sypioides*). The corolla may be narrowly funnelform when fully open (as e.g. *G. aridum, G. somalense, G. triphyllum*), or it may be widely flared (as in *G. turneri* or *G. harknessii*), or it may be fully rotate (as in *G. thurberi* or *G. trilobum*).

The style (relative to the androecium) may be relatively short, virtually buried in the anthers (as in *G. herbaceum* or many strains of *G. hirsutum*) or it may be relatively long, greatly exceeding the androecium and nearly equaling the petals (as in *G. tomentosum, G. darwinii, G. armourianum, G. harknessii*).

The capsules may be more or less round (*G. herbaceum, G. lanceola-tum*), oblong (*G. trilobum*), ovoid (most species), or elongated (*G. barba-dense, G. lobatum*). It is usually glabrous or nearly so but is densely pubes-cent in *G. australe* and *G. triphyllum*. Following dehiscence, the inner suture line bears long fragile hairs in a number of species (e.g. *G. armouri-anum, G. australe, G. bickii, G. davidsonii, G. harknessii, G. klotzschian-um, G. nelsonii, G. somalense, G. thurberi, G. trilobum, G. turneri*) and not in others. The capsules are often prominently glandular, the gossypol glands up to 1 mm diam (*G. anomalum, G. gossypioides, G. lobatum, G. bickii*) but usually smaller. The glands may be raised and warty (e.g. *G. lobatum, G. anomalum, G. bickii, G. capitis-viridis, G. mustelinum, G. sturtianum*) or sunken, making the fruit pitted (e.g. *G. arboreum, G. armourianum, G. barbadense, G. darwinii, G. harknessii, G. lanceolatum, G. populifolium*); the fruits are neither warty nor pitted in other species but are relatively smooth.

Seeds may be as small as 4 to 5 mm long (*G. australe, G. sturtianaum*) and as large as 8 to 10 mm long (*G. harknessii, G. armourianum*), but most of the species have seeds intermediate in size between these extremes. The seed coats have specialized cells of the epidermis that develop as elongated, single-celled hairs or trichomes (Fryxell, 1963, 1964). In this work, these hairs are generally referred to as "fibers" when growing on the plant and as "lint" after they have been harvested, in the case of the cultivated species, and entered commerce. The seed fibers may vary in amount from those species virtually lacking seed fiber (e.g. *G. thurberi, G. davidsonii, G. populifolium*) to the majority of the species, which have densely pubescent seeds. The nature of the fiber differs: two species (*G. australe, G. nelsonii*) have straight fiber, the remaining species have fibers that are variously crimped or curled. Some species have fibers that are tightly appressed to the seeds (e.g., *G. harknessii, G. bickii*) while in other species they are only loosely appressed (e.g. *G. triphyllum, G. raimondii*); the cultivated cottons, of course, have fibers (cotton) that are very loosely fluffed out, in part be-cause of their relatively greater length. The color of the fibers is grayish in some species (*G. lobatum, G. harknessii, G. longicalyx, G. robinsonii, G. gossypioides*) and some shade of brown in others (*G. tomentosum, G. aridum, G. armourianum, G. raimondii*). Most of the cultivated cottons have been selected for white or nearly white fibers.

The preceding survey of the range of variability within *Gossypium* touches lightly on the more obvious characters and does not deal with some of the less obvious but perhaps more important characters, such as disease and pest-resistance, phytochemical characters, fiber properties, and such physiological variables as cold tolerance and photosynthetic rates. For most of these characters data are not yet available or are only imperfectly understood.

2-2 GERMPLASM RESOURCES

2-2.1 History of Plant Improvement

A detailed history of plant breeding and plant improvement is beyond the scope of this chapter and will not be attempted. However, a consideration of this history, sketched in broad outline, is necessary for an understanding of the issues involved in germplasm preservation.

Plant improvement, as a human activity, is as old as agriculture. The first agriculturists could hardly have avoided being selective in obtaining seed for the sowing of the next season's crop. It would be the most natural thing in the world to use the "biggest" or "best" of the available parental material to produce the next generation. Thus, plant selection and improvement originated far earlier than any written record.

Most of history and prehistory has been characterized by limited human mobility. Even though explorers traveled far and wide, this travel had relatively little impact on day-to-day living in local communities and villages. Indeed, even travel between nearby communities was limited by primitive methods of transportation and often by physical barriers of mountains, water, and desert. The consequence, in agricultural terms, is that plant improvement moved forward on a very localized basis. Individual cultivators did their own selection, kept their own planting seed, and developed their own cultivars. Interchange of seedstocks from one farmer to another, from one community to another, was uncommon. The usual result was a patchwork of "landraces," with each farmer, each community, and each river valley having its own cultivars. This isolation resulted in part from the poor communication of the time, but also from an understandable preference for one's own cultivars over one's neighbor's selections.

This patchwork of landraces was characteristic of most crops up to recent times. The pattern served to preserve and maintain (even to produce) a wide range of variability; it was in itself a germplasm preservation system. However, this system began to break down when improved communications facilitated interchange of materials and when "scientific" plant breeding began to produce "super" cultivars. In the present century, programs to foster such cultivars and to improve the agriculture of marginal areas have resulted in displacing the multitude of landraces with a few "super" cultivars. Agricultural productivity has generally been improved, but at the price of more or less destroying this natural germplasm resource. The genetic base of most of our major crop species has been drastically narrowed.

This limitation of the genetic base means a limitation for the plant breeder, who has fewer possibilities for developing new and improved materials for the future. More dramatically, a crop with a narrowed genetic base is much more vulnerable to such challenges as new or more virulent pests or diseases or to new conditions of growth or requirements for growth. These environmental challenges are usually unpredictable, and consequently such "genetic vulnerability" to unforeseen exigencies must be guarded against. With the rapid disappearance of the natural germplasm bank represented by many different landraces and local cultivars, it has become necessary to establish other germplasm banks to substitute for this vanishing resource. To some extent, this effort has been a salvage operation, but it is from such a historical background that modern germplasm collections have developed.

2-2.2 Assembly of Germplasm Resources

The history of cotton gardens, or cotton germplasm collections, goes back at least to 1784, when J. B. P. von Rohr established a cotton garden on St. Croix in the Danish West Indies (now the Virgin Islands) under a commission from the Danish king. He assembled and grew a wide range of cottons, mostly obtained from the West Indies and northern South America. The purpose of von Rohr's work was primarily to evaluate the adaptation of a wide range of cotton germplasm to the climatic conditions that occurred in the Danish West Indies, with a view of improving the agricultural economy of this area. He obtained his materials in a variety of ways, prominent among which was his own travel to many areas within the region, to obtain new types of cotton to include in the St. Croix garden. The results of his studies and observations were published in a small book (two volumes) that described the agricultural methods and problems of the time as well as the various kinds of cotton that he grew in the St. Croix garden (von Rohr, 1791, 1973). Von Rohr prepared herbarium specimens of his cottons, which are still preserved in the Botanical Museum in Copenhagen, so that it is possible to study these specimens and evaluate the cottons he grew (Fryxell, 1969), in spite of the sketchy descriptions in his book. The St. Croix garden was used primarily to describe the cottons and evaluate their adaptation and agronomic potential; it apparently made little contribution to plant improvement.

The next significant cotton garden to be formed was at the Royal Botanical Garden in Palermo, Italy, under the direction of Agostino Todaro. This garden was established around 1860 and was maintained for perhaps two decades. It was brought into being because cotton was introduced into Italy as a major crop at about that time, and there was considerable interest in the botany and agriculture of the plant. Todaro was one of the leading botanists of the day, and the study of *Gossypium* was one of his major contributions (Lanza, 1892; Fryxell and Smith, 1972). The cotton garden he assembled was worldwide in scope and sampled the variability to be found in the genus, with more than 200 accessions included. He obtained the materials in a number of ways, but solicited many of them by correspondence to remote corners of the world, especially through contacts with botanical colleagues and with Italian consulates. The principal purpose of assembling the Palermo garden was to study it, especially from a taxonomic viewpoint, and not to utilize it as a basis for a cotton breeding program. Todaro produced a series of publications based on this study, the most significant *Relazione. . .Monografia* (Todaro, 1877), which deals with cotton agriculture in Italy in the first part, and presents the first taxonomic monograph of *Gossypium* ever published in the second part. His monograph and his cotton garden included both cultivated and wild species of *Gossypium*, although the latter were less well known then than they are today.

In modern times many cotton germplasm collections have been formed in various parts of the world, some of the more significant of which are:

1. The Empire Cotton Growing Corporation collection maintained for many years in Trinidad, but dispersed in 1947 (cf. Hutchinson et al., 1947).

2. A similar collection at Shambat, Sudan, formerly maintained by the Empire Cotton Growing Corporation, now maintained by the Sudanese Department of Agriculture at Wad Medani.

3. A collection at Roque Saenz Pena, Argentina, emphasizing variation in *G. barbadense* and based on the work of Argentine collectors in northern Argentina and adjacent Paraguay and Bolivia.

4. A collection in Tashkent, USSR, worldwide in scope but evidently having its origins in the Central and South American collections of Mauer and Bukasov (Mauer, 1930). It is maintained by Soviet agricultural officials.

5. A collection of cultivars and genetic stocks of *G. arboreum* and *G. herbaceum* maintained at the Cotton Research Station in Surat, Gujarat, India.

6. A collection of the wild species and of cultivars maintained at the Institute of Tropical and Subtropical Agriculture in Prague, Czechoslovakia (Valíček, 1979).

7. A collection in Montpellier, France, maintained by the Institut de Recherches du Coton et des Textiles Exotiques (IRCT), and utilizing a winter garden in Guadeloupe, French West Indies.

8. A collection of the USDA, maintained at several locations in the United States and utilizing a winter garden in tropical Mexico. Cultivars and primitive types of *G. barbadense* and *G. darwinii* are maintained at Phoenix, Arizona; obsolete cultivars of *G. hirsutum* are maintained at Stoneville, Mississippi; primitive types and genetic markers of *G. hirsutum* and all other species are maintained at College Station, Texas. Permanent storage of all these materials is made at the National Seed Storage Laboratory, Fort Collins, Colorado, although "working collections" are also retained at the three locations mentioned.

To a greater or lesser degree, the personnel and the agencies and governments that are responsible for these modern cotton germplasm collections maintain cooperative relationships and exchange cotton germplasm. Plant introduction agencies in the various countries are helpful in facilitating such interchange. From this discussion, it is evident that cotton germplasm resources are assembled by several methods, principally (a) interchange with other germplasm collections, (b) direct solicitation from agricultural officials, botanists, or others, and (c) direct plant exploration. Often the most satisfactory method for acquiring new germplasm, if funds and personnel are available, is to send a capable scientist into the field as a plant explorer.

When a germplasm collection is assembled and mechanisms for its maintenance are set up, the questions of seed storage (Justice and Bass, 1978) and record keeping must be addressed. Since the establishment of the National Seed Storage Laboratory, the matter of seed storage in the United States has not been a problem. Moreover, cottonseed is relatively long-lived, generally lasting for a couple of decades under good (cool and dry) storage conditions. To make such germplasm accessible and useful, satisfactory records must be kept.

The first element in any germplasm record-keeping system is an identification number. This appears to be a straightforward matter but in fact can sometimes be troublesome. The reason it causes problems is that a single germplasm accession frequently has several different identification numbers, because collectors, germplasm curators, plant breeders, and other users of germplasm prefer a particular system of identification. Consequently, when an accession of germplasm (cottonseed sample) changes hands, it frequently has a new identification number assigned to it. A single strain thus accumulates several identification numbers. This practice is deplorable because it causes confusion, but at the same time it is difficult to avoid. Moveover, the literature is already full of papers that cite their germplasm materials sometimes under one system, sometimes another. In U.S. cotton, for example, the following numbering systems may be encountered: P.I. (Plant Introduction), C.B. (Cotton Branch), Tex. (Texas Collection), S.A. (Stoneville Accession), Ft. Collins (National Seed Storage Laboratory), S. G. Stephens (a personal numbering system), and other individual systems less commonly encountered. To minimize this problem for the future it is recommended that the P.I. numbering system be used in preference to the alternative systems, since it is most nearly universal for the U.S. germplasm program.

In addition to an identification number, it is desirable to keep records on the amount of seed in storage and its age, to use as a guide for seed rejuvenation or replenishment. Monitoring the amount of seed in storage involves keeping data on amounts of seed (a) put into storage at initial collection or in subsequent cycles of seed increase and (b) withdrawn from storage for germination tests or for distribution to users.

Additional data of value in germplasm records are the so-called "descriptors," data on the botanical and agronomic characteristics of the items in the collection. These descriptors may concern such traits as plant stature, flower color, fruit size, geographic provenience, resistance to diseases or pests, fiber properties, phytochemical characters, etc. It is important to include data from as many descriptors as possible so that users can select material from the germplasm collection that is most likely to serve their particular needs. A minimum list of descriptors was produced for cotton by an International Board for Plant Genetic Resources, Working Group.

Manipulating records for a germplasm collection, involving monitoring the seed quantity and age as well as keeping track of the descriptors, can become very complex as the size of a collection increases. Moreover, it involves inserting new descriptive data as the collection is screened and evaluated for new characteristics by various specialists. This is most easily accomplished for large collections with computer assistance. Fortunately, computers can assist not only in manipulating this mass of information, but also in producing a printout or "catalog" of the available information at periodic intervals. Such a catalog is essential for the efficient utilization of germplasm materials so that the user can select and request the particular accessions he or she needs. The *Regional Collection of Gossypium Germplasm* (Anonymous, 1974) describes the materials available in the U.S. cotton germplasm collection.

2–2.3 Sources of Variability

Gossypium germplasm can be divided into two categories in three different ways: (a) diploid species vs. tetraploid species, (b) wild vs. cultivated species, or (c) fibered vs. fiberless cottons. I believe the third method of division is the most appropriate in understanding germplasm resources and their utilization.

The nonfiber-producing cottons include all of the (more than 30) wild diploid species of *Gossypium.* Some of these species bear seed hairs and some do not, but none of them bears usable fiber—i.e. the seed fibers that may be present are too short and too firmly adherent to the seed to be of even potential agricultural significance. These species, moreover, are for the most part too distantly related to the cultivated cottons to be considered as usable germplasm in conventional plant breeding. Rather, they are "exotic" sources of germplasm. Yet, as we will see, some of these exotic sources have made significant contributions to cotton improvement by means of special techniques for the utilization of their germplasm.

The fiber-producing cottons include two diploid cultivated species (*G. arboreum* and *G. herbaceum*), three tetraploid cultivated species (*G. hirsutum, G. barbadense,* and *G. lanceolatum*), and three tetraploid species that are more or less wild (*G. tomentosum, G. mustelinum,* and *G. darwinii*), but which are more or less accessible to exploitation in conventional plant breeding programs. Each of the cultivated species (with the exception of the relatively uniform *G. lanceolatum*) shows a wide range of variability in terms of cultivars, strains, feral types, and genetic mutants within each species. *Gossypium darwinii* also shows a large amount of variability that is related to its subdivided distribution among the several islands of the

Galapagos Archipelago. *Gossypium mustelinum* and *G. tomentosum,* on the other hand, exhibit relatively little variability. A large proportion of the *Gossypium* germplasm collection is concerned with sampling the variability within the four principal cultivated species, *G. arboreum, G. herbaceum, G. hirsutum,* and *G. barbadense*, including their semi-wild and feral types.

2-2.4 Using Germplasm Resources

Cotton is a crop native to the tropics and subtropics. Major segments of modern cotton production, however, lie in the temperate zones. A shift in the climatic adaptation of the plant has been necessary to make this geographic shift possible. The most immediate problem concerns the flowering response. Many tropical cottons, including most of the "primitive" types in the germplasm collection, behave photoperiodically and fail to flower under the long-day regime of the temperate-zone growing season. They are therefore not directly available for crossing and incorporation into cotton breeding programs. There are two solutions to this problem: (a) to make the crosses of "primitive" cottons and adapted cultivars at a tropical garden where both will flower simultaneously, and carry out subsequent selection programs in the same environment or (b) to introduce genes for day-neutralism into the germplasm accessions with a back-crossing scheme so as to disturb the genotype as little as possible and thus to have the "primitive" cotton available for utilization directly in the temperate-zone cotton production area. Such "conversion" programs (converting from photoperiodic to non-photoperiodic) are being applied especially to the collection of "primitive" types of *G. hirsutum,* to make available a wider range of germplasm for screening and other utilization by the cotton improvement community.

A second step in germplasm utilization is screening for useful characters. This screening includes observing plants for growth habit; measuring productivity, boll size, or fiber strength; evaluating for resistance to various pests or diseases; or screening for any character deemed to be useful or potentially useful. Screening programs typically are open-ended examinations of all available material to determine the range of variability for the type of resistance or other character in question and to identify those particular accessions with extremes of expression (e.g. resistance to a disease). Sometimes screening programs seek correlations of characters (e.g. resistance to a disease correlated with some phytochemical characteristic) not only to identify potentially valuable germplasm but also to seek to understand the mechanisms of resistance or other character in the plant. Generally speaking, these programs are carried out by specialists using special techniques to inoculate or measure or otherwise evaluate the germplasm being screened.

When desirable germplasm has been located, the cotton breeder then wishes to transfer the desirable traits from the germplasm source to otherwise adapted agricultural cultivars. The ease of doing so depends, in part, on the mode of inheritance of the trait that is to be transferred and, in part, on the closeness of relationship of the donor and donee. A trait that is simply inherited (one or a few genes) can often be transferred effectively even when the two lines are not closely related. Such gene transfer is generally accomplished with back-crossing procedures in a relatively straightforward manner, but sometimes the methods are not so simple if, for example, a gene is being transferred across a ploidy level difference.

The kinds of germplasm utilization described so far are fairly conventional and obvious, but germplasm resources can also be used in more creative ways to produce new kinds of germplasm. One of these is the development of new cytotypes. In these lines the nuclear gene complements of adapted agricultural cultivars (in this case, of *G. hirsutum*) have been transferred into the cytoplasms of *G. barbadense, G. tomentosum, G. arboreum, G. anomalum, G. harknessii,* and *G. longicalyx.* Preliminary data show that these cytoplasms have significant genetic effects and that these novel examples of new germplasm are of potential value in solving agricultural problems. The new types of germplasm are still undergoing screening for cold tolerance, disease and insect resistance, and response to pesticides, among other characters. Many more nuclear-cytoplasmic combinations are potentially available, involving the cytoplasms of the 30-odd wild species and the nuclear complements of any number of adapted agricultural cultivars.

Even though cultivated cottons exist at only two ploidy levels, diploid and tetraploid, it is possible to explore other ploidy levels as a means of simply combining (rather than recombining) germplasm from two sources. This has been done by Muramoto (1969) in crossing the diploid *G. sturtianum* with the tetraploid *G. hirsutum* to produce a sterile triploid hybrid. The chromosomes of this plant were then doubled with the aid of colchicine to the hexaploid level, which restored fertility. In this way the entire genetic complement of one species is added to that of the other, and plant breeding activities can then move forward at a new ploidy level. A new type of germplasm has been created. Clearly, many such new hexaploid types of germplasm can potentially be produced between the 30-odd diploid species and the several tetraploid species. Many, perhaps most, would be little more than novelties, but some combinations of species might well produce superior germplasm. Only one has yet been examined in detail. Presumably, hexaploids involving both diploid and tetraploid cultivated species (rather than wild species) would be of particular agricultural interest, for example *G. arboreum + G. hirsutum.*

In still other cases, relatively conventional methods and materials have produced unconventional results. Transgressive segregation with appropriate selection protocols has given rise to what may be called "exotic" (i.e. new) germplasm. High fiber-strength types have been developed as primary breeding stocks, to some extent involving interspecies hybrids, largely through exploiting transgressive segregation (Culp and Harrell, 1973, 1974). Glandless cottons have been produced in much the same way, achieving a phenotype (glandlessness) that was previously outside the range of variation of the genus *Gossypium.* Transgressive segregation (together with efficient screening) was responsible for achieving a new high level of resistance to the root-knot-nematode *Fusarium*-wilt disease complex (Shepherd, 1974). In all of these cases, germplasm resources were used to create new types of germplasm not previously available.

2–2.5 Examples of Successful Use of Cotton Germplasm Resources

Cotton germplasm has been successfully used in a variety of ways, selected examples of which have been reviewed by Fryxell (1976a).

A disease of localized occurrence is cotton rust, caused by *Puccinia cacabata* Arth. & Holw. (Blank and Leathers, 1963). It is of importance in the southwestern U.S. and northwestern Mexico, where its incidence is variable but its economic impact is significant in some years. Fungicidal control of the disease is costly and not wholly satisfactory. L. M. Blank developed appropriate inoculation techniques, with which he surveyed the available germplasm of *G. hirsutum.* He failed to find any plants exhibiting resistance to the rust organism. He then screened the available diploid species of *Gossypium* and found that *G. anomalum* and a few cultivars of *G. arboreum* were rust resistant, although all the other species were susceptible. Through interspecific hybrids and artificial polyploids plus a backcrossing scheme accompanied by continued screening for resistance, he succeeded in transferring this resistance into agronomically acceptable germplasm of *G. hirsutum.* Now rust resistant cultivars are available for planting in those areas where rust is a significant problem, and the costs of disease control and disease loss are thereby avoided.

Another disease problem resolved by recourse to germplasm resources is the bacterial blight disease (also known as angular leafspot or blackarm) caused by *Xanthomonas malvacearum* (E. F. Smith) Dowson (Brinkerhoff, 1970). Germplasm surveys have, over the years, discovered a whole series of monogenic sources of resistance, coming from *G. hirsutum, G. barbadense, G. herbaceum, G. arboreum,* and *G. anomalum.* Since the disease organism exists in a variety of strains at differing levels of virulence, plant breeders have generally combined two or more genes for resistance and, through selection, raised the effectiveness of the polygenic background in developing new blight-resistance genotypes. Satisfactory blight-resistance cultivars of both *G. hirsutum* and *G. barbadense* have been available for many years.

The most active areas currently attempting to exploit the germplasm collection are the several programs seeking to find host plant resistance to various insects. Researchers are exploring insect response to morphological changes in the plant (nectariless types, modified bract form), chemical changes in the plant (high gossypol, high tannin), and different types of pubescence to affect feeding or egg laying of the insect. Most of these attempts cannot yet be reported as successful, but some programs have achieved their goal, such as the jassid-resistance of extremely hairy cultivars and the nectariless cultivars. Success is anticipated in other programs. Perhaps the most significant instance of past success is the response of the Cotton Belt to the advent of the boll weevil (*Anthonomus grandis* Boh.) around the turn of the century. A measure of host-plant resistance was achieved by developing cotton with new growth habits, specifically by the development of earlier-maturing cottons that could produce a partial crop before the insect populations built up to more damaging levels.

2-3 THE IMPORTANCE OF VOUCHER SPECIMENS AND HERBARIUM DOCUMENTATION

The documentation of plant exploration work and of botanical and agricultural research with voucher specimens (Smith, 1971) could perhaps be considered a form of record keeping and discussed under that heading. But it is a sufficiently specialized form of record keeping to merit separate discussion. A voucher specimen of a plant explorer's find is a part of the record of a germplasm accession, and the care and preservation of the voucher should be considered an integral part of the germplasm preservation process (Fryxell, 1979a).

A voucher specimen is a permanent record of the plant explorer's work. Germplasm accessions from time to time are lost, either by contamination (by inadvertent outbreeding, for example), physical loss, failure of germination, or various other accidents. When such losses occur, as they inevitably do, the herbarium voucher remains as a source of information about the plant. Even when a living accession is not lost, the voucher serves as a source of information when it is not possible or convenient to grow out a seed sample to the mature plant stage for observation. The voucher also serves as a base line to evaluate whether subsequent cycles of seed increase have changed the phenotypic expression of that accession through deliberate or accidental selection or through genetic drift.

Voucher specimens also serve an important role in documenting research, not only in the kinds of research associated with germplasm collections but other kinds as well. In any kind of plant research, it is important that the subject of the research can be unequivocally identified, and this is best accomplished by reference to voucher specimens. In germplasm work it is especially important to preserve vouchers for consultation by other workers or subsequent generations—especially when working with "exotic" materials such as interspecific hybrids, complex hybrids, and derivatives of such materials.

REFERENCES

Anderson, Edgar. 1969. Plants, man and life. University of California Press, Berkeley.

Anonymous. 1974. The regional collection of *Gossypium* germplasm. USDA Rep. ARS-H-2.

Blank, L. M., and C. R. Leathers. 1963. Environmental and other factors influencing development of southwestern cotton rust. Phytopathology 53:921–928.

Brinkerhoff, L. A. 1970. Variation in *Xanthomonas malvacearum* and its relation to control. Ann. Rev. Phytopathol. 8:85–110.

Constance, L. 1964. Systematic botany—an unending synthesis. Taxon 13:157–172.

Culp, T. W., and D. C. Harrell. 1973. Breeding methods for improving yield and fiber quality of Upland cotton (*Gossypium hirsutum* L.). Crop Sci. 13:686–689.

----, and ----. 1974. Breeding quality cotton at the Pee Dee Experiment Station, Florence, S. C. USDA Rep. ARS-S-30.

Fryxell, P. A. 1963. Morphology of the base of seed hairs of *Gossypium*. I. Gross morphology. Bot. Gaz. 124:196–199.

----. 1964. Morphology of the base of seed hairs of *Gossypium*. II. Comparative morphology. Bot. Gaz. 125:108–114.

----. 1969. The West Indian species of *Gossypium* of von Rohr and Rafinesque. Taxon 18: 400–414.

----. 1976a. Germpool utilization: *Gossypium*, a case history. USDA Rep. ARS-S-137.

----. 1976b. A nomenclator of *Gossypium*—the botanical names of cotton. USDA Tech. Bull. 1491.

----. 1979a. Taxonomic research and plant germplasm collections. USDA Rep. ARS-S-4.

----. 1979b. The natural history of the cotton tribe (Malvaceae, tribe Gossypieae). Texas A&M University Press, College Station.

----, and C. E. Smith, Jr. 1972. The contributions of Agostino Todaro to *Gossypium* nomenclature. Taxon 21:139–145.

Hutchinson, J. B., R. A. Silow, and S. G. Stephens. 1947. The evolution of *Gossypium*. Oxford University Press, London.

Justice, O. L., and L. N. Bass. 1978. Principles and practices of seed storage. USDA Handb. No. 506.

Lanza, D. 1892. Agostino Todaro. Malpighia 6:120–132.

Mauer, F. M. 1930. The cottons of Mexico, Guatemala, and Colombia. Bull. Appl. Bot., Genet., Plant Breeding 47 (Suppl.):543–553. [Russian version, with illustrations, on p. 425–464].

----. 1954. Proiskhozhdeniye i Sistematiki Khlopchatnika [Origin and Systematics of Cotton]. Tashkent: Akad. Nauk Uzbek. S.S.R. [in Russian].

Muramoto, H. 1969. Hexaploid cotton: Some plant and fiber properties. Crop Sci. 9:27–29.

Parlatore, F. 1866. Le Specie dei Cotoni. Stamperia Reale, Firenze.

Rohr, J. B. P. von. 1791–1793. Anmerkungen über den Cattunbau, zum Nuzen der Dänischen Westindischen Colonien. 2 Volumes. Hamrich, Altona and Leipzig.

Shepherd, R. L. 1974. Transgressive segregation for root-knot nematode resistance in cotton. Crop Sci. 24:872–875.

Smith, C. Earle, Jr. 1971. Preparing herbarium specimens of vascular plants. USDA Agric. Inform. Bull. No. 348.

Todaro, A. 1877. Relazione sulla cultura dei Cotoni in Italia seguita da una Monografia del genere *Gossypium*. Molina, Rome.

Valicek, P. 1979. Wild and cultivated cottons. Institut de Recherches du Coton et des Textiles Exotiques, Paris. (Published with a 26-page supplement, separately paged, containing a French version ["Cottoniers sauvages et cultive"] without the illustrations or bibliography).

Watt, G. 1970. The wild and cultivated cotton plants of the world. Longmans, Green, and Co., London.

Zaitzev, G. S. 1928. A contribution to the classification of the genus *Gossypium* L. Bull. Appl. Bot., Genet., Plant Breeding 18:1–65. [in Russian, with illustrations, on p. 1–38; in English on p. 39–65].

3 Anatomy and Morphology of Cultivated Cottons

Jack R. Mauney

ARS-USDA
Phoenix, Arizona

To fully comprehend the limitations to the productive capacity of the cotton plant, it is necessary to understand its structures: the vegetative framework, the reproductive branches, flowers, fruit, and seed. These structures in cotton, a typical woody perennial shrub, are botanically relatively simple. The difficulty in analyzing cotton morphology arises from the fact that the plant simultaneously develops both vegetatively and reproductively. Because of its indeterminate growth habit cotton has the most complex morphology of any major field crop grown as an annual.

A complete discussion of the developmental patterns in cotton is given in J. M. Hector's *Introduction to the Botany of Field Crops* (1936). This work by Hector and the Russian monograph, *The Structure and Development of the Cotton Plant* (Baranov and Maltzev, 1937), as well as published reports of Gore (1935), Mauney and Ball (1959), Quintanilha et al. (1962), will, with additional observations made in my laboratory, form the basis for this chapter.

3-1 DEVELOPMENT OF PRIMARY AXIS

The primary axis results from elongation and development of the embryo. In the dormant seed, the primary axis consists of a radicle, a hypocotyl, and a poorly developed epicotyl. The epicotyl contains one true leaf initial and a dome of meristematic cells (Fig. 3 1). Thus, all differentiation

Contribution from ARS-USDA. Published in *Cotton,* Agronomy Monograph No. 24, © ASA-CSSA-SSSA, 677 South Segoe Road, Madison, WI 53711, USA.

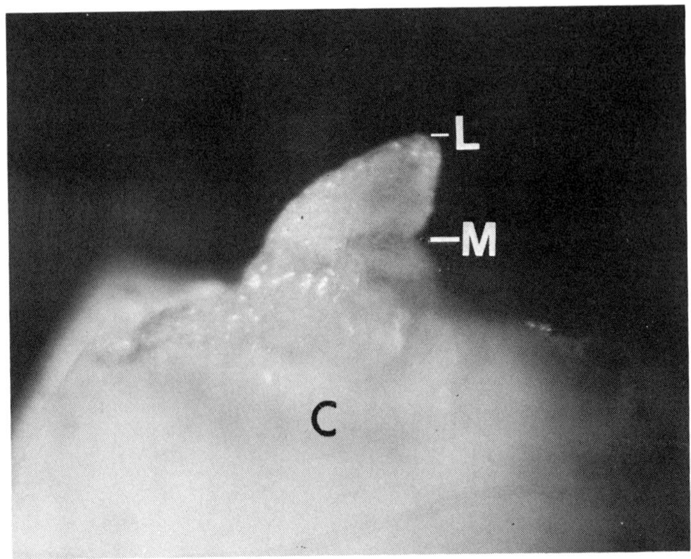

Fig. 3-1. Growing point one day after planting. C—attachment of one cotyledon (attachment of other cotyledon in background behind growing point), L—primordium of first true leaf, M—apical meristem.

of the vegetative framework above the cotyledons must take place subsequent to germination. Cotton remains as a seedling longer than grasses and legumes, because it has no differentiated true leaves, only cotyledons, to expand immediately after germination.

The growing cotton plant actively proliferates new cells on many fronts. To describe the plant completely, each facet of growth should be examined apart from the whole. A picture of both the simplicity of the part and complexity of the whole emerges from this examination.

Four principles of morphological correlation are necessary for a full understanding of the architecture of a cotton plant:

1. Apical meristems make four organs: leaves, stems, roots, and flowers.

2. There are three types of leaves: cotyledons, prophylls, and true leaves. Cotyledons, the kidney shaped seed-leaves, ordinarily reach a breadth of approximately 50 mm. True leaves range from entire to deeply palmately lobed and are sometimes more than 200 mm in breadth. The prophyll is inconspicuous and not ordinarily observed. It is the first leaf on any branch, only 5 mm or less wide and may be mistaken for a stipule, which it resembles (Fig. 3-2).

3. In the axil of each cotyledon, prophyll, or true leaf *one* and *only one branch* apical meristem develops.

4. The only difference between a vegetative branch (monopodium) and a fruiting branch (sympodium) is that the fruiting branch apical meristem, after making a prophyll and a true leaf, terminates in a flower. A vegetative branch continues to make leaves until some stress causes it to cease growth.

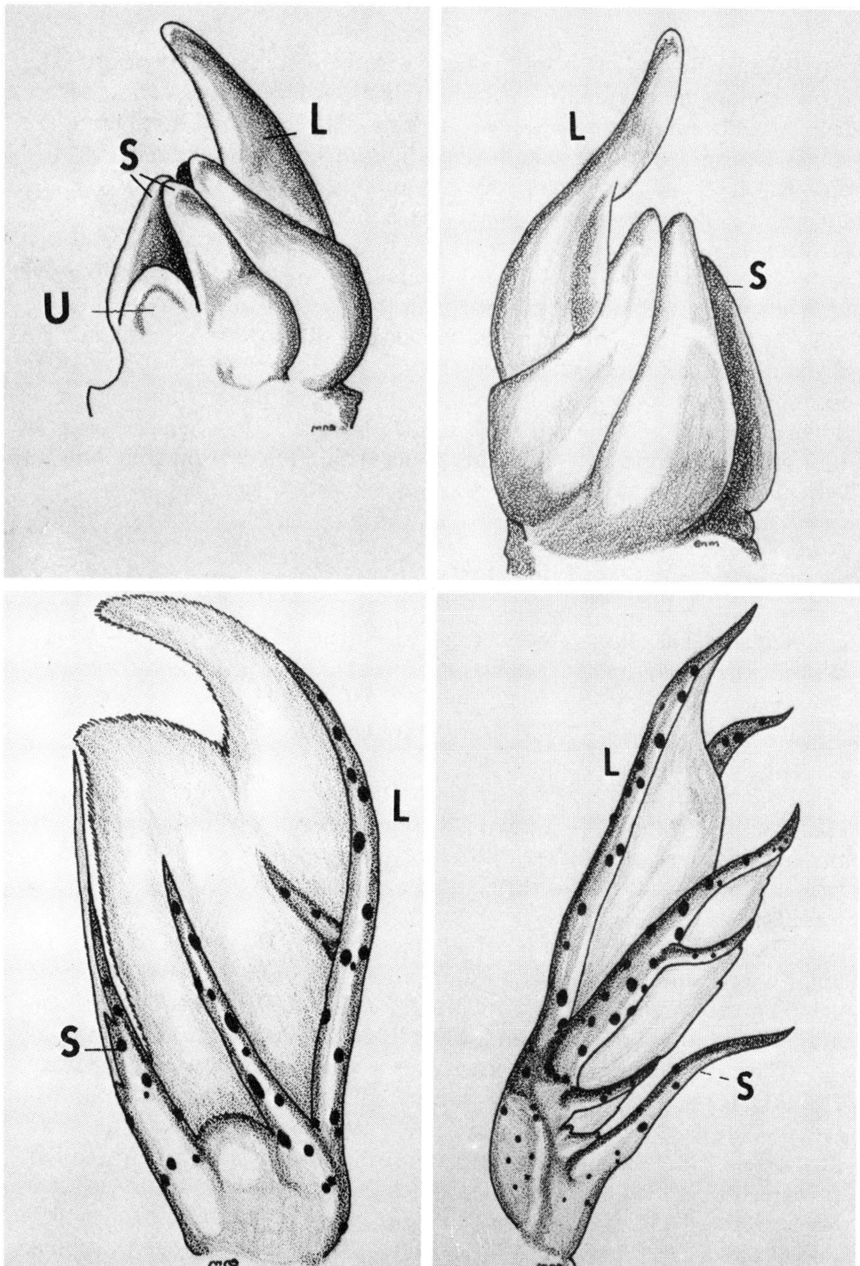

Fig. 3–2 to 3–5. Stages in leaf development. Fig. 3–2. U—unlobed primordium on edge of apical meristem and 0.1 mm L—second youngest leaf with stipules (S) differentiated. Fig. 3–3. 0.2 mm L—third leaf behind apex with 0.2 mm petiole. Fig. 3–4. 1.0 mm L—fifth or sixth leaf behind apex with 1.0 mm petiole. Fig. 3–5. 2.0 mm L—oldest unexpanded leaf. Fig. 3–2 and 3–3, 40X magnification. Fig. 3–4 and 3–5. 10X magnification. (Drawings by Edward Mulrean.)

3-2 ROOTS

Because the radicle is more massive in the seed than the epicotyl, the root develops more rapidly during germination. During the period before true leaves begin expansion, the primary root is penetrating deeply into the soil and branch roots are being formed. The normal root system in cotton is a typical tap root. Secondary root primordia develop endogenously at the juncture of the phloem and the cortex about 12 cm behind the primary growing point. Tertiary roots begin developing about 5 cm behind the secondary growing points. These branch roots are arranged in four rows along the tap root, one row in each quadrant.

Should the apex of the primary root be killed, there is an immediate flush of growth of secondary roots behind the necrotic area. One of these secondary roots may assume a dominant growth pattern and penetrate downward as would the tap root. Generally, however, when the primary root dies, the secondary and tertiary roots proliferate to form a lateral root system which never penetrates as deeply as tap roots.

3-3 AERIAL ORGANS

The aerial portions are more complex. The remaining discussion will describe the developmental pattern and interrelationships of these organs.

3-3.1 Leaf Development

As germination begins, the first true leaf is nothing more than a primordium on the growing point between the two cotyledons (Fig. 3-1). At the end of this stage the first true leaf expands to the sun. By the time the first leaf begins unfolding, the plant has developed six or seven additional leaf initials.

The first evidence of a new leaf is a protuberance on the edges of the apical meristem. The leaf primordium elongates and becomes lobed in developing the stipule (Fig. 3-2). During the early development the stipules may elongate more rapidly than the leaf itself (Fig. 3-3, 3-4, 3-5).

The arrangement of the leaves on a stem is known as phyllotaxy. Cotton has a spiral phyllotaxy with each leaf being 3/8 turn above the last. A stem may have a counter-clockwise (dextrorse) or a clockwise (sinistrorse) phyllotaxy. Half the stems tend to spiral to the right and half to the left. Infrequently, a stem may be seen in which the phyllotaxy has reversed direction.

Ordinarily there are 7 to 11 leaf initials above the youngest unfolded leaf (Fig. 3-6). I have chosen to characterize the developmental stages of these leaves by the length of the petioles, which range from approximately 0.1 mm in the youngest leaf in which lobing can be observed to 2 to 4 mm in the oldest unexpanded leaf. Cotton leaves have several distinctive features. They have two pulvini, one at each end of the petiole, which allow a wide arc of movement of the leaf blade to follow the movement of the sun. They

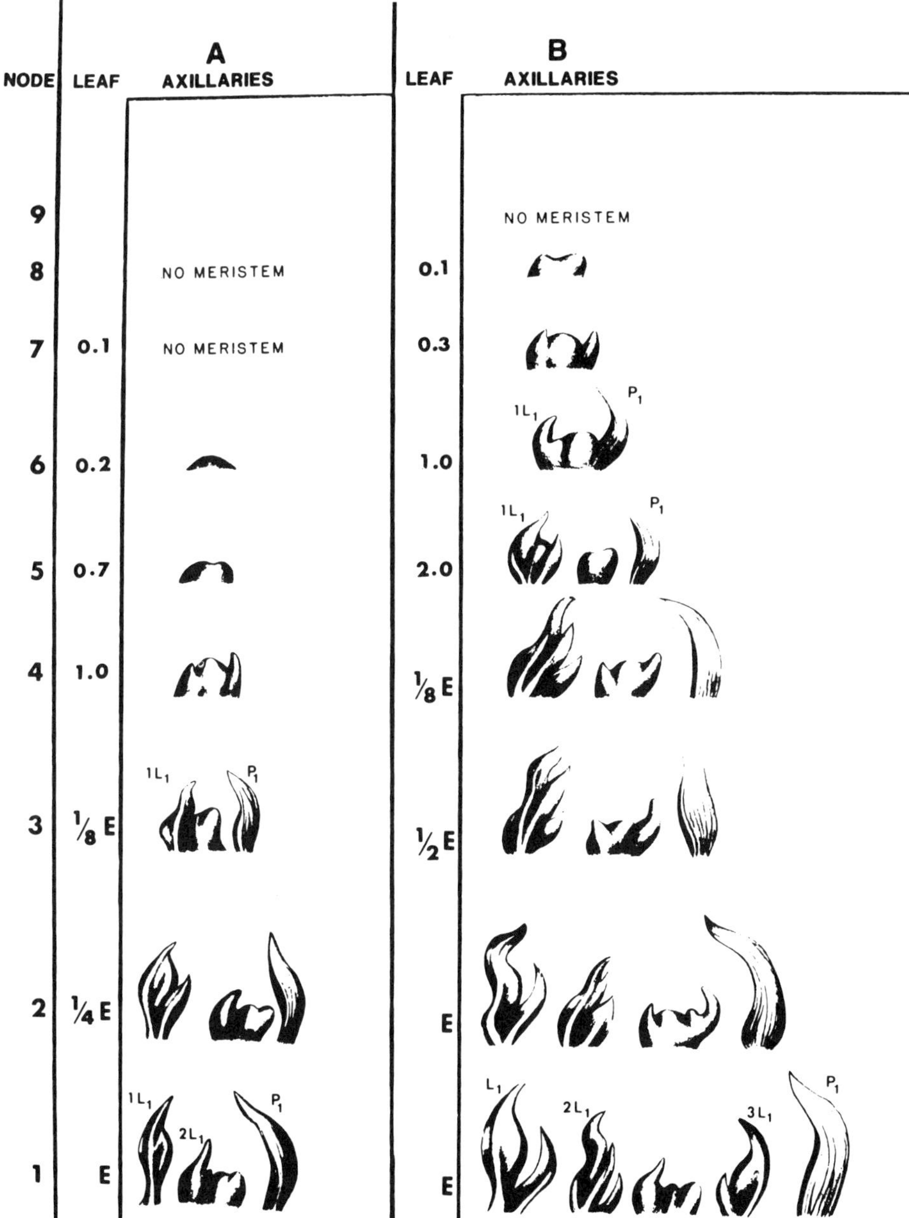

Fig. 3–6. Progressive development of axillary branch axes. Column A represents axillaries of a plant with first true leaf fully expanded, column B represents axillaries of a plant with three leaves expanded. Leaves above unfolded leaves are designated by petiole length (mm). Unfolded leaves are designated by proportion of full expansion (1/8 E, 1/2 E, 3/4 E, E). The progress of the branch at any particular node can be seen by comparing that node on Plant A and B. A similar progression is observed in comparing progressively older branches on each plant. The prophyll (P₁) of the axillary differentiates first, followed by the first true leaf (1L₁). Additional leaves (2L₁, 3L₁) develop as the branch ages. (Drawings by David Moreno.)

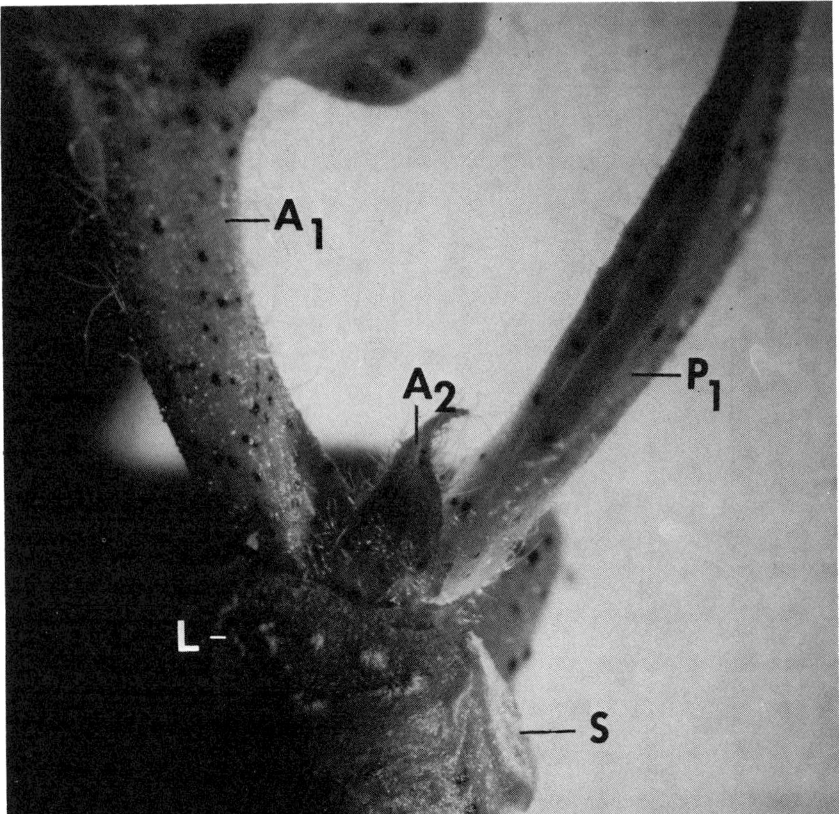

Fig. 3-7. Photograph of vegetative axillary branch bud showing prophyll (P_1), first axillary (A_1) and dormant second axillary (A_2). The main stem leaf has been removed to obtain this view and scars of its leaf (L) and stipules (S) are visible. This arrangement of parts is typical of plants with left-hand phyllotaxy. Other plants may have a reversed arrangement.

have leaf nectaries on the dorsal veins, and pigment glands scattered randomly in all tissues. These glands, which are also present in roots, stems, and flowers, are the repository of gossypol and other toxic compounds distinctive of *Gossypium* and closely related genera. Stomates are located two to three times more frequently on the lower surface than on the upper surface. Dale (1961) observed 37 to 50 stomates per mm^2 on the upper surface and 107 to 121 stomates per mm^2 on the lower leaf surface.

The first true leaf of the primary axis is usually not lobed. Its stipules are small and often abscise early. As the plant matures the leaves formed at the higher nodes are more and more deeply lobed until the climax leaf shape is formed at the 6th to the 10th node.

3-3.2 Typical Branch Development

A branch meristem develops in the axil of each leaf (Fig. 3-6). The initial developmental stages of monopodia and sympodia are indistinguish-

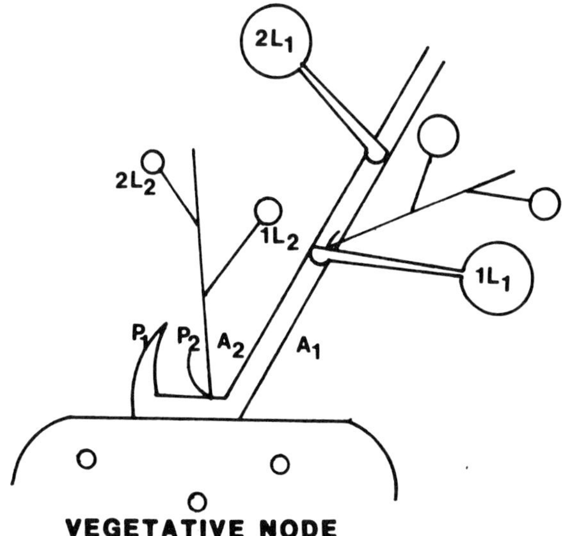

Fig. 3-8. Schematic representation of a vegetative axillary branch. Sawtooths at left—prophylls; open circles—true leaves; P_1—prophyll of first axillary (A_1) axis; P_2—prophyll of second axillary (A_2) axis; $1L_1$—first true leaf of first axillary; $2L_1$—second true leaf of first axillary; $1L_2$—first true leaf of second axillary.

able. The meristem which produces the branch is first visible in the axil of the third leaf below the apical meristem. Esau (1967) states that branch meristems which develop close to the apex are residues of the apical meristem itself. This probably happens in cotton.

The branch meristem begins development by producing a prophyll, an internode and a true leaf. The prophyll is a strap shaped leaf without stipules (Fig. 3-7). Its appearance is so like that of stipules that it can be mistaken for one. The prophyll almost always occupies the position on the branch axis *away* from the next higher node on the main stem. That is, if the plant has a counter-clockwise phyllotaxy, the prophyll will be found on the left side of the branch axis. The node represented in Fig. 3-7 was from a plant with a clockwise phyllotaxy.

After differentiating a prophyll, a branch meristem develops stem tissue of the first internode followed by the second leaf, a true leaf. This leaf is the first conspicuous leaf of the branch, and in the remaining discussion, it will be designated the "first true leaf."

3-3.2.1 Vegetative Branches

At the lower nodes of the primary axis the branch is a monopodium, a vegetative limb. Though the branch is technically a secondary axis the vegetative limb behaves essentially like the primary axis. It assumes an upright orientation and, as long as it is active, produces additional true leaves and internodes. At the time a main stem leaf begins unfolding, the vegetative branch at its base ordinarily has three or four true leaf initials.

Frequently the monopodium becomes dormant at this stage.

The leaves on a vegetative branch arrange themselves with a 3/8 phyllotaxy. There is no correlation between the direction of turn of branch phyllotaxy and that of the main stem.

An axillary meristem develops in the axil of each leaf on the branch. This is true of a prophyll as well as the true leaves. Fig. 3-8 represents schematically the vegetative branch and its axillary meristems. Because there is no internode below the prophyll, the axillary bud of the prophyll seems to lie in the axil of the main stem leaf. Thus, the bud appears to be an additional axillary of the primary leaf.

The naming of the prophyll axillary varies. It has been called the "axillary bud," the "extra axillary bud," and the "accessory bud." Mauney and Ball (1959) have called the major axillary the *first axillary* and the axillary of its prophyll the *second axillary*. The first axillary is responsible for most branching. In its developmental stages the second axillary is identical to the first axillary. It produces a prophyll, internodes, and true leaves. The second axillaries at the base of monopodia are invariably vegetative and usually dormant.

At nodes on which the second axillary elongates into a branch, a third axillary is observed (Fig. 3-9). This branch bud is in the axil of the prophyll of the second axillary. It is dormant unless forced to elongate by pruning of the stems and branches. Presumably, a plant which was forced by pruning to continue to make new branches could develop an indefinite number of axillaries in this manner.

3-3.4.2 Reproductive Branches

Early development of sympodial branches is identical to that described for monopodia (compare nodes 5, 4 and 3, Fig. 3-6B with nodes 11, 10 and 9, Fig. 3-10D). A prophyll, an internode, and a true leaf are differentiated.

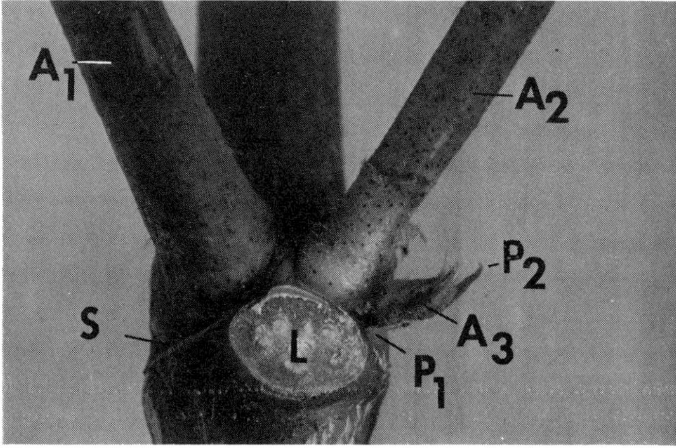

Fig. 3-9. Node showing development of third axillary (A₃) after elongation of first (A₁) and second (A₂) axillary branches. P₁—scare of prophyll of first axillary; P₂—prophyll of second axillary; L—leaf stump; S—scare of stipule.

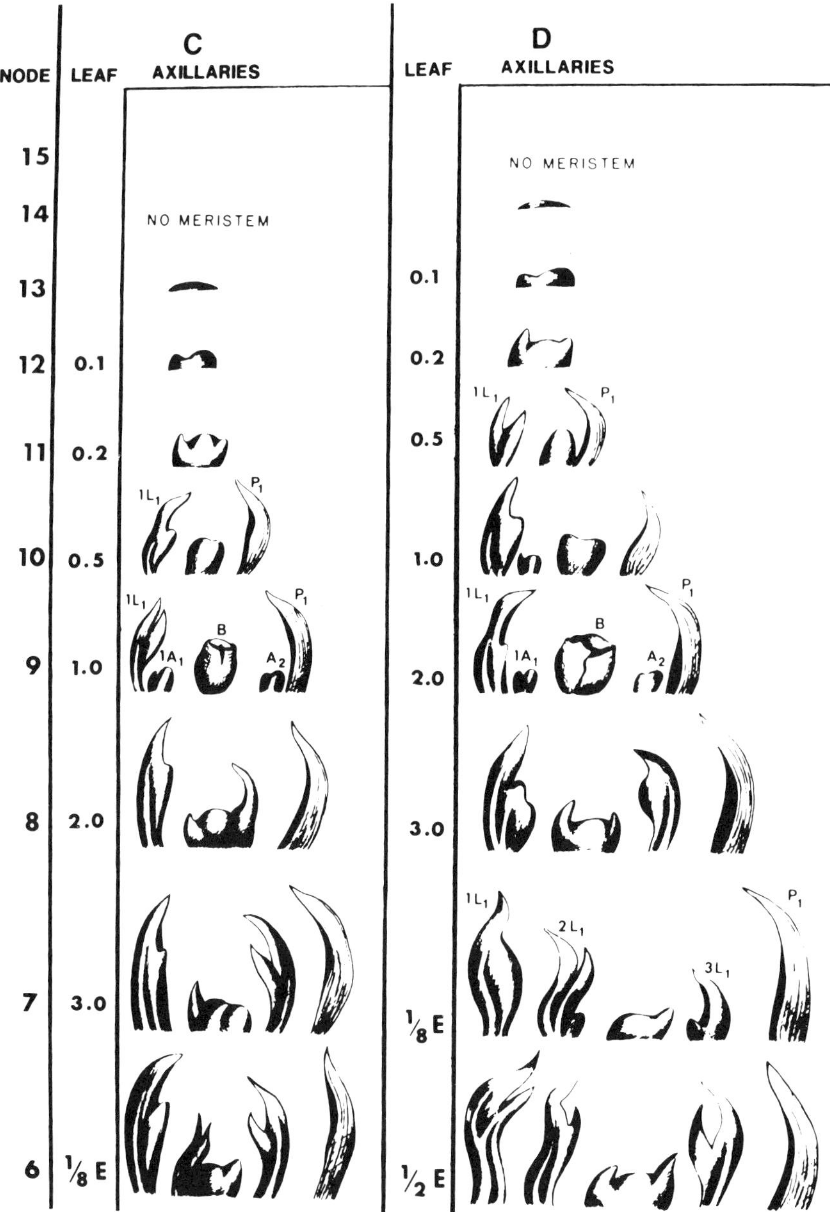

Fig. 3-10. Progressive development of axillary branch axes after floral initiation. Column C represents axillaries of plant with 6 unfolded leaves; column D pictures axillaries of plant with 7 unfolded leaves. Leaf designations and other symbols the same as in Fig. 3-6. Visible development is evident at the base of the second leaf below the apex. After the first axillary (A₁) differentiates the prophyll (P₁) and first true leaf (1L₁), the three bracts (B) of the flower bud develop. The flower bud terminates the first axillary. The second axillary (A₂) and first axillary of the first true leaf (1A₁) develop more actively than on vegetative branches. (Drawings by David Moreno.)

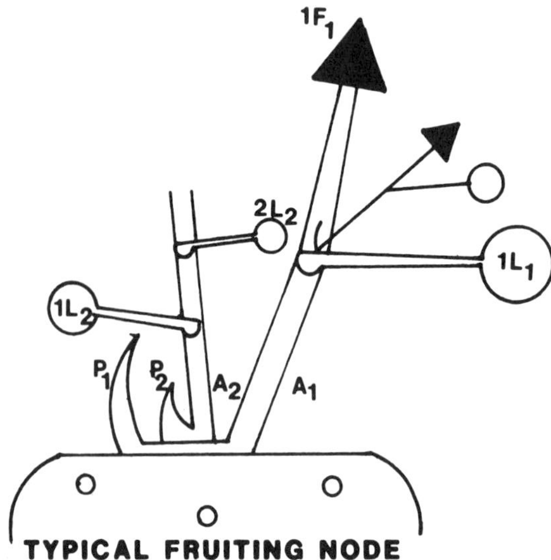

Fig. 3–11. Schematic representation of reproductive axillary branch, sympodium. Black triangles—flower buds; $1F_1$—first flower on first axillary. All other symbols the same as Fig. 3–8.

These are indistinguishable from those of a vegetative axis. All axillary primordia, then, begin development as vegetative meristems. Fruiting branches result from the transformation of these initially vegetative meristems into flower primordium.

The differentiation of the flower begins with formation of a whorl of three bracts after development of the first true leaf. Bract differentiation is first visible on the branch at the fifth node behind the growing point. The floral differentiation consumes the meristem of the axis and thus the axis terminates with the flower. There is no other difference between a vegetative and reproductive axis.

With the extinction of terminal growth by the floral development, the first axillary primordium at the base of the true leaf is stimulated. Very rapidly the primordium develops a prophyll, an internode, a true leaf and a terminal flower. These are the structures associated with the next flower on the sympodium.

Development of the sympodium continues—termination of each new branch by a flower bud and development of the axillaries to produce the next branch (Fig. 3–11). The fact that each internode is a new branch axis gives the sympodium a typical zigzag appearance.

Second axillaries at the base of sympodia may be vegetative or reproductive, dormant or active. The second axillaries at the base of early fruiting branches on the main stem often develop into vigorous vegetative branches. Less frequently, the second axillaries accompanying sympodia develop into fruiting branches.

The second axillaries at the base of the true leaves on the sympodium are ordinarily dormant vegetative buds. Infrequently, they differentiate into floral branches and add additional flowers to the sympodial axis.

The elements of the system described are basically simple. The plant has a complex branching habit because: 1) both the first and second axillary may be vegetative or reproductive, 2) any axillary may be dormant or active, and 3) each segment of a sympodium is a new first axillary axis. Each of the various branches on the plant are the result of an individual apical meristem responding to physiological forces within the plant. These forces determine whether the meristem will continue to produce leaf initials or terminate in a flower and whether the branch axis will elongate or remain dormant.

3-3.2.3 Additional Observations

Some of the implications of the system and some of the frequently observed variations on the general theme are as follows:

1. Because both monopodia and sympodia begin by making a prophyll–internode–true leaf, it is not until the bracts of a flower bud are differentiated that we see differences that would allow diagnosis of the kind of branch, vegetative or reproductive, a given meristem will produce.

In Fig. 3-12, 3-13 and 3-14 the early development of the two branch types is compared. Fig. 3-12 and 3-13 show that the two are almost identical in appearance until the second true leaf becomes lobed or until the second of the floral bracts differentiates.

2. Flowering branches grow more rapidly than monopodia. This difference is often a better characteristic to use in distinguishing the two types of primordia than their appearance. Apical dominance seems to diminish as the plant enters the flowering cycle. Flowering branch meristems approach the primary apex more closely than do the vegetative branch meristems (compare Fig. 3-6A with 3-10D). Likewise the second axillary is more active at the base of a fruiting branch, and the axillary at the base of the first true leaf is ordinarily more active on a fruiting branch than on a vegetative branch (Fig. 3-14).

3. The forces causing the development of fruiting branches seem to act throughout the plant at about the same time. If we examine the fruiting branches that occur on vegetative limbs, we find them progressively on lower nodes as we approach the first fruiting branch of the main stem (Fig. 3-15). From the cotyledonary node we find approximately equal number of nodes (leaf units in Fig. 3-15) to the first fruiting branch on all the vegetative axes. At the time of floral induction, branches at low nodes have more nodes already formed than branches at the higher nodes.

4. The first two fruiting branches are less vigorous than subsequent ones. Frequently the flower at these nodes develops more slowly than is typical later and may blossom on the same day or the day after the flower on the third fruiting branch. This slight inhibition of the early reproductive

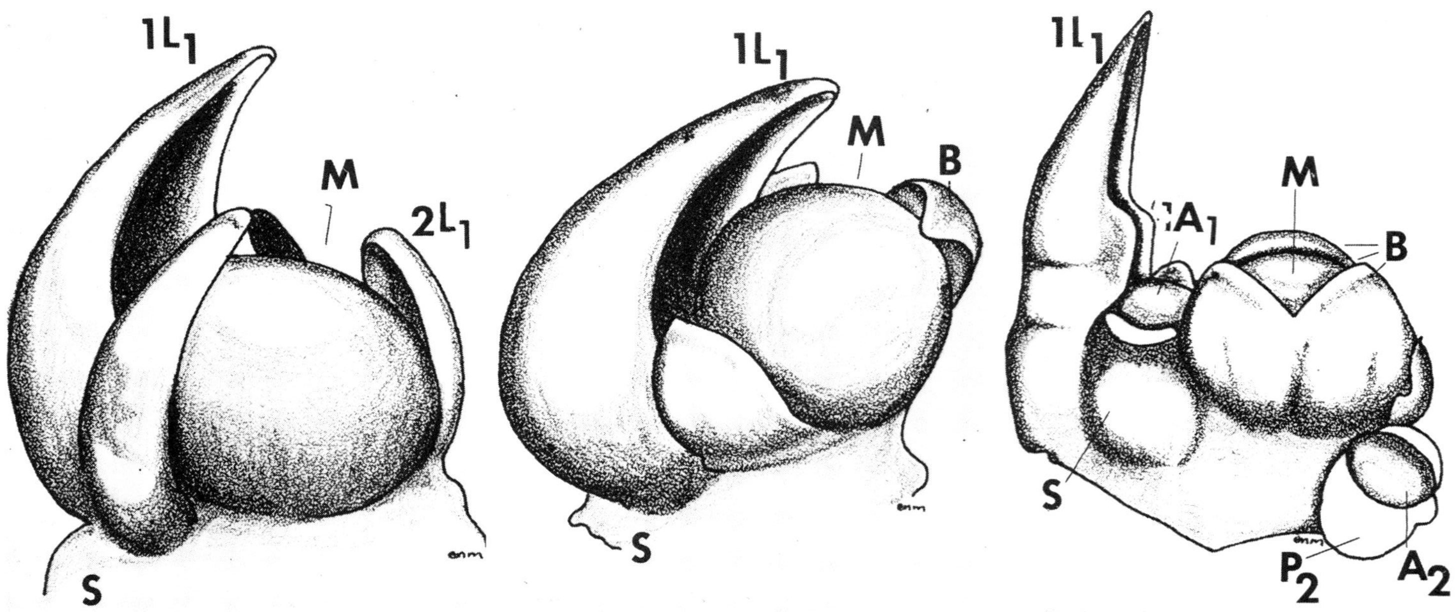

Fig. 3–12, 3–13, 3–14. Comparison of early stages in differentiation of vegetative and reproductive axillary branches. One stipule (S) of the first true leaf ($1L_1$) has been removed from each branch in order to show the apical meristem (M). 40X magnification. Fig. 3–12. Vegetative branch with differentiation of the second true leaf ($2L_1$) beginning. Subsequently, this branch would continue production of leaves in sequence identical to that shown in Fig. 3–2 to 3–5. Fig. 3–13. Reproductive branch with differentiation of the floral primordium beginning. The first of the three bracts (B) has developed on the meristem (M). Fig. 3–14. Subsequent development of the floral branch resulting in the three bracts (B) of the square. Development of the second axillary (A_2) and the first axillary ($1A_1$) of the first true leaf is apparent at this stage. P_1 is the base of the prophyll removed so that A_2 could be seen. (Drawings by Edward Mulrean.)

branches is probably a residual trait based on the dormant fruiting spurs observed in some ancestral desert species such as *G. aridum* (Mauney, 1968).

5. The rate of development of all organs on the plant is a function of temperature. However, the time interval between flowers on successive nodes along an individual sympodium, the horizontal flowering interval (HFI), is several days longer than the interval between first flowers on successive flowering branches, the vertical flowering interval (VFI). Hesketh et al. (1972) calculated the ratio HFI:VFI to be 2.7. Though the ratio was not constant over a wide range of temperature, at 30°C VFI was 3 days and HFI about 8 days.

The reason for this difference can be understood by analyzing the number of leaf units between flowers in the vertical and horizontal dimensions (Fig. 3–15). Because the structures of two successive sympodia can develop simultaneously, vertically there is only one additional leaf, that on the main

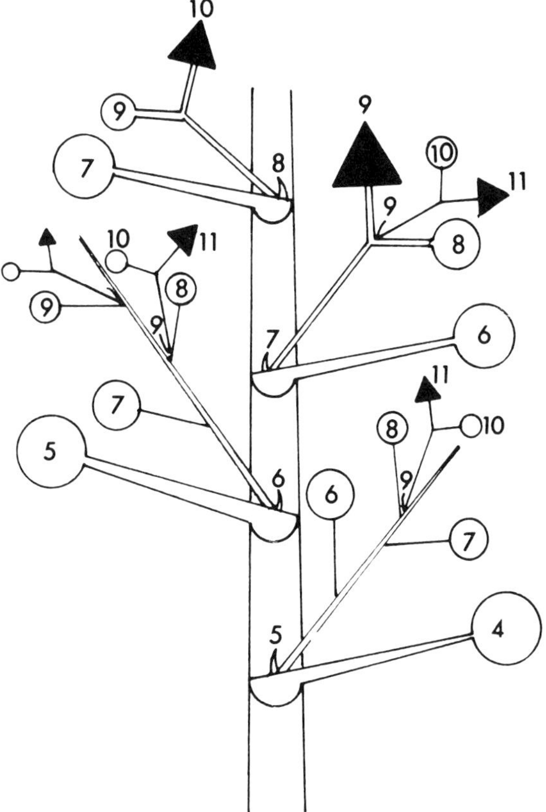

Fig. 3–15. Schematic representation of squares on sympodia on vegetative limbs and those on successive sympodia on the main stem. Numerals represent succession of leaf units (true leaves, prophylls, and bracts of squares). Other symbols same as Fig. 3–8 and 3–11. The plant represented had first flower at node 6. Though all leaf units are not precisely equal amounts of time, the events with equivalent leaf units should happen within a few days.

Fig. 3-16. Node at which two normal sympodia have developed.

stem, between two flowers. On the other hand, the second node on each sympodium cannot develop until after the formation of the first sympodial leaf ($1L_1$). Thus, horizontally there are two leaves, prophyll $2P_1$ and leaf $2L_1$, between successive flowers (Fig. 3-11). First flowers on successive sympodia occur as parallel events separated by one leaf, whereas successive flowers along each sympodium are series events separated by two leaves. Hence the longer time for HFI.

3-3.3   Atypical Branch Development

Certain branching patterns are rarely seen or occur only on certain genetic material or in particular environments. These patterns can be explained as extreme, but consistent, variations of the general scheme.

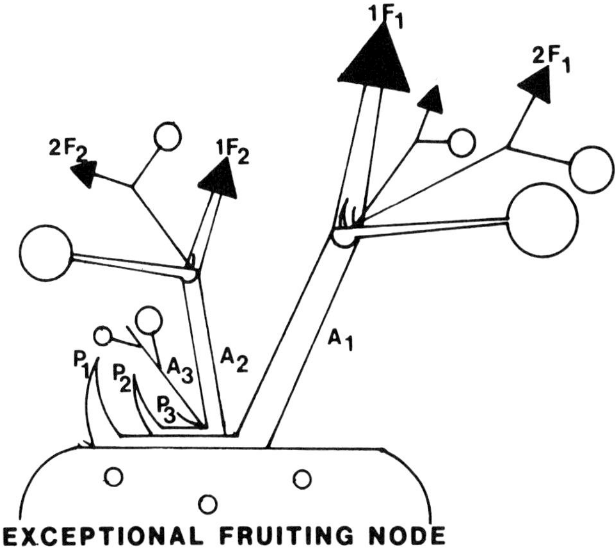

Fig. 3–17. Schematic representation of second axillary sympodium (A₂) and the resulting development of the third axillary (A₃). Symbols same as in Fig. 3–8 and 3–11.

Sympodia may develop from the second axillary position (Fig. 3–16, 3–17). If the forces causing flower differentiation are sufficiently strong to reach the second axillary, the plant can be induced to flower. The tendency to develop sympodia from this location seems to be dependent upon both genetics and environment. Moderate day and cool night temperatures favor development of this trait. *Gossypium barbadense* cultivars show the trait more often than those of *G. hirsutum.*

Though I have not systematically studied branching among cultivars the early fruiting Upland cultivars seem to branch atypically more than others.

When the second axillary is a sympodium it is usually abnormal in that no true leaf is developed. Thus, the branch is single-jointed because of the absence of a leaf axil from which additional nodes can develop. Less frequently the second axillary produces a many-jointed flower branch, an important factor in developing additional fruiting positions in heavily fruited plants.

The fact that second axillary sympodia are uncommon would seem to indicate that floral induction can occur only in a limited area behind vegetative apices. This area includes the three youngest nodes below the meristem (Nodes 14, 13, 12, Fig. 3–10D). Here the first axillary begins differentiation. The second axillary ordinarily differentiates on nodes five and six below the meristem (Nodes 10, 9, Fig. 3–10D). Should the induction area expand to include the differentiation site of the second axillary or should the second axillary differentiate within the narrow limits of induction it could become floral.

Photoperiodic species and cultivars may revert to production of vegetative branches when the inductive condition ceases. This action can be explained from the fact that each flowering node, whether first or last on a sympodium, is the result of a new branch meristem. If induction ceases, the plant must make vegetative axillaries at each new leaf. The primary axis will revert to making vegetative limbs and the sympodia may continue growth as monopodia or become dormant.

While this reversion is ordinarily observed only in selections which are extremely dependent on photoperiod or other environmental features, it is my belief that it also happens occasionally among cultivars.

Several genetically determined, atypical branching habits have been described. The cluster or short branch trait results in the shortening of internodes on the second joint of the sympodia and an accompanying development of the second axillaries into flowers. The foreshortening of the distances between the flower primordia frequently causes them to coalesce into a fasciated, malformed fruit. The whorl mutant of *G. herbaceum* is the only species in the genus in which I have observed a true terminal flower. Sympodia are initiated normally for 6 to 12 nodes, but then the vegetative apex differentiates into the terminal flower. This extinction of the apex permits vegetative branches lower on the stem to develop vigorously but they, in turn, also terminate in flowers.

Gausmann et al. (1977) described a branching trait which they named ''side-by-side.'' As the name implies, on these plants two adjacent branches occur opposed at a single node. Not all leaves and branches on these plants are arranged side-by-side. Most retain the alternate pattern with 3/8 phyllotaxy. Nodes with side-by-side branches usually occur early in the season. Because the two branches develop simultaneously, the first flowers on these branches typically blossom within 1 day of each other. The short flowering interval for these branches contributes significant earliness to genetic stocks that have a high frequency of the side-by-side expression (Namken et al., 1979).

3-4 FLOWER, FRUIT, AND SEED

3-4.1 Flower

The first true differentiation of the flower bud on the branch axis is production of three bracts rather than another true leaf. Subsequently, the other flower parts differentiate from the broad dome meristem that is characteristic of most plant flower buds. Quintanilha et al. (1962) estimated that a flower bud requires 36 days to develop (Fig. 3–18). However, the cross section of a flower bud 36 days prior to anthesis (Fig. 3–18A) shows well developed bracts. Thus, the total time for flower development is more nearly 40 to 50 days from differentiation to anthesis.

Stewart (1982) has outlined the coordination of events during flower morphogenesis and the effects which environment may have on stages of flower development. Formation of the carpels occurs about 35 days before

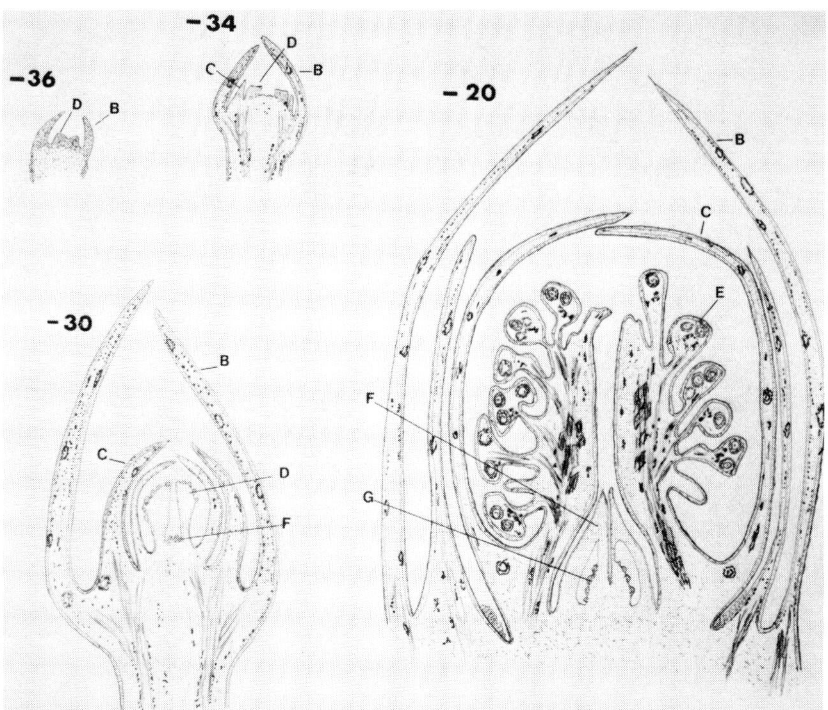

Fig. 3-18. Development of flower bud. Numbers refer to estimated days prior to anthesis (Quintanilha et al., 1962). B—bracts; C—corolla; D—androecium; E—anthers; F—ovary (locules); G—ovules. Drawings adapted from those of Baranov and Maltzev (1937).

anthesis. This is about 10 to 14 days before the flower bud (square) becomes visible. The bud is well protected by the terminal plumule at this stage. Nonetheless, environmental influences apparently determine ovule number at this stage. Pollen mother cells are undergoing meiosis about 22 to 23 days before anthesis. At this stage subsequent pollen viability is particularly sensitive to environmental stresses (Stewart, 1982).

Floral nectaries begin development about 25 days before anthesis. They are located internally on a ring at the inner base of the fused calyx and externally at the base of each of the three bracts.

3-4.2 Fertilization

After anthesis the pollen is deposited on the stigma, germinates, and the pollen tube begins its growth down into the ovary. Growth of the pollen tube takes 12 to 30 hours (Pollock and Jensen, 1964; Stewart, 1982) after which fertilization is accomplished. Cell division of the zygote begins 4 to 5 days after anthesis (Fig. 3-19).

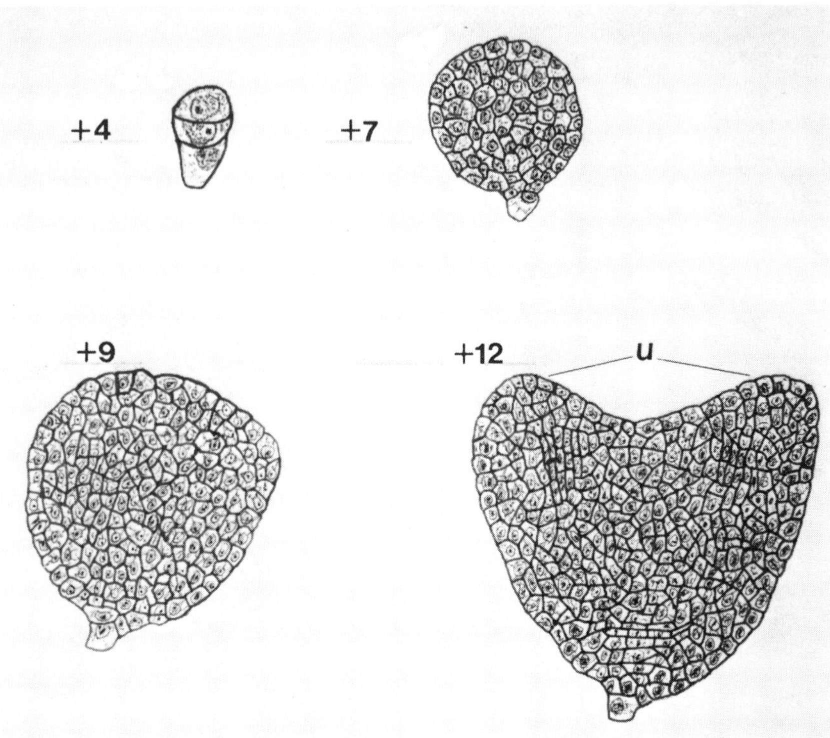

Fig. 3-19. Development of embryo. Numbers refer to estimated days post anthesis. u—differentiating cotyledons of the heart stage embryo. Drawings adapted from Quintanilha et al., 1962.

Enlargement of the young capsule (boll) begins with fertilization. The growth follows a sigmoid curve with the most rapid enlargement during the period 7 to 18 days post anthesis (PA). At 20 days PA the boll will have attained 90% of its final size. The mature size is reached about 25 days PA.

3-4.3 Fiber

Enlargement of the ovules and elongation of the ovule epidermal cells which result in the fiber, parallel closely the growth of the boll. The unfertilized ovule is about 1 mm in length. After fertilization it enlarges to about 8 to 9 mm in 15 to 18 days. The average size of mature seed is about 10 mm.

Initiation and development of the fiber of cultivated cottons have been described and supported by excellent electron photomicrographs by Stewart (1975), Beasley (1975), and Berlin and Woodworth (1983). These observers agree that the seed epidermal hairs, which result in the fiber, begin development within 24 hours PA. The epidermal cells forming these fibers are located randomly over the surface of the ovule, though elongation of those near the chalazal and micopyle ends may lag by a few hours or days (Fig. 3-20). The fiber cell elongates at a linear rate for about 20 days and

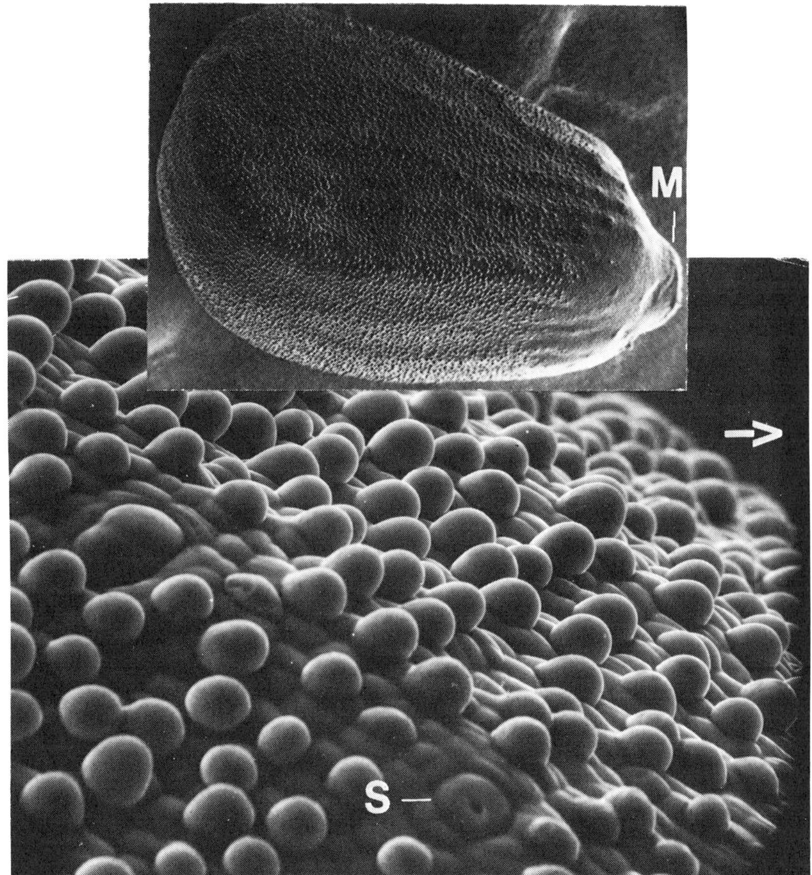

Fig. 3-20. Development epidermal fibers on ovules one day post-anthesis showing tendency of fiber initials to grow toward micropyle (arrow) (X211), inset (X51) showing delay of development near micropyle (M). Note stomata (S). (Photographs by J. McD. Stewart.)

eventually achieves a length of about 25 to 35 mm depending on cultivar and environment (Quisenberry and Kohel, 1975). At about 16 to 19 days PA, secondary thickening of the cell wall begins and continues until fiber maturity at about 35 to 60 days PA. Schubert et al. (1973) showed that secondary wall deposition begins 2 to 5 days prior to cessation of primary wall elongation. Delmar et al. (1980) suggested that a elongation during this period might be limited to the fiber tip or that the secondary wall deposit might be more flexible than the heavy deposits which occur subsequent to 21 days PA. The rate and duration of boll and fiber development are highly influenced by temperature (Gipson, 1982). At maturity the fiber dries, the central vacuole of the cell collapses and the fiber assumes the characteristic spiral ribbon that contributes to its spinability.

At 5 to 10 days PA a second group of epidermal hairs initiate development over the surface of the ovule. These hairs result in very short (about 5 mm or less) fibers that adhere to the seed during ginning and are generally referred to as fuzz or linters. Berlin and Woodworth (1983) observed that the secondary wall thickening of fibers is restricted at the base of the fiber in the epidermis. Fuzz fibers, on the other hand, have a well developed secondary wall at the juncture with the epidermis. This difference in thickening causes the difference in tendency to break away during ginning. Stomata are a feature of the ovule epidermis observed by both Beasley (1975) and Stewart (1975) (Fig. 3-20). The significance of these openings has not been elucidated.

3-4 Embryo

Growth of the embryo lags behind that of the ovule and fruit (Fig. 3-19). The size of the zygote decreases during the early cell division 4 to 5 days after anthesis (Pollock and Jensen, 1964). At 7 to 9 days the embryo is in the globular stage, approximately 0.1 mm in diam. At 10 to 12 days the differentiation of the cotyledon begins.

Pollock and Jensen (1964) found that the pattern of early cell divisions in the zygote is irregular. No definite pattern of cell lineage is established.

Turcotte and Feaster (1967) observed chimeras which can be traced to mixtures of genetically different cells in the young zygote. These cell mixtures behave as a unit in the embryonic stage, and the mature seed develops normally. After germination, sectors develop and indicate the presence of chimera. The fact that cell division patterns are irregular and that genetic mixtures present at the globular stage proceed to differentiate normally suggests that the organization of the embryo develops in response to forces external to the zygote. The physiology of the ovular tissues and the orientation of the embryonic mass of cells probably generate the forces that initiate differentiation.

The development of the cotyledons is the most striking feature of embryo maturation. The cotyledons enlarge rapidly 15 to 25 days after anthesis, fill the embryo sac at about 30 days, then mature and become capable of dormancy. At maturity the boll dehisces and the carpels (burrs) flex open. Subsequent rapid drying of the fibers and seed causes the funiculus to be severed from the placenta and the chalazal cap of the seed to be cemented in place. Christiansen and Moore (1959) confirmed earlier observations that imbibition of water for germination occurs through the chalazal cap. Density and impermeability of the chalazal plug may prevent water entry in certain *Gossypium* species and genetic strains. We have now completely described the life cycle, which began with the germinating seed.

3-5 SUMMARY

Even though the plant is limited to making leaves, stems, roots, and flowers, it can demonstrate a wide array of habits by controlling the arrangement and placement of these organs. Plant habit is determined largely by events at the developing apical meristems. The root system is typically a deeply penetrating tap root. However, if growth restrictions destroy the apical meristem of the primary root, a shallow, fibrous system may result from the proliferation of the secondary roots.

The aerial branching habit is determined by the differentiation and elongation of axillary meristems. The apex of the branch may continue to develop new true leaves or it may differentiate into a flower after the formation of the first true leaf. The branch may elongate rapidly or may be dormant. Flowering branches are less likely to remain dormant than vegetative branches.

Additional variations in branching habit arise from the development of the second axillary position. These positions are in the axils of the prophylls at the base of first axillary branches. The only restriction on the character of a second axillary is that it is more likely to be vegetative and dormant than the first axillary it accompanies.

REFERENCES

Baranov, P. A., and A. M. Maltzev. 1937. The structure and development of the cotton plant: An Atlas. Ogis-Isogis, Moscow-Leningrad.

Beasley, C. A. 1975. Developmental morphology of cotton flowers and seed as seen with the scanning electron microscope. Am. J. Bot. 62:584–592.

Berlin, J. D., and M. J. Woodworth. 1983. Anatomy and morphology of the developing cotton fiber. *In* J. R. Mauney and J. McD. Stewart (ed.) Cotton physiology: A treatise. USDA Handb. (in press).

Christiansen, M. N., and R. P. Moore. 1959. Seed coat structural differences that influence water uptake and seed quality in hard seed cotton. Agron. J. 51:582–584.

Dale, J. E. 1961. Investigation into the stomatal physiology of Upland cotton. Ann. Bot. 25: 38–50.

Delmar, D., N. Carpita, and D. Montezinos. 1980. Cellulose synthesis in fibers of *G. hirsutum* Acala SJ-1. p. 351. *In* Proc. Beltwide Cott. Res. Prod. Conf. St. Louis, Mo.

Esau, Katherine. 1962. Anatomy of seed plants. John Wiley and Sons, N.Y.

Gausmann, H. W., L. N. Namken, M. D. Heilman, and R. H. Dilday. 1977. Histological evidence for a cotton genotype with opposite leaves and fruiting branches at a main stem node. Crop Sci. 17:659–661.

Gipson, J. R. 1982. Integration of events in boll development: Temperature effects. *In* J. R. Mauney and J. McD. Stewart (ed.) Cotton physiology: A treatise. USDA Handb. (in press).

Gore, U. R. 1935. Morphogenetic studies on the inflorescence of cotton. Bot. Gaz. 97:118–138.

Hector, J. M. 1936. Introduction to the botany of field crops. Vol. II. Non-cereals. Central News Agency Ltd., Johannesburg.

Hesketh, J. D., D. N. Baker, and W. G. Duncan. 1972. Simulation of growth and yield of cotton: II. environmental control of morphogenesis. Crop Sci. 12:436–439.

Mauney, J. R. 1968. Morphology. *In* F. C. Elliot, M. Hoover, and W. K. Porter, Jr. (ed.) Advances in production and utilization of quality cotton: Principles and practices. Iowa State Univ. Press, Ames.

–––– and E. Ball. 1959. The axillary buds of *Gossypium*. Bull. Torrey Bot. Club 86:236–244.

Namken, L. N., M. D. Heilmann, and R. H. Dilday. 1979. Arrangement of sympodia and earliness potential of cotton. Crop Sci. 19:620–622.

Pollock, E. G., and W. A. Jensen. 1964. Cell development during early embryogenesis in *Capsella* and *Gossypium*. Am. J. Bot. 51:915–921.

Quintanilha, A. L., Salazar D'Eca, and A. Cabral. 1962. Desenvolvimento de botao floral do algodoeiro em funcao do tempo. Bol. Sco. Brot. 36:189–215.

Quisenberry, J. R., and R. J. Kohel. 1975. Growth and development of fiber and seed in Upland cotton. Crop Sci. 15:463–467.

Schubert, A. M., C. R. Benedict, J. D. Berlin, and R. J. Kohel. 1973. Cotton fiber development—kinetics of cell elongation and secondary wall thickening. Crop Sci. 13:704–709.

Stewart, J. Mc.D. 1975. Fiber initiation on the cotton ovule (*G. hirsutum*). Am. J. Bot. 62:723–730.

––––. 1982. Integrated developmental events and their response to environment in cotton bolls. *In* J. R. Mauney and J. McD. Stewart (ed.) Cotton physiology: A treatise. USDA Handb. (in press).

Turcotte, E. L., and C. V. Feaster. 1967. Semigamy in Pima cotton, *Gossypium barbadense* L. J. Hered. 58:54–57.

4 Qualitative Genetics, Cytology, and Cytogenetics

J. E. Endrizzi
University of Arizona
Tucson, Arizona

E. L. Turcotte
ARS-USDA
Phoenix, Arizona

R. J. Kohel
ARS-USDA and Texas A&M University
College Station, Texas

Cotton was one of the first crops to which the rediscovered mendelian principles of genetics were applied. The results of research on the inheritance of colored lint in *Gossypium barbadense* L. was reported by Balls in 1906; in *G. hirsutum* L. the inheritance of Okra leaf was published by Shoemaker in 1908; and the inheritance of Red leaf was reported by McLendon in 1912. Genetic research was done on the cultivated Asiatic species *G. arboreum* L. and *G. herbaceum* L., and the cultivated New World species *G. hirsutum* and *G. barbadense.* Much of this early research was conducted by personnel of the research organization of the Empire Cotton Growing Corporation.

As research on cotton genetics progressed and as genetic relations were better understood, researchers became aware that *G. arboreum* and *G. herbaceum* were diploid and *G. hirsutum* and *G. barbadense* were allotetraploid. These scientific revelations and concurrent changes in cotton production had their impact on research. Cultivation of the Asiatic species declined while use of the New World species, especially Upland cotton, *G. hirsutum*, increased until it became, by far, the predominant type grown.

Published in *Cotton,* Agronomy Monograph no. 24, © ASA-CSSA-SSSA, 677 South Segoe Road, Madison, WI 53711.

One effect of these changes was a near cessation of qualitative genetic research on the Asiatic species. It is unfortunate that, to a large extent, genetics of the Asiatic diploids has not been effectively related and transferred to the tetraploids. The difficulty associated with interspecific hybridization contributed significantly to this problem. In addition, as research on the genetics of the Asiatic species dwindled, so did the efforts to preserve this germplasm. Although we have the published records, very little of the germplasm was saved. Asiatic cultivars are still grown to a limited extent in India and Pakistan, with the germplasm preserved relating directly to cultivar production and only to incidental preservation of genetic mutant stocks.

The change did not cause a large upsurgence of genetic research in Upland cotton. The complexity of the genetics of an allotetraploid and the uniqueness of the allotetraploid genomic relations shifted emphasis to relations of genomes in the cultivated tetraploids and wild relatives of cotton. This emphasis created increased interest in cotton cytology and cytogenetics, but it centered primarily on studies of genome manipulation, with the evolution of the species taking prominence.

These investigations produced a wealth of knowledge of the cytological and genetical relationships of plant forms between and within species of the genus. However, they generated only limited knowledge of the genetic effects of individual chromosomes on agronomic traits and on overall growth and development of the cotton plant. The lack of large-scale research efforts in this direction was due to two major factors: (i) cytogenetics and qualitative genetics, though generally recognized as very important for genetic improvement, was presumed as not likely to result in immediate or short-term economic returns; and (ii) in the early days, the development of improved breeding, selection, and testing techniques could be relied upon to increase yield.

4-1 GENETIC MUTANTS AND GENE SYMBOLS

The Technical Committee of U.S. Regional Project concerned with cotton genetics and cytogenetics research, S-1, and its successor, S-77, established a subcommittee to establish rules for genetic nomenclature and to preservation of cotton mutant and cytogenetic germplasm. The Technical Committee adopted the subcommittee proposal based on the guidelines of the International Committee on Genetic Symbols and Nomenclature (Anonymous, 1957), and the lists of nomenclature rules and genetic mutants were published (Kohel, 1973b) to provide permanency and widespread distribution. The Technical Committee also established a permanent three-man committee composed of a *G. barbadense* geneticist (currently E. L. Turcotte), A *G. hirsutum* geneticist (currently R. J. Kohel), and a cytogeneticist (currently J. E. Endrizzi) to assist with the implementation and compliance of nomenclature and to insure preservation of genetic stocks.

4–1.1 Asiatic Diploid Species

The change in world cotton production and research emphasis to primarily *G. hirsutum* cultivars created a near cessation of reported qualitative genetic research on the Asiatic cultivated diploids, *G. arboreum* and *G. herbaceum.* The information abstracted by Knight (1954) represents basically the current state of knowledge of Asiatic cotton genetics (Table 4–1).

Table 4–1. List of gene symbols, name, and origin of Asiatic genes.

Gene symbol	Name	*Gossypium* species origin
B_4	Blackarm resistance	*arboreum*
B_{6m}	Blackarm resistance intensifier	*arboreum*
b_8	Blackarm resistance	*arboreum*
b_9	Blackarm resistance	*herbaceum*
chl_1	chlorophyll deficient	*herbaceum*
chl_2	chlorophyll deficient	*arboreum*
cl	short branch	*herbaceum*
Cp_a, Cp_b	Crumpled (complementary)	*arboreum* and *herbaceum*
cr	crinkled	*arboreum*
cu	curly	*arboreum* and *herbaceum*
d	female-sterile dwarf	*arboreum*
d_1	Anakapalle dwarf	*arboreum*
d_2	Cocanada dwarf	*arboreum*
d_3	dwarf bushy	*herbaceum*
de_1, de_2	incomplete boll dehiscence	*herbaceum* and *arboreum*
de_3	closed boll	*herbaceum*
Fz/fz	Tufted seed/fuzzy	*arboreum*
g	no ovary	*arboreum* and *herbaceum*
H_1	Hairy leaves	*herbaceum*
H^{vl}	Super stellate hairs	*herbaceum*
h_a	glabrous-lintless	*arboreum*
h_b	glabrous-lintless	*arboreum* and *herbaceum*
L	Narrow-leaves	*arboreum*
L^B	Mutant broad leaves	*arboreum*
L^J	Mutant intermediate leaves	*arboreum*
L^L	Laciniate leaves	*arboreum*
L^N	Narrow leaves	*arboreum*
Ld_1^K	Khaki lint	*arboreum* and *herbaceum*
lc_2	white lint	*arboreum* and *herbaceum*
Lc_2^B	Light brown lint	*arboreum* and *herbaceum*
Lc_2^K	Khaki lint	*arboreum* and *herbaceum*
Lc_2^M	Medium brown lint	*arboreum* and *herbaceum*
Lc_2^V	Very light brown lint	*arboreum* and *herbaceum*
Lc_3^B	Light brown lint	*arboreum* and *herbaceum*
Lc_4^K	Khaki lint	*arboreum*
li_a	hairy-lintless	*herbaceum*
li_b	hairy-lintless	*herbaceum*
li_c	hairy-lintless (lethal)	*arboreum*
li_d	hairy-lintless (lethal)	*herbaceum*
li_e	hairy-lintless (lethal)	*arboreum*

(continued on next page)

Table 4–1. Continued.

Gene symbol	Name	*Gossypium* species origin
$li_{(sh)}$	short lint	*arboreum*
li_{sp}	sparse lint	*arboreum*
lm	immature lint	*arboreum*
ls	single lobed leaf	*arboreum*
m_1	increased no. of floral parts	*arboreum*
m_2	multibracteolate	*herbaceum*
ne	nectaries absent	*arboreum* and *herbaceum*
P_a, P_b	Pollen color; $P_a P_b$ yellow, $p_a P_b$ cream, $P_a p_b$ pale, $p_a p_b$ cream	*herbaceum*
Ppf	Petalody	*arboreum*
pte	pistillate	*arboreum* and *herbaceum*
$R_2{}^{ASA}$	Sun-red spotted	*arboreum*
$R_2{}^{BO}$	Sun-red, spotless	*arboreum*
$R_2{}^{CS}$	Red calyx, spotted	*herbaceum*
$R_2{}^{DS}$	Thumb-nail red, spotted	*arboreum*
$R_2{}^{DO}$	Thumb-nail red, spotless	*herbaceum*
$R_2{}^{FO}$	Greenstem spotless	*arboreum*
$R_2{}^{GS}$	Tinged petal weak thumbnail red, spotted	*arboreum*
$R_2{}^{HO}$	Greenstem spotless, untinged petal	*arboreum*
$R_2{}^{LO}$	Red leaf, spotless	*arboreum* and *herbaceum*
R^{LS}	Red leaf, spotted	*herbaceum*
$R_2{}^{MS}$	Red margin, spotted	*arboreum*
$R_2{}^{MM}$	Red margin, spotless	*arboreum*
$R_2{}^{OS}$	Greenstem, ghost spot	*arboreum* and *herbaceum*
$R_2{}^{RS}$	Full red, spotted	*arboreum*
$R_2{}^{TS}$	Greenstem, tinged, ghost spot	*arboreum*
$R_2{}^{VS}$	Red vein, spotted	*arboreum*
$R_2{}^{VO}$	Red vein, spotless	*arboreum*
$r_3{}^{oo}$	Green, spotless	*arboreum*
Rl_a	Red lethal, complementary with factor from *G. hirsutum*	*arboreum*
Sr	Spot reducer	*arboreum*
stg	female sterile	*herbaceum*
stp	male-sterile	*arboreum* and *herbaceum*
v_1	virescent yellow	*arboreum* and *herbaceum*
v_2	virescent yellow	*arboreum* and *herbaceum*
v_3	virescent yellow	*arboreum* and *herbaceum*
v_4	virescent yellow	*arboreum* and *herbaceum*
vc	few loculed bolls	*arboreum*
W_1	*Fusarium* resistance	*arboreum* and *herbaceum*
W_2	*Fusarium* resistance	*arboreum* and *herbaceum*
W_3	*Fusarium* resistance	*arboreum* and *herbaceum*
X	Lint color modifier	*arboreum* and *herbaceum*
Y_a	Yellow petal	*arboreum* and *herbaceum*
$Y_a{}^P$	Pale petal	*arboreum* and *herbaceum*
y_a	white petal	*arboreum* and *herbaceum*
Y_b	Chinese yellow petal	*arboreum* and *herbaceum*
$Y_b{}^P$	Chinese pale petal	*arboreum* and *herbaceum*

4-1.2   New World Allotetraploid Species

As mentioned earlier, *G. barbadense* and *G. hirsutum*, now dominate world cotton production. *Gossypium barbadense* cultivars have extra-long, strong, and fine fibers. These cultivars include Sea Island, Egyptian, and Pima cottons as well as extra-long staple cultivars grown in USSR, Sudan, Peru, India, and other countries. Cultivars of *G. hirsutum,* Upland cotton, have shorter and weaker fibers, but are usually higher yielding than those of *G. barbadense.*

Most of the mutants among the allotetraploid species have been spontaneous in origin. Although mutants have been induced by mutagenic agents, only one new mutant in *G. barbadense* and in *G. hirsutum* has been identified from irradiation. Several mutants have been transferred from diploid species and between the tetraploids. A listing of tetraploid mutants, gene symbols, and origins is provided in Table 4–2 and they are discussed in the following sections.

4-1.2.1   Plant Anthocyanin

Anthocyanin pigmentation is expressed in the foliar plant tissues and as an expression of a petal spot at the base of the flower petals. A multiple allelic series of plant anthocyanin expression was established in the Asiatic species (Silow and Yu, 1942). The complexity of anthocyanin pigmentation genetics had produced uncertainty in assigning alleles to specific loci. In *G. hirsutum* three loci are identified, and two, R_1 and R_2, are homoeologous. Located in the D subgenome, R_1, behaves as an incomplete dominant and appears to be the major locus for foliar pigmentation. The wild-type r_1r_1 is a sun-red plant. Some genetic lines of *G. hirsutum* also are called "sun red," compared to the wild-type, and are presumed to carry R_1 alleles, but their precise allelic relations are not documented. It may be a mutation to the R_1^{dar} allelic state found in *G. darwinii* (Stephens, 1974a).

In the *G. hirsutum* A subgenome, R_2 gene conditions petal spot except in race *marie-galante*. Stephens (1974a) reported that some *marie-galante* strains have the R_2^v allele for red stem and veins, apparently introgressed from *G. barbadense*. Cultivated *G. hirsutum* is usually r_2r_2, no petal spot, but the races of *G. hirsutum* frequently have a petal spot. The R_2 allele acts as an incomplete dominant and the size and intensity of petal spot varies widely and is influenced by modifiers, and the inconsistency on individual plants suggest environmental stimuli rather than somatic mutation.

Table 4–2. List of gene symbols, name, and origin of allotetraploid genes.

Gene symbol	Name	*Gossypium* species origin
ac	curly cotton	*hirsutum*
as₁, as₂	asynapsis	*hirsutum*
av₁	albivirescent	*barbadense*
av₂	albivirescent	*barbadense*
B₁	Blight resistance	*hirsutum*
B₂	Blight resistance	*hirsutum*
B₃	Blight resistance	*hirsutum*
B₄	Blight resistance	*arboreum*
B₅	Blight resistance	*barbadense*
B₆	Blight resistance	*arboreum*
B₇	Blight resistance	*hirsutum*
B₈	Blight resistance	*anomalum*
B₉ₖ	Blight resistance	*herbaceum*
B₉ₗ	Blight resistance	*hirsutum*
B₁₀ₖ	Blight resistance	*hirsutum*
B₁₀ₗ	Blight resistance	*hirsutum*
B₁₁	Blight resistance	*herbaceum*
Bᵢₙ	Blight resistance	*hirsutum*
Bₙ	Blight resistance	*hirsutum*
Bₛ	Blight resistance	*hirsutum*
bp	smooth boll	*barbadense*
bw₁, bw₂	withering bracts	*hirsutum*
cg₁, cg₂	cleistogamy	*hirsutum, barbadense*
chl₁, chl₂	chlorophyll deficient	*barbadense, hirsutum*
ckˣ, ckʸ, ckᵒ	corky	*barbadense, hirsutum*
cl₁	cluster	*hirsutum*
cl₂	cluster	*barbadense*
cl₃	cluster	*hirsutum*
cr	crinkled leaf	*barbadense*
Crp	Crumpled	*hirsutum*
cu	cup leaf	*hirsutum*
cy	crinkle yellow	*hirsutum*
de	depauperate	*hirsutum*
Dw	Dirty white lint	*raimondii*
E	Enhancer, fertility	*barbadense*
F	Flowering	*barbadense*
fg, Fgʰ	frego bract	*hirsutum*
fng	fringless nectaries	*barbadense*
fs	female sterility	*hirsutum*
gl₁, gl₆, Gl₁ʸ	glandless boll	*hirsutum, barbadense*
gl₂, gl₃, Gl₂ˢ, Gl₃ʳ	glandless and glanded plants	*hirsutum, barbadense*
(gl₄, gl₅)	(modifiers)	*barbadense, hirsutum*
Gᵒ, Gˣ, Gʸ	Tumorigenesis	*hirsutum, gossypioides*
H₁(H₃, H₄, H₅)	Pubescent (modifiers)	*barbadense, hirsutum*
H₂	Pilose	*hirsutum, tomentosum*
H₆	Pubescent	*raimondii*
hb	hairy boll	*hirsutum, barbadense*
ia	accessory involucre	*hirsutum*
L₁ˡ	Laciniate leaf	*arboreum*
L₂ᵒ, L₂ˢ, L₂ᵘ	Okra leaf	*barbadense, hirsutum*
Lc₁	Brown lint	*hirsutum*
Lc₂	Brown lint	*barbadense, darwinii*, and *tomentosum*
Leᵈᵃᵛ, Le₁, Le₂	Lethal	*barbadense* and *davidsonii*
Lf	Leaf fleck	*hirsutum*
Lf	Filiform	*hirsutum*

(continued on next page)

Table 4–2. Continued.

Gene symbol	Name	*Gossypium* species origin
Lg	Green lint	*hirsutum*
Li	Ligon lintless	*hirsutum*
Lk	Leaky glands	*darwinii*
lp$_1$,lp$_2$	abnormal palisades	*hirsutum*
ltg	light green	*barbadense*
ml	mosaic leaf	*hirsutum*
ms$_1$	male-sterile	*hirsutum*
ms$_2$	male-sterile	*hirsutum*
ms$_3$	male-sterile	*hirsutum*
Ms$_4$	Male-sterile	*hirsutum*
ms$_5$,ms$_6$	male-sterile	*hirsutum*
Ms$_7$	Male-sterile	*hirsutum*
ms$_8$,ms$_9$	indehiscent anthers	*hirsutum*
Ms$_{10}$	Male-sterile	*hirsutum*
Ms$_{11}$	Male-sterile	*barbadense*
mt	mottled leaf	*hirsutum*
N$_1$	Naked seed	*barbadense, hirsutum*
n$_2$	naked seed	*barbadense, hirsutum*
ne$_1$ne$_2$	nectariless	*tomentosum*
ob	open bud	*hirsutum*
P$_1$	Pollen color	*hirsutum*
P$_2$	Pollen color	*barbadense*
pg	pale green	*hirsutum*
R$_1$,R$_1{}^{dar}$	Red plant	*hirsutum*
R$_2$,R$_2{}^v$,R$_2{}^r$	Petal spot	*barbadense, darwinii,* and *hirsutum*
Rc	Cracked root	*harknessii*
Rd	Dwarf red	*hirsutum*
rl$_1$	round leaf	*hirsutum*
Rl$_2$	Round leaf	*hirsutum*
Rf	Fertility restorer	*harknessii*
Rg	Ragged leaf	*hirsutum*
rs	rudimentary stigma	*barbadense*
Ru	Rugate	*barbadense*
s,s^r,s^l,S^h	strap leaf	*hirsutum*
Se	Semigamy	*barbadense*
Sm$_1$,Sm$_1{}^{sl}$	Smooth stem, leaf	*barbadense and armourianum*
Sm$_2$	Smooth leaf	*hirsutum*
Sm$_3$,sm$_3$	Smooth leaf	*hirsutum*
st$_1$	club stigma-style	*hirsutum*
st$_2$	mini stigma-style	*hirsutum*
st$_3$	invert stigma-style	*hirsutum*
v$_1$	virescent	*hirsutum*
v$_2$	virescent (golden crown)	*hirsutum*
v$_3$	virescent	*hirsutum*
v$_4$	virescent	*hirsutum*
v$_5$,v$_6$	virescent	*hirsutum* (irradiation)
v$_7$	virescent	*barbadense*
v$_8$	virescent	*hirsutum*
v$_9$	virescent	*hirsutum*
v$_{10}$	virescent	*hirsutum*
v$_{11}$	virescent	*hirsutum*
vf, Vf^h	veins-fused	*hirsutum*
wr	wrinkled leaf	*barbadense*
Y$_1$	yellow petals	*barbadense, hirsutum*
Y$_2$	Yellow petals	*darwinii*
yg$_1$,yg$_2$	yellow green	*barbadense, hirsutum*
yv	yellow veins	*hirsutum*

A third locus called Red dwarf was identified that produces red plant color (McMichael, 1942). This mutant, *Rd*, is incompletely dominant and causes dwarf red plants with the heterozygote intermediate in coloration and stature. This locus is placed in the D subgenome (Gerstel and Phillips, 1958; and Phillips, 1962), and it is independent of R_1 red plant.

4-1.2.2   Plant Hairs

The wild and cultivated species of *Gossypium* vary widely in the range of plant hairs. Pubescent, H_1, is the major gene used to impart dense pubescence to provide jassid resistance (Knight, 1952b). Saunders (1961b, 1965a, b) described a series of modifier genes that enhance the expression of H_1, H_3-stems, H_4-lower leaf surface, H_5-length. Saunders (1965c) transferred a pubescent character from *G. raimondii* Ulbr. and designated it H_6. He assumed that it was homoeologous to H_1. Pilose, H_2, produces short dense trichomes and short seed fibers (Simpson, 1947), and it is allelic with the native gene of *G. tomentosum* Nutt. ex Seem. (Knight and Sadd, 1953; Knight, 1952b).

A series of genes have been identified that reduce hairiness and one is Sm_2, transferred from *G. hirsutum* race stock (Lee, 1968), Endrizzi (personal communication) has established that Sm_2, H_1, and H_2 are alleles or closely linked loci. A second locus that reduces hairiness in *G. hirsutum,* Smooth stem, Sm_1, was introduced from *G. barbadense* (Lee, 1966) and Smooth leaf, Sm_1^{sl}, was introduced from *G. armourianum* Kearney (Meyer, 1957) and subsequently from *G. barbadense* (Lee, 1976). The locus Sm_1 is on chromosome 25 in the D genome. A third locus harbors an allelic series that reduces the level of hairiness below that of the traditional cultivated Upland cottons. A dominant allele, Sm_3, is found in the *G. hirsutum* races and a recessive allele, sm_3, was found in cultivated Upland cotton (Lee, 1968). The relation of the H_1 modifiers and Sm genes has not been fully investigated, but the H_1-Sm_2 probable allelism suggests further allelic relations may exist. The locus Sm_1 was found on the D subgenome chromosome homoeologous to that of H_1, H_2, and Sm_2 and is presumably the homoeolog of these mutants (Endrizzi, unpublished). It may be that H_6 is an Sm_1 allele.

Hau et al. (1980a) reported a recessive gene, hb_1 that conditions a hairy boll phenotype, and two recessive genes, cg_1 and cg_2, that condition cleistogamic flowers. These genes were isolated in the progeny of an interspecific cross between *G. hirsutum* and *G. barbadense.*

4-1.2.3   Leaf and Bract Shape

The degree of the leaf lobing is highly variable in feral populations of *G. hirsutum,* but the predominant form is the broad leaf-type characteristic of cultivated cottons (Stephens, 1945). Variation in lobing has been identified with alleles at the l_2 locus (D genome). Broad leaf, l_2, is recessive to the most common Okra leaf mutant, L_2^o, (Shoemaker, 1908) found in cultivated forms and certain wild forms such as *G. hirsutum* race *palmeri* (*G. lanceolatum* Tod.) (Stephens, 1945). An intermediate form termed sub-okra, L_2^u, is found in wild *G. hirsutum* and *barbadense*. An extreme type Super-okra, L_2^s, which develops a single leaf blade at maturity was found as a mutant in Okra leaf and as a variant in the Okra leaf wild forms of *G. hirsutum* (Stephens, 1945).

The dominant leaf shape allele has been transferred from the Asiatic cottons. Although a multiple allelic series exists in the Asiatic species (Hutchinson, 1934), only the Laciniate leaf allele, L_1^l, has been transferred to *G. hirsutum* and *G. barbadense. L_1^l and L_2^o are homoeologues.

Hau et al. (1981) reported a filiform leaf mutant, *Lf,* that conditions narrow cotyledons, leaves, and petals in *G. hirsutum.*

Genes at three closely linked loci were identified for their influence on leaf differentiation (Dilday et al., 1975). Crinkle dwarf, *cr*, was first observed as a mutant in *G. barbadense* and was thought to be an allelic series with differential response in *G. barbadense* and *G. hirsutum*. However, subsequent genetic analyses found only the mutant (*cr*) and normal (*Cr*) alleles at the crinkle dwarf locus. The closely linked loci are recessive strap leaf, *s*, the Indore (rugose) alleles, s^r, s^l, and a dominant allele, S^h, and recessive veins-fused mutant, *vf*, and a dominant allele, Vf^h (Dilday et al., 1975).

Flower bract shape, size, and dentation vary widely in the tetraploids, and they vary even wider in the genus. The recessive mutant frego bract, *fg*, has very narrow bracts (Green, 1955). In some genetic backgrounds, a portion of the leaves express an aberrant phenotype to various degrees. The dominant mutant allele, Fg^h, expresses a similar but more extreme phenotype in the heterozygous condition, but it is semilethal when homozygous, similar to *Lf*, S^h, Fg^h, *Rg,* and Rl_2 (Kohel, unpublished).

Knight (1952a) observed a mutant in 'Stoneville 2B' that was originally called deciduous bract. This recessive mutant was subsequently more descriptively named withering bract, bw_1. Homozygous withering bract also can be identified by a downward roll to the leaf edges. Rhyne (1965a) was able to transfer a normal allele from a D genome species to establish that withering bract is inherited as a double recessive, $bw_1bw_1bw_2bw_2$. All tested normal *G. hirsutum* stocks have the genotype $Bw_1Bw_1bw_2bw_2$.

4–1.2.4 *Flowers*

Harland (1936) determined that corolla or petal color in the tetraploid species of *Gossypium* is controlled by duplicate genes. The symbols Y_1 and Y_2 were assigned to these genes by Hutchinson and Silow (1939). The gene Y_1 conditions yellow petals in all allotetraploids except *G. darwinii* in which yellow petals are conditioned by Y_2. In *G. barbadense* yellow petals are the common form, but cream petals have occurred as mutants several times in Sea Island and Pima cottons (Turcotte and Feaster, 1963). Cultivated *G. hirsutum* is predominantly cream petaled with the genotype $y_1y_1y_2y_2$, but both yellow and cream-petaled types are found in the wild forms.

The open bud mutant, *ob*, has shortened corollas that expose the ends of the androecial column and pistil. Open bud was found in a cultivated Upland (Kohel, 1973a) and in a line associated with the transference of Y_2 from *G. darwinii* (Endrizzi, personal communication).

Stephens (1954b) presented data showing that Y_1 is located in the A subgenome. Endrizzi (1975) placed Y_2 on chromosome 18 of the D subgenome.

Pollen color in allotetraploid cotton species varies from cream to deep golden yellow. Harland (1929) used nine color grades to describe pollen color in a study involving strains of *G. barbadense* and *G. hirsutum*. He determined that pollen color was conditioned by one pair of alleles, *P* and *p*, with yellow dominant to cream. Stephens (1954a) determined that the pollen color gene *P* is carried in the A genome, and that *P* is probably homologous with the P_a locus in *G. arboreum*. Turcotte and Feaster (1966) described an orange pollen color mutant in Pima cotton that was conditioned by a second locus for pollen color. They assigned the symbols P_1 and p_1 to the alleles described by Harland, and the symbols P_2 and p_2 to the alleles at the new pollen color locus. Thus, allotetraploid cottons that breed true for yellow pollen color have the genotype $P_1P_1P_2P_2$ and true-breeding cream-pollen strains are $p_1p_1P_2P_2$ or $p_1p_1p_2p_2$, the latter being named neo-cream. The orange-pollen mutant has the genotype $P_1P_1p_2p_2$. The homology of the P_b locus in diploid species has not been determined.

4-1.2.5 Male Sterility

Genes for male sterility have been found with increasing frequency as the search for them has intensified. Eleven loci have been identified and all but the last one, found in *G. barbadense* (Turcotte and Feaster, 1979), are found in *G. hirsutum*. Four dominant genes were found, Ms_4 (Allison and Fisher, 1964), Ms_7 (Weaver and Ashley, 1971), Ms_{10} (Bowman and Weaver, 1979), and Ms_{11}. The recessive male-steriles include two duplicate factors, ms_5ms_6 (Weaver, 1968) and ms_8ms_9 (Rhyne, 1971). These duplicate factors are apparently completely male-sterile but exhibit some anther dehiscence. Among the single recessive factors, only ms_2 (Richmond and Kohel, 1961) is completely male sterile, and ms_1 (Justus and Leinweber, 1960) and ms_3 (Justus et al., 1963) vary from fully fertile to male-sterile.

Meyer (1975) transferred cytoplasms from the wild diploid, *G. harknessii* Brandg., species that interacts with the allotetraploid nuclear genes to produce male sterility, and she identified a source of fertility restoration. Weaver and Weaver (1977) identified this fertility restoration with a single dominant gene, *Rf*, from *G. harknessii* that requires a dominant enhancer, *E*, from *G. barbadense* to enhance pollen fertility of the F_1 hybrid (Sheetz and Weaver, 1980). DaSilva et al. (1981) recorded simple inheritance of Meyer's pollen fertility restoration in a cytogenetic test, but multiple inheritance in a conventional genetic analysis.

When the *Rf* gene was transferred to *G. hirsutum,* Cracked root locus, *Rc*, was linked with it and also transferred (Weaver and Weaver, 1979).

4-1.2.6   Female Sterility

Female sterility results from several loci with different modes of action. Occasionally plants are observed that are female-sterile and on further inspection are also male-sterile. The condition is the result of failure of the chromosomes to form stable bivalents, and it is attributable to duplicate recessive genes, as_1as_2 (Beasley and Brown, 1942). Asynapsis, more properly designated desynapsis, is generally found in certain *G. hirsutum* × *barbadense* hybrid progenies or materials derived from such hybridization. Knight (1954) determined that in those lines that exhibited asynapsis the *G. barbadense* parents were $As_1As_1as_2as_2$ and the *G. hirsutum* parent was $as_1as_1As_2As_2$.

Female-sterile plants in *G. hirsutum* were reported by Stroman (1941) and Justus and Meyer (1963). The mutant segregated in a simple manner and Justus and Meyer assigned the gene symbol fs_1 to it. Description of the mutant plants and their behavior suggests that the mutant may have involved the As_2 locus.

Several other mutants result in female sterility because the stigma-style is physically altered to prevent pollination. A rudimentary stigma and style, *rs*, was found in *G. barbadense* and effects the development of the corolla, stigma, style, and boll (Turcotte and Feaster, 1964). Three recessive *G. hirsutum* mutants have the same effect on female sterility; club stigma-style, st_1, has a club-shaped style with no stigmatic surface, miniature stigma-style, st_2, has a rudimentary style that does not extend beyond the androecial column, and inverted stigma-style, st_3, has a style that is misshapen with "inverted" growth and the corolla ends also fold inward with "inverted" growth (McMichael, 1965).

The homozygous-recessive mutant accessory-involucre, *ia*, also is female-sterile in addition to its more obvious affect on abnormal involucre growth (Kohel, 1965).

4-1.2.7   Lint and Fuzz Color and Pattern

Lint and fuzz color in the allotetraploids vary from white to brown. Cultivated *G. hirsutum* has white lint compared to the creamy colored lint of cultivated *G. barbadense*. Harland (1935b) found that separate loci determine brown lint in *G. hirsutum* and *G. barbadense*. In *G. hirsutum* the incomplete dominant, Lc_1, determines Brown lint while in *G. barbadense* it is Lc_2. Populations of *G. tomentosum* are uniformly homozygous for Lc_2, and various lines of *G. hirsutum* carry Lc_2. A third locus, *Dw*, is the D subgenome homoeolog of Lc_1. The original *Dw* gene was transferred to *G. hirsutum* as the Dirty white gene from *G. raimondii* (Rhyne, 1960), but its expression in *G. hirsutum* is equivalent to Lc_1, whereas Lc_2 is intermediate brown.

In *G. hirsutum* there is an incompletely dominant mutant, *Lg*, that produces Green lint and fuzz at boll opening that fades to a brownish color (Ware, 1932). Stocks from the wild *G. hirsutum* have varying intensities of green fuzz and white lint, and the Florida Green Seed line has dark green fuzz and white lint. The green fuzz mutants have been identified as alleles at the Lg locus (Kohel, unpublished data).

The amount and distribution of fuzz varies widely in the allotetraploids, but the presence or absence of fuzz (naked seed) is determined by two main loci. The dominant Naked seed gene, N_1, and the recessive naked seed gene, n_2, are found in both *G. hirsutum* and *G. barbadense* (Kearney and Harrison, 1927; Ware et al., 1944; Kohel, 1973b). The dominant N_1 is more extreme and consistent in its expression. The n_2 locus appears to have multiple alleles and to be influenced by modified genes (Kohel, unpublished data; Waddle, personal communication). Completely naked seed or partially naked seed expression of n_2n_2 varies with the background genotype, and tufted seed, a naked seeded type except for fuzz at the seed tip, acts as a recessive allele at the n_2 locus. In Pima cotton, bolls produced on lower fruiting branches have fuzzier seeds than bolls produced on higher fruiting branches (Kearney and Harrison, 1928).

The morphological mutant Ligon lintless, Li, also influences fiber and fuzz development. Seed of Ligon lintless plants have fuzz and a uniform layer of very short fibers, about 6 mm in length (Kohel, 1972).

4-1.2.8 *Chlorophyll Deficient and Virescents*

Chlorophyll deficient seedlings are frequently observed in progeny of *G. barbadense* × *G. hirsutum* crosses. Stroman and Mahoney (1925) and Harland (1932) determined that recessive duplicate factors conditioned chlorophyll deficiency. The *G. barbadense* stocks were Chl_1Chl_1 chl_2chl_2 and the *G. hirsutum* stocks were chl_1chl_1 Chl_2Chl_2 that gave rise to the chlorophyll deficient segregants, chl_1chl_1 chl_2chl_2, in the segregating generations (Harland, 1939).

A series of pigment developmental chloroplast mutants have been found in both *G. barbadense* and *G. hirsutum* that include virescent loci v_1 through v_{11}; yellow green, yg_1yg_2, and albivirescents, av_1 and av_2, identified in *G. barbadense* (Turcotte and Feaster, 1978). These mutants are expressed predominantly in the seedling stage; however, they vary widely in the intensity and duration of expression during growth of the plant, each influenced by the environment. Virescent 1, the *G. barbadense* homoeolog, v_7, (Turcotte and Feaster, 1973), and yg_1yg_2 are the most consistent in their expression.

The mutants pale green, pg, found in *G. hirsutum* (Murray and Brinkerhoff, 1966), and light green, ltg, found in *G. barbadense* (Turcotte and Feaster, 1978) are not virescents and retain their expression.

The recessive mosaic leaf mutant, ml, affects the pattern of chlorophyll formation in cotton plants. Mutant plants have a mosaic of chlorophyll and chlorophyll-less sectors that can express as early as the cotyledonary stage and continue through the life of the plant (Lewis, 1958). The abnormal palisade mutant is due to duplicate recessive factors, lp_1lp_2, and can sometimes be confused with mild expression of ml. Abnormal palisades is uniformly expressed in the cotyledons but varies widely in its expression on the leaves. The normal palisade mutant (Kohel, 1964) and the mottled leaf mutant, *mt,* (Lewis, 1960) were identified in progeny from crosses between cultivated and wild races of *G. hirsutum.* Mottled leaf has diffuse yellow areas on the leaves.

Variegated is a plastome mutant that occurs routinely as a spontaneous mutant in cotton plantings (Kohel, 1967; Kohel and Benedict, 1971; Katterman and Endrizzi, 1973; Endrizzi et al., 1974; Kestler et al., 1977). Discrete chlorophyll-less sectors are formed that follow the segregation of mutant chloroplasts in the plant, and they can be transmitted to seed progeny when these sectors are included in the flower.

4-1.2.9 Gossypol Glands

The *Gossypium* species are characterized by the presence of darkly pigmented lysigenous glands that occur throughout the plant. From a cross between cultivated and a primitive race of *G. hirsutum,* McMichael (1960) recovered a mutant, gl_2gl_3, that eliminates all glands on the aerial plant parts and seeds. The normal glanded alleles are incompletely dominant, so except when gl_2 and gl_3 are homozygous, the degree and distribution of glanding depends on the number of recessive and normal alleles, the loci involved, and the presence or absence of the modifiers gl_4 and gl_5 (Lee, 1962). Barrow and Davis (1974) found a new allele, $Gl_2{}^s$, that is intermediate in expression between Gl_2 and gl_2. The glandless genes, gl_2 and gl_3, are used in both *G. hirsutum* and *G. barbadense* improvement programs to produce seed free of glands.

In Egyptian *G. barbadense*, a single dominant gene was discovered following irradiation (Afifi et al., 1966) that results in glandless plants and seeds. This gene is being used in Eypgt to produce gland-free seeds (Bary, personal communication).

McMichael (1954) discovered a mutant, gl_1, that when homozygous recessive causes glands to be absent on the stem and boll. He later described another allele, $Gl_1{}^y$, that has intermediate expression (McMichael, 1970). Murray (1965) found that glandless boll and stem is due to duplicate factors when he identified a second locus in a *G. barbadense* stock, Z101. Normal Upland is Gl_1Gl_1, gl_6gl_6 and Z101 is Gl_1Gl_1, Gl_6Gl_6.

The inheritance of smooth and pitted bolls in Pima cotton was reported by Smith (1942). The smooth-boll trait in Pima is inherited as a monogenic recessive with the gene symbol *bp* (Knight, 1954). Normal, pitted boll Pima has the genotype *BpBp*. Lee (1978) transferred a gene from *G. barbadense* to *G. hirsutum* that he identified as a gland allele, $Gl_3{}^r$, with a rugate boll surface and higher gossypol content.

4-1.2.10 Disease Resistance

Resistance to bacterial blight [*Xanthomonas malvacearum* (E. F. Sm.) Dows.] in cotton is conditioned by a series of genes with the symbol *B* or *b* and appropriate subscript. They are listed in Table 4-2 as qualitative genes as they have been identified with single gene control, but they are not maintained as part of the qualitative genetic mutant germplasm because of the required testing with pathogen differentials for their identification. *Gossypium barbadense* and *G. hirsutum* cotton stocks with bacterial blight resistance genes are maintained by various plant pathology–breeding programs in which they are utilized.

Smith and Dick (1960) determined that a high degree of resistance to Fusarium wilt, *Fusarium oxysporum* f. *vasinfectum* (Atk.) Snyd. & Hans. in Seabrook Sea Island cotton was conditioned by two dominant genes which were additive in effect. One of these was transferred to Upland cotton. No symbols have been assigned to these two genes.

4–1.2.11   General Morphological Mutants

One group of mutants is usually identified on the basis of its expression on leaf development although its expression is not necessarily confined to the leaves. The mutants Ragged leaf, *Rg,* (Kohel and Lewis, 1962), Round leaf, Rl_2, (Percival et al., 1976); and Leaf fleck, *Lf,* (Kohel et al., 1977) are dominant mutants and their names are descriptive of the heterozygote phenotype. The homozygous mutants are semilethal, expressed as small plants with rosetted growth that do not usually survive in culture. There are dominant alleles at the strap leaf, veins fused, and frego bract loci, S^h, Vf^h, and Fg^h, that have similar semilethal behavior when homozygous. The Leaf fleck heterozygote is expressed in the seedling stage, but the remainder of these mutants are not expressed until the plants approach their flowering period.

Rugate leaf, *Ru,* arose by mutation in an experimental strain of Pima cotton (Turcotte and Feaster, 1965), and an identical or similar mutant has been found as chimeral mutants in *G. hirsutum* in Arkansas (Dilday, unpublished) and North Carolina and Texas (Kohel, unpublished observation).

Another Pima leaf mutant, wrinkled leaf, *wr,* was found in the commercial cultivar 'Pima S-4' (Turcotte and Feaster, 1980). Wrinkled leaf is recessive to normal leaf and is expressed only on a portion of the plant, from about node 10 to node 20.

Two leaf mutants described as recessives, cup leaf, *cu,* (Lewis, 1954) and algodon crespo (curly cotton), *ac,* (Tiranti, 1967) have partial expression as heterozygotes. Edges of the homozygous cup leaf mutant roll tightly upward and the heterozygotes have a slight upward curve. Homozygous curly leaf mutants are stunted, have wavy leaf edges, and in general are not vigorous. The heterozygotes can be identified by slight waviness of the leaves.

Round leaf, rl_1, was first described by Brown and Cotton (1937) and also called crenate leaf (Stephens, 1955). The name round leaf describes this mutant, which has a general rounding of the leaves by reduction of leaf lobing. Mutant expression generally begins at flowering.

The recessive mutant, depauperate, *de,* is aptly named (Kohel, 1973d). When homozygous, *dede,* plants express a general depauperate condition from late seedling to peak flowering growth periods. Wide seasonal variation in mutant expression is observed, and in seasons of extreme expression heterozygotes have partial expression.

4-1.2.12 Semigamy

Semigamy is an abnormal type of reproduction in which a sperm nucleus enters an egg cell but does not fuse with the egg nucleus. Subsequently, both nuclei divide independently resulting in F_1 plants that are chimeral for sectors of haploid, maternal tissue and haploid, paternal tissue. Semigamy was found to occur in a line of Pima cotton (Turcotte and Feaster, 1967). It can be utilized to produce at will, haploids of selected parentage. Doubling the chromosome number of haploid, paternal tissue sectors results in pure lines from male gametes (Turcotte and Feaster, 1969). Semigamy in Pima cotton appeared to be conditioned by one dominant gene, *Se,* with dosage effects for frequency of chimeral plants (Turcotte and Feaster, 1975b).

4-1.2.13 Corky

Corky is another trait that is expressed in young cottons. Corky plants apparently have a precocious development of cork cambium (phellogen) resulting in young stems developing a corky surface. Stephens (1950) has reported three alleles in the corky system, ck^o, ck^x, and cy^y. The ck^x allele is found only in *G. hirsutum* var. *marie-galante.* The remaining *G. hirsutum* types carry the allele ck^o. *G. barbadense* cottons are either ck^o or ck^y, with ck^y being the most common; however, Egyptian and Pima cultivars are ck^o. Corky plants have the genotype $ck^x ck^y$. All other combinations of alleles are noncorky. Thus, corky plants are hybrids that result from crosses of *G. hirsutum* var. *marie-galante,* ck^x, and *G. barbadense* plants that are ck^y.

Stephens and Phillips (1972) have shown that primitive cultivars of *G. barbadense* can be separated into two geographical groups. A western group includes the accessions from West of the Andes, and an eastern group includes the accessions from the Amazon and Orinoco basins and the West Indian islands and Central America. The western group is polymorphic for the corky alleles ck^o and ck^y. The eastern group is, with few exceptions, monomorphic for ck^y. Stephens (1974b) found the same geographical pattern for floral nectary fringe hairs. The western group is polymorphic for fringeless and fringed nectaries and the eastern group is monomorphic for fringeless nectaries. Sea Island cultivars had fringeless nectaries. Pima, Egyptian, Russian, and Peruvian cultivars had fringed nectaries. Limited inheritance data showed fringeless dominant to fringed, and determined by one gene pair, *FngFng.*

4-1.2.14 Lethality

Crosses of the diploid species *G. davidsonii* Kell. and *G. barbadense* and *G. hirsutum* result in hybrid embryos that abort, or the seedlings die soon after emergence apparently as a result of a complementary lethal interaction. Lee (1981a, 1981b) determined that all *G. barbadense* and *G. hirsutum* cultivars tested were homozygous for the dominant genes Le_1 and Le_2. He detected a third allele, Le^{dav}, in *G. davidsonii*. He proposed that crosses of cotton with the genotype $Le_1Le_1Le_2Le_2$ and *G. davidsonii* with the genotype $Le^{dav}Le^{dav}$ give triploid, hybrid embryos with the genotype $Le_1Le_2Le^{dav}$ that is lethal. Lee (1981a) isolated a *G. barbadense* stock with the genotype $le_1le_1le_2le_2$ that gave viable triploid hybrids when crossed with *G. davidsonii*. By hexaploid bridging, Lee (1981b) developed stocks with the genotype $le_1le_1le_2le_2Le^{dav}Le^{dav}$ and showed that this stock and *G. barbadense* and *G. hirsutum* stocks with the genotype $Le_1Le_1Le_2Le_2$ could be used for isolating cotton cultivars in seed production schemes. Lee (1981c) determined that Le^{dav} was linked with the Gl_3^{dav} gland determining allele at the Gl_3 locus with 25.9 recombination units.

Stem tumorigenesis in interspecific hybrids involving *G. gossypioides* (Ulbrich) Standley, a D genome species, is governed by a three allele genetic system—G^o, G^x, and G^y (Phillips, 1976). G^xG^y is the only tumorigenetic genotype. *Gossypium gossypioides* carries G^y and *G. raimondii* Ulbrich carries G^o. *G. hirsutum, G. tomentosum,* and *G. mustelinum* Watt carry G^o in one genome and G^x in the other. *Gossypium barbadense* has G^o in both genomes. Hybrids between *G. gossypioides* and G^x diploids develop tumors and die as seedlings. Hybrids between *G. gossypioides* and G^oG^x allotetraploids do not develop tumors until the 10 to 14 leaf stage and can survive several years.

4-1.2.15 Flowering

Flowering response in a short-day noncommercial stock of *G. barbadense* was controlled by one gene pair with the short-day, non-flowering response dominant to flowering. The gene symbols F and f were assigned (Lewis and Richmond, 1960). In *G. barbadense* × *G. hirsutum* interspecific hybrid progenies, Kohel et al. (1965b, 1974) found that flowering response was under multigenic control similar to that found in *G. hirsutum*. Thus, these two species have nonhomologous systems controlling flowering response.

4-1.2.16 Nectaries

Most cottons contain floral, extrafloral, and leaf nectaries. Leake (1911) reported a mutant in the Asiatic cottons that had no leaf nectaries. *Gossypium tomentosum* is unique among cottons in that it is devoid of leaf and extrafloral nectaries (called nectariless). Meyer and Meyer (1961) transferred the nectariless character to *G. hirsutum* and determined that duplicate recessives, ne_1ne_2, control the expression. Holder et al. (1968) established that ne_1 and ne_2 were linked to gl_2 and gl_3, respectively, and formed homoeologous linkage groups. The mutant genes are not completely recessive, and segregation at the Ne_1 or Ne_2 locus can be determined, primarily in testcrosses, by the size and number or absence of nectaries on the leaves and flowers (Holder et al., 1968).

4-1.2.17   Cluster Fruiting

A plant with fruiting branches reduced to one developed internode was found in Pima cotton in 1924 and named "short branch" by Kearney (1930b). He determined that short branch is an A subgenome monogenic recessive trait with the gene symbol cl_2 (Silow, 1946). Kulebyaev (1937) reported the occurrence of a similar trait in F_3 of a cross of two Ashmouni selections. Cluster fruiting habit in *G. hirsutum* is a monogenic recessive, cl_1, (Thadani, 1923), located on chromosome 16 of the D subgenome. Harland (1937) reported that crosses of short branch and cluster segregated as duplicate loci and presumably are homoeologues. Hau et al. (1980b) reported a new cluster gene, cl_3, located in linkage group III on the short arm of chromosome 16. The genes cl_3 and cl_1 are separated by 24.2 recombination units.

4-2   LINKAGES

4-2.1   Asiatic Diploid Species

The last summary of linkage tests in Asiatic cottons was made by Knight (1954). At that time seven linkage groups were identified and several linkage or pleiotropic relations between mutant and morphological or agronomic traits were described. The seven linkage groups are summarized in Table 4-3. Of these seven, linkage groups A II and A VII are identifiable with linkage groups IV and I of the allotetraploids, respectively. Individual loci have suspected homoeology with the allotetraploids, but specific linkage relations with mutants are not known.

4-2.2   New World Allotetraploid Species

The linkage map for the cultivated allotetraploids is presented in Table 4-4. All existing evidence indicates that this map is valid for both *G. barbadense* and *G. hirsutum*. However, most of the linkage analyses and chromosome associations have been determined for *G. hirsutum* and not all of the mutants have been transferred and tested in the *G. barbadense* genome.

Table 4-3. Asiatic linkage map.

Linkage group	Gene order and map units				
I	L	15	cu	19	Lc_1
	L	17.1	Li_d	20.5	Lc_1
II	Ha	7.1	Lc_2		
III	Y_a	24	Lc_3		
IV	P_h	16	Ne		
V	P_a	29.7	Ydp		
VI	R_3	1.2	Y_b		
VII	Chl_2	9	R_2		
	cl	30	R_2		
	b_8	1.4	R_2		

Table 4–4. Allotetraploid linkage map. The symbols sxl are respectively: short arm–centromere–long arm.

Linkage group	Gene order and map units	Chromosome	References
I	R_2 16 cl_2 4 yg_2 32 Lc_1	7	Endrizzi and Taylor, 1968; Harland, 1935a; Kammacher, 1968; Poisson, 1968; Rhyne, 1957; Silow, 1946; Stephens, 1955
II	lp_2 ? v_6 0 $L^o{}_2$ 3 sxl 44 Lg 5 vf 1 s 0.2 cr ←—— 51 ——→ ←—— 6 ——→	15	Dilday et al., 1975; Endrizzi, unpublished data; Endrizzi and Kohel, 1966; Endrizzi and Stith, 1970; Stephens, 1955; Wilson and Kohel, 1970
III	sxl cl_3 24 cl_1 17 R_1 19 yg_1 5 ms_3 33 ac 17 Dw ←— 30 —→ ←——— 26 ———→ ←——— 34 ———→	16	Endrizzi and Kohel, 1966; Harland, 1935a; Kammacher and Schwendiman, 1967; Kohel, unpublished data; Endrizzi, unpublished data; Rhyne, 1957; Hau et al., 1981
IV	Lc_2 10 sxl 4 H_2 (Sm_2,H_1) ←—— 22 ——→	6	Endrizzi, 1975; Endrizzi and Kohel, 1966; Stephens, 1955
V, XIII	gl_2 20 bw_1 39 ne_1 ? ms_8 ←——40——→ gl_2 8 Ms_{11} gl_2 27 Le_1 N_1 14 Ms_{11} N_1 7 Lf sxl 11 N_1	12	Endrizzi, unpublished data; Endrizzi and Ramsay, 1980; Holder et al., 1968; Kohel et al., 1977; Lee, 1965, 1982; Rhyne, 1962a, 1965a, 1965b, Rhyne and Rhyne, 1972; Turcotte and Feaster, 1979

(continued on next page)

Table 4–4. Continued.

Linkage group	Gene order and map units	Chromosome	References
VI	ia 31 fg	3	Endrizzi and Ramsay, 1980; Kohel et al., 1965a
VII	v_5 0 L_1^L 38 sxl 44 lp_1	1	Endrizzi, unpublished data; Endrizzi and Stein, 1975; Endrizzi and Stith, 1970; White and Endrizzi, 1965
VIII	st_1 32 sxl 23 ml_1	4	Endrizzi, unpublished data; Endrizzi and Bray, 1980; White and Endrizzi, 1965
IX	bw_2 5 gl_3 35 ne_2 16 ms_9 ? n_2 gl_3 26 Le_2 Gl_3^{dav} 26 Le_2^{dav}	26	Endrizzi and Ramsay, 1980; Holder et al., 1968; Kohel, 1979; Lee, 1965, 1982; Rhyne, 1962a, 1965a, 1965b; Rhyne and Rhyne, 1972
X	rl_1 20 Rg 15 rx $\longleftarrow$—31—$\longrightarrow$		Kohel, 1972; Percival and Kohel, unpublished data
XI	P_1 4 B_4 ? v_{11} $\longleftarrow$— 36 —$\longrightarrow$	5	Endrizzi and Ramsay, 1980; Percival and Kohel, unpublished data; Tayel et al., 1973
XII	v_{10} 4 Y_1	A	Percival and Kohel, 1976; Stephens, 1954b
XIV	v_8 13 Rd 33 st_3 $\longleftarrow$— 37 —$\longrightarrow$	D	Gerstel and Phillips, 1958; Kohel, unpublished data; Phillips, 1962
XV	v_3 12 Li		Kohel, 1978
XVI	ob_1 sxl 18 Y_2	18	Endrizzi, 1975, unpublished data; Kohel, unpublished data
XVII	Ru 37 yv 30 v_1		Kohel, 1983

The past 25 years has resulted in significant progress with the genetic identification of the allotetraploid chromosomes. Stephens (1955) identified four linkage groups with 12 mutant loci, but one of these loci, N_1, has since been eliminated from its proposed linkage group (Endrizzi and Taylor, 1968). The active development of genetic mutant germplasm with linkage analyses, monosome tests, and other cytological methods for chromosome identification have combined to produce these results.

There are 61 mutant loci identified with 16 linkage groups. Eleven of these linkage groups have been associated with chromosomes, two have been associated with specific subgenomes, and three remain to be associated.

Of the 13 chromosomes that have chromosome or genome identity, four homoeologous pairs have been identified by similar genetic mutants and monosomic phenotypes. Silow (1946) identified the homoeologies between the anthocyanin (R_1 and R_2) and cluster (cl_1 and cl_2) genes of linkage groups I and III, and Rhyne (1958, 1960) extended homoeology to include yellow green (yg_1 and yg_2) and brown lint (Lc_1 and Dw). The proposed homoeology of the leaf shape loci, L_1^l and L_2^o, was established with the chromosome association of lp_1 with L_1^l and lp_2 with L_2^o (Endrizzi and Stein, 1975; Endrizzi and Stith, 1970), and with the linkage of v_5 with L_1^l and v_6 with L_2^o (Kohel, 1973c). Homoeology of linkage group V on chromosome 12 with linkage group IX on chromosome 26 is well established with the identification of subgenome location (Endrizzi and Ramsay, 1979) and the identification of six duplicate loci, gl_2gl_3, bw_1bw_2, ne_1ne_2, ms_8ms_9, N_1n_2, and Le_1Le_2 (Holder et al., 1968; Rhyne, 1962a, 1962b, 1965a; Rhyne and Rhyne, 1972; Endrizzi and Ramsay, 1979; Lee, 1965, 1982; Kohel, 1979; Turcotte and Feaster, 1979). The fourth proposed homoeologous pair, chromosome 6 and 25, is identified by the similar phenotypes of the monosomes (Endrizzi and Ramsay, 1979). Chromsome 6 has linkage group IV, Lc_2 and H_2, and Sm_1 has been associated with chromosome 25.

Of the three homoeologous chromosomes that are genetically marked with 13 homoeologous gene pairs, only five pairs are expressed as completely recessive duplicates (yg_1yg_2, bw_1bw_2, ms_8ms_9, lp_1lp_2, v_5v_6). Gene expression of known or suspected homoeologous loci in the allotetraploids argues for an intermediate level of differentiation. Patterns of gene expression range from the completely recessive duplicates to fully diploidized, e.g., R_1R_2, cl_1cl_2, Lc_1Dw.

4–3 GENETIC MUTANT STOCKS

The lack of qualitative genetic research in the Asiatic species has provided little incentive or immediate need for preservation of the mutant stocks. The limited Asiatic germplasm maintained in the USA is notably devoid of mutant stocks. Although Asiatic cultivars are still grown to a small extent in India and Pakistan, the germplasm preserved relates directly to cultivar production and the preservation of mutants is only incidental because we have not been able to identify or obtain such germplasm.

Current working collections of *G. barbadense* genetic mutants are maintained and developed as part of the Pima genetic and improvement program at Phoenix, Arizona. The older group of *G. hirsutum* mutant stocks is maintained as part of the Stoneville Germplasm Collection, Stoneville, Miss. (Anonymous, 1974). The current working collection of *G. hirsutum* genetic mutants is maintained and developed as part of the Upland genetic program at College Station, Texas.

As a part of the operation of genetic programs, certain basic stocks are developed as biological tools required in the execution of the research activities. Multiple mutant lines were developed to aid linkage analyses. The multiple dominant marker line, Texas 586, includes R_1, H_2, L_2^o, R_2, Y_1, P_1, Lc_1, Lg, and N_1, which genetically mark seven linkage groups. The multiple recessive marker line, Texas 582, includes v_1, cu, gl_1, fg, and cl_1 that genetically mark five linkage groups, four of which are not marked by T586. *G. barbadense* marker lines include the dominant genes R_1, H_2, L_2^o, Lc_1, and the recessive genes y_1, p_1, p_2, v_2, gl_2, gl_3, cl_1, and rs.

To provide a common genetic background for evaluation maintenance, isolines of genetic mutants are developed. The genetic standards Texas Marker 1 of *G. hirsutum* and Pima S-4 of *G. barbadense* are used in the development of isolines (Kohel, 1973b). As part of the maintenance of the mutants, each mutant is routinely backcrossed to the genetic standard. Mutants are backcrossed eight generations to the genetic standard and evaluated in replicated tests (Kohel et al., 1967; Kohel and Richmond, 1971; Turcotte and Feaster, 1975a). Deleterious mutants with no obvious agronomic utility are not backcrossed as long nor are these placed in replicated tests for evaluation.

4-4 UTILIZATION OF GENETIC MUTANTS

Mutant stocks provide basic tools for scientific research into the mechanisms of inheritance, metabolism, and development. Morphological mutants are useful because they have relatively discrete phenotypes and can be readily observed; and because they have simple genetic control, they are readily manipulated. To be useful in any organism, a critical number of mutants is needed to design effective research. We now have enough mutants in cotton to appraise the evolution of the allotetraploids through the evolution of homoeologous loci.

Because the simple genetic mutants have discrete phenotypes, they have major metabolic effects that make them useful in studying the biochemical and developmental pathways of plant characters. The virescent mutants have provided useful tools in the study of photosynthesis in cotton (Benedict and Kohel, 1968; Benedict et al., 1971; Alberte et al., 1974) and the increasing number of these mutants may provide further dimensions to this research. The mutants affecting fiber development provided tools to research questions not answerable with materials possessing the normal range of variability (Kohel et al., 1974).

Genetic mutants play an increasingly important role in cotton improvement as plant breeders work to design and develop cotton plants for newer production systems and increased efficiency. Okra leaf (L_2^o) was recognized as having earliness and a potential to reduce the incidence of boll rot damage in humid areas (Brown and Cotton, 1937), but it was not until chemical herbicides were available to control the increased weed problems that this mutant could be evaluated practically and considered for production. Jones and Andries (1967) found L_2^o to be practical for boll rot control and to enhance early maturity.

The development of host-plant resistance programs resulted in the evaluation of genetic mutants for pest control, and frego bract was found to alter boll weevil behavior (Clower et al., 1970). Hirsute has long been used to control jassid (Knight, 1952b) and the presence or absence of plant hairs was found to interact with specific pests (Lukefahr et al., 1969). The nectariless trait has a significant effect on insects that rely on nectaries as a source of food (Meredith et al., 1973). This trait has been incorporated into the commercial cultivar, 'Stoneville 825,' which was planted on 11% of the U.S. production area in 1981.

The discovery of the glandless genes by McMichael (1960) generated great enthusiasm for producing cotton seeds free of gossypol glands. This enthusiasm was dampened by the observations that many pests prefer glandless to glanded cottons (Jenkins et al., 1966). In glandless cotton, as in most cases of the use of genetic mutants in plant improvement, there is a trade-off of desired benefits with undesirable effects. This situation is not new for plant breeding, but the mutants are definitive in their expression and more readily identifiable with responses. Mutants will continue to be evaluated and used in plant improvement to serve specific use requirements.

4–5 GENOME DIFFERENTIATION AND IDENTIFICATION

During the 1920s and 1930s, chromosome numbers of 2n = 26 and 52 were reported for many *Gossypium* species, including wild and cultivated forms occurring in America, Asia, Africa, and Australia (Youngman and Pande, 1927; Harland, 1928; Banerji, 1929, Baranov, 1930; Kearney, 1930a; Longley, 1933; Skovsted, 1933, 1934a, 1934b, 1935b, 1935c; Webber, 1934a, 1934b, 1935, 1939). Currently, the genus consists of 33 diploid and six allotetraploid taxa (Table 4–5; Fryxell, 1979). All are wild species except for two diploids and two allotetraploids which are grown commercially for their seed and fiber.

The chromosomes of the tetraploid species and the diploid species *G. herbaceum* were noted by early cotton cytologists to vary in size at metaphase I (Beal, 1928; Davie, 1933). These genome size differences were emphasized by Skovsted (1934a, 1934b; 1935a), who reported that the Old World and Australian diploids—*G. arboreum, G. herbaceum, G. stocksii* and *G. sturtianum*—had 13 large chromosomes, whereas the New World diploids—*G. thurberi, G. davidsonii, G. aridum,* and *G. harknessii (G. armourianum,* Skovsted, 1937)—had 13 small chromosomes.

Table 4-5. The Species of *Gossypium*.

Species	Genomic group	Distribution
1. **Diploids** ($2n = 2x = 26$)		
G. herbaceum L.	A_1	Old World cultigen
G. h. var. *africanum* (Watt) Mauer	A_1	Africa
G. arboreum L.	A_2	Old World cultigen
G. anomalum Wawr. & Peyr.	B_1	Africa
G. triphyllum (Harv. & Sand.) Hochr.	B_2	Africa
G. capitis-viridis Mauer	B_3	Cape Verde Islands
G. sturtianum J. H. Willis	C_1	Australia
G. sturtianum var. *nandewarense* (Derera) Fryx.	C_{1-n}	Australia
G. robinsonii F. Muell.	C_2	Australia
G. australe F. Muell.	--†	Australia
G. costulatum Tod.	--	Australia
G. cunninghamii Tod.	--	Australia
G. nelsonii Fryx.	--	Australia
G. pilosum Fryx.	--	Australia
G. populifolium (Benth.) Tod.	--	Australia
G. pulchellum (C. A. Gardn.) Fryx.	--	Australia
G. thurberi Tod.	D_1	Mexico, Arizona
G. armourianum Kearn.	D_{2-1}	Mexico
G. harknessii Brandg.	D_{2-2}	Mexico
G. klotzschianum Anderss.	D_{3-k}	Galapagos Islands
G. davidsonii Kell.	D_{3-d}	Mexico
G. aridum (Rose & Standl.) Skov.	D_4	Mexico
G. raimondii Ulbr.	D_5	Peru
G. gossypioides (Ulbr.) Standl.	D_6	Mexico
G. lobatum Gentry	D_7	Mexico
G. laxum Phillips	D_8	Mexico
G. trilobum (DC.) Skov.	D_9	Mexico
G. turneri Fryx.	--	Mexico
G. stocksii Mast. ex. Hook.	E_1	Arabia
G. somalense (Gurke) Hutch.	E_2	Arabia
G. areysianum (Defl.) Hutch.	E_3	Arabia
G. incanum (Schwartz) Hillc.	E_4	Arabia
G. ellenbeckii (Gurke) Mauer	--	Africa
G. longicalyx Hutch. & Lee	F_1	Africa
G. bickii Prokh.	G_1	Australia
2. *Allotetraploids* ($2n = 4x = 52$)		
G. hirsutum L.	$(AD)_1$	Central America
G. barbadense L.	$(AD)_2$	South America, Central America
G. tomentosum Nutt. ex Seem.	$(AD)_3$	Hawaii
G. mustelinum Miers ex Watt	$(AD)_4$	Brazil
G. darwinii Watt	$(AD)_5$	Galapagos Islands
G. lanceolatum Tod.	(AD)	Mexico

† A dash (--) indicates that the genome designation has not been determined. Species status of *G. lanceolatum* needs experimental verification.

Skovsted's cytological observations of chromosomal differences among the species were extended by Beasley (1940b, 1942), who established a cytological classification of genomes that is closely related to taxonomic affinities. Symbols were devised by Beasley (1940b) to designate the genomes of species. Similar genomes are designated by the same capital letter and closely related genomes are distinguished by a numerical subscript after each letter of that class. Seven diploid and one tetraploid genome groups are now recognized in *Gossypium* (Table 4–5).

Stephens (1947) and Katterman and Ergle (1970) have classified the relative chromosome size of the genome groups. The classification, including the recently discovered G genome is as follows: (i) the C genomes have very large chromosomes; (ii) the E and F genomes have large chromosomes which are slightly larger than those of the A genomes; (iii) the B genomes have large chromosomes, some of which are slightly larger than any in the A genomes (in Stephens's classification, the B and A genome groups are switched in relation to one another); (iv) the A genomes have moderately large chromosomes; (v) the G genome has chromosomes which are moderately large but smaller than those of A genomes (Wilson and Fryxell, 1970; Kadir, 1976; Edwards and Mirza, 1979); and (vi) the D genomes have small chromosomes.

Saunders (1961a) speculates that the center of origin of the genus *Gossypium* is in central Africa because four of the seven diploid genome groups occur on the African continent. Divergence of the genus into the different genome groups is generally assumed to have occurred before or during the separation of the continents in the Cretaceous period (Skovsted, 1934a; Hutchinson et al., 1947; Prokhanov, 1947; Mauer, 1954; Saunders, 1961a; Phillips, 1963; Hawkes and Smith, 1965; Fryxell, 1965, 1979). This event was followed by differentiation within each of the genome groups presumably during the Tertiary period (Fryxell, 1965; Phillips, 1966).

Chromosome homologies (Phillips, 1966), plant morphological characters (Fryxell, 1971; Valicek, 1978), and seed protein patterns (Johnson and Thein, 1970) have been utilized to deduce which one of the genomes is likely to be most representative of the initial genome of the genus.

Data on chromosome pairing in intra- and inter-genomic hybrids of the diploid species were summarized by Phillips (1966, 1974). Phillips hypothesized that reduction of chromosome pairing and chiasma frequency between two genomes can be attributed to structural changes, the degree of which is directly related to the time elapsed since their divergence from a common ancestor. The data showed that the intergenomic hybrids involving the E genome species had the highest frequency of univalents, indicating that the E genome is ancient in origin and therefore closest to the ancestral genome of the genus. Univalent frequency in genomic hybrids of A and B species were very low, suggesting that these two genomes represent a rather recent divergence in which A evolved from the B genome. Univalent frequency in hybrids involving C or D species indicated that the C and D genomes represent evolutionary lines of intermediate age. Divergence of the D genome was assumed to have occurred somewhat before the divergence of the C genome (Phillips, 1963, 1966).

More recently, the univalent frequency of two additional intergenomic combinations, D × F and G(?) × C, has been reported (Phillips, 1974). The F genome is closely related to the A genome and therefore to the B genome (Phillips and Strickland, 1966). The frequency of univalents in the D × F combination is very high and similar to that in B × D. Thus, based on Phillips' hypothesis, it can be assumed that the F genome is of more recent origin similar to that of the A and B genomes.

The C and G genomes occur in Australia, thus it is surmised that the two diverged at a later date, most likely following the isolation of the prototype genome with the Australian land mass.

Genome origin based on intergenomic chromosome affinities as interpreted by Phillips suggests that the original ancestral generic genome was composed of large chromosomes similar to the E genome. The later divergence of genomes with smaller and larger chromosomes would require a downward and upward shift in total nuclear deoxyribonucleic acid (DNA) content.

A comparative study of plant morphological characters among the diploid species by the Wagner Groundplan/Divergence Method was reported by Fryxell (1971). He found that species with the D genome had the least amount of evolutionary advancement, suggesting that the small D genomes are closest to the ancestral or prototype genome. This method of analysis indicates that the other six genomes must have arisen by an upward shift in total nuclear DNA content.

Utilizing numerical taxonomy in evaluating the phenetic relationships of the species of the A through F genomes, Valicek (1978) presented a genome phylogeny similar to that of Johnson and Thein (1970). Valicek's analysis indicates that the B genome is the prototype of the diploid species of the genus. The low correlation of the C and F genomes with other genomic groups, indicated that they may be of more recent origin.

Johnson and Thein (1970) determined the evolutionary affinities of the genomes by evaluating the band patterns of seed proteins. They concluded that the B genome was the closest survivor of the prototype genome, which gave rise to the A genome, one group of D species (D_ϵ), and the C and E genomes. The remaining D species group (D_β) emerged much later from a B-like genome. Thus, these interpretations are similar to Phillips in that they require upward and downward shifts in nuclear DNA content in genome evolution.

4-6 ORIGIN OF THE ALLOTETRAPLOIDS

Skovsted (1934a) was the first to point out that the cytological data indicates that the New World allotetraploids consisted of A and D subgenomes. This proposal was followed by efforts to produce a synthetic hybrid between an A and D species. This feat was accomplished independently in 1940 by Beasley and Harland, who crossed *G. arboreum* with *G. thurberi* and doubled the chromosomes of the F_1 hybrid. Beasley (1940a) observed less than one-half of the chromosomes paired in the diploid F_1 and regular bivalent pairing along with a high degree of fertility in the synthetic amphidiploid, which is what one would expect according to the allotetraploid hypothesis of Skovsted.

Harland (1937), Harland and Atteck (1941), and Silow (1946) provided genetic support for the allopolyploid origin by demonstrating that several loci of the allotetraploids were related (homologous) to those of the diploids by common origin.

The next task was to determine which genomes of the two Asiatic cottons and of the several American wild diploids were most similar to the progenitor forms of the allotetraploids. Stephens (1944) studied leaf developmental patterns in hybrids of Asiatics with various alleles for leaf shape and *G. raimondii,* one of the four American D species, with entire leaves. The results suggested that the D subgenome of the allotetraploids is most closely related to either *G. raimondii* or *G. klotzschianum* and its closely related form *G. davidsonii.*

Hutchinson et al. (1945) discussed a number of factors that should be considered in formation of a natural allotetraploid, which included the kinds of seed hairs occurring in the diploid species and the vigor of the hybrid forms. Among the then-known six D genome species, *G. raimondii* was the most likely candidate as donor of the D subgenome of the allotetraploids which, in combination with the genome of an Asiatic cotton, would provide the genetic system, including convoluted seed hairs, characteristic of the allotetraploids.

Gerstel (1953a) demonstrated that the two Asiatic diploid species, *G. arboreum* and *G. herbaceum* differ by one reciprocal interchange. The pairing data of the hybrids with *G. hirsutum* showed that the genome of *G. herbaceum* rather than that of *G. arboreum* is less structurally differentiated from that of the allotetraploids. In addition, the chromosomes of *G. herbaceum* were shown to be structurally similar to that of the closely related wild diploid *G. anomalum,* whereas *G. arboreum* differed from *G. anomalum* by a simple translocation.

Amphidiploids and allohexaploids involving Asiatic, New World diploids, and the natural allotetraploid species were synthesized and analyzed cytologically for multivalent frequencies and genetically for segregation of mutant genes located in the A and D genomes (Gerstel, 1953b, 1956, 1963; Sarvella, 1958; Gerstel and Phillips, 1957, 1958; Phillips and Gerstel, 1959; Phillips, 1960, 1963, 1963, 1964).

The pairing relationships and genetic ratios observed in the analysis of the hexaploid hybrids involving the D species showed that those involving *G. raimondii* most closely approached autotetraploid behavior, confirming that its genome is indeed most closely related to the contributor of the D subgenome of the tetraploids. Thus, it was concluded that the progenitor genomes of the allotetraploid species were most similar to the genomes extant in *G. herbaceum* and *G. raimondii.* Additional new diploid species have been identified, but they afford no evidence for an origin of the amphidiploids different from that now concluded.

4-7 TIME AND PLACE OF THE ORIGIN
OF THE ALLOTETRAPLOIDS

Skovsted (1934a), Saunders (1961a), and Valicek (1978) speculated that the origin of the allotetraploids goes back to the time when the A and D diploid parental forms overlapped, prior to the origin of the rifting of the continents. Others have also suggested the allotetraploids originated in the area of overlap of the two progenitor genomes, but after the rifting of the continents. Harland (1935a) assumed that the allotetraploids originated in the Polynesian Islands by migration of *G. arboreum* across a trans-Pacific land bridge where it met with a D species in the late Cretaceous or early Tertiary. Stebbins (1947) proposed that an A genome species arrived in North America before the Eocene by way of Behringia. Neither of these hypotheses is now tenable (Stephens, 1947; Gerstel, 1953a; Hutchinson, 1959; Fryxell, 1965, 1979).

Davie (1935) and Mauer (1954) proposed that the allotetraploids arose by hybridization of New World diploids. The former speculated that it occurred in the very recent past, while the latter places the origin in the Eocene.

A recent origin in which a *G. arboreum* cotton was introduced into the New World by humans over a Pacific route was suggested by Hutchinson et al. (1947) and Hutchinson (1959). This idea was discarded when Gerstel (1953a) reported that the A genome of *G. herbaceum,* rather than that of *G. arboreum,* is closer to the A subgenome of the allotetraploids. This suggested the possibility of an Atlantic transfer of an A genome species (Gerstel, 1953a; Phillips, 1963).

The evidence for an Atlantic meeting of the two progenitor diploids suggested to Hutchinson (1962) that the New World origin of the allotetraploids is an ancient event in which the two wild diploids came in contact by natural spread. He postulated that a wild type similar to *G. herbaceum* var. *africanum* existed east of the Andes which hybridized with *G. raimondii.*

A recent and biphyletic or polyphyletic origin of the allotetraploids was proposed by Sherwin (1970) and Johnson (1975). Sherwin's argument is based primarily on the erroneous premise that no wild forms of the New World cottons exist, and on the assumption that a cultivated *G. herbaceum* was transported to northern South America either by people, by abandoned water craft, or sealed in gourds. According to Johnson, man transported *G. herbaceum* to the tropical regions of the New World, where it was cultivated and then hybridized with more than one D genome species.

Comparative cytogenetics and taxonomy of the *Gossypium* species essentially rules out a very recent origin of the allotetraploids involving human transport of the A genome progenitor. Phillips (1974) pointed out that the earliest archeological specimens of cotton in the New World are dated at 4000 to 3000 B.C. which is 2000 years before known archeological cotton in Africa—1000 years before agriculture in Africa—and 3000 years before man made long distance voyages in the African area.

Phillips (1963) interpreted the cytogenetic data as establishing that the allotetraploids are neither ancient nor recent in origin. Cytogenetic data showed that the A and D subgenomes of the allotetraploids, primarily the former, had diverged little during their evolutionary history from the progenitor genomes represented in the Asiatics and *G. raimondii,* which was assumed to be incompatible with ancient origin. The fact that the allotetraploids have differentiated into distinct species, which are widely distributed on islands of the Caribbean and the Pacific, suggested that they are not recent in origin.

Thus, Phillips (1963) concluded as did Fryxell (1965, 1979), who in addition stressed the importance of the several pairs of sibling species and the association of the allotetraploids with littoral habitats, that the allotetraploids arose most likely in the Pleistocene following an Atlantic transfer of the progenitor A genome species to the New World.

4-8 MONO- VS. POLY-PHYLETIC ORIGIN OF THE ALLOTETRAPLOIDS

From the extensive cytological data, most students of *Gossypium* evolution postulate a monophyletic origin for the allotetraploids. However, several individuals have favored a polyphyletic origin (Davie, 1933, 1935; Mauer, 1938, 1954; Kammacher, 1959, 1960; Sherwin, 1970; Parks et al., 1975; Johnson, 1975).

Since there were several diploid species in the New World, Davie (1933, 1935) suggested that they probably intercrossed to produce different amphidiploids. The extreme genetic differences between the allotetraploid species coupled with their geographical distribution led Mauer (1938, 1954) to suggest like Davie, that the allotetraploids had independent origins involving hybridization between different New World diploids. Their proposals were not seriously considered since it was shown that the allotetraploids contained A and D subgenomes (Skovsted, 1934a, b).

Based solely on chromosome pairing behavior in the F_1 hybrids of *G. hirsutum* and *G. barbadense* with *G. raimondii,* Kammacher (1959, 1960) concluded that *G. hirsutum* is less "diploidized" and therefore of more recent origin than *G. barbadense.* The two allotetraploids are assumed to have arisen by separate hybridization events.

Sherwin (1970) speculated that a *G. herbaceum*-like cotton was transported to northern South America, either by rafting or by humans, where it was cultivated and hybridized, possibly with *G. raimondii,* to produce *G. barbadense.* The diploid A genome species was then carried north to Mexico where it hybridized with either *G. gossypioides* or *G. trilobum* to produce

G. hirsutum. However, cytological and taxonomic data excludes the involvement of the D genomes of *G. gossypioides* or *G. trilobum* in the origin of the allotetraploids.

Based on studies of flavonoid compounds of flowers, Parks et al. (1975) assumed that the allotetraploids had separate origins. They speculated that *G. barbadense* and *G. tomentosum* could have arisen possibly from a combination of *G. klotzschianum* and an A species, whereas *G. hirsutum* and *G. mustelinum* originated from a combination of *G. raimondii* and an A species. The cytological data do not support their interpretation since the chromosomes of *G. tomentosum* rather than those of *G. mustelinum* are more similar to the chromosomes of *G. hirsutum* as was demonstrated by Hasenkampf and Menzel (1980). Johnson's (1975) polyphyletic hypothesis was based primarily on the electrophoretic banding patterns of seed proteins, from which he concluded that the allotetraploids arose from possibly three independent hybridization events.

The banding patterns of the two A genome species, *G. herbaceum* and *G. arboreum,* were identical, whereas the D genome species were divided into two groups. One, designated D_β, comprised *G. raimondii, G. lobatum, G. aridum, G. laxum,* and *G. gossypioides* and the other, D_ϵ, consisted of *G. thurberi, G. trilobum, G. davidsonii, G. klotzschianum, G. armourianum,* and *G. harknessii* (Johnson and Thein, 1970).

Except for some accessions of *G. hirsutum* "race" *palmeri,* all accessions of *G. barbadense, G. hirsutum, G. tomentosum, G. darwinii,* and *G. mustelinum* (*G. caicoense* in Johnson) that were examined had identical banding patterns. Their banding patterns were similar to the banding pattern simulating a 1:1 mixture of the protein extract of *G. herbaceum* and any one of the D genome species in the D_β genome group. The exceptional *palmeri* accessions had a different banding pattern which simulated a 1:1 mixture of protein extract of *G. herbaceum* and a species of the D_ϵ genome group. Because of these differences, Johnson concluded that "race" *palmeri* is a distinct species, *G. palmeri* (*G. lanceolatum* in Fryxell, 1979) and that the allotetraploids arose from possibly three independent hybridization events in a center of early American civilization. He presumes that *G. herbaceum* combined with *G. trilobum* of the D_ϵ group to produce *G. palmeri.* This event was followed by the origin of *G. hirsutum* in Southern Mexico and Central America from a *palmeri* prototype with the $AAD_\epsilon D_\epsilon$ genome and from one or more prototypes with the $AAD_\beta D_\beta$ genomes. *G. barbadense* with the $AAD_\beta D_\beta$ genomes is assumed to have originated in South America between *G. herbaceum* and *G. raimondii.* From the Meso-American center of origin, *G. tomentosum* was carried by humans or ocean currents to Hawaii.

Phillips (1963, 1964) points out that if *G. hirsutum* and *G. barbadense* have different D subgenomes due to a biphyletic origin, then the frequency of multivalents and the segregation for a given locus should be different in hexaploid hybrids involving these two species and the same American diploid species. However, both hexaploid hybrids have quite similar multivalent frequencies and segregation ratios, which establishes the sameness or unity of their genomes, and therefore a monophyletic origin.

4-9 NEW EVIDENCE FOR THE ORIGIN OF THE ALLOTETRAPLOIDS

Recent research in *Gossypium* has provided information on the DNA content and on the kinetic components of chromosomal DNA and their relationships to the structural organization of chromatin, which is applicable to any hypothesis on the origin and evolution of species of *Gossypium.*

Size of the chromosomes is a feature of evolutionary significance distinguishing the different genome groups of the diploid species. There is approximately a two-fold difference in size between A and D genome chromosomes (Mikhailova, 1938; Endrizzi and Phillips, 1960; Edwards et al., 1974; Mursal and Endrizzi, 1976, Mursal, 1978). Moreover, this increase in size is distributed rather uniformly throughout the genome, i.e., there is a gradation in size from the longest to the shortest chromosomes of *G. arboreum, G. herbaceum, G. anomalum, G. sturtianum, G. stocksii, G. thurberi, G. bickii,* and *G. hirsutum* (Arutjunova, 1936; Mikhailova, 1938; Abraham, 1940; Jacob, 1942; Wouters, 1948; Afzal, 1949; Bose, 1953–55; Edwards, 1977, 1979; Edwards and Mirza, 1979).

Chromosomes are structurally uninemic; therefore, the differences in genome size are attributable to a longitudinal increase or decrease in the DNA's of individual chromosomes. The eukaryotic chromosome is normally composed of four major sequence classes of DNA: (i) single copy, (ii) moderately repeated (interspersed repetitive), (iii) highly repeated (clustered repetitive), and (iv) foldback or palindromic sequences (Britten and Kohne, 1967, 1968). Recently, the DNA's of *G. arboreum, G. herbaceum, G. thurberi, G. raimondii,* and *G. hirsutum* have been analyzed (Walbot and Dure, 1976; Wilson et al., 1976; Geever, 1980). The fractional components of single copy and different classes of repetitive DNA were analyzed by Walbot and Dure (1976) and Geever (1980). Their results are briefly summarized in Table 4-6 and for the most part, only the data relative to the evolution of the A and D genomes and origin of the allotetraploid will be discussed.

The genome of *G. hirsutum* was analyzed by Walbot and Dure (1976; Table 4-6, part A). In addition to *G. hirsutum,* Geever also analyzed *G. herbaceum* var. *africanum* and *G. raimondii,* the two diploid species with genomes closest to the A and D subgenomes of the allotetraploids (Table 4-6, part B).

Walbot and Dure (1976) estimated that at least 80% of the genome in *G. hirsutum* consists of both a short and long interspersion of single copy and mildly (moderately) repetitive DNA. Approximately 8% of the genome consists of single copy DNA interspersed with much longer sequences of repetitive DNA. It is likely that the genomes of other genomic groups will likewise have high levels of interspersion of these two classes of DNA.

Geever (1980) reported that the genome size of *G. herbaceum* is 1.05 pg while that of *G. raimondii* is 0.68 pg. It is apparent that the former is about one and one-half times greater than the latter. It can be seen in the table that both genomes have basically the same amount of single copy DNA, 0.42 and 0.39 pg, but differ greatly in their quantity of repetitive DNA. *Gossy-*

Table 4-6. Summary data of Genome organization (haploid DNA content) of *Gossypium* species. NT = nucleotides; SC = single copy; Rep. = repetitive; pg = picograms.

A. Interspersion pattern of repetitive and single copy sequence elements of *G. hirsutum* by Walbot and Dure (1976).

Component	Total	Single copy	Repetitive
		% of genome	
1250 NT mildly Rep. element + 1800 NT SC element	61	36	25
1250 NT mildly Rep. element + 4000 NT SC element	21	16	5
SC > 6000 NT from nearest mildly Rep. element	8	8	--
Palindrome and highly Rep. DNA	8	--	8

B. Analysis of SC and Rep. DNA in *G. raimondii* and *G. herbaceum* var. *africanum* by Geever (1980).

Species	Total genome size, pg	Component	Amount in pg	Fraction of genome
G. raimondii	0.68	Fold back	0.054	0.08
		Moderately Rep.	0.204	0.30
		SC	0.422	0.62
G. herbaceum var. *africanum*	1.05	Fold back	0.053	0.05
		Highly Rep.	0.168	0.16
		Moderately Rep.	0.441	0.42
		SC	0.389	0.37

pium herbaceum has 0.44 pg and *G. raimondii* has 0.20 pg of moderately repetitive DNA. The *G. herbaceum* genome also has a significant fraction of highly repetitive DNA which most likely exists also in the genome of *G. raimondii* but is undetectable by the analytical methods used.

It is apparent that the two genomes have essentially the same quantity of single copy DNA, but that the genome of *G. herbaceum* has twice the amount of moderately repetitive DNA and a greater amount of highly repetitive DNA than the genome of *G. raimondii*.

Geever's analysis shows that the only quantitative difference between the chromosomal DNA of the A and D diploid genomes is in their repetitive sequences. This shows that there has been a fairly large shift in this class of DNA in the differentiation of these two genomes, which occurred rather uniformly throughout the chromosome complement of the two species. Similar changes have occurred in the other five genomes since the chromosomes of each show a general graduation in size from the longest to the shortest. Such a uniform increase or decrease in size affecting all 13 chromosomes of each of the seven genome groups is not likely to come about by the gradual accumulation of random cryptic structural changes occurring over a long evolutionary time scale. It seems more likely that such changes where brought about by saltatory fluctuations in the repetitive DNA's.

Amplification of repetitive DNA would differentiate cytologically the chromosomes between the genome groups and alter the regulatory and communicative functions of specific DNA sequences as well as synapsis or chromosome affinity.

The species showing the least evolutionary advancement occur in the D genome group (Fryxell, 1971), indicating that the D genomes are most representative of the ancestral progenitor of the genus in its DNA content and organization. If this is indeed the case, then the six groups of genomes with larger chromosomes evolved from a D genome-like progenitor or a genome derived from the D-like genome by saltatory amplification of the repeated sequences.

Normally in constructing a hypothesis for the origin of the allotetraploids, it has been assumed that the following sequence of evolutionary events must be considered if one accepts the modern synthesis concept of speciation (Stephens, 1947): (i) Long period geographical isolation of the American and Asiatic groups, during which their genomes became cytologically differentiated, (ii) geographical reunion of the two, followed by interspecific hybridization and polyploidy, (iii) geographical reisolation of the two parental species and genetic differentiation of the allotetraploids. The first two are "double-event" rather than "single-event" processes (Stephens, 1947). However, according to our hypotheses, the origin of the allotetraploids could possibly be nearly like a single-event process.

It can be assumed that the rare event of saltatory amplification of repetitive DNA occurred throughout the chromosomes of a D genome-like progenitor and became fixed (homozygous) forming an A genome. The new A genome "species" would have an immediate built-in "isolating" mechanism since the two genomes would differ in their gross structural organization and in gene and chromosome regulation. The only adjustment of any significance that may have been necessary in the newly evolved form would be the development of a more compatible nuclear-cytoplasmic interacting system.

If that is the order of speciation events, it is apparent that the two species would satisfy the three primary requirements for the origin of an allotetraploid: chromosome differentiation, propinquity, and cross compatibility (Hutchinson et al., 1945). Thus, within a relatively short geological time period, a progenitor and the hybrid allotetraploid species could become established as independent, true breeding taxa.

Even though fossil records of *Gossypium* are unknown (Fryxell, 1979), it is presumed that the genus is very old and that continental drift separated the two major genomic chromosome size groups, large chromosomes in the Old World and small chromosomes in the New World, during the Cretaceous period. Thus, according to the present hypothesis, the allotetraploids are ancient in origin occurring perhaps during Cretaceous times.

The events in the origin of the allotetraploid species can be summarized as follows: (i) the origin of an A genome species occurred by macroevolution from a progenitor D genome, (ii) hybridization and polyploidy occurred within a short geological time period following the formation of the A genome parent, (iii) the origin of the amphidiploid probably occurred during the Cretaceous period, (iv) diploidization of the gene regulatory system of the hybrids led to the formation of at least five distinct amphidiploid species, and finally, (v) the generation of genetic diversity that exists primarily in *G. barbadense* and *G. hirsutum*.

4-10 TRANSLOCATIONS

4-10.1 Origin and Identification

Prior to the 1950s, cotton cytologists had observed multivalent chromosome configurations in pollen mother cells of species hybrids. Beasley (1942) referred to the multivalents as translocations without determining the number involved. Later, Gerstel (1953a) reported that the multivalents in hybrids of *G. herbaceum, G. arboreum, G. anomalum,* and the allotetraploids were due to chromosomal interchanges.

During this time, cotton cytologists at the Beasley Laboratory in Texas noticed a high frequency of multivalents at metaphase I in trispecies hybrids of *G. hirsutum* × Asiatic-New World diploid F_1. Upon reanalysis of hybrids between *G. hirsutum* and the Asiatic and New World diploid species, they found that all the viable *G. hirsutum*-New World diploid hybrids almost regularly formed 13 small bivalents and 13 large univalents. However, a different modal pairing behavior was observed in the *hirsutum*-Asiatic hybrids. Like Gerstel (1953a), they observed chromosome structural changes that differentiate the A genomes of *G. arboreum* and *G. hirsutum* from *G. herbaceum,* and they assigned chromosome end arrangements to the first five chromosomes of the three A genomes that would account for the multivalent formation in each hybrid combination. Since only five chromosomes of the three A genomes are involved in the interchanges, this indicates that their remaining eight A genome chromosomes have identical end arrangements (Menzel and Brown, 1954a).

According to Menzel and Brown (1954a), *G. herbaceum* has the standard or primitive chromosome end arrangement for all of its chromosomes, in which the first five are numbered respectively 1-2, 3-4, 5-6, 7-8, and 9-10. *Gossypium arboreum* differs from *G. herbaceum* by a single interchange between chromosomes 1 and 2, thus its end arrangement for the five chromosomes is 1-3, 2-4, 5-6, 7-8 and 9-10. The A subgenome of the allotetraploids differs from the genome of *G. herbaceum* by two independent reciprocal interchanges involving chromosomes 2 and 3, and 4 and 5. Thus, its end arrangement is 1-2, 3-5, 4-6, 7-9, and 8-10. The interchanges in this hybrid were designated respectively, as IV_1 and IV_2. The A genomes of the allotetraploids and *G. arboreum* differ by three interchanges which are apparent in their hybrids by the occurrence of a ring-of-six (VI_1) and a ring-of-four (IV_2). The ring-of-six forms from combining the interchange between chromosomes 1 and 2 of *G. arboreum* with the interchange between chromosomes 2 and 3 of the allotetraploids.

The IV_1, IV_2, and VI_1 chromosome translocations were transferred to *G. hirsutum,* providing the first set of chromosomally identified translocations in *Gossypium* (Brown, 1980).

The first induced translocations reported in cotton were from seed exposed in the atom bomb test at Bikini (Brown, 1950). Later, seeds of various stocks were exposed to radiation (X-rays, gamma, and fast neutron) to induce translocations for isolation and identification in *G. hirsutum* (Menzel and Brown, 1954b; Kammacher, 1958; Brown, 1980). Sixty-two translocations, most of which came from irradiation, have been isolated, made homozygous and identified as to the chromosomes involved (Brown, 1980; Table 4–7). In the table, the A subgenome chromosomes are numbered 1 through 13 and those of the D subgenome, 14 through 26. The 62 translocations identify chromosome 1 through 25 of the 26 chromosomes in the allotetraploid. Chromosome 26 has not been isolated in a translocation, but it is identified as a telocentric chromosome (Table 4–8). Four of the 62 interchanges consist of multiple associations involving interchanges between three or four nonhomologous chromosomes. A tester set selected from the group can be used to identify any one of the 26 chromosomes oc-

Table 4–7. Homozygous translocation lines in *G. hirsutum*. (Data from Seyam and Brown, 1973; Menzel and Brown, 1978a, b; Brown, 1980; Ray, 1981; Brown et al., 1981; Menzel, unpublished data).

Translocation line[†]		Freq dp-df[§]	Translocation line[†]		Freq dp-df[§]
AA Translocations			**AA Translocations**		
IVa	T1L-2Ra[‡]	0.36	8-30-5	T3R-9R	0.47
DP30	T1L-2Lb	0.37	IV$_2$	T4L-5	0.34
2935	T1L-3L	0.41	C14-3	T5-9	0.42
5-4c	T1L-7L	0.57	SL18	T5-12R	0.39
2775b	T1L-8L	0.30	1048	T6-7L	
IV$_1$	T2-3a	0.39	Z9-9	T6L-10R	0.51
1059	T2-3b	0.37	1052	T7R-11R	0.42
7-2b	T2-6	0.33	1043	T7L-12R	0.55
1039	T2-8Ra	0.35	2778	T8-12	0.43
1058b	T2-8Rb	0.30	2785	T10R-11R	0.52
8B-3	T2L-9R	0.46	6-5M	T11R-12L	0.45
8-5Gb	T3R-5	0.51	10-5Kb	T11R-13L	0.50
4010	T3L-6L	0.43			
AD Translocations			**AD Translocations**		
9-5H	T12R-19R	0.55	2780	T1L-14L	0.40
2925	T13R-19R	0.49	2B-1	T2R-14R	0.40
4669	T1L-20R	0.59	AZ-7	T6L-14L	0.51
2772	T9R-20L	0.47	1040	T4R-15L	0.48
2790	T7R-21R	0.50	1058a	T11L-15L	0.43
4675	T10L-21L	0.39	2770	T1R-16Ra	0.49
2775a	T5-23	0.42	4672	T1R-16Rb	0.43
2870	T9R-25	0.43	1036	T9R-17Ra	0.45
6340	T9R-17Rb	0.54	10-5ka	T4L-19R	0.43
1316	T11R-17R	0.58	5-5B	T8R-19R	0.46
4659	T7-18R	0.53	1626	T10R-19R	0.59
E20-7	T3L-19L				
DD Translocations			**DD Translocations**		
2777	T14L-23	0.48	E22-13	T19-21R	0.52
2781	T14R-24R	0.53	2786	T19R-24R	0.57
8-5Ga	T15R-16La	0.50	7-3F	T20R-21L	0.42
2767	T14L-16Lb	0.55	DP4	T20L-22R	0.54
SL15	T15R-20R	0.54	2791	T20R-25	0.68
Complex Translocations			**Complex Translocations**		
VI$_1$	T1-2-3		2779	T5-11-21	
DP6	T2-19-23		4655	T2-11-13-24	

† R and L = Right and Left arms. R arbitrarily assigned to arm with largest number of breaks, not necessarily the longer arm cytologically.
‡ Chromosome designation 1 to 13 = A subgenome, 14 to 26 = D subgenome.
§ dp-df = duplication-deficiency.

curring in a structural or numerical chromosome change (Ray, 1981).

Menzel (1955) has described the cytological procedure for assigning chromosomes in translocations to their respective genomes. This procedure consists of crossing the homozygous translocations to diploid species with either the A genome or the D genome.

The identity of the chromosomes in each of the 62 translocations was accomplished in most cases by intercrossing the homozygous translocations and determining whether the same or different chromosomes were involved (Brown, 1980). In some cases, stocks that were monosomic for known chromosomes were used to identify chromosomes in the translocations and vice versa.

The location of breakpoints in the translocated chromosomes of many translocations has been determined by Menzel and Brown (1978a, b) and Brown et al. (1981). These are given provisionally in Table 4-7 as "right" and "left" arms in which "right" is arbitrarily assigned to the arm involved in the larger number of breaks.

Table 4-8. Monosomes and telosomes for chromosomes of the A and D subgenomes of *Gossypium hirsutum*. (Mono 1 = monosome for chromosome 1; Telo 1L and 1S = telocentric chromosomes for the long and short arms, respectively, of chromosome 1).

Chromosome A subgenome	Chromosome D subgenome
Mono 1	Telo 14L
Telo 1L	
Telo 1S	Telo 15L
Mono 2	Mono 16
Telo 2L	Telo 16L
Telo 2S	
Mono 3	Mono 17
Telo 3L	Telo 17S
Telo 3S	
Mono 4	Mono 18
Telo 4L	Telo 18L
Telo 4S	Telo 18S
Telo 5L	Mono 20
	Telo 20L
Mono 6	Telo 20S
Telo 6L	
Telo 6S	Mono 22
	Telo 22L
Mono 7	Telo 22S
Telo 7L	
Telo 7S	Mono 25
	Telo 25L
	Telo 26S
Mono 9	
Telo 9L	
Mono 10	
Telo 10L	
Telo 10S	
Mono 12	
Telo 12L	

4-10.2   Uses of Translocations

Because *G. hirsutum* is an allotetraploid, viable duplications and deficiencies for segments of chromosomes can be recovered as a result of adjacent disjunction in a heterozygous interchange (Table 4-7; Brown, 1950; Brown et al., 1981; Menzel and Brown, 1952, 1954b, 1978b). It can be seen in Table 4-7 that duplication-deficiency types have been recovered from all the simple interchanges in cotton. Therefore, the translocations probably can be used to assign genes to specific chromosomes.

The translocations are also very useful for detecting incipient genome differentiation between the subgenomes of *G. hirsutum* and the A and D genomes of the diploid species and within the A and D subgenomes of the allotetraploid species (Menzel et al., 1978; Hasenkampf and Menzel, 1980; Menzel et al., 1982). For example, plants homozygous for the AD interchanges are crossed to the New World diploid species to produce F_1's for cytological analysis. The (AD)D hybrid will show a trivalent consisting of the two AD interchanged chromosomes with the normal D chromosome between the two. Therefore, in crosses with two or more D genome species, the chiasma frequencies observed in the D_x chromosome of the different hybrids can be compared and any significant differences would indicate that chromosome differentiation had taken place (Menzel et al., 1978).

Aneuploids can be recovered from heterozygous translocations that may occasionally disjoin unequally at the first meiotic division, producing gametes that are deficient for a chromosome or have an extra chromosome. The extra chromosome and deficient chromosome can be a standard or an interchange chromosome, which can be determined easily when crossed with a standard line.

4-11   MONOSOMES

4-11.1   Origin and Identification

Plants that are deficient for one chromosome of a pair are referred to as monosomics. Unlike diploids, allopolyploids such as *G. hirsutum* can transmit the haplo-deficiency to subsequent generations. Chromosome pairing in monosomic plants of allopolyploid cotton will normally consist of 25 bivalents plus one univalent. Twenty-six different monosomics are theoretically possible.

From 1946 to 1952, a total of 51 monosomic plants of different origin were isolated at the Beasley Laboratory at Texas A&M University and the monosome chromsome of 23 of the lines was identified in cytogenetic tests (Brown and Endrizzi, 1964; Endrizzi and Brown, 1964). The majority of these involved chromosomes 1, 2, 4, and 6 of the A subgenome and 17 and 18 of the D subgenome of *G. hirsutum*. Since then, over 225 additional monosomic plants have been recovered. A vast majority of these involved monosomes for chromosomes 2, 4 and 6 of the A subgenome (Endrizzi and Ramsay, 1979; Edwards et al., 1980).

From this group of monosomic plants, monosomes for 15 of the 26 chromosomes of *G. hirsutum* have been identified (Endrizzi and Ramsay, 1979, 1980). These involve chromosomes 1, 2, 3, 4, 6, 7, 9, 10, and 12 of the A subgenome and 16, 17, 18, 20, 22, and 25 of the D subgenome (Table 4–8). The translocations were used to identify the monosomes.

All of the monosomic lines, including the monotelodisomics to be discussed below, are being transferred to the genetic background of Texas Marker 1 (TM 1), a highly inbred line of 'Deltapine 14,' which serves as the standard reference for plant morphological characters in genetic and cyto-genetic studies (Kohel et al., 1970).

Because of the chromosome imbalance, a syndrome of plant morphological characters is associated with each monosomic type, which can be used to identify easily the monosomic plants from their disomic sibs. Several of the monosomics can be recognized in the seedling stage with a high degree of success (Endrizzi and Ramsay, 1979).

By matching similarities of morphological characters along with the association of duplicate genetic factors, it has been possible tentatively to assign the following A and D subgenome chromosomes as homoeologous sets: 1–15, 4–22, 6–25, 7–16, 10–20, and 12–26. Following the isolation of monosomes for all 26 chromosomes, the chromosomes will be renumbered to identify homoeologous pairs.

4–11.2 Breeding Behavior

Two kinds of cytological progenies are recovered when monosomic plants are self-pollinated or outcrossed as female: disomics and monosomics. The frequency of the monosomic progeny is dependent on the rate of transmission of the haplo-deficient gametes.

Normally, the frequency of monosomic plants obtained from unselected seed will range between 20 to 50%. However, the monosomics for chromosome 3, 9, and 16 usually occur in much lower frequencies of 5 to 10%.

Douglas (1972) and Malek-Hedayat (1981) have found that within a single boll, the seeds having the lowest weight were predominantly monosomic. Malek-Hedayat (1981) also found that the frequency of monosomic progeny can be increased greatly by selecting within a sample of seed those seeds with weights ranging from the minimum weight to minus one standard deviation of the mean weight. This procedure is recommended especially for monosomics 3, 9, and 16 which are normally transmitted in low frequencies.

With the exception of the monosomic for chromosome 9, most of the monosomic chromosomes are fairly stable and misdivide infrequently to give telocentric or isochromosomes (Brown, 1958; Endrizzi and Ramsay, 1979). The monosomic condition of chromosome 9 induces abnormal chromosome separation, apparently during embryo sac development. Monosome 9 plants produce plants that are aneuploid for one or more chromosomes in addition to normal disomes and a few monosomic 9 plants.

The evidence indicates that male transmission of n-1 gametes is rare if it occurs at all although controlled studies have not been made. Of eight monosomes tested, chromosomes 4 and 6 were male-transmitted, but in a very low frequency.

4–11.3   Use of Monosomes

Monosomes provide a quick and simple method of identifying the chromosomes in translocations. They have been used extensively to assign genetic factors to specific chromosomes. The protocol for locating recessive and dominant marker genes on monosomic chromosomes was outlined by Endrizzi (1963). In such tests, the monosomic plant must be used as the female parent for their recovery in the F_1 generation, which will consist of two kinds, the disomic and monosomic F_1's.

Monosomes are ideal for separating duplicate linkage groups so that the linkage value in each group can be determined independently.

Monosomics can also be used in the synthesis of chromosome substitution lines as outlined by Endrizzi et al. (1963b) and White et al. (1967). These lines can be used for the study of the genetic effects of individual chromosomes on plant traits and for estimating the number of genes and their interactions and linkage relationships controlling plant characters of economic importance.

With the development of the monosomic series in two allotetraploid species, more than one alien chromosome may be substituted into one or the other species (White et al., 1967).

4–12   MONOTELODISOMES AND THEIR USE

Telocentric chromosomes arise by misdivision of univalents at the first or second meiotic division and are recovered as monotelodisomics. A monotelodisomic plant has 25 bivalents plus a heteromorphic bivalent, in which the heteromorphic pair consists of a telocentric chromosome plus a complete homologous chromosome.

Twenty-nine monotelodisomic stocks have been identified (Table 4–8) that mark one or both arms of 19 chromosomes. With the exception of monotelodisomic 16L, all other monotelodisomics exhibit a syndrome of morphological characters closely paralleling in one form or another that which is associated with their monosomic counterparts. When either self or cross-pollinated, monotelodisomics produce two kinds of progenies, disomes and monotelodisomes. For most, the female transmission frequency of the telocentric chromosome is generally about 50%. In studies in which the monotelodisomics were used as the male parent in crosses, the telocentrics were transmitted through the pollen, ranging from a few percent for most to a high of 33% for telo 16L. Only the ditelocentric for the long arm of chromosome 16 has been recovered from self-pollination of monotelodisomics.

Following the association of a marker gene and a specific chromosome by the monosome test, the monotelodisomics can then be used for determining the arm location of the marker as well as its linkage relationship with the centromere (Endrizzi and Kohel, 1966).

The monotelodisomics can be employed to identify the individual chromosomes in translocations, and in some cases, to determine the arm in which the translocation had occurred.

4-13 MONOISODISOMES

Isochromosomes are chromosomes with two homologous or identical arms. They arise by misdivision of univalent and telocentric chromosomes. Plants with an isochromosome are referred to as monoisodisomics since they have 25 normal bivalents plus a heteromorphic bivalent consisting of a normal chromosome and an isochromosome.

Isochromosomes can be used for positioning genes in arms of the chromosomes in the same manner as the telocentric chromosomes. However, they may give rates of recombination different from that of mono-telodisomics because of exchanges in the trisomic arms, and thus they may not be as useful as the telocentrics for mapping studies (Endrizzi and Bray, 1980).

4-14 TRISOMES AND ALIEN CHROMOSOME ADDITIONS

Many primary and a few secondary and tertiary trisomics, either spontaneous in origin or, in several cases from unequal disjunction of heterozygous translocations, have been found. Primary trisomic individuals occur occasionally in cultivars of cotton (Endrizzi et al., 1963a).

Alien chromosome addition lines have been bred and studied for their practical use (Kammacher and Poisson, 1964; Poisson, 1967; Kammacher and Schwendiman, 1969; Vieira da Silva and Poisson, 1969; Sharif and Islam, 1970).

Disomic addition lines involving alien chromosomes provide a method for identifying homoeologous chromosomes of different species (Schwendiman, 1974). Schwendiman reported that several *G. anomalum* disomic chromosome addition lines had modified morphological characters that were similar to that of an equal number of *G. stocksii* disomic chromosome addition lines. This finding showed that in each case the *G. anomalum* and *G. stocksii* disomic additions were homoeologs.

REFERENCES

Abraham, P. 1940. Morphology of the somatic chromosomes of three Asiatic cottons. Indian J. Agric. Sci. 10:299–302.

Afzal, M. 1949. Growth and development of the cotton plant and its improvement in the Punjab. Lahore: Government Printing Office, West Punjab, Lahore, India.

Afifi, A., A. A. Bary, S. A. Kamel, and I. Heikal. 1966. Bahtim 110, a new strain of Egyptian cotton free from gossypol. Emp. Cotton Grow. Rev. 43:112–120.

Alberte, R. S., J. D. Hesketh, G. Hofstra, J. P. Thornber. A. W. Naylor, R. L. Bernard, C. Brim, J. E. Endrizzi, and R. J. Kohel. 1974. Composition and activity of the photosynthetic apparatus in temperature-sensitive mutants of higher plants. Proc. Natl. Acad. Sci. USA 71:2414–2418.

Allison, D. C., and W. D. Fisher. 1964. A dominant gene for male-sterility in Upland cotton. Crop Sci. 4:548–549.

Anonymous. 1957. Report of the International Committee on Genetic Symbols and Nomenclature. Intern. Union Biol. Sci., Series B, No. 30.

————. 1974. The regional collection of *Gossypium* germplasm. USDA Rep. ARS-H-2.

Arutjunova, L. G. 1936. An investigation of chromosome morphology in the genus *Gossypium*. C. R. (Doklady) Acad. Sci. USSR. 3:37–40.

Balls, W. L. 1906. Studies in Egyptian cotton. p. 29–89. *In* Yearb. Khediv. Agric. Soc. for 1906. Cairo, Egypt.

Banerji, I. 1929. The chromosome numbers of Indian cottons. Ann. Bot. 43:603–607.

Baranov, P. 1930. Work of the cytoanatomical laboratory of N. I. Kh. I. During the growing period of 1930. Bull. Sci. Res. Cotton Inst. Tashkent 5:7–17 (Russian). From Plant Breed. Abstr. 2:197–198 (1932).

Barrow, J. R., and D. D. Davis. 1974. Gl_2^s—a new allele for pigment glands in cotton. Crop Sci. 14:325–326.

Beal, J. M. 1928. A study of the heterotypic prophases in the microsporogenesis of cotton. La Cellule 38:247–268.

Beasley, J. O. 1940a. The production of polyploids in *Gossypium*. J. Hered. 31:39–48.

————. 1940b. The origin of American tetraploid *Gossypium* species. Am. Nat. 74:285–286.

————. 1942. Meiotic chromosome behavior in species, species hybrids, haploids and induced polyploids of *Gossypium*. Genetics 27:25–54.

————, and M. S. Brown. 1942. Asynaptic *Gossypium* plants and their polyploids. J. Agric. Res. 65:421–427.

Benedict, C. R., and R. J. Kohel. 1968. Characteristics of a virescent cotton mutant. Plant. Physiol. 43:1611–1616.

————, K. J. McCree, and R. J. Kohel. 1971. High photosynthetic rate of a chlorophyll mutant in cotton. Plant. Physiol. 49:968–971.

Bose, S. 1953–1955. VIII. Cytogenetical investigations in cotton, with special reference to chromosome morphology of five types belonging to *Gossypium arboreum* L. Trans. Bose Res. Inst. 19:67–71.

Bowman, D. T., and J. B. Weaver, Jr. 1979. Analyses of a dominant male-sterile character in Upland cotton. II. Genetic studies. Crop Sci. 19:628–630.

Britten, R. J., and D. E. Kohne. 1967. Nucleotide sequence repetition in DNA. Carnegie Inst. Wash. Yearb. 65:78–106.

————, and ————. 1968. Repeated sequences in DNA. Science 161:529–540.

Brown, H. B., and J. R. Cotton. 1937. "Round leaf" cotton. Notes on the appearance and behavior of a peculiar new strain. J. Hered. 28:45–48.

Brown, M. S. 1950. Cotton from Bikini. J. Hered. 41:115–121.

————. 1958. The division of univalent chromosomes in *Gossypium*. Am. J. Bot. 45:24–32.

————. 1980. The identification of chromosomes of *Gossypium hirsutum* L. by means of translocations. J. Hered. 71:266–274.

————, and J. E. Endrizzi. 1964. The origin, fertility and transmission of monosomics in *Gossypium*. Am. J. Bot. 51:108–115.

————, M. Y. Menzel, C. A. Hasenkampf, and S. Naqi. 1981. Chromosome configurations and orientations in 58 heterozygous translocations in *Gossypium hirsutum*. J. Hered. 72:161–168.

Clower, D. F., J. E. Jones, K. B. Benkwith, Jr., and L. W. Sloan. 1970. 'Nonpreference'—a new approach to boll weevil control. La. Agric. 13:10–11.

DaSilva, F. P., J. E. Endrizzi, and L. S. Stith. 1981. Genetic study of restoration of pollen fertility of cytoplasmic male-sterile cotton. Rev. Brasil. Genet. 4:411–426.

Davie, J. H. 1933. Cytological studies in the Malvaceae and certain related families. J. Genet. 28:33–67.

————. 1935. Chromosome studies in the Malvaceae and certain related families. II. Genetica 17:487–498.

Dilday, R. H., R. J. Kohel, and T. R. Richmond. 1975. Genetic analysis of leaf differentiation mutants in Upland cotton. Crop Sci. 15:393–397.

Douglas, C. R. 1972. Relationship of seed weight to cytotype of monosomic progeny in cotton. Crop Sci. 12:530–531.

Edwards, G. A. 1977. The karyotype of *Gossypium herbaceum* L. Caryologia 30:369–374.

————. 1979. Genomes of the Australian wild species of cotton. I. *Gossypium sturtianum,* the standard for the C genome. Can. J. Genet. Cytol. 21:363–366.

————, M. S. Brown, G. A. Niles, and S. A. Naqi. 1980. Monosomics in cotton. Crop Sci. 20:527–528.

————, J. E. Endrizzi, and R. Stein. 1974. Genome DNA content and chromosome organization in *Gossypium.* Chromosoma 47:309–326.

————, and M. A. Mirza. 1979. Genomes of the Australian wild species of cotton. II. The designation of a new G genome for *Gossypium bickii.* Can. J. Genet. Cytol. 21:367–372.

Endrizzi, J. E. 1963. Genetic analysis of six primary monosomes and one tertiary monosome in *Gossypium hirsutum.* Genetics 48:1625–1633.

————. 1975. Monosomic analysis of 23 mutant loci in cotton. J. Hered. 66:163–165.

————, and R. Bray. 1980. Cytogenetics of disomics, monotelo- and monoiso-disomics and ml_1 st_1 mutants of chromosome 4 of cotton. Genetics 94:979–988.

————, and M. S. Brown. 1964. Identification of monosomes for six chromosomes in *Gossypium hirsutum.* Am. J. Bot. 51:117–120.

————, and R. J. Kohel. 1966. Use of telosomes in mapping three chromosomes in cotton. Genetics 54:535–550.

————, F. R. H. Katterman, W. D. Fisher, and L. S. Stith. 1974. Extrachromosomal inheritance of a variegated chlorophyll mutant in cotton. Egypt. J. Genet. Cytol. 3:277–284.

————, S. C. McMichael, and M. S. Brown. 1963a. Chromosomal constitution of "Stag" plants in *Gossypium hirsutum* 'Acala 4-42'. Crop Sci. 3:1–3.

————, and L. L. Phillips. 1960. A hybrid between *Gossypium arboreum* L. and *G. raimondii* Ulbr. Can. J. Genet. Cytol. 2:311–319.

————, and G. Ramsay. 1979. Monosomes and telosomes for 18 of the 26 chromosomes of *Gossypium hirsutum.* Can. J. Genet. Cytol. 21:531–536.

————, and ————. 1980. Identification of ten chromosome deficiencies in cotton. J. Hered. 71:45–48.

————, T. R. Richmond, R. J. Kohel, and M. S. Brown. 1963b. Monosomes—a tool for developing better cottons. Texas Agric. Prog. 9:9–11.

————, and R. Stein. 1975. Association of two marker loci with chromosome 1 in cotton. J. Hered. 66:75–78.

————, and L. S. Stith. 1970. Association of two marker genes with chromosome 1 of cotton. Agron. Abstr., Am. Soc. of Agron., Madison, Wis., p. 9.

————, and T. Taylor. 1968. Cytogenetic studies of N Lc_1 yg_2 R_2 marker genes and chromosome deficiencies in cotton. Genet. Res. 12:295–304.

Fryxell, P. A. 1965. Stages in the evolution of *Gossypium.* Adv. Front. Plant Sci. 10:31–56.

————. 1971. Phenetic analysis and the phylogeny of the diploid species of *Gossypium* L. (Malvaceae). Evolution 25:554–562.

————. 1979. The natural history of the cotton tribe. Texas A&M Univ. Press, College Station.

Geever, R. F. 1980. The evolution of single-copy nucleotide sequences in the genomes of *Gossypium hirsutum* L. Ph.D. thesis. Univ. of Arizona, Tucson. Available from University Microfilms, Ann Arbor, Mich. Publ. No. 8027749 (Dis. Abstr. 41:2056B).

Gerstel, D. U. 1953a. Chromosome translocations in interspecific hybrids of the genus *Gossypium.* Evolution 7:234–244.

----. 1953b. Genetic segregation of allopolyploids in the genus *Gossypium*. Genetics 38:664–665.

----. 1956. Segregation in new allopolyploids of *Gossypium*. I. The R_1 locus in certain New World wild American hexaploids. Genetics 41:31–44.

----. 1963. Evolutionary problems in some polyploid crop plants. Second Int. Wheat Genet. Symp. Hereditas Suppl. 2:481–504.

----, and L. L. Phillips. 1957. Segregation of new allopolyploids of *Gossypium*. II. Tetraploid combinations. Genetics 42:783–797.

----, and ----. 1958. Segregation of synthetic amphidiploids in *Gossypium* and *Nicotiana*. Cold Spring Harbor Symp. Quant. Biol. 23:225–237.

Green, J. M. 1955. Frego bract, a genetic marker in Upland cotton. J. Hered. 46:232.

Harland, S. C. 1928. Cotton notes. Tropic Agric. 5:116–117.

----. 1929. The genetics of cotton. II. The inheritance of pollen colour in New World cottons. J. Genet. 20:387–399.

----. 1932. The genetics of cotton. VI. The inheritance of chlorophyll deficiency in New World cottons. J. Genet. 25:271–281.

----. 1935a. The genetics of cotton. XII. Homologous genes for anthocyanin pigmentation in New and Old World cottons. J. Genet. 30:465–476.

----. 1935b. The genetics of cotton. XIV. The inheritance of brown lint in New World cottons. J. Genet. 31:27–37.

----. 1936. Duplicate genes for corolla colour in *Gossypium barbadense* L. and *G. darwinii* Watt. Z. indukt. Abstamm.-u. Vererb Lehre 71:417–419.

----. 1937. Homologous loci of wild and cultivated American cottons. Nature 140:467–468.

----. 1939. The genetics of cotton. Jonathan Cape, London.

----, and O. M. Atteck. 1941. The genetics of cotton. XVIII. Transference of genes from diploid North American wild cottons (*Gossypium thurberi* Tod., *G. armourianum* Kearney, *G. aridum* comb. nov. Skovsted) to tetraploid New World cottons (*G. barbadense* L. and *G. hirsutum* L.). J. Genet. 42:1–19.

Hasenkampf, C. A., and M. Y. Menzel. 1980. Incipient genome differentiation in *Gossypium*. II. Comparison of 12 chromosomes of *G. hirsutum, G. mustelinum* and *G. tomentosum* using heterozygous translocations. Genetics 95:971–983.

Hau, B., E. Koto, and J. Schwendiman. 1980a. Determinisme genetique de deux mutants du cotonnier capsule pileuse et fleur cleistogame. Cotton Fib. Trop. 35:355–357.

----, ----, ----. 1980b. Examen du groupe de liaison III du cotonnier *Gossypium hirsutum*: Description d'un nouveau phenotype cluster et localisation des genes. Cotton Fib. Trop. 35:359–367.

----, ----, and ----. 1981. Description d'une mutation induisant une feuille filiforme chez le cotonnier *G. hirsutum* L. Cotton Fib. Trop. 36:205–208.

Hawkes, J. G., and P. Smith. 1965. Continental drift and the age of angiosperm genera. Nature 207:48–50.

Holder, D. S., J. N. Jenkins, and F. G. Maxwell. 1968. Duplicate linkage of glandless and nectariless genes in Upland cotton, *Gossypium hirsutum* L. Crop Sci. 8:577–580.

Hutchinson, J. B. 1934. The genetics of cotton. X. The inheritance of leaf shape in Asiatic *Gossypiums*. J. Genet. 28:437–513.

----. 1959. The application of genetics to cotton improvement. Cambridge Univ. Press, London.

----. 1962. The history and relationships of the world's cotton. Endeavour 21:5–15.

----, and R. A. Silow. 1939. Gene symbols for use in cotton genetics. J. Hered. 30:461–464.

----, ----, and S. G. Stephens. 1947. The evolution of Gossypium. Oxford Univ. Press, London.

----, S. G. Stephens, and K. S. Dodd. 1945. The seed hairs of *Gossypium*. Ann. Bot. 9:361–367.

Jacob, K. T. 1942. Studies in cotton IV. Morphology of the somatic chromosomes of eight types of Asiatic cottons. Trans. Base Res. Inst. Calcutta 15:17–27.

Jenkins, J. N., F. G. Maxwell, and H. N. Lafever. 1966. The comparative preference of insects for glanded and glandless cottons. J. Econ. Entomol. 59:352–356.

Johnson, B. L. 1975. *Gossypium palmeri* and a polyphyletic origin of the New World cottons. Bull. Torrey Bot. Club 102:340–349.

Johnson, B. L., and M. M. Thein. 1970. Assessment of evolutionary affinities in *Gossypium* by protein electrophoresis. Am. J. Bot. 57:1081–1092.

Jones, J. E., and J. A. Andries. 1967. Okra-leaf for boll rot control? La. Agric. 10:8–9.

Justus, N. J., and C. L. Leinweber. 1960. A heritable partial male-sterile character in cotton. J. Hered. 51:191–192.

————, and J. R. Meyer. 1963. A female-sterile mutant in a double haploid of Upland cotton. J. Hered. 54:65–66.

————, J. R. Meyer, and J. B. Roux. 1963. A partially male-sterile character in Upland cotton. Crop Sci. 3:428–429.

Kadir, Z. B. Z. 1976. DNA evolution in the genus *Gossypium*. Chromosoma 56:85–95.

Kammacher, P. 1958. La production artificielle d'aberrations chromosomiques chez *Gossypium hirsutum* par les rayons X. Cotton Fib. Trop. 13:1–22.

————. 1959. Relations cytologiques entre l'espece sauvage *Gossypium raimondii* Ulb. et les especes cultivees de cotonniers *G. hirsutum* L. er *G. barbadense* L. Cotton Fib. Trop. 14:1–4.

————. 1960. Observations cytologiques sur deux hybrides F₁ entre especes cultivees tetraploids de cotonniers et l'espece diploid sauvage *Gossypium raimondii* Ulb. Rev. Cytol. Bil. Veg. 22:1–32.

————. 1968. New investigations of linkage group I of *Gossypium hirsutum*. Cotton Fib. Trop. (English Ed.) 23:179–183.

————, and C. Poisson. 1964. Sur les possibilities de transferer du material genetique du cotonnier sauvage *Gossypium anomalum* Waw. et Peyr. A l'espece cultivee *G. hirsutum* L. Cotton Fib. Trop. 19:243–264.

————, and J. Schwendiman. 1967. Etude de la localisation chromosomique du gine ms_3 de sterilite pollinique du cotonnier. Cotton Fib. Trop. 22:417–420.

————, and ————. 1969. Addition au genome de l'espece du cotonnier *Gossypium hirsutum* de deux chromosomes de l'espece sauvage *G. stocksii*. Can. J. Genet. Cytol. 11:169–183.

Katterman, F. R. H., and J. E. Endrizzi. 1973. Studies on the 70S ribosomal content of a plastid mutant in *Gossypium hirsutum* L. Plant Physiol. 51:1138–1139.

————, and D. R. Ergle. 1970. A study of quantitative variations of nucleic acids in *Gossypium*. Phytochemistry 9:2007–2010.

Kearney, T. H. 1930a. Cotton plants, tame and wild and genetics of cotton. J. Hered. 21:325–336, 375–415.

————. 1930b. Short branch, another character of cotton showing monohybrid inheritance. J. Agric. Res. 41:379–387.

————, and G. J. Harrison. 1927. Inheritance of smooth seeds in cotton. J. Agric. Res. 35:193–217.

————, and ————. 1928. Variation in seed fuzziness in individual plants of Pima cotton. J. Agric. Res. 37:465–472.

Kestler, D. P., F. R. H. Katterman, and J. E. Endrizzi. 1977. Buoyant density determination on chloroplast DNA in a variegated cytoplasmic mutant of *Gossypium hirsutum* L. Biochem. Biophys. Res. Commun. 76:720–727.

Knight, R. L. 1952a. The genetics of withering or deciduous bracteoles in cotton. J. Genet. 50:392–395.

————. 1952b. The genetics of jassid resistance in cotton. I. The genes H_1 and H_2. J. Genet. 51:46–66.

————. 1954. Abstract bibliography of cotton breeding and genetics, 1900–1950. Commonw. Agric. Bur., Farnham Royal, Cambridge, England.

————, and J. Sadd. 1953. The genetics of jassid resistance in cotton. II. Pubescent T611. J. Genet. 51:582–585.

Kohel, R. J. 1964. Inheritance of abnormal palisade mutant in American Upland cotton, *Gossypium hirsutum* L. Crop Sci. 4:112–113.

----. 1965. Inheritance of accessory involucre mutant in American Upland cotton, *Gossypium hirsutum* L. Crop Sci. 5:119–120.

----. 1967. Variegated mutants in cotton. Crop Sci. 7:490–492.

----. 1972. Linkage tests in Upland cotton, *Gossypium hirsutum* L. II. Crop Sci. 12:66–69.

----. 1973a. Genetic analysis of the open bud mutant in cotton. J. Hered. 64:237–238.

----. 1973b. Genetic nomenclature in cotton. J. Hered. 64:291–295.

----. 1973c. Analysis of irradiation induced virescent mutants and the identification of a new virescent mutant ($v_5 v_5$, $v_6 v_6$) in *Gossypium hirsutum* L. Crop Sci. 13:86–88.

----. 1973d. Genetic analysis of the depauperate mutant in cotton, *Gossypium hirsutum* L. Crop Sci. 13:427–428.

----. 1978. Linkage tests in Upland cotton. III. Crop Sci. 18:844–847.

----. 1979. Gene arrangement in the duplicate linkage groups V and IX: Nectariless, glandless, and withering bract in cotton (*Gossypium hirsutum* L.). Crop Sci. 19:831–833.

----. 1983. Genetic analyses of the yellow-veins mutant in cotton. Crop Sci. 23:291–293.

----, and C. R. Benedict. 1971. Description and CO_2 metabolism of aberrant and normal chloroplasts in variegated cotton, *Gossypium hirsutum* L. Crop Sci. 11:486–488.

----, and C. F. Lewis. 1962. Inheritance of ragged mutant in American Upland cotton, *G. hirsutum* L. Crop Sci. 2:61–62.

----, and T. R. Richmond. 1971. Isolines in cotton: Effects of nine dominant genes. Crop Sci. 11:287–289.

----, C. F. Lewis, and M. N. Christiansen. 1977. The identification of a new mutant and linkage group in cotton. J. Hered. 68:65–66.

----, ----, and T. R. Richmond. 1965a. Linkage tests in Upland cotton, *Gossypium hirsutum* L. Crop Sci. 5:582–585.

----, ----, and ----. 1965b. The genetics of flowering response in cotton. V. Fruiting behavior of *Gossypium hirsutum* and *Gossypium barbadense* in interspecific hybrids. Genetics 51:601–604.

----, ----, and ----. 1967. Isogenic lines in American Upland cotton, *Gossypium hirsutum* L.: Preliminary evaluation of lint measurements. Crop Sci. 7:67–70.

----, J. E. Quisenberry, and C. R. Benedict. 1974. Fiber elongation and dry weight changes in mutant lines of cotton. Crop Sci. 14:471–474.

----, T. R. Richmond, and C. F. Lewis. 1970. Texas Marker-1. Description of a genetic standard for *Gossypium hirsutum* L. Crop Sci. 10:670–671.

----, ----, and ----. 1974. Genetics of flowering response in cotton. VI. Flowering behavior of *Gossypium hirsutum* L. and *G. barbadense* L. hybrids. Crop Sci. 14:696–699.

Kulebyaev, V. 1937. Something new in Egyptian cotton breeding. (In Russian.) Soc. Reconstr. Agric. 2:227–230. (From Plant Breed. Abstr. 8:376.)

Leake, H. M. 1911. Studies on Indian cotton. J. Genet. 1:202–272.

Lee, J. A. 1962. Genetical studies concerning the distribution of pigment glands in the cotyledons and leaves of Upland cotton. Genetics 47:131–142.

----. 1965. The genomic allocation of the principal foliar-gland loci in *Gossypium hirsutum* and *Gossypium barbadense*. Evolution 19:182–188.

----. 1966. Genetics of the smooth stem character in *Gossypium hirsutum* L. Crop Sci. 6:497–498.

----. 1968. Genetical studies concerning the distribution of trichomes on the leaves of *Gossypium hirsutum* L. Genetics 60:567–575.

----. 1976. Transfer of a smooth leaf allele from *Gossypium barbadense* L. to *Gossypium hirsutum* L. Crop Sci. 16:601–602.

----. 1978. Allele determining rugate fruit surface in cotton. Crop Sci. 18:251–254.

----. 1981a. Genetics of D_3 complementary lethality in *Gossypium hirsutum* and *G. barbadense*. J. Hered. 72:299–300.

----. 1981b. A genetic scheme for isolating cotton cultivars. Crop Sci. 21:339–341.

----. 1981c. A new linkage relationship in cotton. Crop Sci. 21:346–347.

----. 1982. Linkage relationships between *Le* and *Gl* alleles in cotton. Crop Sci. 22:1211–1213.

Lewis, C. F. 1954. The inheritance of cup leaf in cotton. J. Hered. 45:127–128.

----. 1958. Genetic studies of a leaf mosaic mutant causing somatic instability in cotton. J. Hered. 49:267–271.

----. 1960. The inheritance of mottled leaf in cotton. J. Hered. 51:209–212.

----, and T. R. Richmond. 1960. The genetics of flowering response in cotton. Genetics 45: 79–85.

Longley, A. E. 1933. Chromosomes of *Gossypium* and related genera. J. Agric. Res. 46:217–227.

Lukefahr, M. J., T. N. Shaver, and W. L. Parrott. 1969. Sources and nature of resistance in *Gossypium hirsutum* to bollworm and tobacco budworms. p. 81–82. *In* Proc. Beltwide Cott. Prod. Res. Conf., New Orleans, La.

Malek-Hedayat, S. 1981. The relationship between seed weight and 13 monosomic and two monotelodisomic chromosomes in *Gossypium hirsutum*. M.S. Thesis. Univ. of Arizona, Tucson.

Mauer, F. M. 1938. On the origin of cultivated species of cotton. A highly fertile triple hybrid [(*Gossypium barbadense* × *G. thurberi*) × *G. arboreum*] Bull. Acad. Sci. USSR Ser. Biol., p. 695–709. (From Plant Breed. Abstr. 9:318–319, 1939).

----. 1954. Origin and systematics of cotton. Izv. Akad. Nauk. Uzb. SSR, Tashkent. (In Russian.) (From English summary by S. S. Kanash, Tashkent Cotton Station, 1956.)

McMichael, S. C. 1942. Occurrence of the dwarf-red character in Upland cotton. J. Agric. Res. 64:477–481.

----. 1954. Glandless boll in Upland cotton and its use in the study of natural crossing. Agron J. 46:527–528.

----. 1960. Combined effects of the glandless genes gl_2 and gl_3 on pigment glands in the cotton plant. Agron. J. 46:385–386.

----. 1965. Inheritance of aberrant stigmas in the flowers of Upland cotton. J. Hered. 56:21–22.

----. 1970. Yuma glandless, an allelomorph of glandless-one in cotton, *Gossypium hirsutum* L. Crop Sci. 10:202–203.

McLendon, C. A. 1912. Mendelian inheritance in cotton hybrids. Ga. Exp. Stn. Bull. 89, p. 141–228.

Menzel, M. Y. 1955. A cytological method for genome analysis in *Gossypium*. Genetics 40: 214–223.

----, and M. S. Brown. 1952. Viable deficiency-duplications from a translocation in *Gossypium hirsutum*. Genetics 37:678–692.

----, and ----. 1954a. The significance of multivalent formation in three-species *Gossypium* hybrids. Genetics 39:546–557.

----, and ----. 1954b. The tolerance of *Gossypium hirsutum* for deficiencies and duplications. Am. Nat. 88:407–418.

----, and ----. 1978a. Genetic length and breakpoints in twelve chromosomes of *Gossypium hirsutum* involved in 10 reciprocal translocations. Genetics 88:541–558.

----, and ----. 1978b. Reciprocal chromosome translocations in *Gossypium hirsutum*. Arm location of breakpoints and recovery of duplications and deficiencies. J. Hered. 69:383–390.

----, ----, and S. Naqi. 1978. Incipient genome differentiation in *Gossypium*. I. Chromosomes 14, 15, 16, 19 and 20 assessed in *G. hirsutum, G. raimondii,* and *G. lobatum* by means of seven A-D translocations. Genetics 90:133–149.

----, C. A. Hasenkampf, and J. McD. Stewart. 1982. Incipient genome differentiation in *Gossypium*. III. Comparison of chromosomes of *G. hirsutum* and Asiatic diploids using heterozygous translocations. Genetics 100:89–103.

Meredith, W. R., Jr., C. D. Ranny, M. L. Laster, and R. R. Bridge. 1973. Agronomic potential of nectariless cotton (*Gossypium hirsutum* L.). J. Environ. Qual. 2:141–144.

Meyer, J. R. 1957. Origin and inheritance of D_2 smoothness in Upland cotton. J. Hered. 48: 249–250.

––––, and V. G. Meyer. 1961. Origin and inheritance of nectariless cotton. Crop Sci. 1:167–169.

Meyer, V. G. 1975. Male sterility from *Gossypium harknessii.* J. Hered. 66:23–27.

Mikhailova, K. A. 1938. Chromosome morphology of cotton. C. R. (Dokl.) Acad. Sci. URSS 19:181–184.

Murray, J. C. 1965. A new locus for glanded stem in tetraploid cotton. J. Hered. 56:42–44.

––––, and L. A. Brinkerhoff. 1966. Inheritance of pale-green color mutant in cotton. Crop Sci. 6:375–376.

Mursal, I. El Jack. 1978. Chromosome association in haploids and species hybrids of *Gossypium.* J. Hered. 69:413–416.

––––, and J. E. Endrizzi. 1976. A reexamination of the diploid-like meiotic behavior of polyploid cotton. Theoret. Appl. Genet. 47:171–181.

Parks, C. R., W. L. Ezell, D. E. Williams, and D. L. Dreyer. 1975. VII. The application of flavonoid distribution to taxonomic problems in the genus *Gossypium.* Bull. Torrey Bot. Club 102:350–361.

Percival, A. E., and R. J. Kohel. 1976. New virescent cotton mutant linked with the marker gene Yellow petals. Crop Sci. 16:503–505.

––––, R. J. Kohel, and R. H. Dilday. 1976. A dominant round leaf mutant in cotton. Crop Sci. 16:794–796.

Phillips, L. L. 1960. The cytogenetics of speciation in Asiatic cotton. Genetics 46:77–83.

––––. 1962. Segregation in new allopolyploids of *Gossypium.* IV. Segregation in New World × Asiatic and New World × wild American hexaploids. Am. J. Bot. 49:51–57.

––––. 1963. The cytogenetics of *Gossypium* and the origin of New World cottons. Evolution 17:460–469.

––––. 1964. Segregation in new allopolyploids of *Gossypium.* V. Multivalent formation in New World × Asiatic and New World × wild American hexaploids. Am. J. Bot. 51:324–329.

––––. 1966. The cytology and phylogenetics of the diploid species of *Gossypium.* Am. J. Bot. 53:328–335.

––––. 1974. Cotton (*Gossypium*). p. 111–133. *In* R. C. King (ed.) Handbook of genetics 2: Plants, plant viruses, and protists. Plenum Press, N.Y.

––––. 1976. Interspecific incompatibility in *Gossypium.* III. The genetics of tumorigenesis in hybrids of *G. gossypioides.* Can. J. Genet. Cytol. 18:365–369.

––––, and D. U. Gerstel. 1959. Segregation in new allopolyploids of *Gossypium.* III. Leaf shape segregation in hexaploid hybrids of New World cottons. J. Hered. 50:103–108.

––––, and M. A. Strickland. 1966. The cytology of a hybrid between *Gossypium hirsutum* and *G. longicalyx.* Can. J. Genet. Cytol. 8:91–95.

Poisson, C. 1967. Sur les possibilities de transfert de material genetique du cotonnier sauvage *Gossypium anomalum* (Wav. et Peyr.) A l'espece cultivee *G. hirsutum* L. II. Creation de lignees d'addition a 27 paires de chromosomes. Cotton Fib. Trop. 22:401–415.

––––. 1968. Preliminary report on a monosome of *Gossypium hirsutum* corresponding to linkage group I. Cotton Fib. Trop. (English Ed.) 23:183–186.

Prokhanov, Y. I. 1947. The conspectus of a new system of cottons (*Gossypium* L.). Bot Zhur. SSRR 32:61–78. (Cited from Fryxell 1965, ibid., Adv. Front. Plant Sci. 10:31–56).

Ray, D. T. 1981. Identification of the chromosomes in a set of reciprocal translocations in *Gossypium hirsutum* L. Ph.D. thesis. Univ. of Arizona, Tucson. Available from University Microfilms, Ann Arbor, Mich. Publ. No. 8116695 (Dis. Abstr. 42:499B).

Rhyne, C. L. 1957. Duplicated linkage groups in cotton. J. Hered. 48:59–62.

––––. 1958. Linkage studies in *Gossypium.* I. Altered recombination in allotetraploid *G. hirsutum* L. following linkage group transference from related diploid species. Genetics 43: 822–834.

––––. 1960. Linkage studies in *Gossypium.* II. Altered recombination values in a linkage group of allotetraploid *G. hirsutum* L. as a result of transferred diploid species genes. Genetics 45:673–681.

----. 1962a. Inheritance of the glandless leaf phenotype of Upland cotton. J. Hered. 53:115–123.

----. 1962b. Diploidization in *Gossypium hirsutum* as indicated by glandless-stem and boll inheritance. Am. Nat. 96:265–276.

----. 1965a. Duplicate linkage blocks in glandless-leaf cotton. J. Hered. 56:247–252.

----. 1965b. The anomalous behavior of the ghost spot of *Gossypium anomalum* in amphidiploid *Gossypium hirsutum.* Genetics 51:689–698.

----. 1971. Indehiscent anther in cotton. Cotton Gr. Rev. 48:194–199.

----, and P. Rhyne. 1972. Linkage of indehiscent anthers and lack of leaf nectaries in *Gossypium hirsutum* L. Cotton Gr. Rev. 49:57–60.

Richmond, T. R., and R. J. Kohel. 1961. Analysis of a completely male-sterile character in American Upland cotton. Crop Sci. 1:397–401.

Sarvella, P. 1958. Multivalent formation and genetic segregation in some allopolyploid *Gossypium* hybrids. Genetics 43:601–619.

Saunders, J. H. 1961a. The wild species of *Gossypium* and their evolutionary history. Oxford Univ. Press, London.

----. 1961b. The mechanism of hairiness in *Gossypium*. I. *Gossypium hirsutum.* Heredity 16:331–348.

----. 1965a. The mechanism of hairiness in *Gossypium*. III. *Gossypium barbadense*—the inheritance of upper leaf laminar hair. Emp. Cotton Gr. Rev. 42:15–25.

----. 1965b. The mechanism of hairiness in *Gossypium* IV. The inheritance of plant hair length. Emp. Cotton Gr. Rev. 42:26–32.

----. 1965c. Genetics of hairiness transferred from *Gossypium raimondii* to *G. hirsutum.* Euphytica 14:276–282.

Schwendiman, J. A. 1974. Mise en evidence de trois nouvelles homeologies chromosomique entre *Gossypium anomalum* et *G. stocksii.* Can. J. Genet. Cytol. 16:871–881.

Seyam, S. M., and M. S. Brown. 1973. Types of disjunction in interchange heterozygotes of *Gossypium hirsutum* L. Egypt. J. Genet. Cytol. 2:322–330.

Sharif, A., and A. S. Islam. 1970. Cytogenetical studies of some backcross derivatives of *Gossypium hirsutum* × *G. anomalum.* Can. J. Genet. Cytol. 12:454–460.

Sheetz, R. H., and J. B. Weaver, Jr. 1980. Inheritance of a fertility enhancer factor from Pima cotton when transferred into Upland cotton with *Gossypium harknessii* Brandegee cytoplasm. Crop Sci. 20:272–275.

Sherwin, K. H. 1970. Winds across the Atlantic—possible African origins for some Pre-Columbian New World cultigens. Res. Rec. Univ. Mus. S. Ill. Univ. Mesoamerican Studies 6:1–33.

Shoemaker, D. N. 1908. A study of leaf characters in cotton hybrids. Rep. Am. Breed. Assoc. 5:116–119.

Silow, R. A. 1946. Evidence on chromosome homology and gene homology in the amphidiploid New World cottons. J. Genet. 47:213–221.

----, and C. P. Yu. 1942. Anthocyanin pattern in Asiatic cottons. J. Genet. 43:249–284.

Simpson, D. M. 1947. Fuzzy leaf in cotton and its association with short lint. J. Hered. 38:153–156.

Skovsted, A. 1933. Cytological studies in cotton. I. The mitosis and the meiosis in diploid and triploid Asiatic cotton. Ann. Bot. 47:227–258.

----. 1934a. Cytological studies in cotton. II. Two interspecific hybrids between Asiatic and New World cottons. J. Genet. 28:407–424.

----. 1934b. Cytogenetics in relation to plant breeding in cotton. p. 46–49. Emp. Cott. Gr. Corp. Report 2nd Conf.

----. 1935a. Cytological studies in cotton. III. A hybrid between *Gossypium davidsonii* Kell. and *G. sturtii* F. Muell. J. Genet. 30:397–405.

----. 1935b. Some new interspecific hybrids in the genus *Gossypium* L. J. Genet. 30:447–463.

----. 1935c. Chromosome numbers in the Malvaceae I. J. Genet. 31:263–296.

----. 1937. Cytological studies in cotton. IV. Chromosome conjugation in interspecific hybrids. J. Genet. 34:97–134.

Smith, A. L., and J. B. Dick. 1960. Inheritance of resistance to fusarium wilt in Upland and Sea Island cotton as complicated by nematodes under field conditions. Phytopathology 50:44-48.

Smith, E. G. 1942. Inheritance of smooth and pitted bolls in Pima cotton. J. Agric. Res. 64:101-103.

Stebbins, G. L., Jr. 1947. Evidence on rates of evolution from the distribution of existing and fossil plant species. Ecol. Monogr. 17:149-158.

Stephens, S. G. 1944. Phenogenetic evidence for amphidiploid origin of New World cottons. Nature 153:53-54.

----. 1945. A genetic survey of leaf shape in New World cottons—a problem in critical identification of alleles. J. Genet. 46:313-330.

----. 1947. Cytogenetics of *Gossypium* and the problem of the origin of New World cottons. Adv. Genet. 1:431-442.

----. 1950. The genetics of "corky." II. Further studies on its genetic basis in relation to the general problems of interspecific isolating mechanisms. J. Genet. 50:9-20.

----. 1954a. Interspecific homologies between gene loci in *Gossypium*. I. Pollen color. Genetics 39:701-711.

----. 1954b. Interspecific homologies between gene loci in *Gossypium*. II. Corolla color. Genetics 39:712-723.

----. 1955. Linkage in Upland cotton. Genetics 40:903-917.

----. 1974a. Geographic and taxonomic distribution of anthocyanin genes in New World cottons. J. Genet. 61:128-141.

----. 1974b. The use of two polymorphic systems, nectary fringe hairs and corky alleles, as indicators of phylogenetic relationships in New World cottons. Biotropica 6:194-201.

----, and L. L. Phillips. 1972. The history and geographical distribution of a polymorphic system in New World cottons. Biotropica 4:49-60.

Stroman, G. N. 1941. A heritable female-sterile type in cotton. J. Hered. 32:167-168.

----, and C. H. Mahoney. 1925. Heritable chlorophyll deficiencies in seedling cotton. Texas Agric. Exp. Stn. Bull. 333:5-19.

Tayel, M. A. F., L. S. Bird, and J. D. Smith. 1973. Linkage between blight resistance and yellow pollen color genes. J. Hered. 64:208-212.

Thadani, K. I. 1923. Linkage relations in the cotton plant. Agric. J. Ind. 18:572-579.

Tiranti, I. 1967. "Algodon crespo." a new mutation in *Gossypium hirsutum* L. Bol. Genet. Inst. Fito. Castelar 3:29-33.

Turcotte, E. L., and C. V. Feaster. 1963. Inheritance of a cream petal mutant in Pima cotton, *Gossypium barbadense* L. Crop Sci. 3:563-564.

----, and ----. 1964. Inheritance of a mutant with a rudimentary stigma and style in Pima cotton, *Gossypium barbadense* L. Crop Sci. 4:377-378.

----, and ----. 1965. The inheritance of rugate leaf in Pima cotton. J. Hered. 56:234-236.

----, and ----. 1966. A second locus for pollen color in Pima cotton, *Gossypium barbadense* L. Crop Sci. 6:117-119.

----, and ----. 1967. Semigamy in Pima cotton. J. Hered. 58:55-57.

----, and ----. 1969. Semigametic production of haploids in Pima cotton. Crop Sci. 9:653-655.

----, and ----. 1973. The interaction of two genes for yellow foliage in cotton. J. Hered. 64:231-232.

----, and ----. 1975a. Effects of R_1, a gene for red plant color, on American Pima cotton. Crop Sci. 15:875-876.

----, and ----. 1975b. Inheritance of semigamy in American Pima cotton, *Gossypium barbadense* L. Agron. Abstr., Am. Soc. of Agron., Madison, Wis., p. 65.

----, and ----. 1978. Inheritance of three genes for plant color in American Pima cotton. Crop Sci. 18:149-150.

----, and ----. 1979. Linkage tests in American Pima cotton. Crop Sci. 19:119-120.

----, and ----. 1980. Inheritance of a leaf mutant in American Pima cotton. J. Hered. 71: 134–135.

Valicek, P. 1978. Wild and cultivated cottons. Cotton Fib. Trop. 33:363–387.

Vieira da Silva, J. B., and C. Poisson. 1969. Solubilization d'enzymes hydrolytiques chez *Gossypium hirsutum, G. anomalum* et des derives de l'hybridization entre ces deux especes. Can. J. Genet. Cytol. 11:582–586.

Walbot, V., and L. S. Dure, III. 1976. Developmental biochemistry of cotton seed embryogenesis and germination. VIII. Characterization of the cotton genome. J. Mol. Biol. 101: 503–536.

Ware, J. O. 1932. Inheritance of lint colors in Upland cotton. J. Am. Soc. Agron. 24:550–562.

----, W. H. Jenkins, and D. C. Harrell. 1944. Seed characters and lint production. J. Hered. 35:153–160.

Weaver, D. B., and J. B. Weaver, Jr. 1977. Inheritance of pollen fertility restoration in cytoplasmic male-sterile cotton. Crop Sci. 17:497–499.

Weaver, J. B., Jr. 1968. Analysis of a double recessive completely male-sterile cotton. Crop Sci. 8:597–600.

----, and T. Ashley. 1971. Analysis of dominant gene for male sterility in Upland cotton, *Gossypium hirsutum.* Crop Sci. 11:596–598.

----, and D. B. Weaver. 1979. Cracked root mutant in cotton. Inheritance and linkage with fertility restoration. Crop Sci. 19:307–309.

----, and ----. 1979. Cracked root mutant in cotton. Inheritance and linkage with fertility restoration. Crop Sci. 19:307–309.

Webber, J. M. 1934a. Chromosome number and meiotic behavior in *Gossypium.* J. Agric. Res. 49:223–237.

----. 1934b. Cytogenetic notes on cotton and cotton relatives. Science 80:268–269.

----. 1935. Interspecific hybridization in *Gossypium* and the behavior of F_1 plants. J. Agric. Res. 51:1047–1070.

----. 1939. Relationships in the genus *Gossypium* as indicated by cytological data. J. Agric. Res. 58:237–261.

White, T. G., and J. E. Endrizzi. 1965. Tests for the association of marker loci with chromosomes in *Gossypium hirsutum* L. by the use of aneuploids. Genetics 51:605–612.

----, T. R. Richmond, and C. F. Lewis. 1967. Use of cotton monosomes in developing interspecific substitution lines. Crops Res. USDA Rep. ARS 34-91.

Wilson, F. D., and P. A. Fryxell. 1970. Meiotic chromosomes of *Cienfuegosia* species and hybrids and *Hampea* species (Malvaceae). Bull. Torrey Bot. Club 97:367–376.

----, and R. J. Kohel. 1970. Linkage of green-lint and okra-leaf genes in a reciprocal translocation stock of Upland cotton. Can. J. Genet. Cytol. 12:100–104.

Wilson, J. T., F. R. H. Katterman, and J. E. Endrizzi. 1976. Analysis of repetitive DNA in three species of *Gossypium.* Biochem. Genet. 14:1071–1075.

Wouters, W. 1948. Contribution a l'etude taxonomique et caryologique due genre *Gossypium* et application a l'amelioration due cottonier au Congo Belge. Publ. Inst. Nat. Agron. Congo Belge Ser. Sci. No. 34.

Youngman, W., and S. C. Pande. 1927. Occurrence of branched hairs in cotton and upon *Gossypium stocksii.* Nature 119:745.

5   Quantitative Genetics

William R. Meredith, Jr.
ARS-USDA
Stoneville, Mississippi

Most characteristics for improving cotton are inherited as quantitative traits. Factors such as yield, earliness, fiber properties, resistance to pests, stress resistance, and ease of management are conditioned by quantitative genes. Many researchers have been frustrated in attempting to solve their genetic problems by using simple genetic models, wherein few genetic parameters are used to describe complex situations. Quantitative traits are difficult to study because: (i) their expression is modified by environmental and management fluctuations; (ii) a trait, such as yield, is a composite of many other traits, each influenced by many genes, each of which has variable effects; (iii) the expression of an individual gene is often modified by the expression of other genes; (iv) linkage blocks are difficult to breakup; (v) the optimum genotype for a given environment-management system may require gene contributions from many diverse sources; and (vi) the optimum genotype for any one environment-management system is likely to be different from that for another system.

Quantitative inheritance studies should be conducted with the objective of solving or clarifying problems that face breeders. Dudley and Moll (1969) discussed some of the questions breeders must ask to have efficient breeding programs. Some of these questions are listed here:

1. Is genetic variability for the desired characteristic sufficient to warrant a selection program?
2. Which parents or genetic populations are most promising as sources of improved breeding material?
3. How should environments be partitioned into regions or subregions for areas of cultivar adaptation?
4. What are the most efficient methods of testing germplasm (in terms of years, specific locations, management systems, replications, etc.)?

Published in *Cotton,* Agronomy Monograph no. 24, © ASA-CSSA-SSSA, 677 South Segoe Road, Madison, WI 53711.

5. What are the genetic associations among various characteristics?
6. Which breeding and selection procedures will most rapidly and efficiently produce an acceptable level of breeding progress?
7. What type of cultivar (hybrid, synthetic, pure line, etc.) is the most appropriate goal?
8. What is the most efficient selection procedure for combining several characteristics (such as yield and fiber properties)?

This chapter presents a selected review of quantitative inheritance studies. Taken collectively, they should shed some light on the above eight questions. The review will be directed mainly toward yield, the components of yield, and fiber properties. Consequently, it will not attempt to cover thoroughly all cotton characteristics inherited in a quantitative manner.

5–1 BASIC QUANTITATIVE INHERITANCE ANALYSES

Many studies (see References) show that a considerable amount of useful genetic variability is present in cotton. Various mathematical models, using certain basic assumptions, describe and analyze this variability. In most studies not all basic assumptions are met or the number or quality of genotypes and environments are deficient. However, taken collectively, these studies begin to show genetic trends and the utilization of this information allows the geneticist and breeder to employ improved strategies to develop more efficient selection methods and genetic populations. One beginning point to study quantitative inheritance in cotton is to review the various gene action studies.

5–1.1   Gene Action

The gene action and reproductive system of a genetic population provide the information necessary to choose the best selection strategy for that population. For quantitative inheritance purposes, gene action is described in statistical terms as additive, dominance, and epistatic effects and their interactions with environmental factors. The additive effect is defined as the average effect of genes; dominance as the interaction of allelic genes; and epistasis as the interaction of non-allelic genes that influence a particular trait. Gene action effects can be estimated by variance components, by generation means, and by phenotypic distribution methods. The most widely used method in cotton has been the diallel cross, the analyses and assumptions of which have been outlined by Hayman (1954a). Baker (1978) reviewed the usefulness and limitations of this method.

A summary of several gene action studies is given in Table 5–1. All except that by Ramey and Miller (1966) involve the use of diallel crosses. To elucidate the comparisons between traits and studies, the additive component has been set equal to 100 and the dominance component is expressed as a percentage of the additive component. Inspection of the summary indicates 14 significant additive components for the 18 study-trait estimations.

Table 5-1. Relative size† of additive and dominance components and degree of dominance.

Trait	Additive (D)	Dominance (H_1)	$(H_1/D)^{1/2}$	Reference
Lint yield	100**	42**	0.64	Al-Rawi and Kohel (1969)
	100	270**	1.64	Verhalen et al. (1971)
	100**	83	0.91	White and Kohel (1964)
Lint %	100**	35	0.59	Al-Rawi and Kohel (1969)
	100**	3	0.18	Ramey and Miller (1966)
	100**	0	0.00	White and Kohel (1964)
Boll size	100**	53**	0.73	Al-Rawi and Kohel (1969)
	100**	12	0.34	Ramey and Miller (1966)
	100**	12	0.35	White and Kohel (1964)
Fiber length	100**	59**	0.77	Al-Rawi and Kohel (1970)
	100**	20	0.44	Ramey and Miller (1966)
	100	34	0.58	Verhalen and Murray (1969)
Fiber strength	100**	64**	0.80	Al-Rawi and Kohel (1970)
	100**	3	0.17	Ramey and Miller (1966)
	100*	54*	0.73	Verhalen and Murray (1969)
Fiber fineness	100	116**	1.08	Al-Rawi and Kohel (1970)
	100**	5	0.22	Ramey and Miller (1966)
	100	136*	1.16	Verhalen and Murray (1969)

*,** Indicates significance at the 0.05 and 0.01 levels of probability, respectively.

† Additive component set equal 100 and dominance = dominance/additive × 100.

For dominance, eight significant components are indicated. In only three of the cases, one for yield and two for fiber fineness, was the size of the dominance component greater than that of the additive component. These results are similar to those encountered for many other traits in cotton as well as other major crops.

The (dominance/additive)$^{1/2}$ ratio indicates the degree of dominance involved in a study. A ratio of less than 1 indicates partial dominance; a ratio of near 1 indicates complete dominance; and a ratio greater than 1 indicates overdominance. Partial dominance tends to be the general rule, with the exceptions again being yield and fiber fineness which have either dominance or overdominance indicated.

The assumption of no epistasis can also be tested in some diallel crosses. Significant epistasis for plant height, earliness, boll size, and number of bolls per plant were detected by Al-Rawi and Kohel (1969). White (1966) detected significant epistasis both for boll size and for one of its components, seed index. Hayman (1958) reanalyzed Turner's (1953) diallel and found significant complementary epistasis for seed cotton yield. Verhalen et al. (1971), however, detected no epistasis for yield, earliness, lint percentage, or fiber properties.

Generation mean analyses are frequently used to estimate epistatic, additive, and dominance effects. The different generations can be produced easily, and the standard errors are usually small relative to those estimated by variance component analyses. In most generation mean analyses, however, the genetic parameters are correlated and estimates of heritability or selection potential cannot be obtained. Using generation mean analyses, Meredith et al. (1970) reported dominance × dominance epistasis for yield from one of six crosses. In another study also involving six crosses,

Meredith and Bridge (1972) detected additive × additive epistasis for yield from one cross. Others reporting detecting epistasis were Ramey (1963) for lint index, Hosfield et al. (1970) for fiber length, Wilson and Wilson (1976a) for peduncle length, Wilson and Wilson (1976b) for date of first flower, and Kappelman (1971) for resistance to fusarium wilt [*Fusarium oxysporum* f. *vasinfectum* (Atk.) Snyd. & Hans.]. In two 4 × 4 diallels, Lee et al. (1968) controlled the gossypol genotype of the parents so that the various gene effects could be estimated directly. They found that gene action for gossypol content was partitioned such that additive effects accounted for 94%, dominance for 1%, and epistasis for 5%, of the genetic variance.

5-1.2 Reciprocal Differences

Differences in reciprocal crosses may arise from several causes: meiotic irregularities, gamete selection, zygote selection, and cytoplasmic differences. The maternal contributions to an F_1 embryo consist of the environment in which the embryo develops, the substrates available for the developing embryo, the cytoplasm with its inclusions, and the female gamete. The paternal contribution is primarily the male gamete. Most genetic studies in cotton designed to compare reciprocal differences have detected very few differences. However, the exceptions to the general rule are quite important, as they offer special opportunities to breeders.

Kohel (1980) reported a maternal influence on oil content in seed of F_1's. The oil-bearing tissue is embryonic and derives its genotype from both parents. However, the maternal parent provides the substrate and nurtures the developing embryos. Christiansen and Lewis (1973) found reciprocal differences in response to the chilling of germinating seed. The F_1 embryo was susceptible or resistant to chilling depending on the corresponding response of the maternal parent.

Vesta G. Meyer, Mississippi Agric. Exp. Stn., conducted most of the transfer of cytoplasms of exotic species into Upland cotton nuclear backgrounds. A male-sterile cytoplasm and restorer factor from *G. harknesii* Brandagee was detected by Meyer (1975). This development encouraged those who wished to produce cotton hybrids. Cytoplasms from *G. harknesii* and *G. longicalyx* Hutchinson and Lee were found to decrease plant height (Meyer, 1973). *Gossypium arboreum* L., *G. herbaceum* L., *G. anomalum* Wawra ex Wawra & Peyr., and *G. longicalyx* cytoplasms result in decreased anther number and yield. There appears to be little variability among cytoplasms, either within Upland cotton or among the other tetraploid species.

5-1.3 Combining Ability

Combining ability is a term used to describe the breeding value of parental lines used to produce hybrids. Sprague and Tatum (1942) defined combining ability terms as follows: "The term 'general combining ability' is used to designate the average performance of a line in hybrid combination

. . .The term 'specific combining ability' is used to designate those cases in which certain combinations do relatively better or worse than would be expected on the basis of the average performance of the line involved.'' The use of combining ability analyses has greatest use for breeding programs designed to make use of heterosis through hybrids. While there is considerable interest in producing both interspecific and intraspecific cotton hybrids (Davis, 1978) at present very little hybrid cotton is used commercially. Combining ability studies can also produce information about gene action in a base population and thereby aid in selecting parents for producing crosses and segregating populations. The experimental design in cotton studies has usually been the diallel cross without reciprocals. Combining ability studies are not restricted to diallel crosses, but pertain to designs where main effects (general combining ability) and interactions (specific combining ability) can be determined.

The summary of eight combining ability studies for yield given in Table 5-2 shows great variability of results. This result should not be surprising, since the studies were conducted by several researchers, over a range of environments, and with genetically diverse materials. Four of the studies show significance for general combining ability and four show significance for specific combining ability. One study each shows significant general and specific combining ability interactions with environments.

Matzinger (1963) indicated that the general combining ability component consists of additive and additive by additive epistatic variances. Specific combining ability consists of dominance and all types of epistatic variances.

5-1.4 Heterosis

Heterosis is usually defined in one of two ways. For those interested primarily in the F_1 performance per se, the F_1 minus highest-performing parent, expressed as a percentage of that parent, is used. This definition is usually referred to as "useful heterosis." Davis (1978) summarized the re-

Table 5-2. General and specific combining abilities† variance components for yield.

Combining ability		Combining ability × environment		
General	Specific	General	Specific	Reference
0.38	1.02**	--	--	Turner (1953)
3894**	−441	−276	1201	Miller and Marani (1963)
−22	272	503*	−85	Lee et al. (1967)
3393*	−484	−347	2589*	El-Adl and Miller (1971)
87**	3	--	--	Thompson (1971)
1.77	4.60*	0.82	4.97**	Baker and Verhalen (1975)
3481**	1735*	732	683	Miller and Marani (1963)‡
167	1300*	--	--	Meredith and Bridge (1973a)‡

*,** Indicate statistical significance at the 0.05 and 0.01 levels of probability, respectively.

  † Combining abilities given in units designated by the authors.

  ‡ Combining abilities computed from F_2 diallels; all others from F_1 diallels.

search on useful heterosis in cotton. Heterosis is also defined as F_1 minus midparent, expressed as a percent of the midparent. When heterosis defined in this manner is detected, it is an indication of dominance, dominance × dominance epistasis, or both types of gene action.

A summary of heterosis studies, expressed as a percentage of the midparent, is given in Table 5–3. Yield is a complex variable which can be partitioned into smaller more manageable phases or components. Worley et al. (1976) discussed a lint yield model consisting of 12 components. Three components given in Table 5–3, number of bolls/m², seed/boll, and lint/seed [(lint index)/100] were included in their model. Lint percentage, boll size, and seed index were not included in their model directly but were used to determine other components. Also used in their model (Worley et al., 1976) were the fiber properties, length and fineness, to estimate the component average weight per lint fiber.

The trait having highest heterosis was yield with 18%. The summary in Table 5–3 shows that the most important contributors to yield are number of bolls with an average of 13.5%, and boll weight with an average of 8.3%. Lint percentage, seed index, seed/boll, and lint index average heterosis was 1.5, 3.4, 4.7, and 4.2%, respectively. Heterosis for fiber properties was usually smaller than that for yield or its components.

The large amounts of heterosis and specific combining ability for yield can be interpreted as encouragement to those breeders who wish to develop hybrid cotton. Besides this increased yield potential, hybrids have a charismatic effect on growers and allow the developers to control the distribution of hybrid seed. Seed of cultivars developed by conventional breeding methods cannot be controlled as easily as that of hybrids. Factors thus far preventing the wide spread use of hybrids are:

1. The lack of a simply inherited restorer gene that maintains fertility over a wide range of environments.
2. The lack of development of good combiners that have the male sterile cytoplasm or restorer factor.
3. The lack of a dependable and economic method of controlling crosses by insect pollen vectors.

In addition to the above factors, the magnitude of general combining ability and additive variances reported indicate large amounts of genetic variability which are fixable and therefore usable in conventional breeding procedures. Matzinger (1963) indicated, in almost every study of self-pollinated crops, that near-homozygous lines have been developed that equal or surpass the F_1. El-Adl and Miller (1971) conducted breeding experiments that reinforced this conclusion in cotton. An initial F_1 hybrid between 'Coker 100A' and 'Acala 1517' showed significant heterosis over the midparent by 32.6%. After three cycles of recurrent selection six lines were identified and inbred for six generations. The average yield of these six lines exceeded the F_1 hybrid by 55.0%. A diallel cross among these six lines produced hybrids that averaged 9.6% higher yields than the mean of the six parents. This result indicates that even higher yields might be expected from additional cycles of recurrent selection. Lee (1983) reviewed the limitations of the use of hybrids in cotton.

Table 5–3. Heterosis for yield, components of yield, and fiber properties.†

		Yield components					Fiber properties			
Yield	Lint %	No. of bolls	Boll weight	Seed index	Seed/ boll	Lint index	Length	Strength	Fineness	Reference
5.5	2.7	9.0	7.1	9.7	6.0	1.4	2.8	5.6	1.4	Al-Rawi and Kohel (1969, 1970)
3.5	0.6	--	--	--	--	--	0.0	0.0	0.0	Baker and Verhalen (1973)
14.0	1.6	--	--	--	--	--	1.9	0.5	0.2	Baker and Verhalen (1975)
9.6	1.1	4.1	13.4	1.5	2.0	2.8	2.8	0.2	0.0	El-Adl and Miller (1971)
18.8	2.0	33.9	5.4	3.0	--	5.5	1.2§	0.0¶	--	Kime and Tilley (1947)
26.0	1.7	--	8.5	--	--	--	2.8§	−1.0	−0.1	Lee et al. (1967)
20.4	0.7	10.3	7.3	3.9	2.3	4.6	--	--	--	Marani (1963)
19.6	1.6	6.2	8.8	3.8	1.1	6.7	1.1	0.3	1.6	Marani (1968a, 1968b)
22.7	1.1	--	13.4	1.5	--	--	2.8	0.2	0.0	Meredith and Bridge (1972)
16.7	2.0	--	9.1	0.2	--	--	2.4	−1.3	−1.5	Meredith et al. (1970)
19.6	0.0	--	5.6	--	--	--	0.0§	−2.3	1.3	Miller and Lee (1964)
27.5	1.5	--	8.9	--	--	--	3.6	3.3	--	Miller and Marani (1963)
--	--	--	--	--	--	--	2.8	−4.7	−2.9	Quisenberry (1975)
14.9	2.5	6.0	5.7	--	--	--	--	0.0¶	−0.1	Thompson (1971)
33.0‡	--	25.0	7.0	--	12.0	--	--	--	--	Turner (1953)
18.0	1.5	13.5	8.3	3.4	4.7	4.2	2.0	0.1	0.0	Mean
8.21	0.71	11.38	2.69	3.10	4.5	2.12	1.18	2.46	1.30	Standard error

† Heterosis = $100 \times (F_1 - \text{midparent})/\text{midparent}$.
‡ Seed cotton yield, all others lint yield.
§ Length measured as upper half mean, all others 2.5% span length.
¶ Strength reported as Presseley units, all others are T_1 units.

5-2 GENOTYPE × ENVIRONMENT INTERACTIONS

Aside from the gene action the phenotypic performance of a genotype is influenced by its environment. The detection of significant genotype × environment interactions indicates that all phenotypic responses to changes in the environment are not the same for all genotypes. Genotype × environment interactions are important to geneticists and breeders because the magnitude of the interaction component provides information concerning the likely area of adaptation of a given cultivar. The relative magnitudes of the interaction, error, and genotypic components are useful in determining efficient methods of using time and resources in a breeding program.

Because lint yield is considered by many breeders to be the most important single characteristic (Meredith, 1980), yield is a useful reference point from which to examine cotton genotype × environment interactions. While statistical significance of the interaction is important in estimating various genetic and environmental parameters, it is the size of the interaction components relative to those of the genetic components that directs breeders as to the most likely area of adaptation of a successful cultivar. If the interaction components are large relative to the genotypic components, and if they are related to predictable environmental factors (such as geographic areas, elevation, major pest problems, or soil differences), the breeder searches for a cultivar to meet the specific requirements of that environment. If the interaction is small and unpredictable, the breeder searches for a cultivar that has general adaptability and universal performance over the range of environments. Such small and unpredictable environmental factors for cotton would include microclimatic or yearly variations in weather, management practices, and some pest problems.

In 1956 Deltapine 15 came close to being a universally accepted cultivar in the USA, as 34% of the U.S. hectarage was planted to this cultivar (Anonymous, 1956). The idea of a universally accepted cultivar remained at least until the early 1970's when 'Deltapine 16' accounted for 28% of the U.S. hectarage (Anonymous, 1972). However, since then the practice of using a universal cultivar over the USA has gradually given way to using cultivars developed for specific environments.

The ratio of the genotype × environmental components for lint yield from the National Variety Tests (Abou-El-Fittouh et al., 1969), in data relative to the genotypic component (Table 5-4), indicates why cotton breeders are now looking for more local adaptability. The four cultivars used in these 101 tests were developed for the Western, Plains, Delta, and Eastern regions of the USA. The genotypic × year, genotypic × location, and genotypic × year × location are 0.2, 2.3, and 1.3 as large, respectively, as the genotypic component. Inspection of the sizes of the environmental interaction components for yield components and fiber properties (Table 5-4), shows no large interactions except for fiber fineness (micronaire). These factors, yield and fiber fineness, are the two traits that have shown large dominance components in gene action studies. The large genotype × location component for yield suggests a need for developing cultivars for specific regions or

Table 5-4. Ratio of various interaction components with genotypic component for yield, yield components, and fiber properties.†

| | | Yield components | | | Fiber properties | | |
Estimated component‡	Yield	Lint %	Boll size	Seed index	Length	Strength	Fine-ness
Cultivars (σ_g^2)	1.00	1.00	1.00	1.00	1.00	1.00	1.00
Cult. × years (σ_{gy}^2)	0.22	0.22	0.00	0.02	0.01	0.00	0.11
Cult. × location (σ_{gl}^2)	2.32	0.17	0.03	0.02	0.13	0.01	0.26
Cult. × years × location (σ_{gyl}^2)	1.26	0.72	0.10	0.08	0.25	0.04	1.42
Error (σ_e^2)	2.83	0.95	0.08	0.08	0.53	0.12	1.47

† Data from Abou-El-Fittouh et al. (1969).   ‡ Component designation listed in parentheses.

Table 5-5. Ratio of various lint yield interaction components with genotype component.†

| Estimated component | Region | | | | |
	Eastern	Delta	Central	Plains	Western
σ_g^2	1.00**	1.00**	1.00**	1.00**	1.00**
σ_{gy}^2	0.09*	0.21*	0.21*	0.28*	0.11
σ_{gl}^2	0.20**	0.57**	0.65**	0.58**	0.11
σ_{gyl}^2	0.48*	0.57**	0.83*	2.15**	0.22*
σ_e^2	1.97	1.59	2.97	4.77	0.38
n‡	9	15	12	12	4

*,** Designate significant component at the 0.05 and 0.01% levels, respectively.
  † Data taken from Abou-El-Fittouh et al. (1969).
  ‡ n is the number of genotypes in the analysis for each region.

locations. However, for yield components and fiber properties, the analysis indicates genetic differences detected in one year in one part of the USA are likely to be detected in other years in other parts of the country. Again, the only exception appears to be for fiber fineness.

Yield analyses of tests conducted on a regional basis are given in Table 5-5 (Abou-El-Fittouh et al., 1969). While all but two components are statistically significant, the size of the interactions tend to be much less than those reported for the USA as a whole. After regionalization, the genotype × year, genotype × location, and genotype × year × location components, relative to the genotypic component were about 0.2, 0.4, and 0.9, respectively. The summarization in Table 5-5 also reinforces observations concerning the stability of genetic performance in the five regions. The Western region tends to have the most stable environment, the Plains region the least stable, and the other three regions are somewhat intermediate in their likelihood to produce significant genotype × environment interactions.

The results of further reductions in geographic size on genotypic × environment interactions are given in Table 5-6. For yield, the genotype × year and genotype × location interactions tended to be small, except for the Oklahoma tests (Murray and Verhalen, 1970) where a large genotype × location interaction was evident. The interaction in these test locations were related to dryland vs irrigated tests. In the dryland tests, little genetic

Table 5-6. Ratio of variance components with genotype component for yield, component of yield, and fiber properties within states.†

Trait	Genotype	Gen. × yr.	Gen. × loc.	Gen. × yr. × loc.	Error	Reference
Yield	1.00**	−0.342	−0.44	1.57**	8.45	Bridge et al. (1969)
	1.00	−0.035	0.12	1.08**	4.46	Miller et al. (1959)
	1.00	−0.000	3.33*	3.73**	27.29	Murray and Verhalen (1970)
Lint %	1.00**	−0.014	−0.15	0.41**	0.43	Bridge et al. (1969)
	1.00	0.047**	0.07**	0.11**	0.75	Miller et al. (1959)
Boll size	1.00**	0.021	0.16	0.66**	1.24	Bridge et al. (1969)
	1.00	0.021**	0.04**	0.06**	0.70	Miller et al. (1959)
Length	1.00*	0.050	−0.08	0.04	4.50	Bridge et al. (1969)
	1.00	0.125**	−0.06	0.25**	0.98	Miller et al. (1959)
	1.00**	0.104*	0.03	−0.00	2.34	Murray and Verhalen (1970)
Strength	1.00**	−0.304	−0.11	0.55	1.91	Bridge et al. (1969)
	1.00	0.000	−0.03	0.16**	0.48	Miller et al. (1959)
	1.00**	0.144*	0.04	−0.00	2.94	Murray and Verhalen (1970)
Fineness	1.00**	−0.183	−0.10	0.76*	1.53	Bridge et al. (1969)
	1.00	0.035**	0.13**	0.01	0.72	Miller et al. (1959)
	1.00**	0.155	−0.00	0.96**	4.11	Murray and Verhalen (1970)

*,** Component significant at the 0.05 and 0.01 levels, respectively.
  † Data by Bridge et al. (1969) from Mississippi; Miller et al. (1959) from North Carolina; Murray and Verhalen (1970) from Oklahoma.

variability for yield was expressed but in the irrigated tests large genetic differences were evident. Generally, lint yield evaluations for all regional and state tests reveal small genotype × year and genotype × location interactions, but detect a large genotype × year × location interaction. Many researchers believe that these three factor interactions are related to the rather unpredictable differences in yearly weather, crop management, and pest variations interacting with the different fruiting patterns of the various genotypes.

An efficient procedure for sampling environments to evaluate genotypes depends on the genotype × environment interaction components, the cost of testing per environment, and the time desired to have a genotype evaluated. The cost of fiber property determinations is less than that for yield, and as indicated above, the genotype × interaction components for fiber properties are smaller than that for lint yield. The decision about how to sample the environments adequately for cultivar evaluation reduces to what is the best procedure to evaluate yield.

The expected variance of a cultivar mean may be expressed as:

$$V_{\bar{x}} = \sigma_e^2/\text{rly} + \sigma_{gly}^2/\text{ly} + \sigma_{gl}^2/\text{l} + \sigma_{gy}^2/\text{y},$$

where r, l, and y are the number of replications, locations, and years, respectively, and the variance components are those defined in Table 5–4.

Miller et al. (1962) analyzed the results from eight locations east of Texas for a 3-year period. The variance components for the genotype × environment interactions followed the pattern already discussed. Using the above formula, they concluded that the results gave the researcher three choices for sampling: (i) a series of locations in 1 year; (ii) a series of years at one location; and (iii) any combination of years and locations. Since time is very important in decision making, it is likely that most researchers will require no more than 3 years and preferably 2 years for a cultivar evaluation. Several researchers have observed that the genotype × environment interaction produced by different planting dates within the same year produces interaction components equivalent to the unexplained genotype × year × location interaction so prevalent in cotton experiments. Therefore, many researchers include several planting dates within a year to speed-up their genotype evaluations.

5-3 GENETIC ASSOCIATION OF TRAITS

Just as the phenotypic variance of a trait can be partitioned into environmental and genetic components, the covariance between two traits can also be partitioned into environmental and genetic components. Miller and Rawlings (1967a) demonstrated in cotton the method of computing phenotypic and genetic correlations from covariance analyses. If genetic correlations are high, selection for one trait will simultaneously result in changes of other traits. This association may be either harmful or beneficial, depending upon the direction of the genetic correlation and the objectives of the breeders. The genetic mechanisms underlying genetic correlations are pleiotropy or linkage, or both.

The pleiotropic response occurs when the substitution of one gene for another results in phenotypic changes in two or more traits. This result implies a common basic physiological effect shared by the two traits, such as organs arising from a common primordium or the two traits sharing a common biochemical pathway. One example in cotton of probable pleiotropy is the relationship between lint percentage and the smooth leaf genes, Sm_1 and Sm_2 (Lee, 1971). The presence of either of these two dominant genes results in a great reduction of plant and leaf hairs as well as a reduction in lint hairs and, consequently, a lower lint percentage. Genetic correlations caused by linkage arise when the genes conditioning one trait are closely linked to genes conditioning another trait. Miller and Rawlings (1967a) showed that at least part of the negative association of lint yield with fiber strength was caused by linkage in crosses of Upland and exotic germplasm.

Almost every study that detected genetic variation in two or more traits also obtained evidence of genetic associations. A few of the many examples are the following: in the seed, genetic correlations of N (crude protein) with lysine, theonine, glutamic acid, glycine, alanine, and leucine amino acid percentages of -0.51, -0.92, 1.10, -0.73, -0.82, and -0.76, respectively (Meredith et al., 1978), and N with oil percentage -0.64 (Hanny et al., 1978). Hanny et al. (1978) also reported genetic correlations of flowerbud (square) terpenoid percentage with seed terpenoid percentage, terminal terpenoid percentage, and cabbage looper [*Trichopulusia ne* (Hübner)] damage of 0.68, 0.88, and -0.59, respectively. Genetic correlations of lint percentage with lint yield, bolls/plant, seed index, lint index, boll weight, and fiber length, strength, and fineness from three studies averaged 0.80, 0.48, −0.60, 0.36, −0.36, −0.52, −0.12, and −0.31, respectively (Miller et al., 1958). Quisenberry (1975) reported phenotypic correlations of 0.57, 0.88, and 0.38 for plant height with its components, number of main stem nodes, main stem internode length, and node of the first fruiting branch, respectively.

Generally in cotton phenotypic and genetic correlations have been found to be about the same direction and magnitude and themselves highly correlated. Probably no gene, quantitative or qualitative, operates independently of all other genes. Most breeding objectives involve two or more traits.

Genetic associations between qualitative and quantitative traits are very evident in cotton. Kohel (1974) summarized the influence of several morphological characters on yield. The glandless trait results in nearly gossypol-free plant parts and seeds but it also increases insect susceptibility and reduces yield. Narrow parallel-veined, and twisted frego bracts also result in less boll rot, boll weevil (*Anthonomus grandis* Boh.) preference, and greater sensitivity to plant bugs (*Lygus* spp.). Okra leaf creates a 60% reduction in individual leaf area, which results in an open plant canopy that, in turn, is related to increased earliness of yield and reduced boll rot. Smooth leaf genes are associated with a reduction in plant hairs and also seed lint hairs, as mentioned previously; they also reduce trash in the lint and oviposition to *Heliothis* spp. The nectariless trait results in earliness probably because of its reduced susceptibility to plant bugs. The favorable or unfavorable effects that these five traits have on yield varies depending upon environmental fluctuations. These variations are especially significant among those environments in which the traits are exposed to insects. No doubt the same principal of varying correlated responses also applies to quantitative genes, except that the individual quantitative gene effects are not as easy to observe.

As already stated, the trait of most interest to applied breeders is lint yield (Meredith, 1980). Genotypic correlations from the four studies presented in Table 5-7 are typical generally of the results obtained from segregating populations of F_3 or later generations. In general, yield has been positively correlated with lint percentage, fiber elongation, and micronaire values; it is negatively correlated with boll weight, seed weight, fiber length, and fiber strength. As indicated in Table 5-7, lint percentage is a trait that has a high positive correlation with lint yield. Meredith and Bridge (1973b) conducted three cycles of recurrent selection for increased lint percentage within the cultivar 'Deltapine 523.' Mean lint percentages were 33.8, 35.4, 36.6, and 38.0 for the base population, S_1, S_2, and S_3 populations, respectively. Mean lint yields, expressed as a percent of the base populations, were 100.0, 98.8, 107.3, and 106.4%, respectively, for the four respective populations. Increases in lint index, seeds/boll, and micronaire values were detected, and there were simultaneous decreases in both seed index and fiber length. Bridge et al. (1971) reported that the yield improvements of commercial cultivars over the preceding 40 years had been associated closely with increased lint percentage. Most commercial breeders still practice plant

Table 5-7. Genotypic correlations† between lint yield and other traits.

Lint yield vs other traits	Miller and Rawlings (1967a)	Meredith and Bridge (1971)	Fotiadis and Miller (1973)	Scholl and Miller (1976)
Lint %	0.90 (0.12)	0.70 (0.14)	0.79 (0.09)	0.84 (0.08)
Boll weight	0.14 (0.23)	−0.43 (0.22)	−0.14 (0.15)	−0.04 (0.14)
Seed index	--	−0.45 (0.15)	−0.62 (0.12)	−0.28 (0.15)
Fiber length	0.02 (0.20)	−0.47 (0.18)	−0.18 (0.14)	−0.36 (0.13)
Fiber strength	−0.69 (0.14)	−0.54 (0.17)	−0.46 (0.15)	−0.36 (0.14)
Fiber elongation	0.71 (0.19)	0.03 (0.20)	0.02 (0.18)	0.38 (0.16)
Fiber micronaire	0.42 (0.17)	0.42 (0.19)	0.62 (0.12)	0.54 (0.12)

† Standard errors in parentheses calculated by method of Mode and Robinson (1969).

selection for high lint percentage in the F_2 and early generations following a cross. High lint percentage is usually associated with coarser fiber, which has a higher micronaire value.

Since increased fiber strength and lint yield are both major objectives in cotton breeding programs, this negative genetic correlation once presented breeders with a major dilemma. It is, however, one in which considerable genetic and breeding progress has been made. Increased fiber strength has usually been introduced into Upland cotton breeding programs by way of exotic germplasm and interspecific crosses, such as the triple-hybrid cross of (*G. thurberi* Tod × *G. arboreum*) × *G. hirsutum*. Al-Jibouri et al. (1958) and Culp et al. (1979) reported that segregating populations close to the original triple-hybrid cross had yield-fiber strength phenotypic correlations of −0.64 and −0.94, respectively. From such populations, selection theory for increased yield states that correlated negative responses in fiber strength would occur. Miller and Rawlings (1967b) reported the results of such a selection experiment. They conducted three cycles of recurrent selection for increased yield and were successful in increasing yield by 29.7%. Approximately as genetic selection theory predicted, simultaneous increases in lint percentage, fiber elongation, and micronaire values of 8.2, 9.8, and 4.8%, respectively, were obtained. Also as expected, boll weight, seed index, fiber length, and strength decreased by 7.1, 12.5, 3.3, and 6.5%, respectively. Through the middle 1960s, the applied breeders were experiencing similar results.

The correlated responses were caused either by pleiotropy or linkage. Because the initial crosses were interspecific ones, many researchers believed linkage to be responsible for at least part of the correlated responses. Because recombinations of new arrangements of linked genes can only occur between heterozygous chromosome segments, and because cotton is predominately a self-pollinated crop, there is little chance of new recombinants to occur after the F_2 generation. Basing theoretical calculations of linkage associations between parents of diverse genetic backgrounds, Hanson (1959) recommended several generations of random intercrossing to break up undesirable linkage associations. Miller and Rawlings (1967a) and later Meredith and Bridge (1971) (Table 5-8) reported independent intercrossing studies which produced similar results. Genetic correlation of yield and strength were reduced from −0.69 to −0.35 and −0.54 to −0.38, respectively, in these two studies. Using irradiation, Fotiadis and

Table 5-8. Influence of methods of breaking up of linkage blocks on reducing genetic correlations.

	Populations		
Method	Original	Treatment	Reference
Random intercrossing	−0.69	−0.35	Miller and Rawlings (1967a)
Random intercrossing	−0.54	−0.38	Meredith and Bridge (1971)
Irradiation	−0.46	−0.14	Fotiadis and Miller (1973)
Cultivar intercrossing	−0.93†	0.16†	Culp et al. (1979)

† Phenotypic correlations.

Miller (1973) decreased yield-strength genetic correlations of from −0.46 to −0.14.

Simultaneous breeding for increased yield and strength was initiated in 1946 at the Pee Dee Experiment Station, Florence, S.C. (Culp and Harrel, 1974). While the objective of these breeding studies was obviously not planned to test Hanson's 1959 proposal, the results are supportive of it. These researchers used the same principal of intercrossing, except that most of their successes occurred where new yield germplasm was introduced into each new cycle of breeding and selection. Thus, they maintained breeding populations with high amounts of heterozygosity, allowing for breakup of linkage blocks, and also populations with high genetic variances allowing for ample yield selection of new physiological types. In the early stages of their breeding history, the selection of rare yield and fiber strength plants required thousands of evaluations. Gradually, through introgression of new yield germplasm and selection for yield and fiber strength, rare plants occurred in a ratio of about 1:300. With additional crossing (intermating) the frequency of rare plants increased to a ratio of about 1:40. These results suggest that high fiber strength was conditioned by a small number, perhaps as few as two major genes.

The lack of any significant phenotypic correlations of yield and fiber strength in the latest selection cycles indicate that initial deleterious linkage associations may have been broken (Table 5-8). The frequent introduction of new germplasm also allowed for new physiological characteristics for high yield to be developed. The data therefore do not rule out pleiotropic effects being partially responsible for part of the initial negative genetic associations. Scholl and Miller (1976) crossed a low yield and fiber strength cultivar with a high yield and fiber strength strain but still obtained a negative genetic correlation of −0.36 suggesting that some pleiotropy was in-

Table 5-9. Heritability estimates for yield, yield components, and fiber properties.

						Fiber properties			
Yield	Lint %	Boll weight	Seed index	Lint index	Seed/ boll	Length	Strength	Fine- ness	Reference
0.59	0.90	0.77	0.87	0.81	--	0.79	0.90	0.68	Al-Jibouri et al. (1958)
0.66	0.90	0.51	0.87	0.78	0.34	0.90	0.86	0.67	Miller et al. (1958)
0.52	--	0.60	--	--	0.34	0.56	0.86	0.08	Al-Rawi and Kohel (1969, 1970)
0.29	0.28	--	--	--	--	0.46	0.52	0.52	Baker and Verhalen (1975)

volved in this association. Both public and commercial breeding successes indicate that several generations of intercrossing reduce deleterious genetic associations in cotton germplasm of great diversity.

5-4 HERITABILITY AND SELECTION PROGRESS

The percent of the total phenotypic variability attributable to heredity is defined as heritability:

$$\text{heritability} = 100 \; \sigma_g^2 / \sigma_{ph}^2$$

where σ_g^2 = genetic variance
  σ_{ph}^2 = phenotypic variance.

Heritability defined in the broad-sense includes the total genetic variance, but if defined in the narrow-sense includes only the additive genetic variance. Narrow-sense heritabilities are used for making selection progress estimates and selection indexes. The cost of obtaining reliable heritability estimates is very high for most traits, in terms of the time required to develop appropriate genetic populations and to evaluate them with adequate number of years and locations. For this reason, many traits have been only briefly or incompletely studied.

The starting point of most cotton selection studies is with individual plants. Murray and Verhalen (1969) reported broad-sense heritabilities of spaced plants of 75 and 0% for lint yield in 2 successive years. For earliness they found that heritabilities were 55 and 0%; for 2.5% span length, 0 and 20%; for fiber strength (T_0), 0 and 19%; and for micronaire, 46 and 0%. Innes (1973) reported on the basis of plant-to-progeny row data coefficients of determination of 50, 30, and 3% for fiber length, strength, and micronaire. Broad-sense heritabilities of spaced plants for flat squares was low, ranging from 3 to 8% (Khalifa et al., 1974). Selection on an individual plant basis for quantitative traits can be successful. For example, plant selection has resulted in increased lint yield by 5.7% in one generation (Meredith and Bridge, 1973b), lint percentage by 11.2% in three cycles of recurrent selection (Meredith and Bridge, 1973a), and fiber length and strength (Turner et al., 1980).

On a progeny-row basis, Quisenberry (1977) found narrow-sense heritabilities to be 33% for plant height, 76% for fruiting branch internode length, 47% for number of main stem nodes per plant, 51% for node of first fruiting branch, and 62% for mean maturity date. Singh and Weaver (1972) reported broad-sense heritabilities of 73, 75, and 86% for flower bud gossypol. Narrow-sense heritabilities ranged from 75 to 93% for storm-proof boll (Quisenberry et al., 1980). Using standard unit parent-offspring regression, Kohel (1980) estimated the heritability of seed oil to be 35%.

Heritability estimates for lint yield, yield components, and fiber properties from several studies are given in Table 5-9. Heritabilities for yield range from 29 to 66%, those for yield components, with the exception of seed/boll, are generally higher. Fiber properties also usually have high heritabilities.

Most studies with heritability have concentrated on the genetic component to increase heritability. However, heritability may also be more practically enhanced by either reducing or better controlling the environmental variability. Verhalen et al. (1975) subdivided a block of 300 plants of the cultivar 'Westburn' into three grids of 100 plants each. Two selection regimes were used for selecting the upper and lower 10% of the plants, based on fiber length. For one regime, plants were selected from within a grid; for the other regime, plants were selected over the entire block. The phenotypic variation obtained with the grid method was 22% smaller while the narrow-sense heritability was 41 to 52% higher than the values obtained by the non-grid selection procedure.

Another factor under control of the research is what type of environment that is best for evaluating germplasm. Quisenberry et al. (1980) determined the heritabilities and selection responses at three Texas locations by determining lint yield for 2 years for 79 random progenies originating in the F_3 generation. Separate analyses at the three locations indicated heritabilities to be 0% at Lubbock, 14.1% at Big Spring, and 37.8% at College Station. Selection progress at the three respective locations was -0.5, 11.4, and 15.1%. The selections from College Station produced the highest average yields; those selections from Lubbock produced the lowest average yields at all three locations. Feaster et al. (1980) reported similar results with selection studies of Pima (*G. barbadense* L.) cottons at high and low elevations. Selection was effective at low elevations for increased yields at both low and high test environments. However, selections made at the high elevations were effective in increasing yield only for the high elevation test environments. In both Texas and Arizona, it appears that environments deficient in night temperatures limited expression of useful genetic variability.

For many years, cottons selected in the Mississippi Delta have been competitive in yield performance with those developed from the Southwest and West when grown there. The reverse is not true, however; cottons developed in the Southwest and West are not competitive with Delta cottons when grown in the Delta. It is apparent that breeders need to make selections in an environment in which useful genetic variability is best expressed. Genetic progress follows after new useful genetic variability is introduced into the gene pool, after techniques of evaluation are perfected where one can see the useful genetic variability, or after a combination of both situations occurs.

The culmination of quantitative genetic investigations is genetic progress for a trait or combination of traits. Manning (1963) reported yield increases from 12 generations of progeny selection in an African cultivar 'BP52.' Selection was practiced for three components of yield: bolls/plant, seed/boll, and lint/seed. At the end of the tenth generation yield had been increased from 24 to 32%. Kappelman (1980) reported that resistance to fusarium wilt of entries in the regional fusarium wilt test from 1969 through 1978 was significantly higher the last 3 years of his evaluation than in the earlier period. Culp and Harrel (1973) presented the results of about 25 years of breeding for improved yield and fiber quality. In their South Carolina environment, they succeeded in developing strains that consistently equaled the best commercial check in yield, while producing fiber that was 12% higher in yarn strength and 3% longer than that of the commercial check. Meredith and Bridge (1983) analyzed the breeding progress for lint yield from 15 breeders' evaluation tests for the period from 1961 to 1981. They found that yield was increasing from genetic improvements at the rate of 0.7%/year.

5-5 SUMMARY

Large amounts of additive genetic and general combining ability variances indicate continued progress in breeding for yield and fiber properties is likely to continue in cotton. The presence of large amounts of dominance and specific combining ability variances give encouragement to those who wish to attempt to produce high yielding hybrids. Despite regionalization of testing procedures, significant genotype × location × year interactions persist in genotypic evaluation trials and are probably related to small differences related to maturity differences, microclimate, and unpredictable changes in pest populations. Genetic variability of most characteristics is correlated with changes in other characteristics. The negative genetic correlation of yield and fiber strength has been overcome largely by intercrossing both within cross populations and by continuously adding new variability to the gene pool. Heritability of yield is generally lower than that for yield components and fiber properties. Selection progress for many characters has been achieved as evidenced by the steady genetic selection progress for the most important characteristic, yield, or about 0.7% per year, for at least the past 20 years.

REFERENCES

Abou-El-Fittouh, H. A., J. O. Rawlings, and P. A. Miller. 1969. Genotype by environment interactions in cotton—their nature and related environmental variables. Crop Sci. 9:377–381.

Al-Jibouri, H. A., P. A. Miller, and H. F. Robinson. 1958. Genotypic and environmental variances and covariances in an upland cotton cross of interspecific origin. Agron. J. 50:623–636.

Al-Rawi, K. M., and R. J. Kohel. 1969. Diallel analyses of yield and other agronomic characters in *Gossypium hirsutum* L. Crop Sci. 9:779–783.

————, and ————. 1970. Gene action in the inheritance of fiber properties in intervarietal diallel crosses of upland cotton, *Gossypium hirsutum* L. Crop Sci. 10:82–85.

Anonymous. 1956 and 1972. Cotton varieties planted. 1956 and 1972. USDA Agric. Marketing Service, Washington, D.C.

Baker, J. L., and L. M. Verhalen. 1973. The inheritance of several agronomic and fiber properties among selected lines of upland cotton, *Gossypium hirsutum* L. Crop Sci. 13:444–450.

————, and ————. 1975. Heterosis and combining ability for several agronomic and fiber properties among selected lines of upland cotton. Cott. Grow. Res. 52:209–223.

Baker, R. J. 1978. Issues in diallel analysis. Crop Sci. 18:533–536.

Bridge, R. R., W. R. Meredith, Jr., and J. F. Chism. 1969. Variety × environment interaction in cotton variety tests in the Delta of Mississippi. Crop Sci. 9:837–838.

––––, ––––, and ––––. 1971. Comparative performance of obsolete varieties and current varieties of upland cotton. Crop Sci. 11:29–32.

Christiansen, M. N., and C. F. Lewis. 1973. Reciprocal differences in tolerance to seed-hydration chilling in F$_1$ progeny of *Gossypium hirsutum* L. Crop Sci. 13:210–212.

Culp, T. W., and D. C. Harrel. 1973. Breeding methods for improving yield and fiber quality of upland cotton (*Gossypium hirsutum*). Crop Sci. 13:686–689.

––––, and ––––. 1974. Breeding quality cotton on the Pee Dee Experiment Station, Florence, S.C. USDA Publ. ARS-S-30.

––––, ––––, and T. Kerr. 1979. Some genetic implications in the transfer of high fiber strength genes to upland cotton. Crop Sci. 19:481–484.

Davis, D. D. 1978. Hybrid cotton: specific problems and potentials. Adv. Agron. 30:129–147.

Dudley, J. W., and R. H. Moll. 1969. Interpretations and use of estimates of heritability and genetic variances in plant breeding. Crop Sci. 9:257–262.

El-Adl, A. M., and P. A. Miller. 1971. Transgressive segregation and the nature of gene action for yield in an intervarietal cross of upland cotton. Crop Sci. 11:381–384.

Feaster, C. V., E. F. Young, Jr., and E. L. Turcotte. 1980. Comparison of artificial and natural selection in American Pima cotton under different environments. Crop Sci. 20:555–558.

Fotiadis, N. A., and P. A. Miller. 1973. Effects of recurrent seed irradiation on genetic variability and recombination in cotton (*Gossypium hirsutum* L.). Crop Sci. 13:40–44.

Hanny, B. W., W. R. Meredith, Jr., J. C. Bailey, and A. J. Harvey. 1978. Genetic relationships among chemical constituents in seeds, flower buds, terminals, and mature leaves of cotton. Crop Sci. 18:1071–1074.

Hanson, W. D. 1959. The breakup of initial linkage blocks under selected mating systems. Genetics 44:857–868.

Hayman, B. I. 1954a. The analysis of variance of diallel tables. Biometrics 10:235–244.

––––. 1954b. The theory and analysis of diallel crosses. Genetics 39:789–809.

––––. 1958. The theory and analysis of diallel crosses. II. Genetics 43:63–85.

Hosfield, G. L., J. A. Lee, and J. O. Rawlings. 1970. Agronomic properties associated with the glandless alleles in two varieties of upland cotton. Crop Sci. 10:392–396.

Innes, N. L. 1973. Selection for fiber characters in upland cotton. Cott. Grow. Rev. 50:101–102.

Kappelman, A. J. 1971. Inheritance of resistance to fusarium wilt in cotton. Crop Sci. 11:672–674.

––––. 1980. Long-term progress made by cotton breeders in developing Fusarium wilt resistant germplasm. Crop Sci. 20:613–615.

Khalifa, H., W. T. Starmer, W. D. Fisher, and L. S. Stith. 1974. Flat-square in cotton, Acala type. I. Its mode of inheritance and heritability. Crop Sci. 14:519–521.

Kime, P. H., and R. H. Tilley. 1947. Hybrid vigor in upland cotton. Agron. J. 39:308–317.

Kohel, R. J. 1974. Influence of certain morphological characters on yield. Cott. Grow. Rev. 51:281–292.

––––. 1980. Genetic studies of seed oil in cotton. Crop Sci. 784–787.

Lee, J. A. 1971. Some problems in breeding smooth-leaved cottons. Crop Sci. 11:448–450.

––––. 1983. Is there much genetical room left for hybrid cotton? Proc. Beltwide Cott. Prod.-Mech. Conf. San Antonio, Tex.

––––, C. Clark Cockerham, and F. H. Smith. 1968. The inheritance of gossypol level in *Gossypium*. I. Additive, dominance, epistatic and maternal effects associated with seed gossypol in two varieties of *Gossypium hirsutum* L. Genetics 59:285–298.

––––, P. A. Miller, and J. O. Rawlings. 1967. Interaction of combining ability effects with environments in diallel crosses of upland cotton (*Gossypium hirsutum* L.). Crop Sci. 7:477–481.

Manning, H. L. 1963. Realized yield improvement from twelve generations of progeny selection in a variety of upland cotton. *In* W. D. Hanson and H. F. Robinson (eds.) Statistical genetics and plant breeding. NAS-NRC #982.

Marani, A. 1963. Heterosis and combining ability for yield and components of yields in a diallel cross of two species of cotton. Crop Sci. 3:552–555.

————. 1968a. Inheritance of lint quality characteristics in intraspecific crosses among varieties of *Gossypium hirsutum* L. and of *G. barbadense* L. Crop Sci. 8:36–38.

————. 1968b. Heterosis and F_2 performance in intraspecific crosses among varieties of *Gossypium hirsutum* L. and *G. barbadense* L. Crop Sci. 8:111–113.

Matzinger, D. F. 1963. Experimental estimates of genetic parameters and their applications in self-fertilizing plants. *In* W. D. Hanson and H. F. Robinson (eds.) Statistical genetics and plant breeding. NAS-NRC #982.

Meredith, W. R., Jr. 1980. Use of insect resistant germplasm in reducing the cost of production in the 1980's. p. 307–310. Proc. Beltwide Cott. Prod.-Mech. Conf. St. Louis, Mo.

————, and R. R. Bridge. 1971. Breakup of linkage blocks in cotton, *Gossypium hirsutum* L. Crop Sci. 11:695–698.

————, and ————. 1972. Heterosis and gene action in cotton. Crop Sci. 12:304–310.

————, and ————. 1973a. The relationship between F_2 and F_3 progenies in cotton (*Gossypium hirsutum* L.). Crop Sci. 13:354–356.

————, and ————. 1973b. Recurrent selection for lint percent within a cultivar of cotton (*Gossypium hirsutum* L.). Crop Sci. 13:698–701.

————, and ————. 1983. Genetic contributions to yield changes in cotton. *In* W. R. Fehr (ed.) Contribution of genetics to yield gains. Am. Soc. of Agron., Madison, Wis.

————, ————, and J. F. Chism. 1970. Relative performance of F_1 and F_2 hybrids from doubled haploids and their parent varieties in upland cotton, *Gossypium hirsutum* L. Crop Sci. 10: 295–298.

————, C. D. Elmore, and E. E. King. 1978. Genetic variation within cotton for seed N and amino acid content. Crop Sci. 18:577–580.

Meyer, V. G. 1973. A study of reciprocal hybrids between upland cotton (*Gossypium hirsutum* L.) and experimental lines with cytoplasms from seven other species. Crop Sci. 13:439–444.

————. 1975. Male sterility from *Gossypium harknessii*. J. Hered. 66:23–27.

Miller, P. A., and J. A. Lee. 1964. Heterosis and combining ability in varietal top crosses of upland cotton, *Gossypium hirsutum* L. Crop Sci. 4:646–649.

————, and A. Marani. 1963. Heterosis and combining ability in diallel crosses of upland cotton, *Gossypium hirsutum* L. Crop Sci. 3:441–444.

————, and J. O. Rawlings. 1967a. Breakup of initial linkage blocks through intermating in a cotton breeding program. Crop Sci. 7:199–204.

————, and ————. 1967b. Selection for increased lint yield and correlated responses in upland cotton, *Gossypium hirsutum* L. Crop Sci. 7:637–640.

————, H. F. Robinson, and O. A. Pope. 1962. Cotton variety testing: additional information on variety × environment interactions. Crop Sci. 2:349–352.

————, J. C. Williams, and H. F. Robinson. 1959. Variety × environment interactions in cotton variety tests and their implications in testing methods. Agron. J. 51:132–134.

————, ————, ————, and R. E. Comstock. 1958. Estimates of genotypic and environmental variances and covariances in upland cotton and their implications in selection. Agron. J. 50:126–131.

Mode, C. J., and H. F. Robinson. 1969. Pleiotropism and the genetic variance and covariance. Biometrics 15:518–537.

Murray, J. C., and L. M. Verhalen. 1969. Genetics of earliness, yield, and fiber properties in cotton (*Gossypium hirsutum* L.). Crop Sci. 9:752–755.

————, and ————. 1970. Genotype by environment interaction study of cotton in Oklahoma. Crop Sci. 10:197–199.

Quisenberry, J. E. 1975. Inheritance of fiber properties among crosses of Acalas and High Plains cultivars of upland cotton. Crop Sci. 15:202–205.

----. 1977. Inheritance of plant height in cotton. II. Diallel analyses among six semidwarf strains. Crop Sci. 17:347–350.

----, Bruce Roark, D. W. Fryrear, and R. J. Kohel. 1980. Effectiveness of selection in upland cotton in stress environments. Crop Sci. 20:450–453.

Ramey, H. H. 1963. Gene action in the inheritance of lint index in upland cotton. Crop Sci. 3:32–33.

----, and P. A. Miller. 1966. Partitioned genetic variances for several characters in a cotton population of interspecific origin. Crop Sci. 6:123–125.

Scholl, R. L., and P. A. Miller. 1976. Genetic associations between yield and fiber strength in upland cotton. Crop Sci. 16:780–783.

Singh, I. D., and J. B. Weaver, Jr. 1972. Studies on the heritability of gossypol in leaves and flower buds of *Gossypium.* Crop Sci. 12:294–297.

Sprague, G. F., and L. A. Tatum. 1942. General vs. specific combining ability in single crosses of corn. J. Am. Soc. Agron. 34:923–932.

Thompson, N. J. 1971. Heterosis and combining ability of American and African cotton cultivars in a low latitude under high-yield conditions. Aust. J. Agric. Res. 22:759–770.

Turner, J. H., Jr. 1953. A study of heterosis in upland cotton. I. Yield of hybrids compared to varieties. Agron. J. 45:484–486.

----, P. E. Hoskinson, Smith Worley, Jr., and H. H. Ramey, Jr. 1980. Response to selective pressure in early generation progenies of upland cotton (*Gossypium hirsutum* L.). Euphytica 29:615–624.

Verhalen, L. M., and J. C. Murray. 1969. A diallel analyses of several fiber property traits in upland cotton (*Gossypium hirsutum* L.). II. Crop Sci. 9:311–315.

----, J. L. Baker, and R. W. McNew. 1975. Gardner's grid system and plant selection efficiency in cotton. Crop Sci. 15:588–591.

----, R. W. C. Morrison, B. A. Al-Rawi, K. C. Fein, and J. C. Murray. 1971. A. Diallel analysis of several agronomic traits in upland cotton (*Gossypium hirsutum* L.). Crop Sci. 11:92–96.

White, T. G. 1966. Diallel analyses of quantitatively inherited characters in *Gossypium hirsutum* L. Crop Sci. 6:253–255.

----, and R. J. Kohel. 1964. A diallel analyses of agronomic characteristics in selected lines of cotton, *Gossypium hirsutum* L. Crop Sci. 4:254–257.

Wilson, F. D., and R. L. Wilson. 1976a. Breeding potentials of noncultivated cottons. II. Inheritance of peduncle length. Crop Sci. 16:221–224.

----, and ----. 1976b. Breeding potentials of non-cultivated cottons. III. Inheritance of date of first flower. Crop Sci. 16:871–873.

Worley, Smith, Jr., H. H. Ramey, Jr., D. C. Harrell, and T. W. Culp. 1976. Ontogenetic model of cotton yield. Crop Sci. 16:30–34.

6 Physiology

C. R. Benedict

Texas A&M University
College Station, Texas

Past reviews of cotton physiology by Eaton (1950), Eaton (1955), Brown and Ware (1958), Carns and Mauney (1968), Tollervey (1970) and McArthur et al. (1975) have been published. This chapter on cotton physiology deals with the notable progress in seed germination, root physiology, drought tolerance, flowering, flower and fruit abscision, photosynthesis, modeling of yield parameters, water stress, fiber development, cellulose synthesis, ovule and embryo culture, interspecific hybrids of cultivated cotton with wild species of *Gossypium,* harvest-aid chemicals, and chemical control of growth and blooming. The intent of this chapter is to point out the recent advances in the basic research of each of these areas of cotton physiology.

6-1 SEEDLING GROWTH

Good field stands of cotton are affected by imbibition and chilling problems in seed germination, soil microenvironment, soil impedance, and chilling temperatures in seedling emergence (McArthur et al., 1975; Christiansen and Rowland, 1981).

6-1.1 Seed Germination

6-1.1.1 Imbibition

Water uptake is through the chalazal aperture of the dry seed. Chalazal blocks of lignin, pentosans, or waxes can exclude water for seed imbibition for extended periods of time (Christiansen and Rowland, 1981). Cole and

Published in *Cotton,* Agronomy Monograph no. 24, © ASA-CSSA-SSSA, 677 South Segoe Road, Madison, WI 53711.

Christiansen (1975) showed genetic differences in water uptake between varieties of *G. hirsutum* and *G. barbadense.* Cottonseed are normally fully hydrated in approximately 4 to 5 hours at 30°C and the necessary seed moisture for germination is about 55%. Water potential differences between the seed and soil and the resistance to water movement into the seed determines the rate of seed imbibition (Jordan, 1982). Hadas and Russo (1974) showed that the rate of water movement through soils at potentials greater than -15×10^6 Pa should not limit imbibition. Christiansen and Moore (1959) have shown that water entry into the seed may be controlled by hard seed coats prevalent in some wild species of *Gossypium.* The hard seed characteristic is removed by acid delinting with H_2SO_4 to digest the cellulosic fuzz and lint fibers. The hydration of the seed initiates metabolic activity, polysome formation, and transcription and translation necessary for cell division and growth.

6–1.1.2 *Synthesis and Control of Germination Enzymes*

A series of post-germination events in the growth of cottonseed begin with lipolysis of stored triglycerides followed by the β-oxidation of the fatty acids to acetyl-CoA and the conversion of this C_2 unit to succinate in the glyoxysomes. The succinate is converted to oxaloacetate in the mitochondria and the oxaloacetate is converted to phosphoenolpyruvate, phosphoglycerate (PGA), hexose-phosphates and to sucrose by cytoplasmic enzymes (Trelease et al., 1980) as shown in Fig. 6–1.

Ihle and Dure (1972), Dure (1975), and Harris and Dure (1978) describe several developmental events which may be part of preprogramming of cottonseed for successful germination. Ihle and Dure (1972) showed that the de novo synthesis of carboxypeptidase and isocitrate lyase during the germination of cottonseed was insensitive to the transcription inhibitor actinomycin D (ActD)[1] and was concluded to be from pre-existing mRNAs stored in dry seed. Isocitrate lyase and carboxypeptidase were representatives of "germination" enzymes whose synthesis was regulated from stable mRNAs transcribed during embryogenesis. Ihle and Dure (1972) and Dure (1975) described the synthesis of these stable mRNAs at 32 days after anthesis shortly after atrophy of the funiculus. The stable mRNAs were suppressed from translation by the presence of abscisic acid (ABA) synthesized in the ovule and absorbed by the embryo. This suppression prevented vivipary of the embryos in the developing bolls.

Smith et al. (1974) and Radin and Trelease (1976) showed that the de novo synthesis of isocitrate lyase in germinating cottonseed is inhibited by Act D (Fig. 6–2) and thus a stable mRNA synthesized during embryogenesis did not exist. The difference between these results and those of Ihle and Dure (1972) was that the latter authors added Act D to fully imbibed cotton-

---

[1] Abbreviations: actinomycin D, ActD; abscisic acid, ABA; gibberellic acid, GA; indoleacetic acid, IAA; CO_2 exchange rate, CER; ribulose-1,5-bisphosphate carboxylase, RuBPCase; phosphoenolpyruvate carboxylase, PEPCase; phosphoglycerate, PGA; photosynthetic unit, PSU; uridine diphosphate, UDP; guanosine diphosphate, GDP; and 1,1-dimethylpiperididinium chloride, PIX.

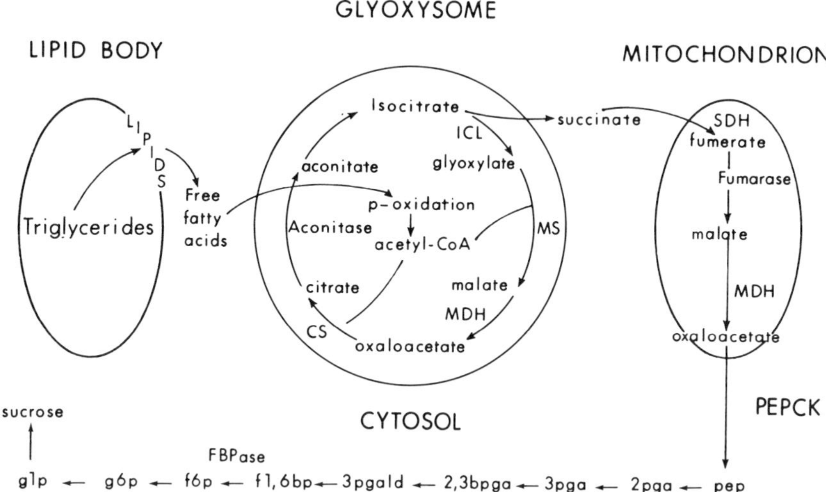

Fig. 6–1. Reactions involved in the conversion of storage lipids to carbohydrate in germinating cottonseed (Trelease et al., 1980).

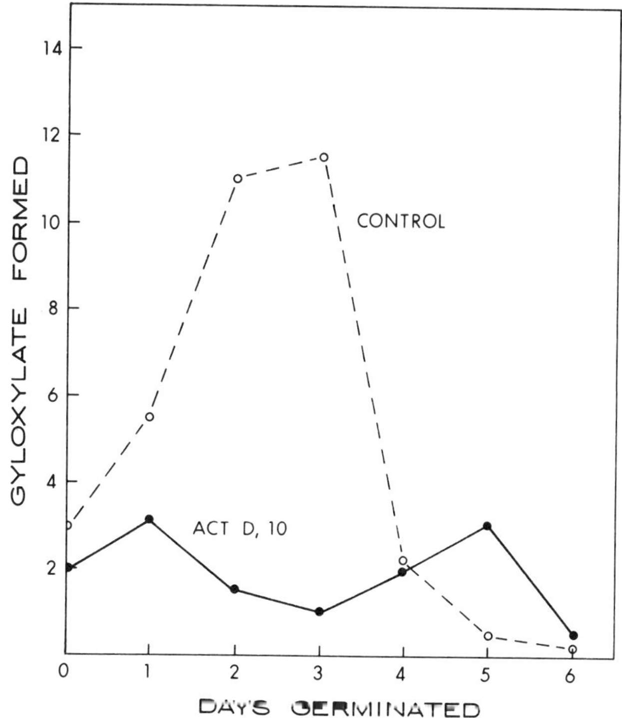

Fig. 6–2. The inhibition of the synthesis of isocitrate lyase by Act D in germinating cottonseed (adapted after Smith et al., 1974).

Table 6-1. Changes in enzyme activity in homogenates of developing embryos and cotyledons of germinated cotton seedlings (Trelease et al., 1980).

Enzyme	Days after anthesis					Hours after germinative growth of mature seeds	
	22	28	38	46	54	24	48
Isocitrate lyase	0	0	0	0	0	0	0
Malate synthase	0	0	0	115	319	711	1 990
Citrate synthase	24	95	142	167	279	447	837
Malate DH	6 660	ND	7 770	7 960	8 390	16 520	19 680
Aconitase	ND†	ND	48	79	57	237	506
Catalase	2	20	25	29	42	218	550
Hydroxylacyl-CoA DH	20	75	265	326	378	1 157	467
Enoyl hydratrase	14	127	1 010	1 235	1 100	4 930	1 870
Thiolase	0	20	99	65	33	174	164
Aspartate AT	ND	ND	314	323	358	1 103	1 601
Alanine AT	ND	ND	341	33	322	616	361
NADP-isocitrate DH	ND	ND	53	63	62	144	270

† ND = not determined.

seed and Smith et al. (1974) added Act D to the imbibing solution. Further, Benedict et al. (1976) concluded that the funiculus does not atrophy at 32 days after anthesis because transportation of photosynthate into the developing ovule continues until the seed has matured 45 to 50 days after anthesis. Dure (1977), Hammett and Katterman (1975), and Harris and Dure (1978) have shown that mRNA pre-exists in dry cottonseed and that this mRNA needs to be processed by polyadenylation for the commencement of normal germination. Trelease et al. (1980) have shown that many enzymes unique to germination of cottonseed are synthesized during embryogenesis and that some or all of the conserved mRNAs may code for the renewed synthesis of these enzymes during germination (Table 6-1). All gluconeogenetic related enzymes examined except isocitrate lyase appear at some time during embryogenesis and are incorporated into glyoxysomes and mitochondria. The glyoxysome-like organelles synthesized during embryogenesis may be related to the postgerminative development of glyoxysomes functioning in gluconeogenesis from storage lipid.

6-1.1.3 Chilling Damage

Cottonseed germination and seedling growth are subject to chilling injury. This injury may completely inhibit seed germination or result in morphological changes which are detrimental to seedling growth. Christiansen et al. (1970) showed that radicles rapidly lost solutes at 5°C in cotton seedlings containing 2 cm long radicles. Chilling temperatures may affect membrane permeability which can be reversed with calcium or magnesium. Cottonseed exposed to chilling temperatures result in a loss of succinate oxidation (Lyons and Raison, 1970) and a depression of isocitrate lyase activity (Mohapatra et al., 1970). Smith et al. (1971) have shown that the low temperature effect is most specific for isocitrate lyase since another glyoxysome enzyme, malate synthetase, and isocitrate dehydrogenase levels are

Table 6–2. Succinate levels in glyoxysomes isolated after 0 and 6 hours exposure of seedlings at 5° C. Cotton and castorbean (*Ricinus communis* L.) seedlings were germinated at 24° C for 72 and 96 hours respectively before chilling. Data are means of three (castorbean) and six (cotton) determinations (Smith and Fites, 1973).

Source	Chilling period	Succinate in gloxysomal protein
	hours	mg mg^{-1}
Castorbean	0	2.9 ± 0.3
Castorbean	6	1.9 ± 0.4
Cotton	0	5.7 ± 0.3
Cotton	6	110.0 ± 14

unaffected by chilling. The depression of isocitrate lyase in cold-sensitive seed is the result of an altered glyoxysomal membrane permeability which leads to a decrease permeability toward succinate (Fites, 1974). The succinate acts as a noncompetitive inhibitor of isocitrate lyase (Godavari et al., 1973) and depresses the lyase activity within the glyosysome. Fites (1974) has shown (Table 6–2) that a 6 hour exposure of cotton seedlings to 5°C results in a 19 fold increase in succinate within the glyoxysomes, and the depressed levels of isocitrate lyase is not due to a conformational change in protein but rather to the buildup of a noncompetitive inhibitor.

6–1.2 Seedling Emergence

The elongation of both hypocotyl and radicle have a temperature optimum of 34.4° C (Wanjura et al., 1972). At optimum temperature the effect of lower soil water potentials ψ_s is to reduce overall growth rates, and the hypocotyl elongation is reduced greater than radicle extension (Jordan, 1982). At ψ_s of -10×10^6 Pa hypocotyl growth was inhibited 87%; whereas, radicle elongation was inhibited 28%. Emergence of seedlings is affected by soil crusting and to emerge through a crust the seedling must exert growth pressure at the hypocotyl hook to penetrate the crust. Jordan (1982) discussed the mechanism by which the hypocotyl senses an external mechanical impedance and transduces this stimulus to growth pressure. Apparently in legumes ethylene production and an increase in radial expansion are proportional to the degree of mechanical impedance thereby adjusting the growth pressure to soil conditions. Wanjura et al. (1971) and Wanjura and Buxton (1972) developed models for the prediction of planting to emergence based on percentage of germination, temperature, soil moisture, and physical impedance of the soil.

6–2 ROOT PHYSIOLOGY

6 2.1 Rhizatron and Water Relations

Browning et al. (1975) studied the water relations of cotton in a rhizatron at Auburn University. The effect of soil compaction, different soil types and pH, and different watering regimes on plant water potentials, root growth, plant height, and stem diameter were measured in order to understand the response of cotton plants to water stress. The data indicated that an acid pH subsoil reduced the growth of cotton. For optimum yields the cotton plant must be continually supplied with water during boll formation. Water stress reduces the growth of new roots and reduces the ability of the root system to supply water to the tops. Their basic data may be useful in simulating the effects of various environmental factors on water uptake by root systems.

6–2.2 Contributions of Heat Tolerance and Root Growth to Drought Resistance

Jordan et al. (1981) studied the relationships of heat tolerance and root growth potential to drought resistance to determine the genetic variability in growth and water use in fifteen exotic strains of cotton grown under contrasting water supplies. The field study was located at Big Spring, Texas on irrigated and dryland plots. The exotic cottons did not flower under the test conditions and heat tolerance was estimated from solute leakage from leaf discs heated at 50°C for 45 min. The data collected from greenhouse growth of plants in 7 × 240 cm acrylic plastic tubes included shoot dry weight, tap root length, and the number of vigorous, downward-growing tap roots at a distance of 1 m below the surface. Water use efficiency calculated on total above ground biomass and soil water use was highly correlated with dry matter production. The correlations of heat tolerance and root growth to field performance are shown in Table 6–3. Cellular damage

Table 6–3. Linear correlation coefficients between all pairs of measured traits from all experiments. Shoot dry weight 86 days after planting was taken as a measure of field performance (Jordan et al., 1981).

Moisture treatment	Field growth, dry weight	Heat tolerance		Greenhouse	
		Total conductivity	% Damaged cells	Root length	Root laterals
Irrigated					
Dry weight	--				
Total conductivity†	0.01	--			
% damaged cells	−0.48*	0.28	--		
Root length	0.14	0.11	0.43	--	
Root laterals	0.19	0.57**	0.14	0.56**	--
Dryland					
Dry weight	--				
Total conductivity	−0.22	--			
% damaged cells	−0.17	−0.09	--		
Root length	0.45*	0.05	0.10	--	
Root laterals	0.57**	0.24	−0.13	0.56**	--

*,** Significantly different from zero at the 0.10 and 0.05 probability levels, respectively.

† Electrical conductivity of bathing solution surrounding 10 leaf discs in which all cells had been killed by autoclaving.

from heat showed consistent but not significant negative correlations with dry matter production from the dryland plots for the last three harvests. At 85 days after planting similar negative correlations were obtained between dry matter yield and cellular damage in irrigated plots. The correlations of tap root length and the number of vigorous laterals with dry matter production were consistent for dryland but not irrigated plots, thus suggesting that root vigor may have allowed superior strains to be better competitors for limited soil water. These results show that significant genetic variability exists among the exotic strains of *G. hirsutum* for dry matter accumulation, heat tolerance, and root growth. Root growth and vigorous growth of root laterals are important to the adaption of cotton to limited supplies of soil water.

6–3 FLOWERING

A common understanding of flowering in plants is that it involves two stages of growth and two chemicals, gibberellic acid (GA) and anthesins (Chailakhyan, 1968). Gibberellic acid is active in the elongation of the stems of flowers, and anthesins are active in the actual differentiation of the floral primordia. Exposure of long-day plants to long photoperiods would increase the synthesis of GA and cause stem elongation from rosette forms. Gibberellic acid would not be limiting in short-day plants, but rather the synthesis of anthesins would be triggered by exposure of the plants to short days. Anthesins have not been identified and another view is that flowering is controlled by an appropriate balance of known phytohormones. The species of *Gossypium* consist of day-neutral, short-day, and long-day plants (Mauney and Phillips, 1963). Commercial Upland cultivars did not exhibit a photoperiodic response. Since most of the cultivated cottons are classified as day-neutral plants, it is not surprising that the application of GA has met with limited success in initiating new floral meristems. Walhood (1958) reported that GA affected fruit set. Dransfield (1961) reported that a single application of GA decreased the number of flowering meristems, although multiple applications of low concentrations of GA did increase the flowering points. Jackson and Fadda (1962) reported a single application of 100 μg of GA retarded flowering of young plants. An increase in flower number following multiple applications (Dransfield, 1961) may have been due to an increase in vegetative and floral branch elongation and not on the differentation of floral meristems. Guinn (1979) points out that preflower water stress has been observed to stimulate subsequent rate of flowering which may be due to increases in hormones. Temperature has an effect on flower initiation. High night temperature (25°C) delays flowering and low night temperatures of 20°C combined with a day temperature of 25°C stimulated flowering (Mauney, 1966).

Mauney (1979) discussed the production of fruiting points on cotton plants. All of the reproductive branches (sympodia) and vegetative branches (monopodia) arise at the base of the leaf on the plant. All of the sympodia originate in the axil of vegetative leaves, and the formation of the reproductive branches is dependent on vegetative growth. Flowering, therefore, is dependent on vegetative growth for additional sympodia, and for the formation of nodes on these sympodia. Hesketh et al. (1972) studied the environmental control of morphogenesis. The concern was the plastochron or the unit of time corresponding to the interval between two successive similar, periodically repeated events. Time intervals between the production of successive squares and flowers on successive fruiting branches on the main stem and between the first two buds or flowers on the same fruiting branch are temperature dependent. At 30°C the production of sympodia equals 2 days and the production of successive nodes on the fruiting branch equals 6 days (Fig. 6-3). The reason why the intervals between events on the main stem and sympodium differ is that events on the

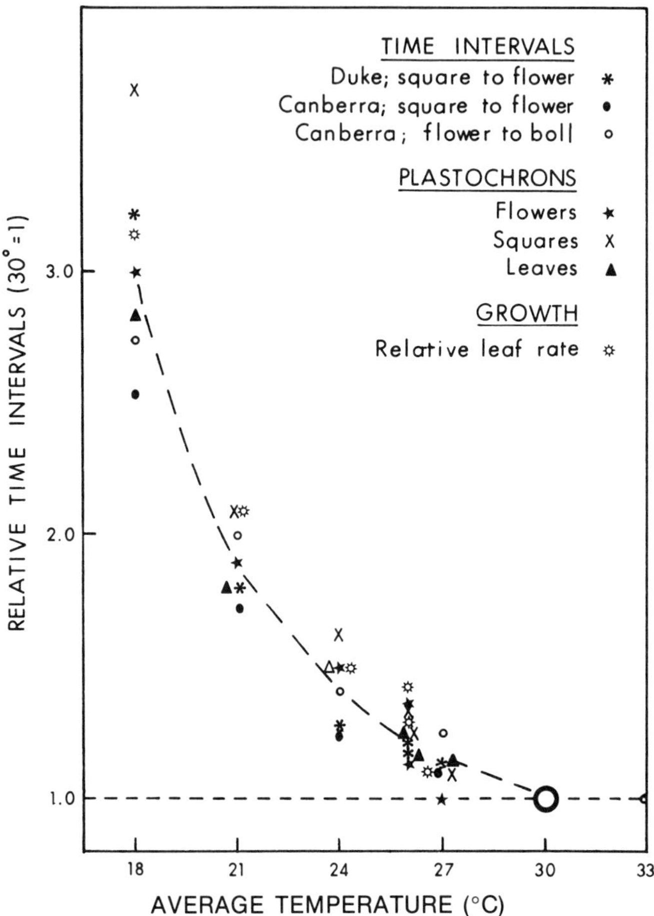

Fig. 6-3. Relative change in isophase and plastochron with decreasing temperature time interval at 30° C for each process plotted being set to 1.0 (Hesketh et al., 1972).

main stem are associated with the initiation of leaf primordia on the shoot apex. Events within a sympodium involve the production of floral bud, flower, open boll, a prophyll, a leaf, two internodes, and new axillary buds in the leaf axil and in the prophyll axis. Guinn (1979) pointed out that as boll load increases, the rate of growth and blooming slow and eventually stops and this phase of growth is known as cut-out.

6-3.1 Earliness

Leffler (1979) states that a commonly used agronomic definition of earliness is the proportion of the total crop that is produced by the time of first picking. Joham (1979) pointed out that from a physiological view the time to first square, open flowers or the nodal position of the first fruiting branch are measures of earliness. Hesketh et al. (1975) list a number of possible earliness factors among genotypes: seedling tolerance to cold associated with planting (Muramoto et al., 1971), shorter apparent plastochrons at the vegetative apex (Hesketh et al., 1975), fruiting branches at a low node on the main shoot (Ray and Richmond, 1966; Low et al., 1969), shorter squaring period (Hesketh and Low, 1968), more flowering sites per fruiting branch and more fruiting branches on vegetative branches (Hesketh et al., 1975), faster squaring or flowering rule (Hearn, 1969), shorter boll periods (Gipson and Ray, 1970), and enhancement of extra earliness in high populations by early cut-out, consistent fruit set at a low main shoot node, and fewer sterile plants (Niles, 1970). Verhalen et al. (1975) concluded that the production of earlier cotton might result from increasing the number of early blooms or from broadening the peak for the number of bolls set. Walhood and Johnson (1976) and Leffler (1979) have presented data which show that seedling vigor and the vigor of the developing canopy influence earliness. Namken et al. (1978) have described in inheritable trait of one or more pairs of sympodia in an opposite configuration on the main stem which leads to earliness by shortening the vertical or main stem flowering interval.

6-3.2 Flower and Fruit Retention

Beginning with the work of Mason (1922) it was thought that a cotton plant retains as many bolls as it can supply with carbohydrates, N, and other nutrients. This became known as the "nutritional theory" of boll shedding. Wadleigh's (1944) experiments on shedding from carbohydrate and N analyses of plants grown on different regimes of light and nitrate support this nutritional theory. Eaton and Ergle (1953) found no differences in the carbohydrate and N concentrations between plants which were shedding and those retaining their bolls. It was concluded (Eaton, 1955) that other explanations of shedding should be sought.

6-3.2.1 Hormones

Eaton and Ergle (1953) proposed that cotton fruit retention was likely to be a hormone-regulated phenomenon. Guinn (1979) reviewed how hormones interact to control abscission. Auxins inhibit, ABA and ethylene promote, and cytokinins and GA have variable effects on abscission. The regulation of abscission by hormones has to be ultimately linked to an increase in the enzymatic activity of pectinase and cellulase which dissolve the middle lamella and weaken the cell wall in the cells of the abscission zone.

Indoleacetic acid (IAA) applied distally to the abscission zone inhibits abscission and IAA applied proximally promotes abscission (Addicott and Lynch, 1955). Abeles (1967, 1969) reasoned that auxin applied distally moves into the abscission zone and inhibits abscission by inhibiting synthesis of cellulase. Morgan and Hall (1962, 1964) showed that 2,4-D and IAA accelerate the release of ethylene in cotton plants. This stimulation in the synthesis of a physiological gas suggests that some of the responses attributed solely to auxins may actually be due to ethylene. Thus, IAA applied dorsally could cause induction of ethylene. Auxins could not move into the abscission zone against polar auxin transport but could cause the induction of ethylene which could migrate to the abscission layer and promote abscission. Evidence supports the view that when auxin promotes abscission it does so by inducing ethylene synthesis, and when it inhibits abscission one important action is to prevent synthesis of hydrolytic enzymes (Abeles, 1967, 1969).

Ethylene has been shown to increase the destruction of auxin and could stimulate abscission, in part, by causing a reduction in auxin levels (Morgan and Hall, 1963). Exogenous ethylene increases the capacity for destruction of IAA in extracts of cotton tissue. However, the major action of ethylene appears to involve its inhibition of auxin transport. Morgan and Gausman (1966) showed that incubation of tissue or intact plant with ethylene reduced the degree of polar auxin transport. Other evidence that auxin transport is important in abscission includes observations that auxin transport inhibitors (Morgan and Durham, 1972, 1973) promote ethylene-induced abscission as does water stress (Jordan et al., 1972) which also reduces auxin transport capacity (Davenport et al., 1977). Beyer and Morgan (1970) have shown that the effect of ethylene on auxin transport is not due to a reduction in auxin uptake or an increase in auxin. These same authors showed that ethylene inhibited the movement of a pulse of auxin in a stem segment (Beyer and Morgan, 1969). Morgan et al. (1970) suggest that ethylene functions in abscission first by lowering the auxin supply to the abscission layer and then by inducing the synthesis of hydrolytic enzyme. In this and a second paper (Beyer and Morgan, 1971) they give evidence that abscissing cotton cotyledons produce physiologically active amounts of ethylene and exhibit declining auxin transport capacity. Further, such levels of ethylene applied to plants induce abscission and inhibit auxin transport. Riov (1974) showed that ethylene induced the synthesis of polygalacturonase in leaf explants of *Citrus simensis*. Abeles (1969) showed that ethylene stimulates cellulase in the abscission zone. Abscisic acid has been found to increase ethylene production and cause an increase in cellulase (Craker and Abeles, 1969). Cytokinins delay senescence and delay abscission. The balance between the production of auxin, ethylene, and ABA probably regulates growth, flowering, fruiting, and abscission (Guinn, 1979).

Abscisic acid was isolated and identified from young fruit of cotton by Ohkuma et al. (1963, 1965). Carns et al. (1955) and Addicott et al. (1964) disclosed the presence of a substance in diffusates of young cotton fruit that could counteract auxin-induced curvature of *Avena* coleotiles. They found that the compound in the diffusates increased with fruit age to a peak in the vicinity of 5 to 7 days then declined. It appears in all likelihood that the compound was ABA. Davis and Addicott (1972) showed that high levels of ABA in cotton fruit occurred 10 and 50 days after anthesis and correlated with abscission of young fruit. Declining or low levels of ABA occurred during the period of most rapid fruit growth. Morgan et al. (1971) and Lipe and Morgan (1972a, 1973) studied the time course of ethylene production in cotton fruits as shown in Fig. 6-4. Significant amounts of ethylene were produced by developing cotton fruits during the period just preceding abscission. After anthesis ethylene production by nonabscising flowers dropped to a very low level in 4 days while production reached a maximum 2 to 4 days after anthesis in those fruits that abscised. The high ethylene production in days near anthesis of the nonabscising fruits probably includes some auxin-induced ethylene production associated with pollination. High levels of auxin in these fruits may have helped prevent abscission. One function of ethylene is the reduction of juvenility auxin reaching the

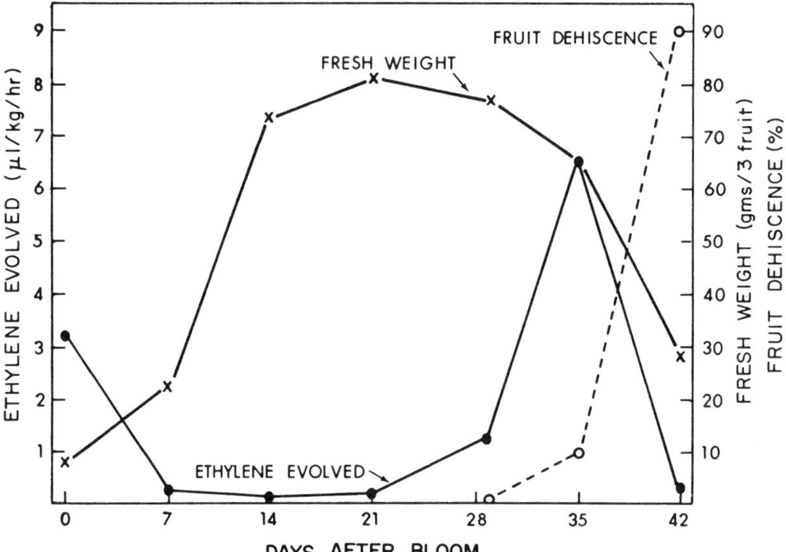

Fig. 6-4. Time course of ethylene production, weight and dehiscence of cotton fruits from anthesis until completion of dehiscence (Morgan et al., 1971).

abscission layer. As the auxin levels drop, the cells in the abscission zone become sensitive to ethylene, which induces the synthesis of hydrolytic enzymes, and separation follows (Morgan et al., 1971). Guinn (1976b, 1978) showed that ethylene increases in cotton fruit during the season and both ethylene production and boll abscission increased to maximum values just prior to irrigation.

There is considerable evidence that ethylene regulates cotton boll dehiscence, a specialized kind of abscission or separation of carpel walls during "ripening" of the cotton fruit (Lipe and Morgan, 1972a, 1972b).

6-3.2.2 Nutrition

Cloudy weather (Mason, 1922) and partial leaf removal (Eaton and Ergle, 1953) can cause shedding of flowers, squares, and bolls. Guinn et al. (1976) have shown that the rate of photosynthesis was increased by 65% immediately after transfer from 330 μl l^{-1} to 630 μl l^{-1} of CO_2 and then declined to 31% higher than in the 330 μl l^{-1} CO_2 atmosphere. As shown in Table 6-4 and Table 6-5 plants growing at the higher CO_2 concentration produced more blooms than those growing at 330 μl l^{-1} CO_2. Also, the lowest abscission rate of bolls occurred on plants growing at the high CO_2 level. The increase in the rate of photosynthesis caused by CO_2 enrichment of the atmosphere increased the number of blooms and decreased boll shedding. Guinn (1974) investigated some environmental factors on abscission of young cotton fruits. Increasing the CO_2 level from 350 μl l^{-1} to 1000 μl l^{-1} decreased the number of bolls abscissed from 30.9% in the control to 16.4% in the high CO_2 (Table 6-6). Increasing the daily photoperiod from 8 to 14 hours also decreased boll abscission (Table 6-7). Shedding was increased by warm nights and low light. These results suggest that a decrease in photosynthesis or an increase in photorespiration may delay fruiting and decrease fruit retention. Guinn (1976a) studied the effect of a nutritional stress on the rate of ethylene production by young cotton bolls. Placing the plants in dim light for as little as 1 day increased ethylene production in 4-day-old cotton bolls about 290% as compared to the control plants (Table 6-8). Bolls from plants exposed to three warm nights produced more ethylene than control plants. The rate of ethylene production by bolls on field grown cotton increased eightfold when the boll load was heavy. These results indicate that a nutritional stress increased ethylene production in young bolls and this ethylene may be a causal factor in boll abscission. These results also indicate that nutritional and hormonal theories of cotton boll shedding are not mutually exclusive because nutritional stress increases ethylene and causes boll abscission.

Guinn (1981) showed that plants placed in dim light for 3 days abscissed nearly all of the young fruit but did not abscise any fruit 15 days or older. Dim light increased ABA concentration in fruits which were 6 to 11 days old by 29 to 155%, but did not increase ABA concentrations in younger or older fruit. Dim light also led to an increase of 28 to 500% in ethylene production in fruits up to 10 days old. The mechanism of how low

Table 6-4. Number of blooms per plant as influenced by CO_2 content of the atmosphere and by concentration of nutrients (Guinn et al., 1976).

	630 μl l^{-1} CO_2	330 μl l^{-1} CO_2	Ratio
	blooms/plant		
2× nutrients	105	69	1.52
Normal nutrients	86	67	1.28
Ratio	1.22	1.03	

Table 6-5. Average numbers of bolls per plant as influenced by CO_2 and nutrient levels (Guinn et al., 1976).

	630 μl l^{-1} CO_2	330 μl l^{-1} CO_2	Ratio
	bolls/plant		
2× nutrients	74.8	41.6	1.80
Normal nutrients	59.3	39.4	1.51
Ratio	1.26	1.06	

Table 6-6. Effects of atmospheric CO_2 level on total fruiting positions (FP) on percentage of the total fruiting positions that abscised their fruiting forms (% abscised) and on number of squares, blooms, and bolls remaining at harvest (No. retained)† (Guinn, 1974).

CO_2 level	Total FP per plant	Percent abscised	No. retained per plant
First test:			
350 μl l^{-1}	14.3 ± 0.52	30.9 ± 2.3	9.9 ± 0.58
1000 μl l^{-1}	16.8 ± 0.60	16.4 ± 2.2	14.2 ± 0.72
Second test:			
350 μl l^{-1}	37.1 ± 1.48	25.6 ± 1.6	28.4 ± 1.54
1000 μl l^{-1}	36.9 ± 1.34	17.1 ± 1.4	30.8 ± 1.44

† Data are averages of 40 plants per treatment. Standard errors of the means are shown.

Table 6-7. Effects of night temperature, photoperiod, and 3.5 days at 650 m^{-2} cd sr on percentage of total fruiting positions that abscised their fruiting forms (% abscised) and on the number of squares, blooms, and bolls remaining at harvest (No. retained)† (Guinn, 1974).

	Growth chamber		Greenhouse	
Treatment	Percent abscised	No. retained per plant	Percent abscised	No. retained per plant
Night temperature:				
30°C	41.0 ± 1.9	17.4 ± 0.91	--	--
20°C	31.8 ± 2.3	18.6 ± 1.03	--	--
Second test:				
8-hour day	48.6 ± 3.0	11.0 ± 0.71	31.8 ± 3.5	24.2 ± 1.55
Long day‡	33.5 ± 2.0	17.2 ± 0.83	20.6 ± 4.0	27.1 ± 2.09
Low light stress:				
650 m^{-2} cd sr	--	--	62.8 ± 2.9	12.8 ± 1.25
Control	--	--	37.7 ± 2.3	21.6 ± 1.19

† Data are averages of 40 plants per treatment. Standard errors of the means are shown.
‡ Long day was 16 hours for the growth chamber test and ranged from 14.35 to 12.53 hours for the greenhouse test.

irradiance increases ABA and ethylene has not been determined, but ethylene responses is related to the availability of photosynthate. Previous sugar and ethylene analysis (Guinn, 1976a) had shown a negative correlation between the sugar content of the bolls and the rate of ethylene production. Guinn (private communication) indicates that photosynthesis is the greatest limiting factor for high yield of cotton and that limited photosynthate causes boll shedding by changing the amount of ethylene and possibly other hormones.

6-3.2.3 Water Stress

Jordan (1979, 1982) discussed the induction of abscission of cotton fruit by a deficiency of water. The resumption of shedding after irrigation may be explained in terms of nutrition. Vegetative growth is highly sensitive to water deficits, but cell division continues at water potentials low enough to inhibit cell enlargement. As a consequence leaf primordia do not enlarge. On rewatering turgor is again established and the new demand for carbohydrate and N by the rapid leaf expansion may be sufficient to trigger shedding (Jordan, 1979). In addition to nutritional stress, water deficits may lead to changes in the concentrations of several plant hormones of magnitude known to induce leaf and boll abscission.

McMichael et al. (1972) studies the effect of water stress on ethylene production in intact cotton petioles. As severe water deficits developed, sharp increases in ethylene production rates of all petioles were noted. The synthesis of ethylene fell rapidly following removal of the deficit by watering. A water deficit of -20×10^6 Pa led to an ethylene production rate of 11 μl/kg fresh weight h^{-1} for the petiole of the third node. The authors con-

Table 6-8. Ethylene evolution by 4-day-old cotton bolls as influenced by 1 or 2 days' exposure of plants to dim light[†] (Guinn, 1976a).

Harvest date	Control	Dim light	% of Control
	μl kg^{-1} h^{-1}		
	First test		
28 January (DL)	3.30 ± 0.84	2.94 ± 0.45	89
29 January (S)	1.31 ± 0.30	3.83 ± 0.44	292
30 January	2.62 ± 0.80	5.56 ± 0.94	212
31 January	2.00 ± 0.24	3.90 ± 1.35	195
1 February	2.41 ± 0.81	1.24 ± 0.46	51
	Second test		
10 September (DL)	1.16 ± 0.21	0.84 ± 0.20	72
11 September (DL)	1.03 ± 0.33	2.07 ± 0.21	201
12 September (S)	0.70 ± 0.11	3.69 ± 0.68	527
13 September	0.67 ± 0.09	2.41 ± 0.42	360
14 September	0.81 ± 0.11	0.86 ± 0.16	106
18 September	0.84 ± 0.07	1.44 ± 0.25	171
19 September	0.75 ± 0.10	0.96 ± 0.22	128

[†] Plants were exposed to dim light, 650 m^{-2} cd sr, on days indicated by (DL) and were returned to full sunlight in the greenhouse on days indicated by (S). Data are averages of 8 to 12 samples/treatment at each harvest date (except for five control bolls on 29 January and seven dim-light bolls on 30 January). Standard errors of the means are shown.

cluded that H_2O stress may directly cause a physical inhibition of auxin transport and stimulation of ethylene synthesis. As discussed previously (Morgan et al., 1971), ethylene can further reduce auxin supply by modifying polar auxin transport and directly induce the synthesis of hydrolytic enzymes in the abscission zone. McMichael et al. (1973) showed that water stress induces abscission in cotton. The extent of both boll and leaf abscission increased in a linear manner as leaf water potentials decreased from -10 to -24×10^6 Pa. Jordan et al. (1972) showed that water stress enhances ethylene-mediated leaf abscission. Treatment with ethylene raised the threshold plant water potential required to induce abscission. The water stress caused the tissue to become predisposed to exogenous ethylene action. Lipe and Morgan (1973) showed that subjecting cotton plants to reduced atmospheric pressure decreased abscission probably because the low pressure removed the endogenous ethylene. Guinn (1976b) has studied the relation between water deficit and ethylene evolution in young cotton bolls. Incubating detached cotton bolls at low humidity increased their rate of ethylene production. Bolls from droughted plants produced more ethylene after the plants wilted compared to control plants. The increased rate of ethylene evolution by young cotton bolls on plants subjected to water stress probably is sufficient to cause abscission of the bolls.

Guinn (1982) studied the ABA levels and rates of abscission in young cotton bolls in relation to availability of water and boll load. Water deficit leads to an increase in ABA concentration of cotton leaves. McMichael and Hanney (1977) and Guinn (1982) show ABA levels and abscission of the bolls increased with water deficits and decreased with relief of water stress by irrigation (Table 6-9). There was a significant correlation of ABA

Table 6-9. Abscisic acid (ABA) concentration in 3-day-old bolls and boll abscision rates as influenced by water deficit. Irrigation and midday leaf water potentials are shown in parentheses (Guinn, 1982).[†]

Date	Irrigation treatment B		Irrigation treatment E		Irrigation treatment F	
	ABA content	Abscision	ABA content	Abscision	ABA content	Abscision
	$\mu g/g$	%	$\mu g/g$	%	$\mu g/g$	%
28 May					(Irrigated, 15 cm)	
4 June	(Irrigated, 15.2 cm)					
17 June	(Irrigated, 12.8 cm)					
24 June			(Irrigated, 12.1 cm)			
30 June	(Irrigated, 10.1 cm)					
2 July	$(-19.1 \times 10^6$ Pa)				$(-27.8 \times 10^6$ Pa)	
3 July	1.23 ± 0.10	19 ± 3			2.05 ± 0.28	68 ± 8
6 July	1.41 ± 0.15	34 ± 8			3.04 ± 0.22	80 ± 7
8 July			$(-29.5 \times 10^6$ Pa)		(Irrigated, 12.2 cm)	
10 July	1.87 ± 0.15	69 ± 9			2.46 ± 0.11	75 ± 8
10 July					$(-22.5 \times 10^6$ Pa)	
14 July	3.02 ± 0.29	89 ± 5	3.76 ± 0.22	92 ± 3		
14 July	$(-28.6 \times 10^6$ Pa)		$(-29.7 \times 10^6$ Pa)			
15 July	(Irrigated, 12.2 cm)		(Irrigated, 12.2 cm)			
17 July	$(-22.8 \times 10^6$ Pa)					

[†] Data are averages of six replications and standard errors of the mean are shown. In relation to bolls, dates refer to time of harvest of 3-day-old bolls.

concentration in 3-day-old bolls with the rate of abscission. The significant correlation provides circumstantial evidence that ABA is involved in boll abscission caused by water deficits.

6–3.3 Mineral Nutrition

Tucker and Tucker (1968), Jones and Bardsley (1968), and Kamprath and Welch (1968) reviewed the role of N, P, and K in the cotton plant and the methods and timing of fertilization. Hinkle and Brown (1968) discussed diagnostic methods, soil properties and availability, and fertility of Ca, Mg, S, Zn, B, and other elements (Mn, Fe, Cu, Mo, Na, and Cl). Emphasis in this section will then be placed on the relation of mineral elements and production of cotton bolls.

6–3.3.1 Fruiting and Fruiting Efficiency

Eaton (1955) reviewed the influence of several mineral elements on fruiting and fruiting efficiency of cotton plants. Early workers found that increasing the percentage of P in fertilizers increased the number of early blooms and the proportion of crop harvested at first harvest, but other workers did not confirm this finding. The S supply to the cotton plant did not alter the number of bolls/100 g fr wt of the plant. Boron-deficient plants are characterized by short sympodia and the failure of young bolls to develop. Increases in B increases relative fruitfulness, i.e. the number of bolls/100 g ft wt of stems and leaves. Since B influences carbohydrate translocation, it is understandable how B effects relative fruitfulness. Molybdenum does not affect relative fruitfulness and relative fruitfulness at several nitrate levels remained unchanged. The uniformity of relative fruitfulness with different amounts of growth under the influence of varied amounts of moisture, N, S, and Mo together with the fact that boll shedding does not occur until a heavy boll load is set, in part, provided the basis for suggesting that boll shedding is controlled by the balance between auxin and anti-auxin.

6–3.3.2 Effects of Elements on Flowering and Fruiting

Joham (1979) reviewed the effect of nutrient elements on fruiting efficiency. The ratio of the dry weight of the bolls/dry weight of stems and leaves is used to describe fruiting efficiency. This seems to be a better measure of the vegetative growth and reproductive growth of cotton plants because it takes into account the differences in boll size between cultivars. Furthermore, fruiting index is similar to the term harvest index commonly used in cereal crops. The elements can be divided into two groups with respect to fruiting index or fruiting efficiency. A deficiency of one group of elements (P, K, Ca, Mg, B and Zn) limits fruit production to a greater extent than vegetative growth; whereas, the deficiency of a second group of elements (N, S, Mo, and Mn) restricts vegetative and fruiting growth to an equal extent. Most of the elements in the first group may affect fruiting efficiency because they function in the control of carbohydrate translocation. A restriction in the flow of carbohydrates out of the leaves could influence the number and size of the bolls. The reason that a deficiency of these elements cause a greater reduction in fruiting than vegetative growth may be due to the proximity of the growing points to the leaves, and the relative polarity of carbohydrate movement to the vegetative and fruiting points. A deficiency of P may limit fruiting due to its role in the conservation of energy. A deficiency of Zn may limit fruiting due to its role in maintaining auxin in an active state.

A deficiency of N, S, Mn, and Mo has little influence on the ratio of vegetative and fruiting growth. A deficiency of N has been found to cause a reduction in both vegetative and fruiting growth. As a result, fruiting efficiency remains unchanged at several levels of N nutrition. Tucker and Tucker (1968) earlier pointed out findings in agreement with this effect of N on fruiting efficiency. Increasing the level of N promotes plant growth and increases the number of axillary positions on the main stem and sympodia. The amount of flowering must be increased with increases in N due to an increased number of sites for flower initiation. However, the total number of flowers in relation to the total number of mature bolls usually is not affected by N over a wide range of N levels in the soil.

Radin and Mauney (1982) reported the effect of N fertilization on several growth parameters of the cotton plant. Nitrogen affects cotton growth rates, leaf area, earliness, boll shedding, lint/boll, and stomatal responses to water stress, but the time to first flower, the time to first open boll, the flowering interval, and the number of seeds/boll were not affected by different levels of N. Different levels of N also affect the fruiting index. Values of 2.7 have been found with low levels of N, and values of 4.1 with high levels of N. High N has been found to increase the number of bolls/plant. Plants grown at low levels of N set fruit earlier, but cut-out also occurs earlier than with plants grown at high levels of N.

6-4 PHOTOSYNTHESIS

Guinn et al. (1976) and Mauney et al. (1979) studied the relationship of photosynthesis and yield of crop plants. Carbon dioxide enrichment to 630 μl l^{-1} resulted in an increase in CO_2 exchange rate (CER) of 15% for cotton, 2% for sorghum [*Sorghum bicolor* (L.) Moench], 41% for soybean [*Glycine max* (L.) Merr.] and 7% for sunflower (*Helianthus annuus* L.) as compared to the CER for these species at 330 μl l^{-1} CO_2 (Table 6-10). Lint yield of cotton increased 180% in plants growing at the high CO_2 concentration (Table 6-11). The selection for increases in CER will not increase the yield of determinate species such as sunflower and sorghum, but may increase the yield of indeterminate species such as cotton and soybeans.

6-4.1   Photosynthetic Characterization of Cotton

6-4.1.1 C_3 Plant

Hesketh (1967) showed that the photosynthetic rate of 'Acala B-54' cotton leaves was enhanced 38% by O_2 free air. This enhancement is due to the presence of photorespiration typical of C_3 plants. The CO_2 compensation point of cotton leaves is 70 μl l^{-1} which indicates the presence of photorespiration (Benedict et al., 1972). The kinetics of CO_2 fixation in cotton leaves has not been reported, but Benedict (1972) and Whelan et al. (1970) showed: the presence of several enzymes of the reductive pentose phosphate cycle in cotton leaves, that there is 25 times more ribulose-1,5-bisphosphate carboxylase (RuBPCase) than phosphoenolpyruvate carboxylase (PEPCase) in cotton leaves, and that the $\delta^{13}C$ value of cotton leaves is -30 $^{o}/_{oo}$. All these results indicate cotton is a C_3 plant.

Table 6–10. Effects of LoCO$_2$ (330 μl l^{-1}) and HiCO$_2$ (630 μl l^{-1}) on CO$_2$ uptake (CER) of Group 1 and Group 2 plants, leaf area, and dry weight accumulation after 12 weeks growth of Group 2 plants. (Adapted after Mauney et al., 1978).

Species	Year	CER LoCO$_2$	HiCO$_2$	Leaf area LoCO$_2$	HiCO$_2$	Dry weight LoCO$_2$	HiCO$_2$
		— nmoles cm^{-2} sec^{-1} —		— dm^2 plant^{-1} —		— g plant^{-1} —	
Cotton	1975	2.96	3.34*				
	1976	2.08	2.39*	153	292**	320	670**
Soybean	1975		3.59				
	1976	1.38	1.95**	100	280**	85	410**
Sunflower	1975		3.40				
	1976	2.64	2.83	120	290**	500	800*
Sorghum	1975	3.90	4.28				
	1976	3.78	3.84	20	23	85	100

*,** Significantly different from LoCO$_2$ measurement at 0.05 and 0.01 levels, respectively.

Table 6–11. Components of yield of Group 1 cotton plants as influenced by LoCO$_2$ (330 μl l^{-1}), HiCO$_2$ (630 μl l^{-1}), and nutrient concentration. (Adapted after Mauney et al., 1978).

Yield component	1975, normal LoCO$_2$	HiCO$_2$	1975, 2 × LoCO$_2$	HiCO$_2$	1976, 2 × LoCO$_2$	HiCO$_2$
Leaf area (dm^2/plant)					195	260
Dry weight (g/plant)					650	950
Blooms/plant (No.)	67	86	69	105	69	110
Bolls/plant (No.)	39	59	42	75	37	80
% retention	66	78	61	71	54	74
Boll weight (g/boll)					4.7	5.5
Lint yield (g/plant)					61	170
Seed yield (g/plant)					114	274
Fraction of yield increase due to:						
Increase bloom/plant		0.67		0.77		0.47
Increase retention		0.40		0.23		0.29
Increase lint/boll						0.24

6-4.1.2 $\delta^{13}C$ Values

Photosynthesis is accompanied by a fractionation of the stable carbon isotopes of CO_2. The discrimination favors the fixation of the $^{12}CO_2$ into plant material, and it is now known that the $\delta^{13}C$ values of plants is highly correlated with the presence of C_3 and C_4 photosynthesis. Wong et al. (1979) showed that the enzymatic fractionation of stable carbon isotypes of CO_2 by RuBPCase is of major importance in determining the $\delta^{13}C$ value of cotton plants. In these experiments RuBPCase was isolated from cotton leaves and purified to electrophoretic homogeneity. The $\delta^{13}C$ values of the substrates (RuBP and CO_2) and product (PGA) of the RuBPCase reaction were determined (Table 6-12). The enzymatic fractionation of the stable carbon isotopes of $^{12}CO_2$, $^{13}CO_2$, by RuPBCase is calculated as follows:

$$5/6\ \delta^{13}C\ RuBP + 1/6\ \delta^{13}C\ CO_2\ \text{fixed} = \delta^{13}C\ PGA$$

substituting $\delta^{13}C$ values

$$5/6\ (-13.9\ ^o/oo) + 1/6\ \delta^{13}C\ CO_2\ \text{fixed} = -24.00\ ^o/oo$$

$$\delta^{13}C\ CO_2\ \text{fixed} = -74.5\ ^o/oo$$

$$\Delta CO_2 = \delta^{13}C\ CO_2\ \text{fixed} - \delta^{13}C\ \text{dissolved}\ CO_2$$

$$\delta^{13}C\ \text{dissolved}\ CO_2 = \delta^{13}C\ HCO_3 - (10.2 \times 0.064 \times T)$$

$$\delta^{13}C\ \text{dissolved}\ CO_2 = -42.0\ ^o/oo$$

substituting

$$\Delta CO_2 = (-74.5\ ^o/oo) - (-42.0\ ^o/oo) = -32.5\ ^o/oo$$

The average ΔCO_2 value calculated in a similar manner for five experiments $= -27.1\ ^o/oo$. The stable carbon isotope fractionation steps in an open system, like the C_3 cotton plant, would be additive and the $\delta^{13}C$ value would be a result of the fractionations associated with: dissolving CO_2 in the cell cytoplasm, isotopic equilibrium of atmospheric CO_2 and dissolved CO_2, and RuBPCase. Fractionations by these steps would give a range of $\delta^{13}C$

Table 6-12. The $\delta^{13}C$ values of the reactants and products of the fractionation of $^{13}CO_2$, $^{12}CO_2$ by RuBPCase (adapted from Wong et al., 1979).

Reactants and products	$\delta^{13}C$ vs. PDB $^o/oo$				
	I	II	III	IV	V
RuBP	-13.9	-13.9	-13.9	-14.3	-14.3
PGA	-24.0	-22.4	-21.9	-23.5	-23.1
HCO_3^-	-34.0	-30.8	-30.8	-33.4	-33.4
CO_2 dissolved	-42.0	-38.8	-38.8	-41.4	-41.4
CO_2 fixed	-74.5	-64.9	-61.9	-69.5	-67.1
$\Delta CO_2\ ^o/oo$	-32.5	-26.1	-23.1	-28.1	-25.8

values for C$_3$ plants between -27.1 to $-44\ ^o/_{oo}$ using a ΔCO_2 of RuBPCase of $-27.1\ ^o/_{oo}$. Therefore, fixation of chloroplast CO_2 by RuBPCase can account for the $\delta^{13}C$ value of cotton plants.

6-4.1.3   Chlorophyll Mutants

Virescent mutants of cotton are among the yellow-green mutants of plants exhibiting a high net CO_2 fixation on a chlorophyll basis (Table 6–13). Benedict et al. (1972) showed that an irradiance of 230 Wm^{-2} the photosynthetic rates of virescent and wild-type cotton leaves were 36.8 mg CO_2 dm^{-2} h^{-1} and 35.5 mg CO_2 dm^{-2} h^{-1}, respectively. The photosynthetic rates on a chlorophyll basis are 36.8 mg CO_2 chl^{-1} h^{-1} for the virescent leaves compared to 12.1 mg CO_2 chl^{-1} h^{-1} for the green wild-type leaves. The relative quantum yield of the virescent leaves is lower than the wild-type from 400 to 500 nm, but similar at 700 nm (Fig. 6–5). The chloroplasts of the mutant leaves lack chloroplast grana (Benedict and Kohel, 1970).

Benedict et al. (1972) showed that the photosynthesis of the virescent leaves is more sensitive to atrazine than the wild-type leaves, a finding that suggests the photosynthetic unit (PSU) of the mutant leaves is smaller than that of wild-type leaves. The high photosynthetic rate on a chlorophyll basis is probably due to a smaller PSU. The composition and activity of the

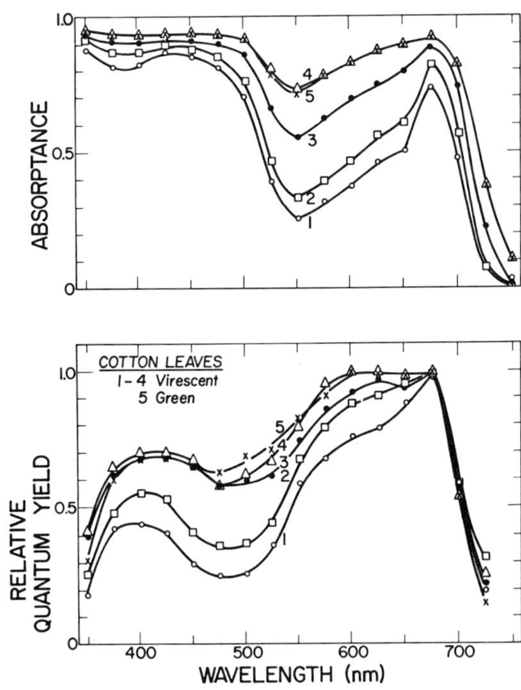

Fig. 6–5. The relative quantum yield of a chlorophyll mutant and wild-type cotton (adapted from Benedict et al., 1972).

Table 6-13. Characteristics of wild-type and mutant cotton leaves
(adapted from Benedict et al., 1972).

	Wild type	Mutant
Photosynthetic rate (mg CO_2 dm^{-2} h^{-1})[†]	39.5	36.8
Photosynthetic rate (mg CO_2 mg^{-1} chl h^{-1})[†]	12.1	36.8
Photosynthetic rate (mg CO_2 dm^{-2} h^{-1})[‡]	7.4	7.1
Photosynthetic rate (mg CO_2 mg^{-1} chl h^{-1})[‡]	1.4	6.0
Chlorophyll content (mg Chl$_{a+b}$ dm^{-2})	3.2	0.65
Carotenoid content (μg dm^{-2})	438	205
Chlorophyll a/b ratio	4.1	3.3
Inhibition of photosynthesis by 10^{-4} M atrazine	12.1%	87.8%
CO_2 compensation point (μl l^{-1} CO_2)	70.0	71.2

[†] Irradiance, 230 W m^{-2}.
[‡] Irradiance, 13 W m^{-2}.

photosynthetic apparatus in virescent mutants of cotton, soybean, and corn have been studied by Alberte et al. (1974). An increase in chl a/b ratio was correlated with the low temperature induced virescent mutation leading to a reduction in total chlorophyll. The alterations in chlorophyll were reflected in the composition and quantity of chlorophyll-protein complexes (Table 6-14) changes in the light-harvesting chlorophyll-protein of photosystem II was the prime consequence of the mutation and resulted in a smaller PSU.

6-4.2   Carbon Dioxide Exchange Rate (CER)

6-4.2.1   Leaf Cross-Section

El-Sharkawy et al. (1965) and Hesketh (1968) have reported CER rates of 45 to 50 mg CO_2 dm^{-2} h^{-1} for cotton plants grown in full sunlight in Arizona. Hesketh (1968) reported lower CER rates for cotton plants grown at low irradiances. This result is probably due to the production of thin leaves with high mesophyll resistances, and leaf anatomical differences among plant species has been studied by El-Sharkawy and Hesketh (1965). Mesophyll resistances (r_m) in cotton leaves is greater than in corn by about 0.3 to

Table 6-14. Analysis of photosystems in virescent cotton plants
(adapted from Alberte et al., 1974).

Strain	Temperature	Ratio[†] CP II/ CP Ia	Leaf Chl/ P700b	CPI Chl/ P700	CPII Chl a/b Ratio
	°C				
Cotton					
T-414 (vir)	20	10	765	80	1.7
T-414	32	6	320	75	1.4
Stoneville					
7-A (w/t)	20	5	275	75	1.4
Stoneville					
7-A (w/t)	32	5	275	75	1.4

[†] CP I and II refer to the photosystem I and the light-harvesting chlorophyll-protein complexes, respectively; b Chl, chlorophyll; c (vir), virescent; d (w/t), wild type.

0.5 cms^{-1} which would account for 10 mg CO_2 dm^{-2} h^{-1}. Gausman et al. (1980a) have shown that applications of mepiquate chloride (1,1-dimethyl-piperididinium chloride, PIX) to cotton leaves increases leaf thickness which could lower resistances to CO_2 uptake. Treated leaves had thinner and larger palisade cells, a thicker spongy parenchyma, and a larger area of air space. After 45 days, PIX-treated leaves had 24 to 37% higher CER rates (Gausman et al., 1980b). Hoffman et al. (1982) reported that PIX applied at blooming stage did not affect photosynthesis and did not affect partitioning of photosynthate into leaf, stem, or fruit. Buckwalter (1982) showed that the time of application of PIX is very important in obtaining yield responses.

6–4.2.2 Levels of CO_2

Growing cotton plants in enriched atmospheres of CO_2 indicate that yield can be regulated by photosynthesis (Mauney et al., 1978). Cotton plants grown at 620 ppm CO_2 resulted in an initial increase in CER of 65% which declined to 31% compared to plants grown at 330 ppm. The number of bolls increased from 41.6 bolls/plant at 330 ppm to 74.8 bolls/plant at 660 ppm fo CO_2, and there was a twofold increase in the dry weight of the bolls. Mauney et al. (1978) showed that average CER increased 15% for cotton and lint yield was increased 180% by high CO_2 treatments on a plant basis. The CER may increase yield in cotton if a sensitive assay for CER can be found for reliable measurements.

6–4.2.3 Cotton Species

El-Sharkawy and Hesketh (1965) studied the CER of 26 species of *Gossypium* grown in Arizona. The CER of the individual leaves of the same species varied between measurements in the winter, spring, summer wet soil, and summer dry soil. The measurements of CER of plants grown in the summer dry soil conditions were thought to be optimal. Under these soil conditions the CER varied from 24 to 51 mg CO_2 dm^{-2} h^{-1}. The high CER was in *G. longicalyx* (F_1 previously designated E_5), and the low CER was in the *G. aridum* (D_4). In general, the CER for the AD genomes ranged from 42 to 45 mg CO_2 dm^{-2} h^{-1}. Benedict et al. (1981) measured the levels of RuBPCase and CER in several species of *Gossypium* grown in the field at College Station, Texas. CER ranged from 16.9 mg CO_2 dm^{-2} h^{-1} for *G. gossypioides* (D_6) to 41 mg CO_2 dm^{-2} h^{-1} for *G. tomentosum* (AD_3). The levels of RuBPCase level in *G. davidsonii* (D_{3-d}) was 44 mg CO_2 dm^{-2} h^{-1}, and the RuPBCase level in the *G. hirsutum* (AD_1) had levels of 99 mg CO_2 dm^{-2} h^{-1}. In general, the species of cotton plants in the D genome had the smallest CER, the lowest RuPBCase activity, and the largest leaf area. These results are similar to those of Hesketh et al. (1980) who measured CER, levels of RuBPCase, leaf area, protein content, stomatal resistance, and chlorophyll content for soybean leaves. The CER and RuBPCase activity varied about twofold in different soybean cultivars. The CER was correlated with leaf area. Results suggested that photoperiod genes and genes for leaf area growth interact with genes controlling photosynthetic CO_2 exchange to produce major differences in CER values among soybean genotypes.

6–4.2.4   Leaf Senescence

Muramoto et al. (1967) studied cotton leaf aging and CER. Young fully expanded leaves had the highest CER. The rate of CO_2 fixation by fully enlarged cotton leaves steadily declines from 50 mg CO_2 dm^{-2} h^{-1} in leaves 1 to 2 days after expansion to a low rate of 10 mg CO_2 dm^{-2} h^{-1} 45 days after expansion (Fig. 6–6). Morris (1964) suggested that the source of photosynthate for developing cotton boll shifts from subtending leaf to boll wall and bract during the decline in CO_2 fixation in the subtending leaves. Benedict and Kohel (1975) studied the amount of ^{14}C-assimilate transported to cotton bolls from the subtending leaves, bracts, and boll walls. The associated bracts furnish only 5 to 10% of the photosynthate for cotton bolls 1 to 24 days post anthesis. In cotton bolls of all ages, the boll wall contributes little photosynthate for their growth. The subtending leaves are the dominant source of assimilate for cotton bolls throughout the boll maturation period. Following a 5 min pulse labeling with $^{14}CO_2$ the amount of ^{14}C-starch synthesized on a leaf area basis does not vary significantly in leaves subtending 10 to 37-day-old cotton bolls.

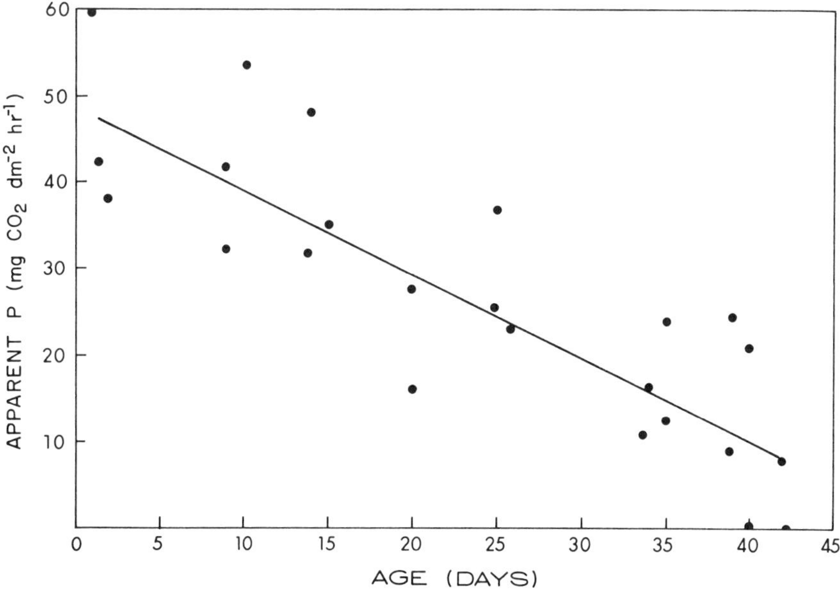

Fig. 6–6.   Rate of apparent photosynthesis of leaves of cotton as a function of leaf age (Muramoto et al., 1967).

6-4.3   Photosynthate Translocation

6-4.3.1   Export of Assimilates

Ashley (1972) noted that 43% of the photosynthate remained in the leaf following 24 hours of translocation. Brown (1973) showed a retention of 35 to 43% of the daily photosynthate in mature cotton leaves. Krieg and Sung (1979) showed that the source of leaf retained a signifiant proportion of the daily photosynthate production for its own use. Twenty-two hours after exposing the leaf to $^{14}CO_2$ as much as 40% of the initially incorporated ^{14}C remained in the source leaf. Other plants retain different amounts of assimilate. Corn translocates assimilate rapidly and completely out of the leaves (Hofstra and Nelson, 1969b). Sugarcane (*Saccharum officinarum* L.) translocates 80% of the ^{14}C-assimilate labeled by an initial pulse of $^{14}CO_2$ in 4 hours (Hartt and Kortschak, 1967). Other tropical grasses export 70% of assimilate within 6 hrs of assimilation, whereas tomato (*Lycopersicon esculentum* Mill.), tobacco (*Nicotiana tabacum* L.), and soybean translocate only 45 to 50% of the assimilate in 6 hours (Hofstra and Nelson, 1969a). Benedict and Kohel (1975) showed a differential rate of export and retention of water soluble and insoluble (starch) assimilate from cotton leaves (Fig. 6-7). There is an initial rapid phase of transport of water soluble compounds from both leaf and bract sources. About 80% of these radioactive compounds disappear 6 hours after exposing the sources to $^{14}CO_2$. After 22 hours, over 90% of this pool is depleted. In contrast there is a retention of ^{14}C in starch up to 6 hours after pulse-labeling the sources with $^{14}CO_2$. After 6 hours there is a slow export of ^{14}C-assimilate from the starch fraction stored in the sources. At the end of 22 hours, 90 to 95% of the ^{14}C-assimilate disappears from mature leaves and bracts. One of the reasons that Ashley (1972) found 43% of the ^{14}C-assimilate was retained in the leaves is that 2 hours elapsed following the labeling before determining the initial amount of ^{14}C-assimilate in the leaves. Benedict and Kohel (1975) concluded that the export of water soluble assimilate from the leaves and bracts of cotton (a C_3 plant) is similar to the export of ^{14}C-assimilate in C_4 plants. Compared to C_4 plants, cotton leaves retain a greater amount of assimilate as starch and export this pool of assimilate slowly.

6-4.4   Photosynthate Partitioning

6-4.4.1   Carbon Partitioning into Leaf Starch and Sugars

Benedict and Kohel (1975) showed there is about two to three times more radioactivity in the water soluble compounds than in starch in both leaves and bracts following a 5 min pulse-labeling with $^{14}CO_2$. Mauney et al. (1979) showed the cotton plants grown at 330 μl l^{-1} of CO_2 accumulated 43 mg sugar g^{-1} of leaves and 74 mg starch g^{-1} of leaves. This contrasted to plants growing at 630 μl l^{-1} of CO_2 where the leaves accumulated 30 mg sugar g^{-1} of leaves and 256 mg starch g^{-1} of leaves.

6-4.4.2   Source-Sink Relationships

Previous studies described the translocation of leaf assimilate in cotton plants. Brown (1973) found that maturing cotton bolls receive assimilate from subtending leaves, associated bracts, boll walls, leaves subtending the sympodia, and leaves higher up on the same side of the stem. Ashley (1972) found the primary source of photosynthate for the cotton boll was the sub-

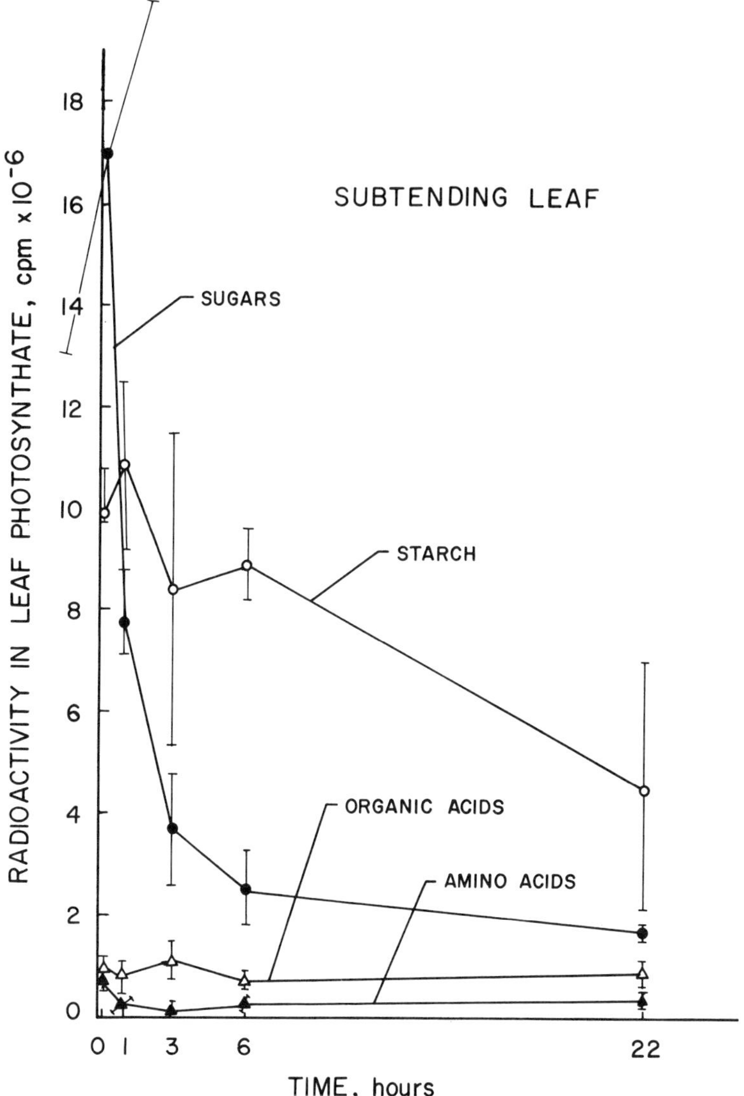

Fig. 6-7. Total disappearance of photosynthate from leaves subtending 30-day-old cotton bolls. The data points are the mean of two replicates and the range for each mean is shown with a line and bar (Benedict and Kohel, 1975).

tending leaf. Sung (1978) showed that leaves in the upper canopy tended to allocate the greatest portion of their assimilate to the vegetation below the treated leaf. The lower canopy leaves directed 50% of the exported assimilate to vegetation below the treated leaf and 30% to the sympodia above the treated leaf. The partitioning of photosynthate within a sympodium showed that the subtending leaf directed photosynthate toward the attached boll. The boll attached to the second position received 14% of the assimilate, and the boll attached to the third position received 1.5% of the assimilate. The data clearly indicate that the direction of assimilate flow is under the influence of sink strength.

6-4.4.3 Partitioning into Boll Wall, Ovule, and Fiber

In an attempt to describe the relative sink strength of different age cotton bolls, Benedict and Kohel (1975) pulse-labeled all of the photosynthetic sources on a single sympodium and determined the relative proportion of ^{14}C-assimilate partitioned into bolls of different ages. About two to three times more ^{14}C-assimilate was partitioned into the developing ovules compared to the boll wall or lint fibers. The maximum ^{14}C-assimilate partitioned to the boll wall occurred in 13-day-old bolls; whereas, the maximum ^{14}C-assimilate partitioned into the ovule and lint fiber occurred in 25 to 30-day-old bolls. Apparently the maximum period of synthetic activity in the different boll components determines the amount of ^{14}C-assimilate partitioned into each component.

6-4.4.4 Functioning Period of the Funiculus

The research by Baranov and Maltzer (1937), Leahy (1948) and Dure (1975) on the anatomy, chemistry, and biochemistry of cottonseed provided information on the developmental period for the endosperm, embryo and fiber maturation, the linear period of oil and protein deposition, and the changing biochemical events necessary for the regulation of ovule maturation, but these studies did not establish if the photosynthetic carbon is deposited in developing ovules throughout the seed maturation period. Studies to determine this active period of transport of assimilates to the ovule are important because Ihle and Dure (1972) and Dure (1975) proposed a temporal map of the biochemical and morphological events in developing cotton cotyledons showing that the funiculus degenerates 32 days after anthesis. A minimum of 50 days after fertilization is required for the maturation of cottonseed. From a yield viewpoint, it is important whether or not leaf assimilates are partitioned into the developing cotton ovules throughout the seed maturation period or whether the increase in the dry weight of the ovule for 40 to 50 days after anthesis occurs at the expense of some other boll component after 32 days. Benedict et al. (1976) showed that the rate of incorporation of ^{14}C-assimilates into fibers of different age cotton bolls was similar to the rate of dry weight increase of the fibers (Fig. 6-8). The ^{14}C-assimilate was incorporated into the fibers during a period of active cellulose deposition in these fibers. Carbon-14 assimilate was incorporated into the ovule wall, sugars, amino acids, oil, and protein in 1 to 45-day-old bolls. This incorporation of photosynthate into the ovules must reflect transport of the ^{14}C-assimilates through the vascular connection. An alternate route of assimilate through the carpel wall, into the locular cavity and absorption by the ovules is not important because the transport of the leaf ^{14}C-assimilate to the carpel wall is complete in 30 to 35-day-old bolls. Thus, the assimilates transported to the cottonseed throughout the maturation period occurs through the vascular connection, and the funiculus does not break 32 days after anthesis.

6-4.5   Carbon Budget

6-4.5.1   GOSSYM Model

Baker et al. (1979) described a computer model, GOSSYM, for the simulation of fruiting and yield predictions of cotton. This is a two-dimensional model including RHIZOS, which describes the location of roots, soil N, soil water, and sink strength of the root system; it simulates the production of fruiting points and fruit abscission, mimics cut-out, and predicts yield.

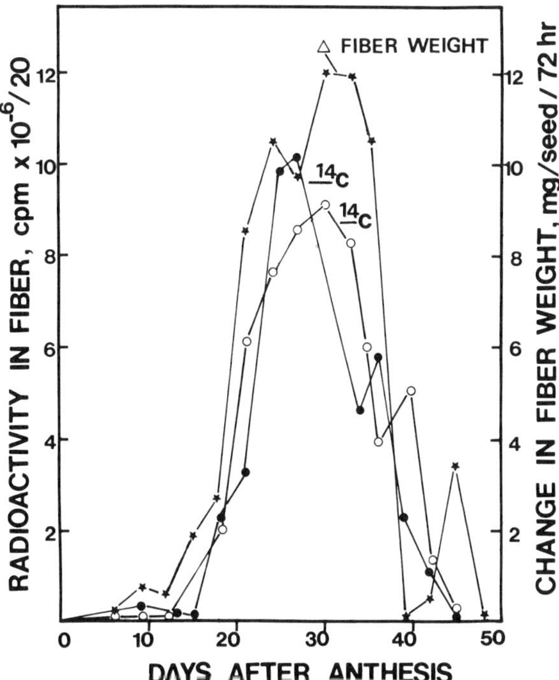

Fig. 6–8. Rate of development of fiber dry weight compared to rate of incorporation of ^{14}C-assimilates into fibers in different age ovules (Benedict et al., 1976).

McKinion and Baker (1982) described the program flow of the cotton model as shown in Fig. 6–9. The growth and yield of a cotton crop are calculated on daily total radiation, minimum and maximum temperature, rainfall, pan evaporation, plant density, row spacing, fertilizer applications, date of crop emergence, and soil characteristics. The CLYMAT routine takes these data and converts them into data which may be called once per simulated day. The TMPSOIL routine calculates soil temperature as a daily average from the daily minimum and maximum air temperature and is called from the CLYMAT routine. FRTLIZ, GRAFLO, ET, UPTAKE, CAPFLO and NITRIF are subprograms of the soil subroutine and are called daily except UPTAKE and CAPFLO, which are called NITER times and NITER two-times, respectively, each day.

The total evapotranspiration from the soil surface and plant canopy is estimated from the model proposed by Ritchie (1972), and it is used to calculate the soil and plant water potential. The amount of moisture is esti-

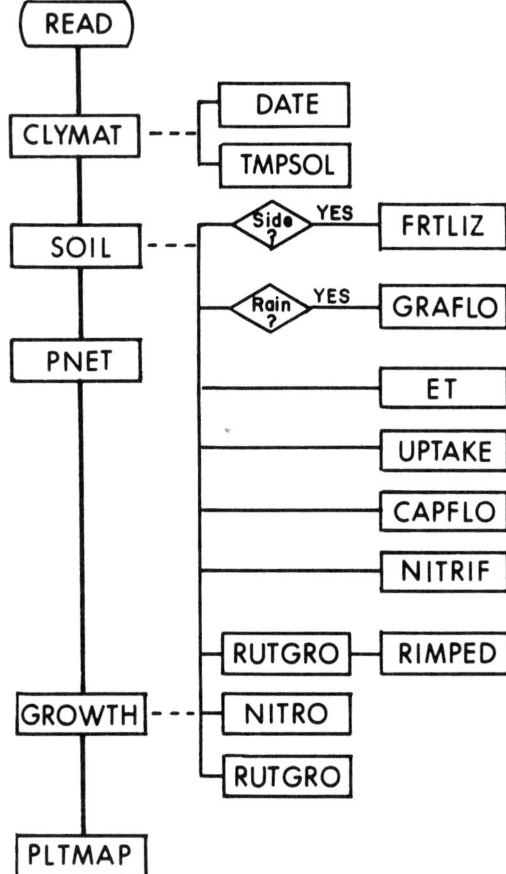

Fig. 6–9. Program flow for GOSSYM (adapted from McKinion and Baker, 1982).

mated from ET and is imposed on the soil profile. Water is withdrawn from the soil profile proportionally to the amount of roots capable of uptake by use of the subroutine UPTAKE. Nitrogen is taken up by passive uptake. Nitrogen is distributed in the soil profile by the subroutine FRTLIZ and moved by GRAFLO and CAPFLO.

The next subroutine is PNET, which calculates canopy photosynthesis and respiration on a daily basis from plant height, row width, and solar radiation. It calculates growth, light, and maintenance respiration, and it uses planting density and row spacing to calculate net photosynthate available per plant. The basis for these latter type of calculations was based, in part, on the work of Hesketh and Baker (1967) on the light and carbon assimilation by plant communities, and Baker and Myhre (1968) on determining the effect of leaf geometry and plant density on seasonal net photosynthesis. Information about respiration needed for model simulation of plant growth was partially based on the work of Hesketh et al. (1971). They discussed the roles of maintenance and growth respiration and carbon balance on the simulation of growth and yield in cotton and on the work of Baker et al. (1972) on the effect of environment on respiration, photosynthesis, and organ development.

The growth of the cotton plant in the model is monitored by the GROWTH subroutine. This growth is a materials balance of carbohydrate and N and the effect of water stress on causing carbohydrate and nitrogen shortages. The basis for this subroutine is found in the work of Hesketh et al. (1972) on the role of temperature and the time interval between the appearance of leaves, floral buds and flowers at successive main stem and fruiting branch nodes of the cotton plant. The analysis of the relationship between photosynthetic efficiency and yield in cotton is based on the work of Baker et al. (1973) which establishes a model where cotton plants adjust to a disparity between real and potential growth and abscises fruit based on the supply-demand ratio rather than on carbohydrate supply per se. It proposes that the trigger behaves like phytochrome in that it activates the auxin mechanisms involved in abscission. The simulation of fruiting under moisture stress, a program that is incorporated into GOSSYM was described by Baker et al. (1979). The stress factors primarily affect the morphogenesis section of the model PLTMAP by causing delays in new node production or by causing square and fruit abscission.

Root growth and N uptake are accounted for in subroutines RUTGO and NITRO called GROWTH. At the end of the simulated growing season the amount of lint produced per hectare is predicted.

6–5 WATER RELATIONS

McArthur et al. (1975) pointed out that cotton is grown under a broad range of atmospheric moisture levels and that the minimum daily leaf water potential can vary from -6 to -28×10^6 Pa depending mainly on the availability of soil moisture. Jordan (1982) presented curves to illustrate the interactions among evaporative demand (E_o), leaf water potential, root xylem water potential (ψ_r, xy), and soil water potential (ψ_s) for typical field growth cotton. Maximum leaf water potentials occur before dawn and range between -1 and -2×10^6 Pa for cotton plants growing in soils with water potentials greater than one bar. Minimum leaf water potentials occur between the hours of 1300 and 1500 and range between -8 to -15×10^6 Pa and -15 to -30×10^6 Pa for cotton plants growing in moist and drying soils.

6–5.1   Photosynthesis and Water Stress

Jordan and Ritchie (1971) found that water stress affects stomatal resistance in potted plants differently than field grown cotton. Stomatal resistance in field cotton was unaffected by leaf water potentials down to -27

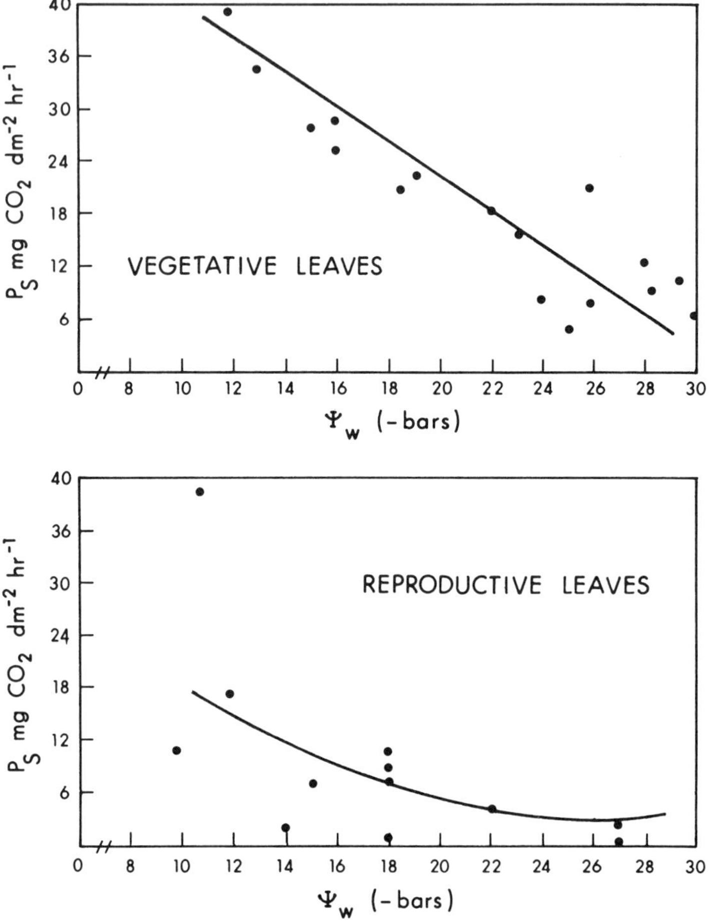

Fig. 6-10. Relationship between photosynthesis (P_S) and leaf water potential (ψ_w) of vegetative (A) and reproductive (B) cotton leaves. Data for vegetative and reproductive leaves are not equivalent radiation (PAR) levels (Ackerson et al., 1977).

$\times$ 10⁶ Pa, but increased to infinity at a leaf water potential of -16×10^6 Pa in potted plants. El-Sharkawy and Hesketh (1965) studied the relation between water deficit and photosynthetic CO_2 fixation in potted cotton plants subject to rapid water deficits. The results suggest that stomatal resistance is a dominant factor in the regulation of photosynthesis. Ackerson and Krieg (1977) and Ackerson et al. (1977) observed severe photosynthetic reduction in field grown cotton at leaf water potentials of -30×10^6 Pa with no stomatal closure (Fig. 6–10). The data indicate that stomata of field grown cotton are relatively insensitive to water stress. Water stress substantially reduced photosynthesis which cannot be due to stomatal closure. Apparently, field grown cotton can osmotically adjust under water stress to minimize stomatal resistance, although photosynthesis is reduced. Hutmacher and Kreig (1982) studied the stomatal and nonstomatal factors controlling photosynthetic rates in cotton. In nonstressed plants the rate of photosynthesis is not correlated with stomatal resistances, and photosynthesis in water-stressed plants is correlated with mesophyll resistances. Krieg (1981) studied the relation between water stress and reduction of photosynthesis and leaf area. A major cause of the reduction of photosynthesis associated with leaf age and water stress was a decline in RuBPCase. Radin and Ackerson (1981) studied stomatal conductance, photosynthesis, and ABA accumulation during drought. Water stressed plants showed a decreased stomatal conductance, increased stomatal sensitivity to elevated CO_2 concentrations, a decreased mesophyll conductance to CO_2, and an increased endogenous ABA content. All of these responses occurred at higher water potentials in nitrogen deficient plants. In both high-N and low-N leaves, the mesophyll conductance to CO_2 declined under water stress more or less in concert with the decline in stomatal conductance to water vapor and an increase in ABA content. Water stress caused by a nonstomatal inhibition in photosynthesis may be the relatively slow development of stress.

6–5.2 Predictions of Evapotranspiration

Many methods have been published for predicting evapotranspiration for field plantings of cotton (McArthur et al., 1975). Ritchie (1972) and Richardson and Ritchie (1973) presented models for predicting evapotranspiration based on net radiations and light interception by canopies.

6–5.3 Genetic Sources for Drought Resistance

Crop maturity is a trait consistently associated with drought tolerance. Once alterations of crop maturity are fully exploited, additional improvements must come from other traits (Jordan, 1982). These traits should allow cotton cultivars to maintain a high internal water potential by increasing water absorption or allow a plant to function even though the plant water potential is lowered. Quisenberry et al. (1981) evaluated the dry matter accumulation of photoperiodic, exotic cotton strains in irrigated and dryland field conditions to estimate water use efficiency and to determine the relationships between field growth parameters and laboratory evaluations of heat tolerance and greenhouse estimates of root growth. The study showed significant variability among the strains for heat tolerance, root growth, dry matter accumulation, and water use efficiency in the dryland regime. Estimates of root and shoot growth conducted in the greenhouse were related to early season field performance when water was not limiting growth. Root morphology and root growth potentials are important to the adaptation of cotton to soil water conditions, a major limitation to growth.

Peterschmidt and Quisenberry (1981) studied drought tolerance mechanisms among cotton genotypes. Genotype T25 had a larger number of lateral roots and the stomates controlled leaf water export during water stress, whereas, genotype T169 did not close its stomates under conditions of zero turgor and visible wilting. Genotype T169 had a high soluble sugar content throughout the day, a condition that may indicate a maintenance of growth during water stress by soluble sugar production. It appears T25 and T169 are adapted to severe environments through different mechanisms. The genotype T169 expresses optimal growth under high water conditions and does not have the drought tolerance mechanisms such as stomatal control and root sink strength of T25. Genotype T25 does not have the production potential of T169 when water is optimal. The lower water potential probably allowed T169 to extract the more tightly bound water held in the limited root zone and would have maintained growth during daily wilting.

6-6 FIBER DEVELOPMENT

The mature cotton fiber is over 25 mm long and 20 μm in diam. The fiber arises by the growth and differentiation of the outer ovule epidermal cells at or near the day of anthesis. Elongation of the fiber occurs for about 15 to 27 days after anthesis. There is an overlap in the formation of the primary and secondary walls and the formation of the secondary wall predominates from 15 to 55 days after anthesis.

6-6.1 Ultrastructure of Early Stages of Cotton Fiber Differentiation

Ramsey and Berlin (1976b) studied the early stages of cotton fiber differentiation. From 16 to 3 days preanthesis both the nucleus and the cytoplasm of the ovule epidermal cells are uniformily electron dense. By 24 hours preanthesis, there was a marked change in appearance of epidermal cells and in some of the cells the vacuoles were filled with electron dense phenolic type substance. At 16 hours preanthesis, certain cells were slightly enlarged and the vacuoles were nearly empty of the electron dense material while the cytoplasm contained electron dense particulate matter. At 8 hours preanthesis, they observed a population of light and dark cells in the epidermal layer of cells. At anthesis the dark type epidermal cells of the ovule give rise to cotton fibers. Ramsey and Berlin (1976a) noted similarities between dictyosome associated vesicles containing fibrils appearing similar in morphology to fibrils in the primary cell wall and plasma membrane associated vesicles which suggested dictyosome involvement in the formation of both the plasma membranes and primary cell wall. Stewart (1975) showed by scanning electron microscopy that fiber density on cotton ovules was about 3300 fibers mm^{-2} and the ratio of fiber initials to total epidermal cells was 1:3.7 at anthesis. Beasley and Ting (1973), Eid et al. (1973), Baert et al. (1975), and Beasley et al. (1974) showed that ovule and fiber growth require Murashige and Skoog medium and the presence of gibberellic acid and auxin. Kinetin is not essential for cotton fiber growth and abscisic acid inhibited fiber growth.

6-6.2 Fiber Growth and Differentiation

The development of fibers of field grown 'Stoneville 213' cotton was studied by Schubert et al. (1973) from anthesis to maturity with reference to fiber elongation, fiber dry weight per seed, and fiber dry weight per seed per unit of length. These variables were plotted against boll age and fitted by computer to appropriate best-fit curves by curvilinear regression analysis. The curve fitting procedures yielded mathematical equations which were differentiated to five accurate growth rate curves (Fig. 6–11). The elongation rate curve showed that lint fiber elongation ceased at 27 days after anthesis. At that time, dry fiber weight had reached 40% of its final value. Spline analysis of the data shows that fiber dry weight per unit length ceased to be constant at 16 to 19 days after anthesis. Theoretically, this is the age at which secondary wall thickening begins. These data suggest that a substantial portion of secondary wall deposition occurs before fiber elongation is completed. Gipson and Joham (1969) and Gipson and Ray (1969) showed that fiber elongation rates, duration of elongation period, and rate of secondary wall deposition vary among cultivars and environmental condition. Benedict et al. (1973) showed that the degree of overlap between the phase of elongation and secondary wall thickening varies among cultivars and environmental condition. Benedict et al. (1973) showed that the degree of overlap between the phases of elongation and secondary wall thickening varies among cultivars with different values of fiber fineness. This knowledge may afford cotton breeders another means of tailoring a cultivar to a particular location. The establishment of an overlap in the primary and secondary wall development suggests the need to examine the mechanism coordinating these two processes.

6-6.3 Cellulose Synthesis

Delmar et al. (1974) examined the utilization of nucleoside diphosphate glucoses in developing cotton fibers. The incorporation of uridine diphosphate (UDP)-glucose and guanosine diphosphate (GDP)-glucose into hot alkali-insoluble products was studied in isolated fibers harvested at various stages of development. The GDP-glucose was active as a substrate only during the period of fiber elongation and the utilization of UDP-glucose steadily increased during secondary wall formation. The reaction product synthesized from GDP-glucose was identified as β-1,4-glucan, but over 90% of the radioactivity incorporated from UDP-glucose was soluble in chloroform-methanol. It was thought that GDP-glucose could be the precursor of primary, but not secondary wall cellulose. Further work (Elbein, 1969), however, showed GDP-glucose probably is only active in plant tissues in synthesizing glucomanan. Delmar (1977) also showed that UDP-glucose synthesis by its pyrophosphorylase is 10,000 times higher than the synthesis of GDP-glucose. No GDP-glucose pyrophosphorylase activity is detectable in cotton fibers. These studies showed there is only one cellulose synthetase in cotton fibers. The finding by Franz (1959) that UDP-glucose is the major soluble nucleoside diphosophate glucose in cotton fibers supports the conclusion that UDP-glucose is the precursor to cellulose in cotton

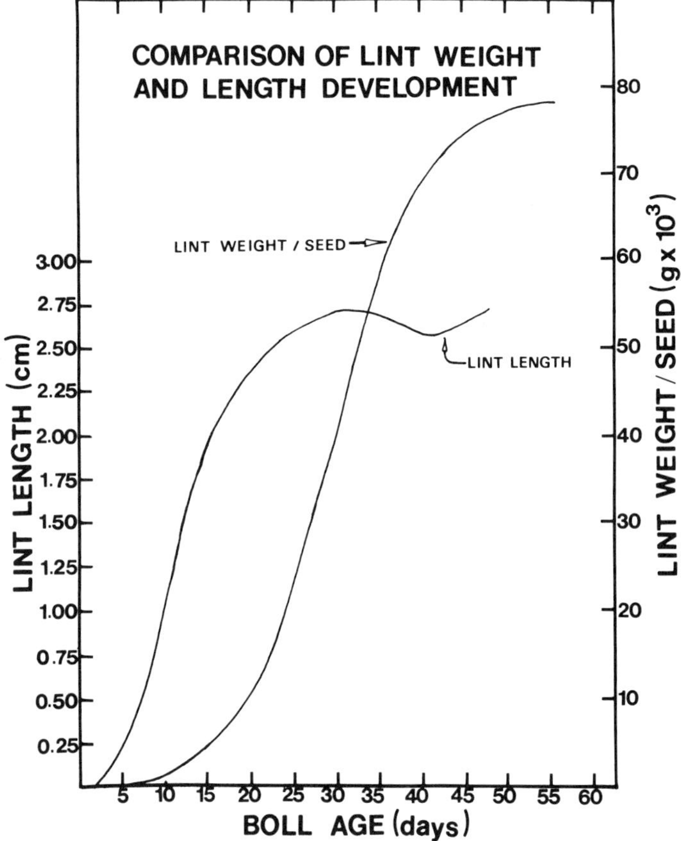

Fig. 6–11. Comparison of the development curves for lint length and lint weight per seed. At 27 days boll age, elongation ceased and lint weight per seed had reached 40% of its final value (Schubert et al., 1973).

fibers. Carpita and Delmar (1981) studied the concentration and metabolic turnover of UDP-glucose in developing cotton fibers. Measurements of pool sizes and rates of labeling of the glucose, glucose-phosphate, and UDP-glucose pools in pulse-chase experiments showed that the rate of synthesis and turnover of UDP-glucose was more than sufficient to account for the accumulation of sucrose, sterylglucosides, β-1,3-glucan and cellulose. Delmar et al. (1977) and Heiniger and Delmar (1977) showed that UDP-(^{14}C)-glucose is incorporated into a hot water-insoluble, chloroform-methanol-insoluble product in the particulate fraction of cotton fibers via a glucan synthetase. The product is not cellulose, but a glucan with predominantly β-1,3-linkages. Huwyler et al. (1978) and Maltby et al. (1979) show that cotton fibers harvested at the time of secondary wall cellulose synthesis contained substantial quantities of a 3-linked glucan, and the glucan synthetase appears to be responsible for catalyzing the in vivo synthesis of this β-1,3-glucan.

Dugger and Palmer (1981) showed that cotton fibers grown on ovules cultured in vitro incorporate UDP-glucose and glucose into cell wall glucan products quite differently. Radioactive UDP-glucose is incorporated into hot water-soluble short-chain glucans with both β,1-4 carbon to carbon linkages, whereas, radioactive glucose is incorporated into cellulose as well as acetic-nitric reagent soluble and water soluble polymers. Within several hours of incubating the ovules with ^{14}C-glucose, cellulose is the most abundant labeled glucan in the cell wall of the fiber (Fig. 6–12).

Birnham et al. (1974) studied boron and phytohormone interactions in unfertilized cotton ovules. Ovules require IAA and/or GA for fiber elongation. Boron is required for maintenance of fiber elongation and normal morphogenesis. Birnham et al. (1977) noted similarities between symptoms of B deficiency and 6-azauracil injury. The ability of uracil to suppress both suggested B deficiency symptoms are related to reduced activity in the pyrimidine biosynthetic pathway. Wainwright et al. (1980) showed a primary event of B deficiency in cotton fiber culture is an alteration in the flow of metabolites through the pyrimidine pathway. In B-deficient cultures, there is less synthesis of UDP-glucose from orotic acid with more label of ^{14}C-orotic acid shunted into RNA. Boron is either involved in the conversion of glucose to UDP-glucose or in the cellulose synthetase reaction, whereby UDP-glucose is utilized in the synthesis of cellulose (Dugger and Palmer, 1981).

These studies do not reveal the mechanism for cellulose synthesis in cotton fibers. Cellulose synthetase has not been isolated. Electron microscopy data do not support the synthesis of cellulose through the endoplasmic reticulum and Golgi apparatus or by large multi-subunit complexes on the plasmalemma. Hopp et al. (1977) have reported data suggesting lipid intermediates are involved in the synthesis of cellulose in *Chlorophyta Prototheca zopfi*. Dolichol derivatives are involved in the synthesis of a glucoprotein containing β,1-4-linked glucoses. This glucoprotein acts as a primer for cellulose (Fig. 6–13) when it is incubated with GDP-glucose. Investigations are exploring the possibility that dolichol-PP-oligosac-

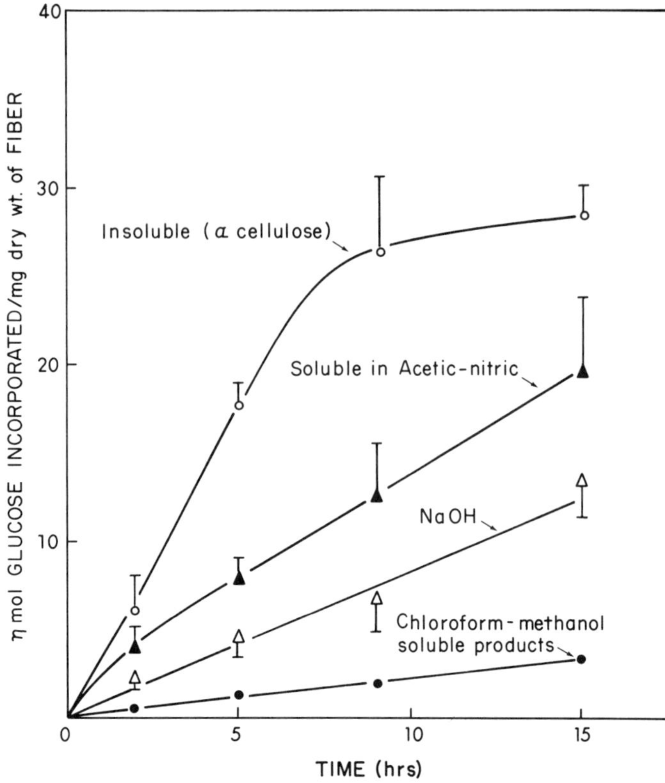

Fig. 6-12. Incorporation of ^{14}C-glucose into intact in vitro-grown cotton fibers (Dugger and Palmer, 1981).

charides are intermediates in cellulose synthesis in cotton fibers (Benedict and Elbein, unpublished results).

6-7 OVULE AND EMBRYO CULTURE

The method of in vitro growth of cotton ovules and their associated fibers developed by Beasley and colleagues (1971, 1973, 1974, 1976, 1977) afforded the developmental physiologist interested in seed formation a new tool to study the effect of plant hormones on ovule and fiber growth. Beasley (1977) pointed out information obtained in the in vitro studies might lead to an improvement of the number, length, and thickness of cotton fibers by field application of specific hormones.

Embryo culture is used to overcome barriers to producing some interspecific hybrids which could not be obtained in situ (Raghavan, 1977). A reliable method for culturing cotton embryos for obtaining interspecific hybrids of *Gossypium* has been developed by Stewart and Hsu (1977).

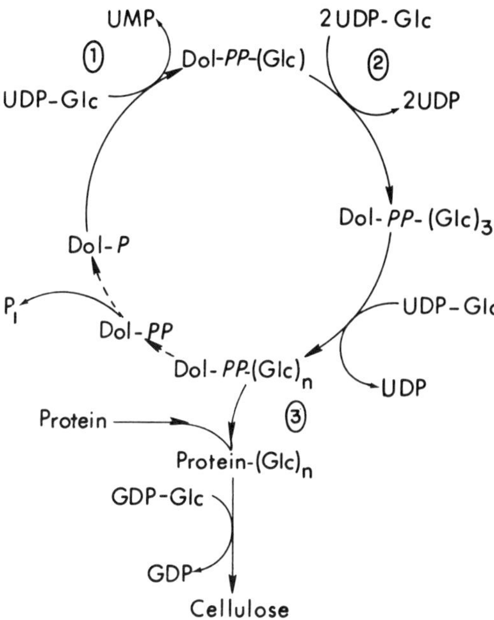

Fig. 6–13. Proposed reaction for the synthesis of cellulose (adapted from Hopp et al., 1977).

6–7.1 Culture of Fertilized and Unfertilized Ovules

Joshi and Johri (1972) expanded the early work on the in vitro culture of cotton ovules and studied the effect of hormones, casein hydrolysate, and yeast extract on the development of embryo and integuments. Early attempts to grow fibers from ovules were unsuccessful until Beasley (1971) showed that extensive fiber development could be obtained by floating fertilized ovules on the high salt Murashige and Skoog's liquid medium. Further studies (Beasley and Ting, 1973) led to a modification of this medium, whereby glucose and fructose replaced sucrose and KNO_3 replaced NH_4NO_3. Growth in this modified medium resulted in less callus and less browning. If fertilization preceded the harvest of the ovaries and transfer of the ovules, fibers continued to develop on the isolated ovules. Growth of the fibers on fertilized ovules, harvested 48 hours postanthesis, was stimulated by GA_3, inhibited by kinetin and ABA, and only slightly stimulated by IAA. Fertilized ovules are deficient in their capacity to synthesize optimum levels of gibberellic acid, but are able to synthesize optimum levels of IAA and cytokinins. Beasley and Egli (1976) showed that the growth of fibers on unfertilized ovules harvested on the morning of anthesis is stimulated by

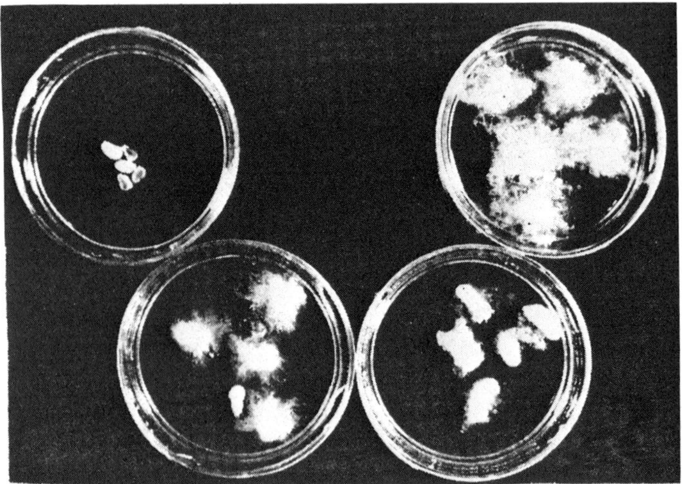

Fig. 6-14. Fiber development (or non-development) in 2 weeks from Acala ovules placed in culture on the morning of anthesis (unfertilized) maintained in basal medium plus no hormones (top left), 5.0 μM IAA (bottom left), 0.5 μM GA$_3$ (bottom right), or 5.0 μM IAA and 0.5 μM GA$_3$ (top right) (Beasley and Egli, 1976).

IAA and GA$_3$ (Fig. 6-14). Unfertilized ovules harvested from the plant on the second day of postanthesis enlarge in the presence of IAA and/or GA$_3$ (Beasley, 1973). Ovules acquire their capacity to respond to phytohormones between the second and third day of preanthsis. Ovules harvested from flowers up to 4 days preanthesis, and younger, produced only callus and root-like appendages in response to IAA and GA$_3$. Unfertilized ovules placed in culture on the morning of anthesis were the experimental choice for all further studies because of the apparent requirement for exogenous IAA and/or GA$_3$ for fiber growth. The in vitro culture of cotton ovules has been used to study not only phytohormone interaction, but also environmental and nutritional factors, and cell wall biogenesis (Beasley, 1974b). These methods have great research potential for studies of cotton embryogenesis and fiber wall biosynthesis.

6-7.2 Embryo Culture

The in-ovule embryo culture technique with in vitro fertilization has been used to by-pass barriers that occur in situ. The procedure offers the potential for formation of new interspecific and intergeneric hybrids. Joshi (1962) first reported the culture of cotton ovules, but embryos reached only 1.8 mm in 80 days. Joshi and Johri (1972) cultured cotton embryos to one-third of their in situ size by using White's medium supplemented with yeast extract, casein hydrolysate, and phytohormones. Stewart and Hsu (1977) re-examined the high salt Murashige and Skoog's medium developed by Beasley and Ting (1973) for the in ovule culture of embryos. If the medium

Table 6-15. Interspecific hybrids of cultivated cottons with wild *Gossypium* species obtained through ovule culture (adapted from Stewart, 1979).

Hybrid	Comment
G. hirsutum × *G. australe*	
G. hirsutum × *G. bickii*	First report
G. hirsutum × *G. somalense*	Semi-lethal; first report
G. barbadense × *G. australe*	First report
G. herbaceum × *G. australe*	
G. herbaceum × *G. somalense*	Appears semi-lethal; first report
G. herbaceum × *G. harknessii*	Synthetic AD; first report
G. arboreum × *G. harknessii*	Synthetic AD
G. arboreum × *G. trilobum*	Synthetic AD; first report
G. arboreum × *G. australe*	

was supplemented with NH_4 salts, more than 50% of the ovules produced embryos and many germinated precociously after 8 to 10 weeks of culture. Stewart (1979) and Stewart and Hsu (1978) have reported that hybrids have been obtained of *G. arboreum* 'Banking,' *G. herbaceum* 'Jayahdar' and *G. herbaceum* var. *africanum* with each of the ten breeding lines of *G. hirsutum* and hybrids of 12 genetic collections of *G. herbaceum* and *G. arboreum* with one genetic line of *G. hirsutum*. Table 6-15 lists the hybrids between commercial cotton and the wild species of *Gossypium* which are difficult to obtain or have never been made. Stewart (1979) pointed out that the relative ease with which hybrids can be obtained makes the *Gossypium* pool available for introgression of desirable traits into commercial cotton.

6-7.3 Somatic Embryogenesis

Price and Smith (1979) showed that somatic embryoids differentiated in suspension cultures of *G. klotzshianum* after 3 to 4 weeks in a liquid medium containing glutamine. The embryoids had an epidermis and meristematic regions, and they formed roots and vestigal leaves. This is a first report of somatic embryogenesis in cell or tissue culture of *Gossypium*.

6-8 HARVEST-AID CHEMICALS

Walhood and Addicott (1968) discussed the principles and practices of harvest-aid programs. These programs are one of the major, beltwide cultural practices, and they involve the use of chemicals to facilitate harvest by reducing foliage either by defoliation or desiccation. The use of the spindle harvester and stripper harvester have created the need to reduce and remove foliage from the cotton plant prior to harvest. The harvest-aid chemical must reduce the moisture content of the foliage for stripper harvesting and leaf moisture must be reduced or leaf fall induced for spindle harvesting. The purposes of the harvest chemicals are as follows:

a. Increase machine harvest efficiency by reducing foliage that interfaces with the operation of the machine.

b. Reduce trash in seed cotton.
c. Reduce moisture in the field.
d. Reduce moisture of the seed cotton.
e. Reduce boll rot.
f. Help lodged plants become more erect.
g. Improve visibility for the machine operator.
h. Reduce populations of destructive insects.
i. Hasten opening of mature bolls.
j. Reduce lint staining.

6–8.1 Defoliation

Defoliation induces premature abscission of the foliage. Under normal field conditions defoliants cause leaf abscission in 7 to 14 days after application. A key limitation to defoliation is the inconsistent responses of the leaves to defoliants. Cathey and Hacskaylo (1971) stated that the results from chemical defoliants are the most unpredictable operation that a farmer performs. The effectiveness of a defoliant is dependent on: uniformity of plant growth, weather conditions, spray coverage, volatilization, photodecomposition, degradation, absorption, and translocation. Chemical additives to harvest-aid chemicals may improve the effectiveness of the defoliant. Certain oils, waxes, or gel-forming carriers added to Def (S,S,S-tributylphosphorotrithioate) increased the effectiveness of the defoliant possibly due to prolonged foliar contact. Barker et al. (1971) evaluated the advantages of bottom and complete defoliation when incorporated in a program of either once-over or twice-over harvesting with two types of cotton pickers. The once-over harvesting method resulted in significantly higher yields and returns in 1967 and in significantly lower yields and returns in 1968. No significant differences were found in 1969 between once-over and two twice-over harvesting methods. Defoliants applied early either as a bottom treatment, when 20% of the bolls were open, or as a total treatment, when 60% of the bolls were open, tended to reduce yield, gross returns, fiber length, and strength. Defoliation increased grade by reducing moisture, yellowness, trash, and incidence of spots in the cotton. Mullins et al. (1972) showed that the use of additives with defoliants under the cool weather conditions of Tennessee was of doubtful value in aiding cotton harvest because of rapid boll opening and reduction in yield. Cathey and Hacskaylo (1974) showed a potential use of glyphosphate in combination with defoliants for regrowth inhibition.

6–8.2 Desiccation

Desiccation is the rapid loss of H_2O from the foliage following the application of a toxic chemical such as arsenic acid (Walhood and Addicott, 1968). Miller and Wilkes (1968) state that desiccation of cotton plants for earlier more efficient harvesting began in the 1950's. Desiccation is the removal of moisture and involves the rapid killing of leaf blades and petioles. The chemical injury prevents the formation of the abscission zone and the leaves are "frozen." Defoliation prior to desiccation has been used in some areas to prepare a cotton crop for mechanical stripping to reduce trash content of the "burr" cotton. Desiccants applied at the proper time can desiccate the foliage and prevent regrowth of the cotton; however, if desiccants are applied too early, such as at 80% open-boll, yield reductions will result. The addition of 2,4-D to desiccants suppresses regrowth, but 2,4-D can accumulate in the seed of the developing boll and reduce the value of seed for planting purposes. The addition of paraquat (1-1'-dimethyl-4,4'-bipyridinium *bis* methylsulfate) to sodium chlorate increases the amount of desiccation and terminal damage.

Brendel and Miller (1979) discussed the possibility that arsenic acid may be removed from the market unless sufficient data are collected showing its benefits and refuting accusations against the chemical. Brendel et al. (1978) and Miller et al. (1978) conducted field trials using alternative chemicals to arsenic acid desiccation consisting of sodium chlorate and endothall (3,6-endoxohexahydrophthalic acid disodium salt) as a "hot" defoliant which does not act as a desiccant. The best results were obtained by the application of the "hot" defoliant when the bolls were 95 to 100% open followed by stripper harvesting immediately after maximum defoliation. Miller and Aldred (1977) found that applying a desiccant with a stalk applicator can reduce the amount of chemical needed for the desiccation of cotton plants.

6-8.3 Chemical Modification of Flowering, Fruiting, and Cut-Out

Cathey and Thomas (1979) noted a considerable amount of success in applying chemicals for cotton growth suppression, abscission of late season fruits, and forced cut-out. The primary aim of the chemical modification of flowering, fruiting, and cut-out has been to obtain 1) shorter plants with better balance between fruiting and vegetative growth, 2) suppression of undesirable late bolls and squares, 3) early maturing cotton and early harvesting, and 4) a reduction of feeding sites for diapausing insects.

Kittock et al. (1973) found the most promising mixture of chemicals in terms of limiting the number of green bolls with a minimum effect on the lint production included 2,4-D and CCC (2-chloroethyl trimethyl ammonium chloride), 2,4-D and CCC, 2,4-D and chlorofluorenol (methyl 2-chloro-9-hydroxy fluorene-9-carboxylate), and CCC alone on Pima cotton (*G. barbadense*), but not on Upland cotton (*G. hirsutum* L.). Kittock et al. (1974) tested growth regulators in the cotton termination program and divided them into two groups. One group, 2,4-D, CCC, and chlorofluorenol, show promise when used alone or in a mixture, and the second group, ethephon (2-chloroethyl-phosphonic acid), cacodylic acid (hydroxydimethylarsine oxide), and dalapon (2,2-dichloropropionic acid), shows no promise. Treatments that reduced green bolls remaining at harvest also reduced pink bollworm larvae with a correlation of 0.89 and 0.91. Kittock et al. (1975) made studies of pink bollworms in plant material, ground trash, and soil. In all comparisons the numbers of pink bollworms were correlated significantly with the number of immature cotton bolls at harvest. The pink bollworms were reduced 92 to 97% by the most effective chemical termination treatment, which included 2,4-D and chlorofluorenol, 2,4-D and CCC, and 2,4-D alone.

Thomas and Hacskaylo (1974) studied cotton fruiting and yield responses to growth retardants used for suppressing late season plant development. Factors that influenced the degree of response, other than compounds used or application rates, were moisture regime and fruit load. Both Cycoel (CCC) and BAS 0660 (n,n-dimethyl-morphalinium chloride) applied near the onset of flowering reduced boll retention. Chlorofluorenol reduced flowering and boll retention. And DPX 1840 (3,3a-dihydro-2(p-methoxyphenyl-8H-pyrozolo(5,1-a) isoindol-8-one) applied early to mid August curtailed late growth and fruit set without reducing yield. The results support the conclusion that late growth and fruit set can be controlled chemically without affecting cotton productivity. Kerr and Royster (1977) studied the biochemical control of growth and maturation of cotton. The growth regulant BAS 08300W (PIX) controlled shoot growth without serious effects on the fruit and leaf tissue. They concluded that the condensing of flowering and fruit maturation into a shorter time span would increase harvest efficiency by once-over harvesting. The better control of flowering and shoot growth in the autumn may promote retention of the existing bolls and the maximum quality of fiber by diverting nutrients to the fruit set during the peak blooming period.

REFERENCES

Abeles, F. B. 1967. Mechanisms of action of abscission accelerators. Physiol. Plant. 20:442–454.

Abeles, F. B. 1969. Abscission: Role of cellulase. Plant Physiol. 44:447–452.

Ackerson, R. C., and D. R. Krieg. 1977. Stomatal and nonstomatal regulation of water use in cotton, corn and sorghum. Plant Physiol. 60:850–853.

----, ----, C. L. Haring and N. Chang. 1977. Effect of plant water status on stomatal activity photosynthesis, and nitrate reductase activity of field grown cotton. Crop Sci. 17:81–84.

Addicott, F. T., and R. S. Lynch. 1955. Physiology of abscission. Ann. Rev. Plant Physiol. 6:211–238.

----, H. R. Carnes, J. L. Lyons, O. E. Smith, and J. L. McMean. 1964. On the physiology of abscissions. p. 687–703. In J. P. Nitsch (ed.) Regulateurs naturels de la crosissance vegetale. Centre National de la Richerche Scientifique, Paris.

Alberte, R. S., J. D. Hesketh, G. Hofstra, J. P. Thornber, A. W. Naylor, R. L. Bernard, C. Rim, J. Endrizzi, and R. J. Kohel. 1974. Composition and activity of the photosynthetic apparatus in temperature-sensitive mutants of higher plants. Proc. Natl. Acad. Sci. USA 71:2414–2418.

Ashley, P. A. 1972. ^{14}C-Labeled photosynthate translocation and utilization in cotton plants. Crop Sci. 12:69–74.

Baert, Th., E. DeLanghe, and L. Waterkeyn. 1975. In vitro culture of cotton ovules. III. Influence of growth hormones upon fiber development. Cellule 71:55–63.

Baker, D. N., and D. L. Myhre. 1968. Leaf shape and photosynthetic potential in cotton. p. 103–109. *In* Proc. Beltwide Cotton Prod. Res. Conf., Hot Springs, Ark.

————, R. R. Bruce, and J. M. McKinion. 1973. An analysis of the relation between photosynthetic efficiency and yield in cotton. p. 110–114. *In* Proc. Beltwide Cotton Prod. Res. Conf., Phoenix, Ariz.

————, J. D. Hesketh, and W. G. Duncan. 1972. Simulation of growth and yield in cotton: I. Gross photosynthesis, respiration and growth. Crop Sci. 12:431–435.

————, J. A. Landivar, and J. R. Lambert. 1979. Model simulation of fruiting. p. 261–264. *In* Proc. Beltwide Cotton Prod. Res. Conf., Phoenix, Ariz.

Baranov, T. A., and A. M. Maltzer. 1937. The structure and development of cotton plants. Laboratory Cytology and Anatomy of the Central Plant Breeding Station, Moscow-Leningrad (English Translation).

Barker, G. L., R. O. Thomas, F. T. Cooke, K. E. Luckett, and E. B. Williamson. 1971. Plant preparation and harvesting system for cotton. p. 33–34. *In* Proc. Beltwide Cotton Prod. Res. Conf., Atlanta, Ga.

Beasley, C. A. 1971. In vitro culture of cotton ovules. BioScience 21:906–907.

————. 1973. Hormonal regulation of growth in unfertilized cotton ovules. Science 179:1003–1005.

————. 1974a. Effects of plant growth substances on in vitro fiber development from unfertilized cotton ovules. Am. J. Bot. 61:188–194.

————. 1974b. Glasshouse production of cotton flowers, harvest procedures, methods of ovule transfer, and in vitro development of immature seed. Cott. Grow. Rev. 51:293–301.

————. 1977. Ovule culture: Fundamental and pragmatic research for the cotton industry. p. 160–248. *In* J. Reinert and Y. P. S. Bajaj (eds.) Applied and fundamental aspects of plant cell, tissue and organ culture. Springer-Verlag, Berlin, Heidelberg.

————, And E. Egli. 1976. Differences in the hormonal regulation of fiber development from ovules of *Gossypium hirsutum*, *G. barbadense* and *G. arboreum* cultured in vitro. p. 39–44. *In* Proc. Beltwide Cotton Prod. Res. Conf., Las Vegas, Nev.

————, A. E. Linkins, and E. H. Birnbaum. 1974. Common ovule culture: A review of progress and review of potential. p. 169–192. *In* H. E. Sheet (ed.) Tissue culture and plant science. Academic Press, London.

————, and I. P. Ting. 1973. The effect of plant growth substances on in vitro fiber development from fertilized cotton ovules. Am. J. Bot. 60:130–139.

Benedict, C. R. 1972. Net CO_2 fixation in yellow-green plants. p. 7–25. *In* C. C. Black (ed.) Net Carbon Dioxide Assimilation in Higher Plants. Symposium of the Southern Section of the American Society of Plant Physiologists sponsored jointly with Cotton Incorporated.

————, and R. J. Kohel. 1970. Photosynthetic rate of a virescent cotton mutant lacking chloroplast grana. Plant Physiol. 45:519–521.

————, and ————. 1975. Export of ^{14}C-assimilate in cotton leaves. Crop Sci. 15:367–372.

————, ————, A. M. Schubert, and J. H. Keithly. 1981. Species variation of photosynthesis in *Gossypium*. Am. Soc. Agron. Abstr. p. 71–80.

————, K. J. McCree, and R. J. Kohel. 1972. High photosynthetic rate of chlorophyll mutant of cotton. Plant Physiol. 49:968–971.

————, R. H. Smith, and R. J. Kohel. 1973. Incorporation of ^{14}C-photosynthate into developing cotton bolls, *Gossypium hirsutum* L. Crop Sci. 13:89–91.

Beyer, E. M., and P. W. Morgan. 1969. Ethylene modification of an auxin pulse in cotton stem sections. Plant Physiol. 44:1690–1694.

----, and ----. 1970. Effect of ethylene on the uptake, distribution and metabolism of in-doleacetic acid-1 ^{14}C and 2-^{14}C and naphthalenacecetic acid-1-^{14}C. Plant Physiol. 46:157–162.

----, and ----. 1971. Abscission: The role of ethylene modification of auxin transport. Plant Physiol. 48:208–212.

Birnham, E. H., C. A. Beasley, and W. M. Dugger. 1974. Boron deficiency in unfertilized cotton (*Gossypium hirsutum*) ovules grown in vitro. Plant Physiol. 54:931–935.

----, W. M. Dugger, and C. A. Beasley. 1977. Interaction of boron with components of nucleic acid metabolism in cotton ovules cultured in vitro. Plant Physiol. 59:1034–1038.

Brendel, T. P., and C. S. Miller. 1979. Defoliation-stripping feasibility tests in Texas. 1978. p. 56–60. *In* Proc. Beltwide Cotton Prod. Res. Conf., Phoenix, Ariz.

----, C. S. Miller, and P. C. Koska. 1978. Alternatives to desiccation. p. 58–60. *In* Proc. Beltwide Cotton Prod. Res. Conf., Dallas, Tex.

Brown, H. B., and J. O. Ware. 1958. Cotton. McGraw-Hill, N.Y.

Brown, K. J. 1973. Factors affecting translocation of carbohydrate in cotton: Movement to the fruiting bodies. Ann. Bot. 32:703–713.

Browning, V. D., H. M. Taylor, M. G. Huck, and B. Klepper. 1975. Water relations of cotton: a rhizotron study. Auburn Univ. Ag. Exp. Stn. Bull. 467.

Buckwalter, H. 1982. Three years of PIX evaluation on cotton in Arizona and Imperial Valley of California. *In* Proc. Beltwide Prod. Cotton Res. Conf., Las Vegas, Nev.

Carns, H. R., J. Hacskalyo, and J. L. Embry. 1955. Relation of indole-3-acetic acid inhibitor to cotton boll development. p. 65–68. *In* Proc. Ninth Annu. Beltwide Cotton Defoliation Conf., Memphis, Tenn.

----, and J. R. Mauney. 1968. Physiology of the cotton plant. p. 41–73. *In* F. C. Elliott, M. Hoover, and W. K. Porter (ed.) Advances in production and utilization of quality cotton: Principles and practices. Iowa State Univ. Press, Ames.

Carpita, N. A., and D. P. Delmar. 1981. Concentration and metabolic turnover of UDP-glucose in developing cotton fibers. J. Biol. Chem. 256:308–315.

Cathey, G. W., and J. Hacskaylo. 1971. Increasing the effectiveness of defoliants. p. 31–33. *In* Proc. Beltwide Cotton Prod. Res. Conf., Atlanta, Georgia.

----, and ----. 1974. Evaluation of glyphosphate as a harvest aid chemical on cotton. p. 57. *In* Proc. Beltwide Cotton Prod. Res. Conf., Dallas, Tex.

----, and R. O. Thomas. 1979. Exogenous modification of flowering, fruiting and cut-out. p. 277–279. *In* Proc. Beltwide Cotton Prod. Res. Conf., Phoenix, Ariz.

Chailakhyan, M. Kh. 1968. Internal factors of plant flowering. Annu. Rev. Plant Physiol. 19:1–36.

Christiansen, M. N., H. R. Carns, and D. J. Slayter. 1970. Stimulation of soluble loss from radicles of *Gossypium hirsutum* by chilling, anaerobiosis, and low pH. Plant Physiol. 46:53–56.

----, and R. P. Moore. 1959. Seed coat structural differences that influence water uptake and seed quality in hard seed cotton. Agron. J. 51:582–584.

----, and R. Rowland. 1981. Cotton physiology—Seed and germination. Symposium on Cotton Physiology. III. Seed and Germination. p. 315–318. *In* Proc. Beltwide Cotton Prod. Res. Conf., New Oreleans, La.

Cole, D. F., and M. N. Christiansen. 1975. Effect of chilling duration on germination of cottonseed. Crop Sci. 15:410–412.

Craker, L. E., and F. B. Abeles. 1969. Abscission: Role of abscisic acid. Plant Physiol. 44:1144–1149.

Davenport, T. L., W. R. Jordan, and P. W. Morgan. 1977. Movement and endogenous levels of abscisic acid during water-stress-induced abscission in cotton seedlings. Plant Physiol. 59:1165–1168.

Davis, L. A., and F. T. Addicott. 1972. Abscisic acid: Correlation with abscission and with development in the cotton fruit. Plant Physiol. 49:644–648.

Delmar, D. P. 1977. The biosynthesis of cellulose and other plant cell wall polysaccharides. *In* F. A. Loewus and V. C. Runeckles (ed.) Recent Adv. in Phytochem. 11:45–77.

----, C. A. Beasley, and L. Ordin. 1974. Utilization of nucleoside diphosphate glucoses in developing cotton fibers. Plant Physiol. 53:149–153.

----, U. Heiniger, and C. Kulow. 1977. UDP-glucose: Glucan synthetase in developing cotton fibers. I. Kinetic and physiological properties. Plant Physiol. 59:713–718.

Dransfield, M. 1961. Some effects of gibberellic acid on cotton. Emp. Cotton. Grow. Rev. 38: 3–16.

Dugger, W. M., and R. L. Palmer. 1981. ^{14}C-Glucose incorporation into cell wall glucans of in vitro grown fibers. p. 45. XIII Int. Bot. Cong. Abstr., Sydney, Australia.

Dure, L. S. III. 1975. Seed formation. Annu. Rev. Plant Physiol. 26:259–278.

----. 1977. Stored messenger ribonucleic acid and seed germination. p. 335–346. *In* A. A. Khan (ed.) The physiology and biochemistry of seed dormancy and germination. Elsevier North Holland Press, N.Y.

Eaton, F. M. 1950. Physiology of the cotton plant. Adv. Agron. 2:11–25.

----. 1955. Physiology of the cotton plant. Annu. Rev. Plant Physiol. 6:299–328.

----, and D. R. Ergle. 1953. Relationship of seasonal trends in carbohydrate and nitrogen levels and effects of girdling and spraying with sucrose and urea to the nutritional interpretation of boll shedding in cotton. Plant Physiol. 28:503–520.

Eid, A. A., H. De Langhe, and L. Waterkeyn. 1973. In vitro culture of fertilized cotton ovules. I. The growth embryos. Cellule 69:361–371.

Elbein, A. D. 1969. Biosynthesis of a cell wall glucomannan in mung bean seedlings. J. Biol. Chem. 244:1608–1616.

El-Sharkawy, M., and J. Hesketh. 1965. Photosynthesis among species in relation to characteristics of leaf anatomy and CO_2 diffusion resistances. Crop Sci. 5:517–521.

----, M. J. Hesketh, and H. Muramoto. 1965. Leaf photosynthetic rates and other growth characteristics among 26 species of *Gossypium.* Crop Sci. 5:173–175.

Fites, R. C. 1974. Temperature-induced physical-phase transition in the permeability of glyoxysomes to succinate. p. 49–51. *In* Proc. Beltwide Cotton Prod. Res. Conf., Dallas, Tex.

Franz, G. 1959. Soluble nucleotides in developing cotton hair. Phytochemistry 8:737–741.

Gausman, H. W., J. Stabenow, F. R. Rittig, D. E. Escarbar, and M. V. Garza. 1980a. Mepiquat chloride effects on cotton leaf anatomy. Proc. 7th Annu. Plant Growth Regulator Working Group, Dallas, Tex.

----, H. Walter, F. R. Rittig, D. E. Escobar, and R. R. Rodriques. 1980b. Effect of mepiquat chloride (PIX) on CO_2 uptake of cotton plant leaves. p. 106. *In* Proc. 7th Annu. Plant Growth Regulator Working Group, Dallas, Tex.

Gipson, J. R., and H. E. Joham. 1969. Influence of night temperature on growth and development of cotton (*Gossypium hirsutum* L.). III. Fiber elongation. Crop Sci. 9:127–129.

----, and L. L. Ray. 1969. Fiber elongation rates in five varieties of cotton (*Gossypium hirsutum* L.) as influenced by night temperatures. Crop Sci. 9:339–341.

----, and L. L. Ray. 1970. Temperaure variety interrelationship in cotton. I. Boll and fiber development. Cotton. Grow. Rev. 47:257–271.

Godavari, H. R., S. S. Badour, and E. R. Waygood. 1973. Isocitrate lyase in green leaves. Plant Physiol. 51:863–867.

Guinn, G. 1974. Abscission of cotton floral buds and bolls as influenced by factors affecting photosynthesis and respiration. Crop Sci. 14:291–293.

----. 1976a. Nutritional stress and ethylene evolution by young cotton bolls. Crop Sci. 16:89–91.

----. 1976b. Water deficit and ethylene evolution by young cotton bolls. Plant Physiol. 57: 403–405.

----. 1978. Bloom production, ethylene evolution, and boll abscission as affected by water deficit. p. 53. *In* Proc. Beltwide Cotton Prod. Res. Conf., Dallas, Tex.

----. 1979. Hormonal relations in flowering, fruiting and cut-out. p. 265–276. Proc. Beltwide Cotton Prod. Res. Conf., Phoenix, Ariz.

----. 1982. Fruit age and changes in abscisic acid content, ethylene production, and abscission rate of cotton fruit. Plant Physiol. 69:349–352.

————. 1982. Abscisic acid and abscission of young cotton bolls in relation to water availability and boll load. Crop Sci. 22:580–583.

————, J. D. Hesketh, K. E. Fry, J. R. Mauney, and J. W. Radin. 1976. Evidence that photosynthesis limits yield of cotton. p. 60–61. *In* Proc. Beltwide Cotton Prod. Res. Conf., Las Vegas, Nev.

Hadas, A., and D. Russo. 1974. Water uptake by seeds as affected by water stress, capillary conductivity, and seed-soil water contact. II. Analysis of experimental data. Agron. J. 65: 647–652.

Hammett, J. R., and F. F. Katterman. 1975. Storage and metabolism of poly (adenylic acid)-mRNA in germinating cotton seeds. Biochemistry 14:4375–4379.

Harris, B., and L. S. Dure III. 1978. Developmental regulation in cotton seed germination: polyadenylation of stored messenger RNA. Biochemistry 17:3250–3256.

Hartt, C. E., and J. P. Kortschak. 1967. Translocation of ^{14}C in the sugarcane plant during the day and night. Plant Physiol. 42:82–94.

Hearn, A. B. 1969. Growth and performance of cotton in a desert environment. J. Agric. Sci. (Camb.) 73:65–97.

Heiniger, U., and D. P. Delmar. 1977. USP-glucose: Glucan synthetase in developing cotton fibers. II. Structure of the reaction product. Plant Physiol. 59:719–723.

Hesketh, J. 1967. Enhancement of photosynthetic CO_2 assimilation in the absence of oxygen, as dependent upon species and temperature. Planta (Berl.) 76:371–374.

————. 1968. Effect of light and temperature during plant growth on subsequent leaf CO_2 assimilation rates under standard conditions. Aust. J. Biol. Sci. 21:235–241.

————, and D. Baker. 1967. Light and carbon assimilation by plant communities. Crop Sci. 7: 285–293.

————, ————, and W. G. Duncan. 1971. Simulation of growth and yield in cotton: Respiration and carbon balance. Crop Sci. 11:394–398.

————, ————, and ————. 1972. Simulation of growth and yield in cotton: II. Environmental control of morphogenesis. Crop Sci. 12:436–439.

————, H. C. Lane, R. S. Alberte, and S. Fox. 1975. Earliness factors in cotton: New comparisons among genotypes. Cott. Grow. Rev. 52:126–133.

————, and A. Low. 1968. The effect of temperature on components of yield and fiber quality of cotton varieties of diverse origin. Cotton Grow. Rev. 45:243–257.

————, W. L. Ogren, M. E. Hageman, and D. B. Peters. 1980. Correlations among leaf CO_2-exchange rates, areas and enzyme activities among soybean cultivars. Photosyn. Res. 2:21–30.

Hinkle, D. A., and A. L. Brown. 1968. Secondary nutrients and micronutrients. p. 281–320. *In* F. C. Elliot, M. Hoover, and W. K. Porter, Jr. (ed.) Advances in production and utilization of quality cotton: Principles and practices. The Iowa State University Press, Ames.

Hoffman, W. C., A. K. Dobrenz, and R. E. Briggs. 1982. Photosynthesis and assimilate partitioning in PIX treated cotton. *In* Proc. Beltwide Cotton Prod. Res. Conf., Las Vegas, Nev.

Hofstra, G., and C. D. Nelson. 1969a. A comparative study of translocation of assimilated ^{14}C from leaves of different species. Planta (Berl.) 88:103–112.

————, and ————. 1969b. The translocation of photosynthetically assimilated ^{14}C in corn. Can. J. Bot. 47:1435–1442.

Hopp, H. E., P. A. Romero, G. E. Daleo, and R. Pont Lezica. 1977. Synthesis of cellulose precursors. The involvement of lipid-linked sugars. Eur. J. Biochem. 84:561–571.

Hutmacher, R. B., and D. R. Kreig. 1982. Stomatal and non-stomatal factors controlling photosynthic rates in cotton. Proc. Beltwide Cotton Prod. Res. Conf., Las Vegas, Nev.

Huwyler, H. R., G. Franz, and H. Meier. 1978. β-1,3-Glucans in the cell wall of cotton fibers (*Gossypium arboreum* L.). Plant Sci. Lett. 12:55–62.

Ihle, J. N., and L. S. Dure III. 1972. The development biochemistry of cotton seed embryogenesis and germination. III. Regulation of the biosynthesis of enzymes utilized in germination. J. Biol. Chem. 247:5048–5055.

Jackson, J. E., and N. R. Fadda. 1962. Effects of gibberellic acid on the flowering and fruiting of *Gossypium barbadense.* Emp. Cott. Grow. Rev. 8:125–130.

Joham, H. E. 1979. The effect of nutrient elements on fruiting efficiency. p. 306–311. *In* Proc. Beltwide Cotton Prod. Res. Conf., Phoenix, Ariz.

Jones, U. S., and C. E. Bardsley. 1968. Phosphorous nutrition. p. 213–254. *In* F. C. Elliot, M. Hoover, and W. K. Porter, Jr. (ed.) Advances in production and utilization of quality cotton: Principles and practices. The Iowa State University Press, Ames.

Jordan, W. R. 1979. The influence of edaphic parameters on flowering, fruiting and cut-out. A. Role of plant water deficit. p. 297–301. *In* Proc. Beltwide Cotton Prod. Res. Conf., Phoenix, Ariz.

----. 1982. Water relations in cotton. *In* J. D. Teare and M. M. Peet (ed.) Crop water relations. Wiley Interscience, New York.

----, P. W. Morgan, and T. L. Davenport. 1972. Water stress enhances ethylene-mediated leaf abscission in cotton. Plant Physiol. 50:756–758.

----, J. E. Quisenberry, and B. Roark. 1981. Contributions of heat tolerance and root growth potential to drought resistance. p. 40–41. *In* Proc. Beltwide Cotton Prod. Res. Conf., New Orleans, La.

----, and J. T. Ritchie. 1971. Influence of soil water stress on evaporation, root absorption and internal water status of cotton. Plant Physiol. (Lancaster) 48:783–788.

Joshi, P. C. 1962. In vitro growth of cotton ovules. p. 199–204. *In* Plant embryology—A symposium. Council of Scientific and Industrial Res. Sangam Press Private-Ltd., New Delhi.

----, and B. M. Johri. 1972. In vitro growth of ovules of *Gossypium hirsutum.* Phytomorphology 22:195–209.

Kamprath, E. J., and C. E. Welch. 1968. Potassium nutrition. p. 256–280. *In* C. Elliott, M. Hoover, and W. K. Porter, Jr. (ed.) Advances in production and utilization of quality cotton: Principles and practices. The Iowa State University Press, Ames.

Kerr, H. D., and C. M. Royster. 1977. Biochemical control of growth and maturation of cotton. p. 65. *In* Proc. Beltwide Cotton Prod. Res. Conf., Atlanta, Ga.

Kittock, D. L., H. F. Arle, and L. A. Bariola. 1974. Current status of chemical termination of cotton fruiting. p. 55–56. *In* Proc. Beltwide Cotton Prod. Res. Conf., Dallas, Tex.

----, ----, and ----. 1975. Chemical termination of cotton fruiting in Arizona in 1974. p. 71. *In* Proc. Beltwide Cotton Prod. Res. Conf., New Orleans, La.

----, ----, J. R. Mauney, and L. A. Bariola. 1973. Effect of several chemical treatments on early termination of cotton fruiting. p. 33. *In* Proc. Beltwide Cotton Prod. Res. Conf., Phoenix, Ariz.

Krieg, D. 1981. Leaf development and function as related to water stress. p. 41–42. *In* Proc. Beltwide Cotton Prod. Res. Conf., New Orleans, La.

----, and F. J. M. Sung. 1979. Source-sink relations of cotton as affected by water stress during boll development. p. 302–305. *In* Proc. Beltwide Cotton Prod. Res. Conf., Phoenix, Ariz.

Leahy, J. 1948. Structure of the cottonseed. p. 105–116. *In* A. E. Bailey (ed.) Cottonseed and cotton seed products. Interscience Publishers, Inc., N.Y.

Leffler, H. R. 1979. Physiology of earliness. p. 264–265. *In* Proc. Beltwide Cotton Prod. Res. Conf., Phoenix, Ariz.

Lipe, J. A., and P. W. Morgan. 1972a. Ethylene: Role in fruit abscission processes. Plant Physiol. 50:759–764.

----, and ----. 1972b. Ethylene: Response of fruit dehiscence to CO_2 and reduced pressure. Plant Physiol. 50:765–768.

----, and ----. 1973. Ethylene regulator of young fruit abscission. Plant Physiol. 51:949–953.

Low, A., J. D. Hesketh, and H. Muramoto. 1969. Some environmental effects on the varietal node number of the first fruiting branch. Cotton Grow. Rev. 46:181 188.

Lyons, J. M., and J. K. Raison. 1970. Oxidative activity of mitochondria isolated from plant tissues sensitive and resistant to chilling injury. Plant Physiol. 45:386–389.

Maltby, D., N. C. Carpita, D. Montezinos, C. Kulow, and D. P. Delmar. 1979. β-1,3-Glucan in developing cotton fibers. Structure, localization, and relationship of synthesis to that of secondary wall cellulose. Plant Physiol. 63:1158–1164.

Mason, T. G. 1922. Growth and abscission in sea island cotton. Ann. Bot. 36:457–483.

Mauney, J. R. 1966. Floral initiation of upland cotton *Gossypium hirsutum* L. in response to temperature. J. Exp. Bot. 17:452–459.

––––. 1979. Production of fruiting points. p. 256–260. *In* Proc. Beltwide Cotton Prod. Res. Conf., Phoenix, Ariz.

––––, K. E. Fry, and G. Guinn. 1978. Relationship of photosynthetic rate of growth and fruiting of cotton, soybean, sorghum, and sunflower. Crop Sci. 18:259–263.

––––, G. Guinn, K. E. Fry, and J. D. Hesketh. 1979. Correlation of photosynthetic carbon dioxide uptake and carbohydrate accumulation in cotton, soybean, sunflower, and sorghum. Photosynthetica 13:260–266.

––––, and L. L. Phillips. 1963. Influence of day length and night temperature on flowering of *Gossypium*. Bot. Gaz. 124:278–283.

McArthur, J. A., J. D. Hesketh, and D. N. Baker. 1975. Cotton. p. 297–325. *In* L. T. Evans (ed.) Crop physiology. Cambridge Univ. Press, N.Y.

McKinion, J. M., and D. N. Baker. 1982. Modeling experimentation, verification, and validation: Closing the feedback loop. Trans. ASAE (In press).

McMichael, B. L., and B. W. Hanny. 1977. Endogenous levels of abscisic acid in water-stressed cotton leaves. Agron. J. 69:979–982.

––––, W. R. Jordan, and R. D. Powell. 1972. An effect of water stress on ethylene production by intact cotton petioles. Plant Physiol. 49:658–660.

––––, ––––, and ––––. 1973. Abscission processes in cotton: Induction by plant water deficits. Agron. J. 65:202–204.

Miller, C. S., and W. H. Alfred. 1977. Determination of the efficiency of stalk application of desiccants. p. 69–72. *In* Proc. Beltwide Cotton Prod. Res. Conf., Atlanta, Ga.

––––, T. P. Brendel, and P. C. Koska. 1978. Search for a new cotton desiccant. p. 60–64. *In* Proc. Beltwide Cotton Prod. Res. Conf., Dallas, Tex.

––––, and L. H. Wilkes. 1968. Cotton desiccation practices and experimental results in Texas. MP-307. Texas Agric. Exp. Stn., College Station, Tex.

Mohapaptra, N., E. W. Smith, R. C. Fites, and G. R. Noggle. 1970. Chilling temperature depression of isocitratase activity from cotyledons of germinating cotton. Biochem. Biophys. Res. Commun. 40:1253–1258.

Morgan, P. W., E. M. Beyer, Jr., J. A. Lipe, and J. A. McAfee. 1971. Ethylene, a regulator of cotton leaf shed and boll opening. p. 42–44. *In* Proc. Beltwide Cott. Prod. Res. Conf., Atlanta, Ga.

––––, and J. I. Durham. 1972. Abscission: Potentiating action of auxin transport inhibitors. Plant Physiol. 50:313–318.

––––, and ––––. 1973. Morphactins enhance ethylene-induced leaf abscission. Planta 110:91–93.

––––, and H. W. Gausman. 1966. Effects of ethylene on auxin transport. Plant Physiol. 41:45–52.

––––, and W. C. Hall. 1962. Effect of 2,4-dichlorophenoxyacetic acid on the production of ethylene by cotton and grain sorghum. Physiol. Plant 15:420–427.

––––, and ––––. 1963. The effect of ethylene on the IAA oxidase system in cotton. p. 32–33. Cott. Defol. Physiol. Conf., Memphis, Tennessee.

––––, and ––––. 1964. Accelerated release of ethylene by cotton following application of indolyl-3-acetic acid. Nature 201:91.

––––, D. L. Ketring, E. M. Beyer, and J. A. Lipe. 1970. Functions of naturally produced ethylene in abscission, dehiscence and seed germination. p. 502–509. *In* D. J. Carr (ed.) Plant growth substances. Springer-Verlag Berlin, Heidelberg, New York.

Morris, D. A. 1964. Photosynthesis by the capsule wall and bracteoles of the cotton plant. Emp. Cotton Rev. 41:49–51.

Mullins, J. A., Jr., R. Overton, and T. McCuthen. 1972. Cotton defoliants and additives. Results under cool weather conditions of Tennessee. p. 50-51. *In* Proc. Beltwide Cott. Prod. Res. Conf., Memphis, Tenn.

Muramoto, H., J. D. Hesketh, and D. N. Baker. 1971. Cold tolerance in a hexaploid cotton. Crop Sci. 11:589-591.

----, ----, and C. D. Elmore. 1967. Leaf growth, leaf aging, and leaf photosynthetic rates of cotton plants. p. 161-165. *In* Proc. Beltwide Cott. Prod. Res. Conf., Dallas, Tex.

Namken, L. N., M. D. Heilman, and R. H. Dilday. 1978. The relationship of fruiting character in earliness and yield of selected genotypes. p. 93. *In* Proc. Beltwide Cotton Prod. Res. Conf., Dallas, Tex.

Niles, G. A. 1970. Development of plant types with special adaption to narrow row culture. p. 63-65. *In* Proc. Beltwide Cotton Prod. Res. Conf., Memphis, Tenn.

Ohkuma, K., F. T. Addicott, O. E. Smith, and W. E. Thiessen. 1965. The structure of abscisin. II. Tetrahedron Lett. 29:2529-2535.

----, J. L. Lyon, F. T. Addicott, and O. E. Smith. 1963. Abscisin II, and abscission-accelerating substance from young cotton fruit. Science 142:1592-1593.

Peterschmidt, N. A., and J. E. Quisenberry. 1981. Plant water status among cotton genotypes. p. 43-44. *In* Proc. Beltwide Cott. Prod. Res. Conf., New Orleans, La.

Price, H. J., and R. H. Smith. 1979. Somatic embryogenesis in suspension cultures of *Gossypium klotzschianum* Amderss. Planta 145:305-307.

Quisenberry, J. E., W. R. Jordan, B. A. Roark, and D. W. Fryrear. 1982. Exotic cottons as genetic sources for drought resistance. Crop Sci. 21:889-895.

Radin, J. W., and R. C. Ackerson. 1981. Water relations of cotton plants under nitrogen deficiency. III. Stomatol conductance, photosynthesis, and abscisic acid accumulation during drought. Plant Physiol. 67:115-119.

----, and J. R. Mauney. 1982. The nitrogen stress syndrome in cotton. *In* J. R. Mauney and J. McD. Stewart (ed.) Cotton physiology: A treatise. USDA Handb. (in press).

----, and R. N. Trelease. 1976. Control of enzyme activities in cotton cotyledons during maturation and germination. I. Nitrate reductase and isocitric lyase. Plant Physiol. 57: 902-905.

Ragahavan, V. 1977. Applied aspects of embryo culture. p. 375-397. *In* J. Reinert and Y. P. S. Bajaj (ed.) Applied and fundamental aspects of plant cell, tissue, and organ culture. Springer-Verlag, Berlin, Heidelberg.

Ramsey, J. C., and J. D. Berlin. 1976a. Ultrastructural aspects of early stages in cotton fiber elongation. Am. J. Bot. 63:868-876.

----, and ----. 1976b. Ultrastructure of early stages of cotton fiber differentiation. Bot. Gaz. 137:11-19.

Ray, L. L., and T. R. Richmond. 1966. Morphological measures of earliness of crop maturity in cotton. Crop Sci. 6:527-531.

Richardson, C. W., and J. T. Ritchie. 1973. Soil water balance of small watersheds. Trans. ASAE 16:72-77.

Riov, J. 1974. A polygalacturonase from citrus leaf explants. Role in abscission. Plant Physiol. 53:312-316.

Ritchie, J. T. 1972. Model for predicting evaporation from a row crop with incomplete cover. Water Resources Res. 8:1204-1213.

Schubert, A. M., C. R. Benedict, J. D. Berlin, and R. J. Kohel. 1973. Cotton fiber development—Kinetics of cell elongation and secondary wall thickening. Crop Sci. 13:704-709.

Smith, E. W., and R. C. Fites. 1973. The influence of chilling temperature alteration of glyoxysomal succinate levels on isocitritase activity from germinating seedlings. Biochem. Biophys. Res. Comm. 55:647-654.

----, ----, and G. R. Noggle. 1971. Effects of chilling temperature on isocitratase and malate synthetase levels during cottonseed germination. p. 45-47. *In* Proc. Beltwide Cotton Prod. Res. Conf., Atlanta, Ga.

Smith, R. H., A. M. Schubert, and C. R. Benedict. 1974. Development of isocitric lyase activity in germinating cotton seed. Plant Physiol. 54:197–200.

Stewart, J. M. 1975. Fiber initiation on the cotton ovule (*Gossypium hirsutum*). Am. J. Bot. 62:723–730.

————. 1979. Use of ovule cultures to obtain interspecific hybrids of *Gossypium*. p. 44–56. *In* J. T. Barber (ed.) Plant tissue culture. Southern Section, Am. Soc. Plant Physiol., New Orleans, La.

————, and C. L. Hsu. 1977. In ovule embryo culture and seedling development of cotton (*Gossypium hirsutum* L.). Planta 137:113–117.

————, and ————. Hybridization of diploids and tetraploid cottons (*Gossypium* sp.) through in ovule embryo culture. J. Hered. 69:404–408.

Sung, F. J. M. 1978. Source-sink relationships of sorghum and cotton as effected by water stress. Ph.D. thesis. Texas Tech. Univ., Lubbock. Univ. Microfilms. Ann Arbor, Mich. (Diss. Abstr. 39:7904974).

Thomas, R. O., and J. Hacskaylo. 1974. Cotton fruiting and yield responses to growth retardants used for suppressing late-season plant development. p. 61. *In* Proc. Beltwide Cotton Prod. Res. Conf., Memphis, Tenn.

Tollervey, P. W. 1970. Physiology of the cotton plant. Cott. Grow. Rev. 47:245–256.

Trelease, R. N., J. A. Miernyk, J. S. Choinski, Jr., and S. J. Bortman. 1980. Enzyme synthesis in developing cotton embryos. p. 355–366. *In* Symposium on cotton boll development II. Carbon metabolism in developing cottonseed. Proc. Beltwide Cotton Prod. Res. Conf., St. Louis, Mo.

Tucker, T. C., and B. B. Tucker. 1968. Nitrogen nutrition. p. 183–211. *In* F. C. Elliott, M. Hoover, and W. K. Porter, Jr. (ed.) Advances in production and utilization of quality cotton: Principles and practices. The Iowa State University Press, Ames.

Verhalen, L. M., R. Mamaghani, W. C. Morrison, and R. W. McNew. 1975. Effect of blooming date on boll retention and fiber properties in cotton. Crop Sci. 15:47–52.

Wadleigh, C. H. 1944. Growth status of the cotton plants as influenced by the supply of nitrogen. Arkansas Agric. Exp. Stn. Bull. No. 466. p. 138.

Wainwright, I. M., R. L. Palmer, and W. M. Dugger. 1980. Pyrimidine pathway in boron-deficient cotton fiber. Plant Physiol. 65:893–896.

Walhood, V. T. 1958. Effect of gibberellins on yield and growth of cotton. p. 27–30. Proc. 13th Annu. Beltwide Cotton Defoliation Physiol. Conf., Houston, Tex.

————, and F. T. Addicott. 1968. Harvest-Aid programs: Principles and Practices. p. 407–431. *In* F. C. Elliott, M. Hoover, and W. K. Porter, Jr. (ed.) Advances in production and utilization of quality cotton: Principles and practices. The Iowa State University Press, Ames.

————, and R. E. Johnson. 1976. Inception of solar radiation by a constant population of cotton plants. p. 70–71. *In* Proc. Beltwide Cotton Prod. Res. Conf., Memphis, Tenn.

Wanjura, D. F., and D. R. Buxton. 1972. Water uptake and radicle emergence of cottonseed as affected by soil moisture and temperature. Agron. J. 64:427–431.

————, ————, and H. N. Stapleton. 1971. A model for describing cotton growth during emergence. Transactions ASAE 16:227–231.

Whelan, T., W. M. Sackett, and C. R. Benedict. 1970. Carbon isotope discrimination of a plant possessing the C_4 dicarboxylic acid pathway. Biochem. Biophys. Res. Commun. 41:1205–1210.

Wong, W. W., C. R. Benedict, and R. J. Kohel. 1979. Enzymic fractionation of the stable carbon isotopes of carbon dioxide by ribulose-1,5-bisphosphate carboxylase. Plant Physiol. 63:852–856.

7 Breeding

G. A. Niles
Texas A&M University
College Station, Texas

C. V. Feaster
ARS-USDA
Phoenix, Arizona

Man has attempted to improve cotton since the time that the plant was domesticated. Cursory and empirical as those efforts were, they nevertheless affected the differentiation of a multitude of forms adapted to varying environmental and cultural conditions. These selections undoubtedly constituted much of the bedrock germplasm from which contemporary cultivars have descended.

The modern era in cotton improvement might be considered as beginning in the early 1900s, coincidentally with the advent of the boll weevil (*Anthonomus grandis* Boh.) into the USA and the rediscovery of Mendel's laws of heredity. The advent of Mendelism did not prompt drastic changes in selection and hybridization techniques, but it did provide plant scientists with improved understanding of breeding systems and how they might be employed to greatest effect. The incursion of the boll weevil around the turn of the century mandated a decidedly new approach to cotton improvement, as cotton breeders were called upon to restructure the plant into new, fast-fruiting and early-maturing forms, while retaining acceptable levels of yield and quality.

Breeders of the early post-boll weevil period did an admirable job making the necessary modifications in plant growth and fruiting habit, and cotton was able to survive an event that some people viewed as catastrophic and likely destructive of the cotton industry in the USA. Since that earlier breeding accommodation to the boll weevil, breeders have successfully weathered numerous crises by developing new cultivars that substantially

Published in *Cotton,* Agronomy Monograph no. 24, © ASA-CSSA-SSSA, 677 South Segoe Road, Madison, WI 53711.

resolve problems in disease resistance, fiber quality, stormproofness, seed quality, maturity, and yield. The following sections provide a historical perspective of the elaboration of cotton cultivars in the USA over the past 300 years or so.

7-1 AREAS OF CULTIVATION IN THE WORLD

Cotton is harvested from almost 32.4 million ha in more than 40 nations of the temperate and tropic regions of the world (Anonymous, 1981a). The crop is grown as far north as 47° N Lat in the Ukraine and 37° N Lat in the USA; in the Southern Hemisphere production extends to about 32° S Lat in South America and Australia. World production during the 1977–1980 period averaged 63.5 million bales (standard bale is 218 kg or 480 lb each); 80% came from nine countries in which average annual production exceeded 1 million bales. Cotton is a warm weather plant and the cultivated species are not tolerant of freezing temperatures. Even so, production is not limited to the tropics, as the development of adapted cultivars and effective production techniques permit successful culture in regions where the frost-free period is less than 180 days.

7-1.1   Asiatic Cottons

Production of the Asiatic cottons, either *Gossypium arboreum* or *G. herbaceum,* is limited almost exclusively to Southeast Asia, primarily on dry and unproductive areas of India and Pakistan not suited for *G. hirsutum* or *G. barbadense.* Scant data are available to document the extent of this production, but the latest available information indicates that 10% or less of the total production in both India and Pakistan is planted to Asiatic types (Anonymous, 1979). Worldwide, less than 4% of production is of the Asiatic types.

7-1.2   *Gossypium barbadense*

Firm data on the production of *G. barbadense* types are sparse. Data tabulated by the International Cotton Advisory Committee (Anonymous, 1979) indicate that extra-long staple types (34.9 mm or longer) account for approximately 5.1% of the world's production of cotton. These are grown because of their uniquely long, fine, and strong fiber. Major producers of the *G. barbadense* cultivars are Egypt, Sudan, USSR, Peru, and USA; relatively small production occurs in Israel, Morocco, Colombia, China, and Yemen.

7-1.3 *Gossypium hirsutum*

Among the cultivated species, *G. hirsutum* is, by far, the predominant
form of cotton grown in the world at the present time. It is grown primarily
because of its relatively high productivity and wide adaptability. Approximately 91% or more of the world production is planted to cultivars of *G.
hirsutum* types, many of which were derived from American Upland cultivars.

7-2 AREAS OF CULTIVATION IN THE USA

Cotton production is essentially confined to a tier of 15 states stretching from North Carolina to California. Production from 1974 to 1980 averaged about 11.6 million bales from 4.8 million ha, distributed as shown in
Table 7-1.

During the past two decades certain perceptible shifts have occurred in
the beltwide distribution of cotton production. These are demonstrated in
Figure 7-1, which shows the general trends in four regions of the U.S. Cotton Belt. Most obvious is the reduced production of the southeastern states
and the attendant increases in the southwestern and western regions.

7-2.1 *Gossypium barbadense*

Sea Island production on the mainland and island areas of South Carolina and Georgia and on the coastal plain areas of southern Georgia and
northern Florida usually amounted to less than 100 000 bales annually from
about 1790 until 1920, when production of the crop became unprofitable
because of the boll weevil.

The first commercial production of Pima in the Southwest was in 1912
(McGowan, 1961). A total of 98 000 ha was planted in 1920—81 000 in
Arizona and 17 000 in California. Since 1920, production has seldom ex-
ceeded 40 000 ha, with production confined primarily to Arizona, New
Mexico, and west Texas.

7-2.2 *Gossypium hirsutum*

The Upland cottons grown in the USA can be classified, roughly, into
four major types—Acala, Delta, Plains, and Eastern. Although this classification is rather arbitrary, it reflects the major variations in adaptation,
growth characteristics, and fiber properties that exist among contemporary
Upland cultivars.

Table 7-1. Average annual harvested production of cotton in the USA, 1974 through 1980.†

State	Hectares (1000s)	Bales (1000s)
Alabama	156.4	335.7
Arizona	202.1	1 080.0
Arkansas	324.9	727.7
California	532.4	2 617.0
Florida	2.4	6.4
Georgia	81.3	171.0
Louisiana	207.1	534.0
Mississippi	520.7	1 342.3
Missouri	93.6	192.7
New Mexico	49.1	117.0
North Carolina	29.1	63.3
Oklahoma	194.5	312.0
South Carolina	58.6	133.3
Tennessee	128.7	231.3
Texas	2 263.2	3 765.1
Other States	1.5	2.8
Totals	4 845.4	11 631.6

† Adapted from Anonymous (1981a).

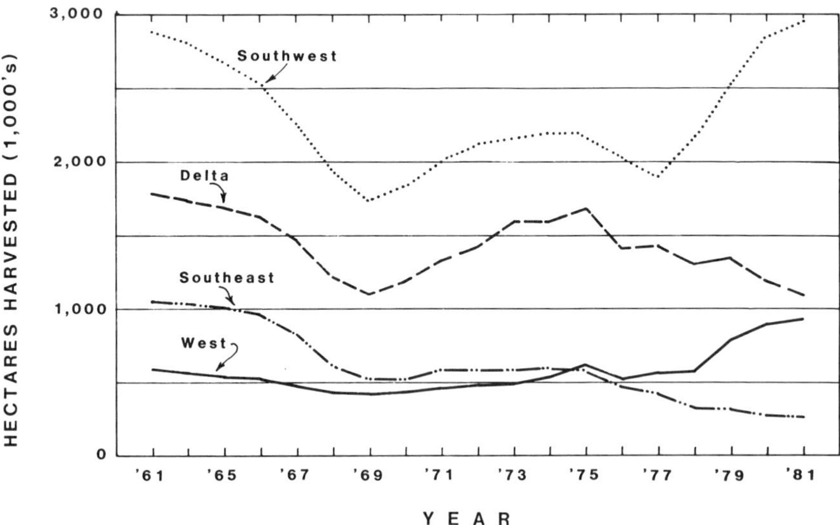

Fig. 7-1. Three-year moving averages of harvested production of cotton in four regions of the USA, 1960 through 1980. Regions defined as follows: West = Calif., Ariz., N.M., Nev.; Southwest = Tex., Okla., Kan.; Delta = Mo., Ark., Tenn., Miss., La., Ill., Ky.; Southeast = Va., N.C., S.C., Ga., Fla., Ala. (Adapted from Anonymous, 1981a.)

7–2.2.1   Acala Types

Production of Acala-type cultivars is confined to irrigated areas in West Texas, New Mexico, Arizona, and California (Anonymous, 1980). In the first three of these states, the Acala cultivars grown are predominantly of the Acala 1517 family, whereas production in California is confined to cultivars of the Acala SJ series. The Acalas accounted for an estimated 11% of U.S. production in the period 1976 through 1980 (Anonymous, 1981b).

7–2.2.2   Delta Types

Approximately one-third of the U.S. cotton production is of the Delta type, primarily of the Deltapine and Stoneville series. Adaptation of Delta-type cultivars, generally, is quite broad and representative cultivars are grown in nearly every cotton-producing state. Nationally, Deltapine and Stoneville cultivars accounted for 30% of the planted area from 1976 through 1980 (Anonymous, 1981b).

7–2.2.3   Plains Types

The Plains type comprises a rather heterogeneous group of cultivars essentially confined to Texas and Oklahoma, with limited production in eastern New Mexico. In the aggregate, they account for more than 40% of USA production area, with Lankart (12%) and Paymaster (7%) cultivars being the predominant examples of this type (Anonymous, 1980).

7–2.2.4   Eastern Types

This type is represented mainly by the Coker and McNair cultivars that account for the bulk of the production in North Carolina, South Carolina, and Georgia, with minor production in other states.

7–3   BREEDING ACCOMPLISHMENTS AND CULTIVAR SHIFTS IN THE USA

The accomplishments in cotton cultivar development in the USA and the relationships of contemporary types can best be put into perspective by reviewing, briefly, the history of cotton cultivar evolution over the past 250 years. This history has been documented by numerous authors. The account that follows here is a synthesis of information related by Watts (1907), Ramey (1966), Handy (1896), Moore (1956), and Ware (1937, 1951).

When European explorers visited the New World in the 15th and 16th centuries, they found the natives using cotton as a textile fiber for clothing and household goods. This basic cotton industry utilized fibers obtained from indigenous cottons that were independent of the cottons of the Old World. Cotton goods, and presumably cottonseed, were carried back to Europe for introduction into the cotton-growing areas around the Mediterranean, from which seed later were reintroduced into the new colonies of North America.

Cotton culture in the USA apparently began with the Jamestown colonists, but systematic cultivation was not established until after 1620. Origin of the seedstocks used in the early colonies is unclear, but it appears that the earliest cottons introduced into the USA came from the Levant. Some of these importations probably were returned New World stocks, as well as forms of *G. herbaceum* that then were the common cotton in the Near East and Central Asia. Subsequently, White Siam and Chinese Nankeen, both undoubtedly forms of *G. arboreum*, were introduced into Louisiana in the mid-18th century. Subsequently introductions of New World cottons from the West Indies, Mexico, Central America, and South America gradually replaced the Asiatic types. Establishment of these tropical cottons to the higher latitudes of the American colonies undoubtedly was a difficult process because of the perennial habit and photoperiod requirements that prevented them from fruiting in the long-day growing season of the temperate zone and from maturing their fruit before the advent of killing frost in the fall. The immediate and stringent selection pressure for the day-neutral habit facilitated the isolation and establishment of annual or near-annual forms, which became the first authentic American cottons of the early colonial period.

The early established annual forms, apparently of *G. hirsutum,* were of two principal types—naked or black-seeded and fuzzy green-seeded. By the time that Sea Island cotton, *G. barbadense,* was introduced about 1785, the green-seeded cottons had become the dominant types grown in the Colonies, and remained the most important *G. hirsutum* type until the Mexican highland stocks were introduced in the next century.

Early colonists in the coastal areas did little to develop a cotton industry, but relied mostly on tobacco (*Nicotiana* spp.), rice (*Oryza* spp.), and indigo (*Indigofera* spp.) enterprises. These agricultural industries provided the purchasing power with which textile requirements could be satisfied through importation of either raw materials or finished textiles. In the interior areas of the Carolinas and Georgia, the colonists had to be more self-sufficient in satisfying their textile needs. Adapted stocks of *G. hirsutum* provided raw cotton for development of a home textile industry, and these cottons became known as "Upland" types. The designation Upland later was used in contrast to the terms "Lowland" or "Sea Island" after the latter was established in cultivation.

Settlement of the interior regions was slow until about 1760, following the French and Indian War, when Pennsylvania Dutch, Tide-water people, and European immigrants moved into the up-country regions. Cotton

culture also spread among these newly settled regions, but it did not become
a commercial crop of consequence until the Revolutionary War when textiles and raw cotton supplies from other countries were curtailed. Demand
for cotton textiles became so great that cotton production moved into
coastal areas of Maryland, Delaware, and New Jersey, and even into
eastern Pennsylvania.

Arkwright's successful development of a water-powered spinning
frame in 1769 provided a strong impetus for the development of the factory
system of textile manufacturing in England. This and other technical advances helped make England the dominant country in the spinning of high
count yarns from which fine cotton fabrics could be woven. England's
burgeoning textile industry required increasing supplies of raw cotton that
could not be satisfied by the conventional growths of the Mediterranean
area and India. For a time, cotton was provided by the West Indies, Brazil,
and Peru, but these supplies became inadequate. Following the Revolutionary War, a regular cotton trade was begun with the USA. The invention of
the cotton gin by Eli Whitney in 1793 greatly facilitated the processing of
cotton in the USA and enabled the country to satisfy the growing English
demand for cotton.

Introduction of Sea Island cotton in the mid-1780's had special significance for the U.S. cotton industry. While the Sea Island types were not
grown extensively, they produced lint of special value in the manufacture of
thread, lace, and fine cloth by English and European spinners. By the end
of the 18th century, production of both Upland and Sea Island cottons was
firmly established in the USA, and in the first half of the 19th century the
USA became the predominant supplier of raw cotton for the English textile
industry.

The foregoing historical sketch provides a background for considering
two post-colonial events that significantly influenced the differentiation of
cotton cultivars in the USA. These two events, the introduction of Sea
Island cottons and of the Mexican highland stocks, will be considered in the
following sections.

7-3.1 *Gossypium barbadense*

South America is the center of origin of *G. barbadense* (Hutchinson,
1959). Until the 16th century, it was the fiber crop of South America (Kerr,
1960). These cottons were photoperiodic, and the fiber was medium staple
(23.8 to 27.0 mm in length) and coarse, as typified by current Tanguis
cottons of Peru.

7-3.1.1 *Sea Island*

Gossypium barbadense first appeared in the USA about 1785, where it
became known as Sea Island (Hutchinson, 1959). The Sea Island cotton was
day-neutral in flowering habit and had strikingly different fiber properties

from the *G. barbadense* of South America. The exact origin of Sea Island is unknown, but Kerr (1960) and Stephens (1975, 1976) suggested that the most logical hypothesis to explain the long-staple Sea Island type was that it developed by transgressive inheritance through the introgression of length genes from outside the species, possibly from *G. hirsutum.* The high quality fiber of Sea Island demanded a high price that compensated for its low yields. Its excellent quality was maintained by isolation of seed production on islands off the coast of South Carolina and Georgia, where it was kept from mixing with Uplands and photoperiod forms of *G. barbadense.* Many of the planters conducted their own selection programs for maintaining or improving fiber quality. They practiced annual selection of superior individual plants and increased the seed of these plants to provide for planting of future crops (Kearney, 1930). These breeding programs resulted in many cultivars being developed with fiber lengths up to 50.8 mm or longer. The Sea Island industry became unprofitable because of the boll weevil, and production ceased in 1920.

Attempts to revive the Sea Island industry in the 1930s were unsuccessful. By 1940, it was evident that a cotton that would be competitive with the Uplands of that era could not be developed solely by selection within Sea Island. A program of intercrossing among different cultivars of *G. barbadense* was initiated by the USDA and the Georgia Agricultural Experiment Station. The program resulted in the release of 'Coastland' (Jenkins, 1953). Coastland was developed from intraspecies crossing of three *G. barbadense* types—Sea Island, Egyptian, and Tanguis. Coastland combined the best characters of the parent material, giving improved agronomic characteristics as well as high quality fiber. Coastland represented a striking agronomic advance over Sea Island, but it did not become a commercial success. The *G. barbadense* improvement program in the Southeast was terminated in 1962.

7–3.1.2 Pima (American-Egyptian)

In the early 20th century, when *G. barbadense* was first grown in the southwestern USA, it was known as American-Egyptian cotton. The germplasm was reintroduced from Egypt. The introduced Egyptian cotton is believed to have been derived from hybridization of Sea Island and a *G. barbadense* tree cotton called Jumel. Mr. Jumel, a French engineer, observed this tree cotton in a garden near Cairo. It supposedly originated from seed obtained in the upper Nile region of Sudan. Selection from the heterogeneous population resulting from mixing and crossing Jumel with Sea Island resulted in the development of 'Ashmouni' about 1860. The next several Egyptian cultivars were derived either by selecting within Ashmouni or from crosses of Ashmouni with Sea Island (Kearney, 1943). The early developed Egyptian cottons were of lower quality than Sea Island. W. L. Balls, working in Egypt from 1910 to 1940, introduced the concept of inbreeding, utilized measurements of various fiber properties, spinning tests, and gradually developed Egyptian cottons that could compete with the quality of the Sea Islands (Kerr, 1960). During the development of these cottons no germplasm from the outside was introduced into the Egyptian program.

The first Pima cultivar released by the USDA was 'Yuma', in 1908. It was selected from 'Mitafifi', an Egyptian cultivar developed in 1887 from a cross of Ashmouni and Sea Island and introduced into the southwestern USA about 1900. The fiber of Yuma was about 38 mm in length and more productive than Mitafifi (Hathorn, 1951). Between 1908 and 1949, four additional cultivars, 'Pima', 'SXP', 'Amsak', and 'Pima 32' were developed from the Egyptian germplasm base, either by direct selection or through hybridization followed by selection (Feaster and Turcotte, 1962). These five cultivars (Yuma, Pima, SXP, Amsak, and Pima 32) had a fairly narrow genetic base from introduced Egyptian cotton. They were developed and maintained by a strict pedigree system that resulted in a loss of genetic variability. They were similar in many respects, having long fiber and being generally late and relatively unproductive.

The 'Pima S-1' cultivar released in 1951 differed considerably from the previous cultivars and was developed in a much different manner (Bryan, 1955). Pima S-1 was developed by selection from a complex series of crosses involving Sea Island, Pima, Tanguis, and 'Stoneville'. The first three were *G. barbadense,* while the latter was *G. hirsutum,* which probably contributed to cultivar development through limited introgression. Crossing for this development was begun in 1934 and the single plant selection that gave rise to the Pima S-1 cultivar was made in 1947. Pima S-1 was considered as a new base point in the Pima improvement program. Compared with previously developed Pima cottons, Pima S-1 had larger bolls, higher lint percentage, and small but highly productive plants (Waddle, 1953). The fiber of Pima S-1 was shorter and coarser than that of previous Pima cultivars, but it had a yarn strength greater than might be expected on the basis of its fiber properties (Feaster and Turcotte, 1962).

With the development of Pima S-1, the genetic base was broadened much beyond the germplasm base from which Pima 32 was developed. Pima S-1 was crossed with 3-79, a productive experimental strain of the same background as Pima 32. An F_3 selection from this cross became 'Pima S-2' and was released for commercial production in 1960 from the cooperative USDA improvement program at Phoenix, Arizona (Feaster and Turcotte, 1962). Compared with Pima S-1, Pima S-2 was more productive, shorter in stature, earlier maturing, higher in lint percentage and similar for fiber properties, except for a slight increase in micronaire. Yields of Pima S-2 averaged 25% higher than those of Pima S-1 at elevations below 426 m and 10% higher at elevations above 609 m. This difference was attributable to the greater heat tolerance of Pima S-2 during the fruiting period.

'Pima S-4', a selection from a cross of experimental strain (P32 × S1 10-8) and Pima S-2, was released in 1966. This cross combined a lower yielding, higher quality strain (P32 × S1 10-8) with a higher yielding, lower quality cultivar (Pima S-2) (Feaster et al., 1967). At low elevations, Pima S-4 averaged 17% greater yield than Pima S-2. At elevations above 457 m, the yields from Pima S-4 and Pima S-2 were similar. Pima S-4 had more tolerance than Pima S-2 to high night temperatures, since it was capable of setting more fruit than Pima S-2 during July and August at low elevations where night temperatures are high.

In 1975, 'Pima S-5' was released from the improvement program at Phoenix, Arizona (Feaster et al., 1976). Pima S-5 was derived from an F_4 selection from a cross of experimental strain (3-79 × S1 × S1 78-567-228-325) and Pima S-4. The major advantages of Pima S-5 were higher yield and earlier maturity. Yield advantages for Pima S-5 over Pima S-4 ranged from 10 to 17%, depending on elevation. The relatively greater adaptability of Pima S-5 at the lower elevations appeared to be related to its greater heat tolerance during the fruiting period.

Pima S-5 was the third of a series of cultivars released after the germplasm had been broadened in the development of Pima S-1. Each succeeding cultivar showed increased heat tolerance and had increasingly greater adaptability at low elevation than at high elevation. The germplasm pool currently shows considerable variability and potential for continued cultivar improvement for short plant stature and greater heat tolerance. Two non-commercial strains, 79–103 and 79–106, were released from this germplasm pool in 1980 (Feaster and Turcotte, 1980). These strains were early-maturing and short-statured, and they were released for potential use in combination with *G. hirsutum* cultivars for interspecific hybrid production. The short stature and height stability of these strains offer the potential of reducing the excessive vegetative growth that has been a persistent problem with *G. hirsutum* × *G. barbadense* hybrids.

At the same time that Pima S-4 was released for commercial production, 'Pima S-3' also was released (Feaster et al., 1967). Pima S-3 and Pima S-4 appeared adapted to a two-cultivar system. Pima S-3 was recommended for the high elevations, and Pima S-4 for the low elevations and for high elevations where conditions of high productivity caused Pima S-3 to be late-maturing and rank. Pima S-3 was a selection from Hybrid B. The Hybrid B material was developed at the Far West Texas Research Station, El Paso, Texas, by mass crossing of Pima S-1, Pima experimental strain 1-71, Tanguis, Ashmouni, 'Giza 12', Pima 32, various Coastland strains, and the Upland strain C-1. Pima S-3 was late-maturing, had no heat tolerance, and had limited production for a short time at high elevations.

7–3.2 *Gossypium hirsutum*

As the American cotton industry expanded in the late 18th and early 19th centuries, the green-seeded Upland stocks that were in general use proved to be inadequate for providing the quantities and qualities of fiber needed to sustain the industry in the USA (Ware, 1951). According to this author, the first recorded importation of Mexican highland stocks was made in 1806 by Walter Burling of Mississippi. This stock eventually became the parent stock of a large proportion of the short and medium staple cultivars that were developed in the eastern portion of the Cotton Belt. It is beyond the scope of this chapter to document the development of the multiplicity of Upland cultivars developed up to the boll weevil era. Suffice it to say that Burling's Mexican and other introductions from Mexico, either directly or indirectly, contributed greatly to cultivar development in the USA in the pre-boll weevil era.

7–3.2.1 Acala Types

The Acala types of Upland cotton are unique in that they were derived from introductions made into the USA after the advent of the boll weevil in the early 1900s and their development has been well documented. The initial stock was collected by G. N. Collins and C. B. Doyle in 1906 near the village of Acala, Chiapas, Mexico. Plants of this accession were grown at San Antonio, Texas, from 1907 to 1911, and selections were made to improve adaptation and uniformity. Subsequent selection in the Acala materials in Texas and Oklahoma gave rise to three Acala stocks that were primary progenitors of the Acala families developed in the USA. Development of the Acalas proceeded along three fairly distinct paths, beginning with the No. 5 and No. 8, and later with the No. 9 selections.

The No. 5 selection proved to be more determinate, more compact and earlier fruiting than No. 8. The seedstock was increased in Oklahoma, and in 1918 C. N. Nunn released 'Acala 5', which was distributed in Oklahoma, Arkansas, and North Texas. The later-released 'Nucala' was the source of breeding stocks for Acala breeding programs of the Oklahoma, Arkansas, and Tennessee Agricultural Experiment Stations.

Stocks of the No. 8 selection were maintained and increased in Texas to provide seed for distribution. Interest in the Acala and other relatively indeterminate types was limited by the fact that the boll weevil was forcing breeders and producers to look with greater favor on the faster-fruiting, early-maturing types. Seed of the No. 8 selections were sent to California, where the first commercial planting of Acala was made near Bakersfield in 1919 (Turner, 1974). 'Acala 8' proved to be very popular with California growers, not only for its yielding ability, but also for its vigorous growth, upright plant conformation and excellent fiber properties. Cotton interests in California recognized that the Acalas were well-adapted, produced high yields and fiber of preferred qualities (Gilmore, 1918), and that the integrity and marketing advantage of California cotton should be protected. In 1925 the California legislature enacted a One-Variety Law that prohibited the growth of any cotton cultivar other than Acala in the San Joaquin Valley. The California breeders, utilizing a type selection scheme, established a number of successive Acala stocks in which staple lengths of 27.0 to 30.2 mm provided a substantial marketing advantage. In the 1930s, two new problems arose to disturb the California cotton industry; these were the spread of Verticillium wilt (*Verticillium dahliae* Kleb.) and fiber quality. The wilt prompted an extensive screening on the part of breeders and pathologists to locate sources of tolerance and to develop techniques for field inoculation and wilt evaluation. Wilt tolerance was identified in 'Missdel', 'Cook', 'Mexican Big Boll,' 'Kekchi', 'Tuxtla', and Pima, and these germplasms were integrated into the wilt-breeding program (Turner, 1974).

With the development of new fiber instrumentation in the late 1930s, the relationships of fiber fineness and fiber strength, as well as length and length uniformity, to yarn quality and spinning performance were recognized. The marketing situation for California cotton became disturbing when domestic textile mills labeled the irrigated Acala growths as "soft" and inferior in spinning performance to Delta-grown cottons of the same length. In 1939, 'Acala P18C', an inbred selection from Acala 8, was released and became the standard cultivar for the San Joaquin Valley from 1944 to 1948. This cultivar did not remedy either the wilt or the fiber quality problem, but it provided growers with a higher yielding cultivar until a better one could be developed. California breeders, intent on developing wilt tolerance and higher fiber quality in the San Joaquin cottons, introduced stocks from Mississippi, New Mexico, and Arizona for use in a hybridization program. The first significant achievement occurred with release of 'Acala 4-42' about 1943. Acala 4-42 originated from a single plant selection in 'Acala 1517', a New Mexico cultivar. This new cultivar was not considered a better yielder than Acala P18C, but it produced a shorter plant, began fruiting earlier, and showed a greater degree of storm resistance in the open boll. The biggest advantage was its improvement in fiber quality. Compared to Acala P18C, Acala 4-42 was 20% stronger and produced 42% fewer neps in the spinning process (Turner, 1974). Sufficient seed were produced to plant the entire San Joaquin Valley to Acala 4-42 in 1949, and within a few years San Joaquin Valley cotton regained its reputation for superior fiber quality.

In the 1950's an extensive crossing program was pursued, utilizing Acala stocks and selected strains from the southeastern Cotton Belt, to combine fiber quality with yield, early maturity, and high bloom retention. Turner (1974) describes the frustration of this effort:

"Many disappointments were encountered in this new breeding endeavor from 1954 to 1960. For example, a number of F_3 and F_4 progenies with earliness and productivity were derived from Early Fluff × Hopi Acala 46-124 crosses. The fiber properties approached the Acala 4-42 standard, but by 1955 they were found to be highly susceptible to Verticillium wilt. . . . Other "east-west" crosses provided segregates that could withstand moderate wilt conditions, but in most cases earlier maturity and/or fiber properties were sacrificed. When selective pressure in the F_2 to F_4 generations was primarily for early maturity, poor taproots were present. This resulted in a high water requirement. Frequent irrigation, in turn, accelerated Verticillium damage in late summer. Furthermore, attempts to recapture the desired tolerance to wilt resulted in late-maturing progenies that would defeat our early harvest objective."

The extensive hybridization program begun in the 1950s gave rise to AXTE-1, a strain of complex origin, which was subsequently crossed with New Mexico 2302 (2302 later was released in New Mexico as 'Acala 1517D'), giving rise to strain 12302, which was released in 1967 as 'Acala SJ-1'. Acala SJ-1 was the inaugural release of a series of "SJ" cultivars that have maintained the high quality reputation of San Joaquin Valley cotton and also established improved levels of performance for yield, wilt resistance, and earliness. The latest release in the series is 'Acala SJ-5', which became commercially available in 1978.

The third avenue for development of the Acala type centered in New Mexico (Staten, 1971; Ware, 1951). In 1922, W. T. Young in the El Paso Valley obtained seed of the Acala No. 9 stock from Ferris Watson, a Texas cotton breeder. Young mass selected the stock and developed 'Young's Improved Acala'. G. N. Stroman at the New Mexico Agricultural Experiment Station practiced plant-to-row selection in Young's Acala to produce 'Acala 1064', which was released in 1937. Reselection within Acala 1064 gave rise to Acala 1517 and 'Acala 1517A'. The Acala 1517 cultivar was popular because of its yielding potential, earliness, and fiber strength, but it was not resistant to the Verticillium wilt that became an increasing problem in the few years following its release. Breeders at the New Mexico Station and the USDA Field Station at Las Cruces, New Mexico, continued to screen for wilt tolerance within the Acala 1517 family; this effort resulted in release of 'Acala 1517WR' in 1946 and 'Acala 1517B' in 1949. Release of 'Acala 1517C' in 1951 represented a notable advance in yield over Acala 1517A and 1517B. Acala 1517C was well adapted throughout the El Paso marketing territory, and it was recognized as a high quality cotton especially well-suited for combed yarns. The initial release of Acala 1517C was lacking in wilt resistance, but substitution of two subsequent reselections effectively increased the degree of wilt tolerance of the Acala 1517C cultivar. Chronic appearance of bacterial blight [*Xanthomonas campistris* pv *malvacearum* (Smith) Dye] eastern New Mexico prompted the development of blight-resistant 1517 types, the first of which ('Acala 1517BR-1') was released in 1957. Race 1 resistance was transferred from Stoneville 20, and later development of the race 2 biotype rendered the cultivar susceptible. 'Acala 1517BR-2', released in 1961, carried dual-race resistance. In 1960, 'Acala 1517D' was released as a replacement for Acala 1517C. Exact parentage of Acala 1517D is not known, but one of the parental strains was thought to represent a degree of introgression of *G. barbadense* germplasm. In recent years the New Mexico program has produced several other new cultivars that retain the essential features of the Acala 1517 series, with earlier maturity and improved storm resistance. These include 'Acala 1517-70', 'Acala 1517-75', 'Acala 1517-77', 'Acala 1517E-1', and 'Acala 1517E-2'.

The Acala type occupies about 10% of the U.S. production area, concentrated in California and New Mexico, with minor production in Arizona and Texas. Although the California one-variety law has been rescinded, the existing "one quality" law effectively limits production in the San Joaquin Valley to the Acala type, mostly the Acala SJ-2 and Acala SJ-5 cultivars. In Arizona, Acala production is confined to the high altitude valleys in the eastern part of the state; in Texas, Acala cultivars are essentially confined to counties west of the Pecos River in Far West Texas.

7–3.2.2 Delta Types

As was noted previously, cotton cultivar development in the post-colonial, pre-boll weevil period resulted in an extensive diversity of cultivars, with many having a thread of common parentage. With the invasion of the boll weevil around 1900, a majority of the older cultivars became unsuited for production under weevil conditions. Ewing (1948) noted that cultivars previously grown in the Mississippi Delta, such as 'Black Rattler', 'Keeno', 'Allen Long Staple', and other Upland cultivars were slow in fruiting, late in maturity, and vulnerable to severe fruit damage by the boll weevil. Breeders began placing strong emphasis on development of faster-fruiting, earlier-maturing genotypes in which damage from the boll weevil might be minimized. Not all of the pre-boll weevil cultivars were discarded; a few were retained and used as parent stocks for either selection or hybridization.

Cultivars that primarily represent the contemporary Delta type were derived, at least in part, from germplasm of the older pre-boll weevil cultivars. The Deltapine and Stoneville series are prominent and successful examples of Delta-type cultivars.

The Deltapine series provides a good example of how the older, pre-boll weevil germplasm was utilized in fashioning of newer, more precocious cultivars adapted to the boll weevil conditions. Progenitors of the Deltapine series included the 'Express', 'Mebane Triumph', 'Sunflower', and 'Polk' cultivars. Both Express and Mebane Triumph are traceable to the 'Bohemian' stock that originated from an introduction into Texas from Mexico about 1860. Sunflower and Polk were among the latest of the Upland long-staple cultivars developed in Mississippi in the pre-boll weevil era. The Mebane Triumph × Sunflower combination that entered into the derivation of the Deltapines also established the 'Foster' cotton from which the Delfos series was derived.

The other major cultivar group in the Delta type includes the Stoneville, Empire, and Bobshaw families. Stoneville originated in Mississippi from a straight selection from 'Lone Star', a Texas cultivar that traces back through 'Jackson Round Boll' to an unknown "Stormproof" introduction from Mexico somewhere around the mid-1880s. Both the Empire and Bobshaw families originated from selections within Stoneville stocks. The Stoneville germplasm also gave rise to a number of other cultivar families, including 'Stonewilt', 'Dortch', 'Arkot', and 'Paula'.

The Delta type has been grown very successfully in the USA over the past 50 years or so, and cultivars of this type have occupied major portions of the rainbelt area from southern Texas to Alabama. Delta-type cultivars also are dominant in Arizona and the Imperial Valley of California.

7–3.2.3 Plains Type

The Plains type developed almost totally in the Texas and Oklahoma region of the Cotton Belt. Most of the contemporary cultivars of this type are characterized by relatively compact plant habit, relatively determinant fruiting habit, and either storm-resistant or stormproof bolls. Determinacy is considered necessary because of moisture and temperature factors that limit the effective growing season; storm resistance or stormproofness provides protection to open bolls until the entire crop is matured and ready for once-over harvest with stripper machines.

Cultivars of the Plains type predominantly trace back to the Big Boll Stormproof stocks introduced from Mexico about 1850, and most especially to the 'Texas Stormproof', 'Meyer', and Bohemian cultivars that were established within the next few years. Three Big Boll Stormproof stocks survived the boll weevil crisis and figured prominently in the development of early-fruiting cultivars suitable for production under boll weevil conditions. These were the 'Rowden', Jackson Round Boll and Mebane Triumph cultivars. Lone Star (selected from Jackson Round Boll), Mebane Triumph, and 'Half and Half' (selected from 'Cook') germplasms have been especially important progenitors of the present-day Plains type. Half and Half was the parent stock from which the 'Macha' cultivar was selected; the Macha stormproof boll has figured prominently in evolution of several series of Plains cultivars. The Paymaster family, developed in the High Plains of Texas, is partly traceable to Kekchi, an introduction from Guatemala in 1904. Today, the Plains type includes more than 40 named cultivars that are grown almost exclusively in Texas, Oklahoma, and eastern New Mexico, on about 2.8 million hectares, or about one-half of the total USA production area.

7-3.2.4 *Eastern Type*

Of the four major cultivar types, the Eastern possibly represents the widest diversity of germplasm. Lines of descent are traceable to certain prominent post-boll weevil cultivars (Lone Star, Delfos, Foster, and Express), with probable hybridization with various derivatives of pre-boll weevil stocks and introgression from Sea Island. Derivation of cultivars of this type has been influenced considerably by the need for resistance to Fusarium wilt (*Fusarium oxysporum* f. sp. *vasinfectum* (Atk.) Snyd. & Hans.), especially in the southeastern part of the Cotton Belt. Current or recent cultivar groups included in the Eastern type include Coker, Auburn, McNair, and Dixie King issues.

7-3.3 Exotic Germplasm

Fryxell (1976) reviewed the utilization of germplasm from the wild species and primitive forms for improvement of cultivated cotton. A notable example is the trispecies hybrid material derived from crossing *G. thurberi, G. arboreum,* and *G. hirsutum.* The Triple Hybrid germplasm has been used effectively as a source of high fiber strength in the Acala and Pee Dee breeding programs. Diploid species of *Gossypium* have contributed genes for bacterial blight resistance, resistance to cotton rust (*Puccinia cacabata* Arth. & Holw.), smooth leaf, cytoplasmic male sterility and fertility restoration. Germplasm from various tetraploid forms have been sources of blight resistance, nectariless, glandless, and high bug gossypol.

7–4 BREEDING APPROACHES

Cotton is considered to be a normally self-pollinating crop, although natural crossing may range from zero to more than 50%. Cotton pollen is heavy and sticky, and its transfer by wind is essentially nil. Transfer of pollen is effected by insects, primarily various wild bees, bumblebees (*Bombus* spp.), and honeybees (*Apis mellifera*), and it seems probable that other insects also may function as pollen vectors. Populations of pollen-vectoring insects are influenced by proximity of other vegetation, climate, and amount of insecticide spraying in the vicinity.

Control of pollination in cotton is relatively simple. Self-pollination can be done with use of cloth or paper bags, malleable wire, string, paper clips, cellulose acetate, or other devices that will effectively seal the flower bud and prevent entry of insects. Hybridization is also relatively easy to accomplish. Cotton has a complete flower in which emasculation is done by removing the staminal column, usually the day before flowering or in early morning before pollen is shed. Pollen from the male plant is placed on the stigma of the emasculated flower, and the pollinated stigma or flower is covered to prevent contamination and desiccation. With good emasculation and pollination technique, a crossed boll will yield 18 to 25 seed.

In a review of cotton breeding procedures and methods, Richmond (1951) discussed various considerations in cotton improvement and outlined certain breeding systems that had been utilized. He noted that cotton breeders do not follow rigid systems, but rather that considerable flexibility is dictated by breeding goals, finances, time, facilities, and the state of genetic knowledge about the species. The following section provides a general discussion of the principal breeding systems used in cotton improvement.

7–4.1 Methods

7–4.1.1 *Selection Within Existing Cultivars*

Intra-cultivar selection undoubtedly was the principal technique utilized in establishing adapted, productive genotypes from the accessions introduced into the U.S. Cotton Belt. The history of cotton cultivar development in the USA indicates that these introductions must have possessed considerable genetic variability, and plant selectors were successful in isolating a wide array of cultivar types through straight selection over many succeeding generations. Certainly, many of the introductions must have represented "land races" that harbored considerable genetic heterogeneity to support their adaptation to differing climatic and cultural situations.

ability was reduced, intra-cultivar selection became less effective, and breeders employed hybridization more extensively to generate new populations amenable to selection. The increasing complexity of breeding objectives and improved understanding of genetic mechanisms over the past few decades have encouraged breeders to employ more imaginative approaches in the task of crop improvement.

Excellent examples are available to illustrate the effective use of intra-cultivar selection in cotton cultivar development. The first two Pima cultivars grown in the USA were derived from selections within existing cultivars. Yuma was a selection from Mitafifi, an introduction from Egypt, and Pima, in turn, was selected from Yuma. In the Uplands, an excellent example of successful intra-cultivar selection is the development of the Stoneville series from the Lone Star progenitor, beginning with the primary plant selection in 1916. Over the succeeding 50 years or so, the Stoneville breeders have isolated a succession of cultivars, some of which became commercially significant. Obviously, the utilization of intra-cultivar selection does not preclude the possibility that a degree of hybridization has taken place by natural means.

7-4.1.2 Hybridization and Selection

Today, hybridization within and among types, followed by selection, is the predominant method utilized for improvement of cotton. The appeal of this breeding approach stems from the fact that intercrossing, generally, expands the range of genetic variability within which a breeder can practice selection. In most instances, a pedigree system of selection is used, beginning with the primary selections in the F_2 generation. Under certain circumstances, a breeder will use a bulk breeding method in which plant selection is delayed for a few generations, allowing for natural selection to act on successive populations.

The hybridization-selection method has been utilized effectively in many instances in the development of cotton cultivars. In breeding of the Pima types, hybridization within a type followed by selection resulted in the development of SXP, Amsak, and Pima 32. These three cultivars differed little from Yuma and Pima, which had been developed from selection within existing cultivars. This result reflected the relatively narrow germplasm base of the introduced Egyptian cottons from which the five Pima cultivars were derived (Feaster and Turcotte, 1962). Hybridization among types of *G. barbadense* and with *G. hirsutum* broadened the germplasm base sufficiently for significant improvement in plant type, production and maturity, while maintaining fiber quality. This improvement was realized with the development of Pima S-1, and subsequently in the development of Pima S-2, Pima S-4, and Pima S-5 (Feaster and Turcotte, 1970). Understandably, as the process of selection proceeded and genetic vari-

In *G. hirsutum* numerous examples exist to illustrate the effective use of hybridization and selection for the establishment of new cultivars; one such example is the development of the Acala SJ-1 cultivar in California (Turner, 1974), as outlined in Fig. 7-2. The Acala SJ-1 cultivar represented a successful synthesis of excellent adaptation and production (from the Acala P18C, Acala 29, and New Mexico 2302 stocks), tolerance to Verticillium wilt (from Missdel), excellent fiber quality (from New Mexico 2302), extra high fiber strength (from Triple Hybrid), and early maturity and high fruit retention (from 'Early Fluff').

7-4.1.3   Recurrent Selection and Gene Pools

Recurrent selection has not been utilized to any great extent for the improvement of cotton. Limited studies have demonstrated the effectiveness of recurrent selection techniques for altering population means while retaining genetic variability (Bilbro, 1961; Lewis, 1956), and for breaking up of linkage blocks and increasing genetic recombination (Miller and Rawlings, 1967a; Meredith and Bridge, 1971). The use of recurrent selection for yield improvement in cotton was reported by Miller and Rawlings (1967b), and Meredith and Bridge (1973) used a form of recurrent selection to improve lint percent. Richmond (1954) discussed the use of recurrent selection for fiber strength improvement in a trispecies hybrid.

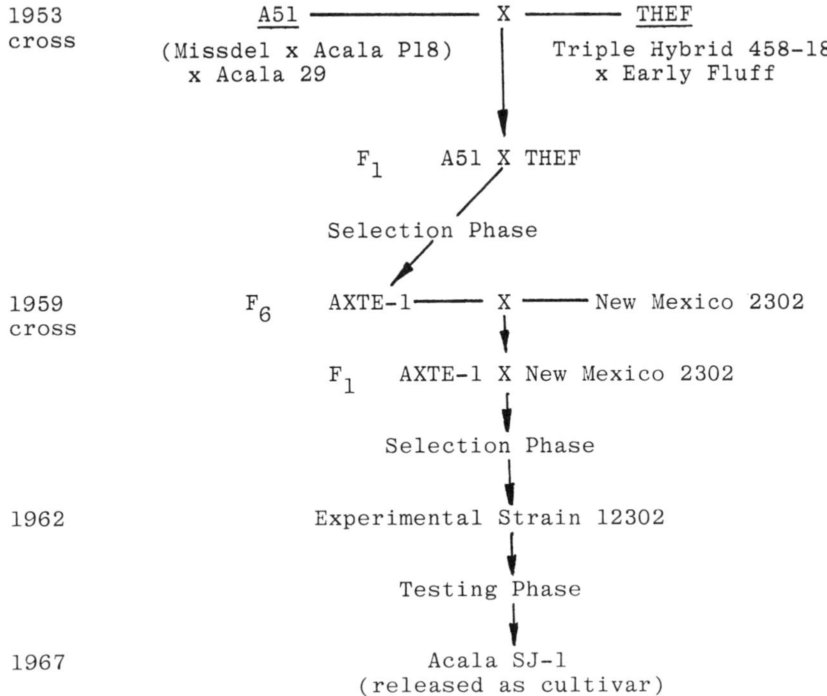

Fig. 7-2. Pedigree of Acala SJ-1. (Adapted from Turner, 1974.)

Gene pools have been established by various breeders working in both *G. barbadense* and *G. hirsutum*. These may be formed by various means, such as growing parental stocks under conditions of high natural crossing, through mass intercrossing of parentals, compositing, etc. Workers at the Lubbock Center of the Texas Agricultural Experiment Station for years have maintained populations designated as "composites." These were established through use of genetic male sterility to facilitate intercrossing. After several generations of intercrossing, the populations presumably represent an extensive reassortment of genetic variability, and the populations are ready for selection. Cotton breeders at the Pee Dee Experiment Station in South Carolina have utilized the germplasm pool approach for developing a series of breeding lines that represent significant improvement in the association of high fiber strength with lint yield (Culp and Harrell, 1974, 1977, 1981).

For *G. barbadense,* two major gene pools exist in the USA. One is maintained at a low elevation breeding site and the other at a high elevation breeding site. At the lower elevation, the pool includes a higher degree of heat tolerance in genotypes that begin fruiting relatively low on the plant and continue fruiting throughout the season. The pool maintained at high elevation does not have a high degree of heat tolerance and does not perform well at low elevation. The pool maintained at low elevation contains genotypes with general adaptability throughout the Pima belt. Fruiting height response has provided a breeding tool for evaluating adaptability at the various elevations (Feaster and Turcotte, 1965). The yield of Pima cotton is highly negatively correlated with fruiting height response under conditions of high night temperatures (low elevation), but the two characters are not correlated under moderate night temperatures (high elevation). The magnified expression of fruiting height, plant height, and production at the low elevation breeding site permits a selection of genotypes that are productive at each elevation (Feaster et al., 1980).

7-4.1.4 Hybrid Cotton

Heterosis in crosses within *G. hirsutum* (Hawkins et al., 1965; Jones and Loden, 1951) and in crosses of *G. hirsutum* and *G. barbadense* (Fryxell et al., 1958; Marani, 1967) has been well documented. Utilization of the heterosis through hybrid cotton was, at that time, impractical because hand crossing was required. Recently, the potential for commercial hybrid cotton production has received considerable attention since Meyer (1973a, b, 1975) detailed the mechanism for utilizing cytoplasmic-genetic sterility for hybrid cotton production and released germplasm necessary to utilize the mechanism. Meyer showed that the transfer of the *G. hirsutum* nuclear genome into *G. harknesii* cytoplasm produces complete male sterility ("A" line); fertility can be restored to male-sterile lines with *G. harknesii* cytoplasm by introducing either one dominant gene or a homozygous recessive gene from *G. harknesii* ("R" line). Pure-breeding male-fertile lines, which can restore male fertility to completely male-sterile lines, can be isolated from certain stocks with *G. harknesii* cytoplasm. Male-sterile lines are maintained by crossing with normal, fertile *G. hirsutum* stocks ("B" lines).

Weaver and Weaver (1977) modified Meyer's theory on genes involved in restoration. They found fertility restoration to be controlled by a single gene pair showing incomplete dominance, or, in some cases, a completely dominant gene, depending on the genotypic background of the parents. Considerable effort has been devoted to the extraction of good restorer lines from the originial material released by Meyer. Weaver and Weaver (1977) found that Pima cotton (*G. barbadense*) possesses a factor or factors that greatly increase the expression of the fertility restorer gene. Sheetz and Weaver (1980) suggested that the enhancer factor is controlled by a single gene expressing dominance. They found some indication that the enhancer may, in some cases, express incomplete dominance. They noted fertility in hybrid populations heterozygous for the restorer gene that possess the enhancer factor is comparable, if not superior, to that of the homozygous restorer lines that do not have the enhancer factor. Da Silva et al. (1981), through a conventional genetic study, indicated that the restoration of pollen fertility may be controlled by three genes. Weaver and Weaver (1979) observed a deleterious character, "cracked root," in populations carrying genes for fertility. They indicated that cracked root is controlled by a single gene pair expressing complete dominance in Upland stocks. The major genetic problem in producing cotton hybrids with the *G. harknesii* cytoplasmic-genetic male sterility system still appears to be absence of satisfactory restorer lines, especially under high temperatures.

Pollen transfer in the production of hybrid cotton also is an obstacle in many areas of the Cotton Belt (Moffett and Stith, 1972; Waller et al., 1981). Natural pollinators often are lacking and honeybees may not provide adequate pollination. The application of insecticides to control harmful insects reduces the effectiveness of native pollinators or introduced honeybees.

7-4.1.4.1 Interspecific Hybrids. Davis (1974, 1979a, 1979b) and Davis and Palomo (1980) evaluated interspecific hybrids involving *G. hirsutum* and *G. barbadense*. The hybrid vigor in *G. hirsutum* × *G. barbadense* F$_1$ hybrids is usually associated with late maturity and excessive vegetative growth, relative to reproductive development. Davis (1979b) determined lint yield and fiber properties of two interspecific hybrids between *G. hirsutum* and *G. barbadense*. Both hybrids showed significant heterosis for yield of seed cotton. The hybrids produced 48 and 42% more seed cotton and 33 and 26% more lint cotton, respectively, than the Upland parent. The adaptation of the hybrid was somewhat limited. Both hybrids had fiber that was slightly longer, finer, and about equal in strength to their *G. barbadense* parents. The extrafine fiber may be a problem in spinning performance.

Germplasm released by Feaster and Turcotte (1980) may help reduce the persistent problem of excessive vegetative growth in *G. hirsutum* × *G. barbadense* hybrids. Until recently, *G. barbadense* cultivars have been generally late-maturing and rank, but the 1980 germplasm release provided early-maturing and short-statured strains of *G. barbadense* that could be used in interspecific hybrids.

7-4.1.4.2 Intraspecific Hybrids. In a review of the hybrid cotton situation, Davis (1978) cited several authors that had reported heterosis for yield in intra-*hirsutum* hybrids. He noted one report from India that indicated a 138% increase in net yield of an F_1 hybrid over a commercial check cultivar. This was the greatest degree of heterosis noted; several other reports indicated heterosis for yield on the order of 17%. In a later report of cotton hybrids, Davis (1981) commented that there are problems in obtaining heterosis for yield in Upland × Upland hybrids, and Hess (1981) expressed the opinion that consistently superior yielding Upland × Upland hybrids do not exist at the present time. The obvious question that arises is whether intra-*hirsutum* hybrids offer enough heterosis for yield to make their production and use economically feasible.

It is likely that hybrids would offer advantages other than yield increases, and these possibilities enhance the potential attractiveness of the hybrid approach. For example, hybrids may offer an effective approach to problems of insect and disease resistance, plant and boll conformation, earliness, seedling vigor, etc. Unfortunately, little data are available to support this supposition. Generally, fiber properties in Upland × Upland hybrids are quite stable, comparable to mid-parent values, and do not show appreciable heterosis (Davis, 1978). Inter-cultivar *G. hirsutum* crosses appear to afford little potential for improvement of fiber properties.

Davis (1981) reported that at least 13 commercial firms were working on development of hybrid cottons, but that immediate prospects for a commercial F_1 hybrid are not optimistic. Restoration of fertility in the hybrid remains a problem; the matter of pollen transfer has not been adequately resolved, and there is an obvious need to develop information on combining ability of prospective parent stocks.

7-4.2   Current Trends in Cultivar Improvement

7-4.2.1   Plant Type

In general, the trend in cotton breeding is toward a diminution of plant size. The degree to which this size modification is acceptable is influenced both by yield potential and harvest methods. In regions where machine-stripping is the dominant method of harvest, Upland cotton producers tend to prefer relatively small plants with a fairly compact zone of fruiting. In certain instances, stripper-type cultivars show a semi-cluster fruiting habit. Where spindle machine picking is utilized, there appears to be less interest on the part of breeders to modify plant type to any significant degree. Machine-picking is the usual method of harvest in the higher rainfall regions of the South and Southeast, and in the irrigated areas of California, Arizona, and portions of New Mexico and South Texas. Production potential in these areas is greater than in the Plains region and relatively large plants with dispersed fruiting habit are needed to achieve this potential. In the Pima series, plant type has been modified appreciably in recent years. Since 1951, each succeeding cultivar release of the Pima series has been shorter and more compact, and they begin fruiting at a lower node on the main stem.

7–4.2.2 Earliness Complex

The introgression of Upland into Pima germplasm during the development of Pima S-1 apparently provided the basis for obtaining a continuing transgressive segregation for earliness of maturity. Each succeeding cultivar release of Pima has been earlier-maturing, with the most recent germplasm releases, 79-103 and 79-106 in 1980, being about 2 weeks earlier than Pima S-5, which was released in 1975.

In many Upland breeding programs, the earliness complex has acquired an increasing priority in recent years. This emphasis on earliness has been prompted, in part, by pest management considerations and a growing effort to increase production efficiencies by decreasing inputs of fertilizer, water, crop protectants, and energy. Properly, "earliness" should be viewed in respect to its components, which include seed germination and stand establishment, onset of squaring, onset of blooming, rate of blooming, boll retention, boll maturation period, and time of crop maturity. Each of these components apparently is subject to genetic modification, and they can be manipulated separately or concurrently, depending on which aspect of earliness merits the greatest priority in a breeding program. For example, Walker et al. (1977) showed that vulnerability of green bolls to boll weevil damage declines markedly after the boll reaches about 13 days of age. Thus, rate of blooming is an important determinant of how rapidly fruit can be set and how quickly the developing fruit crop becomes less subject to weevil attack.

Research conducted in Texas (Niles, unpublished data) demonstrates that selection for "earliness" can be overemphasized, in the sense that highly precocious genotypes can be selected in which inherent potential for yield and fiber quality development are seriously reduced.

7–4.2.3 Fiber and Seed Properties

Fiber quality is a major consideration in development of Pima cultivars. The trend is toward increased fiber length uniformity, higher fiber strength, and slightly coarser fiber. These properties are important in the spinning of fine yarns of high strength and good appearance. A high percentage of Pima cotton classes 36.5 mm in length.

Quality of fiber likewise is an important consideration in the breeding of Upland cultivars, although not to the same degree as in the Pima cottons. Among the Upland cultivars produced in the USA today, the Acalas produce the highest "quality" fiber, in that they are the longest and strongest (Anonymous, 1981b). Obviously, fiber properties receive strong emphasis in breeding Acala cultivars, as their "quality" reputation generally earns certain marketing advantages. In the medium-staple Upland cultivars, fiber quality factors are receiving renewed attention, especially for fiber strength. The emphasis on strength is prompted mostly by changes in textile technology which allow for more rapid processing that places greater physical stress on the cotton fiber. The increasing utilization of open-end spinning calls for certain modifications in raw fiber properties, especially increased strength and length uniformity. For the most part, only modest improvements of fiber length are desired in new cultivars, the principal concern being to retain an optimum balance between productivity and fiber quality.

Traditionally, seed have been a by-product of the production of lint, and their market value is only about one-fourth or one-third that of the lint. Considerable research has been done over the past 2 decades to develop productive cultivars in which seed are glandless, or low in gossypol. Gossypol is a natural plant compound that is localized in pigment glands throughout the plant, including the seed. The compound is toxic to non-ruminant animals and imparts undesirable properties to cottonseed oil. Removal of these glands by breeding essentially eliminates the gossypol, thereby increasing the potential use of cottonseed for animal feed and human food products. Glandless stocks have been developed by a number of cotton breeders, and many of these stocks are as productive as their normally glanded counterparts. Thus far, production of glandless cultivars has been very limited, and likely will remain so until adequate incentives are provided to prompt producers to switch from conventional cultivars to the glandless types.

Essentially nothing has been done to genetically improve other chemical properties of cottonseed, including oil, proteins, or amino acids, although limited research indicates adequate genetic variability exists to permit improvement (Kohel, 1978).

7–4.2.4 Environmental Stress Tolerance

Increasing attention has been given within the past few years to the matter of environmental stress and to the possible effectiveness of breeding for stress tolerance. Evidence is growing to demonstrate that genetic variability exists among cotton stocks for salinity tolerance, heat tolerance, and tolerance of moisture stress (Lauchli et al., 1981; Jordan et al., 1981; Peterschmidt and Quisenberry, 1981). Tolerance to high temperatures during the fruiting period has been increased with each succeeding commercial Pima cultivar release since 1960 (Feaster and Turcotte, 1962; Feaster et al., 1967, 1976).

7–4.2.5 Boll Size and Prolificacy

It seems unlikely that future cultivar development will involve any appreciable modification of boll size, since the range of boll size found in current cultivar types generally is quite compatible with harvest by machine. In circumstances where hand-harvesting prevails, boll size is a moderately important character to consider in cultivar development, and, within limits, increased boll size is a legitimate breeding consideration.

Viewed in a functional sense, the boll is a unit package of yield; thus, high yield is achieved when the size and number of bolls per unit area is maximized. Kerr's opinion (1966) that variation in boll size has played a relatively minor role in yield improvement in recent years suggests that prolificacy (i.e., number of bolls per plant or per unit area) is an important factor to consider in selection for improved yield. This view is supported by the yield models described by Maner et al. (1971) and Worley et al. (1976). Kerr (1966) also considered that, in respect to yield, there is an optimum range for boll components (including boll size), and that the optimum established in different cultivar types or groups of cultivars varies with environment, cultural conditions, and yield levels.

In the Pima types, boll size was increased with the development of Pima S-1. There was a slight decrease in boll size with development of Pima S-2, and boll size has remained relatively constant through the development of Pima S-4 and Pima S-5. Boll size for Pima S-5 averages about 3.5 g. The yield potential of Pima, particularly at lower elevations, has been increased significantly in cultivar releases since 1950.

7–4.2.6 Pest Resistance

Pathogens, nematodes, insects, and spider mites cause significant crop losses in U.S. cotton. For the 1980 season, diseases and nematodes caused an estimated yield loss of 10.3% (Anonymous, 1981c); yield reductions due to insect and mite pests was estimated at 8.7% (Anonymous, 1981d).

Breeding for host plant resistance is an important consideration in many public and private breeding programs. Meredith (1980) reported results of a survey of breeding priorities, which indicated that breeding for disease resistance is considered a more important objective than is breeding for insect resistance. Among diseases, Verticillium wilt was perceived as most important; *Heliothis* spp. was considered most important among the insect pests. The status of breeding for pest resistance has been reviewed by Bird (1980), Niles (1980), and Schuster (1980).

7–5 APPLIED BREEDING

7–5.1 Commercial

Of the specified cultivars listed in the 1980 USDA report on cultivars planted (Anonymous, 1980), more than 80% of the U.S. production area was planted to cultivars developed by commercial companies. The majority of private seed breeding companies employ competent scientists to direct their breeding programs, as well as to keep abreast of technical developments in plant breeding and its associated disciplines.

7-5.2 State

Involvement of state agricultural experiment stations in the development of commercial cultivars traditionally has been limited. In general, the state agencies do not view the development of commercial cultivars as a prime objective of cotton breeding programs, but rather that experiment station releases are justified when a potential new cultivar represents a distinct and significant improvement or satisfies a need that is not adequately addressed by the private sector.

7-5.3 Federal

Activities of the federal agencies in applied cotton breeding are quite restricted, as the responsibilities of these agencies are considered to be regional or national in nature. At the present time, the only applied cotton breeding program conducted by a federal agency in the USA is the Pima breeding program maintained by the Agricultural Research Service, USDA, at Phoenix, Arizona, in cooperation with the State Experiment Stations of Arizona, New Mexico, and Texas. The production of Pima is too small in any one state to justify an individual state Pima improvement program, and the total production is too small to attract commercial breeding efforts.

7-5.4 Plant Variety Protection

The Plant Variety Protection Act of 1970 was enacted to provide legal protection against unauthorized reproduction and sale of seed of sexually reproducing plants. Under the Act, a breeder can maintain exclusive rights and obtain royalties on seed sales. The breeder also may invoke the ''Title V'' provision that requires that only certified seed can be offered for sale. A special exemption is provided for farmers to save planting seed for their own use.

7-6 DEVELOPMENTAL BREEDING

Developmental breeding, as considered here, includes those activities concerned with the establishment of primary breeding stocks that possess new, improved or unique morphological, biochemical, or physiological properties of potential value in the construction of new cultivars. In effect, developmental breeding bridges the gap between genetics and applied breeding; it is to some extent ''basic'' in orientation, and at the same time somewhat ''applied'' in nature. Developmental breeding has been a very productive endeavor in cotton, and has contributed much to the evolution of cultivars that possess new or improved characteristics of importance to the cotton industry. In fact, many valuable plant traits that represented significant new developments in commercial cultivars were initially established, by developmental breeding, in primary breeding stocks that formed the parental germplasm for new cultivars. Examples include Stoneville 20 (resistance for bacterial blight, Race 1), Coquette (resistance to Verticillium wilt and, a parent stock of 'Acala 1517V'), Arkugo 4 (very early maturity, derived from an introduced Yugoslavian stock), CA 491 (a semi-dwarf, very compact and determinate strain utilized in breeding stormproof, stripper cultivars) and Pima 3-79 (a parent stock in the development of Pima S-5).

7-6.1 State

Developmental breeding activities of the state agencies are primarily aimed at establishment of breeding stocks with growth and performance traits that satisfy relatively local, or state needs. As an example, the developmental breeding program conducted at the Lubbock Center of the Texas Agricultural Experiment Station provides for the selection and evaluation of improved stormproof genotypes suitable for once-over machine harvest, with earlier maturity, improved fiber properties, and tolerance to Verticillium wilt. As such types are established and evaluated for their performance potential, they may be made available to other public and private breeders as developmental breeding stocks. Generally, these developmental breeding stocks are lacking in one or more attributes, and are not suitable for release as commercial cultivars.

7-6.2 Federal

A distinction between state and federal developmental breeding is not altogether discrete; it lies primarily in the relative importance and breadth of a specific problem area. Federal research in developmental breeding focuses mainly on problem areas of general interest across one or more states, or of beltwide importance. Breeding activities at the Boll Weevil Research Laboratory, for example, are oriented to the identification and establishment of weevil-resistance traits in relatively normal and productive genetic backgrounds that are suitable for refinement, selection, or hybridization by other breeders. In a similar way, the host plant resistance program of the USDA Entomology Research Center at Brownsville, Texas, is focused on the identification, incorporation, and evaluation of plant traits that condition resistance to the bollworm-budworm complex, as well as to other economically important insect pests of cotton throughout the U.S. Cotton Belt. Developmental breeding programs presently are conducted at USDA units located at Florence, South Carolina; Raleigh, North Carolina; Auburn, Alabama; Stoneville and State College, Mississippi; College Station and Lubbock, Texas; Phoenix, Arizona; and Shafter, California.

7-7 CULTIVAR MAINTENANCE

When a cotton cultivar is released for production, consideration must be given to the system by which it is to be propagated during succeeding years, without deterioration in its performance characteristics. Cultivar maintenance is an important operation in the process of cultivar development and distribution, and usually it is a responsibility of the originating breeder. The maintenance procedure may involve a substantial expense, in that it diverts personnel and facility resources away from the cultivar development activity. Cultivar maintenance also poses an element of risk, since poorly conceived or executed procedures may lead to cultivar deterioration. In the strictest sense, cultivar maintenance involves retention of the cultivar without genetic change. However, certain programs are designed to improve the cultivar during the maintenance process. An excellent discussion of cultivar maintenance concepts in cotton is presented by Lewis (1970).

7-7.1 Methods

The procedures to be employed for cultivar maintenance depend on the breeder's objectives, i.e. prevention of deterioration or improvement of the cultivar. If a cultivar is to be maintained, as is, with the least chance for genetic change, the most effective procedure is storage of breeder seed. At the same time that a cultivar is released, a designated amount of breeder seed is put into storage under conditions that minimize loss in seed viability. A portion of seed is removed periodically for use in starting another cycle of increase in the seed multiplication procedure.

Other procedures employed for cultivar maintenance involve some form of plant selection. Both roguing of off-type plants and type selection are utilized to preserve a cultivar when growers have little or no intention of improving it. In the first instance, undesirable or off-type plants are removed from the field; in the second, plants that best conform to the cultivar type are selected and their seed bulked to form a new supply of breeder seed. The modal bulk procedure outlined by Manning (1955) is a form of mass selection in which individual plants are selected and evaluated successively for certain yield components and fiber properties. At each level of selection, established criteria are used to justify retention of the individual selections. Seeds of plants that survive the successive selection levels are massed into a modal bulk. A plant-to-row system may be used in which individual plant selections are screened initially for desirable properties, and seed of the better ones is planted to produce progeny rows the following year. Rows that satisfy established standards set by the breeder for the cultivar are massed for a new supply of nucleus breeder seed. The plant-to-row system may be extended to provide for at least one cycle of replicated testing, after which seed of the superior strains are massed. Various forms of pedigree system are used for cultivar maintenance in cotton. In the first year, a large number of plants are selected; in the second year, progeny rows are grown from the individual plants that met established performance standards. The better progeny rows are identified, and these are advanced to multi-year replicated testing and gradual increase. Finally, a single strain, or a composite of strains, is used to form a new breeder seed source. The pedigree system has been utilized in various modified forms, with essentially the same ultimate practice of preserving the cultivar from one or more tested strains.

7–7.2 Implications of Maintenance Methods

The various procedures of cultivar maintenance should be evaluated in respect to their practical feasibilities (cost, labor, space, time, etc.) and how well they preserve the genetic integrity of a cultivar.

If the objective of a cultivar maintenance procedure is to preserve the cultivar without genetic alteration, the seed storage or seed bank system is the obvious choice. It would require the availability of suitable space in which the environment, either natural or controlled, adequately ensures seed longevity. The causes of cultivar deterioration would be eliminated, although a shift of gene frequency could occur if deterioration and loss of seed viability during storage does not occur at random in the population of seed. Apart from the costs associated with storage facilities, and perhaps periodic testing of seed viability, the seed bank system probably is the least expensive method of cultivar maintenance and the one that offers the least possibilities for adversely altering the genetic constitution of a cultivar during its maintenance. At the time that seed are placed into storage, a system of withdrawal must be programmed to assure a continuing supply of breeder seed over the expected life span of the cultivar.

Systems of maintenance that involve selection allow for considerable modification in procedure, as well as effect. Lewis (1970) noted that selection in the maintenance program is based on the assumption that the population derived from plants that are allowed to reproduce will be equal or superior to the population that might be generated if all plants were permitted to contribute to the ensuing generation. To some extent, any system of selection will result in changes of gene frequency in a population; the magnitude of this change will depend on the intensity of selection pressure. Presumably, roguing of off-type plants and type selection will result in least change of gene frequencies. Both methods should tend to increase frequencies of those genes that contribute the most to development of the phenotype that the breeder considers to be the ideal representation of the cultivar under maintenance.

As selection criteria become more objective and specific in governing the retention (or discarding) of plant selections or progeny rows, shifts in gene frequencies become greater. Further, as population size diminishes, inbreeding increases and genetic variability decreases. This loss of genetic variability may reduce the buffering capacity of a population and narrow its range of adaptability. Strict adherence to the pedigreed system will result in loss of heterozygosity and ultimately result in a population of homozygous plants; it severely limits the possibilities for further genetic improvement and increases genetic vulnerability.

REFERENCES

Anonymous. 1979. ICAC cotton production survey. Doc. 4 of the Secretariat, International Cotton Advisory Committee.

————. 1980. Cotton varieties planted, 1980 crop. USDA, Agricultural Marketing Service.

————. 1981a. Supplement for 1981 to statistics on cotton and related data, 1960–78. USDA, Economics and Statistics Service, Stat. Bull. 617.

————. 1981b. U.S. cotton handbook—1981. Cotton Council International and National Cotton Council of America, Memphis, Tenn.

————. 1981c. Cotton disease loss estimate committee report. p. 3. Proc. Beltwide Cotton Prod. Res. Conf., New Orleans, La.

————. 1981d. Report of the cotton insect loss committee of the thirty-fourth annual conference on cotton insect research and control. p. 136. Proc. Beltwide Cotton Prod. Res. Conf., New Orleans, La.

Bilbro, J. D. 1961. Comparative effectiveness of three breeding methods in modifying coarseness of cotton fiber. Crop Sci. 1:313–316.

Bird, L. S. 1980. Breeding for disease and nematode resistance in cotton. p. 86–100. *In* M. K. Harris (ed.) Biology and breeding for resistance to arthropods and pathogens in agricultural plants. Texas Agric. Exp. Stn. MP-1451.

Bryan, W. E. 1955. Prospects for Pima S-1 cotton. Progr. Agric. Ariz. 7:6.

Culp, T. W., and D. C. Harrell. 1974. Breeding quality cotton at the Pee Dee Experiment Station, Florence, S.C. USDA Res. Rep. ARS-S-30.

————, and ————. 1977. Yield and fiber quality improvements in Upland cotton (*Gossypium hirsutum* L.). South Carolina Agric. Exp. Stn. Tech. Bull. 1061.

————, and ————. 1981. Lint yield and fiber quality improvements in PD lines of Upland cotton. South Carolina Agric. Exp. Stn. Tech. Bull. 1081.

Da Silva, F. P., J. E. Endrizzi, and L. S. Stith. 1981. Genetic study of restoration of pollen fertility of cytoplasmic male-sterile cotton. Rev. Brasil. Genet. 4:411–426.

Davis, D. D. 1974. Synthesis of commercial F_1 hybrids in cotton. I. Genetic control of vegetative and reproductive vigor in *Gossypium hirsutum* L. × *G. barbadense* L. crosses. Crop Sci. 14:745–749.

————. 1978. Hybrid cotton: specific problems and potentials. Adv. Agron. 30:129–157.

————. 1979a. Yield and fiber properties of selected F_1 hybrids. p. 71–72. Proc. Beltwide Cotton Prod. Res. Conf., Phoenix, Ariz.

————. 1979b. Synthesis of commercial F_1 hybrids in cotton. II. Long, strong-fibered *G. hirsutum* L. × *G. barbadense* L. hybrids with superior agronomic properties. Crop Sci. 19:115–116.

————. 1981. Update on cotton hybrids. p. 26–27. Proc. Western Cotton Prod. Conf., Lubbock, Tex.

————, and A. Palomo. 1980. Yield stability of interspecific hybrids. p. 81–82. Proc. Beltwide Cotton Prod. Res. Conf., St. Louis, Mo.

Ewing, E. C. 1948. History of cotton varieties. Proc. Fifth Annu. Spinner-Breeder Conf., Greenville and Stoneville, Miss.

Feaster, C. V., and E. L. Turcotte. 1962. Genetic basis for varietal improvement of Pima cottons. USDA-ARS 34-31.

————, and ————. 1965. Fruiting height response: a consideration in varietal improvement of Pima cotton, *Gossypium barbadense* L. Crop Sci. 5:460–464.

————, and ————. 1970. Breeding methods for improving Pima cotton and their implications on variety maintenance. Crop Sci. 10:707–709.

————, and ————. 1980. Registration of American Pima cotton germplasm. Crop Sci. 20:831–832.

————, E. L. Turcotte, and E. F. Young, Jr. 1967. Pima cotton varieties for low and high elevations. USDA-ARS 34-90.

————, ————, and ————. 1976. Registration of Pima S-5 cotton. Crop Sci. 16:604.

----, ----, and ----. 1980. Comparison of artificial and natural selection in American Pima cotton under different environments. Crop Sci. 20:555–558.

Fryxell, P. A. 1976. Germpool utilization: *Gossypium*, a case history. USDA, ARS-S-137.

----, G. Staten, and J. H. Porter. 1958. Performance of some wide crosses in *Gossypium*. New Mexico Agric. Exp. Stn. Bull. 419.

Gilmore, J. 1918. Cotton in the San Joaquin Valley. Univ. of California Circ. 192.

Handy, R. B. 1896. History and general statistics of cotton. p. 17–66. *In* The cotton plant; its history, botany, chemistry, culture, enemies, and uses. USDA Exp. Stn. Bull. 33.

Hathorn, Scott, Jr. 1951. American-Egyptian cotton—an economic analysis. Arizona Agric. Exp. Stn. Bull. 238.

Hawkins, B. S., H. A. Peacock, and W. W. Ballard. 1965. Heterosis and combining ability in upland cotton—effect on yield. Crop Sci. 5:543–546.

Hess, D. C. 1981. Hybrid cotton development. p. 28–29. Proc. Beltwide Cotton Prod. Mech. Conf., New Orleans, La.

Hutchinson, J. 1959. The application of genetics to cotton improvement. Cambridge University Press, London.

Jenkins, J. G. 1953. Coastland—A new long staple cotton for the Southeast. Georgia Coastal Plain Exp. Stn. Bull. 53.

Jones, J. E., and H. D. Loden. 1951. Heterosis and combining ability in Upland cotton. Agron. J. 43:514–516.

Jordan, W. R., J. E. Quisenberry, and B. Roark. 1981. Contributions of heat tolerance and root growth potential to drought resistance. p. 40–41. Proc. Beltwide Cotton Prod. Res. Conf., New Orleans, La.

Kearney, T. H. 1930. Cotton breeding today works with main types known in remote past. U.S. Government Printing Office, Washington, D.C. p. 182–190.

----. 1943. Egyptian-type cottons: their origin and characteristics. Report of Div. of Cotton and other Fiber Crops and Diseases, USDA. Mimeo.

Kerr, T. 1960. The potentials of barbadense cottons. p. 57–60. Proc. 12th Cotton Improvement Conf., Memphis, Tenn.

----. 1966. Yield components in cotton and their interrelations with fiber quality. p. 276–283. Proc. 18th Cotton Improvement Conf., Memphis, Tenn.

Kohel, R. J. 1978. Survey of *Gossypium hirsutum* L. germplasm collections for seed-oil percentage and seed characteristics. USDA, ARS-S-187.

Lauchli, A., L. M. Kent, and J. C. Turner. 1981. Physiological responses of cotton genotypes to salinity. p. 40. Proc. Beltwide Cotton Prod. Res. Conf., New Orleans, La.

Lewis, C. F. 1956. Cotton Breeding Methods. Proc. Cotton Improvement Conf., Atlanta, Ga.

----. 1970. Concepts of varietal maintenance in cotton. Cott. Gr. Rev. 47:272–284.

Maner, B. A., S. Worley, D. C. Harrell, and T. W. Culp. 1971. A geometrical approach to yield models in Upland cotton. Crop Sci. 11:904–908.

Manning, H. L. 1955. Response to selection for yield in cotton. Cold Spring Harbor Symp. Quant. Biol. 20:103–110.

Marani, A. 1967. Heterosis and combining ability in intraspecific and interspecific crosses of cotton. Crop Sci. 7:519–522.

McGowan, J. C. 1961. History of extra-long staple cottons. Hill Printing Company, El Paso, Tex.

Meredith, W. R. 1980. Use of insect resistant germplasm in reducing the cost of production in the 1980's. p. 307–310. Proc. Beltwide Cotton Prod. Res. Conf., St. Louis, Mo.

----, and R. R. Bridge. 1971. Breakup of linkage blocks in cotton, *Gossypium hirsutum* L. Crop Sci. 11:695–697.

----, and R. R. Bridge. 1973. Recurrent selection for lint percent within a cultivar of cotton (*Gossypium hirsutum* L.). Crop Sci. 13:698–701.

Meyer, V. G. 1973a. Fertility restorer genes for cytoplasmic male sterility from *Gossypium harknessii*. p. 65. Proc. Beltwide Cotton Prod. Res. Conf., Phoenix, Ariz.

----. 1973b. Registration of sixteen germplasm lines of Upland cotton (Reg. No. G.P. 3 to G.P. 18). Crop Sci. 13:778.

----. 1975. Male sterility from *Gossypium harknessii*. J. Hered. 66:23–27.

Miller, P. A., and J. O. Rawlings. 1967a. Breakup of initial linkage blocks through intermating in a cotton breeding population. Crop Sci. 7:199–204.

----, and J. O. Rawlings. 1967b. Selection for increased lint yield and correlated responses in Upland cotton. Crop Sci. 7:637–640.

Moffett, J. O., and L. S. Stith. 1972. Pollination by honeybees of male-sterile cotton in cages. Crop Sci. 14:476–478.

Moore, J. H. 1956. Cotton breeding in the Old South. Agric. History 30:95–104.

Niles, G. A. 1980. Breeding cotton for resistance to insect pests. p. 337–369. *In* F. G. Maxwell and P. R. Jennings (ed.) Breeding plants resistant to insects. John Wiley and Sons, N.Y.

Peterschmidt, N. A., and J. E. Quisenberry. 1981. Plant water status among cotton genotypes. p. 43–44. Proc. Beltwide Cotton Prod. Res. Conf., New Orleans, La.

Ramey, H. H. 1966. Historical review of cotton variety development. p. 310–326. Proc. 18th Cotton Improvement Conf., Memphis, Tenn.

Richmond, T. R. 1951. Procedures and methods of cotton breeding with special reference to American cultivated species. Adv. Agron. 4:213–245.

----. 1954. Recurrent selection in a three-species cotton hybrid. Proc. Seventh Cotton Improvement Conf., Greenville-Stoneville, Miss.

Schuster, M. F. 1980. Insect resistance in cotton. p. 101–112. *In* M. K. Harris (ed.) Biology and breeding for resistance to anthropods and pathogens in agricultural plants. Texas Agric. Exp. Stn. MP-1451.

Sheetz, R. H., and J. B. Weaver, Jr. 1980. Inheritance of a fertility enhancer factor from Pima cotton when transferred into Upland cotton with *Gossypium harknessii* Brandegee cytoplasm. Crop Sci. 20:272–275.

Staten, Glen. 1971. Breeding Acala 1517 cottons, 1926 to 1970. New Mexico State Univ. College of Agric. and Home Econ. Memoir Series No. 4.

Stephens, S. G. 1975. Some observations on photoperiodism and the development of annual forms of domesticated cottons. Econ. Bot. 30:409–418.

----. 1976. The origin of Sea Island cotton. Agric. History 50:391–399.

Turner, J. H. 1974. History of Acala cotton varieties bred for San Joaquin Valley, California. USDA, ARS W-16.

Waddle, B. M. 1953. Progress in American-Egyptian cotton breeding. Proc. Spinner-Breeder Conf., Spartanburg, S.C.

Walker, J. K., J. R. Gannaway, and G. A. Niles. 1977. Age distribution of cotton bolls and damage from the boll weevil. J. Econ. Entomol. 70:5–8.

Waller, G. B., F. D. Wilson, and J. H. Martin. 1981. Influence of phenotype, season and time-of-day on nectar production in cotton. Crop Sci. 21:507–511.

Ware, J. O. 1937. Plant breeding and the cotton industry. p. 657–744. USDA Yearb., U.S. Government Printing Office, Washington, D.C.

----. 1951. Origin, rise and development of American Upland cotton varieties and their status at present. Univ. of Arkansas, Fayetteville. Mimeo.

Watts, G. 1907. The wild and cultivated cotton plants of the world. Longmans, Green and Co., N.Y.

Weaver, D. B., and J. B. Weaver, Jr. 1977. Inheritance of pollen fertility restoration in cytoplasmic male-sterile upland cotton. Crop Sci. 17:497–499.

Weaver, J. B., Jr., and D. B. Weaver. 1979. Cracked root mutant in cotton: Inheritance and linkage with fertility restoration. Crop Sci. 19:307–309.

Worley, Smith Jr., H. H. Ramey, Jr., D. C. Harrell, and T. W. Culp. 1976. Ontogenetic model of cotton yield. Crop Sci. 16:30–34.

8   Crop Growing Practices

B. A. Waddle
University of Arkansas
Fayetteville, Arkansas

Cotton has a reservoir of technology available to those who grow and harvest the plant as a commercial crop. How to selectively apply technology among diverse climates and soils by people of diverse skills and equipment is intended as the thrust of this chapter. What follows provides information to support the many decisions made or to be made by growers, and it is not an inventory of current practices or state of the art of growing cotton. A 20 000 ha corporate farm operation in California, USA, grows the same general type of cotton (*Gossypium hirsutum* L.) as a 1 ha family farm in remote parts of Africa. Appropriate technology is available to both farms. Application skills and available equipment, however, differ such that contrasting production practices obviously are followed.

Production practices on a large corporate farm and on a small family unit both reflect the grower's effort to fit his crop to the climate and soils in his area. Climate generally sets the geographic limits for cotton across the world. Cotton originated in the tropics, and the plant becomes inactive at temperatures below 15°C. Activity slows as this temperature is approached. The crop needs more than 160 days above 15°C. Almost any soil used for commercial production of any other field crop can be used to grow cotton if temperature and moisture are favorable. Soils used for cotton may require special preplant preparation and may need nutritional supplements. If temperature and soil are acceptable for cotton, the crop still may not be grown unless sufficient moisture is available. An acceptable cotton crop requires at least 50 cm of water during its growing season. This water must be resident in the soil or be provided by rainfall or irrigation. More than half the world's cotton crop is produced with essentially all its moisture provided by irrigation. Temperature, sunlight, rainfall, soil, supplemental nutrients, and water for irrigation are the major resources available to cotton growers. These are the key elements used by cotton growers to obtain control of the crop; that is, to have the crop develop as a planned progression.

Published in *Cotton,* Agronomy Monograph no. 24, © ASA-CSSA-SSSA, 677 South Segoe Road, Madison, WI 53711.

8-1 CLIMATIC RANGES ACCEPTABLE FOR COTTON

8-1.1 In the USA

Cotton in the USA is grown south of the isotherms giving 200 frost-free days. The area lies south of 36° N Lat except for a small area below the junction of the Ohio and Mississippi rivers, a small area in central Oklahoma, and the northern portion of the San Joaquin Valley in California where cotton is grown close to 37° N. There is a direct relationship between potential yield and the number of growing days. Recently, Young et al. (1980) equated growing days to the "heat units" or "day-degrees," but for this discussion the term growing day will be used to represent the number of cumulative hours above 15°C expressed in units of 24, assuming adequate growth requirements being met otherwise. A working relationship between growing days and yield is shown in Fig. 8-1.

Different production regions of the USA vary in production costs and in other crop options (Anonymous, 1966). Minimum acceptable yields can be computed for cotton in the several production regions. These values vary from 0.75 bales/ha in parts of Texas and Oklahoma to 10.0 bales/ha in parts of California. Therefore, the minimum number of growing days needed for cotton depends upon minimum levels of acceptable yield. The rough relationship shown in Fig. 8-1 indicates as few as 120 growing days

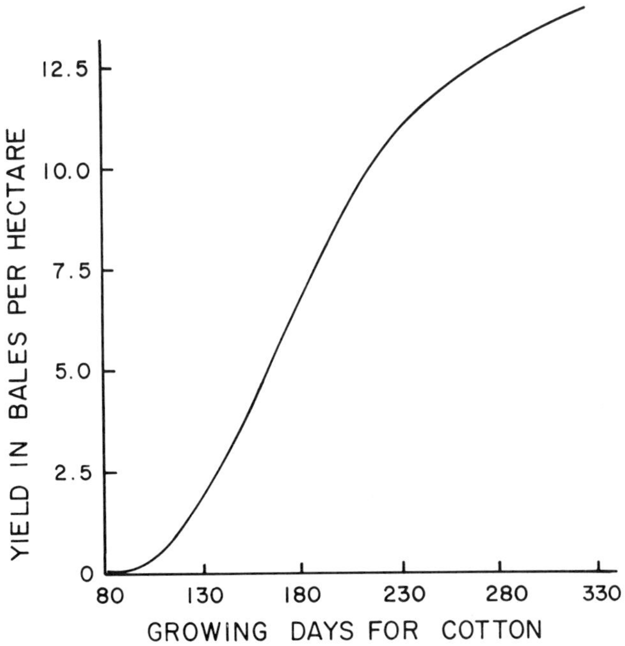

Fig. 8-1. Relationship of yield and growing days, assuming neither moisture nor nutrients as limiting.

may be acceptable in some regions while more than 200 may be required elsewhere. A fallacy in giving primacy to growing days is the disparity of impact on yield between growing days acquired after peak bloom and those acquired before peak bloom. All cottons require approximately the same number of growing days to reach first bloom regardless of where the cotton is grown.

Another factor limiting cotton yield is the total amount and distribution of rainfall during the growing season. A minimum of approximately 50 cm of moisture is needed to mature the crop even at the low acceptable yields of 0.75 bales/ha. Normal rainfall approaches 150 cm in the Southeast; 120 cm in the Delta; 50 cm in the High Plains of Texas and Oklahoma; and 15 cm in California and Arizona. Normal rainfall provides the total moisture needs of cotton generally east of 100° W Long with supplemental irrigation used effectively where available. This area frequently is referred to as the Rain Belt of the USA. As suggested by Raney and Cooper (1968), there is a water deficit sufficient to reduce yields at some time each season wherever cotton is grown east of 100° W Long. The risk becomes greater as one moves from east to west. Essentially no cotton is grown west of 100° W Long without irrigation.

In some coastal areas of the Mid-South and the Southeast there is too much rain, in general, for cotton. All cottons grown in these regions must be harvested ahead of excessive rains that occur in late autumn along the Texas coast and in the late fall elsewhere. Most rainfall during the growing season across the Cotton Belt occurs in thunderstorms, sometimes accompanied by violent winds and damaging hail. Runoff during thunderstorms makes total rainfall less meaningful. Extended periods of relative humidity above 90% adversely affect fruit set and potentially reduce yields. Coastal areas and the lower valleys of Arizona and California (Yuma–Brawley–Indio) have annual periods of extended high humidities in the late summer. Regularly occurring periods of adverse moisture have been neutralized with crop management decisions on cultivar to be planted, date of planting, and irrigation schedules.

Sunshine is vital to cotton, according to Doyle (1941), and areas with more than 50% cloudiness are not suited for cotton, regardless of temperature and moisture. Areas growing cotton in the USA vary in cloudiness from less than 10% to a little more than 30%. Solar energy is essential for a rapid warming and drying of soil in the spring. According to Raney and Cooper (1968) the heat required to raise the temperature 1°C is five times greater for water than for an equal weight of an average soil. This point relates to the need for effective field drainage systems and seedbed preparations that extend the number of cotton growing days.

Rapidity and consistency of spring warming determines where cotton can be grown successfully. Within the past 20 years new areas with water for irrigation have been developed in high valleys of southwest Texas and Arizona only to fail because of unexpected late frosts in the spring and early freezes in the fall.

Considerable evidence is accumulating to assign "smog" a significant role in reducing yield, especially in parts of the San Joaquin Valley of California. Smog can be considered as an interaction of man with climate and may have a wider impact on cotton in the future.

8–1.2 In Other Countries

Outside the USA cotton is grown under the same climatic constraints as described above. Areas of cotton production extend up to 42° N Lat in the USSR (Uzbek, Tadjik, and Kazakh Republics), but prior to 1965 extended to 47° N Lat in the area above the Black Sea. After 1965 all cotton in the USSR has been produced in the east under irrigation. Essentially all of the cotton grown in the USSR is grown between 37° and 42° N Lat. This region approaches 180 growing days annually being warmed by what some call the Mediterranean or Sahara winds. The only other world cotton area approaching 40° N Lat is above Peking, PRC. Relatively little of the world's cotton is produced in the southern hemisphere. Cotton is grown near 30° S Lat near Narrabri in Australia and in northern Argentina. All cotton is grown under similar climatic and soil constraints. Cotton is a "heat loving" plant, but more than 50% of the world crop is grown in temperate zones above 30° N Lat.

Large regions of rain-grown cotton are to be found in the PRC, India, Pakistan, African Republics, and Brazil. Rainfall varies from 50 to 150 cm. Parts of India, Pakistan, and Africa experience periodic droughts, and yields of cotton are very low in those years when droughts occur.

In some areas cotton is grown successfully under excessive rainfall. In Central America, for example, cotton is planted during the rainy season that may accumulate more than 250 cm of rain after planting. These plantings are in soils of recent volcanic origin and are well drained internally. In other areas cotton is grown with less than 50 cm of rainfall by utilizing cultivars and production practices that effectively use all the rain that falls to produce a yield of as little as 0.5 bales/ha. Yield and market potentials of other crop options for these regions are such that cotton is the more common land use even at these low levels of yield.

8–2 SOILS ACCEPTABLE FOR COTTON

8–2.1 In the USA

Any soil used effectively to grow any row crop in the USA below 37° N Lat can also be used to grow cotton, except where flooding restricts the number of days that equipment can have excess to the field. The basic needs of cotton from the soil are water, oxygen, available nutrients, and anchorage for roots.

Soils provide the cotton crop's only source of water between rains or between irrigations. Soils vary greatly in water holding capacity and in water movement potentials. The most desirable soils, in terms of moisture relationships with the growing plant, are the silt loams. Sands and clays must have special attention. A preplant irrigation for cotton on a sandy soil, for example, will require a second irrigation by the second week after first flower buds (squares) are visible. In contrast, a silt loam soil should hold until after first bloom, or 2 to 3 weeks later. Clay soils require special water delivery systems to avoid saturation and oxygen stress. The amount of water available for cotton in any soil is a function of rooting depth, soil texture (storage indices), and resistance to evaporation losses.

According to Brown and Ware (1958), cotton has been grown on more than half the 36 great soil groups recognized at that time in the USA with a major part of the crop being grown on six of these groups: The Red-Yellow Podzolic, Rendzina, Alleuvial, Reddish Chestnut, Reddish Prairie, and the Reddish Brown Lateritic soils. The group classification as cited by Brown and Ware (1958) has been revised and described in Southern Cooperative Series Bulletin No. 174 as edited by Buol (1973). Under the new classification, cotton is now being grown on seven major orders or taxa: Entisols and Aridosols in California, Arizona, and the upper Rio Grande cotton areas; Mollisols and Alfisols in the plains areas of Texas and Oklahoma (Reddish Prairie and Reddish Chestnut and some Rendzina of the old classification); Inceptisols along the major river systems in the Rain Belt (Alleuvials); Ultisols in the coastal plains from Texas to Virginia (Red-Yellow Podzolic); and Vertisols as prairie islands in the coastal plain forests (Rendzina).

The Entisols and Aridosols are relics of the Ice Ages. These are very fertile soils with a high organic matter content preserved through the centuries by their desert ecosystem. They require irrigation to grow cotton.

The Mollisols and Alfisols are an uplifted ocean floor that support short and tall grass prairie ecosystems with enough rainfall to give some degree of weathering, especially toward their eastern borders.

The Inceptisols reflect the geology of their watersheds. Those along the lower Red, Arkansas, and Mississippi rivers are especially productive for cotton. Alleuvials along rivers arising in regions of the Ultisols, however, are not generally suited for cotton.

The Ultisols are of marine origin with rain sufficient to support hardwood and pine forest ecosystems. These soils readily erode and leach if left unprotected and generally are low enough in pH to require liming to avoid toxic levels of manganese and aluminum. These soils respond well to supplemental fertilization.

Vertisols are also of marine origin but have chalk outcroppings or underlay. These soils supported a prairie ecosystem as large islands in an otherwise forested land.

Detailed descriptions, properties, and response potentials of all major soil orders and suborders are to be found in Handbook No. 436, published by USDA, SCS in 1975. In considering any specific soil as a medium for growing cotton, these descriptions may alert growers to potential problems associated with internal drainage, erosion, pH, and attending toxic materials, salt potentials, etc. Detailed soil maps are now available that show specific series and associations. Certain of these soil series are preferred for cotton as indicated in Table 8–1.

Cotton can be grown on most soils that lie below 37° N Lat in the USA, but the question of economic feasibility rests largely with the properties of the specific soil in question. These basic properties largely determine how well the soil in question can meet the nutritional needs of the cotton plant.

8–2.2   Outside the USA

Cotton is grown outside the USA on soils similar to those in the USA. Major cotton areas are found in USSR, PRC, India, Pakistan, the emerging nations of Africa, the Eastern Desert countries of Syria, Turkey, Lebanon, and Israel, UAR (Egypt), Brazil, other South American countries, Central America, Mexico, Australia, Malaysia, and Indonesia.

Table 8–1. A listing of the more productive soil series used for cotton in the USA by production regions.

Western†	Code#	Tex-Okla‡	Code#	Mid-south§	Code#	Southeast¶	Code#
Merced	14	Yahola	5	Commerce	7	Marlboro	13
Milham	15	Willacy	9	Norwood	8	Norfolk	13
San Emigdio	16	Abilene	9	Convent	6	Varina	13
Hanford	17	Tillman	11	Gallion	2	Faceville	13
Pond	18	Amarillo	1	Dubbs	2	Iredale	2
Panoche	19	Port	10	Dundee	3	Cecil	12
Dinuba	18	Canadian	10	Tutwiler	2	Orangeburg	13
Chino	14	Tipton	9	Rilla	2	Congaree	8
Temple	20	Clairemont	5	Memphis	2	Lloyd	3
Foster	14	Harkey	7	Loring	4		
		Pullman	11	Herbert	3		
				Bosket	2		

† Kent Brittan and Tom Kerby, California.
‡ Bob Metzer, Texas and Laval Verhalen, Oklahoma.
§ Tom Burch, Louisiana, Stan Chapman, Arkansas, Phil Hoskinson, Tennessee.
¶ Lawrence Harvey, South Carolina and Alan York, North Carolina.
# Codes refer to following Soil Orders and Subgroups:

Alfisols	Entinsols	Mollisols	Ultisols
1. Paleustalfs	5. Ustifluvents	9. Argiustolls	12. Hapludults
2. Hapludalfs	6. Fluvaquents	10. Haplustolls	13. Paleudults
3. Ochraqualfs	7. Torrifluvents	11. Paleustolls	
4. Fragiudalfs	8. Udifluvents		

Aridosols			
14. Haploxerolls	17. Xerortheuts	20. Haplaquepts	
15. Haplargids	18. Haploxeralfs		
16. Xerofluvents	19. Torriorthents		

One major order of soil in terms of bales produced is Aridosols (Sierozems) found in the USSR and western PRC. These soils require irrigation, are subject to wind erosion, and have drainage and salinity problems. Other major soil orders in terms of land area where cotton is grown involves Ultisols, Alfisols, and Vertisols found in eastern PRC, India, and Pakistan, the emerging nations of Africa, Brazil, and Australia. These soils are similar to U.S. counterparts and have varying degrees of weathering. Cotton is grown on Entisols in the high valleys of southeastern USSR, eastern PRC, and desert countries of the Near East. Inceptisols are used for cotton production in river valleys of eastern PRC and India-Pakistan. The old classification of "Oxisols" as described by Brown and Ware (1958), are considered generally as Mollisols (Argiustools, Haplaustalfs, Ustics, and Udics). These are the excellent cotton soils of the Pacific Coastal Plain in Central America, and parts of Africa. They are pseudo-volcanic in origin, have excellent internal drainage and tolerate excessive rainfall but are droughty when the rainy season ends.

Lint yields from the soils above vary from more than 2000 kg/ha to less than 150 kg/ha. These yields reflect the basic soil properties that provide the growing cotton plant with moisture, oxygen, available nutrients, and root anchorage. Moisture from rainfall can be supplemented with irrigation. Oxygen concentration can be improved by tillage and maintained with adequate field drainage. Nutrients can be added in commercial forms. In this context, the soils available for cotton grown outside the USA attract the same kinds of cultural practices as those within.

8-3 SOIL ORGANIC MATTER

8-3.1 Natural Levels

Soil organic matter is portrayed by Albrecht (1938) as one of our most important national resources. It is a renewable resource but is easily consumed by exploitation. It is a natural component in soils used for cotton and has uniquely supported high levels of cotton production in the past and, more significantly, will directly affect levels of cotton yields and those of alternate crops in the future.

The soil orders named above as being developed under prairie and desert ecosystems are relatively high in organic matter (2 to 4%). Those originating under a forest ecosystem are, today, generally low in organic matter content (0.50 to 1.25%). When the native land cover is removed and the soil is tilled, each successive tillage accelerates the breakdown of the resident soil organic matter, especially where soil moisture is adequate to sustain the microflora in an active phase. In a majority of cases an equilibrium is finally reached. Cotton is grown below the 200 frost-free day isotherm, and this factor relates to organic matter. Jenny (1930) reported that, other things being equal, each drop of 10° C in average annual temperature was associated with a two- to three-fold increase in native soil organic matter. Cotton, therefore, has been grown on soils lower in organic matter than corn and wheat. This relationship of organic matter in soils and temperature has often led to the erroneous conclusion that cotton is the cause of the lower organic matter soils of the South.

The cotton area historically described as The Old South in the USA has experienced serious soil erosion and leaching, and today its lands are used rarely for cotton or any other row crop. Cotton frequently has been blamed for the soil's deterioration; the most serious loss being in organic matter. These soils are in the Ultisol group and would have experienced this deterioration with any row crop once the forest cover was removed and the plow inserted (Brown and Ware, 1958). According to Raney and Cooper (1968), these soils can be recovered for effective cotton production with liming, and other mineral supplements, crop rotation, and other soil management practices.

The level of organic matter in the soil is more highly correlated with yield, according to Albrecht (1938), other things being equal, than is water holding capacity, soil texture, and mineralization potential. New importance has been assigned to soil organic matter as cotton production has moved toward totally mechanized production. Pesticides have become an essential part of the emerging production system. The degradation of applied pesticides is often a function of soil organic matter. Data compiled by Meredith (1982) leaves little doubt that regional average cotton yields in the Mid-South have declined slightly since 1965, the year generally accepted as the beginning of total use of herbicides across the area. Part of this decline in average yield has been assigned by Frans et al. (1982) to soil accumulations of herbicides.

Another function of soil organic matter is to provide feedstocks for specific microflora components that are capable of suppressing epidemics of certain pathogens capable of inciting serious losses in cotton. A common grower experience, for example, is to see efforts to increase yields by building up soil organic matter with a green manure crop nullified in the subsequent cotton crop by seedling diseases. If the level of organic matter in the soil is high enough, antagonist against several major soil inhabiting pathogens are able to maintain such a microflora balance that no economic loss is caused by these pathogens. Unfortunately, technology has not defined adequate levels of organic matter for the several soil types used for cotton.

Traditional benefits of soil organic matter such as soil tilth or favorable soil aggregation, buffering against low pH, slow and balanced soil mineralization, favorable water transmission potentials, resistance to erosion, etc. as covered by Nikiforoff (1938), remain as values in cotton growing areas. Soils used for rain-grown cotton are experiencing a serious decline in soil organic matter. It is well to consider grower options designed to arrest the decline, to hold the status quo, or, in a few situations, to increase soil organic matter.

8-3.2 Organic Matter Management

Organic matter content of soil can be temporarily increased by physical additions to the soil, by green manure crops, by cover crops, and by specific crop rotations.[1] As emphasized by Raney and Cooper (1968), it is not practical to consider procedures that temporarily increase organic matter in a soil if the only concern is available nitrogen and other nutrients. The dominance of the short-term economic evaluation being forced upon U.S. cotton growers prevents any serious consideration of procedures listed briefly below.

Manure from domestic animals is a proven procedure for increasing soil organic matter, according to White (1896). The Tarahumara Indians of the Sonoran Desert of Mexico still consider a person's wealth as a function of the number of animals owned because of the manure supply that animals generate. Manure remains a viable soil-building activity in parts of the PRC, India, Pakistan, and certain of the emerging Republics of Africa. Vegetable compost, gin trash, rotted sawdust, etc. also have proven values as soil builders but are not used at the present time because of costs of transport and spreading. Cotton disease and weed problems sometimes are caused by spreading gin trash directly from the gin, but not if the trash is properly composted before spreading (Griffis and Mote, 1978). Other kinds of problems are sometimes caused by the spreading of such materials as sawdust and rice hulls. Generally, the use of manure and vegetative materials are not among the economically viable options employed by U.S. cotton growers.

8-3.2.1 *Green Manure and Cover Crops*

Green manure crops grown for the purpose of being plowed under as a soil building practice have long been used effectively by cotton growers (White, 1896). Government subsidies in the USA at various times have encouraged growers to utilize green manure crops, with grower use terminating often when the subsidy is terminated. Cotton grown on Ultisols, having less than 1% organic matter, can utilize green manure crops effectively in the current economic situation if a 3- to 5-year land use agreement can be negotiated with land owners. One Arkansas demonstration field, approximately 10 ha in size, was used for vetch and sudan green manure double crop in 1978 and again in 1979. In 1980, when planted to cotton, this field made significantly more cotton than the cumulative yields of a comparable control field that was continued in cotton with recommended practices (C. Denver, personal communication). The field contained 0.45% OM when the demonstration began, and this was essentially unchanged (0.50% OM in 1981) the second year in cotton, again with increased yield.

Green manure crops may be more compatible with the current economic situation when used with other crops in rotations with cotton. However, there are problems when vetch as a green manure crop followed by soybeans is used as a rotation ahead of cotton. Apparently, on low organic matter soils, the vetch crop increases the hazard from nematodes on soybeans, and the cotton crop following soybeans does not reflect a residual gain (Hinkle and Fulton, 1963).

In Texas (Mollisols and Alfisols) legumes such as guar, annual sweetclover, biennial sweet clover, and cowpeas have proven records as green manure crops. In other parts of the Rain Belt (Ultisols) the legumes vetch, cowpea, mung bean, crimson clover, lespedeza, and Crotalaria, and some non-legumes have been successful green manure crops as described by Pieters and McKee (1938). Crotalaria is so well adapted to these soils that it has become a noxious weed in soybeans. Nonlegumes, mainly sorghums, have also been used as green manure crops in all areas east of 100° W Long.

Cover crops are grown in all areas of the Rain Belt primarily to provide ground cover to reduce soil erosion during the winter months. They can be shredded and the resulting residue left on the soil surface to give protection beyond spring planting. Many crops considered for green manure also can be considered as a cover crop. Winter grains and grasses can serve as cover crops equally as well as others listed above. Cover crops are fall seeded in standing cotton stalks and disked lightly, or they may be drilled with no other tillage. They may be seeded ahead of defoliation and harvest to let dropping leaves provide sufficient cover to obtain an acceptable stand. One effective means of spring disposal of cover crops is to shred the standing cover crop and allow cut vegetation to wilt fully. Then disk lightly ahead of disk bedders and allow beds or ridges 3 to 5 days to "set". Beds can then be prepared for planting, and a preplant herbicide can be applied and incorporated lightly. In some cases a desiccant may be needed at planting time.

Essentially the same crop disposal procedure described for cover crops can also be used for green manure crops since the value of these crops in restoring or maintaining organic matter should be the same as if the crops truly were plowed under. If either standing cover crops or green manure crops are plowed under, a 2-week or longer waiting period is required before planting. Even with waiting there is a risk of stand loss in the subsequent cotton plantings (Hinkle, 1969).

More than 25 years of continuous cotton preceded annually by vetch as a green manure crop was reported from Louisiana by Melville and Rasburg (1980) with only a small but significant change in the level of soil organic matter but with sizeable increases in cotton yields. Nine successive cotton crops preceded by vetch and rye as a cover crop in Arkansas have now given a second successive year of advanced yields over a comparable control (B. A. Waddle, unpublished data). No appreciable change has been recorded in soil organic matter and the yield increases may have been related to increased nitrogen availability.

Changes in costs of applied forms of nitrogen and in federal legislative developments (U.S. Clean Water Act) concerning field run-off may shift interests again to cover crops, especially in the Ultisols. A number of ongoing evaluation sites have been established in the cotton part of the Rain Belt. The energy situation is also undergoing change, and this will influence the costs of nitrogen and may also encourage use of green manure crops.

8-3.2.2 *Crop Rotation*

Crop rotation has been recommended for all U.S. cotton growing areas, including the Entisols and Aridisols of western states. Reasons by Leighty (1938) to justify crop rotation were (a) better maintenance of soil fertility, (b) preventing soil erosion, and (c) improved pest control, including insect, disease, and weed pests. To these can be added (d) alternating families of herbicides to reduce potential soil residues. In some rotation studies significant yield gains have been reported for cotton. In others no significant yield differences were recorded (Spurgeon and Grisson, 1965; Hinkle, 1969). Many of the reasons for crop rotation given above have been supported. Cotton-corn-cotton, cotton-soybean-rice, and alfalfa-alfalfa-alfalfa-cotton-cotton, and others are proven rotations for Ultisols and Inceptisols. In Mollisols, Alfisols, and Vertisols proven rotations are cotton-sorghum-wheat and, where irrigation can be utilized, cotton-corn-wheat and alfalfa-alfalfa-cotton-cotton. Rotations for the Entisols and Aridisols of the western USA are cotton-safflower-cotton, cotton-barley or wheat-cotton.

Crop rotation can be justified in most cotton growing areas, but U.S. cotton growers are reluctant to rotate cotton because rotation complicates cotton production and poses an extra challenge to management. Crop rotation is a cost-effective soil management practice.

8-4 SOIL CHEMISTRY AND MANAGEMENT

8-4.1 Soil Tests

Soils used for cotton vary in native fertility. Most cotton growers have confidence in soil test indications of needed nutrients and in industry's capacity to deliver soil supplements to fill that need. A soil test indicates nutrient levels estimated for the sample involved. To have a soil sample predict the subsequent cotton crop's need for supplemental nutrients requires more than the direct chemical analyses.

Geologic origin, age, and degree of weathering delineate major soil groups. Local variations set limits that are called soil types and are given names. The soil series from which a sample is taken indicates characteristics of the several vertical strata that subsequent cotton roots will penetrate. Soil series also allow us to set expectations for organic matter content, water retention, movement, release potentials, and exchange capacities. All of these factors are considered in interpretations offered for a soil test result.

Cropping history and anticipated yields if nutrients are not limiting are jointly considered with soil series to develop a fertilizer recommendation for a specific field. Each recommendation is backed by numerous tests designed to give response correlations. Consequently, an experienced soil testing laboratory recommendation can be followed with considerable confidence (Milsted and Peck, 1973).

8-4.2 Amounts Needed

The three major supplements for cotton are nitrogen; phosphorus (P_2O_5); and potassium (K_2O). Calcium (CaO) and magnesium (MgO) are used by cotton in lesser amounts and normally are resident in adequate quantities in the soil. Cotton yielding 2.5 bales/ha removes approximately 40 kg/ha of nitrogen, 16 kg/ha of P_2O_5, 17 kg/ha of K_2O, 7 kg/ha of MgO, and 4 kg/ha of CaO (Berger, 1969). With 3.75 bales/ha yields, 62, 25, 26, 11, and 6 kg/ha of nitrogen, P_2O_5, K_2O, MgO, and CaO, respectively, are removed. A yield of 7.5 bales/ha removes approximately 125, 50, 52, 22, and 13 kg/ha of the same nutrients. Berger's (1969) estimates indicate that 260% of nitrogen and P_2O_5 removed is needed in the soil to grow the crop as compared to 470% of the K_2O removed is needed to grow the crop. For a minimal yield of 2.5 bales/ha, the soil must deliver approximately 100 kg nitrogen, 50 kg P_2O_5, and 80 kg K_2O as well as adequate amounts of other nutrients. These amounts must be resident in the soil or must be added as supplements.

The nutrients intercepted by cotton roots are derived from weathering of primary minerals, decomposition of organic matter, seepages, depositions from the atmosphere, and from grower applications. Roots of cotton occupy only a 1 to 2% of the soil volume of the root zone according to Corey and Schulte (1973). Factors affecting nutrient uptake by cotton roots are oxygen, temperature, ionic antagonism, and toxic substances. Cotton roots slow their uptake of nutrients and moisture when oxygen content of the soil falls below 10% or when soil temperatures in the root zone drop below 15°C. Huck (1970) places the terminal point for activity for cotton roots at 3 to 5% oxygen. The soil's capacity to supply oxygen increases as the soil dries, giving emphasis to field drainage and hard pan removals in areas receiving summer rainfall. Ionic antagonism and toxic substances become involved when growers fail to correct drainage problems or continually apply nitrogen such as anhydrous ammonia or ammonia sulfate.

Soil orders, suborders, groups, and series differ in levels of resident nutrients. A blanket recommendation cannot be made for each series because of too much field-to-field variation resulting from variable cropping histories and previous fertilizer applications. Soil testing is the preferred basis for estimating the soil's reservoir of nutrients. All cotton-growing states in the USA and essentially all cotton-growing countries of the world, provide soil testing services for cotton growers (Milsted and Peck, 1973).

Grower application of fertilizers usually are based on soil test results. An average application of fertilizer for cotton is approximately 100, 59, and 50 kg/ha of nitrogen, P_2O_5, and K_2O in the USA (Berger, 1969). In western U.S. cotton regions, the rate of nitrogen applied ranges from 160 to 260 kg/ha. East of 100° W Long, the rates used range from 30 to 150 kg/ha. Nitrogen is stressed in this context because cotton is grown in the USA on soils relatively low in organic matter. Similar variations are to be found with other nutrient supplements applied by cotton growers.

8-4.3 Plant Tissue Tests

Nutrients applied and those naturally available in the soil need only to meet the growth and fruiting requirements of the cotton plant. Concepts of sufficiency and adequacy have meaning in this sense. Differential nutrient uptake related to cotton's growth stage was first documented by McBryde (1891). He suggested that almost 90% of the total nutrient uptake was completed by peak bloom (3rd or 4th week of blooming). This finding, supported consistently by others over a half-century according to Sabbe and MacKenzie (1973), encouraged heavy preplant applications of fertilizer to assure sufficiency. In recent years growth and fruiting problems, complicated by unusual insect and disease damage levels when excess nitrogen was taken up by the cotton plant, have stimulated the use of plant analyses to monitor the nutritional status of the plant throughout the fruiting season. This concept was implemented by Tucker (1965) in Arizona and used effectively by Maples et al. (1977) in Arkansas. The end result has been to give cotton growers an element of crop control that previously was not available.

The nitrate monitoring system as developed in Arkansas by Maples et al. (1977) and implemented by Miley (1982) begins with a nitrogen inventory. The amount of nitrogen in the 0 to 1 m soil profile is estimated, usually accompanied by a direct estimate of the amount and location in the soil of nitrate-nitrogen (NO_3-N). Thirty percent of the total anticipated deficiency for nitrogen is applied preplant. Petiole collections and analyses begin the second week after first flower buds are visible and continue systematically. Analyses are made for NO_3 and phosphorus from 25 to 30 petioles of the most recent mature leaves, usually the fourth from the top of the plant. Desired levels of NO_3-N for the several stages of growth in cotton vary by region according to yield expectations. The key is to maintain sufficiency but avoid excess, especially in early flower bud and first open boll stages of development. If nitrogen is needed, it may be supplied in irrigation water, or in or on the soil as a side or topdressing. These methods of application may not be feasible at some point during the growing season. Supplemental nitrogen, in special situations, may be foliarly applied as urea at rates up to 8 to 10 kg/ha.

8-4.4 Salinity

A side effect of continued use of commercial fertilizers is the salinity potential, especially in specific situations. Cotton grown in areas receiving 75 cm or more rainfall annually is exposed to a fluctuating ground water table. Leaching takes salts, primarily nitrates, into the lower soil horizons. Periods with no rain accelerates movement back toward the soil surface. A "wicking" effect is described by Tucker and Tucker (1968) as a process that may concentrate excess salts in the root zone of cotton. Also, if anhydrous ammonia is used as the nitrogen source the additional increase in salt will give stand problems in the subsequent cotton crop. In the Ultisols carbonates move out of the subsoil and accumulate at the soil surface to create potential problems with excess salinity. It has been suggested that an electrical conductivity reading of 500 dS m^{-1} is a warning level with root activity and growth slowing rapidly at readings about 500 dS m^{-1} (W. Miley, personal communication).

Salinity problems are intensified wherever hard pans or fragipans are present, and cotton is grown continuously. One procedure for reducing a salinity problem is subsoiling to break soil pans.

Field flooding to move salts back into lower soil horizons is a proven salinity correction procedure described by Scofield (1938). Broadcasting the fertilizers that are applied rather than banding is another proven salinity reduction practice. The use of cover crops has worked effectively to reduce salinity problems in some areas.

Salinity interferes with seedling growth, and this interference frequently extends to late June in many U.S. cotton growing regions. This loss of 20 to 40 growing days cannot be tolerated in tight economic situations in which the most logical means of reducing the production costs per bale is with increased yield per unit of land.

8-5 MOISTURE AVAILABILITY AND CONTROL

Cotton requires at least 50 cm of water to grow a crop of minimum acceptable yield. Yield level, other things being equal, becomes a function of available water supply at successive stages of growth (Table 8-2).

The water requirements of cotton are met largely from the soil with small amounts being taken in through stem lenticels and leaf stomates. Soil water available for cotton includes that present when the crop was planted plus that which is added by rainfall and irrigation after planting. As indi-

Table 8-2. Hypothetical water demand by cotton at sequential growth stages for increasing lint yields.

Water needs by stages of growth	Yields in bales/ha			Approximate no. days
	0.75	3.75	7.50	
		cm		range
Seedling (Planting to flower bud initiation)	8	8	10	40–60
Fruiting (through 4th week of blooming)	12‡	14	20	40–50
Maturing (5th week of blooming) to 1st week of open boll)	18	23	32	15–25
Opening (1st week open to all open)	12	27	38	35–60
Totals	50	72	100	

† Approximated for mid-South (USA) cotton with lower and higher yields estimated accordingly.

‡ Lower yields were assumed to have shorter fruiting, maturation, and opening periods; higher yields, longer periods.

cated earlier, U.S. cotton grown west of 100° W Long generally must be irrigated and that grown east of this line generally is described as rain-grown, although between 15% to 20% of U.S. rain-grown cotton is given supplemental irrigation.

In both eastern and western regions, surface and internal drainage problems may have caused more localized yield reductions than has lack of sufficient available water. As emphasized by Grissom (1957), moisture control is a prerequisite to sustained high yields wherever cotton is grown.

8–5.1 Drainage

The drainage requirements of any field to be used for cotton are a function of quantity and intensity of rainfall, rate of soil water intake, and topography (Edminster and Reeve, 1957). Rainfall probabilities for any region can be estimated from previous rainfall records if the data base is adequate. Each soil series has a predictable rate of intake, assuming the absence of soil restrictions such as tillage-related pans. Topographical characteristics of a field give slopes down the row and across, and also give indications of external surface water that may move onto and across the field. A drainage consultant considers all facets in developing a drainage plan. Unfortunately, as stated by Grissom (1957), "The prevailing philosophy is to move raindrops from point of impact to main drainage channels in shortest time possible". Increases in soil temperature in the spring are inversely related to soil moisture. This faster soil warm-up and the level of soil oxygen required by cotton stimulate interests in having adequate surface drainage for cotton.

Cotton growers in the Rain Belt of the USA prefer to have surface drainage such that equipment can be supported within 3 days after the end of any rain during the growing season. This encourages the plowing out of water furrows as the last step of each tillage operation. These water furrows are not total drainage solutions and are generally considered to be erosion accelerators (Grissom, 1957).

Rhizotron studies at Auburn, Alabama (Huck, 1970), clearly demonstrate the role of soil oxygen in stimulating root growth and in promoting nutrient and water uptake by cotton. Cotton roots, like those of many plants, stop functioning when soil oxygen goes below 10%. Soil oxygen content is related to air spaces. Patrick et al. (1973) found that oxygen was limiting in the subsoil of the Bruin soil series, one of Louisiana's most productive soil series for cotton, until late June and continued to be limiting until mid-July for the heavier Tunica series. They reported more than 10% oxygen in the surface 15 cm in all soils checked. Browning et al. (1975) reported root deterioration with as little as 30 min exposure to a saturated soil (less than 3% oxygen). In this context surface drainage alone does not correct oxygen stress in the subsoil.

The development of a pan or restriction in the soil profile accelerates surface soil saturation when rain falls or irrigation water is applied. An oxygen stress that occurs when soil oxygen falls below 10%, and is prolonged more than a few hours in the fruiting stage of growth, will induce shedding of small flower buds and small bolls. Poor fruit set, according to Patrick et al. (1973), is commonly found in cotton grown on soils with poor internal drainage. The presence of a pan or restricted zone in the soil profile presents a great challenge for the surface drainage plan of any field.

In the USA several government agencies are involved in planning drainage systems. The U.S. Soil Conservation Service develops surface drainage plans for fields and farms. The U.S. Bureau of Land Reclamation and the U.S. Army Corps of Engineers build laterals, ditches, and canals to service field systems. These are all too frequently localized, responding to the "squeaking-wheel-gets-the-grease" philosophy, and suffer from the absence of firm regional drainage policy. The original Levee Districts organized along major river systems in the USA (Harrison, 1951) have assumed some drainage responsibilities. Muck and wet lands have encouraged the formation of drainage district organizations. There is still a need for regional operational structures that address drainage problems specifically.

Surface drainage requires some degree of land forming initially to remove low spots in a field, often encompassing as much as 20% of the field area (W. E. Woodall, personal communication). Removal of low spots, especially in the mid-South should increase field productivity sufficiently to pay for removal costs in one season according to Grissom (1957). Upland soils of the southeastern USA can be formed or terraced to facilitate irrigation and drainage simultaneously, since these Ultisols (Cecil, Norfolk, Russton, Memphis, and Loring series) are fairly well drained internally. The Mollisols and Alfisols of the drier southwestern regions of the USA are exposed to infrequent but highly intense rains. Surface drainage plans must avoid erosion. Level, closed-end, and broad-base terraces are used on the Amarillo, Brownfield, Portales, Abilene, Vernon, San Saba, Denton, and other series in the northern and western reaches of these soil orders (Johnston, 1957).

Some terrace slope and grassed waterways are used not to implement drainage but to minimize erosion on the Houston, Austin, Wilson, Victoria, Katy, and other series as found in the eastern and southern reaches of the Mollisol and Alfisol soil orders.

Drainage problems west of 100° W Long are largely internal. Continued use of surface irrigation can raise water tables and accelerate salt problems, especially in certain Aridisols, such as the Arlington (coarse-loamy, mixed, thermic Haplic Durixeralfs) series. As emphasized recently by Brown (1982) prime agricultural lands in California and along the Tigris-Euphrates rivers in Iraq have been removed from agricultural use because of elevated water tables following long-time surface irrigation.

8–5.2 Irrigation

More than 60% of the world's cotton crop is grown under irrigation. Surface irrigation may be done by flood or furrow irrigation or by overhead

Table 8-3. Available water stored in soils varying in texture according to Longenecker and Erie (1968).

Soil texture	Available water
	cm/0.3 m (in./ft)
Coarse sands	2-3†
Fine sandy loams	3-4
Loams	4-5
Clay loams	4-6
Clays	5-7

† Approximate values.

sprinkler. Subsurface irrigation is obtained by methods creating an artificial water table at some usable depth below the soil surface. Water for irrigation comes from streams, surface reservoirs, canals, and wells (Criddle and Haise, 1957). The amount of irrigation water needed during the growing season depends upon how much may be stored in the soil when the crop is planted, how much water may come as rainfall, and the length of growing season.

The amount of available water that can be stored in a soil varies according to soil series, soil depth, recently induced soil profile restrictions, cropping history, and other factors. Bloodworth, as cited by Longenecker and Erie (1968), estimated available water storage differentials according to soil texture, as shown in Table 8-3. Criddle and Haise (1957) reported similar values. Aldrich (1957) suggested a working soil depth of 2.0 m in much of California's cotton area, and Guinn et al. (1981) suggested a working depth of 1.5 to 2.0 m in Arizona. This range corresponds to 0.3 m to 0.6 m in the southern High Plains and mid-South cotton areas (Johnston, 1957) and only 0.3 to 0.5 m in the Southeast. In U.S. cotton production areas, variable amounts of water can be stored in the soil prior to planting.

8-5.3 Growing Cotton Where Irrigation Is Required

Soils used for cotton grown west of 100° W Long in the USA can be assumed to have little water stored in the soil and no usable rainfall during the growing season. Water to grow cotton comes from irrigation. California growers, for example, apply between 60 to 120 cm of water to cotton (Grimes et al., 1978).

Preplant irrigations are commonly made in the USA west of 100° W Long, especially where surplus stream and canal water is available at little or no cost. In California and Arizona the amount applied preplant varies from 20 to 40 cm, depending upon depth of soil to be watered. In the southern High Plains the amount applied preplant may be as little as 10 cm or as much as 20 cm depending upon summer rainfall expectation and the number of growing days available for the crop. Minimal amounts generally are applied because the water originates from wells in the declining Ogallala, a limited aquifer (Bilbro, 1974) with excessive pumping costs.

In other parts of the world a preplant irrigation is made because of water availability and, in some cases, this may be the only irrigation applied to cotton. In southern Arabia (Hearn, 1980), Turkestan (Christides and Harrison, 1955), parts of Africa (Farbrother, 1974), and India (Berger, 1969) it is a common practice to spread off-season flood waters over fields to be used for cotton. Where this is the only irrigation, cotton yields are related to the amount of water that can be stored in the potential root zone.

Surplus water can be used to flush accumulated salts from the potential root zone if the lower soil profile will accept inflow. This practice should be considered as a reclamation procedure with the land used for 1 year to grow a crop other than cotton. Barley has been used for this in the western USA.

Irrigations made after the preplant watering begin normally between the second week after first flower buds are visible and the second week of blooming and are repeated at 10- to 20-day intervals depending upon soil characterisics, water supply, and yield expectations. Limited water supply and high costs in all irrigated cotton growing areas have enoucraged the use of conservative moisture regimes. The elimination of one or two irrigations has been accomplished with no yield reduction by planting a faster fruiting or more prolific cultivar, by delaying the first irrigation after planting, and by deleting the last irrigation normally applied (Guinn et al., 1981).

In areas where irrigation is required to grow cotton, water source and availability largely determine the moisture regime to be used for cotton. Water districts have been formed in the USA and in other parts of the world to serve agricultural needs. Control devices built on watersheds retain snow and rain waters for later delivery via canals or ditches. The California State Water Project, the Gila River Project in Arizona, and the Lower Colorado River Project serving both states are major water districts in the USA; the Gezira Scheme of the upper Nile and the lower Nile in Africa; and the Soviet Water Districts based on the Syrdar'ya and Amurdar'ya rivers are the major water districts serving cotton. In these partly or totally government-controlled water districts, users receive a total allotment and a use schedule. In some cases there are variable charges, i.e. costs are tied to supply and distance between outlet and source. In 1981, for example, a cotton grower at the south end of the California aqueduct near Bakersfield paid approximately \$3.75 ha cm^{-1} as compared with a rice grower near the water source above Sacramento who paid approximately \$1.85 ha cm^{-1}. Canal water costs in the USA and restrictions on wells have encouraged the building of holding reservoirs to stockpile surplus and low-cost stream water. One large corporate cotton ranch near Bakersfield, California, has reduced water costs to less than \$1.50 ha cm^{-1} by utilizing a combination of California aqueduct water, reservoirs, and a system of pumps and canals to reuse all tail water discharge, (Larry J. Chrisco, Ranch Manager, Buena Vista Ranch, J. G. Boswell Company, personal communication).

Water for irrigation is released in cotton either at the soil surface or above the soil surface. Release above the soil surface may be within 1 m of the soil or as much as 3 to 5 m, depending upon the pressured system used (Criddle and Haise, 1957). Release at the soil surface may be into furrows, the more common method, or as a flood between levees, dikes, or borders, sometimes called basins. Surface release is 50 to 90% efficient, depending upon slopes, intake rates, head build-up, tailing losses, etc. Sprinkler systems are 70 to 80% efficient, depending upon droplet size, wind velocities, evaporative loss, etc.

Surface release as furrow irrigation may have water delivered as open header ditches, by surface piping, or by buried pipe or conduit with risers and valves at regular intervals. Ultimate release is with gated pipe having adjustable gates, siphon tubes, or other methods. A furrow system uses gravity. Slopes down furrows should neither exceed 0.30% nor be less than 0.05%. Cross slopes up to 2.0% can be controlled with beds maintained at least 15 cm in height (Quackenbush and Thorne, 1957). Furrow irrigation should not be attempted if water intake is less than 0.75 cm/hour or more than 7.50 cm/hour.

Furrow irrigation frequently gives poor distribution of water down the row. Row ends near the header ditch or pipe source may become stressed for oxygen before the row outlet receives adequate water. Alternate furrow runs, stream cutbacks, and resets are sometimes needed to assure a more even water distribution. An overhead release gives much better water control.

Basin irrigation is used with the same soil intake precautions as with furrow irrigation. Usually the land is formed with zero side slope and a 0.05% to 0.20% downslope as described by Criddle and Haise (1957). One large corporate cotton ranch at Buena Vista, Calif., levels the land with laser equipment to a precision of zero slide slope and only a 0.01% down slope. Basins are 24 ha each. Each hour two pumps deliver approximately 80 ha cm^{-1} from bordering canals. During the season, 10 to 15 ha cm^{-1} are applied each irrigation and left for approximately 6 hours and then the standing water is removed by the same type of pumps used to deliver the water (Larry J. Chrisco, personal communication).

The time required to irrigate cotton is a function of delivery stream flow and the amount of water to be applied (Criddle and Haise, 1957). A delivery rate of 1.7 m^3 min^{-1} approximates an hourly rate of 1 ha cm^{-1} (1 ft^3 sec^{-1} approximates 1 acre-inch/hour). If a well is to be used for irrigation, a minimal delivery should be approximately 6.8 m^3 hour^{-1} ha^{-1} to be irrigated from that well (12 gal min^{-1} acre^{-1}).

Overhead release may be with hand moved pressured pipe, risers, and sprinkler heads set 9 to 12 m apart and sprinkler lines set approximately 18 m apart. In some areas a blank row is dedicated to sprinkler lines and the pipes are never moved. Sprinkler lines that move, some more than 2 km in length, are a more recent development. These may be center pivot types or laterally moving types. A few single release units that release in a circle up to 60 m or more in diameter are also used west of 100° W Long.

8–5.4 Growing Cotton Where Irrigation Is A Supplement to Rainfall

Cotton grown east of 100° W Long in the USA may suffer from moisture stress at some time in each growing season, but this stress does not always result in a direct loss in yield, according to Clower and Patrick (1965). All areas have a summer rain potential. This potential separates irrigation philosophy between west and east.

Rain-grown cotton rarely receives a preplant irrigation, and the common concern of growers having supplemental irrigation capabilities is when to irrigate the first time. A few growers will irrigate essentially by calendar date, but this method requires a rare combination of drainage and crop control. The water requirement of the cotton plant during its several stages of growth and development and the available water in the soil determines when to irrigate.

The amount of water in the soil that is available to growing plants involves absolute water content of the soil and the energy, called suction, required to remove water from the soil (Richards and Richards, 1957). A number of devices are used to estimate the amount of available water in any soil. One is the tensiometer, a suction-estimating device that gives readings in units related to atmospheric pressure. Many tensiometers are calibrated in centibars of pressure. The "bar" is a metric unit that represents 0.987 atmospheres of pressure or 10^6 Pa in SI units. The traditional constants that set limits to the volume of soil water available to plants were field capacity and wilting percentage. Their difference was called available water. A more acceptable term in extractable water. Each soil has an upper and a lower limit of extractable water. The integrated difference is the amount of extractable water in centimeters available to the plant. Field capacity varies in volume of water according to soil type and rooting depth. The permanent wilting percentage is that percentage of soil water held against 15×10^6 Pa (15 bars) of pressure in the upper part of the soil profile, but this holding strengths at the wilting point may differ greatly in a lower horizon. The dynamic nature of soil water makes all estimating devices crude approximators. The tensiometer estimates from 25 to 75% of the strength or tension at which water is held in the soil. In most soils irrigation is justified at a reading of 55 to 60%. Tensiometer readings do not consider the influence of osmotic pressure on moisture tensions, and there are precautions associated with installation and activation. Tensiometers are more effective if placed at 22 to 25 cm and 55 to 65 cm at sites located in each quadrant of the field. The upper depth must be above any residual pan. The lower depth reinforces the irrigation decision and helps determine the amount of water needed. A reading 24 hours after irrigation is very valuable. Tensiometers should be serviced (bled to remove air from tube, refilled, and vacuum pulled) after each reading. This procedure takes less than 5 min. Each instrument must be protected against freezing, removed each fall, and reset each spring at time of planting.

Other devices are available for estimating soil water status. Examples of these are gypsom blocks and neutron probes. Both require specific calibrations before their readings are reliable. These devices also have inter-active potentials with specific soil conditions that bias their estimates of the soil water status.

Growers east of 100° W Long who have irrigation capabilities generally have used plant indicators of moisture stress to determine when irrigation is needed. Wilting by noon or by mid-morning has been used effectively as an indicator of need. A reddening of the upper stem areas and the darkening of the green in uppermost leaves have been used effectively to show need for irrigation, but modern cotton cultivars having heritable reddish tendencies have negated the use of plant color changes as water sensors. The so-called "pressure bomb" device that measures the water status of the plant has proven efficiency in showing the need for irrigation (Grimes et al., 1978). This device currently is well suited as an irrigation research tool where re-peatable results are needed. Other measures of plant water status, such as stem diameter differentials, exudate flow from excised stems or roots, the infrared "gun" that reads leaf surface temperatures, and other devices are being used as plant indicators of water needs.

Supplemental irrigation is used to avoid water stress in cotton. Hearn (1980) suggested that the shoot of the cotton plant continues to develop in-dependent of water supply until three quarters of the available water has been used and then growth ceases rather abruptly. Supplemental irrigation gives cotton growers an element of control over the fruiting period. Spooner et al. (1958) demonstrated the significant suppression of cultivar-by-year interactions for fiber properties when supplemental irrigation is used on rain-grown cotton.

If limited irrigation water is available for rain-grown cotton, a con-sensus recommendation is that a single irrigation about 3 weeks after first bloom should give the maximum return to water applied. If water for sup-plemental irrigation is not limited, a common concern of growers of rain-grown cotton is when to stop irrigating. In soils having a water-holding capacity of 20 cm in the cotton root zone the last water should be applied no later than 1 to 2 weeks before the last effective flower where flower ef-fectiveness is a function of decreasing temperatures (Hearn, 1980). As water storage potential decreased below 20 cm, the last watering date approaches the last effective date of flowering.

8–5.5 Multiple Cropping

Water control encourages multiple cropping to extend land use. In the western USA and other parts of the world where probable yields exceed five bales/ha, so much of the total growing season is consumed by cotton that a second crop is not normally considered. Safflower is sometimes used in the western USA as a winter crop in rotation with cotton. Safflower can be fol-lowed with wheat or barley, and these cereals can be followed with cotton in the southernmost reaches of the western USA production areas. Radishes and lettuce and other short-term crops can be inserted between cotton crops, but this is not generally done.

A new interest in double-cropping cotton after wheat has been demonstrated in the southern parts of the mid-South and the southeastern USA. Moisture control is the key to obtaining an acceptable stand of cotton following wheat. Growers in a region composed of southeast Alabama, southern Georgia, and the Panhandle of Florida in the USA double-cropped cotton successfully after wheat in 1981 on approximately 35 000 ha with supplemental irrigation. Their cotton yields varied from 2.00 to 3.75 bales/ha (Gary A. Herzog, personal communication).

Little double-cropping of cotton is attempted outside the USA. A few countries require a food crop from the land each year. In Peru, for example, a cotton grower is required to plant enough food crops to feed all resident workers and to place any surplus into normal public food channels (Berger, 1969). Cotton growers in Peru traditionally plant beans into standing cotton, and these plantings provide food for all the workers of the specific cotton unit. Cotton may be planted in each of the 12 months in parts of India, Pakistan, Peru, and other countries; a system of continuous cropping rather than double-cropping is followed (Berger, 1969; Afzal, 1977).

8–6 PRODUCTION SYSTEMS

Cotton has many common production practices, but it has a wide array of production systems employed by growers in diverse parts of the world. Cotton production in the USA, Mexico, Central America, the Middle East, Australia, and in bits and pieces elsewhere is labor extensive or mechanically oriented. The emerging republics of Africa, UAR (Egypt), India, Pakistan, and others grow cotton largely with man and animal power. Machines are used to prepare the land for planting and to plant in USSR, PRC, and parts of Brazil; weed control and harvesting are done primarily with hand labor. In this context it should be noted that mechanical harvesting is used generally on the large state farms in USSR and PRC. The production system used to grow cotton is related to the local labor situation.

A second factor determining the production system used for cotton is size of the production unit. A mechanical harvester, picker or stripper, has a specific minimum volume to justify its operation. The same can be said of tillage and planting machines.

Production systems are the integration of crop residue disposal, subsoiling, seedbed preparation, planting, and cultivation up to harvest. Integrated systems of production are designed for production efficiency even when grown under diverse soil and climatic conditions. Both USA and foreign production systems will be described separately as (a) irrigated, (b) rain-grown with supplemental irrigation, and (c) as dryland.

8–6.1   Irrigated Cotton Production Practices

Crop residues from crops grown ahead of cotton in the western USA more or less decompose except where cotton was the previous crop. Old cotton stalks are shredded mechanically, and the debris essentially covers the soil. In much of the USSR, PRC, Middle East, and parts of Africa and India-Pakistan, cotton stalks are pulled by hand and burned as fuel. Pulling and burning also is justified in specific areas as measure of disease control (Smirnova, 1980). In areas where cotton stalks rarely grow taller than 75 cm, the growers ignore the old stalks.

Subsoiling is a standard practice ahead of cotton in the irrigated west. In some instances, the soil is shattered to a depth of 1 m with 80 to 100 kg/ha of nitrogen as anhydrous ammonia released at that depth (M. Hoover, personal communication). The more common depth in the west is 40 to 50 cm, using the so-called "parabolic" subsoiler. The general use of subsoiling ahead of irrigation improves internal soil drainage. Subsoiling is a practice of mechanized production areas rather than labor intensive areas. A notable exception is in the USSR and PRC where heavy machinery can be rented by cotton communes and used as needed (Anonymous, 1974).

Seedbed preparations depend, to some extent, upon the water regime to be followed. If a basin system of surface irrigation is to be used as a pre-plant irrigation in the western USA, the field is left flat after subsoiling; then it is watered; and as soon as equipment can be supported it is disked or harrowed; and finally ridged or left flat for planting. The same proce-dure generally is followed in the USSR and in other areas growing cotton under irrigation.

If preplant irrigation is applied in furrows, ridges are formed, and the water is released. The ridges (beds) settle and firm up awaiting the preplant and planting operations.

In Egypt and parts of Sudan the ridges are formed, usually in an east-west orientation, and the fields are flooded. Or, the ridges are formed and seed are worked by hand, foot, and stick into holes on the south side of the ridges and then irrigated. In other irrigated areas, a preplant irrigation is simply a spate flood covering fields with little or no seedbed preparation ahead of planting.

Planting cotton that is to be grown under irrigation differs little from planting cotton that is to be grown under natural rainfall with or without supplemental irrigation. As shown in Fig. 8–2 cotton is planted each of the 12 months of the year in some part of the world. Irrigated cotton is gener-ally planted when soil temperatures at a depth of 20 cm reach 18° C for 3 to 5 days in succession (Berger, 1969; Wilkes and Corley, 1968). Where daily soil temperatures at the 20 cm depth are available a preferred planting date is the day the 10-day average temperature reaches 18° C. At this tempera-ture germination and emergence from the soil will require approximately 100 hours.

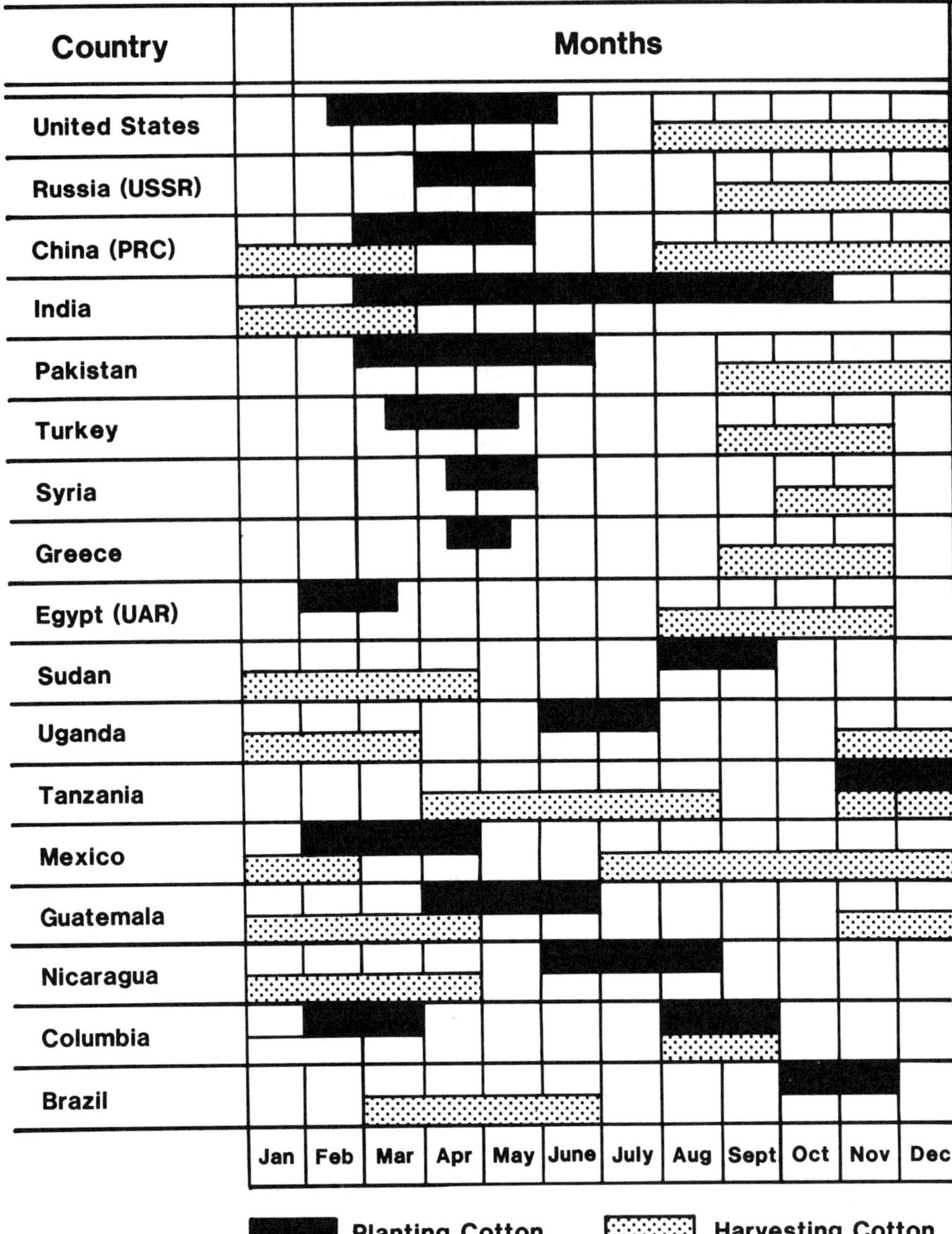

Fig. 8-2. Dates of planting and harvesting in major cotton countries according to Berger (1969).

Cottonseed separated from the lint in the ginning process are difficult to plant precisely with mechanical devices because the short seed fibers cause the seed to clump and flow improperly in planters. Seed for planting in irrigated and non-irrigated areas alike are delinted to some degree. Acid delinted seed have all the lineters removed and are free flowing in bulk. They can be precisely metered with modern mechanical planters. Re-ginning the seed cuts away a portion of the seed linters and, especially when these partially delinted seed also are passed across a flame, these "machine de-linted" seed can also be planted with most mechanical planters. Acid-delinted seed will emerge quicker than machine-delinted seed and are preferred for planting fields to be irrigated. This preference reflects both the advance in emergence and the greater precision in metering.

Planting rates for irrigated cotton are designed to give 10 to 15 plants per drilled meter of row which converts to 100 000 to 140 000 plants/ha if rows are 1 m apart. More plants per hectare may cause excessive growth at the expense of yield (Wilkes and Corley, 1968). Uniformity of emergence influences weed control programs and harvesting efficiency. Good quality in planting seed assures a uniform emergence where temperature or moisture are not limitations and a firm seedbed is used.

Irrigation water with a relatively low salt content can cause salt problems at planting on some soils because of the size of the preplant irrigation and the residual salts in the subsoil. Frequently this problem can be overcome either by using double-wide beds and planting in the relatively salt-free shoulders or covering the seed in the planting operation with a peak of excessive soil to "wick-up" excess salts and then dragging off the peak 3 to 4 days after planting (Wilkes and Corley, 1968).

8–6.2 Production Practices in Rain-grown Cotton

Most of the soils used for cotton in the USA form traffic pans or hard pans when cotton is grown continuously. The tendency to form hard pans is more common in rain-grown cotton areas. Subsoiling 25 to 30 cm deep is a general preplant practice, with some going down to 45 cm. Some growers subsoil every year and others in alternate years. Brown et al. (1955) failed to obtain any benefits from subsoiling a Grenada silt loam soil for 5 consecutive years, but many others obtained positive responses to subsoiling (Wilkes and Corley, 1968).

Seasonal variations in rainfall generally have given significant year-by-cultural practice interactions in the rain-grown area. Data accumulations have failed to support any one set of production practices followed from seedbed preparation to crop maturity. It is generally understood, however, that in the temperate zones of the world a lower survival rate (30 to 50%) should be expected of germinable seed planted early in the season. In this same climatic area, the soils used for cotton will be more productive if 200 000 to 250 000 plants/ha are obtained with early plantings.

Growers in the rain-grown region of the USA generally plant more seed (14 to 18 kg/ha acid-delinted) if planting is considered as being earlier than normal and less seed (7 to 11 kg/ha acid-delinted) if planting is considered as being later than normal.

In tropical zones of the world, planting rates do not vary to the extent suggested above because seedling survival expectations are essentially the same regardless of date planted. Most tropical production areas are subjected to alternating periods (4 to 6 months) of rain and no rain. Planting is made during gaps in the wet period, and the crop matures during the dry period. Desired plant populations range from 100 000 to 150 000 plants/ha.

Preferred planting dates in rain-grown production areas of the USA can be anticipated by developing rainfall and temperature deviation probabilities based on long-term weather records. By using 30 years of weather data established 28 April to 9 May as a preferred planting time in Arkansas (Downey, personal communication). This can be considered as high probability planting. The temperature constraint imposed on planting as cited by Wilkes and Corley (1968) is met in Downey's preferred planting date. Fielding (1975) developed a similar recommendation for high probability planting date based on rainfall expectations in northern and eastern Uganda.

8–6.3 Production Practices in Dryland Cotton

Dryland cotton is cotton grown with 30 to 50 cm of rainfall in areas where cropping options other than cotton cannot be justified. In the USA this dryland area is confined to Texas and Oklahoma but constitutes almost 20% of the total U.S. hectares used for cotton. Growers plan for dry weather by minimal seedbed disturbance before planting, by using lister planters to plant in water furrows rather than on ridges, and by planting two rows and skipping two or four as a moisture conservation practice (Wilkes and Corley, 1968).

Planting rates for dryland cotton in the USA are designed to obtain 150 000 to 200 000 plants/ha of actual cotton; i.e. 12 to 18 plants per meter of row in rows 1 m apart. Planting date is highly variable. A general guide has been to plant at any time sufficient soil moisture is available to complete germination after 1 May. In this area, planting date can be as late as 15 June with yield expectations as low as 200 kg/ha of lint being acceptable.

8–7 CROP DEVELOPMENT SCHEDULES

The cotton plant grows with a biological rhythm that gives a regularity of appearance of leaves, branches, and fruit. The development schedule is highly predictable and is influenced by temperature, soil moisture, nutrition, and cultivar (Gipson and Ray, 1970). For example, if a preferred harvest period can be defined in terms of the local climatic constraints imposed on cotton, a preferred blooming period can be defined regardless of cultivar planted. With blooming period defined as a critical point on the development schedule, a preferred planting period can be identified for cotton in general terms. Crop control requires that cultivar and cultural practices be selected so that this preferred planting period will coincide with a favorable local climatic period.

Downey (personal communication) has established 1 to 20 October as a preferred harvest date for cotton in Arkansas. With lint yield expectations in the range of 500 to 800 kg/ha the Arkansas crop should begin flowering 10 to 15 July and reach peak flowering 3 weeks later to have matured the crop within this preferred harvest period of 1 to 20 October. Downey also established two very hazardous harvest periods for Arkansas as 25 August to 25 September and after 25 October. The hazards are primarily rainfall and resulting field deterioration of the bolls of cotton. Crop control, therefore, must be exercised to fit the crop to the harvest window.

Planting date and cultivar are critical choices of crop control in the Arkansas example. If planting is successful at the preferred 1 May period, a full season cultivar, as represented currently by 'Stoneville 825' or 'Deltapine 62', must be planted to meet the preferred harvest window. Planting a "short season" cultivar will bring the crop into maturity during the early high-hazard harvest period. On the other hand if planting is delayed or if replanting is necessary such that plantings are made in late May, using the full season cultivars brings the crop into maturity during the late high-hazard harvest period. Using a short season cultivar, as represented currently by 'Tamcot CAMD-E' or 'DES-56', for the late planting should bring the crop into maturity during the preferred harvest period.

Crop control exploits development schedules in cotton and the loose system of control developed in Arkansas can be developed for any region where cotton is grown, and especially so where the crop is irrigated.

McMahon and Low (1972) suggested a normal crop development schedule based on growing degree-day (GDD) accumulations. They defined GDD as the sum of daily mean temperatures above a base temperature of 10° C. They computed an average of 2640 GDD for the total cotton crop based on data from four diverse cotton-growing countries. Young et al. (1980) computed day-degree (DD) by subtracting a constant (12.8° C) from daily maximum temperaures and accumulating the degrees on an hourly basis. They computed the Arizona crop as expending approximately 75 000 DD in a 200-day growing season. They also computed "heat units" as the average daily temperature minus 12.8° C. The difference is referred to as "heat units." These were accumulated for the growing season. They estimated Arizona cotton crop as using more than 2000 heat units in their 200-day growing season.

The number of GDD, DD, or any kind of thermal unit or index related to the cotton plant's change from one developmental stage to another depends upon how these units are computed. No standard procedure has emerged from the revival of interest in the "heat-loving" tendencies of the cotton plant. The relationship can be seen in Table 8–4.

Table 8-4. A comparison of thermal unit measurements required for several stages of growth and development in cotton.

Unit means as	Emergence	From planting to:				
		1st true leaf	First flo. bud	First bloom	First open boll	All bolls open
Calendar days[†]	5–15	13–30	30–70	51–94	100–175	120–230
GDD[‡]	--	--	--	900	--	2 640
DD[§]	--	6 700	25 000	33 000	58 000	--
Heat units[§]	--	154	600	860	1 600	--
Hours above 17° C[¶]	100	200	700	1 200	2 200	2 700

† Arbitrary range under Arkansas' conditions, 2.5 bales/ha.
‡ Base temperature is 10° C as used by McMahon and Low (1972).
§ Base temperature is 12.8° C as used by Young et al. (1980).
¶ Used in Arkansas by Waddle.

Thermal units measured by McMahon and Low (1972), Young et al. (1980), and others were in terms of daily maximum or average temperatures. Other units in Table 8-4 were arbitrarily assigned. In any case, thermal unit accumulations are used to predict or to anticipate a crop development schedule. New electronic recorders that continuously record thermal units should improve the normal crop development predictions, but this supposition remains to be demonstrated.

Another significant element of crop control relates to the cotton phenomenon of "cut-out." Cut-out describes the tendency of plants to abort young bolls and pinhead flower buds after an indefinite number of bolls have been set. Many factors are capable of inducing cut-out. A suggestion of approaching cut-out is the appearance of uppermost white flower four or five nodes below the plant terminal. When this white flower reaches node number 3 below the terminal, the cut-out process generally procedes without interruption until the terminal is dormant and most bolls in the uppermost four nodes have aborted. When cut-out is complete, an interval of about 4 weeks is required for the plant to initiate new flower buds. This means 7 weeks are needed between cut-out and a new bloom which will require an additional 70 days to make an open boll.

Early maturing or determinate cultivars appear to have an accelerated cut-out process. These cultivars yield up to 25% less lint per hectare than full season cultivars when grown under a full season growth schedule in terms of available water and nutrients. Accelerated cut-out or excessive shedding was evaluated by Tugwell and Waddle (1964) with measured inputs of hours of sunlight, soil moisture, temperatures, and bolls already set on the plant. They found that bolls already set were easily the strongest stimulant to shedding, and this net relationship was stronger in the determinate than in the full season cultivar.

Crop control begins with subsoiling if needed, and with a soil profile inventory for nitrates. The cultivar chosen should suit the requirements of the grower. Planting good seed on a firm seedbed with carefully monitored or calibrated herbicide delivery systems should minimize loss of growing days between planting and first flower bud. Moisture needs must be met and cut-out delayed as long as possible with timely irrigations if crop control is to be maintained. When any grower loses control of the cotton crop, i.e., when the developmental schedule is interrupted such that crop and climatic development are out of synchronization, the cotton grower faces a salvage situation.

REFERENCES

Anonymous. 1966. Cotton: Supply, demand, and farm resource use. Southern Coop. Series bull. 110.

----. 1974. USSR agriculture atlas. U.S. Central Intelligence Agency, Washington, D.C. Special Report.

Afzal, Muhammad. 1977. The yield of cotton. Proc. Pakistan Acad. Sci. 14(1&2):1–24.

Albrecht, W. A. 1938. Loss of soil organic matter and its restoration, p. 347–360. *In* Soils and men. USDA Yearb. Agric., U.S. Government Printing Office, Washington, D.C.

Aldrich, D. G. 1957. The dry mild-winter region. p. 467–474. *In* Soil. USDA Yearb. Agric., U.S. Government Printing Office, Washington, D.C.

Berger, Josef. 1969. The world's major fiber crops, their cultivation and manuring. Centre d'Etude de l'Azote, Zurich, Switzerland.

Bilbro, J. D. 1974. Effect of preplant-only irrigation on cotton yields. Agron. J. 66:833–834.

Brown, D. A., R. H. Benedict, and B. B. Bryan. 1955. Irrigation of cotton in Arkansas. Arkansas Agric. Exp. Stn. Bull. 552.

Brown, H. B., and J. O. Ware. 1958. Climate and soils for cotton. p. 264–290. *In* Cotton. McGraw-Hill Book Co. Inc., N.Y.

Brown, L. R. 1982. Soils and civilization. Audubon 84:18–24.

Browning, V. Douglas, H. M. Taylor, M. G. Huck, and Betty Klepper. 1975. Water relations of cotton: A rhizotron study. Auburn Univ. Agric. Exp. Stn. Bull. 467.

Buol, S. W. (ed.) 1973. Soils of the Southern States and Puerto Rico. Southern Coop. Series Bull. 174.

Christides, Basil G., and G. J. Harrison. 1955. Cotton growing problems. McGraw-Hill Book co., N.Y.

Clower, K. N., and W. H. Patrick, Jr. 1965. Soil moisture extraction and physiological wilting of cotton on Mississippi River alleuvial soils. Louisiana Agric. Exp. Stn. Bull. 598.

Corey, R. B., and E. E. Schulte. 1973. Factors affecting the availability of nutrients in plants. p. 23–33. *In* Soil testing and plant analysis. (Rev. Ed.) Soil Sci. Soc. Am., Madison, Wis.

Criddle, W. D., and H. R. Haise. 1957. Irrigation in arid regions. p. 359–367. *In* Soil. USDA Yearb. Agric., U.S. Government Printing Office, Washington, D.C.

Doyle, C. B. 1941. Climate and cotton. p. 348–363. *In* Climate and man. USDA Yearb. Agric. U.S. Government Printing Office, Washington, D.C.

Edminster, T. W., and R. C. Reeve. 1957. Drainage problems and methods. p. 378–385. *In* Soil. USDA Yearb. Agric., U.S. Government Printing Office, Washington, D.C.

Farbrother, H. G. 1974. Irrigation practices in the Gezira. Cotton Res. Corp. Res. Mem. No. 89.

Fielding, J. 1975. Problem of Optimum sowing dates for cotton in Northern and Eastern Uganda. Cotton Grow. Rev. 52:103–111.

Frans, R. E., R. E. Talbert, and Brent Rogers. 1982. Influence of long term herbicide programs on continuous cotton. p. 228–229. *In* Proc. Beltwide Cotton Prod. Res. Conf., Las Vegas, Nev.

Gipoon, J. R., and I. I. Ray. 1970. Temperature-variety interrelationships in cotton. Cotton Grow. Rev. 47:257–271.

Griffis, C. L., and C. R. Mote. 1978. Weed seed viability as affected by the composting of cotton gin trash. Arkansas Farm Res. 27:8.

Grimes, D. W., W. L. Dickens, and H. Yamadu. 1978. Early season water management for cotton. Agron. J. 70:1009–1012.

Grissom, P. H. 1957. The Mississippi Delta Region. p. 524–531. *In* Soil. USDA Yearb Agric., U.S. Government Printing Office, Washington, D.C.

Guinn, G., J. R. Mauney, and K. E. Fry. 1981. Irrigation scheduling and plant population effects on growth, bloom rates, boll abscission, and yield of cotton. Agron. J. 73:529–534.

Harrison, R. W. 1951. Levee districts and levee building in Mississippi. Delta Council and Mississippi Agric. Exp. Stn. Cooperating with USDA.

Hearn, A. B. 1980. Water relations in cotton. Outlook Agric. 10:159–166.

Hinkle, D. A. 1969. Crop rotation studies on Sharkey Clay soil. Arkansas Agric. Exp. Stn. Rep. Series 176.

----, and N. D. Fulton. 1963. Yield and Verticillium wilt incidence in cotton as affected by crop rotation. Arkansas Agric. Exp. Stn. Bull. 674.

Huck, M. G. 1970. Variation in taproot elongation rate as influenced by composition of the soil air. Agron. J. 62:815–818.

Jenny, Hans. 1930. A study on the influence of climate upon the nitrogen and organic matter content of the soil. Missouri Agric. Exp. Stn. Res. Bull. 152.

Johnston, J. R. 1957. Southern Plains. p. 516–523. *In* Soil. USDA Yearb. Agric., U.S. Government Printing Office, Washington, D.C.

Leighty, C. E. 1938. Crop rotation. p. 406–430. *In* Soils and men. USDA Yearb. Agric., U.S. Government Printing Office, Washington. D.C.

Longenecker, D. E., and L. J. Erie. 1968. Irrigation water management. p. 321–345. *In* W. J. Porter, M. Hoover, and F. C. Elliot (ed.) Advances in production and utilization of quality cotton. Iowa State Univ. Press, Ames.

Maples, Richard, J. G. Keogh, and W. E. Sabbe. 1977. Nitrate monitoring for cotton production in Loring-Calloway Silt Loam. Arkansas Agric. Exp. Stn. Bull. 825.

McBryde, J. B. 1891. A chemical study of the cotton plant. Tennessee Agric. Exp. Stn. Bull. 120–145.

McMahon, J., and A. Low. 1972. Growing degree days as a measure of temperature effects on cotton. Cotton Grow. Rev. 52:103–111.

Melville, D. R., and G. E. Rasburg. 1980. The effects of winter cover crops on the production of cotton grown on Norwood Fine Sandy Loam. p. 38–48. *In* Ann. Res. Rep. Red River Valley Agric. Exp. Stn. Louisiana State Univ., Bossier City, La.

Meredith, W. R., Jr. 1982. Changes in cotton yields since 1950. p. 35–38. Proc. Beltwide Cotton Prod.-Mech. Conf., Las Vegas, Nev.

Miley, W. N. 1982. Plant tissue analysis as a cotton production management guide. p. 56–60. Proc. Beltwide Cotton Prod.-Mech. Conf. Las Vegas, Nev.

Milsted, S. W., and T. R. Peck. 1973. Principles of soil testing. p. 13–21. *In* Soil testing and plant analysis. (Rev. Ed.) Soil Sci. Soc. Am., Madison, Wis.

Nififoroff, C. C. 1938. Soil organic matter and soil humus. p. 929–939. *In* Soils and men. USDA Yearb. Agric., U.S. Government Printing Office, Washington, D.C.

Patrick, W. H., Jr., R. D. Delaune, and R. M. Engler. 1973. Soil oxygen content and root development of cotton in Mississippi River alleuvial soils. Louisiana State Univ. & Agric. Exp. Stn. Bull. 673.

Pieters, A. J., and Roland McKee. 1938. The use of cover and green-manure crops. p. 431–444. *In* Soils and men. USDA Yearb. Agric., U.S. Government Printing Office, Washington, D.C.

Quackenbush, T. H., and M. D. Thorne. 1957. Irrigation in the East. p. 368–378. *In* Soil. USDA Yearb. Agric., U.S. Government Printing Office, Washington, D.C.

Raney, W. A., and A. W. Cooper. 1968. Soil adaptation. p. 77–115. *In* W. J. Porter, M. Hoover, and F. C. Elliot (ed.) Advances in production and utilization of quality cotton. Iowa State Univ. Press, Ames.

Richards, L. A., and S. J. Richards. 1957. Soil moisture. p. 49–60. *In* Soil. USDA Yearb. Agric., U.S. Government Printing Office, Washington, D.C.

Sabbe, W. E., and A. J. MacKenzie. 1973. Plant analysis as an aid to cotton fertilization. p. 299–313. *In* Soil testing and plant analysis. (Rev. Ed.) Soil Sci. Soc. Am., Madison, Wis.

Scofield, C. S. 1938. Soil, water supply, and soil solution in irrigation agriculture. p. 704–716. *In* Soils and men. USDA Yearb. Agric., U.S. Government Printing Office, Washington, D.C.

Smirnova, A. A. 1980. Protection of cotton in the USSR. Outlook Agric. 10:206–207.

Spooner, A. E., D. A. Brown, and B. A. Waddle. 1958. Effects of irrigation on cotton fiber properties. Arkansas Agric. Exp. Stn. Bull. 601.

Spurgeon, W. I., and P. H. Grissom. 1965. Influence of cropping systems on soil properties and crop production. Mississippi Agric. Exp. Stn. Bull. 710.

Tucker, T. C. 1965. The cotton petiole, guide to better fertilization. Plant Food Rev. 11:9–11.

----, and B. B. Tucker. 1968. Nitrogen nutrition. p. 184–211. *In* W. J. Porter, M. Hoover, and F. C. Elliot (ed.) Advances in production and utilization of quality cotton. Iowa State Univ. Press, Ames.

Tugwell, N. P., and B. A. Waddle. 1964. Yield and lint quality of cotton as affected by varied production practices. Arkansas Agric. Exp. Stn. Bull. 682.

Wilkes, L. H., and T. E. Corley. 1968. Planting and cultivation. p. 117–149. *In* W. J. Porter, M. Hoover, and F. C. Elliot (ed.) Advances in production and utilization of quality cotton. Iowa State Univ. Press, Ames.

White, H. C. 1896. The manuring of cotton. p. 169–196. *In* The cotton plant. Bulletin No. 33. USDA Office of Exp. Stn. U.S. Government Printing Office, Washington, D.C.

Young, E. F., Jr., R. M. Taylor, and H. D. Petersen. 1980. Day-degree units and time in relation to vegetative development and fruiting for three cultivars of cotton. Agron. J. 20: 370–274.

9 Cotton Protection Practices in the USA and World

Section A: Insects

R. L. Ridgway
ARS-USDA
Beltsville, Maryland

Section B: Diseases

A. A. Bell
ARS-USDA and Texas A&M University
College Station, Texas

Section C: Nematodes

J. A. Veech
ARS-USDA and Texas A&M University
College Station, Texas

Section D: Weeds

J. M. Chandler
Texas A&M University
College Station, Texas

Published in *Cotton*, Agronomy Monograph no. 24, © ASA-CSSA-SSSA, 677 South Segoe Road, Madison, WI 53711.

A. Insects

Management practices specifically designed to prevent or reduce damage to the cotton plant by insect and mite pests are necessary to produce a profitable crop almost everywhere in the world. Such protection of the cotton crop can best be accomplished through the integration of the most appropriate control measures into insect pest management systems.

A number of previous reviews are available on cotton insects (Newsom and Brazzel, 1968; Ridgway, 1972; Newsom, 1974; Bottrell and Adkisson, 1977; Reynolds et al., 1975; Phillips et al., 1980; Ridgway et al., 1983). This paper will focus on the importance of cotton insects and on the development of improved programs for insect pest management. Emphasis will be placed on examples using *Heliothis* spp. because of the worldwide importance of these insects.

9A-1 IMPORTANCE OF COTTON INSECTS

Aston and Winfield (1972) identified 46 major cotton insect pests in 32 countries. However, the principal insect pests in nine major cotton-producing countries may not exceed 10 or 15 species (Table 9A-1), and perhaps the majority of losses can be attributed to the following six species, among which are included three closely related bollworms: the American bollworm, *Heliothis armigera* (Hübner); the cotton bollworm, *H. zea* (Boddie); the tobacco budworm, *H. virescens* (F.); the pink bollworm, *Pectinophera gossypiella* (Saunders); the Egyptian cotton leafworm, *Spodoptera literoralis* (Boisduval); and the boll weevil, *Anthonomous grandis* Boheman. However, a number of important pests sporadically cause serious losses or may consistently cause serious losses in certain geographical areas (Table 9A-2). Additional information is available on the insect pests of cotton in various parts of the world, such as the USSR (Alimukhamedov and Khodjaev, 1978), the PRC (Anonymous, 1977a; Chu, 1981), Egypt (Isa, 1981), Africa (Pearson and Darling, 1958), Mexico (Sifuentes, 1978), and the USA (Anonymous, 1982a).

Some insight into the losses in yield caused by insects and mites can be obtained, at least for the USA, by examining some of the experimental data obtained in controlled experiments. Data summarized by Schwartz (1983) indicated that losses due to the cotton bollworm, tobacco budworm, and boll weevil exceeded 50% of the crop when control measures were not applied and ranged between 15 and 20% when control measures were applied (Table 9A-3). However, most of these data were probably collected in areas where insect infestations were known to be high; therefore, the results probably would not be representative of all cotton-growing areas in the USA.

Table 9A-1. Principal insect and mite pests in some of the major cotton producing countries. Modified from Aston and Winfield (1972).

Common name	Scientific name	No. of countries where control measures required	Most injurious insect and mite pests in some major cotton producing countries								
			USA	Mexico	Brazil	Egypt	Turkey	USSR	Pakistan	India	PRC
Pink bollworm	*Pectinophora gossypiella* (Saunders)	26	x	x	x	x	x		x	x	x
American bollworm	*Heliothis armigera* (Hübner)	24					x	x	x	x	x
Cotton bollworm	*Heliothis zea* (Boddie)	7	x	x							
Tobacco budworm	*Heliothis virescens* (F.)	4	x	x	x						
Spiny bollworm	*Earias* spp.	19					x		x	x	
Egyptian cotton leafworm	*Spodoptera literoralis* (Boisduval)	6				x					
Boll weevil	*Anthonomous grandis* (Boheman)	4	x	x							
Cotton aphid	*Aphis gossypii* (Glover)	28									x
Spider mites	*Tetranychus* spp.	20						x			
Cutworms	*Agrotis* spp. and others	16						x			

Table 9A-2. Some additional important insects attacking cotton.
Modified from Aston and Winfield (1972).

Common name	Scientific name
White flies	*Bemisia* spp.
Lygus bugs	*Lygus* spp.
Jassids	*Empoasca* spp.
Cotton stainers	*Dysdercus* spp.
Cotton fleahopper	*Pseudatomoscelis seriatus* (Reuter)
Thrips	*Frankliniella* spp. and others
Cotton leaf perforator	*Bucculatrix thurberiella* (Busck)
Cotton leafworm	*Alabama argillacea* (Hübner)
Cabbage looper	*Trichoplusia ni* (Hübner)
Red bollworm	*Diaparopsis castanea* (Hampson)
Egyptian bollworm	*Earias insulana* (Boisduval)
Spiny bollworm	*Earias biplaga* Wlk.
Spotted bollworm	*Earias fabia* (Stoll)

Table 9A-3. Estimated percentage of yield loss caused by some pests of cotton calculated
from experimental data obtained in the USA from 1945 to 1980 (Schwartz, 1983).

			Calculated losses	
Pest	Data sets	Observations per data set	With best control	Without control (untreated check)
		no.	%	
Cotton bollworm and tobacco budworm	67	7.3	14.7	63.0
Pink bollworm	1	3.0	9.2	61.0
Boll weevil	37	5.7	20.6	50.7
Spider mites	6	7.2	0.5	21.3
Cotton aphid	17	5.6	7.9	18.9

Losses in yield caused by insects to the worldwide cotton crop have
been estimated to average 16% of the potential crop (Cramer, 1967). These
loss estimates, when compared with data on cotton production in some of
the major cotton-producing countries of the world (Anonymous, 1982b),
indicate that insects are responsible each year for the loss of over 10 million
bales of cotton (Table 9A-4). The cost of control should also be included in
estimates of losses caused by insects. In 1980 nearly $1 billion was spent
worldwide for insecticides applied to cotton (Anonymous, 1981a). This estimate may be low since certain countries produce and apply insecticides that
are not likely to be reported in international market surveys. Other direct
costs of insect control include those associated with applying insecticides,
monitoring insect levels, and the application of noninsecticidal control
measures.

In addition to the losses in yields and the costs of insect control, substantial indirect losses occur as a result of the destruction of beneficial
insects and the development of insecticide resistance that result from the use
of insecticides on cotton. For instance, Pimental et al. (1980) estimated that
the annual costs in the USA from the destruction of natural enemies was
$119 million and the cost of increased insecticide resistance was $36 million.

Table 9A-4. Preliminary estimates of cotton production in some major cotton producing countries for 1981/82 and estimates of percent loss due to insects.

Country	Area†	Yield†	Production†	Loss from insects‡
	million ha	kg/ha	million bales	%
USA	5.6	608	15.6	17.0
Mexico	0.3	885	1.4	16.0
Brazil	2.0	294	2.7	15.0
Egypt	0.5	1007	2.3	13.0
Turkey	0.7	745	2.2	12.0
USSR	3.2	948	13.8	--
Pakistan	2.2	351	3.5	16.0
India	8.1	170	6.4	16.0
PRC	5.1	582	13.6	--
All other	5.7	--	9.8	--
World	33.4	465	71.3	16.1

† Anonymous (1982b).
‡ Cramer (1967).

Many estimates of losses caused by insects and mites attacking cotton are subjective and perhaps contain some bias. Nevertheless, considerable evidence exists which indicates that losses are substantial and that considerable effort is warranted to reduce these losses.

9A-2 DEVELOPMENT OF COTTON PROTECTION PROGRAMS

The development of effective programs to protect the cotton plant from insect pests requires a broadly based approach. Fundamental knowledge of the physiology, genetics, behavior, and quantitative population ecology is needed for each major insect pest. Such knowledge reveals the most vulnerable part of the insect's life system and produces the information needed to successfully utilize the preferred control measures. Control measures can be placed into three general categories: (i) cultural control, including plant growth regulators and host plant resistance; (ii) biological controls, including parasites, predators, microbial agents, and autocidal methods; and (iii) chemical controls, including pheromones, insect growth regulators, and insecticides. The various controls that need to be integrated into a management system and the strategy or combination of strategies to be employed in dealing with a specific situation should be determined.

9A-2.1 Biology and Ecology

Two of the more important control measures currently used in the USA evolved from fundamental knowledge of the physiology of the pink bollworm and boll weevil. Knowledge of the factors that cause the induction of diapause in the pink bollworm provided the basis for cultural control of the pink bollworm (Adkisson and Gaines, 1960); and knowledge of diapause in

the boll weevil led to the development of the concept of using autumn applications of insecticides or diapause control of the boll weevil (Brazzel et al.,
1961). More recently, knowledge of diapause in the bollworm and tobacco
budworm combined with monitoring of adult populations with pheromone
traps and computer modeling has provided new insights into insect migration (Hartstack et al., 1982). Also, physiological studies directed towards
growth regulation have led to the synthesis of ecodysone analogues which
may be useful in preventing reproduction in the boll weevil (Villivaso and
Thompson (1983).

As more consideration is given to management of a total insect population (Knipling, 1979), knowledge of genetic diversity within a species and relationships between closely related species will become more important. For
instance, genetic analyses of different boll weevil biotypes is providing useful insights into the diversity within the boll weevil complex (Bartlett, 1981;
Terranova, 1982). Association of genetic differences with specific biological
parameters such as diapause, pheromone response, and mating behavior are
needed to design improved insect management programs.

Quantitative population ecology as affected by biotic and abiotic
factors can also provide an important framework for development of improved pest control programs. A number of population models are available that provide a convenient basis for reviewing this important area.
Particularly useful models are available for analyzing populations of
Heliothis spp. (Fig. 9A–1; Hartstack and Witz, 1983) and the boll weevil
(Brown et al., 1983). However, considerable additional biological data is
needed to validate and refine these and other models.

9A–2.2 Cultural Controls and Host Plant Resistance

Manipulation of the cotton plant by management and by genetic means
are ways to provide varying degrees of protection from insect attack. Cultural control of the pink bollworm through stalk destruction has been used
successfully in the southwestern USA (Noble, 1969), where destroying the
plants prior to the induction of diapause is compatible with other management practices and with the yield potential for that geographical area. Conversely, in southern California and Arizona, where a longer growing season
is available, cotton producers are extremely reluctant to limit the growing
season and reduce yields. In this situation, the use of plant growth
regulators that will terminate fruiting but not significantly affect fruit
maturation shows considerable promise for providing cultural control of
the pink bollworm without reducing yield (Bariola et al., 1981).

The development of rapid fruiting cultivars and cotton production
systems that will produce adequate yields in a short period of time provides
an opportunity to grow profitable cotton crops and avoid late season pest
attack. Cotton cultivars for use in such a system have been developed for

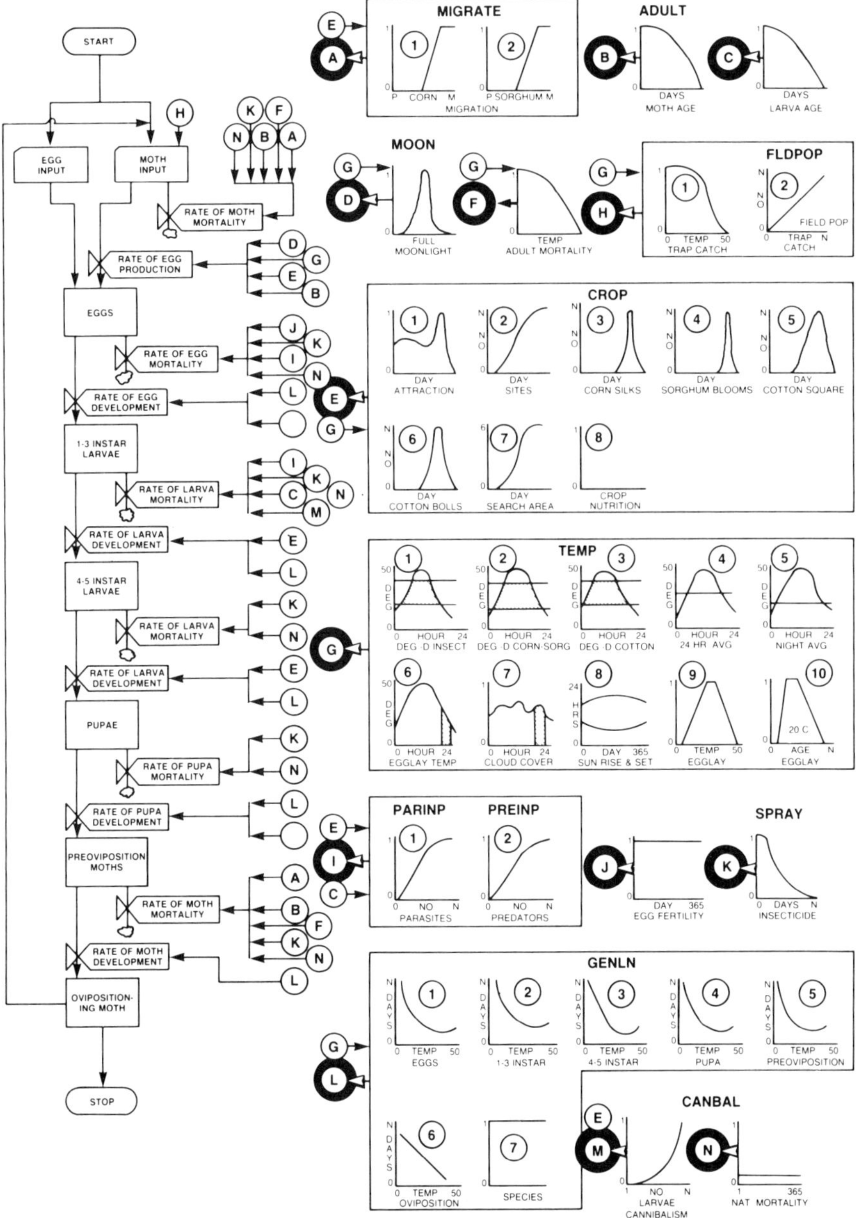

Fig. 9A-1. Flowchart illustrating the components of a population model for *Heliothis zea* and *H. virescens* (Hartstack et al., 1976).

Table 9A–5. Yields of cotton lint for three cotton genotypes, Weslaco, Texas, 1978 and 1979 (Namkin et al., 1981).

| | Cotton lint yields | | | |
| | 1978 | | 1979 | |
Cultivar or strain	130 days	Final	130 days	Final
	kg/ha			
GH-11-9	1358 a*	1567 a	513 a	853 a
Tamcot SP-37	1222 b	1373 ab	229 b	446 b
Stoneville 213	754 b	1103 c	68 b	378 b

* Means followed by the same letters are not significantly different at the 0.05% level according to Duncan's multiple range test.

selected areas within the semi-arid cotton growing regions of the south-western USA (Table 9A–5; Namkin et al., 1983). Similar systems that will provide economical yields in the higher rainfall areas of the south and southeastern USA are not yet available.

A wide range of cotton cultivars that are resistant to insect attack are available (Table 9A–6; Niles, 1980). Although many promising resistant characters are known, practical use is somewhat limited. The pubescence character has been used for a number of years in Africa and in the East to protect cotton from jassids (*Empoasca* spp.). More recently, commercial cultivars with the nectarless character have been introduced in the USA, where their use is increasing. These cultivars are capable of reducing populations of *Lygus* spp. and of the pink bollworm by up to 50% (Meredith et al., 1973; Wilson, 1980).

9A–2.3 Biological Controls

Naturally occurring parasites and predators are major regulators of insect populations (Ridgway and Lingren, 1972; Table 9A–7). Knowledge of important natural enemies of *Heliothis* spp. in the USA was reviewed and efficacy indexes estimated for some of the more important predators (Ables et al., 1983). Recent attempts to determine the effects of a complex of predators on a pest population not only emphasize the importance of conserving natural enemies, but also provide a basis for integrating natural enemies with other suppression methods (Fig. 9A–2). Substantial progress has also been made in the development of augmentation programs for control of *Heliothis* spp. Augmentations have been successful with a number of species (Ridgway and Vinson, 1977), but perhaps efforts with the egg parasite, *Trichogramma* spp., are the most advanced (Ridgway et al., 1981a; Voronin and Grinberg, 1981).

The development and use of pathogens or microbial agents for control of cotton insects have been of particular interest because they can be used to control pests without having significant effects on natural enemies. Many scientists have contributed to the development of viruses, bacterial agents, and adjuvants for use on cotton (Ignoffo et al., 1965; Dulmage, 1970; Bell

Table 9A–6. Insect and mite resistance characters in cotton† (Niles, 1980).

Trait	Boll Weevil	Heliothis spp.	Lygus spp.	Cotton fleahopper	Spider mites	Pink bollworm	Empoasca spp.	Thrips	Aphids	Cotton leaf perforator	Whitefly
Frego bract	R	N	S	S	N		S	N	N		
Nectarilessness	N	R	R	R		R	R	N	N		
Glabrousness	N	R	(?)	R(?)		R	S	S	(?)		
Terpenoids (high square gossypol, heliocides)	N	R	R	R			R	S	(?)		S
Heavy pubescence	R	S	(?)	R	N	R	R	(?)	(?)	R(?)	(?)
Red plant color	R	N	N	N					N		
Okra leaf	N										R
Oviposition-suppression factor	R										
Plant bug suppression factor			R	R							
Early-rap d fruiting	E	E				E					

† R = resistance; S = susceptible; E = escape; N = neutral; (?) = conflicting evidence or not verified.

Table 9A-7. Some common predators or groups of predators known to attack *Heliothis* spp. Adapted from Ridgway and Lingren (1972).

| | Stage of *Heliothis* spp. primarily attacked | | | |
| | | Larvae (by instar) | | |
Predator	Egg	1–2	3	4–5
Big-eyed bugs				
Lygaeidae				
Geocoris punctipes (Say)	x	x		
Damsel bugs				
Nabidae				
Reduviolus americoferus (Carayon)	x	x	x	
R. alternatus (Parshley)	x	x	x	
Minute pirate bugs				
Anthocoridae				
Orius insidiosus (Say)	x	x		
O. tristiocolor (White)	x	x		
Green lacewings				
Chrysopidae				
Chrysopa carnea (Stephens)	x	x	x	
C. oculata (Say)	x	x	x	
C. rufilabris Burmeister	x	x	x	
Lady beetles				
Coccinellidae				
Coleomegilla maculata (De Greer)	x	x		
Hippodamia convergens Guérin-Meneville	x	x		
Scymnus spp.	x	x		
Orb weaver spiders				
Araneidae		x	x	
Lynx spiders				
Oxyopidae				
Oxyopes salticus		x	x	
Jumping spiders				
Salticidae	x	x	x	x
Crab spiders				
Thomisidae		x	x	

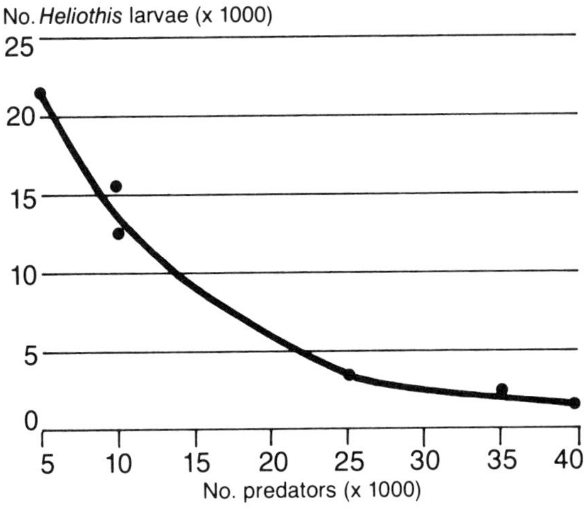

Fig. 9A-2. The relationship between the number of predators and the number of *Heliothis* spp. larvae per 0.4 ha (1 acre) (Ables et al., 1983).

and Kanavel, 1977). Although these materials will not provide adequate control of very high densities of such pests as *Heliothis* spp., some of them are in practical use, and their potential value has been clearly demonstrated (Table 9A-8; Coppedge et al., 1972; Kinzer et al., 1976; Bell, 1982). Also, the effectiveness of the microbial agents may be enhanced by the addition of low dosages of chlordimeform (Pieters et al., 1978) without drastically reducing the population of natural enemies.

Research on autocidal methods, particularly sterility, has indicated that this approach shows promise for control of such insects as the pink bollworm (Henneberry, 1980), the boll weevil (Wright and Villivaso, 1983), and the tobacco budworm (Ridgway et al., 1981b). Perhaps the most intriguing of the sterility methods is the backcross sterility that is produced by crossing the tobacco budworm with a closely related species, *Heliothis subflexa* (Guenée) (Laster, 1972). The resulting females, when mated with fertile tobacco budworm males, continue to produce fertile females and sterile males. Releases of these backcross insects on St. Croix, the Virgin Islands, have resulted in the infusion of a very high rate of sterility into the native population (Proshold, in press).

9A-2.4 Chemical Controls

Chemical controls have in the past been associated primarily with conventional and systemic insecticides. However, with the advances that have been made in the identification of naturally occurring compounds such as pheromones, hormones, and deterrents, and with the synthesis of these natural products and their mimics, the adoption of a broader concept of chemical control seems appropriate.

Table 9A-8. Yields of cotton lint after application of the bacterial agent, *Bacillus thuringiensis* Berliner, and the nuclear polyhedrosis virus, *Baculovirus heliothis,* at College Station, Texas, and Phoenix, Ariz.

	Dosage	Cotton lint yields
	———— kg/ha ————	
Experiment 1 (1967)†		
Untreated control		233 a*
B. thuringiensis	4.14	579 b
Monocrotophos	1.68	901 c
Experiment 2 (1970)‡		
Untreated control		150 a
B. heliothis	80 larval equil.	407 b
Monocrotophos	1.68	616 c
Experiment 3 (1976)§		
Untreated control	0.56	109 a
B. thuringiensis	0.56	279 b
B. thuringiensis + adjuvant	3.36	369 c

* Means followed by the same letter are not significantly different at the 5% level according to Duncan's Multiple Range Test.
† Kinzer et al. (1976).
‡ Coppedge et al. (1972).
§ Bell (1982).

The identification of sex or aggregating pheromones for the major insect pests of cotton and advances in the development of delivery systems provides very unusual opportunities and complex challenges for developing practical uses for these materials (Mitchell, 1981; Kydonieus and Beroza, 1982). The most common approaches being considered are mass trapping, mating disruption, and destruction of males with the use of attracticides. Mass trapping has been demonstrated to be effective in suppressing low density populations of the boll weevil (Table 9A-9; Lloyd et al., 1983). This suppression method has been used successfully as a component of a boll weevil eradication program (Anonymous, 1981c). Mating has been disrupted in the pink bollworm by the use of its sex pheromone, gossyplure, in field experiments resulting in a reduction of pink bollworms in bolls (Table 9A-10; Henneberry et al., 1981; Doane and Brooks, 1981). Research is also underway to explore mating disruption to control *Heliothis* spp. and the Egyptian cotton leafworm (Rothchild, 1981; Campion et al., 1981), but additional information is needed before practical control programs can be designed for these pests.

The use of an attracticide in which an insecticide is combined with pheromones to disrupt mating shows promise for insect suppression. For instance, the addition of a very small amount of a pyrethroid insecticide

Table 9A-9. Suppression of a simulated first generation population of boll weevils in cotton fields 1.6 ha in size, 1979. Adapted from Lloyd et al. (1983).

Traps per 0.4 ha	Female weevils captured†	Reduction of weevils
	Avg. no.	%
1.9	2.0	27
3.6	3.0	41
4.2	5.5	75
5.7	6.7	92

† Assuming 50% of the emerging boll weevils were females and an estimated average of 7.3 female weevils emerged in each field.

Table 9A-10. Mean number of pink bollworm larvae per 100 cotton bolls from pheromone-treated and control fields at Rainbow Valley, Arizona, 1979 (Henneberry et al., 1981).

Sampling date	Gossyplure‡	Control
9 July–16 August	0.2	0.7
21 August	1.8 a*	6.0 b
23 August	1.0	4.0
27 August	1.8	5.8
30 August	4.3 a	15.8 b
6 September	3.0 a	20.8 b
10 September	2.0	5.3
13 September	0.5	6.8
17 September	0.8 a	14.8 b
21 September	0.3	3.8
25 September	2.5 a	10.5 b
27 September	0.8	3.3

* Means followed by different letters are significantly different at the 5% level according to Duncan's multiple range test.

† These data were taken from fields that also received treatments of a two-component mixture of pheromones from the tobacco budworm.

to the adhesive used to apply gossyplure formulations to cotton has resulted in substantial increases in the level of control of the pink bollworm obtained when compared to use of gossyplure formulations alone (Staten and Haworth, 1981).

Insect growth regulators are highly selective chemicals that interfere in different ways with the normal development of certain life stages of treated insects. Although many growth regulators have been evaluated in laboratory and field experiments for control of pests of cotton, only a few substituted benzoylphenyl ureas have shown sufficient promise for practical use. Of these, diflurbenzuron is being utilized on a limited basis to control the boll weevil on cotton in the USA. This compound is effective against the boll weevil, if used in an area-wide management program, and it appears to have little adverse effect on populations of beneficial species associated with cotton (Bull et al., 1983). Diflurbenzuron has also proven to be particularly effective when used in combination with methomyl for control of the Egyptian cotton leafworm (Anonymous, 1980; Anonymous, undated).

The development and use of synthetic organic insecticides has made possible the production of higher yields of cotton in areas where insects would have otherwise prevented the production of an economical cotton crop. The major classes of synthetic organic insecticides that have or that are currently being used on cotton include the chlorinated hydrocarbons, the organophosphates, the carbamates, the chlorformandines, and the synthetic pyrethroids. The common and chemical names of these toxicants currently used on cotton in the USA are given in Table 9A–11. Current suggested rates of applications and annual revisions are available in the Annual Report of the Cotton Insect Research and Control Conference, which is published each year (Anonymous, 1982a).

Table 9A–11. Common and chemical names of insecticides used for cotton pest control in the USA (Anonymous, 1982a).

Common name	Chemical name	Other designation†
acephate	O,S-dimethyl acetylphosphoramidothioate	Ortho 12,420; xOrthene
aldicarb	2-methyl-2-(methylthio)propionaldehyde O-(methylcarbamoyl)oxime	Union Carbide; 21149; UC 21149; xTemik
azinphosmethyl	O,O-dimeyl S-[(4-oxo-1,2,3-benzotriazin-3(4H)-yl)methyl]phosphorodithioate	*Guthion
carbaryl	1-naphthyl methylcarbamate	*Sevin
carbophenothion	S-[[(p-chlorophenyl)thio]methyl] O-O-diethyl phosphorodithioate	*Trithion
chlordimeform	N'-(4-chloro-o-tolyl)-N-N-dimethyl-formamidine	*Galecron; xFundal
chlorpyrifos	O,O-diethyl O-(3,5,6-trichloro-2-pyridyl) phosphorothioate	*Lorsban
demeton	O,O-diethyl O (and S)-[2-(ethylthio)ethyl] phosphorothioate	*Systox; mercaptophos
diazinon	O,O-diethyl O-(2-isopropyl-6-methyl-4-pyrimidinyl) phosphorothioate	*Spectracide
dicofol	4,4′ dichloro (trichloromethyl)benzhydrol	*Kelthane
dicrotophos	dimethyl phosphate ester of (E)-3-hydroxy-N,N-dimethylcrotonamide	*Bidrin

(continued on next page)

Table 9A–11. Continued.

Common name	Chemical name	Other designation†
dimethoate	*O,O*-dimethyl *S*-(methylcarbamoylmethyl) phosphorodithioate	*Rogor; *Cygon
disulfoton	*O,O*-diethyl *S*-[2-(ethylthio)ethyl] phosphorodithioate	*Di-Syston; thiodemeton
endosulfan	6,7,8,9,10,10-hexachloro-1,5,5a,6,9,9a-hexahydro-6,9-methano-2,4,3-benzodioxathiepin 3-oxide	*Thiodan
endrin	1,2,3,4,10,10-hexachloro-6,7-epoxy-1,4,4a,5,6,7,8,8a-octahydro-1,4-*endo-endo*-5,8-dimethanonaphthalene	Compound 269
EPN	*O*-ethyl *O*-(*p*-nitrophenyl) phenylphosphonothioate	EPN 300
ethion	*O,O,O′,O′*-tetraethyl *S,S′*-methylene bis(phosphorodithioate)	*Nialate
fenvalerate	cyano(3-phenoxyphenyl)methyl 4-chloro--(methylethyl) = benzeneacetate	*Pydrin
malathion	*O,O*-dimethyl phosphorodithioate of diethyl mercaptosuccinate	*Cythion
methamidophos	*O,S*-dimethyl phosphoramidothioate	*Monitor; *Tamaron
methidathion	*O,O*-dimethyl phosphorodithioate S-ester with 4-(mercaptomethyl)-2-methoxy- -1,3,4-thiadiazolin-5-one	*Supracide; Ultracide
methomyl	*S*-methyl *N*-[(methylcarbamoyl)oxy]thioacetimidate	*Lannate; *Nudrin
methyl parathion	*O,O*-dimethyl O-(*p*-nitrophenyl) phosphorothioate	*Metacide; 'Wofatox
monocrotophos	dimethyl phosphate ester with (*E*)-3-hydroxy-*N*-methylcrotonamide	*Azodrin
naled	1,2-dibromo-2,2-dichloroethyl dimethyl phosphate	*Dibrom
oxydemeton-methyl	*S*-[2(ethylsulfinyl)ethyl] *O,O*-dimethyl phosphorothioate	*Metasystox-R
parathion	*O,O*-diethyl *O*-(p-nitrophenyl) phosphorothioate	*Thiophos; *Niran
permethrin	(3-phenoxyphenyl)methyl 3-(2,2-dichloroéthenyl)-2,2-dimethylcyclopropanecarboxylate	*Ambush PP557
phenamiphos	ethyl 4-(methylthio)-*m*-tolyl isopropylphosphoramidate	*Nemacur
phorate	*O,O*-diethyl *S*-diethyl *S*-[(ethylthio)methyl] phosphorodithioate	*Thimet
phosphamidon	2-chloro-2-diethylcarbamoyl-1-methylvinyl dimethyl phosphate	*Dimecron
propargite	2-(*p-tert*-butylphenoxy)cyclohexyl 2-propynyl sulfite	*Comite; *Omite
sulfur	sulfur	
sulprofos	*O*-ethyl *O*-[4-methylthio)]phenyl] *S*-propyl phosphorodithioate	*Bolstar; Bay NTN 9306
toxaphene	chlorinated comphene containing 67% to 69% chlorine	Camphechlor
trichlorfon	dimethyl (2,2,2-trichloro-1-hydroxyethyl) phosphonate	*Dipterex; *Dylox

† * Indicates a proprietary name.

Table 9A-12. Ten major insecticides used on cotton in the USA, 1979 (McDowell et al., 1982).

	Hectare treatments[†]		
Insecticide	Single material applications	Tank mix applications	Total
		1000	
Chlordimeform	749	1478	2227
Dimethoate	489	15	504
Dicrotophos	847	66	913
EPN	12	1948	1960
Fenvalerate	1390	122	1512
Methomyl	220	320	540
Methyl parathion	423	2789	3212
Parathion	70	234	304
Permethrin	1344	723	2067
Toxaphene	15	383	398

† Hectare treatments are the product of the number of hectares treated × the number of treatments applied.

Worldwide information on the specific insecticides currently used on cotton is limited. However, in the USA, where nearly 25%, based on dollar value, of all cotton insecticides are used, the major insecticides include methyl parathion, chlordimeform, permethrin, EPN, and fenvalerate (Table 9A-12; McDowell et al., 1982).

Even though synthetic organic insecticides will continue to be essential to combat insect pests of cotton, reliance upon these materials has resulted in the development of resistance to insecticides in many species of insects and mites and in the resurgence of secondary insect pests as a result of the destruction of natural enemies. Specifically, resistance to one or more classes of insecticides has been reported for all of the major insect and mite pests of cotton (Georghiou, 1980; Wolfenbarger et al., 1982; Anonymous, 1982a). Perhaps the greatest concern is the development of resistance to chlorinated hydrocarbon, organophosphate, and carbamate insecticides by such pests as *Heliothis* spp. and the Egyptian cotton leafworm (Georghiou, 1980). The resurgence of secondary pests resulting from destruction of natural enemies by insecticides includes the use of both foliar and systemic insecticides to control such insects as the boll weevil and the cotton flea-hopper, *Pseudatomoscelis seriatus* (Reuter) (Ridgway et al., 1967; Ridgway, 1969; and Ridgway and Lingren, 1972) and foliar applications to control *Lygus* spp. (van den Bosch et al., 1971). More recently, the whitefly, *Bemisia tabaci* (Gennadius), has emerged as a serious pest of cotton in parts of Africa (Anonymous, 1981b) and in western USA (Anonymous, 1982a). Although the causes of the outbreaks are not yet known, insecticide resistance and secondary pest resurgence associated with insecticide use are potential causes. Because of the adverse side effects of using insecticides, judicious use must be emphasized, and decisions on their use should be made within the context of a knowledge of all available suppression methods.

9A-2.5   Insect Management Systems

Knowledge of the biology and ecology of pest and beneficial insects and of the available cultural, biological, and chemical control measures provides the basis for designing and implementing improved insect management systems. Likewise, the strategies to be utilized should be considered when designing an integrated system. Three of the four strategies suggested by the USDA (Anonymous, 1977b) are particularly pertinent to the scope of this discussion: (i) local population management, (ii) area-wide management, and (iii) elimination of a population from a geographically defined area.

Efficient insect management systems will rely on various decision-making technologies in the selection and use of control measures. A wide range of survey methods designed to assess insect populations and plant conditions are available. Of the methods available, manual inspection to estimate numbers of pests, beneficial species, and fruiting forms is the most commonly utilized in implementing the local population management strategy. Descriptions of a number of sampling methods and treatment thresholds for the *Heliothis* spp. are available (Sterling, 1979).

The use of manual surveys can be illustrated by the following recommendation developed by the Arkansas Agricultural Extension Service (Barnes et al., 1978): Treatment thresholds for *Heliothis* spp. in Arkansas are tied closely to the phenology of the cotton plant. No control recommendations are made for *Heliothis* spp. until first flowering. After appearance of the first flowers through about the 6th week of squaring, treatment is recommended when one damaged square is found per 0.3 m of row and eggs and larvae are present. During the period of high squaring rate, 7th through the 10th week, treatment is suggested when one square per 0.3 m of row is damaged. Eggs and larvae should also be present during this period. During the boll maturation period, treatment is suggested when one damaged square per 0.3 m of row or one damaged boll per 1.0 m of row is found. No treatment for *Heliothis* spp. is normally made after 15 to 20 September.

Substantial progress has also been made in utilizing quantitative assessments of the number of predators when deciding whether or not to apply control measures for *Heliothis* spp. A decision-making chart is available that provides estimates of the number of predators required to prevent a treatment threshold from being surpassed when varying numbers of pest eggs are present (Fig. 9A-4) (Ables et al., 1983).

Area-wide surveillance of insects can be used to improve local population management. For example, light and pheromone traps can be used to determine the need for and the timing of manual surveys. Pheromone trap catches of *Heliothis* spp. and weather data can be used in conjunction with a computer model to predict the timing of oviposition by *Heliothis* spp. (Fig. 9A-3; Hartstack et al., 1976; Hartstack and Witz, 1983). After 3 years of using this technique, the Texas Agricultural Extension Service reported that Texas cotton producers realized in a combined net benefit of about $5 million in 1979 (Hartstack and Witz, 1983).

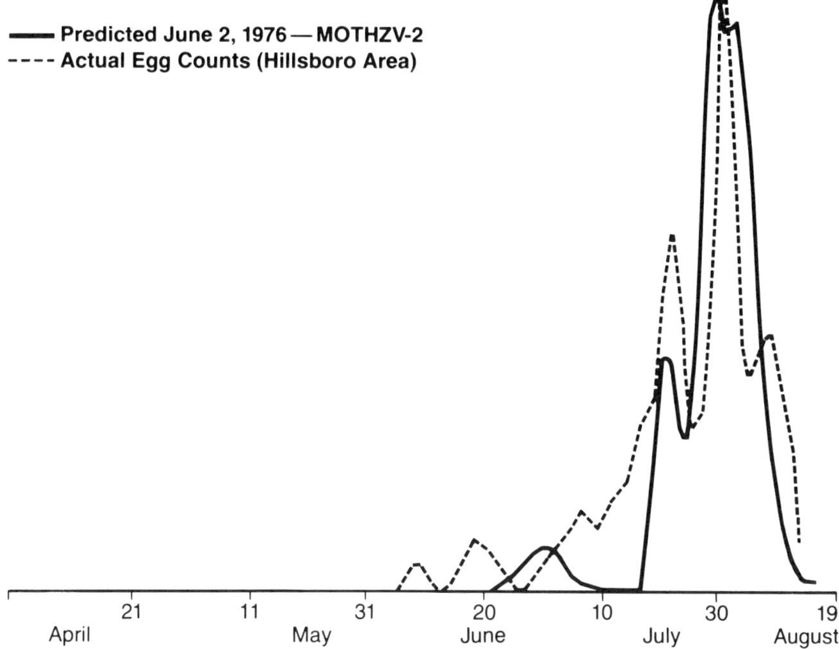

Fig. 9A-3. A comparison of the forecast of *Heliothis zea* oviposition with the actual field egg counts, Hillsboro, Texas, 1976. The two curves were adjusted to the same maximum number of eggs to emphasize the comparisons of the time of oviposition (Hartsack and Witz, 1983).

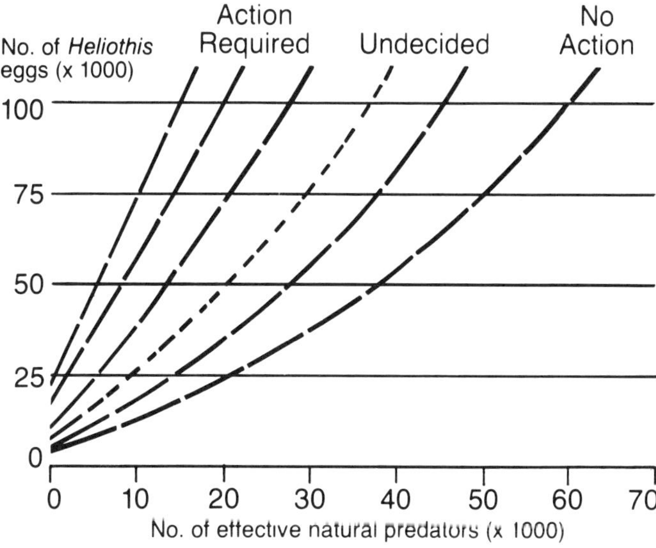

Fig. 9A-4. A decision-making chart showing the number of naturally occurring predators and *Heliothis* spp. eggs per 0.4 ha of cotton. A required corrective action is expected when the number of large *Heliothis* spp. larvae reaches or exceeds 0.4 ha (Ables et al., 1983).

The area-wide population management strategy may evolve from the local population management strategy (Rabb, 1972). A number of control measures will either benefit from or will require use on an area-wide basis. Cultural methods such as stalk destruction for control of the pink bollworm have been used on an area-wide basis for several years (Adkisson and Gaines, 1960), but as new technologies are developed, such as the use of pheromones for mass trapping, mating disruption, or attracticides and improved biological controls, area-wide management of insect populations will become an increasingly important control strategy. Very recently a pink bollworm abatement district has been formed to implement the use of pink bollworm pheromone on all cotton land in the northern part of the Imperial Valley in California (Hendrichs, 1982), and mass trapping on an area-wide basis has been suggested for control of the Egyptian cotton leafworm (Gubbins and Campion, 1982).

The elimination of reproduction or the actual eradication of an insect population from a geographically defined area may be an appropriate step in the evolution of insect control actions under some conditions (Ridgway and Lloyd, 1983). For instance, the boll weevil is an exotic insect in the USA that has a very limited host range, and a number of suppression technologies are available for use against a total population (Ridgway et al., 1983). The application of selected technologies against the boll weevil in northern North Carolina apparently has eliminated reproduction of the boll weevil (Anonymous, 1981c) on about 16 000 ha of cotton.

A number of local, area-wide, and elimination insect management systems have been presented previously: (i) a short-season management system for cotton production for Texas (Frisbie et al., 1982); (ii) a community-wide bollworm management program in Arkansas (Phillips et al., 1980); (iii) a cotton insect management program for the San Joaquin Valley of California (Phillips et al., 1980); (iv) an optimum pest management program in Mississippi (Hamer et al., 1983); and (v) a boll weevil eradication program in North Carolina (Ganyard et al., 1981).

These examples, together with specific information from a local area, can provide the basis for development of an improved insect management system for other areas. Moreover, economic evaluation of a specific insect management system is highly desirable and can provide a sound basis for continuing the use and improvement of a particular system (Lacewell et al., 1977; Simpson and Parvin, 1983; Carlson and Suguiyama, 1983). For instance, Carlson and Suguiyama (1982) calculated that the return on investment was about 20% from an insect elimination management system implemented in North Carolina. These results would indicate that this particular system was economically sound and its operation should be continued. Also, this economic analysis provides a basis for identifying ways of reducing costs of similar programs in the future.

9A-3 FUTURE PROSPECTS

The world cotton crop will continue to be threatened by insect pests, and insecticides will continue to be essential for protecting the crop from attack by insect pests. However, integrated insect management systems for the future are likely to utilize new technologies, such as the use of pheromones and insect growth regulators, and are likely to include increased use of resistant cultivars and modified cotton cultural methods designed to avoid insect attack. Increased use of biological controls, such as microbial agents and augmentation of natural enemies, is also likely. Because of the complexity associated with some of the new technologies, emphasis is likely to be concentrated on using them against certain key pests such as *Heliothis* spp., the pink bollworm, the boll weevil, and perhaps the Egyptian cotton leafworm. Also, with more use of methods designed for suppression of specific insect pests and with the development of improved decision-making methods, insecticides are likely to be used more selectively in the future than they have been in the past.

Mention of pesticides does not constitute a recommendation for use. The use of trade names does not constitute a guarantee, warranty, or endorsement of the products by the USDA.

REFERENCES

Ables, J. R., J. L. Goodenough, A. W. Hartstack, and R. L. Ridgway. 1983. Entomophagous arthropods. p. 103-128. Chap. 5. *In* R. L. Ridgway, E. P. Lloyd, and W. H. Cross (ed.) Cotton insect management with special reference to the boll weevil. Agric. Handb. 589, USDA, Washington, D.C.

Adkisson, P. L., and J. C. Gaines. 1960. Pink bollworm control. Texas Agric. Exp. Stn. Misc. Publ. 444.

Alimukhamedov, S., and S. H. Khodjaev (ed.). 1978. Pests of cotton and methods of their control. *In* Major pests of cotton. [In Russian]. Publ. House, Vsbeebirta, Tashkent.

Anonymous, Undated. Official Egyptian pest control manual. p. 244-247.

----. 1977a. Insect control in the People's Republic of China. Committee on Scholarly Communication with the People's Republic of China, Rept. No. 2. National Academy of Sciences, Washington, D.C.

----. 1977b. U.S.D.A. policy on management of pest problems. Secretary's Memorandum No. 1929.

----. 1980. Label for Deenate (Registered trademark for a mixture of diflurbenzuron and methomyl). E. I. DuPont de Nemours & Co., Inc., Wilmington, Del.

----. 1981a. A look at world pesticide markets. Farm Chem. September p. 55, 58.

----. 1981b. Integrated pest control in agriculture. Rep. Tenth Session FAO/UNEP Panel of Experts Mtg., FAO Hdqtrs., Rome.

----. 1981c. Biological evaluation of alternative beltwide boll weevil/cotton insect management programs. Overall Evaluation, App. A. SEA Staff Rep. USDA, Washington, D.C.

----. 1982a. Annual (35th) Conf. Rep. on Cotton-Insect Research and Control, Las Vegas, Nev. Southern Reg., ARS, USDA, New Orleans, La.

----. 1982b. World crop production. Foreign Agric. Circ. WCP-7-82. FAS/ERS, USDA, Washington, D.C.

Aston, J. L., and B. Winfield. 1972. Insect section introduction. p. 44–45. *In* Cotton, CIBA-Geigy Agrochemicals Tech. Monogr. No. 3.

Bariola, L. A., T. J. Henneberry, and D. L. Kittock. 1981. Chemical termination and irrigation cutoff to reduce overwintering populations of pink bollworms. J. Econ. Entomol. 74:106–109.

Barnes, G., J. J. Kimbrough, and M. L. Wall. 1978. Cotton insect management program. Arkansas Coop. Ext. Lflt. No. 52.

Bartlett, A. C. 1981. Isozyme polymorphisms in boll weevils and thurberia weevils from Arizona. Ann. Entomol. Soc. Am. 74:359–362.

Bell, M. R. 1983. Microbial agents. p. 129–152. Chap. 6. *In* R. L. Ridgway, E. P. Lloyd, and W. H. Cross (ed.) Cotton insect management with special reference to the boll weevil. Agric. Handb. 589, USDA, Washington, D.C.

----, and R. F. Kanavel. 1977. Field tests of a nuclear polyhedrosis virus in a bait formulation for control of pink bollworms and *Heliothis* spp. on cotton in Arizona. J. Econ. Entomol. 70:625–629.

Bottrell, D. G., and P. L. Adkisson. 1977. Cotton insect pest management. Annu. Rev. Entomol. 22:451–481.

Brazzel, J. R., T. B. Davich, and L. D. Harris. 1961. A new approach to boll weevil control. J. Econ. Entomol. 54:723–730.

Brown, L. G., R. W. McClendon, and J. W. Jones. 1983. Cotton and insect management simulation model. p. 437–479. Chap. 17, *In* R. L. Ridgway, E. P. Lloyd, and W. H. Cross (ed.) Cotton insect management with special reference to the boll weevil. Agric. Handb. 589, USDA, Washington, D.C.

Bull, D. L., J. R. Ables, and E. P. Lloyd. 1983. Insect growth regulators with emphasis on the use of benzoylphenyl ureas. p. 207–236. Chap. 9. *In* R. L. Ridgway, E. P. Lloyd, and W. H. Cross (ed.) Cotton insect management with special reference to the boll weevil. Agric. Handb. 589, USDA, Washington, D.C.

Campion, D. G., L. J. McVeigh, P. Hunter-Jones, D. R. Hall, R. Lester, B. F. Nesbitt, G. J. Marrs, and M. R. Alder. 1981. Evaluation of microencapsulated formulations of pheromone components of the Egyptian cotton leafworm in Crete. p. 253–265. *In* E. R. Mitchell (ed.) Management of insect pests with semiochemicals—Concepts and practice. Plenum Press, N.Y.

Carlson, G. A., and L. F. Suguiyama. 1983. Economic evaluation of the boll weevil eradication trial in North Carolina, 1978-80. p. 497–517. Chapt. 19. *In* R. L. Ridgway, E. P. Lloyd, and W. H. Cross (ed.) Cotton insect management with special reference to the boll weevil. Agric. Handb. 589, USDA, Washington, D.C.

Chu, H. F. 1981. Integrated pest management of cotton in China. p. 537–539. *In* T. Kommedahl (ed.) Proc. Symp. IX Intl. Congr. Plant Protect. II: Integrated plant protection for agricultural crops and forest trees. Washington, D.C. Burgess Publ. Co., Minneapolis, Minn.

Coppedge, J. R., R. E. Kinzer, and R. L. Ridgway. 1972. Field evaluations of *Bacillus thuringiensis* and *Heliothis* nuclear polyhedrosis virus for control of *Heliothis* on cotton. p. 43–48. *In* Consol. Prog. Rept. PR-3086. Texas A&M Univ., College Station, Tex.

Cramer, H. H. 1967. Plant protection and world crop production. Farbenfabriken Bayer Ag., Leverkusen, W. Germany.

Doane, C. G., and T. W. Brooks. 1981. Research and development of pheromones for insect control. p. 285–304. *In* E. R. Mitchell (ed.) Management of insect pests with semiochemicals—Concepts and practice. Plenum Press, N.Y.

Dulmage, H. T. 1970. Insecticidal activity of HD-1, a new isolate of Bacillus thuringiensis. J. Invertebr. Pathol. 15:232–239.

Frisbie, R. E., J. R. Phillips, W. R. A. Lambert, and H. B. Jackson. 1983. Opportunities for improving cotton insect management programs and some constraints on beltwide implementation. p. 521–558. Chap. 20. *In* R. L. Ridgway, E. P. Lloyd, and W. H. Cross (ed.) Cotton insect management with special reference to the boll weevil. Agric. Handb. 589, USDA, Washington, D.C.

Ganyard, M. C., J. R. Brazzel, J. H. Dillier, and A. E. Miller. 1981. Boll weevil eradication trial. p. 38–40. Proc. Cotton Prod. Mech. Conf., New Orleans, La.

Georghiou, G. P. 1980. Insecticide resistance and prospects for its management. Residue Rev. 76:131–145.

Gubbins, K. E., and D. G. Campion. 1982. Economic aspects of pheromone trapping techniques for the control of Egyptian cotton leafworm. Outlook Agric. 11:62–66.

Hamer, J. L., G. L. Andrews, R. W. Seward, D. F. Young, Jr., and R. B. Head. 1983. Optimum pest management trial in Mississippi. p. 385–408. Chapt. 15. *In* R. L. Ridgway, E. P. Lloyd, and W. H. Cross (ed.) Cotton insect management with special reference to the boll weevil. Agric. Handb. 589, USDA, Washington, D.C.

Hartstack, A. W., and J. A. Witz. 1983. Models for cotton insect pest management. p. 359–381. Chap. 14. *In* R. L. Ridgway, E. P. Lloyd, and W. H. Cross (ed.) Cotton insect management with special reference to the boll weevil. Agric. Handb. 589, USDA, Washington, D.C.

----, J. D. Lopez, R. A. Muller, E. G. King, W. Sterling, J. Witz, and A. Eversol. 1982. Evidence of long-range spring migration of *Heliothis zea* (Boddie) into Texas and Arkansas. Southwest. Entomol. 7:188–201.

----, ----, J. P. Hollingsworth, R. L. Ridgway, and J. D. Lopez. 1976. MOTHZV-2: a computer simulation of *Heliothis zea* and *Heliothis virescens* population dynamics. Users manual. ARS-S-127. USDA, Washington, D.C.

Hendrichs, T. 1982. Imperial cotton pest program cranks up. Southwest Farm Press, 12 June. p. 17, 31.

Henneberry, T. J. 1980. Potential for sterile moth releases for pink bollworm management. p. 52–66. *In* H. M. Graham (ed.) Pink bollworm control in the western United States. Agric. Rev. Mans. ARM-W-16. SEA-USDA, Washington, D.C.

----, L. A. Bariola, H. M. Flint, P. D. Lingren, J. M. Gillespie, and A. F. Kydonieus. 1981. Pink bollworm and tobacco budworm mating disruption studies on cotton. p. 267–284. *In* E. R. Mitchell (ed.) Management of insect pests with semiochemicals—Concepts and practice. Plenum Press, N.Y.

Ignoffo, C. M., A. J. Chapman, and D. F. Martin. 1965. The nuclear-polyhedrosis virus of *Heliothis zea* (Boddie) and *Heliothis virescens* (Fabricius). III. The effectiveness of the virus against field populations of *Heliothis* on cotton, corn, and grain sorghum. J. Invertebr. Pathol. 7:227–235.

Isa, A. L. 1981. Cotton pest problems in Egypt. p. 545–546. *In* T. Kommedahl (ed.). Proc. Symp. IX Intl. Congr. Plant Protect., Vol. II: Integrated plant protection for agricultural crops and forest trees. Washington, D.C. Burgess Publ. Co., Minneapolis, Minn.

Kinzer, R. E., L. A. Bariola, R. L. Ridgway, and S. L. Jones. 1976. Systemic insecticides and a nuclear polyhedrosis virus for control of the bollworm and tobacco budworm on cotton. J. Econ. Entomol. 69:697–701.

Knipling, E. F. 1979. The basic principles of insect population suppression and management. Agric. Handb. No. 512, USDA, Washington, D.C.

Kydonieus, A. F., and M. Beroza (ed.). 1982. Insect suppression with controlled release pheromone systems. Vol. I. CRC Press, Inc., Boca Raton, Fla.

Lacewell, R. D., J. E. Casey, and R. E. Frisbie. 1977. An evaluation of integrated pest management programs in Texas: 1964–1974. Texas Agric. Exp. Stn., Tech. Rep. No. 77-4.

Laster, M. L. 1972. Interspecific hybridization of *Heliothis virescens* and *H. subflexa*. Environ. Entomol. 1:682–687.

Lloyd, E. P., G. H. McKibben, J. E. Leggett, and A. W. Hartstack. 1983. Pheromones for survey, detection, and control. p. 179–205. Chap. 8. *In* R. L. Ridgway, E. P. Lloyd, and W. H. Cross (ed.) Cotton insect management with special reference to the boll weevil. Agric. Handb. 589, USDA, Washington, D.C.

McDowell, R., C. Marsh, and C. Osteen. 1982. Insecticide use on cotton in 1979. Econ. Res. Serv. Staff Rep. No. AGES 820519. USDA, Washington, D.C.

Meredith, W. R., Jr., C. D. Ranney, M. L. Laster, and R. R. Bridge. 1973. Agronomic potential of nectarless cotton. J. Environ. Qual. 2:141–144.

Mitchell, E. R. (ed.). 1981. Management of insect pests with semiochemicals—Concepts and practice. Plenum Press, N.Y.

Namkin, L. N., M. D. Heilman, N. J. Jenkins, and P. A. Miller. 1983. Plant resistance and modified cotton culture. p. 73–101. Chap. 4. *In* R. L. Ridgway, E. P. Lloyd, and W. H. Cross (ed.) Cotton insect management with special reference to the boll weevil. Agric. Handb. 589, USDA, Washington, D.C.

Newsom, L. D. 1974. Pest management: History, current status and future progress. p. 1–18. *In* Proc. Summ. Inst. Biol. Contr. Plant Insects and Dis. F. G. Maxwell and F. A. Harris (ed.). Univ. of Mississippi, Jackson, Miss.

————, and J. R. Brazzel. 1968. Pests and their control. p. 367–405. *In* F. C. Elliott et al. (ed.) Advances in production and utilization of cotton. Iowa State University Press, Ames.

Niles, G. A. 1980. Breeding cotton for resistance to insect pests. p. 337–369. *In* G. F. Maxwell and P. R. Jennings (ed.) Breeding plants resistant to insects. John Wiley and Sons, N.Y.

Noble, L. W. 1969. Fifty years of research on the pink bollworm in the United States. Agric. Handb. No. 357, USDA, Washington, D.C.

Pearson, E. O., and R. C. M. Darling. 1958. The insect pests of cotton in tropical Africa. Empire Cotton Growing Corp. and Commonwealth. Inst. Entomol., London.

Phillips, J. R., A. P. Gutierrez, and P. L. Adkisson. 1980. General accomplishments toward better insect control in cotton. p. 123–154. *In* C. B. Huffaker (ed.) Technology of pest control. John Wiley and Sons, N.Y.

Pieters, E. P., S. Y. Young, W. C. Yearian, W. L. Sterling, D. R. Melville, and F. R. Gilliland. 1978. Efficacy of Elcar-chlordimeform and Dipel-chlordimeform mixtures for control of *Heliothis* spp. on cotton. Southwest. Entomol. 3:237–240.

Pimentel, D., D. Andow, D. Gallahan, I. Schreiner, T. E. Thompson, R. Dyson-Hudson, S. N. Jacobson, M. A. Irish, S. F. Kroop, A. M. Moss, M. D. Shepard, and B. G. Vinzant. 1980. Pesticides: Environmental and social costs. p. 99–158. *In* D. Pimentel and J. H. Perkins (ed.) Pest control: Cultural and environmental aspects, AAAS Selected Symp. 43, Westview Press, Boulder, Colo.

Proshold, F. I. Release of backcross insects on St. Croix, U.S. Virgin Islands, to suppress the tobacco budworm (*Lepidopter: Noctuidae*): Infusion of sterility into a native population of tobacco budworm. J. Econ. Entomol. (in press).

Rabb, R. L. 1972. Principles and concepts of pest management. p. 6–29. *In* Proc. Natl. Insect Mgmt. Wkshp., Purdue Univ., West Lafayette, Ind.

Reynolds, H. T., P. L. Adkisson, and R. F. Smith. 1975. Cotton insect pest management. p. 379–444. *In* R. L. Metcalf and W. Luchmann (ed.) The introduction to pest management. John Wiley and Sons, N.Y.

Ridgway, R. L. 1969. Control of the bollworm and tobacco budworm through conservation and augmentation of predaceous insects. p. 127–144. *In* Proc. Tall Timbers Conf. Ecol. Animal Control by Habitat Mgmt. Tallahassee, Fla.

————. 1972. Integrated control of cotton insects in the United States. p. 190–195. *In* Methods of integrated insect control in cotton, Doc. 8, Intl. Cotton Advis. Comm., Managua, Nicaragua.

————, J. R. Ables, C. Goodpasture, and A. W. Hartstack. 1981a. *Trichogramma* and its utilization for crop protection in the U.S.A. p. 41–48. *In* J. R. Coulson (ed.) Use of beneficial organisms in the control of crop pests. Proc. Joint Am.-Soviet Conf. Entomological Society of America, College Park, Md.

————, and P. D. Lingren. 1972. Predaceous and parasitic arthropods as regulators of *Heliothis* populations. p. 48–56. *In* Distribution abundance and control of *Heliothis* species in cotton and other host plants. South. Coop. Ser. Bull. No. 169.

————, ————, C. B. Cowan, Jr., and J. W. Davis. 1967. Populations of arthropod predators and *Heliothis* spp. after application of systemic insecticides. J. Econ. Entomol. 60:1012–1016.

----, and E. P. Lloyd. 1983. Evolution of cotton insect management in the United States. p. 3-28. Chap. 1. *In* R. L. Ridgway, E. P. Lloyd, and W. H. Cross (ed.) Cotton insect management with special reference to the boll weevil. Agric. Handb. 589, USDA, Washington, D.C.

----, ----, and W. H. Cross (ed.). 1983. Cotton insect management with special reference to the boll weevil. Agric. Handb. No. 589, USDA, Washington, D.C.

----, F. I. Proshold, T. J. Henneberry, D. F. Martin, and D. Keaveny. 1981b. St. Croix studies with sterile pink bollworms and backcross tobacco budworms. p. 152-153. Proc. Beltwide Cotton Prod. Res. Conf., New Orleans, La.

----, and S. B. Vinson (ed.). 1977. Biological control by augmentation of natural enemies. Plenum Press, N.Y.

Rothchild, G. H. L. 1981. Mating disruption of lepidopterous pests: Current status and future prospects. p. 207-228. *In* E. R. Mitchell (ed.) Management of insect pests with semiochemicals—Concepts and practice. Plenum Press, N.Y.

Schwartz, P. H. 1983. Losses in yield of cotton due to insects. p. 329-358. Chap. 13. *In* R. L. Ridgway, E. P. Lloyd, and W. H. Cross (ed.) Cotton insect management with special reference to the boll weevil. Agric. Handb. 589, USDA, Washington, D.C.

Sifuentes, J. A. 1978. Cotton pests in Mexico. (In Spanish). Inst. Nac. Investig. Agric., Mexico City, Mexico.

Simpson, E. H., III, and D. W. Parvin, Jr. 1983. Impact of alternative cotton insect management strategies on producer income in Mississippi. p. 481-496. Chap. 18. *In* R. L. Ridgway, E. P. Lloyd, and W. H. Cross (ed.) Cotton insect management with special reference to the boll weevil. Agric. Handb. 589, USDA, Washington, D.C.

Staten, R. T., and J. K. Haworth. 1981. Device for insect control. U.S. Pat. Appl. Serial No. 252,992.

Sterling, W. L. (ed.). 1979. Economic thresholds and sampling of *Heliothis* species on cotton, corn, soybeans, and other host plants. South. Coop. Ser. Bull. No. 231.

Terranova, A. C. 1982. Inheritance of estarases in *Anthonomus grandis.* Ann. Entomol. Soc. Am. 75:261-265.

van den Bosch, R., T. F. Leigh, L. A. Falcon, V. M. Stern, D. Gonzalez, and K. S. Hagen. 1971. The development program of integrated pest management of cotton insects in California. p. 377-394. *In* C. B. Huffaker (ed.) Biological control. Plenum Press, N.Y.

Villivaso, E., and M. J. Thompson. 1983. Synthetic ecdysteroids plus acute irradiation: method for sterilizing the boll weevil (*Cleoptera: Curculionidae*). J. Econ. Entomol. 76:63-68.

Voronin, K. E., and A. M. Grinberg. 1981. The current status and prospects of *Trichogramma* utilization in the U.S.S.R. p. 49-51. *In* J. R. Coulson (ed.) Use of beneficial organisms in the control of crop pests. Proc. Joint Am.-Soviet Conf. Entomol. Soc. Am., College Park, Md.

Wilson, F. D. 1980. Cotton cultivars resistant to the pink bollworm. p. 46-51. *In* H. M. Graham (ed.) Pink bollworm control in the western United States. Agric. Revs. Manuals ARM-W-16, SEA-USDA, Washington, D.C.

Wolfenbarger, D. A., V. P. R. Bodegas, and G. R. Flores. 1982. Development of resistance in *Heliothis* spp. in the Americas, Australia, Africa, and Asia. Bull. Entomol. Soc. Am. 27:181-185.

Wright, J. E., and E. J. Villavaso. 1983. Boll weevil sterility. p. 153-178. Chap. 7. *In* R. L. Ridgway, E. P. Lloyd, and W. H. Cross (ed.) Cotton insect management with special reference to the boll weevil. Agric. Handb. 589, USDA, Washington, D.C.

B. Diseases

Diseases cause losses in cotton production wherever the crop is grown. The types and relative importance of diseases vary considerably from one area to another depending on the *Gossypium* species, the environment, cultural practices, and availability of alternate hosts of pathogens. Estimates of losses from the major infectious diseases and total disease losses in the United States are shown in Table 9B-1.

Cotton disease losses are very difficult to assess and are probably greatly underestimated, because many of the most important ones are caused by soilborne pathogens. These pathogens often debilitate plants, reducing boll set and yield without appreciably affecting the size or outward appearance of the plant. Losses from individual diseases are also difficult to assess because of the complex interactions among fungal pathogens, nematodes, and biological antagonists—all of which occur together in soils. Fungicides invariably destroy some beneficial microorganisms (biological antagonists of pathogens, nitrogen-fixing bacteria, mycorrhizal fungi, etc.) as well as pathogens; thus estimates of losses are conservative. Resistant cultivars, likewise, rarely control diseases completely and may be overcome by disease complexes, again causing conservative estimates. Recent experiments with soil solarization (Katan, 1980; Pullman et al., 1981) and rotation with paddy rice (Pullman and DeVay, 1981), which preserve beneficial organisms while destroying complexes of pathogens, have indicated disease losses of 15 to 65% from soilborne pathogens alone, even on lands yielding more than 857 kg/ha.

The major causes of disease losses in most temperate climates, such as in USSR (Smirnova, 1980) and PRC (Ford et al., 1981), are similar to those in the USA, except that *Phymatotrichum* root rot occurs only in North America. *Verticillium* wilt is extremely severe in USSR and Peru, and has forced a switch in culture from *G. hirsutum* L. to the more resistant *G. barbadense* L. in Peru. *Chalara* (*Thielaviopsis*) root rot and macrosporiosis (probably caused by *Alternaria macrospora* Zimm) also are major diseases in USSR. Both seedling diseases and boll rots caused by *Phytophthora* species are important in PRC.

Several diseases are of major importance only in tropical or subtropical areas. These include the viral diseases, leaf curl, mosaic, and blue disease in Africa; and fungal foliar diseases, such as escobilla and powdery mildew in South America, areolate mildew in Africa and Asia, *Alternaria* leaf spot in Africa and Asia, tropical rust in Brazil, India, and Jamaica, and *Myrothecium* leaf blight in India and Pakistan. *Macrophomina* root rot (also referred to as *Rhizoctonia* (*baticola*) root rot) is a major disease in the hot dry areas of Asia and central Africa. This disease is more severe on *G. herbaceum* L. than on other cultivated species. Internal boll rots caused by *Nematospora* spp. are most important in central and eastern Africa, where the vectors of these fungi, *Dysdercus* species, occur abundantly. Bacterial blight can be devastating in the tropics and at one time destroyed an estimated 28% of the

Table 9B-1. Percentage losses from potential cotton production caused by infectious diseases in the USA from 1956 through 1980.†

Diseases	Time interval (years)					
	1956–1961	1961–1966	1966–1971	1971–1976	1976–1981	Avg.
	%					
Seedling diseases	2.68	2.79	3.77	2.78	2.65	2.93
Verticillium wilt	1.79	2.54	3.38	2.51	2.57	2.56
Fusarium wilt	1.13	0.97	0.81	0.73	0.42	0.81
Phymatotrichum root rot	1.10	1.53	0.70	0.59	1.38	1.06
Bacterial blight	1.90	1.10	1.15	0.92	0.39	1.09
Boll rots	2.10	2.18	3.14	2.97	2.18	2.51
Total	14.07	15.74	12.94	11.30	9.99	12.81

† Data taken from the cotton disease loss estimates of the Cotton Disease Council, published annually in the *Proceedings* of the Beltwide Cotton Production Research Conferences.

African cotton crop (Ebbels, 1980). This disease generally is more severe on *G. barbadense* than on *G. hirsutum* but causes little damage to *G. arboreum* or *G. herbaceum. Corticium* spp. and *Alternaria macrospora* are most important as seedling pathogens in moist tropical climates.

9B-1 SYMPTOMS AND EPIDEMIOLOGY OF DISEASES

Excellent colored photographs of cotton disease symptoms and listings of key references are available in the *Compendium of Cotton Diseases* (Watkins, 1981). *Fungal Wilt Diseases of Plants* (Mace et al., 1981) also contains extensive recent information on the wilt diseases of cotton.

9B-1.1 Seed and Seedling Diseases

During their germination and initial establishment cotton seeds and seedlings are extremely vulnerable to attack by a wide range of pathogens. This is a transient period of susceptibility in the life of the plant, and under normal planting conditions vigorous seedlings progressively become more resistant to seedling pathogens beginning 4 to 7 days after planting (Roncadori and McCarter, 1972). At 2 to 3 weeks after planting seedlings usually resist the attack of seedling pathogens enough to prevent serious damage unless unusually cold or wet conditions are present. The period of susceptibility coincides with very low tannin and terpenoid contents in hypocotyl and root tissues (Hunter, 1978; Hunter et al., 1978). As these compounds increase in tissue, parallel increases in resistance also occur.

Seed and seedling diseases may be caused by seedborne pathogens that occur on or in seed before planting or by soilborne pathogens that reside in soil, usually as dormant propagules embedded in organic matter. The soilborne pathogens are usually the most important causes of seedling diseases and most difficult to control. Several surveys of seedborne and soilborne seedling pathogens have been published (Davis, 1975; DeVay et al., 1982;

Johnson et al., 1978; Simpson et al., 1973). *Fusarium* spp., *Alternaria* spp. and *Aspergillus niger* Van Tiegham are the most widespread and common pathogens found in seed at harvest; the same fungi cause boll rots and deteriorate moist fiber. In local areas, high percentages of seed also may be infested wth the boll-rotting pathogens *Glomerella gossypii* Edg., *Xanthomonas campestris* pv *malvacearum* (Smith) Dye, *Aspergillus flavus* Lk. ex Fr., *Rhizopus* spp., or *Diplodia gossypina* Cke. Moist seed also may be invaded during storage in trailers or modules. Microbial activity in moist seedcotton causes heating; thus, thermophilic organisms prevail under these conditions. *Aspergillus* spp. (especially *A. flavus*) and *Rhizopus* spp. are the main thermophilic pathogens to invade seed.

The most important soilborne pathogens are *Thanatephorus cucumeris* (Frank) Donk (syn. *Rhizoctonia solani* Kuehn), *Pythium* spp., *Chalara elegans* Nag Raj and Kendrick (syn. *Thielaviopsis basicola* (Berk and Br.) Ferr.), *Fusarium* spp., and *Corticium rolfsii* Curzi (syn. *Sclerotium rolfsii* Sacc.). *Pythium* is most important in the northern and southern extremes of the cotton growing areas of the world, where cool, moist conditions prevail at planting. *Corticium,* in contrast, occurs mostly under tropical conditions. The other pathogens occur throughout cotton growing regions.

Most symptoms of seedling diseases are similar regardless of cause. These include a soft watery rot of the seed, root, shoot, or entire plant, damping-off (a rot of the hypocotyl near the soil line causing the seedling to fall over and die), soreshin (sunken brown to reddish brown lesions on the hypocotyl and upper root), and necrotic lesions on cotyledons and leaves. *Chalara* causes a dark discoloration of roots which is somewhat distinct from symptoms of other seedling diseases; hence, the disease often is referred to as black root rot. The rot and lesions caused by *Thanatephorus* are attributed at least in part to endopolygalacturonase enzymes produced by the pathogen (Brookhouser et al., 1980). Adequate tannin concentrations, such as those occurring in older seedlings, cause inactivation of these enzymes as well as inhibit growth of the pathogen, whereas low levels of tannin precursors actually stimulate enzyme production by highly virulent strains (Hunter, 1978).

The severity of seedling diseases depends on the duration of the susceptible stage in seedling development, the amounts of nutrients lost from seeds and seedlings to stimulate external growth of pathogens, the genetic potential of the pathogen for pathogenesis, and the inoculum concentration of the pathogen. Factors that may both prolong the susceptible stage and increase nutrient leakage from seed and seedlings include seed deterioration prior to planting; i.e., low vigor seed (Bishnoi and Delouche, 1980; Minton and Supak, 1980), cool temperatures (Hunter and Guinn, 1968; McCarter and Roncadori, 1971; Wanjura and Minton, 1981), pesticide chemicals (Lewis and Papavizas, 1979; Miller et al., 1979), nematodes (Carter, 1975a; Cauquil and Shepherd, 1970) and anaerobic conditions (usually due to excess soil moisture). A few studies have compared the virulence of different species of seedling pathogens, but little attention has been given to pathotypes within species. *Pythium ultimum* Trow. is generally more virulent than other *Pythium* species, particularly at 20 to 27°C. The few isolates of *Thanatephorus* from cotton that have been identified belonged to anastomoses group 4; however, no direct comparisons of virulence of isolates from different anastomoses groups are known to the author. Effects of inoculum concentration on seedling disease severity have been described for *Chalara* (Tabachnik et al., 1979), *Pythium* (Mitchell, 1978) and *Rhizoctonia* (Carter, 1975a). Increases in inoculum invariably cause increases in disease incidence and severity; however, the relationship is not linear and depends on the specific changes involved. Inoculum concentrations in soils depend on rotational sequences, cultivar susceptibility in previous years, and soil type and texture (Carter, 1975b; Lewis, 1979). Many of these factors probably indirectly affect inoculum concentrations by changing the populations of biological antagonists of the pathogen (Howell, 1982; Howell and Stipanovic, 1980). Disease control measures either minimize those factors that contribute to disease severity or provide chemical protection of seedlings during their susceptible stage.

9B-1.2   Root and Vascular Diseases

9B-1.2.1   Wilts

Wilts caused by the fungi *Verticillium dahliae* Kleb. and *Fusarium oxysporum* f. sp. *vasinfectum* (Atk.) Snyd. & Hans. are major causes of disease losses almost everywhere that cotton is grown. These diseases have been reviewed in detail (Ebbels, 1975; Mace et al., 1981; Ranney, 1973).

The two wilt pathogens only rarely occur together because of the different conditions favoring their growth and disease development. *Verticillium* wilt shows an extremely strong response to temperatures; all cottons are highly resistant to the disease when mean temperatures exceed 29°C, but disease resistance decreases rapidly as temperatures are decreased (Bell, 1982). At mean temperatures of 25 to 27°C all cottons are colonized by the fungus, but show variable levels of resistance. The disease is most prevalent in neutral to alkaline loam and clay soils, where there is limited rainfall, and irrigation is normally used to grow the crop. In contrast, *Fusarium* wilt occurs mostly in somewhat acid, coarse-textured soils, probably because of its synergistic relationship with root knot nematodes. About 100 times more *Fusarium* inoculum is required without nematodes than with them to cause the same amount and severity of disease (Garber et al., 1979b). *Fusarium* wilt is favored by warm temperatures and is common in areas of high rainfall. Because of these different requirements, *Fusarium* usually attacks plants in the seedling or early development stages, whereas *Verticillium* usually attacks plants late in the growing season after fruiting has begun.

Symptoms of wilt diseases are sufficiently similar to one another and to those of root rots that the causal fungus should be isolated to confirm the cause. For example, early reports of *Fusarium* wilt from Peru (Abbott, 1931), in fact, probably involved *Verticillium* wilt, but the causal organism was not diagnosed correctly because the pathogen was not isolated and identified. The causal fungi are usually isolated readily from infested petioles, upper stems, or small slightly discolored wedges of xylem tissue.

Epinasty of upper leaves is often the first symptom of wilt diseases. Plants then generally become stunted, show chlorosis on leaves along the margins and between veins, and subsequently show excessive defoliation that may become complete with certain strains of *V. dahliae.* The involvement of a wilt fungus is confirmed by brown streaks in the vascular system when the stem is sectioned.

Both *V. dahliae* and *F. oxysporum* f. sp. *vasinfectum* include a number of different pathotypes that attack cotton. In *Fusarium* these are distinguished as races 1 through 6 and are identified by their virulence to different species and cultivars of cotton and other crops as shown in Table 9B–2 (Armstrong and Armstrong, 1980). In *Verticillium*, genetically different groups of isolates are distinguished as anastomosis compatability groups (Puhalla, 1979, personal communication). Sixteen groups are known and isolates belonging to groups 1, 2, 3, 6, 7, and 8 have been recovered from diseased cotton in various parts of the world. Isolates in group 1 cause a severe defoliating-type disease and are more virulent than those in other groups. Isolates of group 1 currently are known only in Peru, Mexico, and North America, and have been isolated from cotton, sesame, safflower, peanut, velvetleaf weed, rose, elm, olive, and maple.[1]

Verticillium and *Fusarium* can survive for many years in the soil as dormant microsclerotia and chlamydospores, respectively. Disease frequency and intensity are usually directly related to the concentration of these propagules (Ashworth et al., 1979), which are stimulated by root exudate to germinate and then invade the root (Fitzell et al., 1980). A new crop of infectious propagules normally is not formed until tissues are dead. Thus, only a single cycle of reproduction of the overwintering infectious propagules occurs each year. Consequently, factors affecting concentrations of the surviving dormant propagules in soils have similar effects on disease severity, and may be manipulated for disease control. Species and cultivars vary considerably in their suscpetibility to wilt fungi, and resistant cultivars provide an important means of controlling disease.

9B–1.2.2   Root Rots

Three fungi, *Phymatotrichum omnivorum* (Shear) Dug., *Chalara elegans,* and *Macrophomina phaseolina* (Tassi) Goid. (syn. *Sclerotium baticola* Taub. or *Rhizoctonia baticola* (Taub.) Briton-Jones) cause major root rots on old as well as young plants. *Phymatotrichum* is indigenous to many alkaline, calcarious soils of the southwestern USA and northern Mexico,

---

[1] Scientific names of other plants mentioned in this chapter are as follows: sesame, *Sesamum indicum* L.; safflower, *Carthamus tinctorius* L.; peanut, *Arachis hypogaea* L.; velvetleaf weed, *Abutilon theophrastii* Medic.; rose, *Rosa* spp.; elm, *Ulmus americana* L.; olive, *Olea europaea* L.; maple, *Acer* spp.; barley, *Hordeum vulgare* L.; wheat, *Triticum aestivum* L.; rice, *Oryza sativa* L.; sorghum, *Sorghum vulgare* L.

Table 9B-2. Reaction of plant species and cultivars used to differentiate races
of *Fusarium oxysporum* f. sp. *vasinfectum*.

Species and cultivar	Race					
	1	2	3	4	5	6
G. hirsutum						
Acala	S†	S	R	R	R	--
Rowden	S	S	R	R	--	S
G. barbadense						
Ashmouni	R	R	R	R	S	--
Coastland	R	R	S	--	--	R
Sakel	S	S	S	R	S	--
G. arboreum						
Rozi	R	R	S	S	--	R
Abelmoschus elegans						
Clemson Spineless	S	S	R	R	--	S
Nicotiana tabacum						
Burley 5	S	S	R	R	--	R
Gold Dollar	R	S	R	R	--	R
Medicago sativa						
Grimm	S	S	R	R	--	R

† S = susceptible, R = resistant, -- = not tested. Results are those reported by Armstrong and
Armstrong (1980).

and annually destroys 3% or more of the crop in this area during warm,
moist periods in early summer (Lyda, 1978). Losses would be much greater,
except cotton production has been abandoned on several hundred thousand
hectares infested with this pathogen. *Macrophomina* root rot is a major
problem where cotton is grown under hot, dry conditions in Pakistan,
India, Greece, and central Africa, but not normally in the United States. At
College Station, Texas, this pathogen is highly virulent to *G. herbaceum*,
completely killing all plants of many accessions, but it rarely attacks other
Gossypium species. A local cultivar, but not U.S. cultivars, of *G. hirsutum*
was heavily diseased by *Macrophomina* in Greece (Watkins, 1981). *Chalara*
destroys mostly secondary and tertiary roots, resulting in stunting or wilting
but rarely death. Consequently, it is more serious on seedlings than on older
plants. The pathogen is favored by cool, wet conditions and alkaline soils,
such as found in southwestern and western USA and USSR. Because the
pathogen usually only debilitates plants, losses from this disease may be
greater than is normally estimated.

Root rot fungi, other than *Chalara,* generally cause rapid wilting and
death of plants, and the brown necrotic leaves remain attached to the plant.
This reaction is unlike the wilt diseases which usually result in defoliation of
infected leaves. Also, unlike the wilt fungi, *Phymatotrichum* and *Macrophomina* cause extensive decay of the root bark and lower stem bark. The
olive-tan mycelial strands (rhizomorphs) of *Phymatotrichum* appear super-
ficially on the rotted root, whereas the small black sclerotia of *Macrophomina* are scattered throughout the rotted bark and wood. The wood below bark rotted by *Phymatotrichum* appears red to wine in color near the
soil line; the color distinguishes this disease from other root rots. The red
color probably is caused by phlobaphenes or anthocyanidins formed by the

action of phytotoxic oxalic acid from the fungus on condensed proanthocyanidins of the cotton plant.

Chalara normally rots only epidermal and cortical cells and does not penetrate the endodermis or periderm. However, rot of the internal wood tissues sometimes occurs in *G. bardadense* apparently from reactivation of residual infections that originally occurred in seedlings. Such infections may result in marked swelling of the lower stem and upper root and occasionally rapid wilting and death. All infections by *Chalara* show a characteristic blackening of infected tissue. This blackening may be caused by the action of methylacetate and other fusel oils produced by the pathogen (Tabachnik and DeVay, 1980).

Both *Phymatotrichum* and *Macrophomina* have unusual ecological adaptions that contribute to their survival and importance as pathogens (Lyda, 1978; Watkins, 1981). Most fungi form sclerotia in or on organic debris, but *Phymatotrichum* mycelial strands can form sclerotia apart from organic debris down to 2 m in the soil. *Phymatotricum* sclerotia are most concentrated between 10 and 40 cm deep, which allows escape from most natural antagonists and makes chemical treatments or tillage treatments extremely difficult and expensive. The germination of sclerotia are favored by temperatures above 80°C and by high CO_2 levels such as result from summer rains or irrigation. Thus, the fungus normally becomes active in late June to mid-August when cotton plants have reached early fruiting stage and have roots permeating the soil where the fungus is present.

Macrophomina is perhaps the best adapted of all plant pathogens to combinations of high temperature and low water potential; maximum growth occurs at 35 to 40°C and -10 to -20 Pa $\times 10^6$ (Bell, 1982), conditions that severely limit growth and active defense reactions in the plant and growth of most microbial antagonists of *Macrophomina*. Under usual temperatures and moistures, this pathogen is a poor competitor with antagonists and is unable to attack plants; sclerotia usually remain dormant until favorable conditions for the fungus return.

Chalara forms dark chlamydospores, in characteristic chains of three to six cells, which allow the fungus to survive for prolonged periods in the soil. Chlamydospores are stimulated to germinate by exudates and directly penetrate the young root tissue. Pathogenesis is favored by temperatures below 16°C. New chlamydospores are formed in the rotted tissue.

Control practices for root rots are directed at destroying inoculum in the soil or managing the crop so as to avoid or limit the duration of environmental conditions favoring the diseases.

9B–1.3 Foliar Diseases

The pathogens in this group normally infect leaves, stems, and bolls, and occasionally seedling roots. These pathogens normally survive in or on infested seed and infested plant residues that have not been buried to decompose. Alternate hosts can maintain the pathogen in the absence of cotton. Thus, these pathogens usually can be controlled by using high quality, pathogen-free seed combined with clean tillage practices that eliminate cotton plant residues and alternate hosts. High levels of resistance to some of these pathogens are available in some cultivars, and these can be used when other approaches are not feasible.

9B-1.3.1 Bacterial Blight

Before the development of resistant cultivars, bacterial blight, caused by *X. campestris* pv *malvacearum* and also known as angular leaf spot and blackarm, was the most important cotton disease in the world. The disease is still one of the most important bacterial diseases of plants in the USA and the world, but it is rapidly declining in importance. Variation occurs in the pathogen, and 18 races can be distinguished as shown in Table 9B-3 (Hussain and Brinkerhoff, 1978). Race 18 is the most virulent and has been found in both Texas and Pakistan. However, several combinations of blight resistance genes available in numerous American and African cultivars confer immunity even to race 18 (Brinkerhoff and Hussain, 1978).

The initial symptom of bacterial infection is a waxy, water-soaked appearance of infected leaf, stem, or boll tissue. Later tissues turn into brown to black necrotic areas, which on leaves may be confined by small veins to give an angular appearance. Chlorosis develops around the necrotic areas, and the leaves defoliate when extensive infection occurs. In highly susceptible cultivars, necrotic areas along major veins of the leaf and on stems may become dark brown to black, and are often covered with a light-colored crust composed of dried bacterial slime and cells. When infected leaves are crushed in water and placed on potato–carrot–detrose agar, numerous smooth yellow colonies of the bacterium can be found after 2 to 3 days.

The bacterium survives in infested seed and in dry residues of diseased plants left on the field surface. Infection occurs when bacteria are introduced into stomata by splashing rain, blowing sand, hail, and insects. Thus, the use of pathogen-free seed or seed treatments and clean plowing are effective in eliminating early season infections and possible subsequent epidemics.

Table 9B-3. Susceptible (S) and resistant (R) disease reactions of nine of lines cotton to 18 races of *Xanthomonas campestris* pv *malvacearum* (Hussain and Brinkerhoff, 1978).

Cultivar or line	Race																	
	1	2	3	4	5	6	7	8	9	10	11	12	13	14	15	16	17	18
Acala 44	S	S	S	S	S	S	S	S	S	S	S	S	S	S	S	S	S	S
Stoneville 2B-S9	S	S	S	S	S	S	S	S	R	S	S	S	R	S	S	S	S	S
Stoneville 20	R	S	R	R	R	R	R	S	S	S	R	S	R	S	S	S	S	S
Mebane B1	R	R	R	R	R	S	S	S	R	S	R	R	R	R	R	S	R	S
1-10 B	R	R	S	R	S	S	S	S	R	S	R	R	R	S	S	R	R	S
20-3	R	R	R	S	S	R	S	R	S	S	R	R	R	S	R	S	S	S
101-102 B	R	R	R	R	R	R	R	R	R	R	R	R	R	R	R	R	R	R
Gregg	R	R	--	--	--	--	--	--	--	--	S	S	--	--	--	--	--	S
DP × P-4	R	R	--	--	--	R	R	--	--	R	R	R	--	--	--	--	--	S

9B-1.3.2   Fungal Diseases

Numerous fungi can cause infections of stems and leaves of very young cotton plants exposed to prolonged cool, cloudy, wet weather or of old leaves and bolls approaching maturity and senescence. These fungi include necrotrophic pathogens that cause death and necrosis of invaded tissues, and biotrophic pathogens that depend on live tissues for their nutrition and well-being.

Examples of the necrotrophic fungi include *Alternaria tenuis* Auct., *A. macrospora, Aschochyta gossypii* Woronichin, *G. gossypii, Helminthosporium gossypii* Tucker, *Myrothecium roridum* Tode, *Cercospora gossypii* Cooke, and *T. cucumeris.* The centers of necrotic lesions produced by these fungi vary from a light ash-grey to brown color and the borders are red-brown to purple brown. The exact colors and sizes of lesions depend on the specific pathogen, the cultivar, age of tissue, and environmental conditions. Symptoms are too similar to consistently make accurate diagnoses without isolating and confirming the causal fungus.

Examples of biotrophic fungi include the rust fungi *Phakopsora gossypii* (Arth.) Hirat., *Puccinia cacabata* Arth. & Holw., and *Puccinia schedonnardi* Kellerm. and the powdery mildew fungi *Salmonia malachrae* (Seaver) Blumer & Muller and *Leveillula taurica* (Lev.) Arn. *Mycosphaerella areola* Ehrlich & Wolf is a partial biotroph that does not cause death of invaded cells until just before defoliation, but later makes saphrophytic growth in fallen leaves similar to necrotrophic pathogens. The biotrophs and *M. areola* are best recognized by the superficial appearance of the fungal structures on leaves and stems. *Phakopsora* uredia are yellowish brown, varying from 0.5 to 3.0 mm in diam, and are surrounded by purplish borders, 1 to 5 mm. Uredia first appear as oval, corky pustules, and then become round. They open by an apical pore, and uredospores emerge through a cirrus. The pycnial and aecial stages of *Puccinia* occur on cotton, whereas uredial and telial stages occur on grasses, most of which are species of *Bouteloua.* Pycnial pustules occur mostly on upper leaf surfaces and are bright yellow to orange in color; aecia of similar color occur on lower surfaces. The powdery mildew fungi occur generally over leaves giving them an appearance of being covered by white dust or snow. Conidiophores of *Leviellula* emerge through stomata to the leaf surface and abundantly bear hyaline conidia (14 to 25 μm $\times$ 43 to 79 μm) on the leaf surface, whereas *Salmonia* forms abundant hyphae and cleistothecea (52 to 67 μm in diam; light yellow brown) but few if any conidia on the leaf surface. *M. areola* also forms hyaline conidiophores through stomates, but conidia have one to four cells, and the fungus is restricted to angular lesions bounded by small leaf veins (areoli).

The necrotrophic pathogens and *M. areola* survive as dormant hyphae or spores in infected plant debris and in and on seed. Emerging seedlings may become infected by the fungi on the seed coat or by spores carried by wind, water, or insects from the infected debris to the young plants. Environmental conditions desirable for planting cotton also may be conducive to production of new spore crops on debris from the previous year. Several additional cycles of infection may occur during the growing season as new spores are formed in necrotic lesions and spread to other plants. Control procedures are directed at disinfection of seed, elimination of infested debris, and finally fungicides to control established infections.

The biotrophs depend on alternate hosts or volunteer cotton plants to maintain the pathogen in the absence of the cultivated crop. The alternate hosts for *Phakopsora* and powdery mildews are malvaceous plants, often weeds, whereas those of *Puccinia* spp. are grasses which may occur abundantly in range lands next to cotton fields. New infections arise each year from these alternate or volunteer hosts. Control procedures depend on eliminating alternate or volunteer hosts or using fungicides if extensive infection appears imminent.

9B-1.3.3   *Viral and Mycoplasmal Diseases*

At least 16 viral and 2 microplasma-like (MLO) diseases have been reported on cotton (Tarr, 1964; Watkins, 1981). However, the causal agents have not been isolated and characterized; thus, many of these diseases may be caused by the same or similar infectious agents. These diseases are mostly of importance where cotton is grown in small plantings in tropical or subtropical climates in Africa, Asia, Central America, and South America. Under these conditions cotton is often ratooned and alternate malvaceous hosts and volunteer cotton are present all year to act as reservoirs of the virus. Also, the insect vectors that are required to transmit cotton viruses and MLO often thrive under these conditions.

Viral diseases may be divided into four groups based on symptoms and vectors. The first, leaf curl, causes thickening and enations of leaf veins, leaf curling, and twisting and curving of stems and petioles. The affected veins appear abnormally dark green and opaque. The virus is transmitted by whitefly, *Bemisia tabaci* Genn., and is of most importance in Africa. Diseases referred to as mosaics, leaf crumple, and leaf mottle have similar symptoms and all are transmitted by *B. tabaci,* indicating involvement of similar viruses. Plants are often stunted, sterile, and have crinkled, mottled, and blistered leaves. The green blisters or chlorotic patches occur in a mosaic pattern over the leaves. Diseases in this group have been reported in Africa, South America, and areas of the western United States where cotton is ratooned. Blue disease is a major viral cause of disease losses in several African countries. This virus is transmitted by *Aphis gossypii* Glover. Affected plants are dwarfed and have thick, dark blue-green leaves with light areas next to veins and edges that curl down; stems often grow in a zigzag pattern. Diseases referred to as leafroll, vein clearing disease, and veinal mosaic in the USSR, Thailand, Philippines, and South America have symptoms similar to blue disease and also are transmitted by *A. gossypii,* which indicates similar causal viruses. Vein clearing reported in Texas, USA, apparently is caused by a fourth type of virus, because it is transmitted by the leafhopper, *Scaphytopius albifrons* Hepner. Virus diseases are controlled by eliminating reservoir hosts and insect vectors. Resistant cultivars are available for certain viral diseases.

9B-1.4 Boll Diseases

Boll rot diseases have been reviewed in detail by Cauquil (1975). These diseases are most severe where moist conditions prevail in the canopy of the plant for most of the day. The incidence of boll rot is increased by increases in rainfall, humidity, and plant size and density. Fiber and seed are often rotted when they are exposed by wounds before bolls are mature. Insect injury, spindle picker injury, hail injury, or genetically faulty bolls may contribute to the incidence of fiber and seed deterioration. Young intact bolls are generally resistant to infection by any pathogen, but as bolls age, especially beyond 40 days, they become progressively more susceptible to attack.

The first organisms able to attack the aging boll directly are *X. campestris* pv *malvacearum, Colletotrichum* spp., and *Diplodia gossypina.* As bolls near dehiscence, *Fusarium* spp. and *Alternaria* spp. also may penetrate bolls directly, although these pathogens more commonly enter through wounds or lesions caused by *Xanthomonas* or *Colletotrichum.*

The greatest boll rot losses in the USA occur in the lower Mississippi River delta area of Louisiana and Mississippi, where over half the crop is lost some years. Most of the rots are caused by *Diplodia* or *Fusarium*; *Colletotrichum* is a major cause only in very wet years (Sanders and Snow, 1978). The same three fungi are also responsible for most of the boll rots in other humid, high-rainfall areas of the world. In the western United States and other arid regions, true boll rots occur rarely, but a number of fiber and seed-rotting organisms may cause problems if extensive boll injury occurs from insects or if bolls open slowly with extensive dew. *Erwinia* species of bacteria and the fungi *A. niger, A. flavus, Rhizopus,* and *Nigrospora* often rot fiber and seed in injured bolls in hot, dry climates (Watkins, 1981).

Most boll rot pathogens produce airborne spores that come to rest on the bolls, bracts, or exposed fibers. These spores germinate and procede to ramify through boll parts. Bract tissue often dies before other boll parts and may serve as an important means of entry to the boll. The boll rot fungi apparently grow and sporulate prolifically on flowers, squares, and bolls that are shed from the plant and fall on the ground. An average of more than 1 million conidia of *Fusarium* were found per shed flower and square in Louisiana, and large numbers of both *Diplodia* and *Fusarium* were found in air samples collected over cotton fields (Sanders and Snow, 1978). *Colletotrichum,* likewise, sporulates prolifically on shed plant parts and cotton debris.

Insects occasionally transmit boll rot pathogens as well as provide wounds for them (Watkins, 1981). Several insects can carry *A. flavus,* and viable spores are found in frass. Pink bollworms also are important in providing wounds for *Aspergillus.* The mite *Siteroptes reniformis* Krantz carries spores of *Nigrospora oryzae* into bolls and deposits them on fiber where they cause a rot. The stinkbug, *Euschistus impictiventris* Stal. appears to be involved with boll rots caused by *Erwinia.* Other studies have shown that boll weevils and boll worms also result in greater levels of boll rots.

The control of boll rots begins with eliminating early season sources of inoculum by seed treatment and clean tillage. Plant size and density should be kept to a minimum by minimum nitrogen fertilization, minimum irrigation, wide row-spacing and use of open-canopy cultivars such as those with okra leaf, frego bract, and dwarf stature. Insecticides may decrease boll rots by preventing insect wounds necessary for the entrance of certain pathogens. Fungicides generally have not been effective because of the huge amounts of inoculum produced on debris located under the foliage canopy.

9B-2 CONTROL STRATEGIES

Disease control practices are designed to reduce pathogen inoculum or restrict development of the pathogen in host tissue. This reduction is best accomplished through integration of a number of approaches. The most economical means of controlling diseases is often through modification of cultural practices. Sometimes this is the only approach needed, but if inadequate, it may be supplemented by using cultivars with increased resistant to the disease. Under disease free conditions resistant cultivars sometimes do not yield as well as more susceptible cultivars and may have lower fiber quality. Thus, a small cost as reduced yield or price is often involved with their use. Chemical controls are normally used only as a last resort when other controls are inadequate. Chemical control is generally the most expensive, and its use must be weighed against the cost of disease losses. Biological controls may be a less expensive and safer substitute for chemical controls, but are not yet fully developed for commercial use in most countries. Soil solarization also has given dramatic disease control, but is still relatively expensive.

9B-2.1 Cultural Practices

Some of the procedures that can be modified to control diseases include planting, fertilization, irrigation, crop rotations, tillage, use of pesticides, and harvest. Modifications that alleviate diseases generally are also beneficial for insect and nematode control.

The use of high quality, vigorous seed is important both for decreasing seedling diseases and late season diseases such as *Verticillium* wilt (Green and Minton, 1980; Minton and Supak, 1980). Several ways for measuring seed quality have been suggested; the simplest is measurement of radicle length after seeds are incubated in germination towels for 72 hr at 25°C (Bishnoi and Delouche, 1980). Delaying planting until soil temperatures are

18°C or higher speeds seedling development, shortening the period of susceptibility to seedling diseases. However, it is important to plant as early as possible in most areas to escape late season diseases such as *Phymatotrichum* root rot, *Verticillium* wilt, and boll rots. Excessive soil moisture favors *Pythium, Ascochyta,* and *Chalara* infections and, therefore, should be avoided at planting.

Excessive nitrogen fertilization aggrevates most cotton diseases but is particularly damaging for wilt diseases (Mace et al., 1981) and boll rots (Caquil, 1975). Thus, nitrogen fertilization should be limited to levels for optimal yields. Ammonium ions generally increase severity of *Fusarium* wilt, whereas nitrate ions increase *Verticillium* wilt. Russian farmers use urea sprays on young plants for wilt control (Smirnova, 1980), but this practice has not been adopted elsewhere. Potassium deficiencies greatly increase the severity of both *Verticillium* and *Fusarium* wilt. Host resistance to these diseases increases progressively with increasing doses of potassium fertilizer. It is important that potassium be maintained at levels required for optimal growth or higher. Calcium fertilization and adjustment of soil pH· to 6.0 or higher are effective in reducing both seedling diseases and *Fusarium* wilt, as well as reducing aluminum and manganese toxicity (Bell, 1982; Foy et al., 1981). Generally, variable effects on diseases have been reported for phosphorus and minor element fertilization (Bell, 1982; Ranney, 1973). However, plants generally recover better from disease when adequate levels of all nutrients are maintained.

Either inadequate or excess irrigation may aggrevate disease (Bell, 1982). *Macrophomina* only attacks water-stressed plants. Timely, adequate irrigation can be used to prevent this disease. Most other soilborne pathogens are favored by excessive irrigation, particularly when irrigation lowers soil temperature. Outbreaks of *Verticillium* wilt and *Chalara* root rot often follow irrigation late in the growing season. Thus, it is particularly important to keep irrigation to a minimum once mean temperatures begin to decrease appreciably late in the growing season.

Rotations with barley, wheat, rice, sorghum, or legumes generally reduce the amount of inoculum of soilborne pathogens of cotton and decrease incidence and severity of disease. Paddy rice for a single year has given tremendous increases in cotton yields along with marked reductions in most soilborne pathogens (Pullman and DeVay, 1981). Flooding alone also reduces diseases but is not as effective as paddy rice. Repeated planting of resistant cotton cultivars also reduces nematode populations (Hyer et al., 1979) and *Verticillium* microsclerotia (Minton, 1980) and decreases subsequent damage by *Fusarium* wilt and *Verticillium* wilt, respectively, when susceptible cultivars are planted. In contrast, pathogens such as *Pythium, Fusarium,* and *Macrophomina* may actually be increased by rotations with crops such as corn and soybeans. Thus, rotational crops usually are beneficial, but may be detrimental, depending on the specific pathogen.

Both the timing and type of tillage practices may effect disease control. Clean tillage practices that completely bury cotton debris as soon as possible after harvest and destroy volunteer plants and alternate hosts are very important for preventing build up of inoculum and early season infections by seedling pathogens, foliar pathogens, and viruses. Organisms such as *Xanthomonas* and *Verticillium* survive much better in dry plant debris left on the soil surface than in that buried in the soil. Plowing or chiseling down to 46 to 61 cm have been the most effective means of destroying *Phymatotrichum* sclerotia in soil. Completely turning over green manure crops with a deep moldboard plow is the most effective control of all. This procedure apparently places sclerotia of the pathogen into the upper soil layers allowing their destruction by antagonists.

Various herbicides and insecticides commonly increase but occasionally decrease seedling and wilt diseases, depending on the specific chemical, the cultivar, and environmental conditions (Bell, 1982). Minimal effective doses of herbicides should be used to prevent phytotoxicity and accompanying increases in disease. Some herbicides apparently aggrevate diseases less than others (Lewis and Papavizas, 1979; Miller et al., 1979). Machinery used for cultivating or spraying should not be run through fields before the foliage is dry; otherwise, bacterial blight might be spread.

Several harvest practices influence diseases. Picking and storing moist cotton in trailers or modules can stimulate growth of *Aspergillus* species and aflatoxin production, and cause deterioration of seed. This organism requires only 15% moisture for growth; thus, picking and storing moist cotton should be prevented. Spindle-picking cotton before plants are dry also may increase incidence of boll rots. Improperly adjusted machinery may injure unopened bolls and increase boll rots. Gin trash often is a source of inoculum of *Verticillium* and *Xanthomonas,* if not treated properly (Sterne et al., 1979). Composting trash eliminates some of the inoculum, and early plowdown to allow thorough rotting before planting further reduces inoculum in trash. Infested gin trash should be destroyed rather than spread on fields that are free of the pathogens.

9B–2.2 Host Plant Resistance

Many gains in disease control over the past 15 years are directly attributable to improved disease resistance and escape in most cultivars. Numerous cultivars with complete immunity to all races of bacterial blight have been developed in both the United States and Africa and have given remarkable control of this pathogen (Brinkerhoff and Hussain, 1978). Levels of resistance to *Fusarium* wilt in commercial cultivars have increased progressively in recent years with a concurrent decrease in losses from this disease (Kappelman, 1980b). New cultivars resistant to nematodes effectively control *Fusarium* wilt, as well as nematodes, by disrupting the synergism between these organisms (Hyer et al., 1979). The MAR cultivars developed by Bird (1982) have improved resistance to seed deterioration and shortened duration of susceptibility to seedling diseases, and thus, suffer less damage from seedling pathogens. Many of the open-canopy cultivars, having okra leaf and frego bract, and short-season cultivars decrease boll rots both by facilitating drying of foliage and allowing better coverage with insecticides to control boll-damaging insects.

Disease resistance genes have often been found in *Gossypium* species other than the ones being cultivated (Bird, 1973; Mace et al., 1981; Watkins, 1981). Immunity to bacterial blight in Upland cotton, for example, has been obtained by introducing genes from *G. barbadense* and *G. arboreum* and using these in combination with minor genes in *G. hirsutum.* The same materials have yielded high levels of resistance to areolate mildew. High levels of resistance to *Fusarium, Verticillium,* and nematodes have been obtained from *G. barbadense* and *G. darwinii.* Immunity to *Puccinia* was found among progeny from the triple species hybrid *G. hirsutum* × (*G. arboreum* × *G. thurberi*), and resistance to blue virus was found among progeny from the triple species hybrid *G. hirsutum* × (*G. arboreum* × *G. raimondii*). The male sterile cytoplasm from *G. harknessi* may be beneficial for increasing resistance to bacterial blight (Mahill and Davis, 1978). Interspecific hybrids and cytoplasm from other species may also be sources of resistance to other pests and need expanded study.

Several strategies have been developed for incorporating multiple resistance to diseases and nematodes into cotton cultivars. All of these seek to combine immunity to bacterial blight and resistance to both *Fusarium* and *Verticllium* wilt. In addition, Bird (1982), uses primary screens for resistance to seed deterioration at 13°C and escape from seedling diseases at 18°C, and Sappenfield et al. (1980) plant seeds in nematode-infested soils. More emphasis needs to be placed on resistance to nematodes, because these pests may negate resistance genes to fungal pathogens unless resistance to the nematode is also present (Bell, 1982).

Sources of appreciable resistance still are not known for many important pathogens such as *Pythium, Thanatephorus, Chalara, Fusarium roseum, Fusarium moniliforme, Diplodia, Phymatotrichum, Alternaria, Ascochyta,* and *Cercospora.* Only low levels of resistance to defoliating strains of *Verticillium* are available, and considerable loss still occurs in the most resistant cultivars when cool temperatures are present. Chemical controls are often needed to help control these pathogens.

9B–2.3 Chemical Controls

Fungicidal chemicals are currently used to protect seedlings during their susceptible period of development, to destroy microsclerotia and chlamydospores of wilt and root rot fungi in soil, and to contain foliar diseases when they approach damaging levels. Several growth regulators can be used to enhance the resistance of cotton plants to *Verticillium* (Erwin et al., 1979). Research data showing the performance of specific chemicals can be found in the annual reports of the seed treatment committee and soil fungicide committee of the Cotton Disease Council, published in the annual *Proceedings* of the Beltwide Cotton Production Research Conferences. Recommendations for seed treatments by state for the USA in 1980 are also included in the *Compendium of Cotton Diseases* (Watkins, 1981) and a thorough discussion of the use of soil fungicides for controlling wilt diseases is included in *Fungal Wilt Diseases of Plants* (Mace et al., 1981).

Treatment of seeds and incovering soil with fungicides has generally increased seedling vigor and usually improved the stand. Better seedling performance often results in less late season diseases, such as *Verticillium* wilt. Seed and soil treatments are particularly beneficial for mechanically damaged or partially deteriorated seed (Green and Minton, 1980; Minton and Sapak, 1980). Different seed treatments usually give similar results on acid and machine-delinted seed and in different locations (Minton and Quisenberry, 1981). Seed treatments rarely show beneficial results when severe weather follows planting.

Various attempts have been made to increase the efficacy of seed treatments. Adding acid neutralizers to treatment mixtures has increased effectiveness slightly (Minton, 1978). Chemicals applied as organic infusions have given control similar to that obtained with slurries (Halloin et al., 1978). Mixtures of fungicides that show separate potent activity against *Pythium, Rhizoctonia,* and *Chalara* give much better improvement of seedling vigor and stand than single fungicides which have narrower activity spectra (Garber et al., 1979a; Papavizas et al., 1980). Likewise, including insecticides with fungicides and adding incovering soil treatments to seed treatments have given further increases in seedling vigor (Kappelman, 1980a). In general, increases in seedling vigor appear more important than increases in stand percentage for increasing yield (Pinckard and Melville, 1975). Combining insect and nematode control with disease protection for seedlings is an important consideration, because insect and nematode damage often increase disease severity and prolong the period of seedling susceptibility to diseases.

Several chemical fumigants are effective in eradicating *Verticillium* and *Phymatotrichum* in soils (Heilman et al., 1978; Mace et al., 1981). These include methyl bromide, mixtures of methyl bromide and chloropicrin, and dichloropropenes (telone). Effective treatments, however, frequently require large doses of the chemical and deep placement, and thus are very expensive. These chemicals may have adverse effects on beneficial organisms and may decrease yields because of such effects. Drenches with the systemic fungicide benomyl at 100 ppm are very effective in eradicating wilt fungi, even in infected plants, in greenhouse pots, but doses and placement required for effective control in the field are prohibitively expensive (Erwin et al., 1978).

Benomyl has been very effective in controlling foliar diseases such as areolate mildew (Watkins, 1981), but is not particularly toxic to other pathogens like *Alternaria macrospora* or *Helminthosporium gossypii.* The triphenyltin derivatives appear to have the best promise for controlling the latter foliage pathogens (Padaganur and Siddaramaiah, 1979). Several different fungicides may be used to control rust and mildew diseases (Watkins, 1981).

9B–2.4 Biological Control

Biological control, as used here, refers to the suppression of plant pathogens by the manipulation of nonhost organisms, usually microorganisms. Three approaches may be used: i) the selective stimulation of beneficial organisms antagonistic to pathogens, ii) enrichment of soils or seed with microorganisms that produce antibiotics effective against pathogens, and iii) enrichment of soils with hyperparasites of pathogens. An example of the first approach is treatment of cottonseed with sorbose. This sugar inhibits mycelial extension in *Rhizoctonia* and *Verticillium*, but stimulates growth of beneficial fungi such as *Trichoderma.* Treatment of cottonseed with L-sorbose has given significant increases in emergence and survival in *Rhizoctonia*-infested soils (Howell, 1978). An example of the second approach is the addition of cell suspensions of *Pseudomonas fluorescens* to soils or cottonseed. This treatment increases stands and seedling vigor in *Rhizoctonia-* and *Pythium*-infested soils, apparently because of production of the antibiotics pyrrolnitrin and pyoluteorin, respectively (Howell and Stipanovic, 1980). The fungus *Gliocladium virens* gives biocontrol by both the second and third approach. This fungus produces the antibiotics heptelidic acid and gliovirin which are toxic to *Rhizoctonia* and *Pythium,* respectively; in addition, *G. virens* is a hyperparasite of *Rhizoctonia* (Howell, 1982).

Several other microorganisms also are antagonistic to cotton pathogens and have promise for biological control. *Trichoderma harzianum* significantly decreased cotton seedling diseases caused by *Sclerotium rolfsii* and *Rhizoctonia solani* in the field in Israel (Elad et al., 1980). *Laetisaria arvalis* may be a biocontrol organism for both *Rhizoctonia* and *Pythium* (Burdsall et al., 1980). Many soils are naturally suppressive to certain diseases, and recent evidence indicates that such suppressiveness is due to enhanced populations of biological antagonists (Scher and Baker, 1980). Thus, suppressive soils may be a rich source of new biological antagonists, and in addition, ecological studies of such soils may facilitate the successful deployment of biological control agents on a commercial scale.

9B–2.5 Solar Pasteurization

Recent studies have shown that covering moist soils with clear plastic in midsummer for several weeks causes temperatures to rise sufficiently to kill propagules of many important pathogens, such as *Pythium, Verticillium, Fusarium,* and *Chalara,* without killing many of the beneficial microorganisms. Such treatments, referred to as solar pasteurization (Katan, 1980) or soil solarization (Pullman et al., 1981), have given dramatic increases in cotton yields concomittant with control of most soilborne pathogens. The major limitation of this control is that it must be applied during midsummer, thus elimating cotton culture for 1 year. However, benefits persist for several years, so that the treatment still may be economical over several years, especially in soils heavily infested with complexes of soilborne pathogens.

9B-3 FUTURE PERSPECTIVES AND RESEARCH NEEDS

Tremendous progress has been made over the past 40 years in controlling diseases caused by *Colletotrichum, Xanthomonas,* and *Fusarium.* Each of these diseases at one time was a major threat to the cotton industry, destroying 40% or more of the crop in certain states and years. *Colletotrichum,* for example, was responsible for nearly 80% of the seedling diseases and boll rots in the southeast USA in the early 1940s, but is not of little importance in any area. These tremendous gains in disease control have been made because of the development of effective seed treatments, improved cultural practices, and the discovery and incorporation of various disease resistance genes into commercial cultivars. Additional sources of resistance to all three organisms are available for use if new pathotypes should arise. The disease control practices now available probably will virtually eliminate these important pathogens as pests of cotton, if the practices are used. However, it must be recognized that all three pathogens are universally present in cotton-growing areas and are highly variable with numerous races already known. It is possible that new highly virulent races might arise at anytime. For this reason new sources of resistance to these pathogens should be discovered and transfered into breeding stocks, which can be quickly used in the event of new more virulent races. At the same time we need to learn more about the genetics and physiology of these pathogens and their potential for increasing in virulence.

Most of the pathogens responsible for current disease losses in the USA cannot be controlled readily by seed treatment, cultural practices, or known sources of host plant resistance. Current chemical controls for these pathogens are prohibitively expensive unless major losses are incurred. There is a tremendous need for new and innovative approaches to economically control these organisms. Such approaches must be based on a thorough knowledge of the ecology, biochemistry, and genetic variability of these pathogens. Biological control would appear to have the greatest promise as an economical control for these pathogens. Ideal biological control agents should grow and thrive in the rhizosphere of the plant and protect against a broad spectrum of soilborne pathogens. Considerable effort should be made to discover and develop biocontrol agents that are well suited to cotton cropping systems and different soil types, but still give antagonism to the target pathogens.

Increased attention needs to be given to discovering and utilizing new sources of host resistance. Such resistant most likely will come from species or genera related to cultivated cotton, and may require unique genetic techniques for transfer. This means that greater knowledge is needed of the biochemical mechanisms of resistance in these related plants and of the biochemical and genetic nature of interspecific and intergeneric incompatibility in *Gossypium*. Such information is essential to devise bridging crosses or genetic engineering techniques for transfer of specific resistance genes that alter specific biochemical mechanisms of resistance.

Solar pasteurization should be pursued especially as a technique to get better estimates of soilborne disease losses. It also appears to have great promise for controlling soilborne disease complexes, if economical obstacles can be overcome.

REFERENCES

Abbott, E. V. 1931. Further notes on plant diseases in Peru. Phytopathology 21:1061–1071.

Armstrong, G. M., and J. K. Armstrong. 1980. Race 6 of the cotton-wilt *Fusarium* from Paraguay. Plant Dis. 64:596.

Ashworth, L. J., Jr., O. C. Huisman, D. M. Harper, L. K. Stromberg, and D. M. Bassett. 1979. Verticillium wilt disease of cotton: Influence of inoculum density in the field. Phytopathology 69:483–489.

Bell, A. A. 1982. Plant pest interaction with environmental stress and breeding for pest resistance: Plant diseases. p. 335–363. *In* M. E. Christiansen and C. F. Lewis (eds.) Breeding plants for less favorable environments. John Wiley and Sons, Inc., N.Y.

Bird, L. S. 1973. Cotton. p. 181–198. *In* R. R. Nelson (ed.) Breeding plants for disease resistance, Concepts and applications. The Pennsylvania State University Press, University Park.

————. 1982. The MAR (Multi-adversity resistance) system. Plant Dis. 66:172–176.

Bishnoi, U. R., and J. C. Delouche. 1980. Relationship of vigour tests and seed lots to cotton seedling establishment. Seed Sci. Technol. 8:341–346.

Brinkerhoff, L. A., and T. Hussain. 1978. Evaluation of American and African bacterial blight resistant Upland cotton lines in Pakistan in 1977. Plant Dis. Rep. 62:1082–1085.

Brookerhouser, L. W., J. G. Hancock, and A. R. Weinhold. 1980. Characterization of endopolygalacturonase produced by *Rhizoctonia solani* in culture and during infection of cotton seedlings. Phytopathology 70:1039–1042.

Burdsall, H. H., Jr., H. C. Hoch, M. G. Boosalis, and E. C. Setliff. 1980. *Laetisaria arvalis* (Aphyllophorales, Corticiaceae): a possible biological control agent for *Rhizoctonia solani* and *Pythium* species. Mycologia 72:728–736.

Carter, W. W. 1975a. Effects of soil temperatures and inoculum levels of *Meloidogyne incognita* and *Rhizoctonia solani* on seedling disease of cotton. J. Nematol. 7:230–233.

————. 1975b. Effects of soil texture on the interaction between *Rhizoctonia solani* and *Meloidogyne incognita* on cotton seedlings. J. Nematol. 7:234–236.

Cauquil, J. 1975. Cotton boll rot: Laying out a trial of a method of control. Amerind Publishing Co. Pvt. Ltd., New Delhi.

————, and R. L. Shepherd. 1970. Effect of rootknot nematode-fungi combinations on cotton seedling disease. Phytopathology 60:448–451.

Davis, R. G. 1975. Microorganisms associated with diseased cotton seedlings in Mississippi. Plant Dis. Rep. 59:277–280.

DeVay, J. E., R. H. Garber, and D. Matheron. 1982. Role of *Pythium* species in the seedling disease complex of cotton in California. Plant Dis. 66:151–154.

Ebbels, D. L. 1975. Fusarium wilt of cotton: A review, with special reference to Tanzania. Cott. Grow. Rev. 52:295–339.

————. 1980. Cotton diseases. Outlook Agric. 10:176–183.

Elad, Y., I. Chet, and J. Katan. 1980. *Trichoderma harzianum*: A biocontrol agent effective against *Sclerotium rolfsii* and *Rhizoctonia solani*. Phytopathology 70:119–121.

Erwin, D. C., S. D. Tsai, and R. A. Khan. 1978. Reduced number of microsclerotia formed by *Verticillium dahliae* in cotton tissue exposed to systemic benzimidazole fungicides and desiccation. Phytopathology 68:1488–1494.

----, ----, and ----. 1979. Growth retardants mitigate Verticillium wilt and influence yield of cotton. Phytopathology 69:283–287.

Fitzell, R., G. Evans, and P. C. Fahy. 1980. Studies on the colonization of plant roots by *Verticillium dahliae* Klebahn with use of immunofluorescent staining. Aust. J. Bot. 28: 357–368.

Ford, R. E., H. L. Bissonnette, J. G. Horsfall, R. L. Miller, D. Schlegel, B. G. Tweedy, and L. G. Weathers. 1981. Plant pathology in China, 1980. Plant Dis. Rep. 65:706–714.

Foy, C. D., H. W. Webb, and J. E. Jones. 1981. Adaptation of cotton genotypes to an acid, manganese toxic soil. Agron. J. 73:107–111.

Garber, R. H., J. E. DeVay, A. R. Weinhold, and D. Matheron. 1979a. Relationship of pathogen inoculum to cotton seedling disease control with fungicides. Plant Dis. Rep. 63:246–250.

----, E. C. Jorgenson, S. Smith, and A. H. Hyer. 1979b. Interaction of population levels of *Fusarium oxysporum* f. sp. *vasinfectum* and *Meloidogyne incognita* on cotton. J. Nematol. 11:133–137.

Green, J. A., and E. B. Minton. 1980. Influence of mechanical damage and fungicide seed treatments on germination and stand with cottonseed. Crop Sci. 20:535–539.

Halloin, J. M., E. B. Minton, and H. D. Peterson. 1978. Fungicide application to cottonseed using methylene chloride carrier. Crop Sci. 18:909–910.

Heilman, M. D., P. R. Nixon, R. V. Cantu, and R. E. Smithey. 1978. Use of soil-applied liquid fumigant for Phymatotrichum root rot control in cotton. Plant Dis. Rep. 62:609–612.

Howell, C. R. 1978. Seed treatment with L-sorbose to control damping-off of cotton seedlings by *Rhizoctonia solani.* Phytopathology 68:1096–1098.

----. 1982. Effect of *Gliocladium virens* on *Pythium ultimum, Rhizoctonia solani,* and damping-off of cotton seedlings. Phytopathology 72:496–498.

----, and R. D. Stipanovic. 1980. Suppression of *Pythium ultimum*-induced damping-off of cotton seedlings by *Pseudomonas fluorescens* and its antibiotic, pyoluteorin. Phytopathology 70:712–715.

Hunter, R. E. 1978. Effects of catechin in culture and in cotton seedlings on the growth and polygalacturonase activity of *Rhizoctonia solani.* Phytopathology 68:1032–1036.

----, and G. Guinn. 1968. Effect of root temperature on hypocotyls of cotton seedlings as a source of nutrition for *Rhizoctonia solani.* Phytopathology 58:981–984.

----, J. M. Halloin, J. A. Veech, and W. W. Carter. 1978. Terpenoid accumulation in hypocotyls of cotton seedlings during aging and after infection by *Rhizoctonia solani.* Phytopathology 68:347–350.

Hussain, T., and L. A. Brinkerhoff. 1978. Race 18 of the cotton bacterial blight pathogen, *Xanthomonas malvacearum,* identified in Pakistan in 1977. Plant Dis. Rep. 62:1085–1087.

Hyer, A. H., E. C. Jorgenson, R. H. Garber, and S. Smith. 1979. Resistance to root-knot nematode in control of root-knot nematode-Fusarium wilt disease complex in cotton. Crop Sci. 19:898–901.

Johnson, L. F., D. D. Baird, A. Y. Chambers, and N. B. Shamiyeh. 1978. Fungi associated with postemergence seedling disease of cotton in three soils. Phytopathology 68:917–920.

Kappelman, A. J., Jr. 1980a. Effect of fungicides, insecticides, and their combinations on stand establishment and yield of cotton. Plant Dis. 64:1076–1078.

----. 1980b. Long-term progress made by cotton breeders in developing Fusarium wilt resistant germplasm. Crop Sci. 20:613–615.

Katan, J. 1980. Solar pasteurization of soils for disease control: Status and prospects. Plant Dis. 64:450–454.

Lewis, J. A. 1979. Influence of soil texture on survival and saprophytic activity of *Rhizoctonia solani* in soils. Can. J. Microbiol. 25:1310–1314.

----, and G. C. Papavizas. 1979. Influence of the herbicides EPTC and prometryn on cotton seedling disease caused by *Thielaviopsis basicola*. Can. J. Plant Pathol. 1:100–106.

Lyda, S. D. 1978. Ecology of *Phymatotrichum omnivorum*. Ann. Rev. Phytopathology 16: 193–209.

Mace, M. E., A. A. Bell, and C. H. Beckman (ed.) 1981. Fungal wilt diseases of plants. Academic Press, New York.

McCarter, S. M., and R. W. Roncadori. 1971. Influence of low temperature during cottonseed germination on growth and disease susceptibility. Phytopathology 61:1426–1429.

Mahill, J. F., and D. D. Davis. 1978. Influence of male sterile and normal cytoplasms on the expression of bacterial blight in cotton hybrids. Crop Sci. 18:440–443.

Miller, J. H., C. H. Carter, R. H. Garber, and J. E. DeVay. 1979. Weed and disease responses to herbicides in single- and double-row cotton (*Gossypium hirsutum*). Weed Sci. 27:444–449.

Minton, E. B. 1978. Neutralizers and pesticides and their sequence of application to acid-delinted cottonseed: Effects on germination, stand, and Verticillium wilt of cotton. Crop Sci. 18:831–835.

----. 1980. Effects of row spacing and cotton cultivars on seedling diseases, Verticillium wilt and yield. Crop Sci. 20:347–350.

----, and J. E. Quisenberry. 1981. Beltwide evaluation of fungicides applied to acid-delinted cottonseed. Agron. J. 73:684–686.

----, and J. R. Supak. 1980. Effects of seed density on stand, Verticillium wilt, and seed and fiber characters of cotton. Crop Sci. 20:345–347.

Mitchell, D. J. 1978. Relationships of inoculum levels of several soilborne species of *Phytophthora* and *Pythium* to infection of several hosts. Phytopathology 68:1754–1759.

Padaganur, G. M., and A. L. Siddaramaiah. 1979. Laboratory evaluation of fungicides against two cotton pathogens: *Alternaria macrospora* Zimm and *Helminthosporium gossypii* Tucker. Pesticides 13:35–36.

Papavizas, G. C., J. A. Lewis, E. B. Minton, and N. R. O'Neill. 1980. New systemic fungicides for the control of cotton seedling disease. Phytopathology 70:113–118.

Pinckard, J. A., and D. R. Melville. 1975. Some new developments in cottonseed dressings. Plant Dis. Rep. 59:262–266.

Puhalla, J. E. 1979. Classification of isolates of *Verticillium dahliae* based on heterokaryon incompatibility. Phytopathology 69:1186–1189.

Pullman, G. S., and J. E. DeVay. 1981. Effect of soil flooding and rice culture on the survival of *Verticillium dahliae* and incidence of Verticillium wilt in cotton. Phytopathology 71: 1285–1289.

----, ----, R. H. Garber, and A. R. Weinhold. 1981. Soil solarization: Effects on Verticillium wilt of cotton and soilborne populations of *Verticillium dahliae, Pythium* spp., and *Thielaviopsis basicola*. Phytopathology 71:954–959.

Ranny, C. D. (ed.) 1973. Verticillium wilt of cotton. USDA Misc. Publ. ARS-S-19.

Roncadori, R. W., and S. M. McCarter. 1972. Effect of soil treatment, soil temperature, and plant age on Pythium root rot of cotton. Phytopathology 62:373–376.

Sanders, D. E., and J. P. Snow. 1978. Dispersal of airborne spores of boll-rotting fungi and the incidence of cotton boll rot. Phytopathology 68:1438–1441.

Sappenfield, W. P., C. H. Baldwin, J. A. Wrather, and W. M. Bugbee. 1980. Breeding multiple disease resistant cottons for the North Delta. p. 280–283. Proc. Beltwide Cotton Prod. Res. Conf., St. Louis, Mo.

Scher, F. M., and R. Baker. 1980. Mechanism of biological control in a *Fusarium*-suppressive soil. Phytopathology 70:412–417.

Simpson, M. E., P. B. Marsh, G. V. Merola, R. J. Ferretti, and E. C. Filsinger. 1973. Fungi that infect cottonseeds before harvest. Appl. Microbiol. 26:608–613.

Smirnova, A. A. 1980. Protection of cotton in the USSR. Outlook Agric. 10:206–207.

Sterne, R. E., T. H. McCarver, and M. L. Courtney. 1979. Survival of plant pathogens in composted cotton gin trash. Ark. Farm Res. 28:9.

Tabachnik, M., and J. E. DeVay. 1980. Black root rot development in cotton roots caused by *Thielaviopsis basicola* and the possible role of methyl acetate in pathogenesis. Physiol. Plant Pathol. 16:109–117.

––––, ––––, R. H. Garber, and R. J. Wakeman. 1979. Influence of soil inoculum cncentrations on host range and disease reactions caused by isolates of *Thielaviopsis basicola* and comparison of soil assay methods. Phytopathology 69:974–977.

Tarr, S. A. J. 1964. Virus diseases of cotton. Commonw. Mycol. Inst., Kew, Surrey, England Misc. Publ. 18.

Wanjura, D. F., and E. B. Minton. 1981. Delayed emergence and temperature influences on cotton seedling vigor. Agron. J. 73:594–597.

Watkins, G. M. (ed.) 1981. Compendium of cotton diseases. The American Phytopathological Society, St. Paul, Minn.

C. Nematodes

Nematodes (nema = Greek for thread) are elongate, unsegmented, invertebrate animals that range in size from microscopic to up to 1 m in length. They occupy a variety of ecological niches and have various feeding habits. Most are free living in the soil and feed on various types of flora, fauna, and organic debris; others are animal parasites, and many are plant parasites. The plant parasitic nematodes are either endo- or ecto- (a few are semiendo) parasites, but all are obligate parasites. Endoparasitic nematodes are characterized by complete penetration of the host, i.e., the entire body of the nematode enters the host; ectoparasitic nematodes feed on cells without entering the host. Since nematodes are obligate parasites the cells on which they feed must be living.

This chapter will be concerned with those nematodes that are parasites on cotton (*Gossypium* sp.). However, it should be kept in mind when devising protection strategies, that some nematodes may be beneficial to cotton production [e.g. *Orrina phylobia* Brzeski (syn. *Nothanguina phylobia,* Thorne), a parasite of the silver leaf nightshade (*Solanum elaeagnifolium,* Cav.) weed (Orr et al., 1975), and the mermithid parasite of boll weevil *Anthonomus grandis,* (Cleveland, 1980)]. Although several species of nematodes attack cotton, the Southern root-knot nematode, *Meloidogyne incognita* (Kofoid & White) Chitwood; reniform nematode, *Rotylenchulus reniformis* Linford & Oliviera; sting nematode, *Belonolaimus longicaudatus* Rau; and lance nematodes, *Hoplolaimus columbus* Sher., and *H. galeatus* (Cobbs) Thorne, are recognized as the most serious threats to production (Heald and Orr, 1982; Sasser, 1972; Watkins, 1981).

According to loss estimates released annually by the Cotton Disease Council (CDC) (see the 1957 through 1982 *Proceedings* of the Beltwide Cotton Production Research Conferences, published by the National Cotton Council, Memphis, Tenn.) nearly 2% of the total U.S. cotton crop was lost to nematodes. This estimate is misleading since losses often exceed 50% of the crop (Heald and Orr, 1982). Loss averages calculated in 5-year increments from 1956 through 1980 (Table 9C-1) show a steady increase in the amount of the crop that is destroyed by nematodes. This increase may indicate that i) no progress has been made in protecting cotton from nematodes, ii) new problems are rapidly developing, or iii) the loss estimates have not accurately reflected the effects nematodes have on cotton. I believe that the last two possibilities are mostly responsible for the gradual increase in cotton losses due to nematodes. Recently, Orr et al. (1982) took exception to the commonly excepted loss estimates and attempted to arrive at more accurate figures. Test plots established to specifically measure yield loss to nematodes on the Texas High Plains indicated an average loss of 25%. Considering that the High Plains account for about 50% of the state's cotton production, the losses in Texas from nematode could be much higher than the CDC estimate (Table 9C-2). If losses in the other cotton producing states (Table 9C-2) are similarly underestimated, a very serious threat to cotton production is being largely overlooked and neglected.

In addition to being serious pests in their own right, nematodes are also involved in interactions with fungi: the most notable interaction is the root-knot nematode—fusarium wilt complex. Although we know little about the subject, nematodes likely exacerbate damage done to cotton by insects and environmental stresses and reduce the capacity of the cotton plant to compete with weeds.

Table 9C-1. Average losses of cotton from nematodes in the USA from 1956 through 1980.

Years	Average loss
	%
1956–1960	1.3
1961–1965	1.7
1966–1970	2.1
1971–1975	2.3
1976–1980	2.2
1956–1980	1.9

Table 9C-2. Average losses (%) of cotton from nematodes in 14 cotton-producing states in the 1980 growing season.

Ala.	Ariz.	Ark.	Calif.	Ga.	La.	Miss.	Mo.
2.0	2.0	0.5	0.6	4.8	4.5	0.8	0.1

N.Mex.	N.C.	Okla.	S.C.	Tenn.	Tex.		$\bar{x}$
3.0	8.0	0.8	2.0	0.2	1.2		1.4

9C-1 NEMATODES PARASITIC ON COTTON

Nematodes parasitic on cotton have many features in common with one another and with other plant parasitic nematodes. Preadult stages are commonly referred to as juveniles or larvae. Except for some adult stages nematodes are too small to be easily seen with the unaided eye; with magnification they can be seen to be eel-shaped. They all inhabit the soil, or more precisely, the film of water that surrounds the soil particle, and they all possess a hollow spear shaped structure—a stylet—that can be extended from an opening in the head and inserted into a host cell when the nematode feeds. Salivary gland secretions, which dissolve cell contents, are injected into host cells through the stylet; host cell contents are extracted via the same structure. Finally, being obligate parasites all nematodes parasitic on cotton must feed on living host cells to complete their life cycles.

The common names of the four major types of nematode parasites of cotton are diagnostically descriptive and helpful in determining which nematode is causing a problem. Table 9C-3 is intended as an empirical diagnostic tool for such determinations. All that is required for the determination is a hand lens and careful lifting of the plant from the soil.

9C-1.1 Root-Knot Nematode

The root-knot nematode, *Meloidogyne incognita,* is a sedentary endoparasite. It is the most important nematode parasite of cotton because it is endemic wherever cotton is grown, and it has an extremely broad host range. The nematode is capable of severely limiting yields either alone or in concert with other microorganisms or ecological situations.

Table 9C-3. Empirical identification of nematode parasites on cotton.

Nematode	Diagnostic characteristic	Nematode characteristic
Root-Knot (*Meloidogyne incognita*)	Roots knotted, i.e., distorted by small to medium sized galls.	Pear-shaped, mature female is embedded in root, gelatinous matrix containing ovoid eggs envelops posterior part of nematode and is often exposed to the root surface.
Reniform (*Rotylenchulus reniformis*)	The appearance of the egg bearing female nematodes; no pronounced gall on root.	Mature egg bearing female is kidney shaped; usually only head and neck embeded in root; gelatinous matrix containing eggs exposed.
Lance (*Hoplolaimus columbus* or *H. galeatus*)	Brown-yellow discoloration of epidermis at the site of infection.	May not be present on roots (ecto- or endo-parasite).
Sting (*Belonolaimus longicadatus*)	Dark, sunken lesions, stubby roots	Usually not present on roots, (ecto-parasite).

9C-1.1.1   Taxonomy

The taxonomy of *M. incognita* has been reviewed by Sasser (1972), and a synopsis of that report is presented here. The cotton root-knot nematode was originally named *Heterodera radicola* Atkinson, but the name was changed to *H. marioni* Chit. when all the root-knot nematodes were classified under the same binomial with the cyst nematodes. Chitwood partitioned the root-knot nematodes into the genus *Meloidogyne*, and Buhrer later relegated the cotton root-knot nematode to the subspecies *M. incognita* var. *acrita.* Since Triantaphyllou and Sasser could not morphologically distinguish *M. incognita acrita* from *M. incognita incognita,* they proposed, and Whitehead adopted, the binomial *M. incognita* to include both subspecies.

9C-1.1.2   Life Cycle

Root-knot nematode eggs are ovoid in shape (32×77 μm) and may be found free in the soil, enclosed in a gelatinous matrix on the surface of the host root, or embedded in the root tissues. The juvenile nematode (L_1) within the egg undergoes its first cuticular molt while still in the egg; three post-hatch molts occur later. The second stage juvenile (L_2) escapes from the egg by using its stylet to cut a slit in the eggshell. The released L_2, which is the infective stage of the nematode, migrates through the soil until it encounters a host root. It may survive in the soil for several weeks before locating a host. The encounter between the migrating nematode and the growing plant root may occur by chance, or the L_2 may be chemically attracted to cotton roots (McClure, 1972). After infection, the nematode establishes a permanent feeding site, which develops into a multinucleate giant cell. Throughout its development, the nematode feeds on the giant cell. Sex differentiation of the nematode is greatly affected by the parasitic burden of the host; fewer nematodes develop into females when the host is heavily parasitized (Triantaphyllou and Hirschmann, 1964; Triantaphyllou, 1973). When a nematode develops into a female, its body becomes pear-shaped. Reproduction is parthenogenetic and males are not necessary (Triantaphyllou and Hirschmann, 1964). Up to 500 eggs are deposited in a gelatinous matrix exuded at the time of egg laying. In cotton, eggs may be produced in as few as 22 days depending on the level of cultivar susceptibility and environmental conditions (Minton, 1962).

More detailed accounts of the life cycle of *M. incognita* have been presented elsewhere (Heald and Orr, 1982; Taylor and Sasser, 1978; Watkins, 1981).

9C-1.1.3   Infection and Histopathology

Second stage juvenile nematodes penetrate host roots near the root tip (McClure and Robertson, 1973). The nematodes migrate intercellularly through the cortex toward the stele. Generally, little damage is done to the cortex during the migration. A permanent feeding site is established in the developing stele and the nematode soon becomes sedentary. The permanent feeding site develops into a hypertrophied, thick-walled, multinucleate giant cell. The induction of a giant cell seems to be absolutely necessary for the subsequent development of the nematode. The host vascular system at the site of giant cell induction becomes distorted and extensive hypertrophy and hyperplasia results in the formation of a gall or knot on the roots. In some early work on the nature of resistance of cotton to the root-knot nematode, Brodie et al. (1960) reported that roots of susceptible and resistant cultivars were equally penetrated by the nematode. Similar studies on other cultivars confirmed this finding (McClure et al., 1974a; Carter et al., 1977; Minton, 1962; Veech and McClure, 1977). Brodie et al. (1960) reported that at inoculum densities of 900 larvae/root tip a greater percentage (ca. 40%) of the larvae succeed in infecting plants than at inoculum densities of 150 larvae/root tip (ca. 3% infection). However, McClure et al. (1974a,b), using lower inoculum densities (10 or 200 larvae/seedling), reported that the percentage of infection was nearly constant (ca. 20%).

Various host responses have been reported in resistant cotton infected by the root-knot nematode. In addition to resistant responses, Brodie et al. (1960), McClure et al. (1974a, b), and Veech (1979) observed typical susceptible responses in resistant hosts. That is, some infective larvae initiate giant cells in resistant hosts and develop to gravid females, albeit it slower than in susceptible hosts. Most studies on the nature of resistance of cotton to the root-knot nematode have concluded that resistance is based on the physiological ability of the plant to respond to infection by producing a chemical(s) that inhibits or retards the development of the nematode (Minton, 1962; Brodie et al., 1960; McClure et al., 1974a, b; Veech and McClure, 1977; Veech, 1978, 1979).

Among resistant host responses, Brodie et al. (1960) reported three types of reactions: root necrosis, retarded gall development, and failure of the majority of the nematodes to reach maturity. McClure et al. (1974a, b), recognized essentially only inhibition of larval development, but this response could occur either before or after, giant cell initiation. Brodie et al. (1960), but not McClure et al. (1974a, b) observed some host necrosis at nematode feeding sites. On the other hand, McClure et al. (1974b), but not Brodie et al. (1960) observed lysed nematodes and galls without nematodes. As different as these resistant responses seem to be, they may all be manifestations of the same biochemical mechanism of resistance. I suggest that the differences are brought about by the time the resistant response is initiated, or by the rate at which it occurs.

Veech and McClure (1977) and Veech (1978, 1979) proposed a biochemical mechanism of resistance predicated on the accumulation of terpenoid aldehyde phytoalexins at the site of nematode feeding. This mechanism could account for the various observed resistant responses. The terpenoid aldehydes are initially absent from host tissues at the site of nematode feeding, but they accumulate there in response to feeding. At low concentrations, terpenoid aldehydes reversibly inhibit nematode motility but at high concentrations the terpenoid aldehydes are toxic to the nematode (Veech, 1979). Although both susceptible and resistant plants may accumulate terpenoid aldehydes at the site of nematode feeding, the accumulation is more rapid and occurs at more nematode feeding sites in resistant hosts than in susceptible hosts. Furthermore, there is a good positive correlation between the level of host resistance and the ability of the host to accumulate terpenoid aldehydes in response to infection (Veech, 1978). Unfortunately from the standpoint of a rapid method of screening for nematode resistance, there is no relationship between the level of host resistance and the concentration of constitutive (present prior to infection) terpenoids aldehydes.

In a field study using five cotton cultivars with differing levels of resistance, Minton (1962) found the soil population of *M. incognita* increased more slowly under resistant plants than under susceptible plants. However, as plants reached the harvest stage in September, soil populations of the nematode were at their highest levels and were nearly equal under susceptible and resistant plants.

9C–1.1.4   Races

In a greenhouse study on four cultivars of *Gossypium hirsutum* L. and one *G. barbadense* L. cotton, larvae in a population identified as *M. incognita acrita* reached the adult female stage of development 22 to 29 days after inoculation, respectively, but larvae in a population identified as *M. incognita* did not reach this stage of development in 121 days on *G. barbadense* or *G. hirsutum* 'Mexican' (Minton, 1962). If the current practice of grouping both of these nematode populations under one binomial is legitimate, then host specific races of *M. incognita* exist. Taylor and Sasser (1978) analyzed the host preferences of many *M. incognita* populations from a worldwide collection. They concluded that there were four races of *M. incognita*. Races 1 and 2 do not reproduce well on cotton, but Races 3 and 4 do. Usually it is a simple task, using host differentials, to identify the nematode race infesting a field. Table 9C–4 shows the responses of the host differentials to the four races of *M. incognita*.

Table 9C–4. Ability of the four races of *Meloidogyne incognita* to reproduce on tobacco and cotton (Sasser, 1972).

	Host differential	
M. incognita Race	*Nicotiana tabacum* 'NC95'	*Gossypium hirsutum* 'Deltapine 16'
1	−	−
2	+	−
3	−	+
4	+	+

Meloidogyne acronea Coetzee appears to be an extremely devastating nematode parasite of cotton in Malawi (Bridge et al., 1976). In many ways it is very unlike *M. incognita.* It is a semi-endoparasite that does not induce the galls or knots which typify *M. incognita* infection. Like *M. incognita,* *M. acronea* females produce an egg sac. When females are congregated at an infection site, the egg sacs may collectively appear as a small gall. Unlike *M. incognita,* mature *M. acronea* females become filled with eggs and are transformed into cyst-like structures. Because the nematode attacks the very fine feeder roots, evidence of its presence may be lost when lifting the plants for examination. This fact may account for the lack of detection of the nematode in other parts of Africa.

9C-1.2   Reniform Nematode

The reniform nematode (*Rotylenchulus reniformis*) was first reported as a parasite on cotton in Louisiana by Smith and Taylor (1941). It may have been observed on cotton in Georgia even earlier (Smith, 1940). The nematode is a sedentary endoparasite that feeds in the pericycle and induces the formation of a uninucleate giant cell. It does not induce root galling as does the root-knot nematode. This nematode adversely effects cotton yields (Minton and Hooper, 1959; Minton et al., 1964) and can compound other problems by delaying crop maturity. According to Yik (1981) all commercial cultivars of cotton are susceptible to the reniform nematode, but Sasser (1972) lists several cultivars that appear resistant to the reniform nematode–fusarium wilt complex.

9C-1.2.1   Life Cycle

The life cycle of the reniform nematode has been described by Sivakumar and Sishadri (1971). It differs from that of the root-knot nematode in that the L_2, L_3, L_4 juveniles, and males do not feed and are not infective. The infective stage is the immature female (Birchfield, 1962). Since sex is determined prior to infection, the sex ratio is not regulated by post infection factors as it is in the root-knot nematode. Sexual reproduction occurs and non-parasitic males are often seen coiled around parasitic females. On cotton, the life cycle lasts 17 to 23 days (Birchfield, 1962).

9C-1.2.2   Host Range

The reniform nematode is a severe parasite on soybeans (*Glycine max*), sweet potato (*Ipomoea batatas*), and other crops commonly rotated with cotton. Of 43 crops tested by Birchfield and Brister (1962) only 11 were rated as non-hosts. Although commercial cottons are susceptible, resistance has been identified in several *Gossypium* spp. (Yik, 1981; Muralidharan and Sivakumar, 1977).

9C-1.2.3   Geographical Distribution

The reniform nematode is geographically more restricted than the root-knot nematode (Sasser, 1972). In general it prefers tropical to subtropical climates (Heald and Orr, 1982), and it is a serious pest in all of the U.S. coastal cotton-producing states (Minton and Hooper, 1959; Neal, 1954; Fassuliotis and Rau, 1967; Lambe and Horne, 1963). In Egypt it causes serious problems on *G. barbadense.* The nematode has also been reported south of the Sahara (Luc and de Guiran, 1960; Martin, 1955; Peacock, 1956) and in South America.

9C-1.2.4   Infection and Histopathology

The histopathology of the susceptible and resistant responses of cotton to the reniform nematode have been described (Cohn, 1974, 1976; Carter, 1981; Yik, 1981). Susceptible and resistant plants are equally penetrated (Carter, 1981), and penetration occurs at any place along the root axis. However, small feeder roots are preferred over larger roots (Heald and Orr, 1982). Although the nematode does not feed on the epidermis or cortex, some damage is done to those tissues during migration of the nematode toward the stele (Birchfield, 1962). Differences between susceptible and resistant plants are manifest within 36 hours after penetration; fewer nematodes can be observed in resistant plants inoculated with the same number of nematodes as susceptible plants (Carter, 1981). In resistant plants, egress is greater or more nematodes are lysed than in susceptible plants (Carter, 1981). For the nematodes remaining in resistant roots the host response may resemble the susceptible response for a short time. That is, feeding is initiated in the endodermis and may extend into the pericycle; some hypertrophy may also occur (Birchfield, 1972; Carter, 1981; Cohn, 1974, 1976; Yik, 1981). The resistant response is characterized by what appears to be a hypersensitive reaction (Carter, 1981; Yik, 1981). The host cell walls adjacent to the pericycle and endodermis collapse and a thick deposit of safranin-positive material (indicative of necrosis) forms between the pericycle and the cortex and extends about half the circumference of the stele (Carter, 1981). Such a reaction could effectively isolate the nematode in a zone of dead cells. Since the nematode is sedentary and cannot migrate to living cells, it will die from toxic metabolites of the host, or from starvation. The resistant response is also characterized by dead or degenerated female nematodes and small or no egg masses; large healthy females and large egg masses characterize the susceptible response (Yik, 1981).

9C-1.3   Lance Nematodes

The lance nematodes *Hoplolaimus columbus* and *H. galeatus* are ecto-, endo-, and semiendo parasites of cotton. These nematodes have been of economic concern for a number of years in the Atlantic coastal areas of

North Carolina, South Carolina, and Georgia (Fassuliotis, 1974; Schmidt-Kraus, 1979; Schmidt-Kraus and Lewis, 1981). *Hoplolaimus galeatus* is more widely distributed than *H. columbus* in the southeastern Coastal Plain. Both species cause severe stunting of cotton (Wakins, 1981). *Hoplolaimus* spp. have been reported parasitizing cotton in Arkansas (Riggs, 1974) and Alabama (Rodriguez-Kabana and Pearson, 1972), but the incidences there were localized. Outside of the USA, lance nematodes are a problem in Asia and northern Africa (Schmidt-Kraus, 1979).

Bird et al. (1974) think that *H. columbus* is a neospecies, and as such it will become increasingly abundant over a broader geographical range than it currently inhabits; concommitantly, it will increase in aggressiveness as a parasite. If this is an accurate assessment, we should, in the future, expect more severe and widespread problems from this nematode.

Lewis et al. (1976) described the symptomology of infected cotton. Typically, infected roots display shallow surface depressions at the point of nematode penetration of the root. Slight cellular discoloration may occur at this point and along the path of the parasite through the cortex. Dark, necrotic, cortical lesions are formed at the feeding site of the nematode. Cotton grown in severely infested fields will often be stunted, chlorotic, and partially to totally defoliated (Fassuliotis, 1974). The extensive cortical damage caused by this nematode leads one to suspect that the nematode may be a predisposing factor for wilt or seedling disease problems. However, there is no evidence to support this conclusion, and Yang et al. (1976a) reported *H. galeatus* did not increase wilt problems.

Fassuliotis (1974) compiled a partial host-range list for *H. columbus* and reported that the common weeds [bermudagrass (*Cynodon* L.), pigweed (*Chenopodium* L.), sicklepod (*Arabis* L.), and nutsedge (*Cyperus* L.)] were good or excellent hosts for the nematode. Lewis and Smith (1976) extended the host list with several agronomic crops, some of which are commonly rotated with cotton.

Among the weed host of *H. columbus,* nutsedge presents an interesting problem in that it may enhance the potential for economic losses caused by the nematode (Bird and Hogger, 1973; Hogger, 1975). Nutsedge, which is an excellent host for the nematode, initiates growth in the early spring when soil population densities of the nematode are low. Because the nematode has a generation time of only 7 to 8 days (Lewis et al., 1976) large populations soon develop and by the time cotton is planted damaging population densities of nematodes may occur.

Hoplolaimus columbus has been reported to affect population densities of other nematodes parasitic on cotton. *Scutellonema brachyurum* (Steiner) Andrassy generally does not reproduce well on cotton (Schmidt-Kraus and Lewis, 1979), but in the presence of *H. columbus* its reproduction is markedly increased (Schmidt-Kraus and Lewis, 1981). Conversely, *H. columbus* tends to inhibit the reproduction of the root-knot nematode, *M. incognita* (Schmidt-Kraus and Lewis, 1981), and in some cases the lance nematode may replace the root-knot nematode as the predominant parasitic nematode (Bird et al., 1974).

9C-1.4   Sting Nematode

The sting nematode, *Belonolaimus longicautatus* Rau, was first reported on cotton in Virginia (Owens, 1951). It is a migratory ectoparasite that feeds on the epidermis and cortex. All stages of the nematode feed. According to Sasser (1972) the sting nematode can be the most devastating of all nematodes on cotton. However, because its geographic distribution in the USA is largely limited to the east, and cotton production is moving west, the nematode is probably the least important of the four major nematode parasites. The geographic distribution is limited by soil type (Heald and Orr, 1982), and in general the nematode prefers a light sandy soil.

A complex of the sting nematode and fusarium wilt is devastating to cotton production. A wilt resistant cotton (e.g., 'Coker 100') grown in *Fusarium* infested soil shows little or no wilt. However, if the sting nematode is present, more than 60% of the plants succumb to the fungus (Holdeman and Graham, 1954).

9C-1.5   Nematode-Fungus Interactions

Detailed reviews of the literature on the interaction of nematodes and disease-causing organisms have been presented by Powell (1971a, b). Sasser (1972) presented similar information specifically related to cotton.

9C-1.5.1   Wilts

The interaction between the root-knot nematode and *Fusarium oxysporum* f. sp. *vasinfectum* (Atk.) Snyd. & Hans., on cotton has been recognized since 1892 (Sasser, 1972). The most important element in this interaction is the nematode. When nematodes are controlled, or nematode resistance is incorporated into cotton, problems with fusarium wilt are greatly diminished. Resistance to *Fusarium* alone, however, does not offer adequate protection. Cottons developed for resistance to *Fusarium* on soils infested with *M. incognita* usually maintain their resistance when simultaneously challenged by both organisms. But wilt resistant cottons developed on soils not infested with the nematode often wilt in the presence of the fungus and nematode (Sasser, 1972; Heald and Orr, 1982). Garber et al. (1979) demonstrated the effect the nematode can exert on susceptibility to fusarium wilt. They obtained wilt in the presence of 50 *M. incognita* larvae and 650 propagules of the fungus. In the absence of the nematode it required 77 000 fungus propagules to produce the same level of wilt.

The role of the nematode in enhancing wilt is not well understood. Since mechanical wounding of roots increases the amount of wilt, it would seem that the nematode's role is simply to create a portal of entry. However, if that is the case, one would expect an accumulation of fungus at the points of nematode penetration; but this is usually not observed (Perry, 1963). It is more likely that a physiological interaction occurs between the fungus, nematode, and plant.

According to Martin et al. (1956), different populations of the nematode vary in their ability to increase the incidence of wilt. These differences may be related to what we now recognize as the different races of *M. incognita*.

Recently there has been interest in determining the effect on wilt of multiple interactions. Since nematodes often occur in mixed populations, Yang et al. (1976a) investigated the role of concomitant inoculations of *M. incognita* and *H. galeatus* on fusarium wilt. Wilt was not enhanced when *H. galeatus* was the only nematode present, but when *H. galeatus* and *M. incognita* occurred together the incidence of wilt was greater than that produced by the fungus alone, or when the fungus and *M. incognita* occurred together. Similarly, fungi that normally do not enhance fusarium wilt may do so in the presence of *M. incognita*. The incidence of fusarium wilt is greater in the presence of concomitant populations of *Trichoderma harzianum* Rifai and *M. incognita* than when only *M. incognita* and the wilt fungus interact; but *T. harzianum* alone does not enhance wilt (Yang et al., 1976b). This suggests that organisms commonly considered antagonists to plant pathogens (e.g. *T. viride* Pers. ex S. F. Gray) may exacerbate diseases, in the presence of nematodes.

The severity of wilt induced by *Verticillium dahlie* Kleb., is enhanced by *M. incognita* (Bazan de Segura and Aguilar, 1955; Khoury, 1980), *R. reniformis* (Prasad and Padaganur, 1980), and *Belonolaimus gracilis* and *Pratylenchus* spp. Cobb, (Heald and Orr, 1982). Also by delaying the maturation of a crop, nematodes enhance losses by verticillium wilt; *V. dahlie* is a cool season pathogen and delayed crop maturation would favor the pathogen. Sappenfield (1963) reported that root-knot—fusarium resistant cottons also have some resistance to verticillium wilt.

9C–1.5.2   Seedling Diseases

The interaction between *Rhizoctonia solani* Kuehn and *M. incognita* in the incitement of soreshin was reported by Reynolds and Hanson (1957). They noted a substantial decrease in disease severity in field plots fumigated with ethylene dibromide for nematode control. Others (Brodie and Cooper, 1964; Cauquil and Shepherd, 1970; Carter, 1975) later showed that both organisms together had a greater adverse effect on cotton than either organism alone. White (1962) showed that fumigation to control nematodes was as effective in reducing seedling disease as fumigation to control fungi (*R. solani* or *Thielaviopsis basicola* [Berk. & Br.] Ferr.).

The interaction of *M. incognita* and *R. solani* on cotton is synergistic (Carter, 1975; Cauquil and Shepherd, 1970). Synergistic effects also occur when *M. incognita* interacts with *Alternaria tennuis* Auct., *Fusarium oxysporum* f. sp. *vasinfectum,* and *Glomerella gossypii* Edg. (Cauquil and Shepherd, 1970). The mechanism by which synergism occurs is not understood. Since root pruning does not enhance the disease (Brodie and Cooper, 1964; Carter, 1981), it is not likely that wounding of the plant roots by the nematode is responsible for the enhancement of the disease. This result could be expected because the fungus primarily attacks the hypocotyl and the nematode attacks the roots. Mechanical punctures of the hypocotyl in the presence of one or both organisms does increase disease severity. Carter (1975) reported that the optimum temperature for disease development in the presence of *M. incognita* and *R. solani* was 18 and 25°C respectively, and that 2500 larvae per seedling were required to increase disease above that which occurs in the presence of the fungus alone. Additionally he showed a direct correlation between coarse particle content of the soil and the synergistic effect between nematode and fungus.

Soreshin is also enhanced by *M. arenaria,* (Neal) Chit., *R. reniformis,* and *Hoplolaimus* sp. (Brodie and Cooper, 1964).

9C-2   CONTROL PRACTICES

Strategies for protecting cotton from losses by nematodes can be divided into four major categories; chemical, cultural, biological, and host plant resistance. These may be practiced individually or in various combinations. In developed countries chemical control is the most widely practiced method of reducing losses. In less developed countries, cultural practices are relied upon most heavily. The limited availability of nematode-resistant commercial cottons has kept host plant resistance techniques from gaining wide use. Nevertheless, the report of the U.S. Office of Technology Assessment (Anonymous, 1979) concluded that host plant resistance offered the best hope for long range control of crop pests. Biological control of nematode pests of cotton is relatively new, and it would be premature to assess its potential.

The objective of all control strategies is to reduce the soil population density of parasitic nematodes low enough to prevent economic losses. Barker and Olthof (1976) wrote a provocative report on the relationship between nematode population densities and their effect on crops. In that report they discussed the concept of economic thresholds. Generally, the economic threshold population is based on the nematode population density at the time of planting. If the population is below that threshold, control practices are usually not economically worthwhile. Accurate methods of sampling nematode population densities (Barker and Campbell, 1981) and effective diagnostic and advisory service (Nusbaum and Barker, 1971) are usually requisite to determining economic thresholds. Economic thresholds for cotton have not been developed for most ecological and cultural situations, and the practice of nematode control is usually empirical.

9C-2.1   Chemical Control

There is no paucity of literature on the chemical control of nematodes. The principles of such control and the modes of action of nematicides have been reviewed in detail by others (Wright, 1981; Decker, 1972; Van Gundy and McKenry, 1977; Whitehead, 1978). Essentially there are two types of

Table 9C-5. Vapor pressures and Henry's constants of the common fumigant nematicides.

Fumigant	Vapor pressure	Henry's constant[†]
	mm Hg	
Methyl bromide	1380.0	4.1
Dichloropropane	21.0	20.2
Methyl isothiocyanate	21.0	88.0
Trichloronitromethane	20.0	10.8
Dibromoethane	8.0	42.7
Dibromochloropropane	0.6	163.8

[†] Ratio of the concentration of fumigant in the soil water to the concentration in the soil atmosphere.

chemical control agents—fumigants and non-fumigants. Both are generally used for early season protection, but some may be applied as post-plant treatments. By reducing the initial nematode population, preplant treatment allows the plants to establish good root systems and reduces the number of nematode generations that can occur in a growing season.

The fumigant nematicides are generally applied as preplant treatments and they must be dissipated from the soil prior to planting because many are phytotoxic. The alkyl halides—methyl bromide, dibromochloropropane, dibromoethane, dichloropropane, and trichloronitromethane—are the most common chemical fumigants. Methyisothiocyanate is also an effective nematode fumigant.

The fumigants are usually injected into the soil as a gas or liquid, but they may also be applied as a drench and watered in. When injected, they are ideally placed at the deepest level at which control is desired. Usually, however, injection depth is determined by economics and available equipment. The fumigants are volatile and soluble in water. Because nematodes inhabit the soil-water film, the chemicals are effective as nematicides when they are dissolved in the water film. Thus, soil moisture at the time of application is critical to successful treatment. If the soil is too dry the fumigant will diffuse to the atmosphere too rapidly; if it is too wet, diffusion through the soil is limited. Generally if the soil moisture is right for tilling, it is acceptable for fumigation. Soil temperature is also an important factor affecting successful fumigation. As the soil temperature increases, the concentration of fumigant in the soil atmosphere increases. But the increase in the concentration in the soil atmosphere is at the expense of the concentration in the soil water. At colder temperatures more fumigant will be dissolved in the soil moisture than at warmer temperatures. Henry's Law holds that the ratio of the concentration of fumigant in the soil atmosphere to the concentration in the soil water is constant at any temperature. Henry's constants, therefore are useful for determining the concentration of nematicide required for the prevailing soil temperatures; the constants for the common fumigants are shown in Table 9C-5. Generally, soil temperatures between 10 to 20°C at 20 cm depth are acceptable for fumigation.

Three factors that affect the success of fumigation treatments, but over which we generally have less control, are soil texture, pH, and organic matter content. Fumigants diffuse rapidly in soils with large pores and may not be retained long enough to be effective. This diffusion can be remedied by tarping the soil; but tarping is expensive. Soil pH may have a profound effect on the efficacy of the nematicide and fumigants may also be rendered ineffective by absorption onto organic matter.

The nonfumigant nematicides are generally organophosphates or carbamates and virtually all were designed as insecticides. Although they can be formulated as liquids they are generally applied as granules. Compared to the fumigants they have low phytotoxicity and most can be applied at planting or after planting. Many of the nonfumigants are systemic; some are transported in the phloem, but most are transported in the xylem. Generally their adverse effects on nematodes are reversible; when the compounds dissipate from the soil the nematodes often recover, hence, the nonfumigants are usually considered nematistatic. The modes of action of nematicides have been reviewed by Wright (1981) and Van Gundy and McKenry (1977).

Unlike the fumigants that move through the soil primarily in the gaseous phase, the nonfumigants are dispersed through the soil in the soil water. Their movement is usually slower than that of the fumigants but all of the nematicide is in the soil fraction (soil water) where it can be effective, and little is lost to the atmosphere. Solubility of a nematicide in water and its absorption by organic matter are determined by the polarity of the chemical; there is a positive relationship between polarity and solubility, but a negative relationship between polarity and absorption. A very polar nematicide will be very soluble in water and not extensively absorbed by soil organic matter. The same physiochemical processes that reduce absorption of polar nematicides by soil organic matter also reduce the penetration of the nematicide through the cuticle of the nematode. Nematicides are, therefore, synthesized to accommodate these opposing conditions; they are neither very polar nor very nonpolar.

Unlike the fumigant nematicides that are usually applied deep in the soil to accommodate their upward diffusion, the nonfumigant nematicides are usually applied at or near the soil surface since their general direction of dispersion is downward. Because of their low phytotoxicity the nonfumigants are often applied to the soil during the planting operation or added to irrigation water.

9C–2.2 Biological Control

At present there are no biotic agents commercially available for use for the control of nematodes (Mankau, 1981). Theoretically, biotic agents could offer the specificity for target organisms that is generally lacking in chemical control agents. *Bacillus penetrans* Mankau is a bacterial pathogen that is well suited to parasitism of the root-knot nematode. Infected nematodes are often killed by the bacterium, and those surviving do not reproduce (Sayre, 1980). Unfortunately, the inability to culture the bacterium in vitro has hampered its development as a biocontrol agent (Mankau, 1980a). Mankau (1980b) assessed prospects of the biocontrol of nematodes with fungi. He was not impressed with the potential of the nematode-trapping fungi because of their nonspecificity and facultative saprophytism. The main disadvantage of these fungi is that their populations decline under annual cropping situations. He was impressed, however, with the potential of several fungal parasites of nematode eggs (e.g. *Paecilomyces lilacinus* and *Dactyella oviparasitica*).

9C–2.3 Cultural Control

The most widely practiced cultural methods of nematode control are crop rotation and fallowing. Brown (1978) and Decker (1972) have reviewed the principles of cultural control. Fertilization and subsoiling (Hussey, 1977) can be effective tools for nematode management, but they are seldom employed specifically for nematode control.

9C–2.3.1 Crop Rotation

The general objective of crop rotation is to reduce the soil population density of the target nematode to a level below the economic threshold for the base crop. Rotations work best when only a single species of parasitic nematode occurs in a field and a high value resistant crop is available. Unfortunately, soils seldom contain only one species of parasitic nematode and suitable second crops may be difficult to find. However, even when several nematode species exist in a field, rotations can be very effective management tools.

Numerous rotation schedules that include cotton have been reported and a few examples are described here. Because the selection of a suitable second crop is largely determined by economics and local conditions, it is not possible to construct rotation schedules acceptable to all situations. Johnson et al. (1975) found that a population of *M. incognita* increased yearly under monoculture of cotton or corn (*Zea mays* L.) but the nematode population declined under peanut (*Arachis hypogaea* L.) and soybean. Rotating these crops kept the nematode population to an acceptable level. If the dominant population of *M. incognita* were a race especially virulent on soybean, however, such a rotation might not have been successful. The reniform nematode was kept to acceptable levels in a cotton field by rotating with soybean resistant to the nematode (Gilman et al., 1978). Cotton yields were better following 2 years of resistant soybean than after 1 year, but even 1 year in resistant soybean substantially increased yield over cotton monoculture. Thames and Heald (1974) showed that cotton yields were significantly increased in reniform nematode infested soil following a single rotation with *Sorghum*. Furthermore, to illustrate the effectiveness of rotation, preplant fumigation of soil under the rotation did not increase cotton yield over that of the rotation alone. In fields infested with *M. incognita,* cotton yields following 1 year of Bahiagrass were double those obtained from cotton monoculture (Rodriquez-Kabana and Pearson, 1972).

Carter and Nieto (1975) rotated barley (*Hordeum vulgare* L.) as a winter crop in a desert cotton field and effectively reduced the soil population of *M. incognita*. But since the soil temperatures were below those required for nematode reproduction the barley may have functioned somewhat as a trap crop.

9C–2.3.2   Fallowing

Loss of immediate economic return is the obvious disadvantage to fallowing land, but it can be effective in reducing nematode populations. Reynolds and O'Bannon (1966) reported that clean fallowing was as effective as soil fumigation in reducing the population of *M. incognita*. Hogger (1975) reported a number of common weeds were good to excellent host for *H. columbus* and that the nematode over wintered very effectively in *Trifolium incarnatum*. Thus, weeds in a fallowed field could be very detrimental to reducing nematode populations.

9C–2.4   Host Plant Resistance

These studies can be divided into two aspects: the development of resistant cottons and the elucidation of the mechanisms of resistance. A cursory review of our knowledge concerning the mechanisms of resistance was presented under the specific nematode headings.

One problem plant breeders and others interested in host resistance must resolve is a definition of the relevent terms: resistance, susceptibility, tolerance, and intolerance. Tolerance is generally accepted to mean no appreciable loss in yield at a level of infection that causes an economic loss on other cultivars of the same crop (Schafer, 1971). Resistance to nematodes can be defined in two ways. In one context, resistance is a measure of host sensitivity to the nematode, e.g., an index to the degree of damage caused to the host by the nematode; but in another context, resistance is a measure of host efficiency, e.g., an index to the degree of nematode reproduction on the host (Barker and Olthof, 1976). It is possible to have a nematode do extensive damage to a host and still be relatively incapable of reproducing on that host; or vice versa, causing little damage but achieving relatively good reproduction. According to the definitions above, the host would be classified as tolerant or intolerant according to its yield in comparison with other cultivars sustaining equal infection. The level of host resistance, however, remains in doubt. Barker and Olthof (1976) reviewed the concepts of resistance and tolerance, and their antonyms, and strongly suggested that when the term resistance is used it should be made clear if resistance is predicated on nematode reproduction or host sensitivity.

Most efforts have been made to introduce resistance to the root-knot nematode into *G. hirsutum* cottons. Sasser (1972) reviewed the achievements in this area up to the late 1960s and noted that until that time there were no programs for breeding for resistance to other nematodes. Since then, efforts have been made to identify sources of resistance to the reniform nematode and additional progress has been made on introducing root-knot nematode resistance into commercial Upland cottons.

Among the more notable recent advances in breeding for resistance to the root-knot nematode are the efforts by Shepherd (1970; 1974a, b, c; 1979), Hyer et al. (1979), and others (Anonymous, 1978). The increases in yield achieved by the MAR (multi-adversity resistance) breeding technique (Bird, 1982; Kappleman and Bird, 1981) have been considerable but the actual level of resistance attained is difficult to evaluate.

In Shepherd's (1974a, b, c; 1979) breeding program resistance to nematodes is based on host efficiency. That is, the level of resistance is inversely proportional to the extent of nematode reproduction. Initially he measured this level indirectly by estimating the extent of root galling (relative root-knot indices). Although galling is an index of host sensitivity, it may also have a direct relationship to the level of nematode reproduction and hence to resistance. But that relationship is not absolute. For instance, 'Auburn 623 RNR' and 'M8' cottons have galling indicies of 1.1 and 4.9, respectively (1 = none to very slight galling, 5 = very heavy galling), and they support mean egg numbers of 200 and 122 000 eggs per gram of root, respectively (Shepherd, 1979). For these cultivars the relationship holds. However, 'La. Mexican Wild' and 1029 have galling indicies of 2.3 and 3.7, respectively, but they each support about 12 000 eggs per gram of root. For these cultivars the relationship does not hold. The most dramatic departure from the relationship occurs with *G. barbadense* where the galling index is 1.3, indicating good resistance, but nearly 43 000 eggs are recovered per gram of root, indicating moderate susceptibility. Recently, Shepherd (1979) adopted a direct measure of resistance predicated on host efficiency (nematode reproduction) and now makes selections based on the number of nematode eggs recovered from infected roots.

The importance of the root-knot nematode in the establishment of fusarium wilt has been known for many years. For most of this time efforts to control the disease centered on breeding for resistance to the fungus. However, the reports by Hyer et al. (1979) and Shepherd (1970) show that resistance to the nematode is more important than resistance to the fungus in controlling this nematode disease complex. For instance, 'N 6072' is susceptible to fusarium wilt but resistant to the root-knot nematode. 'Auburn 56' is tolerant to the wilt but less resistant to the nematode than N 6072. 'Delcot 277' is tolerant to the wilt but susceptible to the nematode, and 'Acala SJ-1' is susceptible to wilt and the nematode. In tests on nematode and fungus infested plots the nematode resistant–wilt susceptible N 6072 yielded 60% more than the doubly susceptible Acala SJ-1 and the nematode susceptible–wilt tolerant Delcot 277; it yielded 18% more than the less resistant–wilt tolerant Auburn 56 (Hyer et al., 1979). Shepherd (1970) obtained similar results and also concluded that resistance to the nematode is a major factor in resistance to the complex.

Few efforts have been made to develop resistance in cotton to the reniform nematode. An excellent bases for such a program has been established by Carter (1981) and Yik (1981). Yik evaluated numerous *Gossypium* spp. and cotton collections for resistance based on nematode reproduction. According to her findings *G. longicalyx* Hutch. & Lee, is immune to the reniform nematode; *G. stocksii* Mast. in Hook., *G. somalense* (Gurke) Hutch., and *G. barbadense* Texas 110 are highly resistant; among *G. hirsutum*, 'Texas 893' and 'La RB 15702' are resistant. Carter reported that several *G. arboreum* L. are also resistant to *R. reniformis*.

REFERENCES

Anonymous. 1978. Notice of release of a resistant cotton, non-commercial breeding stock of cotton N6072. USDA, SEA, and California Agric. Exp. Stn., Univ. California, Davis.

————. 1979. Pest management strategies in crop protection. Vol. I. Office of Technology Assessment. U.S. Government Printing Office, Washington, D.C.

Barker, K. A., and T. H. A. Olthof. 1976. Relationships between nematode population densities and crop responses. Ann. Rev. Phytopathol. 14:327–353.

————, and C. L. Campbell. 1981. Sampling nematode populations. p. 451–474. *In* B. M. Zuckerman and R. A. Rhode (ed.) Plant parasitic nematodes. Vol. III. Academic Press, N.Y.

Bazan de Segura, C., and F. P. Aguilar. 1955. Nematodes and root rot disease of Peruvian cotton. Plant Dis. Rep. 39:12.

Birchfield, W. 1972. Differences in host-cell responses to the reniform nematode. Phytopathology 62:747.

————. 1962. Host-parasite relations of *Rotylenchulus reniformis* on *Gossypium hirsutum*. Phytopathology 52:862–865.

————, and L. R. Brister. 1962. New hosts and nonhosts of reniform nematodes. Plant Dis. Rep. 46:683–685.

Bird, G. W., O. L. Brooks, and C. E. Perry. 1974. Dynamics of concomitant field populations of *Hoplolaimus columbus* and *Meloidogyne incognita*. J. Nematol. 6:190–194.

————, and C. W. Hogger. 1973. Nutsedges as hosts of plant parasitic nematodes in Georgia cotton fields. Plant Dis. Rep. 58:402.

Bird, L. S. 1982. The MAR (Multi-adversity Resistance) system. Plant Dis. Rep. 66:172–176.

Bridge, J., E. Jones, and L. J. Page. 1976. *Meloidogyne acronea* associated with reduced growth of cotton in Malawi. Plant Dis. Rep. 60:5–7.

Brodie, B. B., L. A. Brinkerhoff, and F. B. Struble. 1960. Resistance to the root-knot nematode, *Meloidogyne incognita acrita,* in upland cotton seedlings. Phytopathology 50:673–677.

————, and W. E. Cooper. 1964. Relation of parasitic nematodes to postemergence damping-off of cotton. Phytopathology 54:1023–1027.

Brown, E. B. 1978. Cultural and biological control methods. p. 269–282. *In* J. F. Southey (ed.) Plant nematology. Her Majesty's Stationery Office, London.

Carter, W. W. 1975. Effects of soil temperatures and inoculum levels of *Meloidogyne incognita* and *Rhizoctonia solani* on seedling disease of cotton. J. Nematol. 7:229–233.

————. 1981. Resistance and resistant reaction of *Gossypium arboreum* to the reniform nematode *Rotylenchulus reniformis*. J. Nematol. 13:368–374.

————, and S. Nieto, Jr. 1975. Population development of *Meloidogyne incognita* as influenced by crop rotation and fallow. Plant Dis. Rep. 59:402–403.

----, ----, and J. A. Veech. 1977. A comparison of two methods of synchronous inoculation of cotton seedlings with *Meloidogyne incognita*. J. Nematol. 9:251–253.

Cauquil, J., and R. L. Shepherd. 1970. Effect of root-knot nematode-fungi combinations on cotton seedling disease. Phytopathology 60:448–451.

Cleveland, T. C. 1980. Boll weevil: A parasitic nematode of the boll weevil. J. Ga. Entomol. Soc. 16:122–125.

Cohn, E. 1976. Cellular changes induced in roots by two species of the genus *Rotylenchulus*. Nematologica 22:169–173.

----. 1974. Histology of the feeding site of *Rotylenchulus reniformis*. Nematologica 19:445–458.

Decker, H. 1972. Plant nematodes and their control. 1981 English translation, USDA, Amerind, New Delhi.

Fassuliotis, George. 1974. Host range of the Columbia lance nematode, *Hoplolaimus columbus*. Plant Dis. Rep. 58:1000–1002.

----, and G. J. Rau. 1967. The reniform nematode in South Carolina. Plant Dis. Rep. 51:557.

Garber, R. H., E. C. Jorgenson, S. Smith, and A. H. Hyer. 1979. Interaction of population levels of *Fusarium oxysporum* f. sp. *vasinfectum* and *Meloidogyne incognita* in cotton. J. Nematol. 11:133–137.

Gilman, D. F., J. E. Jones, C. Williams, and W. Birchfield. 1978. Cotton-soybean rotation for control of reniform nematodes. La. Agric. 21:10–11.

Heald, C. M., and C. C. Orr. 1982. Nematode parasites of cotton. *In* W. R. Nickle (ed.) Plant and insect nematodes. Marcell Dekker, N.Y.

Hogger, C. H. 1975. Plant-parasitic nematodes associated with weeds and agronomic crops in Georgia. Ph.D. Thesis. Univ. of Georgia, Athens. Univ. Microfilm No. DCJ-76-02236, Ann Arbor, Mich.

Holdeman, Q. L., and T. W. Graham. 1954. Effect of the sting nematode on expression of Fusarium wilt in cotton. Phytopathology 44:683–685.

Hussey, R. S. 1977. Effects of subsoiling and nematicides on *Hoplolaimus columbus* populations and cotton yields. J. Nematol. 9:83–85.

Hyer, A. H., E. C. Jorgenson, R. H. Garber, and S. Smith. 1979. Resistance to root-knot nematode in control of root-knot nematode-fusarium wilt disease complex in cotton. Crop Sci. 19:898–901.

Johnson, A. W., C. C. Dowler, and E. W. Hauser. 1975. Crop rotation and herbicide effects on population densities of plant-parasitic nematodes. J. Nematol. 7:158–168.

Kappelman, A. J., Jr., and S. L. Bird. 1981. Indirect selection for resistance to the fusarium wilt-root-knot nematode complex in cotton. Crop Sci. 21:66–68.

Khoury, F. Y. 1980. The influence of *Rhizoctonia solani* (Kuhn) and of *Meloidogyne incognita acrita* Chitwood on the infection of cotton plants by *Verticillium albo atrum* Reinke and Berth. Ph.D. Thesis. Univ. Arizona, Tucson, Ariz.

Lambe, R. C., and W. Horne. 1963. The reniform nematode in cotton in the lower Rio Grande valley of Texas. Plant Dis. Rep. 47:941.

Lewis, S. A., and F. H. Smith. 1976. Host plants, distribution, and ecological associations of *Hoplolaimus columbus*. J. Nematol. 8:264–270.

----, ----, and W. M. Powell. 1976. Host-parasite relationships of *Hoplolaimus columbus* on cotton and soybean. J. Nematol. 8:141–145.

Luc, M., and G. de Guiran. 1960. Les nematodes associes aux plants de L'Onest Africaine liste prelimenaire. L'Agron. Trop. 15:434–449.

Mankau, R. 1980a. Biological control of nematode pests by natural enemies. Ann. Rev. Phytopathol. 18:415–440.

----. 1980b. Biocontrol: Fungi as nematode control agents. J. Nematol. 12:244–252.

----. 1981. Biological control of nematode root parasites. p. 37–38. Proc. Beltwide Cotton Prod. Res. Conf., New Orleans, La.

Martin, G. C. 1955. Plant and soil nematodes of the Federation of Rhodesia and Nyssaland. J. Rhodesia Agric. 52:346–361.

Martin, W. J., L. D. Newsom, and J. E. Jones. 1956. Relationship of nematodes to the development of Fusarium wilt in cotton. Phytopathology 46:285–289.

McClure, M. A. 1972. Comparative biochemistry of cotton resistant and susceptible to the root-knot nematode. Phytochemistry 11:2209–2212.

----, and J. Robertson. 1973. Infection of cotton seedlings by *Meloidogyne incognita* and a method of producing uniformly infected root segments. Nematologica 19:428–434.

----, K. C. Ellis, and E. L. Nigh. 1974a. Resistance of cotton to the root-knot nematode, *Meloidogyne incognita*. J. Nematol. 6:17–10.

----, ----, and ----. 1974b. Post-infection development and histopathology of *Meloidogyne incognita* in resistant cotton. J. Nematol. 6:21–26.

Minton, E. B., A. L. Smith, and E. J. Cairns. 1964. Effects of 7 nematode species on 10 cotton selections. Phytopathology 54:625 (Abstr).

Minton, N. A. 1962. Factors influencing resistance of cotton to root-knot nematodes (*Meloidogyne* spp.). Phytopathology 52:272–279.

----, and B. E. Hooper. 1959. The reniform nematode and sting nematode in Alabama. Plant Dis. Rep. 43:47.

Muralidharan, R., and C. V. Sivakumar. 1977. Susceptibility of certain Indian varieties of cotton and wild species of *Gossypium* to the reniform nematode, *Rotylenchulus reniformis*. Indian J. Nematol. 5:116–118.

Neal, D. C. 1954. The reniform nematode and its relationship to the incidence of *Fusarium* wilt of cotton at Baton Rouge, Louisiana. Phytopathology 44:447–450.

Nusbaum, C. J., and K. A. Barker. 1971. Diagnostic and advisory programs. p. 281–301. *In* B. M. Zuckerman, W. F. Mai, and R. A. Rhode (ed.) Plant parasitic nematodes Vol. I. Academic Press, N.Y.

Orr, C. C., J. R. Abernathy, and E. B. Hudspeth. 1975. *Nothanguina phyllobia,* a nematode parasite of silverleaf nightshade. Plant Dis. Rep. 59:416–418.

----, A. R. Robinson, C. M. Heald, J. A. Veech, and W. W. Carter. 1982. Estimating cotton losses to nematodes. p. 22. *In* Proc. Beltwide Cotton Prod. Res. Conf., Las Vegas, Nev.

Owens, J. V. 1951. The pathological effects of *Belonolaimus gracilis* on peanuts in Virginia. Phytopathology 41:29.

Peacock, F. C. 1956. The reniform nematode in the Gold Coast. Nematologica 1:307–310.

Perry, D. A. 1963. Interaction of root-knot and fusarium wilt of cotton. Emp. Cott. Grow. Rev. 40:41–47.

Powell, N. T. 1971a. Interactions between nematodes and fungi in disease complexes. Ann. Rev. Phytopathol. 9:253–274.

----. 1971b. Interaction of plant parasitic nematodes with other disease causing agents. p. 119–136. *In* B. M. Zuckerman, W. F. Mai, and R. A. Rhode (ed.) Plant parasitic nematodes. Vol. II. Academic Press, N.Y.

Prasad, K. S. K., and G. M. Padaganur. 1980. Observations on the association of *Rotylenchulus reniformis* with *Verticillium wilt* of cotton. Indian J. Nematol. 10:91–92.

Reynolds, H. W., and R. G. Hanson. 1957. Rhizoctonia disease of cotton in presence or absence of the cotton root-knot nematode in Arizona. Phytopathology 47:256–261.

----, and J. H. O'Bannon. 1966. The efficacy and residual nature of experimental chemicals for controlling *Meloidogyne incognita acrita* on Deltapine cotton in Arizona. Plant Dis. Rep. 50:512–514.

Riggs, R. D. 1974. Lance nematode, a new problem in Arkansas cotton and soybeans. Ark. Farm Res. 23:8.

Rodriguez-Kabana, R., and R. W. Pearson. 1972. Changes in populations of lance and stunt nematodes in cotton following a Bahiagrass sod. J. Nematol. 4:233.

Sappenfield, W. P. 1963. *Fusarium* wilt, root knot nematode and *Verticillium* wilt resistance in cotton: Possible relationships and influence on cotton breeding methods. Crop Sci. 3:133–135.

Sasser, J. N. 1972. Nematode diseases of cotton. p. 187–214. *In* J. M. Webster (ed.) Economic nematology. Academic Press, N.Y.

Sayre, R. M. 1980. Biocontrol: *Bacillus penetrans* and related parasites of nematodes. J. Nematol. 12:260–270.

Schafer, J. F. 1971. Tolerance to plant disease. Ann. Rev. Phytopathol. 9:235–252.

Schmidt-Kraus Helmuth. 1979. Pathogenicity, population dynamics and nematode-nematode interactions of *Hoplolaimus columbus* Sher, *Scutellonema brachyurum* Andrassy and *Meloidogyne incognita* Chitwood on cotton in South Carolina. Ph.D. Thesis. Clemson University, Clemson, S.C. Univ. Microfilms No. 79-25088, Ann Arbor, Mich.

––––, and S. A. Lewis. 1979. *Scutellonema brachyurum*: Host plants and pathogenicity on cotton. Plant Dis. Rep. 63:688–691.

––––, and ––––. 1981. Dynamics of concomitant populations of *Hoplolaimus columbus, Scutellonema brachyurum,* and *Meloidogyne incognita* on cotton. J. Nematol. 13:41–48.

Shepherd, R. L. 1970. Breeding for resistance to the root knot fusarium wilt complex in cotton. p. 68. Proc. Beltwide Cotton Prod. Res. Conf., Houston, Tex.

Shepherd, R. L. 1974a. Breeding root-knot resistant *Gossypium hirsutum* L. using a resistant wild *G. barbadense* L. Crop Sci. 14:687–691.

––––. 1974b. Transgressive segregation for root-knot nematode resistance in cotton. Crop Sci. 14:872–875.

––––. 1974c. Registration of Auburn 623 RNR cotton germplasm. Crop Sci. 14:911.

––––. 1979. A quantitative technique for evaluating cotton for root-knot nematode resistance. Phytopathology 69:427–430.

Sivakumar, C. V., and A. R. Shihadri. 1971. Life history of the reniform nematode, *Rotylenchulus reniformis* Linford and Oliveira, 1940. Indian J. Nematol. 1:7–20.

Smith, A. L. 1940. Distribution and relation of *Meadow Nematode, Pratylenchus pratensis,* to *Fusarium Wilt* of cotton in Georgia. Phytopathology 30:710.

––––, and A. L. Taylor. 1941. Nematode distribution in the 1940 regional cottonwilt plots. Phytopathology 31:771.

Taylor, A. L., and J. N. Sasser. 1978. Biology, identification and control of root-knot nematodes (*Meloidogyne* species). North Carolina State Univ. Graphics, Raleigh.

Thames, W. H., and C. M. Heald. 1974. Chemical and cultural control of *Rotylenchulus reniformis* on cotton. Plant Dis. Rep. 58:337–341.

Triantaphyllou, A. C. 1973. Environmental sex determination of nematodes in relation to pest management. Ann. Rev. Phytopathol. 11:441–462.

––––, and H. Hirschmann. 1964. Reproduction in plant and soil nematodes. Ann. Rev. Phytopathol. 2:57–80.

Van Gundy, S. D., and M. V. McKenry. 1977. Action of nematicides. p. 263–283. *In* J. G. Horsfall and F. B. Cowling (ed.) Plant disease: An advanced treatise. Vol. 1. Academic Press, N.Y.

Veech, J. A. 1978. An apparent relationship between methoxy-substituted terpenoid aldehydes and the resistance of cotton to *Meloidogyne incognita.* Nematologica 24:81–87.

––––. 1979. Histochemical localization and nematoxicity of terpenoid aldehydes in cotton. J. Nematol. 11:240–246.

––––, and M. A. McClure. 1977. Terpenoid aldehydes in cotton roots susceptible and resistant to the root-knot nematode, *Meloidogyne incognita.* J. Nematol 9:225–229.

Watkins, G. M. 1981. Compendium of cotton diseases. Am. Phytopathol. Soc., St. Paul, Minn.

White, L. V. 1962. Root knot and the seedling disease complex of cotton. Plant Dis. Rep. 46:501–504.

Whitehead, A. C. 1978. Chemical control (a) Soil treatment. p. 283–296. *In* J. F. Southey (ed.) Plant nematology. Her Majesty's Stationery Office, London.

Wright, D. J. 1981. Nematicides: Mode of action and new approaches to chemical control. p. 421–449. *In* B. M. Zuckerman and R. A. Rohde (ed.) Plant parasitic nematodes. Vol. III. Academic Press, N.Y.

Yang, H., N. T. Powell, and K. R. Barker. 1976a. Interactions of concomitant species of nematodes and *Fusarium oxysporum* f. sp. *vasinfectum* on cotton. J. Nematol. 8:74–80.

----, ----, and ----. 1976b. The influence of *Trichoderma hazianum* on the root-knot-*Fusarium wilt* complex in cotton. J. Nematol. 8:81–86.

Yik, Choi-Pheng. 1981. Resistant germplasm in *Gossypium* species and related plants to the reniform nematode. Ph.D. Thesis. Louisiana State University and Agricultural and Mechanical College, Baton Rouge, La. Univ. Microfilms No. DEN81-17652, Ann Arbor, Mich.

D. Weed Control

In the United States prior to the early 1900s, weeds were removed from cotton (*Gossypium hirsutum* L.) fields by hand-hoeing. In the early 1900s, a typical weed control program in Ellis County, Texas, was described by Cates (1917):

"After planting, all the cultivation is done with a 2-horse, 1-row, 2-shovel cultivator equipped with buzzard-wing sweeps instead of shovels. For the first cultivation, which is given 10 days or 2 weeks after planting, small 15 cm (6 in) or 20 cm (8 in) sweeps are used next to the cotton and 25 cm (10 in) or 30 cm (12 in) sweeps on the outside. The cotton is chopped to a stand after the first cultivation and gone over with a hoe once or twice later in the season, to chop any extra stalks and weeds. For later cultivations 30, 36, and 41 cm (12, 14, and 16 in) sweeps are used on the same cultivator. During the season, four or five cultivations are given, and level cultivation is always practiced. The most prevalent and troublesome weeds of this country are careless weed (*Amaranthus* sp.), hurraw grass (*Panicum texanum* Buckl.), cocklebur (*Xanthium pensylvanicum* Wallr.), morning-glory (*Ipomoea* sp.), and johnsongrass [*Sorghum halepense* (L.) Pers]."
In similar information from Washington County, Miss., it was noted that cotton was cultivated every week or 10 days during the growing season for a total of 9 or 10 cultivations (Cates, 1917).

In the late 1930s and early 1940s, a dramatic shift from the use of animal-drawn cultivators to tractor-mounted cultivators occurred. By 1946, approximately 44% of the cotton grown was cultivated with tractor-mounted cultivators (Brown and Ware, 1958). From 1945 to 1954, weed control techniques and equipment developed included cross-plowing, rotary weeders, flame cultivators, and mechanical choppers (Brown and Ware, 1958; Christidis and Harrison, 1955). These control techniques were very helpful, but did not provide adequate control within the cotton drill.

The era of selective chemical weed control in cotton began to emerge in the late 1940s and early 1950s with the discovery and development of several herbicides. Prior to this time, non-selective chemicals such as sodium chlorate, borax, carbon bisulfide, and sodium arsenite were used as soil sterilants, but had very little potential for use in cotton. The herbicides 2,4-D [(2,4-dichlorophenoxy)acetic acid] and 2,4,5-T [(2,4,5-trichlorophenoxy) acetic acid] were evaluated for weed control in cotton, but cotton was very sensitive to these hormone-type herbicides.

In 1950, dinoseb (2-*sec*-butyl-4,6-dinitrophenol) and a herbicidal oil, LHO-1, were evaluated as preemergence and postemergence weed control agents, respectively (Harris, 1960). Dinoseb caused moderate to severe stunting and burning of the hypocotyls, and consequently never gained acceptance as a preemergence. Postemergence application of dinoseb plus LHO-1 resulted in good to excellent weed control.

The most satisfactory herbicidal oils for cotton were found to be petroleum fractions, those commonly referred to as naphthas. The naphthas used had a boiling range of approximately 150 to 200°C, contained 23 to 25% aromatic compounds and were not fortified (Colwick, 1960).

In 1954, additional herbicides were evaluated. Chlorpropham (isopropyl *m*-chlorocarbanilate) provided excellent crabgrass (*Digitaria* sp.) control but poor pigweed (*Amaranthus* sp.) control (Harris, 1960). Diuron [3-(3,4-dichlorophenyl)-1,1-dimethylurea] applied preemergence plus postemergence provided excellent weed control. It was concluded from studies conducted between 1950 and 1960 that herbicides effectively controlled weeds in cotton without reducing cotton yield. The herbicides did not accumulate in the soil, and in adverse weather they greatly reduced weed losses and made it possible to save the crop (Harris, 1960).

In the early 1960s, several new selective herbicides were introduced and used extensively to control weeds in cotton. New selective herbicides continue to be developed, and currently, most commercially planted cotton receives at least one application of some herbicide with a substantial portion of the crop receiving two or more applications of herbicides.

9D-1 ECONOMICALLY IMPORTANT WEED SPECIES INFESTING UNITED STATES COTTON

Before we tilled the soil, the concept of plants out of place had not been conceived. As we developed cultivated plants, weeds seem to have evolved, so that now we have weeds that are adapted to the same environment as our major crops. Holm (1978) states that across the world about 200 weed species are involved in 95% of the weed problems related to food production. He lists 80 species as primary weeds and 120 species as secondary weeds. To date, we have named about 0.25 million flowering plants; thus, these 200 weed species comprise less than 1% of all the world's plant species. Worldwide, over 100 species of plants have been reported to be weeds in cotton (Holm et al., 1977). Several of the world's worst weeds are serious problems in cotton. They include purple nutsedge (*Cyperus rotundus* L.), johnsongrass, barnyardgrass [*Echinochloa crus-galli* (L.) Beauv.], and crowfootgrass [*Dactyloctenium aegyptium* (L.) Richter]. The economically important weed species infesting cotton fields in the United States are listed in Table 9D-1. This list covers the 10 most important weeds in cotton in each state of 15 cotton-producing states. Approximately two-thirds of the species are broadleaf weeds and one-third grasses or sedges. Of these weeds, 24% are perennials and 75% are annuals. Six of the species are

Table 9D-1. Economically important weed species in U.S. cotton, 1981 (Whitwell et al., 1981)

Common name	Number of states
Annual broadleaf species	
Anoda, spurred	6
Buffalobur	5
Cocklebur, common	13
Crotons	4
Groundcherry, smooth	3
Lambsquarter, common	1
Marigold, wild	1
Morningglories	15
Pigweeds	11
Ragweed, common	3
Sesbania, hemp	3
Sicklepod	3
Sidas	10
Smartweed, Pennsylvania	1
Spurges	8
Starbur, bristly	1
Sunflower	1
Thistle, Russian	1
Velvetleaf	3
Annual grasses	
Barnyardgrass	7
Crabgrass	8
Goosegrass	7
Junglerice	7
Panicums (browntop, fall, Texas)	8
Signalgrass, broadleaf	3
Perennial broadleaf species	
Bindweed, field	3
Blueweed, Texas	1
Bursage, woolyleaf	1
Horsenettle	2
Nightshade, silverleaf	3
Perennial grasses	
Bermudagrass	9
Johnsongrass	13
Perennial sedges	
Nutsedges	13

found in 10 or more states. Nine species are troublesome in 5 to 9 states. Eighteen species are troublesome in four or fewer states. Morningglories were listed as troublesome in every state.

Proper identification of weeds infesting cotton fields plus a thorough understanding of their life cycles can aid in the development of effective and economically sound control programs. In-depth summaries of distribution, competitiveness, and control have been prepared for morningglories (Crowley, 1977), prickly sida (*Sida spinosa* L.) (Baker, 1977), spurred anoda [*Anoda cristata* (L.) Schlecht.] (Chandler, 1977a), Wright groundcherry (*Physalis wrightii* Gray) (Taylor and Heathman, 1978), common cocklebur (Whitwell, 1980), sicklepod (*Cassia obtusifolia* L.) (French, 1980), fall panicum (*Panicum dichotomiflorum* Michx.) (Coble, 1980), field bindweed (*Convolvulus arvensis* L.) (Wiese, 1978), silverleaf nightshade (*Solanum elaeagnifolium* Cav.) (Abernathy, 1978), lanceleaf sage (*Salvia reflexa* Hornem.) (Supak, 1978), bermudagrass [*Cynodon dactylon* (L.) Pers.] (Jordan, 1977), johnsongrass (Gates, 1977), purple nutsedge (Wills, 1977b), and yellow nutsedge (*Cyperus esculentus* L.) (Keeley, 1978).

Within a geographic region of the U.S. Cotton Belt, the frequency of individual weed species and the shifts in population composition over time are very important in developing appropriate control programs. Both the numbers and viability of seed or vegetative propagules can play a vital role in the survival of weed species within a plant community. The number of seed produced by a single weed varies from a few hundred to several thousand. A single prickly sida plant in a non-competitive environment can produce 45 000 seed, while individual spurred anoda and velvetleaf (*Abutilon theophrasti* Medic.) plants can produce 19 000 and 17 000 seed, respectively. Seed production data alone would indicate that prickly sida would probably be the dominant species. However, the viability of weed seed over an extended period of time varies widely according to species. After 5 years in the soil, prickly sida seed viability declines to less than 5%, while spurred anoda and velvetleaf seed viability declines to only 50% (Egley and Chandler, 1982). With a perennial such as johnsongrass, the level of seed production for a single plant may be only 3200 seed, but the seed viability may be 50% after 6 years in the soil (McWhorter, 1972). Also, the level of rhizome production can range from 1 kg/m^2 in clay soil to 2.4 kg/m^2 in loam soil. Texas panicum (*Panicum texanum* Buckl.), an annual grass, can produce 23 000 seed per plant but less than 10% of the seed will be viable after 6 years in the soil (Egley and Chandler, 1982; Chandler and Santelmann, 1969).

Weed populations are not static but change in responses to changing edaphic factors, control procedures, and cropping sequence (Buchanan et al., 1975b; Hoveland et al., 1976; Weber et al., 1974). Repeated applications of fluometuron [1,1-dimethyl-3-(α,α,α-trifluoro-*m*-tolyl) urea] have been associated with decreased broadleaf weed populations and increased prominence of yellow nutsedge and annual grass species in Georgia (Dowler and Hauser, 1974). The widespread and long-term use of trifluralin (α,α,α-trifluoro-2,6-dinitro-N,N-dipropyl-p-toluidine) has decreased populations of large crabgrass [*Digitaria sanguinalis* (L.) Scop.] and other annual grasses, but has increased populations of prickly sida and other broadleaf weeds (Frans, 1969).

9D-2 WEED INTERFERENCE IN COTTON

Where weeds are allowed to grow unchecked, yield is drastically reduced. The growth stage of the crop at which weed competition occurs is most critical. Weed-crop competition studies indicate that weed competition during the first 2 months of the crop's life is more injurious than during the second 2 months. The critical period of weed competition depends on: (i) the maximum period for which weeds can be tolerated without affecting final yield, and (ii) the period after which weed growth does not affect final yield. Buchanan and Burns (1970) showed that cotton competing with a mixed population of annual grasses or broadleaf weeds requires weed-free maintenance for up to 9 weeks after emergence (Fig. 9D–1). Studies in furrow-irrigated cotton show that yellow nutsedge competition for more than 4 weeks reduced seed cotton yield (Keeley and Thullen, 1975). Season-long yellow nutsedge competition reduced yield 34% as compared with 30% when it competed for 6 weeks. The duration of joint competition of browntop panicum [*Panicum fasciculatum* Sw. var. *reticulatum* (Torr.) Beal], jungle-rice [*Echinochloa colonum* (L.) Link], red sprangletop [*Leptochloa filiformis* (Lam.) Beauv.], Wright groundcherry, and Palmer amaranth (*Amaranthus palmeri* S. Wats.) on irrigated cotton was studied (Arle and Hamilton, 1973). It was found that weeds allowed to compete after the first (6 weeks) or second (9 weeks) irrigation reduced cotton yield by 16% and

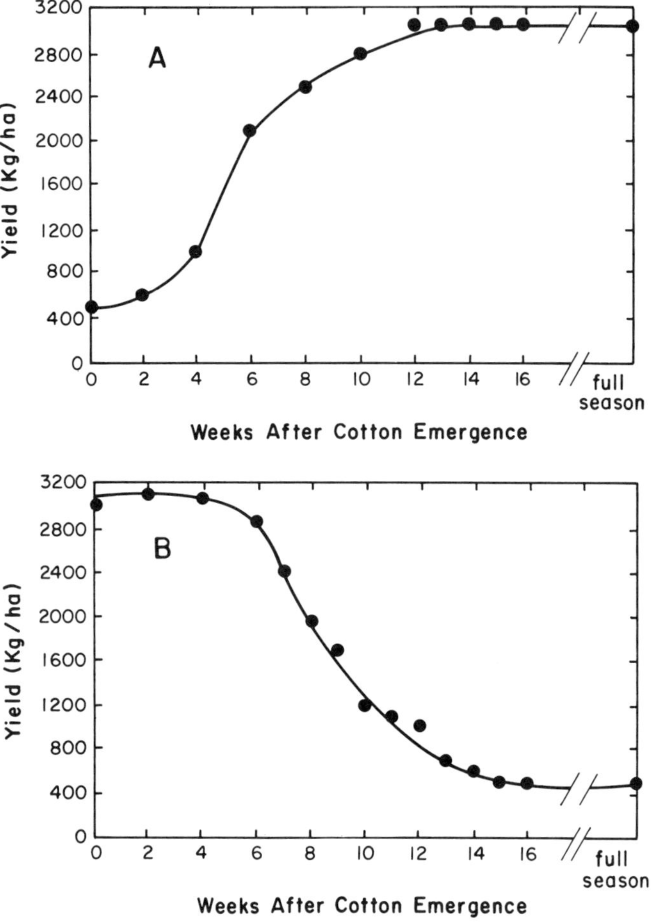

Fig. 9D–1. Yield of seed cotton as affected by period of weed-free maintenance or weed competition. Data are averages of three experiments at two different locations. (A) Weeks of weed-free maintenance; (B) weeks of weed competition (Buchanan, 1981).

12%, respectively. Yield were unaffected by competition that began after the third (11 weeks) or fourth (13 weeks) irrigation, or competition that ended after the first, second, or third irrigation. Prickly sida did not reduce yields when a weed-free period of 5 to 6 weeks was maintained or competition was terminated by 7 weeks after emergence (Buchanan et al., 1977a).

Several studies have been conducted to evaluate the weed density—crop yield relationship or threshold effect. Most of this research has been with single broadleaf annual weed species at sveral densities competing with cotton. These weed density studies show that the degree of competition varies between species, years and location (Chandler, 1977b; Crowley and Buchanan, 1978; Buchanan et al., 1980; Buchanan and Burns, 1971a and 1971b; Smith and Tseng, 1970). Within a range of densities for each specific weed, an economic threshold level exists. It is very difficult to develop a specific economic threshold level for a given weed species due to changes in the environment and cultural practices. Buchanan and Burns (1971a, 1971b) showed that identical densities of sicklepod, tall morningglory [*Ipomoea purpurea* (L.) Roth], common cocklebur, and redroot pigweed (*Amaranthus retroflexus* L.) vary in competitiveness depending on soil type. The studies showed that the weeds competed more aggressively with cotton on a sandy clay loam soil than on a sandy loam soil. They suggested that more favorable fertility and moisture relationships in the sandy clay loam soil could have accounted for the observed differences. In other field studies, nitrogen applications at planting did not change the competitive relationship between cotton and the weeds (Buchanan and McLaughlin, 1975). Weed competition in cotton was not influenced by cotton row widths of 53, 79, or 106 cm, but the length of the weed-free period was reduced with a reduction in row width (Rogers et al., 1976). The density of cotton did not influence the competitive effect of sicklepod or pigweed (Street et al., 1981).

To develop useful economic threshold data for a specific weed, both weed density and duration of infestation must be considered. Studies have shown that full-season competition by spurred anoda, velvetleaf, and prickly sida at 8, 16, and 64 plants per 12 m of crop row respectively, resulted in statistically significant reductions in seed cotton yield. Competition by spurred anoda at 16 plants per 12 m, and by velvetleaf or prickly sida at 64 plants per 12 m, from 2 weeks after cotton emergence until harvest, statistically reduced yields (Chandler, 1977b).

From the above discussion, it is apparent that the severity of weed competition is dependent on several variables, and these variables must be considered in the development of economic threshold data for a given weed species.

9D-3 COTTON LOSSES DUE TO WEEDS

Cotton losses due to the presence of weeds may be severe although the damage caused is not always as obvious as losses caused by other pests. These losses occur at various stages in the cotton production cycle. Weeds compete with cotton for water, nutrients, light, and other growth factors.

Weeds (a) reduce the yield of cotton; (b) impair the quality of cotton; (c) increase costs of hand tillage, mechanical tillage, fertilizer, and herbicides; (d) prevent efficient irrigation and water management; and (e) serve as hosts and habitats for insects, nematodes, disease-causing organisms, and rodents (Shaw, 1964).

With the advent of the mechanical cotton spindle harvester only 30 years ago, minimizing weed populations became extremely importnt in cotton fields. Poor weed control can reduce cotton harvesting efficiency, lint quality, and yield. Work with tall morningglory in Alabama showed that cotton harvesting efficiency can be reduced 3 to 31%, depending on morningglory density and seasonal growth rate (Crowley and Buchanan, 1978). A higher percentage of trash was found in seed cotton when weeds were present. Work in North Carolina showed that one mature grass plant per 6 m of crop row can reduce the quality of cotton one grade (Colwick, 1960). Grade reduction, however, does not reflect the total impact of weeds on quality reduction. Grass and other foreign matter in machine-harvested cotton requires more elaborate cleaning in the gin, which in turn can further adversely affect the spinning quality of the fiber.

9D-3.1 Estimated Cotton Losses Without Herbicides

To fully understand the importance of herbicides in the production of cotton, it is beneficial to consider the losses that could be expected if weeds were controlled with only non-chemical management. Abernathy (1981) showed that cotton production without herbicides in the USA would be very expensive and laborious. In 1976 the total value of the U.S. cotton crop was $3.3 billion. The total expenditures for herbicides was $120 million. Without these herbicides, a 32% yield reduction could have been expected from competition of weeds. Another 3% would be lost to indirect causes, such as grade reductions from weed contamination. A 5% reduction could be attributed to harvest loss due to interference with the harvest operation. Thus, the total direct cost of weeds would be $1.3 billion. Without herbicides, three additional tillage operations would be required at a cost of $62.9 million. The cost of hand weeding a cotton crop is economically unfeasible but a reasonable cost of at least $50/ha could be considered. A net loss of $1.5 billion could be anticipated in cotton production without herbicides, or a loss of $317/ha.

9D-3.2 Estimated Cotton Losses Utilizing Current Technology

The use of herbicides over the past 3 decades has increased in cotton production systems and has replaced most of the hand labor associated with controlling weeds. Even with this shift, weeds are still major pests in cotton fields. In 1980 it was estimated that weed interference in cotton fields resulted in an average yield reduction of 7.4% or 686 877 bales (Whitwell et al., 1981). This was a $242 million loss (Table 9D-2). The number of different weed species responsible for these losses were relatively few. In 1980 johnsongrass, morningglories, nutsedges, and bermudagrass reduced cotton yield in all cotton-growing regions of the USA (Table 9D-3). Common cocklebur and prickly sida were major problems in the Southeast and mid-South regions, while pigweeds were extremely important in the Southwest and West regions. Approximately two-thirds of the losses from weeds in cotton can be attributed to these seven species.

9D-3.3 Estimated Production Costs Associated with Controlling Weeds in Cotton

Since 1950 the use of hand labor for controlling weeds has been drastically reduced. Herbicides applied with mechanized equipment have been substituted. When evaluating losses from weeds, the monetary investment in equipment, labor, and herbicides used to provide economical control must be considered.

Table 9D-2. Estimated average annual losses due to weed in cotton, by state, USA, 1980 (Whitwell et al., 1981).

| State | Estimated loss from potential production | | |
	Reduction	Quantity	Value†
	%	bales	$ US (× 1 000)
Alabama	7	18 500	6 509
Arizona	9	120 000	42 255
Arkansas	10	51 106	17 981
California	2	65 000	23 080
Florida	10	816	287
Georgia	10	8 102	2 850
Louisiana	9	47 221	16 614
Mississippi	6	60 526	21 295
North Carolina	8	3 888	1 367
New Mexico	7	9 028	3 176
Oklahoma	6	18 040	6 347
South Carolina	4	3 452	1 214
Tennessee	9	18 000	6 332
Texas	7	262 500	92 358
Total		686 877	241 665

† Calculation of value was based on the average 1980 U.S. price of $352/bale. Ginned bale weight used was 218 kg.

Table 9D-3. Estimated percent of cotton yield losses caused by the seven most frequently reported weeds in U.S. cotton; 1980 (Whitwell et al., 1981).

| Weed species | Cotton production regions | | | |
	Southeast	Midsouth	Southwest	West
	%			
Johnsongrass	5	13	15	16
Cocklebur, common	22	12	4	--
Morningglories	11	13	13	8
Pigweeds	5	2	31	22
Sida, prickly	9	17	1	--
Nutsedges	9	8	7	9
Bermudagrass	2	5	2	6

Table 9D–4. Estimated cost of equipment, labor, and herbicides used to control weeds in U.S. cotton; 1980.

Production regions	Planted hectares ($\times$ 1 000)	Avg. total cost ($U.S. $\times$ 1 000)		
		Equipment	Labor	Herbicides
Southeast	266	9 475	2 599	14 899
Midsouth	1 168	69 268	48 266	64 912
Southwest	2 928	65 079	88 579	75 925
West	828	43 213	45 109	19 739
Total	5 190	187 035	184 553	175 475

In 1980 cotton was planted on 5.2 million ha in the USA. The total cost for equipment, labor, and herbicides used to control weeds was estimated to be $547 million (Table 9D–4). Herbicides account for 32% of the total cost while equipment and labor costs were 34% each.

9D–4 WEEDS AS HOSTS FOR PESTS

Weeds serve as hosts for numerous pests, therefore, weed control can reduce losses caused by diseases and insects as well as losses due to competition.

9D–4.1 Pathogens

The fungus *Phymatotrichum omnivorum* (Shear) Dugger occurs as a plant pathogen throughout a region extending from the Gulf of Mexico to southern California, and from Utah into Mexico. In this region, root rot caused by this pathogen is the most destructive plant disease (Taubenhaus and Ezakiel, 1936). Weeds aid the survival of the fungus in cotton fields. The fungus can overwinter on roots of weeds that grow through the winter, and spread from these roots to roots of cotton (Taubenhaus, 1936). Weed species with complete resistance to root rot are all monocots, and the susceptible weed species that are of economic importance in cotton are all dicots. The moderately susceptible species are "passive carriers" and as such may survive for some time after portions of their root systems are infected. Velvetleaf, a member of the Malvaceae, is very susceptible to this disease.

According to Ray and McLaughlin (1952) *Rhizoctonia solani* Kuehn is the most important fungus involved in diseases of cotton seedlings because of its high degree of virulence and its frequency. Considerable variation exists in the pathogenicity of *R. solani* isolates (Kermkamp et al., 1952). The degree of their host specificity is variable. The fungus has a wide host range, attacking over 250 plants of economic importance, including annual and perennial broadleaf, grass, and sedge species (Anonymous, 1960). Weeds of economic importance in cotton that serve as a host for *R. solani* are listed in Table 9D–5.

Table 9D–5. Economically important weed species found in cotton that serve as hosts for pathogens and insects that are injurious to cotton.

	Phymatotrichum root rot†	Rhizoctonia solani	Verticillium albo-atrum	Verticillium dahliae	Root-knot nematodes	Bollworm	Tobacco budworm
Annual broadleaf species							
Beggarweed, Florida						X	X
Buffalobur	+					X	X
Cocklebur, common	+				X	X	X
Croton, tropic	X						X
Croton, woolly	X				X		
Groundcherry, smooth	+					X	X
Jimsonweed			X		X	X	
Lambsquarter, common	+	X			X	X	
Mallow, Venice				X			
Morningglory smallflowered	X					X	X
Morningglory, tall	+		X				
Pigweed, rough		X	X				
Pigweed, spiny		X			X	X	X
Pigweed, tumble		X					
Purslane, common	+	X		X	X		
Ragweed, common	X	X				X	X
Sida, prickly	X		X			X	X
Smartweed, Pennsylvania	+				X	X	X
Spurge, spotted	X	X					
Sunflower	X					X	
Velvetleaf	X		X			X	X
Annual grasses							
Barnyardgrass		X		X			
Crabgrass, large						X	
Crabgrass, smooth		X					
Junglerice		X					
Panicum, Texas						X	
Perennial broadleaf species							
Bindweed, field	X	X				X	
Blueweed, Texas	+						
Dock, curly							X
Horsenettle, Carolina	+	X	X			X	X
Morningglory, bigroot							X
Nightshade, silverleaf	+			X	X		
Perennial grasses							
Bermudagrass		X				X	
Johnsongrass						X	X
Perennial sedges							
Nutsedge, purple		X					
Nutsedge, yellow		X		X	X		

† Moderately susceptible species = + and highly susceptible species = X.

Verticillium wilt (*Verticillium albo-atrum* Reinke and Berth, or *Verticillium dahliae* Kleb.) is so widespread that it is considered a destructive disease in most of the major cotton-producing areas of the USA. Although some cultivars of cotton exhibit more tolerance than others to the causal organism, none are completely resistant (Johnson and Brinkerhoff, 1976). Crop rotation of cotton with sorghum is used to control Verticillium wilt. Minton (1972) found that the incidence of Verticillium wilt in cotton planted after 4 years of grain sorghum [*Sorghum bicolor* (L.) Moench] infested with *Amaranthus* sp. was higher than in cotton planted after 4 years of weed-free grain sorghum. Weeds of economic importance that serve as hosts for *Verticillium albo-atrum* are annual and perennial broadleaf species (Table 9D–5), while hosts for *Verticillium dahliae* include barnyardgrass, yellow nutsedge, and broadleaf species (Johnson and Brinkerhoff, 1976; Brown and Wiles, 1970; Johnson and Brinkerhoff, 1977).

The root-knot nematode, *Meloidogyne* spp., formerly classified as *Heteroidia marioni* (Corner), has spread to nearly every country in the world and flourishes under conditions of cultivation (Tyler, 1937). Nearly 900 species of plants have been reported as hosts of the root-knot nematode. Weed species of economic importance found in cotton that serve as hosts for root-knot nematodes are mainly annual broadleaf species (Tyler, 1937; Hogger and Bird, 1976). Weeds as host plants may not show serious symptoms of disease, but nematodes feed and multiply on them, and the soil is thus kept infected (Table 9D–5).

9D–4.2   Influence of Herbicides on Pathogens

Recent reviewers of the literature discussed the influence of herbicides on plant pathogens (Katan and Eshel, 1973; Altman and Campbell, 1977). The responses can range from positive to negative depending upon the herbicide and the pathogen and, with a single herbicide, the effect may depend on the dosage.

An increase in seedling damage caused by *Rhizoctonia solani* in cotton treated with trifluralin was reported (Pinckard and Standifer, 1966). Neubauer and Avizohar-Hershenson (1973) found that trifluralin increased susceptibility of cotton to *R. solani* and also increased the saprophytic activity of the fungus in soil. Chandler and Santelmann (1968) found that trifluralin in the field interacted with *R. solani* to create excessive injury to cotton seedlings, but fluometuron or prometryn [2,4-bis(isopropylamino)-6-(methylthio)-*s*-triazine] did not interact with the pathogen. Miller et al. (1979) showed that neither trifluralin nor diuron influenced the occurrence of *Pythium ultimum* Trow or *R. solani* on cotton. The presence of *Thielaviopsis basicola* (Berk, and Br.) Ferr. on cotton seedlings was increased by trifluralin after three annual applications. The level of Verticillium wilt was not influenced by trifluralin or diuron.

9D–4.3   Insects

The bollworm [*Heliothis zea* (Boddie)] is one of the major cotton insect pests in the USA. The bollworm is of economic importance throughout the world and is often referred to as the corn earworm or tomato fruitworm (Quaintance and Brues, 1905). It has a wide range of host species that includes several economically important crops and numerous weeds. Weeds of economic importance in cotton that serve as hosts for the bollworm are listed in Table 9D–5. Johnsongrass and velvetleaf serve as minor early season hosts while Carolina geranium (*Geranium carolinianaum* L.), and cutleaf geranium, (*Geranium dissectum* L.), serve as important early season hosts for the first generation of the bollworm (Snow et al., 1966; Stadelbacher, 1979). Florida beggarweed, [*Desmodium tortuosum* (Sw.) DC], small-flower morningglory [*Jacquemontia tamnifolia* (L.) Griseb.], and prickly sida are economically important weeds that serve as late season hosts for substantial buildup of over-wintering diapausing pupae (Roach, 1975).

The tobacco budworm [*Heliothis virescens* (F.)], like the bollworm, feeds on many annual broadleaf species but can also use perennial grass and broadleaf species as hosts (Table 9D–5). Important weedy host species of the tobacco budworm are Florida beggarweed, Carolina geranium, cutleaf geranium, and small-flower morningglory (Stadelbacher, 1981; Brazzel et al., 1953; Barber, 1937).

In the USA, cotton is the preferred host of the pink bollworm [*Pectinophora gossypiella* (Saunders)]. Species of *Abutilon, Althea, Hibiscus,* and *Malvastrum* have been reported to be infested but should be disregarded as serious hosts for the pink bollworm in the USA (Little and Martin, 1942).

The cotton fleahopper [*Pseudatomoscelis seriatus* (Reuter)], an insect that damages the young fruit forms of cotton, is a major insect pest of cotton in the South (Little and Martin, 1942). The cotton fleahopper feeds and deposits its eggs on many species. Some of these species, such as pigweeds, common lambsquarter, (*Chenopodium album* L.), sunflower, (*Helianthus* sp.), morningglories, and horsenettle, (*Solanum carolinense* L.), infest cotton fields. Others grow outside the cotton field, but still serve as important hosts. For example, several species of *Oenothera* and *Monarda* serve as hosts for building up early season populations. Members of the genus *Croton* are the principal plants in which overwintering eggs are deposited.

The Lygus bugs (*Lygus* sp.) are important cotton insect pests throughout the U.S. Cotton Belt. They attack a wide range of hosts and will go readily from native plants to cultivated crops. Redroot pigweed is the most preferred native host (Little and Martin, 1942).

In cotton the fall armyworm [*Spodoptera frugiperda* (J. E. Smith)] is a pest of only minor importance. This insect prefers to feed on species belonging to the family *Poaceae,* such as crabgrass, corn, sorghum, and bermudagrass, and would probably confine its attack entirely to these plants if they were always available (Lugenbill, 1928). Cotton is seldom seriously injured unless it has been poorly cultivated and the field has been overrun with grass. The larvae, not finding a sufficient supply of food on the grass to reach maturity, go to cotton and do considerable damage to the plants. Weeds infesting cotton that serve as host species for fall armyworm include common cocklebur, large crabgrass, johnsongrass, common lambsquarter, smooth pigweed (*Amaranthus hybridus* L.), purslane (*Portulaca* sp.), and purple nutsedge.

9D–4.4   Influence of Herbicides on Insects

Information on the influence of herbicides on insects is very limited. Stevens (1967) reported that DSMA (disodium methanearsonate) or MSMA (monosodium methanearsonate) applied to cotton reduced the infestation of fleahoppers for a number of days after application. Stam et al. (1978) observed no significant effects with dinoseb, diuron, or MSMA on insects and spider populations when these herbicides were applied as post-directed sprays. However, foliar applications of MSMA reduced populations of fleahoppers. Miller and Miller (1979) found that plant bug populations in cotton may be reduced through the application of dinoseb as a directed spray for weed control. Beneficial insects are also reduced, but not as greatly as plant bug populations. They concluded that applications of dinoseb for weed control may eliminate the need for insecticidal sprays to control plant bug populations.

9D–5   METHODS OF CONTROLLING WEEDS IN COTTON

Currently in the USA, weeds are controlled by the judicious integration of cultural, mechanical, and chemical control procedures. Less frequently used weed control methods include hand labor, flame cultivation, and biological agents.

9D–5.1   Hand Labor

In 1950, hand labor for weed control in cotton accounted for 60 to 70% of the total labor required to produce a cotton crop (Crowe and Holstun, 1953). Between 1950 and 1960, hand hoeing changed from a primary to a supplementary method of weed control. This was made possible with the introduction of herbicides. Currently, hand labor is still used to some extent to supplement other weed control practices, but it is used primarily to remove only those weeds that have survived herbicidal treatment. The use of hand labor as a major component of future weed control programs is questionable due to the scarcity and cost of hand labor (Wiese and Chandler, 1979).

9D–5.2   Flame Cultivation

Flame cultivation began with the introduction of butane and propane in 1945, and it can be used effectively to control small weeds in cotton (Edwards, 1964). The effectiveness of flame cultivation in destroying weeds depends upon the growth characteristics of the weeds exposed to the flame (Williford et al., 1973).

Flame cultivation does not leave harmful residues in the soil or in the crop plants. Flaming does not disturb the soil, and often flaming is possible when soil is too wet for cultivation. A disadvantage is that flame kills only emerged tissue. Plants such as nutsedges, johnsongrass, or even annual grasses that have well-established root systems, will generate new tissues within several days after flaming. Flame cultivation, to be effective on established weeds, must be repeated several times. From one to five applications of flame effectively control most weeds less than 7 or 10 cm tall (Holstun and Wooten, 1966).

The availability of selective herbicides and the increased cost of fuel have contributed to the reduced use of flame cultivation for weed control in cotton. It is unlikely that flame cultivation will be an important part of integrated weed management systems in the future.

9D–5.3 Biological Control

Geese have been used as biological control agents of certain weeds in cotton (Johnson, 1960; Mayton et al., 1945). Geese prefer johnsongrass and bermudagrass, but will eat seedlings of nutsedges, puncturevine (*Tribulus terrestris* L.), crabgrass, and most other grasses. They will not eat pigweed, lambsquarter, cocklebur, or most other broadleaf weeds. Geese as weed control agents present numerous management problems, such as supplemental feeding, protection from predators, and careful use of insecticides. Owing to these disadvantages, they are not widely used to control grasses in cotton.

Presently, biological weed control with insects, plant pathogens, or nematodes has not developed sufficiently to be used commercially in the production of agronomic crops in the USA. The control of purple nutsedge with early season release of a native moth, *Bactra verutana* Zeller, was found to be technologically feasible but not economically practical because of the large numbers of larva required for adequate purple nutsedge control in cotton (Frick and Chandler, 1978).

A host-specific nematode, *Nothanguina phyllobia*, is currently being studied as a possible biological control agent for silverleaf nightshade, an economically important weed in cotton production in the southwest USA (Orr et al., 1975). The leaf-feeding nematode produces galls on the leaves and stems of the weed. Heavily infested plants are stunted, and some are killed. In view of the cost effectiveness of biological control techniques and the widespread importance of several major weeds in cotton, continued research in this area should be emphasized.

9D–5.4 Cultural Control

Cultural practices that favor the growth of the cotton plant over weed flora are of primary importance in successful weed management. Starting a cotton crop in a well-prepared seedbed without established weeds allows the cotton seedling to compete favorably with germinating weed seedlings and also facilitates later weed control operations.

Crop competition is one of the most economical weed control methods available to cotton producers. The choice of a cotton cultivar can significantly influence the degree of season-long weed control obtained through crop suppression of weeds. Later season weed control with normal and okra leaf type cotton is better than with super okra leaf type cotton (Andries et al., 1974; Weaver et al., 1978). The use of narrow row spacing in cotton, such as 53 cm as compared to the conventional spacing of 106 cm, decreases the weed-free maintenance period, but row spacing did not influence the length of time weeds emerging with the crop could compete (Rogers et al., 1976).

The rapid increase in the importance of purple nutsedge as a weed in cotton is largely explained by changes in cultural practices. Up until the late 1950s or early 1960s, nutsedge was controlled with extensive cross-cultivation and hand-hoeing. When skip row cotton was introduced in the early 1960s, cross-cultivation was largely eliminated. Not only did this favor nutsedge, but the skips between rows were ideal for nutsedge growth since there were no cotton plants to provide shade competition (Wills, 1977a).

Crop rotations and fallowing are not common practices in cotton production, but they sometimes offer the most practical solutions to problems of weed control (Hauser and Arle, 1958). Rotating from one crop to another, or fallowing for a season, often permits the use of additional herbicides or tillage practices on weeds that are difficult to control in any single crop. Spurred anoda in cotton is very competitive and difficult to control, but in soybeans [*Glycine max* (L.) Merr.], spurred anoda is much less competitive and can be controlled with bentazon [3-isopropyl-1H-2,1,3-benzothiadiazin-4(3H)-one 2,2-dioxide] (Chandler and Oliver, 1979).

9D–5.5 Mechanical Control

Mechanical tillage can be divided into preplanting tillage and postplanting tillage. It has been estimated that 50% of all preplant tillage is done to control weeds (Wiese and Chandler, 1979). Proper seedbed formation and maintenance of a uniform row profile do much to promote uniform stands of cotton and effective control of weeds.

Postplanting tillage is extensively used to control weeds growing between rows but is infrequently used to control weeds within the rows. Cultivation alone as a means of weed control in cotton is ineffective. Cultivation is, however, valuable in complementing other weed control programs (Holstun, 1963; Foy, 1959; Upchurch and Selman, 1968). Generally, two to five cultivations are used in cotton production systems.

An extensive series of recent experiments revealed that the total benefit gained from cultivation of cotton can be accounted for by control of weeds (Buchanan and Hiltbold, 1977). They found no consistent, favorable response of cotton to either deep or shallow cultivations. Consequently, under most conditions, mechanical cultivation should be considered primarily on the merits of controlling weeds, to prevent soil erosion, or to provide furrows for irrigation.

9D-5.6 Chemical Control

The use of herbicides for control of weeds in cotton grew rapidly in the 1950s and herbicides were extensively used in the early 1960s. In 1976, approximately 8.3 million kg of herbicides were applied to 3.96 million ha (Eichers et al., 1976). Currently, there are about 21 herbicides labeled for weed control in cotton. The common name, structural formula, and other pertinent information is given for each of these herbicides in Table 9D-6. Herbicides were readily accepted for the control of weeds in cotton because they were effective on a wide range of weed species, and they were efficient and relatively economical.

Producers apply herbicides at six critical stages during the annual production of cotton. These include treatments made preplant to foliage of existing weeds, preplant soil incorporated, preemergence, directed postemergence, over-the-top postemergence, and at late postemergence. Across the Cotton Belt, the amount of herbicides used within these six stages of the production cycle varies widely. Information from four geographic regions of the U.S. Cotton Belt illustrates the wide variation in herbicides utilized within the production cycle. From 1974 to 1976, the percentage of cotton treated preplant to existing weed foliage ranged from 5% in Mississippi to 20% in central Texas, with 10% of the area treated in west Texas and California (DeBord, 1977). Preplant soil incorporated herbicides were utilized on 88% of the planted area in California and Mississippi, with 60% of it treated in west Texas and 40% in central Texas. Preemergence treatments were applied to 1, 15, 45, and 85% of the cotton planted in California, west Texas, central Texas, and Mississippi, respectively. The percentages of harvested area treated with a directed postemergence herbicide were 6, 20, 52, and 75% respectively, for these same regions. Mississippi treated 35% of its harvested area with an over-the-top herbicide application, but less than 1% was treated in the other regions. In California, 14% of the harvested area was treated with a late postemergence herbicide application, whereas 40% was treated in Mississippi.

Although an effective weed control program must be designed to control those weed species that currently infest a cotton field, attention must also be given to the appearance of new weed species (Buchanan et al., 1975a; Weber et al., 1974). A thorough knowledge of the weeds infesting cotton fields and their response to available herbicides is essential for economical production of cotton. The general level of control by individual herbicides on broad groups of economically important weeds infesting cotton are presented in Table 9D-6.

Foliar herbicides are applied to existing weed vegetation before planting (preplant) the crop. The advantage of preplant foliar treatments is that they are not used unless a problem exists; in the absence of a crop, there is no problem with crop safety when using non-residual herbicides. The main disadvantage is that planting must be delayed until the problem weed develops sufficiently to permit treatment.

Herbicides labeled for weed control in cotton as preplant foliar treatments include dalapon (2,2-dichloropropionic acid), paraquat (1,1'-di-

Table 9D-6. Herbicides labeled for use in cotton in the USA; 1981.

Common name, U.S. producer, and trade name	Structural formula	Rate	Year of introduction, control, and remarks†
		kg ai/ha	
Preplant-foliage treatments			
Dalapon Dow Dowpon	$CH_3-\overset{\displaystyle Cl}{\underset{\displaystyle Cl}{C}}-COOH$	4.4–8.2	1953 AB = P, AG = E, PB = P, PG = E. Treat when grasses are actively growing. Warm weather needed before and after spraying. Spray when johnsongrass is 20 to 30 cm or after bermudagrass stolons are 10 cm long. Plow or disk after 3 days.
Glyphosate Monsanto Roundup	$HOOC-CH_2-NH-CH_2-\overset{\displaystyle O}{\underset{\displaystyle OH}{P}}-OH$	1.1–1.5	1974 AB = E, AG = E, PB = E, PG = E. Apply to actively growing weeds that have reached early head or early bud growth stage. Allow at least 7 days after application before tillage.
Paraquat Chevron Ortho Paraquat CL	$CH_3-\overset{+}{N}\!\!=\!\!=\!\!\overset{+}{N}-CH_3\ 2Cl^-$	0.5–1.1	1965 AB = E, AG = E, PB = P, PG = P. Apply when weeds and grasses are succulent and growth is from 3 to 15 cm tall. Not dependent upon temperature for activity.
Preplant-incorporated treatments			
Fluchloralin BASF Wyandotte Basalin	F_3C (dinitro-benzene ring) $-N\big({-}CH_2{-}CH_2{-}CH_3,\ {-}CH_2{-}CH_2{-}Cl\big)$; NO_2 groups	0.5–1.6	1976 AB = P, AG = E, PB = P, PG = P. Incorporate 3 to 5 cm deep. Herbicide losses of 20% or greater can occur if incorporation is delayed 24 hours. Apply the high rates only on areas heavily infested with rhizome johnsongrass.

(continued on next page)

Table 9D-6. Continued.

Common name, U.S. producer, and trade name	Structural formula	Rate	Year of introduction, control, and remarks†
		kg ai/ha	
Pendimethalin American Cyanamid Prowl	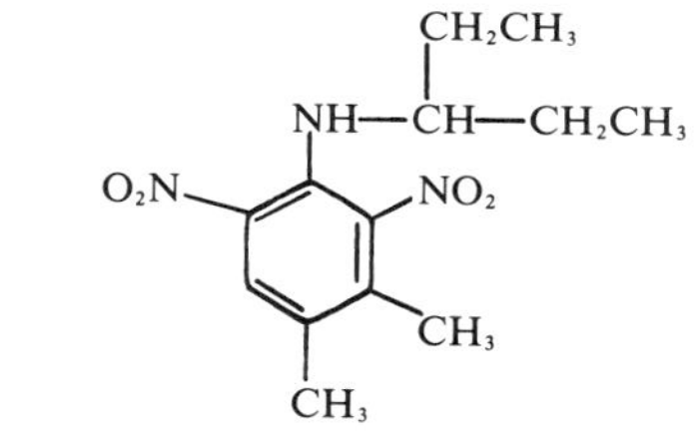	0.5–1.6	1976 AB = P, AG = E, PB = P, PG = P Incorporate 3 to 5 cm deep. Herbicide losses of 15% or greater can occur if incorporation is delayed 24 hours. Apply the high rates if heavy weed populations are anticipated.
Profluralin Ciba-Geigy Tolban	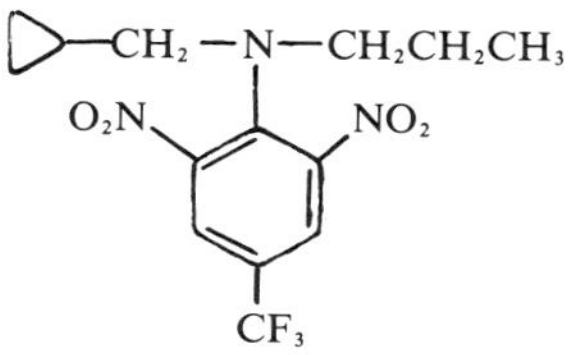	0.5–3.3	1975 AB = P, AG = E, PB = P, PG = P Immediate incorporation is strongly recommended. Herbicides losses of 40% of greater can occur if incorporation is delayed 24 hours. Apply the high rates only on areas heavily infested with rhizome johnsongrass.
Trifluralin Elanco Treflan	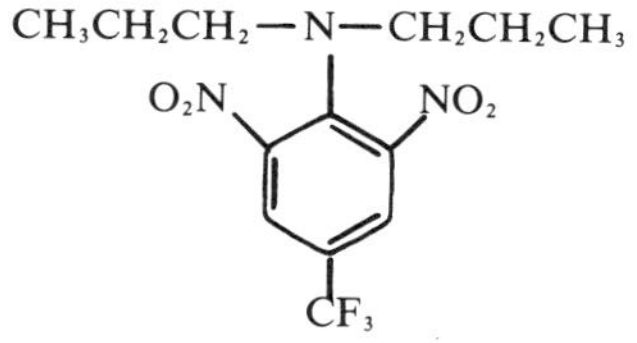	0.5–2.2	1963 AB = P, AG = E, PB = P, PG = P Immediate shallow incorporation is strongly recommended. Herbicide losses of 30% or greater can occur if incorporation is delayed 24 hours. Apply the high rate only on areas heavily infested with rhizome johnsongrass.

(continued on next page)

Table 9D-6. Continued.

Common name, U.S. producer, and trade name	Structural formula	Rate	Year of introduction, control, and remarks†
		kg ai/ha	
EPTC Stauffer Eptam	$CH_3CH_2-S-\overset{O}{\overset{\|}{C}}-N(CH_2CH_2CH_3)_2$	2.2	1961	AB = F, AG = E, PB = P, PG = P Effective nutsedge control can be obtained. Apply after cotton has 2 to 4 leaves and injected no closer than 10 cm on each side of drill. Cotton has minimum tolerance to this herbicide.
Preemergence treatments Alachlor Monsanto Lasso		1.1–2.2	1969	AB = P, AG = G, PB = P, PG = P Provides good annual grass control. Labeled for use in parts of Oklahoma and Texas. It should not be applied on sands or loamy sand soils. If sprinkler irrigation is used, apply after pre-irrigation.
Bensulide Stauffer Prefar	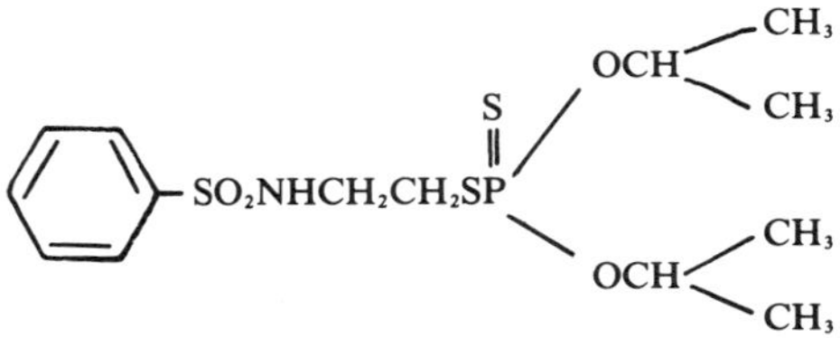	1.1–2.2	1968	AB = P, AG = G, PB = P, PG = P Provides good annual grass control. Use only on cotton that will be irrigated. Apply to soil before furrowing for the preplant irrigation. Disking before furrowing may increase herbicide effectiveness.

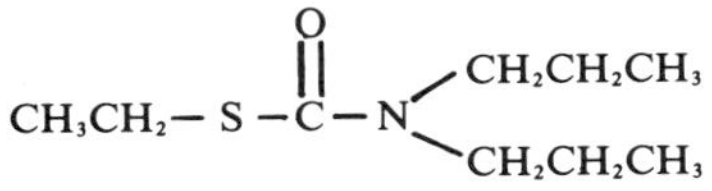

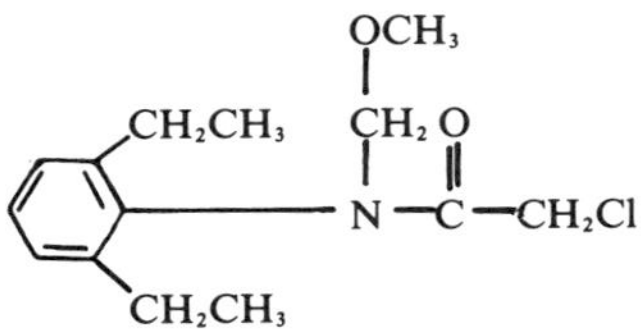

(continued on next page)

Table 9D-6. Continued.

Common name, U.S. producer, and trade name	Structural formula	Rate	Year of introduction, control, and remarks†
		kg ai/ha	
Cyanazine Shell Bladex		0.9–2.2	1972
Diuron DuPont Karmex		0.5–2.2	1954
DCPA Diamond Shamrock Dacthal		4.9–11.5	1956
Dipropetryn Ciba-Geigy Sancap		1.3–2.2	1973

AB = G, AG = F, PB = P, PG = P. Should cotton stand be lost due to adverse weather conditions, the field can be replanted to cotton or soybeans. Do not use on coarse or medium textured soils or on soils with less than 1% organic matter.

AB = G, AG = F, PB = P, PG = P. Do not use on sand as crop injury may result. Do not use with diuron when systemic insecticides are used on seed.

AB = P, AG = G, PB = P, PG = P. Exceptionally safe on cotton. May be used on sandy soil at the lower rates. Shallow incorporation given better control on irrigated land. Many vegetables can follow cotton treated with DCPA during the same year.

AB = G, AG = P, PB = P, PG = P. For use in sand, loamy sand, and fine sandy loam soils of Arizona, New Mexico, Oklahoma, and Texas. Do not make broadcast application to furrow planted cotton.

(continued on next page)

Table 9D-6. Continued.

Common name, U.S. producer, and trade name	Structural formula	Rate	Year of introduction, control, and remarks†
		kg ai/ha	
Fluometuron Ciba-Geigy Cotoran	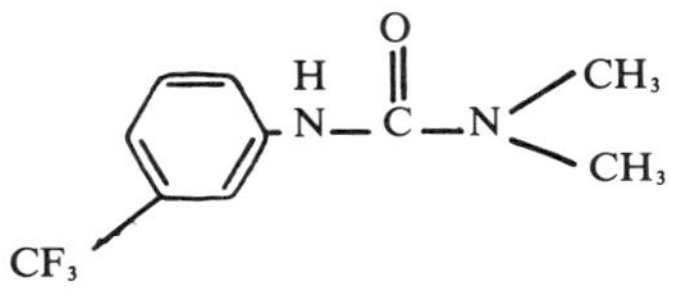	0.9–2.2 1964	AB = G, AG = G, PB = P, PG = P Where dry weather conditions prevail, the herbicidal activity may be delayed or reduced. Crop injury may occur with fluometuron when systemic insecticides are used on seed.
Norflurazon Sandoz Zorial	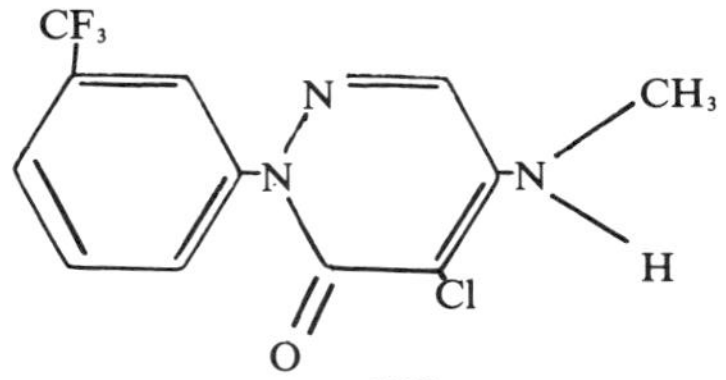	1.1–2.2 1975	AB = G, AG = G, PB = P, PG = P For use in the cotton belt where 100 cm of rain annually is expected. Norflurazon suppresses nutsedge but is highly injurious to glandless strains of cotton. See label for rotation restrictions.
Oryzalin Elanco Surflan	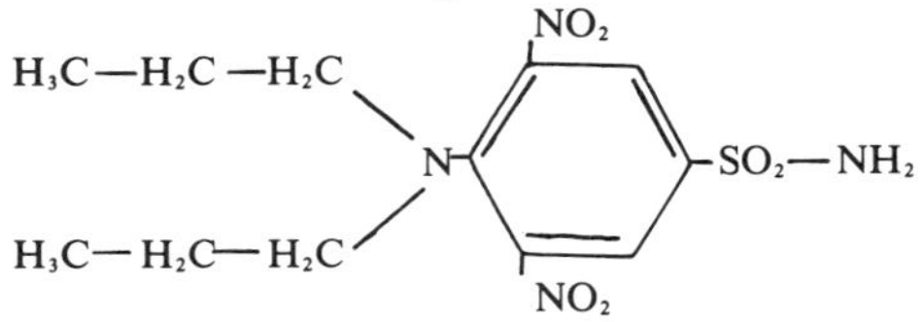	1.5–1.6 1976	AB = F, AG = G, PB = P, PG = P A one-half inch rain or more is necessary to activate the herbicide. Not rcommended for soils containing more than 5% organic matter.

(continued on next page)

Table 9D-6. Continued.

Common name, U.S. producer, and trade name	Structural formula	Rate	Year of introduction, control, and remarks†
		kg ai/ha	
Prometryr Ciba-Geigy Caparol		1.3–3.0	1964 AB = G, AG = F, PB = P, PG = P Do not use on sand or loamy sand soils. Rainfall or irrigation is needed following application to obtain good weed control. Avoid broadcast application to cotton planted in furrows over 6 cm deep. Do not use on glandless varieties.

Postemergence directed and postemergence over-the-top treatments

Cyanazine Shell Bladex		0.6–1.1	1972 AB = G, AG = F, PB = P, PG = P Apply as well directed basal spray to cotton before weeds exceed 5 cm in height. Apply after cotton is 15 cm tall and before first bloom. The inclusion of an arsenical often improves activity on certain weed species.
Dinoseb Dow Premerge 3		1.6–2.4	1947 AB = G, AG = G, PB = P, PG = P Apply as a carefully directed spray when cotton is at least 15 cm tall. Do not apply when soil surface is wet, when cotton plants are extremely succulent, or when small cotton plants are affected by seedling disease.

(continued on next page)

Table 9D–6. Continued.

Common name, U.S. producer, and trade name	Structural formula	Rate	Year of introduction, control, and remarks†
		kg ai/ha	
Diuron DuPont Karmex	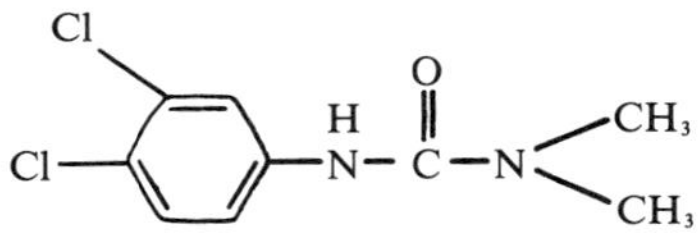	0.4	1954 — AB = G, AG = G, PB = P, PG = P Apply as a directed spray after cotton is 15 cm tall up to first bloom to weeds less than 5 cm tall. Do not treat drought-stressed weeds. The inclusion of an arsenical often improves activity on certain species.
Fluometuron Ciba-Geigy Cotoran	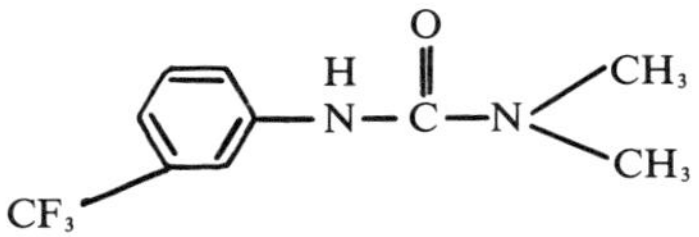	0.9–2.2	1964 — AB = F, AG = G, PB = P, PG = P Apply as directed, semi-directed or over-the-top spray when cotton is 7 to 15 cm tall. Relatively safe on young cotton and provides residual preemergence control. The inclusion of an arsenical often improves activity on certain species.
Methazole Velsicol Probe	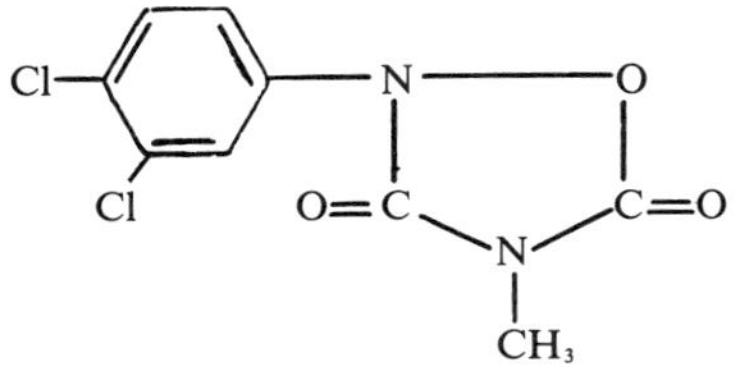	0.5–1.6	1976 — AB = F, AG = P, PB = P, PG = P Apply as directed spray to the base of cotton plants, use only lower rates on 7 cm cotton and higher rates on 15 cm cotton. The inclusion of an arsenical often improves activity on certain weed species.
MSMA Diamond Shamrock Daconate	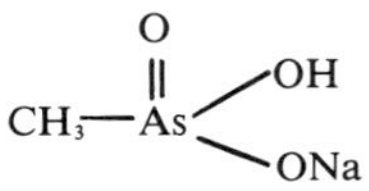	2.2	1963 — AB = P, AG = G, PB = P, PG = F Apply as a directed spray after cotton is at least 7 cm tall and before first bloom. Usually 2 applications needed for nutsedge control. Combination with other herbicides more effective—if a broad spectrum of weeds is present.

(continued on next page)

Table 9D-6. Continued.

Common name, U.S. producer, and trade name	Structural formula	Rate	Year of introduction, control, and remarks†
		kg ai/ha	
Prometryn Ciba-Geigy Caparol	$S-CH_3$ ring structure (CH_3)$_2$CHNH— triazine —NHCH(CH_3)$_2$	0.5–0.6	1964 — AB = G, AG = G, PB = P, PG = P Apply as directed spray to the base of cotton plants 51 to 20 cm tall. Do not apply to furrow planted cotton until furrows are leveled. Avoid spraying drought-stressed weeds. Do not use on coarse sand or loamy sand soils.
Glyphosate Monsanto Roundup	$HOOC-CH_2-NH-CH_2-P(=O)(OH)-OH$	0.9–2.7 kg ai in 75 L water	1974 — AB = F, AG = G, PB = F, PG = G Recirculating Sprayers—Apply to weeds growing above the crop canopy. Keep spray stream at least 5 cm above crop. Avoid droplets, mist, or splatter on to cotton.
		33% solution	Ropewick—Apply to weeds growing above the crop canopy. Keep wiper at least 5 cm above crop. Avoid dripping and do not exceed 5 km. Performing a second wiping in opposite direction to the first may improve control.
		15% solution	Recirculating Wiper—Apply to weeds growing above the crop canopy. Keep wiper at 5 cm above crop and do not exceed 8 km.
		2.2–5.5	Shielded sprayers—Apply to low growing perennials growing between crop rows and avoid contact with desirable vegetation.
		1–2% solution	Spot treatment—Spray at 1% to wet actively growing johnsongrass in the boot to head stage of growth or bermudagrass at 2% in the seed head stage of growth. Avoid contact with cotton.

(continued on next page)

Table 9D-6. Continued.

Common name, U.S. producer, and trade name	Structural formula	Rate	Year of introduction, control, and remarks†
Late postemergence treatments			
Cyanazine Shell Bladex	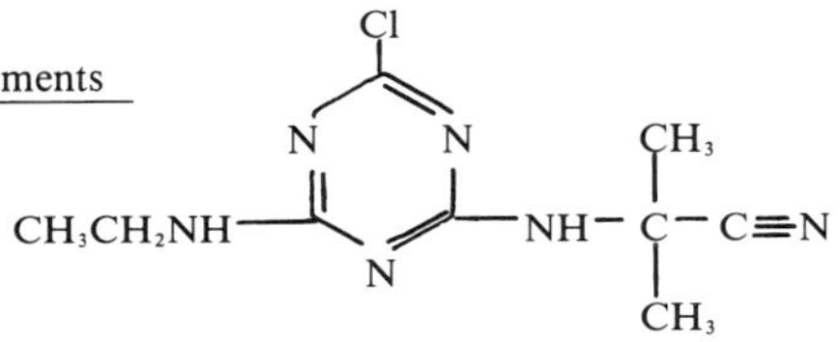	0.9–1.7	1972	AB = G, AG = F, PB = P, PG = P. Apply as a directed spray after cotton is at least 30 cm tall. Do not apply to sandy soils. Use on soils with over 1% organic matter. Use cyanazine where shorter residual effect is desired.
Diuron DuPont Karmex	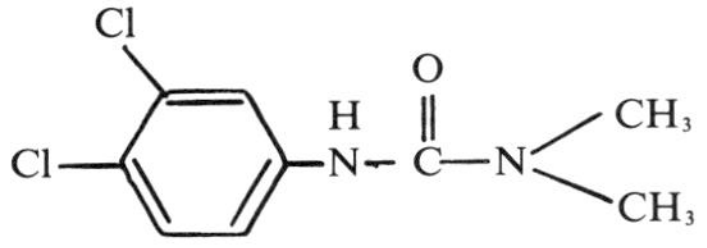	0.9–1.3	1954	AB = G, AG = G, PB = P, PG = P. Apply as a broadcast spray directed to the base of plants after cotton is 30 cm tall. Do not use on sands. In irrigated areas, water management is essential to insure thorough wetting of beds after application.
Linuron DuPont Lorox	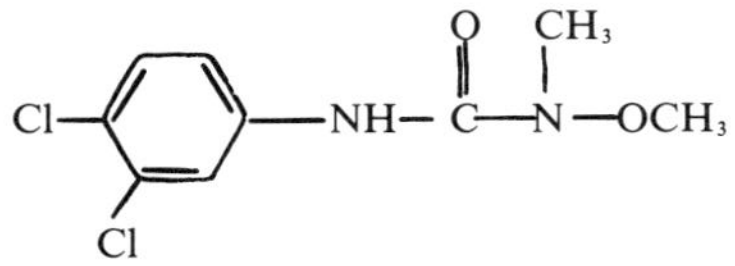	0.6–1.1	1962	AB = G, AG = G, PB = P, PG = P. Apply as a directed spray after cotton is at least 50 cm tall. Do not use on sand or sandy loam soils. Do not treat Pima varieties. In irrigated areas, water management is essential to insure thorough wetting of beds after application.

(Anonymous, 1981a; Anonymous, 1981b; Ashburn, 1981; French and Swann, 1981; Harvey et al., 1981; Miller et al., 1981; Weaver, 1981; WSSA Herbicide Handbook Committee, 1979; USDA, 1980.)

† Estimates of control are based on appropriate lable rate of each herbicide for specific conditions of application. Weed type identification: AB = annual broadleaf species, AG = annual grasses, PB = perennial broadleaf species, PG = perennial grasses. Level of control: E = excellent (90 to 100%), G = good (80 to 90%), F = fair (70 to 80%), P = poor (0 to 70%).

methyl-4,4′-bipyridinium ion), MSMA, and glyphosate [N-(phosphono-methyl)glycine] (Table 9D–6). Dalapon is active on perennial grasses such as bermudagrass and johnsongrass. Unfortunately, these grasses are very slow to initiate growth in the spring, which often limits the level of control obtained with dalapon. Paraquat is very effective in the control of small annuals, and is not dependent on temperature for activity; MSMA controls annual grass and topkills johnsongrass and nutsedge. Temperatures should be above 21°C for best activity. Glyphosate provides excellent control of johnsongrass, bermudagrass, and most other emerged annual and perennial weeds. Best johnsongrass control with glyphosate is obtained when application is made at the heading stage of growth.

Preplant incorporated treatments usually provide excellent annual grass control (Table 9D–6). The major advantage of these treatments is that they can be implemented along with seedbed preparation. Generally the producer applies these treatments as broadcast applications, making them relatively expensive. The disadvantage to preplant-incorporated treatments is that they must be used prior to the development of a weed problem.

Herbicides labeled for weed control in cotton as preplant incorporated treatments include trifluralin, profluralin [N-(cyclopropyl-methyl)-α,α,α-trifluoro-2,6-dinitro-N-propyl-p-toluidine], fluchloralin [N-(2-chloroethyl)-2,6-dinitro-N-propyl-4-(trifluoromethyl)aniline], and pendimethalin [N-(1-ethylpropyl)-3,4-dimethyl-2,6-dinitrobenzenamine] (Table 9D–6). These herbicides have excellent activity on many species of annual grasses and some small-seeded broadleaf weeds. They are generally ineffective in control of large-seeded broadleaf weeds and many perennial weed species. These herbicides are seldom applied as a single treatment for a total weed control program.

A slight variation of the preplant incorporated application technique is the preplant-injected application of EPTC (S-ethyl dipropylthiocarbamate) (Holstun et al., 1963). This herbicide is used in cotton primarily because of its excellent activity against nutsedges. However, care must be taken to avoid injecting this herbicide directly in the drill row, because cotton will be injured.

Preemergence applications are made as the crop seeds reach the soil, and anytime thereafter until the crop plants emerge from the soil. The major advantages of preemergence treatments are that they can be applied as part of the planting operation. They can be conveniently used as band applications, and result in a highly economical treatment. Like the preplant incorporated treatment, they are applied prior to the development of a weed problem.

Several herbicides are labeled for preemergence use in cotton (Table 9D–6). They represent several chemical families and range of herbicidal activity. Herbicides applied as preemergence applications include diuron, DCPA (dimethyl tetrachloroterephthalate), fluometuron, prometryn, bensulide [0,0-diisopropyl phosphorodithioate S-ester with N-(2-mercapto-ethyl)benzenesulfonamide], alachlor [2-chloro-2′,6′-diethyl-N-(methoxy-methyl)acetanilide], cyanazine <2-{[4-chloro-6-(ethylamino)-s-triazin-2-yl]amino}-2-methylpropionitrile>, dipropetryn [2-(ethylthio)-4,6-bis(isopropylamino)-*s*-triazine], norflurazon [4-chloro-5-(methylamino)-2-(α,α,α-trifluoro-*m*-tolyl)-3(2H)-pyridazinone], and oryzalin (3,5-dinitro-N^4,N^4-dipropylsulfanilamide). Remarks in Table 9D–6 need to be considered since each herbicide has unique characteristics that are matched with specific weed species to be controlled, geographic locations, and environmental conditions.

Postemergence treatments are applied after emergence of the crop plants from the soil, but before harvest is completed. Three types of postemergence applications are generally used in cotton. They are directed postemergence, over-the-top postemergence, and late post-emergence. The most extensive use of postemergence treatments in cotton is to control weeds in the drill-row (Baker et al., 1967). Limited use is made of postemergence applications in the row middles. The major advantage of the postemergence technique is that applications of herbicides can be delayed until the weed problem has developed.

Directed postemergence treatments require that the weeds be shorter in height than the cotton. Special equipment is required to make directed postemergence treatments, and, for best results, the cotton seedbed must be relatively smooth and free of debris. This uniformity is important so that the spray nozzles are kept at a constant height above the soil surface to avoid injury to cotton. Herbicides labeled for directed post-emergence application in cotton include dinoseb, diuron, MSMA, prometryn, fluometuron, cyanazine, and methazole [2-(3,4-dichlorophenyl)-4-methyl-1,2,4-oxadiazolidine-3,5-dione] (Table 9D–6). For broad spectrum weed control, MSMA plus one of the other postemergence herbicides listed are applied as tank mixtures.

Over-the-top postemergence treatments are easy to apply with ground or air equipment, but it is difficult to obtain adequate coverage of weeds, particularly when they are smaller and are partially covered by the cotton canopy. Fluometuron and DSMA are the only herbicides labeled for over-the-top application.

Late postemergence treatments are made after the cotton is 35 to 60 cm tall. Herbicides labeled for late postemergence application are diuron, linuron [3-(3,4-dichlorophenyl)-1-methoxy-1-methylurea], prometryn, fluometuron, and cyanazine. Applications are made either directly to freshly cultivated soil or to weeds growing in uncultivated soil. Late postemergence treatments are not usually required, but poor cotton stands, presence of perennial weed species, and particularly viny growing weeds, can occasionally justify applications of herbicides at this time.

Until recently, the only means of controlling weeds growing taller than the crop late in the season was by hand. The development of the recirculating sprayer provided a useful technique for controlling late-season weeds that are taller than the cotton, but it was not widely used until glyphosate became available (McWhorter, 1977). Other equipment developed to apply glyphosate selectively to weeds above the crop canopy include the ropewick applicator and recirculating wiper (Dale, 1979; Chandler, 1981). The major disadvantage of these late season techniques is that the weeds have caused crop losses through competition before they are controlled. On the other hand, these procedures are exceedingly efficient in that herbicides are applied only to the weeds. Consequently, only a minimum amount of herbicide is released to the environment.

9D–6 EQUIPMENT FOR WEED CONTROL IN COTTON

In the early 1950s, the discovery and development of selective herbicides made it necessary to give special attention to application equipment and methods of application. Early herbicide applications were accomplished primarily with knapsack sprayers. The introduction of the low-volume sprayer with agricultural spray nozzles was a major breakthrough in herbicide application equipment. The sprayer can be used to apply spray volumes of 8 to 60 L/ha. The low-volume spray system is still the basic component of most herbicide application equipment being used today (Williford, 1981).

In cotton, most herbicides that are applied preplant must be incorporated into the soil. Herbicide application is made with a boom-type sprayer and herbicide incorporation can be accomplished with several devices. By far, the most popular incorporating tool is the pulverizing tandem disk harrow. Other commonly used implements are the Triple K harrow® , Do-All® bed conditioner, rotary hoe, field cultivator, rolling cultivator, power-driven rotary tiller, and the Lely-Roterra® . A wide variation in mixing efficiency of the various incorporation implements has been found. The pattern of incorporation depends on soil texture, depth of penetration, design of the device, forward speed, and rotor speed in the case of power-driven devices (Barrentine and Jordan, 1976).

Conventional preemergence applications may be made with single nozzles in treating bands of less width than the row or they may be applied with conventional boom sprayers if the entire area is being treated (Wooten and Holstun, 1959). If a band is being treated, vertical gauging of the nozzle to maintain a constant height is necessary. A single nozzle which varies in height while spraying delivers a variable rate per unit of area actually sprayed.

While herbicides have revolutionized cotton production and weed control, mechanical postemergence weed control is still practiced on almost all cotton farms. The row-crop cultivator is the principal tool for mechanical weed control. Both fixed and spring-tine cultivators are available. Proper adjustment and operation speed are key factors in the efficiency obtained with row crop cultivators (Williford, 1981).

Postemergence application of herbicides is an integral part of the total weed control program in cotton. Directed postemergence sprays are applied to a band under the crop plants, with the spray nozzles directed under the foliage to reach the target area without hitting the crop. This spray is applied with a skid-mounted or wheel-mounted applicator or with shield applicators mounted to cultivators. Directed sprays are used when cotton and weeds are small, often before cultivation begins, or even later so long as the cotton plants are larger than weeds and weeds are small enough to be susceptible to the herbicide (Jordan and Barrentine, 1976).

A few herbicides are labeled for over-the-top, postemergence application in cotton. These applications are made with the same type of spray boom that is used to broadcast preemergence herbicide treatments. Aerial application is widely used for this type of treatment.

With the development of glyphosate, several applicators have become available that will apply a non-selective herbicide to weeds growing above the crop canopy. The box-type recirculating sprayer was first marketed in the mid-1970s. With this applicator, the herbicide solution is applied through solid stream nozzles above and at right angles to the crop row. Herbicide spray not deposited on weeds is intercepted and recovered in a trap. The solution is returned to the original supply tank by a venturi system (McWhorter, 1970).

The rope wick applicator was also developed to apply systemic herbicides to weeds taller than crop plants (Dale, 1979). The rope wick applicator was first marketed in 1978. The rope wick applicator has no pumps or moving parts and is constructed using polyvinyl chloride or aluminum pipe and nylon rope. The rope-wick principle is based upon capillary and gravitational movement of the herbicide solution from the reservoir through the ropes. The herbicide solution is transferred from the ropes to weeds by a wiping action. Contact of the herbicide-laden rope with crop plants is avoided by positioning the applicator 5 to 10 cm above the crop.

Like the recirculating sprayer and the rope wick applicator, the recirculating wiper or ultimate Stoneville applicator, applies herbicide only to the weeds growing taller than the crop plants (Chandler, 1981). The recirculating wiper was first marketed in 1981. The basic components are a length of 20 cm-diam aluminum irrigation tubing, a pad of nylon carpet, a section of flat expanded metal, and cone spray nozzles. The applicator is constructed by removing a longitudinal section of the tubing from the 0100 (1 o'clock) to the 0400 (4 o'clock) position and replacing it with a hinged section of expanded metal which supports the carpet. The spray nozzles are mounted on the tubing at the 0800 (8 o'clock) position with the tips inside the tubing. The nozzles spray onto the back of the carpet pad that covers the opening in the tube. The spray solution moves from the back of the carpet through the carpet fibers by capillary action. Excess spray solution flows from the bottom of the tubing into a shallow reservoir and is returned to the supply tank. The recirculating wiper is between the recirculating sprayer and the rope wick in terms of simplicity of operation, cost of construction, and cost of herbicide application.

Many weeds are localized in the field and, for maximum economy, only the infested areas should be treated with spot application of herbicides. Although hand spraying is labor-intensive, walking the field and spraying with a knapsack sprayer is used by some farmers because the cost of herbicide is low.

9D-7 ECONOMICS OF WEED CONTROL SYSTEMS IN COTTON

The optimum weed control program for a specific cotton field can be developed only by carefully evaluating the total weed problem, and then diligently planning an economical control program. In developing an effective weed control program for cotton, careful attention should be given to: (a) selection of the most effective herbicides available; (b) consideration of proper application equipment and techniques; (c) application of herbicides at the proper time; and (d) the potential use of spot treatments and supplemental hand hoeing. The cost of hand labor is highly variable and depends largely on the effectiveness of the other control measures used in the system. As the relative cost of labor increases, the effectiveness of a weed control system becomes more important (Holstun and Wooten, 1966).

Weed control systems using individual preemergence or postemergence treatments may be highly effective in reducing production costs but when combined they reduce costs still further (Holstun, 1963; Miller et al., 1961). The effectiveness of single systems is limited because no herbicide or other method of control will economically control enough species of weeds under all conditions. When persistent weeds are present, the use of multiple control systems often fails to provide economical control of all species, but such systems greatly reduce the possibility of failure.

The production system used by a producer to produce cotton and the management of weeds in that system is unique for the various production regions of the Cotton Belt (Tables 9D-7 to 9D-10). The production environment, especially the amount of available moisture influences the weed control strategy and the costs incurred. The total cost for full-season weed control in four separate geographic locations across the U.S. Cotton Belt in

Table 9D-7. Estimated cost of full-season weed control in solid cotton. Based on Georgia practices and four-row equipment in 1980 (C. M. French and G. Westberry, personal communication).

| | 1980 cost ($ U.S./ha) | | | |
Operation	Equipment	Labor	Material	Total
Disk twice (6.4 m); broadcast and incorporate 0.84 kg trifluralin per ha	$21.94	$1.94	$12.97	$36.85
Plant; apply 0.66 kg fluometuron per ha (35 cm band on 95 cm row spacing)	0.00	0.00	10.64	10.64
Cultivate; postdirect 0.66 kg fluometuron + 0.83 kg MSMA per ha (45 cm band on 95 cm row spacing)	3.56	2.01	15.35	20.92
Cultivate; postdirect 0.41 kg cyanazine + 0.83 kg MSMA per ha (45 cm band on 95 cm row spacing)	3.56	2.01	8.51	14.08
Cultivate; postdirect 0.41 kg cyanazine + 0.83 kg MSMA per ha (45 cm band on 95 cm row spacing)	3.56	2.01	8.51	14.08
Cultivate	3.00	1.81	0.00	4.81
Total	$35.62	$9.78	$55.98	$101.38

Table 9D–8. Estimated cost of full-season weed control in solid cotton grown on sandy soil, based on Mississippi Delta practices and six-row equipment in 1980 (Parvin et al., 1980).

	1980 cost ($ U.S./ha)			
Operation	Equipment	Labor	Material	Total
Disk (6.4 m); broadcast and incorporate 0.56 kg trifluralin per ha	$ 9.14	$ 1.38	$ 7.78	$ 18.30
Plant; apply 0.56 kg fluometuron per ha (50-cm band on 100-cm row spacing)	0.00	0.00	6.50	6.50
Cultivate	7.68	1.61	0.00	9.29
Cultivate; postdirect 0.9 kg MSMA per ha (50-cm band on 100-cm row spacing)	10.00	1.83	2.70	14.62
Cultivate; postdirect 0.33 kg cyanazine + 1.12 kg MSMA per ha (50-cm band on 100-cm row spacing)	10.00	1.83	5.71	17.54
Hand hoe (4.9 hours per ha)	0.00	15.32	0.00	15.32
Cultivate; postdirect 1.12 kg of MSMA + 0.45 kg fluometuron per ha (50-cm on 100-cm row spacing)	7.51	1.35	8.67	17.53
Cultivate; postdirect 1.12 kg of MSMA + 0.22 kg diuron per ha (50-cm band on 100-cm row spacing)	7.51	1.35	4.87	13.73
Hand hoe (4.9 hours per ha)	0.00	15.32	0.00	15.32
Cultivate; postdirect 1.12 kg linuron per ha	7.51	1.35	19.27	28.13
Total	$59.35	$41.34	$55.59	$156.28

Table 9D–9. Estimated cost of full season weed control in solid cotton. Based on Texas practices and eight-row equipment in 1980 (Abernathy, 1981; personal communication).

	1980 cost ($ U.S./ha)			
Operation	Equipment	Labor	Materials	Total
Spring tooth harrow (9.1 m); broadcast and incorporate 0.56 kg trifluralin per ha	$ 8.65	$ 0.62	$ 9.27	$18.54
Plant; apply 0.27 kg prometryn per ha (25-cm band on 100-cm row spacing)	0.00	0.00	3.09	3.09
Cultivate; wiper application of 1:2 solution of glyphosate	6.18	1.24	3.71	11.13
Spot spray; 2% solution of glyphosate	2.47	2.47	9.88	14.82
Cultivate	6.18	1.24	0.00	7.42
Hand hoe (4.9 hours/ha)	0.00	21.17	0.00	21.71
Total	$23.48	$27.28	$25.95	$76.71

Table 9D–10. Estimated cost of full-season weed control in solid cotton. Based on Kern County, California practices and four-row equipment in 1980 (P. E. Keeley and K. D. Olson, personal communication).

	1980 cost ($ U.S./ha)			
Operation	Equipment	Labor	Materials	Total
Disk (4.2 m); broadcast and incorporate 0.67 kg trifluralin per ha	$15.71	$ 6.55	$ 9.14	$31.40
Cultivate	11.36	5.68	0.00	17.04
Hand hoe (9.8 hours per ha)	0.00	30.64	0.00	30.64
Cultivate	11.36	5.68	0.00	17.04
Cultivate; postdirect 1.55 kg prometryn per ha	13.76	5.93	14.70	34.39
Total	$52.19	$54.48	$23.84	$130.51

1980 has been estimated to range from \$76 to \$156/ha. In the eastern half of the U.S. Cotton Belt, the number of herbicide applications and total herbicide cost is about twice that used in the western half. Equipment and labor coss in the mid-South and western states was substantially higher than the Southeast and Southwestern regions. Expenditures for labor were noticeably low in the southeastern region and markedly high in the western region. With the exception of the southeast, hand hoeing is still used on a limited basis to control those weeds that escape other control methods.

With the introduction of new chemicals and technology at substantial cost, the management of weeds in a totally integrated production system will be mandatory for future economical production of cotton in the USA.

Mention of pesticides does not constitute a recommendation for use, nor does it imply that the pesticides are registered under the U.S. Federal Insecticide, Fungicide, and Rodenticide Act as amended. The use of trade names in this publication does not constitute a guarantee, warranty, or endorsement of the products by the Texas Agricultural Experiment Station.

REFERENCES

Abernathy, J. R. 1978. Pernicious weeds in cotton—silverleaf nightshade. p. 130. Proc. Beltwide Cotton. Prod. Res. Conf., Dallas, Tex.

----. 1981. Estimated crop losses due to weeds with nonchemical management. p. 159-167. *In* D. Pimental (ed.) Handbook of pest management in agriculture. Vol. I. CRC Press, Boca Raton, Fla.

Altman, J., and C. L. Campbell. 1977. Effect of herbicides on plant diseases. Annu. Rev. Phytopathol. 15:361-385.

Andries, J. A., A. G. Douglas, and A. W. Cole. 1974. Herbicides, leaf type, and row spacing response in cotton. Weed Sci. 22:496-499.

Anonymous. 1960. Index of plant diseases in the United States. Agric. Handb. No. 165.

----. 1981a. Recommended chemicals for weed and brush control. Arkansas Agric. Exp. Stn., MP-44.

----. 1981b. Weed control guidelines for Mississippi. Mississippi Agric. Exp. Stn.

Arle, H. F., and K. C. Hamilton. 1973. Effect of annual weeds on furrow-irrigated cotton. Weed Sci. 21:325-327.

Ashburn, E. L. 1981. Chemical weed control in cotton. Tennessee Agric. Exp. Stn., Publ. 652.

Baker, R. S. 1977. Pernicious weeds in cotton—prickly sida. p. 163. Proc. Beltwide Cotton Prod. Res. Conf., Atlanta, Ga.

----, J. T. Holstun, Jr., W. L. Barrentine, and E. E. Schweizer. 1967. Directed and overhead sprays of postemergence herbicides on cotton. Proc. South. Weed Sci. Soc. 22:59.

Barber, G. W. 1937. Seasonal availability of food plants of two species of *Heliothis* in eastern Georgia. J. Econ. Entomol. 30:150-158.

Barrentine, W. L., and T. N. Jordan. 1976. Equipment for incorporating herbicides. Weeds Today 7:20-21.

Brazzel, J. R., L. O. Newsom, J. S. Roussel, C. Lincoln, F. J. Williams, and G. Barnes. 1953. Bollworm and tobacco budworm as cotton pests in Louisiana and Arkansas. Louisiana Agric. Exp. Stn. Bull. 482.

Brown, F. H., and A. B. Wiles. 1970. Reaction of certain cultivars and weeds to a pathogenic isolate of *Verticillium albo-atrum* from cotton. Plant Dis. Rep. 54:508-512.

Brown, H. B., and J. O. Ware. 1958. Cotton. Chapter 14, p. 313-341. McGraw-Hill Book co., N.Y.

Buchanan, G. A. 1981. Management of weeds in cotton. p. 215-242. *In* D. Pimental (ed.) Handbook of pest management in agriculture. Vol. III. CRC Press, Boca Raton, Fla.

----, and E. R. Burns. 1970. Influence of weed competition on cotton. Weed Sci. 18:149–154.

----, and ----. 1971a. Weed competition in cotton. I. Sicklepod and tall morningglory. Weed Sci. 19:576–579.

----, and ----. 1971b. Weed competition in cotton. II. Cocklebur and redroot pigweed. Weed Sci. 19:580–582.

----, and A. E. Hiltbold. 1977. Response of cotton to cultivation. Weed Sci. 25:130–134.

----, and R. D. McLaughlin. 1975. Influence of nitrogen on weed competition in cotton. Weed Sci. 23:324–328.

----, R. H. Crowley, and R. D. McLaughlin. 1977. Competition of prickly sida with cotton. Weed Sci. 25:106–110.

----, ----, J. E. Stree, and J. A. McGuire. 1980. Competition of sicklepod (*Cassia obtusifolia*) and redroot pigweed (*Amaranthus retroflexus*) with cotton (*Gossypium hirsutum*). Weed Sci. 28:258–262.

----, C. S. Hoveland, and M. C. Harris. 1975. Response of weeds to soil pH. Weed Sci. 23:473.

----, ----, V. L. Brown, and R. N. Wade. 1975. Weed population shifts influenced by crop rotations and weed control programs. Proc. South. Weed Sci. Soc. 28:60.

Cates, H. R. 1917. Farm practices in the cultivation of cotton. USDA Bull. 511.

Chandler, J. M. 1977a. Pernicious weeds in cotton—spurred anoda. p. 162. Proc. Beltwide Cotton Prod. Res. Conf., Atlanta, Ga.

----. 1977b. Competition of spurred anoda, velvetleaf, prickly sida, and Venic mallow in cotton. Weed Sci. 25:151–158.

----. 1981. The ultimate Stoneville applicator for postemergence weed control. Proc. South. Weed Sci. Soc. 34:294.

----, and L. R. Oliver. 1979. Spurred anoda: A potential weed in southern crops. USDA, SEA. ARM-S-2.

----, and P. W. Santelmann. 1968. Interaction of four herbicides with *Rhizoctonia solani* on seedling cotton. Weed Sci. 16:453–456.

----, and ----. 1969. Growth characteristics and herbicide susceptibility of Texas panicum. Weed Sci. 17:91–93.

Christidis, B. G., and G. J. Harrison. 1955. Cotton growing problems. Chapter 16, p. 435–462. McGraw-Hill Book Co., N.Y.

Coble, H. D. 1980. Pernicious weeds in cotton—fall panicum. p. 183. Proc. Beltwide Cotton Prod. Res. Conf., St. Louis, Mo.

Colwick, R. F. and Technical Committee Members, S-2 and W-24. 1960. Weed control equipment in mechanized cotton production. South. Coop. Ser. Bull. No. 71.

Crowe, G. B., and J. T. Holstun, Jr. 1953. The economics of weed control in cotton. Mississippi Agric. Exp. Stn., Cir. 179.

Crowley, R. H. 1977. Pernicious weeds in cotton—morningglories. p. 161. Proc. Beltwide Cotton Prod. Res. Conf., Atlanta, Ga.

----, and G. A. Buchanan. 1978. Competition of four morningglory (*Ipomoea* sp.) species with cotton (*Gossypium hirsutum*). Weed Sci. 26:484–488.

Dale, J. E. 1979. A non-mechanical system of herbicide application with a rope wick. PANS 25:431–436.

DeBord, D. O. 1977. Cotton insect and weed loss analysis. The Cotton Foundation.

Dowler, C. C., and E. W. Hauser. 1974. The effect of cultivation on weeds controlled by fluometuron in cotton. Proc. South. Weed Sci. Soc. 27:112.

Edwards, F. E. 1964. History and progress of flame cultivation. p. 3–6. First Annu. Symp. Res. on Flame Weed Control, Natural Gas Processors Assoc.

Egley, G. H., and J. M. Chandler. 1982. Longevity of weed seeds after 5.5 years in the Stoneville 50-year buried seed study. Weed Sci. 30:190–197.

Eichers, T. R., P. A. Andrilenas, and T. W. Anderson. 1976. USDA, Agric. Econ. Rep. No. 418.

Foy, C. L. 1959. Combined use of preemergence herbicide and cross-cultivation in cotton. Weeds 7:459–462.

Frans, R. W. 1969. Changing ecology of weeds in cotton fields. p. 29. Proc. Beltwide Cotton Prod. Mech. Conf., New Orleans, La.

French, C. M. 1980. Pernicious weeds in cotton—sicklepod. p. 184. Proc. Beltwide Cotton Prod. Res. Conf., St. Louis, Mo.

French, M. C., and C. W. Swann. 1981. Chemical weed control in cotton. Georgia Agric. Exp. Stn., Bull. 826.

Frick, K. W., and J. M. Chandler. 1978. Augmenting the moth (*Bactra verutana*) in field plots for early-season suppression of purple nutsedge (*Cyperus rotundus*). Weed Sci. 26: 703–710.

Gates, D. W. 1977. Pernicious weeds in cotton—johnsongrass. p. 164. Proc. Beltwide Cotton Prod. Res. Conf., Atlanta, Ga.

Harris, V. C. 1960. Weed control in cotton over a 10-year period by use of the more promising materials and techniques. Weeds 8:616–624.

Harvey, L. H., C. L. Parks, B. J. Gossett, F. H. Smith, and E. C. Murdock. 1981. Produce high yields of quality cotton. Clemson Univ. (S.C.) Cir. 589.

Hauser, E. W., and H. F. Arle. 1958. Johnsongrass as a weed. USDA Farmer's Bull. 1537.

Hogger, C. H., and G. W. Bird. 1976. Weed and indicator hosts of plant-parasitic nematodes in Georgia cotton and soybean fields. Plant Dis. Rep. 60:233–226.

Holm, L. G. 1978. Some characteristics of weed problems in two worlds. Proc. West. Soc. Weed Sci. 31:3.

————, D. L. Plucknett, J. V. Pancho, and J. P. Herberger. 1977. The world's worst weeds, distribution, and biology. p. 495–499. University Press of Hawaii, Honolulu.

Holstun, Jr., J. T. 1963. Cultivation techniques in combination with chemical weed control in cotton. Weeds 11:190–194.

————, and O. B. Wooten. 1966. Weeds and their control. p. 152–181. *In* F. C. Elliot, M. Hoover, and W. K. Porter, Jr. (ed.) Advances in production and utilization of quality cotton: Principles and practices. Iowa State University Press, Ames.

————, ————, R. A. Parker, and E. E. Schweizer. 1963. Triband weed control, a new concept for weed control in cotton. USDA, ARS Rep., 34.

Hoveland, C. S., G. A. Buchanan, and M. C. Harris. 1976. Response of weeds to soil phosphorus and potassium. Weed Sci. 24:194.

Johnson, C. 1960. Management of weeder geese in commercial fields. Calif. Agric. 14:5.

Johnson, W. M., and L. A. Brinkerhoff. 1976. Susceptibility of some crops and weeds to *Verticillium dahliae* Kleb. isolated from cotton. p. 21–22. Proc. Beltwide Cotton Prod. Res. Conf., Las Vegas, Nev.

————, and L. A. Brinkerhoff. 1977. Symptoms and formation of microsclerotia in weeds inoculated with the Verticillium wilt pathogen. p. 183. Proc. Beltwide Cotton Prod. Res. conf., Atlanta, Ga.

Jordan, T. N. 1977. Pernicious weeds in cotton—bermudagrass. p. 163. Proc. Beltwide Cotton Prod. Res. Conf., Atlanta, Ga.

————, and W. L. Barrentine. 1976. Equipment for applying herbicides. Weeds Today 7:14–16.

Katan, J., and Y. Eshel. 1973. Interactions between herbicides and plant pathogen. Residue Rev. 45:145–177.

Keeley, P. E. 1978. Pernicious weeds in cotton—yellow nutsedge. p. 128. Proc. Beltwide Cotton Prod. Res. Conf., Dallas, Tex.

————, an R. J. Thullen. 1975. Influence of yellow nutsedge competition of furrow-irrigated cotton. Weed Sci. 23:171–175.

Kermkamp, M. F., S. M. Chen, B. C. Ortega, C. T. Tsiang, and A. M. Khan. 1952. Investigations of physiologic specialization and parasitism of *Rhizoctonia solani*. Univ. of Minnesota Agric. Exp. Stn. Bull. 200.

Little, V. R., and D. F. Martin. 1942. Cotton insects of the United States. Burgess Publishing Co., Minneapolis, Minn.

Lugenbill, P. 1928. The fall army worm. USDA Tech. Bull. 34.

Mayton, E. L., E. U. Smith, and D. King. 1945. Nutgrass eradication studies: IV. Use of chickens and geese in the control of nutgrass (*Cyperus rotundus* L.). Agron. J. 37:785–789.

McWhorter, C. G. 1970. A recirculating spray system for postemergence weed control in row crops. Weed Sci. 18:285–287.

————. 1972. Factors affecting johnsongrass rhizome production and germination. Weed Sci. 20:41–45.

————. 1977. Weed control in soybeans with glyphosate applied in the recirculating sprayer. Weed Sci. 22:584.

Miller, J. H., C. H. Carter, R. H. Garber, and J. E. DeVay. 1979. Weed and disease response to herbicides in single- and double-row cotton. Weed Sci. 27:444–449.

————, H. M. Kemper, D. W. Cudney, B. B. Fisher, and P. E. Keeley. 1981. Weed control in cotton. Univ. of California Leaflet 2991.

————, H. M. Kemper, J. A. Wilkerson, and C. L. Foy. 1961. Control of barnyardgrass (*Echinochloa crus-galli*) in western irrigated cotton. Weeds 9:273–281.

————, and C. E. Miller. 1979. Plant bug reduction through the use of Preemerge 3 Dinitro Amine herbicide as a directed spray in cotton. Down Earth 35:14–15.

Minton, E. B. 1972. Effect of weed control in grain sorghum on subsequent incidence of Verticillium wilt in cotton. Phytopathology 62:582–583.

Neubauer, R., and Z. Avizohar-Hershenson. 1973. Effect of the herbicide, trifluralin, on Rhizoctonia disease in cotton. Phytopathology 63:651–652.

Orr, C. C., J. R. Abernathy, and E. B. Hudspeth. 1975. *Northanguina phyllobia* a nematode parasite of silverleaf nightshade. Plant Dis. Rep. 59:416–418.

Parvin, Jr., D. W., J. G. Hamill, and F. T. Cooke, Jr. 1980. Budget for major crops, Delta of Mississippi. Mississippi Agric. Exp. Stn., AECMR No. 95.

Pinckard, J. A., and L. C. Standifer. 1966. An apparent interaction between cotton herbicidal injury and seedling blight. Plant Dis. Rep. 50:172–174.

Quaintance, A. L., and C. T. Brues. 1905. The cotton bollworm. USDA, Bur. Entomol. Bull. 50.

Ray, W. W., and J. H. McLaughlin. 1952. Isolation and infection tests with seed and soil-borne pathogens. Phytopathology 32:233–238.

Roach, S. H. 1975. *Heliothis* sp.: Larvae and associated parasites and diseases on wild host plants in the Pee Dee area of South Carolina. Environ. Entomol. 4:725–728.

Rogers, N. K., G. A. Buchanan, and W. C. Johnson. 1976. Influence of row spacing on weed competition with cotton. Weed Sci. 24:410–413.

Shaw, W. C. 1964. Weed science-revolution in agricultural technology. Weed Sci. 12:153.

Smith, D. T., and U. H. Tseng. 1970. Cotton development and yield as related to pigweed (*Amaranthus* sp.) density. p. 37. Proc. Beltwide Cotton Prod. Res. Conf., Houston, Tex.

Snow, J. W., J. J. Hamm, and J. R. Brazzel. 1966. *Geranium carolinianaum* as an early host for *Heliothis zea* and *H. virescins* in the southeastern United States, with notes on associated parasites. Ann. Entomol. Soc. Am. 59:506–509.

Stadelbacher, E. A. 1979. *Geranium dissectum*: an unreported host of the tobacco budworm and bollworm and its role in their seasonal and long term population dynamics in the Delta of Mississippi. Environ. Entomol. 8:1153–1156.

————. 1981. Role of early-season wild and naturalized host plants in the buildup of the F_1 generation of *Heliothis zea* and the *H. virescens* in the Delta of Mississippi. Environ. Entomol. 10:766–770.

Stam, P. A., D. F. Clower, J. B. Graves, and P. E. Schilling. 1978. Effects of certain herbicides on some insects and spiders found in Louisiana cotton fields. J. Econ. Entomol. 71:477–480.

Stevens, M. 1967. The effect of DSMA and MSMA applied to cotton for weed control on the infestation of fleahoppers and tarnish plant bugs. Proc. South. Weed Conf. 20:405–409.

Street, J. E., G. A. Buchanan, R. H. Crowley, and J. A. McGuire. 1981. Influence of cotton (*Gossypium hirsutum*) densities on competitiveness of pigweed (*Amaranthus* spp.) and sicklepod (*Cassia obtusifolia*). Weed Sci. 29:253–256.

Supak, J. R. 1978. Pernicious weeds in cotton—lanceleaf sage. p. 130. Proc. Beltwide Cotton Prod. Res. Conf., Dallas, Tex.

Taylor, B. B., and E. S. Heatherman. 1978. Pernicious weeds in cotton—Wright ground-cherry. p. 131. Proc. Beltwide Cotton Prod. Res. Conf., Dallas, Tex.

Taubenhaus, J. J. 1936. Phymatotrichum root-rot on winter and spring weeds of South Central Texas. Am. J. Bot. 23:167–168.

----, and W. N. Ezekiel. 1936. A rating of plants with reference to their relative resistance or susceptibility of Phymatotrichum root-rot. Texas Agric. Exp. Stn. Bull. 527.

Tyler, J. 1937. The root-knot nematode. Univ. of Calif. Cir. 330.

Upchurch, R. P., and F. L. Selman. 1968. Compatibility of chemical and mechanical weed control methods. Weed Sci. 16:121–130.

USDA. 1980. Suggested guidelines for weed control. USDA, Agric. Handb. No. 565. U.S. Government Printing Office, Washington, D.C.

Weaver, D. 1981. Suggestions for weed control with chemicals in cotton. Texas Agric. Exp. Stn. Supplement to MP-1059.

Weaver, Jr., J. B., J. Miller, and D. Weaver. 1978. Tolerance and weed control response of three cotton genotypes to three levels of two dinitroaniline herbicides. Georgia Agric. Exp. Stn. Res. Bull. 213.

Weber, J. B., J. A. Best, and W. W. Witt. 1974. Herbicide residues in weed species shifts on modified soil field plots. Weed Sci. 22:427.

Whitwell, T. 1980. Pernicious weeds in cotton—common cocklebur. p. 182. Proc. Beltwide Cotton Prod. Res. Conf., St. Louis, Mo.

----, L. W. Wells, and J. M. Chandler. 1981. Report of the 1980 cotton weed loss committee. p. 175–184. Proc. Beltwide Cotton Prod. Res. Conf., St. Louis, Mo.

Wiese, A. F. 1978. Pernicious weeds in cotton—field bindweed. p. 129. Proc. Beltwide Cotton Prod. Res. Conf., Dallas, Tex.

----, and J. M. Chandler. 1979. Weeds. p. 232–238. *In* W. B. Ennis, Jr. (ed.) Introduction to crop protection. Am. Soc. Agron., Madison, Wis.

Williford, J. R. 1981. A review of postemergence equipment available to the grower. p. 170–171. Proc. Beltwide Cotton Prod. Res. Conf., New Orleans, La.

----, O. B. Wooten, R. S. Baker, and W. L. Barrentine. 1973. Water-shield flame cultivation for cotton and soybeans. Mississippi Agric. Exp. Stn. Info. Sheet 1217.

Wills, G. D. 1977a. Nutsedge deals misery to cotton growers. Weeds Today 8:16–17.

----. 1977b. Pernicious weeds in cotton—purple nutsedge. p. 164. Proc. Beltwide Cotton Prod. Res. Conf., Atlanta, Ga.

Wooten, O. B., and J. T. Holstun, Jr. 1959. Special equipment for chemical weed control. Mississippi Agric. Exp. Stn. Bull. 576.

WSSA Herbicide Handbook Committee. 1979. Herbicide handbook of the Weed Science Society of America. Fourth Edition. Weed Sci. Soc. Am., Champaign, Ill.

10 Harvesting

Rex F. Colwick
ARS-USDA
Mississippi State, Mississippi

William F. Lalor
Cotton Incorporated
Raleigh, North Carolina

Lambert H. Wilkes
Texas A&M University
College Station, Texas

The recently developed technology of harvesting cotton mechanically probably had a greater effect on the cotton industry in the USA than any other since the invention of the cotton gin. Although a patent was issued for a cotton picker as early as 1850, a breakthrough that started the complete mechanization of cotton production did not occur until a workable cotton picker was introduced by International Harvester Co. in 1942. A stripper harvester was marketed by Deere and Co. in Texas in the early 1930s, but its use did not become widespread until the 1940s (Smith, 1938; Hagen, 1951). Progress in mechanizing the U.S. cotton crop can be measured by recording the percentage of cotton mechanically harvested in the 10-year intervals following the introduction of the spindle harvester in 1942. We found some firm estimates of mechanical harvesting collected by the National Cotton Council in 1954 which indicated that about 22% of the U.S. crop was harvested by machine that year. Recalling the rapid proliferation of mechanical harvesting, we extrapolated that in 1952 only about 10% of the U.S. cotton crop was mechanically harvested. Then, only 10 years later, 59% of the U.S. crop was mechanically harvested. Only 10 years later, 99% of the U.S. crop was mechanically harvested (USDA, 1962, 1972). Since that time, harvesting by hand has been less than 1% in the USA. By comparison, the USSR harvested by hand about half of its crop in 1976 (USDA, 1977).

Harvesting technology, as well as advances in pest control and other phases of production, has reduced the man-hour requirements for producing a bale of cotton from about 140 to a range of 7 to 20 man-hours per bale, depending on the geographical area and type of production. Man-hours per hectare range from about 12 to 15 in Alabama and Georgia to about 45 for irrigated cotton in Arizona (McArthur and Cooke, 1980).

10-1 MACHINES

10-1.1 Types and Principles of Operation

Mechnical cotton harvesters in use in 1980 were of two basic types, pickers and strippers. They harvested about 62 and 37%, respectively, of the U.S. crop in 1980. A third machine was used in gleaning seed cotton dropped on the ground by machines or blown there by storms. Gleaning has remained at about the 1% level of the crop for the past 10 years (Glade and Cole, 1980).

Harvesters of the picker type harvest cotton from open bolls and leave incompletely opened bolls and empty burs on the plant. Harvest operations can be started with these machines before all of the crop is open, and they can be used more than once per season to harvest the crop as it progressively matures. The picking is accomplished by revolving spindles. Spindle harvesters on the market today use either tapered, barbed spindles or straight spindles to remove the seed cotton from the bur. The tapered spindle has three rows of machined barbs or teeth designed to engage and hold the cotton fiber while the seed cotton is wrapped around the spindle. The straight spindle may have one or two rows of machined teeth. In some cases square straight spindles are used to harvest extra-long staple cotton. Multiple columns of rotating spindles are arranged around a rotating drum which projects the spindles into the plant. The drum mechanism is synchronized to forward travel such that spindles possess no forward motion with respect to the plant until they have been withdrawn. As the spindles are withdrawn from the plants with the seed cotton wrapped around them, a rotary or stripper doffer brushes the seed cotton from the spindle into a pneumatic system which conveys the seed cotton into a basket. When the basket is full, the seed cotton is dumped into a transporting vehicle or other receptacle for transport to the gin or for temporary storage. Spindle harvesters on the market today are all self-propelled and mostly two-row machines; however, four-row machines have been introduced into the market recently.

Stripper harvesters strip the entire plant of both open and unopened bolls. They must, therefore, be used in a once-over operation after all of the crop is mature. This type of harvesting was first referred to as "sledding," because one of the first stripping devices was a sled with fixed stripping fingers positioned in front, parallel to the direction of travel and inclined upward to the rear. As the sled moved forward, the stripping fingers applied an upward combing action to remove the cotton bolls from the plant while the plants were permitted to pass between adjacent fingers. Some of these horse-drawn sleds with stripping fingers made of fence pickets were used in Texas as early as 1914 (Smith, 1964).

Modern strippers can be classified according to two general stripping principles—the finger principle, similar to the old sled type, and the roll whereby the plants pass between two inclined rotating rollers. The stripping action of the roll is similar to that of the snapping action of a corn picker. Stripping rollers are made of steel and, more often, have flights of resilient materials such as nylon brushes and rubber strips. Most strippers being sold today are self-propelled; however, tractor-mounted models are still available. Two-row units are the most common but four-row strippers have been introduced into the market in the past several years. Most of the modern strippers have some type of boll separator or burr extractor as part of the conveying system.

Ground retrievers, or gleaning machines, first appeared in Arizona in the late 1950s, and are still being used to some extent in the irrigated areas of the Far West. These machines usually use notched belts which open notches as they rotate around a roller at ground level and pinch the seed cotton as they close when they leave the roller. Conversely, the seed cotton is discharged as the notches open again over a top-roller. Seed cotton harvested by these machines is refered to as "ground cotton" in the trade.

10–1.2 Application by Location and Culture

Use of harvest machines varies by regions. The spindle harvester is used predominantly in the medium to high rainfall areas of the South and Southeast, and in the irrigated cotton of the Far West. Fifty to 75% of the spindle picked crop is harvested twice (McArthur and Cooke, 1980). Spindle harvesters can be used on cotton plants of practically any size, but stripper harvesters are usually better adapted to plants less than 1.2 m tall. Stripper harvesters are usually used on the kinds of cotton generally grown in the Black Lands and Rolling Plains of Texas and in the High Plains of Texas and Oklahoma. Stripping can be used in other areas of the Cotton Belt when conditions are such that plant size is not excessive and the crop matures uniformly and early, and where satisfactory desiccation or defoliation can be achieved either by chemicals or frost (Colwick et al., 1979).

10–1.3 Performance Characteristics

Field losses from the spindle picker normally range from 5 to 15%. The average field losses are usually estimated to be about 11%. This represents 121 kg of lint per hectare left in the field in cotton yielding 1100 kg/ha (two bales/acre). Losses from the stripper harvester range from 2 to 5% (Colwick and Williamson, 1968). The trash content of seed cotton harvested by spindle pickers normally averages about 6%. Trash and bur content of stripped cotton averages about 25% (Pendleton and Moore, 1968).

Performance rate of both pickers and strippers varies from area to area and by the method of estimation. Parvin et al. (1981) used a performance rate for the two-row spindle harvester of 1.85 hour/ha (0.75 hour/acre) for the first harvest and 1.35 hour/ha (0.55 hour/acre) for the second harvest for the mid-South. The USDA Economic Statistics and Cooperative Service Farm Enterprises Data System (FEDS) uses an average figure of 2.30 hour/ha (0.932 hour/acre) throughout the South and Southwest (Krenz, 1981). The FEDS calculates performance rates based on the mid-points of the ranges in speed and field efficiency given in the American Society of Agricultural Engineers Yearbook. The FEDS gives the performance rate of 1.69 hour/ha (0.684 hour/acre) for a self-propelled, two-row stripper in the Texas High Plains and the same for two-row tractor mounted strippers in the Texas Black Lands and Edwards Plateau areas. For the four-row stripper, they give a figure of 0.84 hour/ha (0.342 hour/acre) in the Texas-Oklahoma area (Krenz, 1981). Actual performance will have wide variations from these average figures caused by the age and make of the machine, the yield of the crop, terrain, row length, weather, field condition, and operator control.

10-1.4   Cost Elements

Cost elements also vary from farm to farm depending upon the annual machine use, the area harvested, the yield of the crop, the age and condition of the machine, etc. Average figures based on the wide data gathering of FEDS are close to more localized studies such as those given in experiments in Mississippi (Parvin et al., 1980). Typical cost figures collected by FEDS and the Mississippi Experiment Station are given in Table 10-1. The "second pick" costs from Mississippi are based on the estimate that 65% of the field is harvested twice. In a more recent study Parvin et al. (1981) estimate harvesting costs about 20% higher in 1981 than the 1979 figures given in the table.

10-2   CULTURE AND MANAGEMENT PRACTICES AFFECTING MECHANICAL HARVESTING

10-2..1   Land Selection and Field Configuration

Mechanical harvesters, as well as other modern mechanized equipment, owe their success in large measure to more precision and timeliness in performing their functions. Timeliness and precision are not possible unless the land is adequately prepared. The first requirement is water control.

The steps necessary for water control are different but equally important whether for irrigation, drainage, or conservation. In the Piedmont, Coastal Plains, and hill areas of the South, where some of the cotton land is rolling or hilly, soil and water conservation is the primary objective of water control, but drainage and irrigation are also considerations in some areas.

Table 10–1. Performance and cost estimates of cotton harvesting machines; 1979.

Area	Machine type	Performance rate	Cost	
		hours/ha	$/hour	$/ha
Texas High Plains†	Stripper 2-row	1.689	56.90	95.88
Texas Rolling Plains†	Stripper 4-row	0.844	61.35	76.52
Texas Blackland and Edwards Plateau†	Stripper 2-row§	1.689	17.78	30.03
Rio Grande Valley dryland†	Picker 2-row	2.304	56.88	131.05
	Stripper 4-row	0.844	71.34	60.22
Oklahoma irrigated†	Picker 2-row	2.304	57.10	131.58
	Stripper 4-row	0.844	71.93	60.76
Arkansas Delta†	Picker 2-row	2.304	61.31	141.31
Mississippi Delta†	Picker 2-row	2.304	62.13	152.40
Mississippi Delta‡	Picker 2-row			
	1st harvest	1.852	68.49	126.86
	2nd harvest	1.358	44.91	61.00

† Excerpted from Krenz (1981).
‡ Excerpted from Parvin et al. (1980).
§ Tractor mounted, remainder are self-propelled.

Mechanical harvesters can be used on sloping land provided suitable drainage and erosion control systems are used. Terracing systems that lend themselves to mechanization and conservation have been improved, but as the size of row-crop machinery has increased, the problems of correlating efficient machine use and conservation practices are more difficult. The improved systems are designed to include water disposal outlets in the major draws, and straighter and more evenly spaced terraces and rows. The water disposal outlets should be flat and shallow to permit crossing with the harvester. The terracing system should be designed to provide the greatest possible number of long, straight, or gradually curving rows. The elimination of point rows, the layout of continuous rows made possible by crossing the outlets, and elimination of sharp turns in the rows all contribute to better operating and harvesting efficiency.

The number of turns at row ends per hectare is inversely proportional to the length of rows. At a speed of 4.8 km/hour turning time as a percentage of total operating time decreases from about 18 to 4% when row length is increased from 100 to 300 m (Renoll, 1979).

In the mid-South Delta area, most of the cotton land is relatively flat. Drainage is the most important consideration there. Since 1950 great strides have been made in shaping the surface of fields to eliminate low spots where water can accumulate. Land forming for irrigation has been a standard practice in the western areas for many years. Initially, most of the land forming in the mid-South was done to facilitate furrow irrigation, but many farmers soon discovered that the benefits from good drainage were of more value than those from supplemental irrigation. Land forming speeds drying of the entire field and permits more timely operation of all mechanized equipment.

Land forming often includes changes in field layout to give long, straight rows with adequate turnrows. Turnrows should be smooth and wide enough to permit the harvester to turn without stopping and to enter the row straightforward. The time to make each turn can decrease 33% on a smooth turnrow as compared to a rough one. Turnrow width for two-row pickers and strippers should be at least twice the length of the machine (Renoll, 1979).

Fields should be cleared of stumps, large rocks, bricks, scrap iron, and other debris which can damage the spindles and spindle bars of cotton pickers or the conveying systems of strippers. Most pickers are equipped with safety clutches that disengage the drums when obstructions enter the picking zone; however, these clutches often can only minimize the damage rather than prevent it, as it is difficult to devise a system to stop instantaneously the momentum of the picker head. This problem is diminishing as land leveling and more efficient production practices result in cleaner fields.

10–2.2 Crop Residue Disposal and Seedbed Preparation

Proper disposal of crop residue will reduce clogging of mechanical harvesters and production equipment. In some areas early residue disposal is required for pest control. As soon as the crop is harvested, stalks should be cut close to the ground and into small pieces. If this residue is then buried by tillage, it usually decays and causes no further trouble. In some areas where plants are large and decay is slow, the stubble and root section does not decay but remains on top of the ground after tillage. Later it clogs planters and pest control equipment and may be picked up by the harvester. A few machines are available to shred completely the entire plant including the root, but a simple and economical solution to this problem has not been found.

Seedbed preparation affects stand and weed control, thereby indirectly affecting harvester performance. A smooth seedbed at planting time enhances the development of evenly shaped row profiles for efficient harvesting. Since harvesting efficiency has been shown to vary directly with yield, another indirect benefit of good seedbed preparation is sturdy plants and deep root systems that contribute to high yield. In soils where compacted zones or other deficiencies contribute to small, shallow root development, plants are often pulled up by the harvester or fall over after a heavy rain. In these cases subsoiling has been directly beneficial to harvesting and has produced higher yields.

10–2.3 Planting Practices

For mechanical harvesting the most important aspect of planting is uniformity of spacing in the row and between rows, uniformity of plant size and shape, and uniformity of the shape of the seedbed. Many factors are involved in obtaining a good uniform stand of cotton (Wilkes and Corley, 1968).

Accurate row spacing is important for multiple-row harvesters. For most pickers, strippers, and cultural equipment currently available, 96.5 or

101.6 cm (38 or 40 in.) row spacing is most convenient. Experiments with close-row and "broadcast" type spacing have been conducted in several areas, and stripper-type machines have been developed for harvesting these various patterns of plant arrangement (Kirk et al., 1964). The potential of this type of culture varies with climate and the availability of cultivars specifically adapted for this purpose.

As early as 1954, USDA and Texas researchers began to experiment with cotton planted in narrow rows with a grain drill. The limiting factor at that time was difficulty of weed control and the lack of an efficient harvester for gathering the crop grown in narrow rows. A harvester was developed by that group by 1963 (Kirk et al., 1964). In recent years considerable research has been done in all areas of the U.S. Cotton Belt on row spacings closer than the standard 96.5 or 101.6 cm (38 or 40 in.) spacing to influence the size of the plants and in an effort to mature the crop earlier. This activity has led to a number of descriptive labels for this type of planting, such as "narrow row," "broadcast," "short season," and "high pop" planting systems. Most of these systems rely on the finger-type stripper for harvesting, but recent developments in brush type harvesters have produced machines that will harvest multiple 76.2 cm (30 in.) rows with individual brush type row units. In some cases, double rows have been planted on beds spaced 101.6 cm apart and harvested with a spindle-type picker.

Recent work in Arizona and California showed some advantages for narrow-row spacing, particularly in early maturity and yield. This practice has been one method to control excessive vegetative growth by introducing mild water stress and setting the crop sooner. Experimentation with growth regulating chemicals has also been done. Buxton et al. (1977) reported that the number of vegetative branches per plant was reduced an average of one branch per plant for each 74 100/ha (30 000/acre) increase in plant density. The vegetative branches that developed at the higher densities were smaller and contained fewer nodes. This practice placed more mature bolls on the main stem instead of on vegetative branches and resulted in more efficient stripper harvesting.

Research in California indicates improved photosynthetic efficiency in narrow rows providing fiber yields up to 25% greater than yields from conventional row plantings. The crop can be harvested 15 to 60 days earlier. Scheduling the crop in an insect pest management system gives an advantage to the narrow row system (Walhood et al., 1979). Harvesting tests comparing a finger type stripper to a spindle picker gave a 10% yield advantage and a 20% overall production cost reduction to the narrow-row, stripper-harvesting system. There were no practical differences in fiber or yarn properties, and the extra costs of mill processing were considered small compared with the cost savings associated with changing to a narrow-row stripping system (Kirk et al., 1981).

Narrow-row, short-season production systems have generally given favorable yield results in widespread experiments in Texas (Whiteley and Rinn, 1976; Whiteley et al., 1976; Sprott et al., 1976). These experiments were conducted in the Rio Grande Valley, coastal bend area, Brazos River

Valley, and the black land area as well as previously mentioned experiments on the Texas High Plains.

A small amount of research has been done on harvesting narrow-row, "short-season" cotton in the more humid areas of the U.S. Cotton Belt. Trials with an experimental cotton combine were generally unsuccessful in Mississippi (Tupper et al., 1974). Other experiments in the Mississippi Delta with narrow-row cotton have indicated that the adapted cultivars and soil conditions are such that consistently good results are difficult to obtain with stripper harvesters (Tupper et al., 1974; Luckett et al., 1977). Experiments with narrow row production systems in the hill section of Mississippi have been somewhat more successful (Colwick et al., 1979). In a silty clay loam soil, narrow row production hastened maturity in 2 years of a 4-year test and gave yield advantages in 3 years. These tests were harvested with a finger-type stripper. Tests in the Mississippi Delta with a brush-roll stripper have shown that stripping is practical with standard row spacings on the heavier soils in years of medium to light rainfall (Luckett et al., 1977).

Research in Georgia (Baker, 1972) found that planting presently available cotton cultivars in narrow rows or high populations does not offer the farmer any advantages over conventional methods of production.

Plant population and spacing are two of the variables in planting that can have a great effect on mechanical harvesting. Studies were conducted on the effects of plant population and spacing on mechanical harvesting in the 1940s and 1950s. Results of these tests were summarized by Colwick et al. (1965). Some of these results specifically applicable to harvesting conditions will be cited here because they have an important influence on machine efficiency.

In eight tests in Alabama over a 3-year period, the average harvest results showed no significant differences in efficiency among drilled plantings of 36 000, 50 000, 97 000, and 143 000 plants per hectare. However, there were indications in individual tests that plant population had an effect on mechanical harvesting, usually in favor of the higher plant populations. As plant population increased, plant height and limb length decreased, while height of fruiting increased.

Cotton stand investigations at Stoneville, Mississippi, indicated that mechanical pickers can operate efficiently over a wide range of different spacings and populations (Williamson and Fulgham, 1956). In drier years both yield and picking efficiency were reduced when plant populations were above 185 000 plants per hectare.

Plant population studies in California showed that populations in the range of 74 000 to 128 000 plants per hectare were satisfactory for yield and mechanical harvesting. Plant characteristics were more suitable for mechanical picking in the upper part of this range, but where lodging was a problem, the lower end of the range (74 000 to 90 000) was more desirable (Colwick et al., 1965).

These data indicate that a plant population of 99 000 per hectare (about 9 plants/m in 1 m rows) is best for mechanical picking under most conditions. This conclusion is made recognizing that a variation of plus or minus 50 000/ha in the final stand from the suggested figure will not seriously endanger the picking qualities or the yield of the crop.

Plant-population, harvesting studies in Texas and Oklahoma have been chiefly confined to the effects on stripping. Results from Chickasha, Lubbock, and College Station have all shown the same trends in the effects of populations on plant characteristics and stripper performance. Work with drill planting in 1 m rows showed that satisfactory stands for stripping range from 74 000 to 124 000 plants per hectare. It is not necessary to thin a stand that has been drilled to produce a population within this recommended range (Colwick and Williamson, 1968).

Studies of populations for both stripping and picking resulted in recommendations that are surprisingly similar for all areas of the U.S. Cotton Belt. Perhaps of greater significance is the establishment of the fact that, under most conditions, population control is a method of influencing plant conformation within a particular cultivar of cotton. Many of the plant growth characteristics can be predicted for various population ranges. The plant characteristics desirable for machine harvesting, smaller plants and higher first fruiting branches, generally improve as the population increases to certain limits that will vary to some extent under differing soil and climatic conditions. When the upper limits are exceeded, the unfavorable results are often a reduction in yield, later maturity, smaller bolls, and more trash in the harvested cotton (Wilkes and Corley, 1968).

10-2.4 Pest Control

The control of weeds, insects, and diseases has a major influence on mechanical harvesting. This relation is particularly true with weed control. Poor weed control can reduce harvesting efficiency, cotton quality, and yield. Researchers in North Carolina reported that one mature grass plant in 6 m of row could reduce the quality of cotton one grade (Colwick et al., 1960). Studies in Mississippi showed that modern gin machinery including two lint cleaners can remove most of certain types of grass (Williamson, 1963). Classer's grade of cotton harvested from light grass-infested plots was equal to that of the clean check plots in a 2-year test at the Mississippi Delta Branch Station. Recent statistics, however, have indicated significant losses in cotton quality because of grass in the lint. In the 1980–81 U.S. crop year, for example, 281 752 bales were reduced in grade because of grass. This amount was about 2.6% of the crop classed by the USDA (USDA, 1981a). Based on an average discount rate of 379 points, this represented a loss to the farmer of $18/bale, and a total loss of $5 million to U.S. producers (USDA, 1981b).

Classer's grade does not always reflect the total impact of quality reduction. Grass and other foreign matter in machine-harvested cotton require more elaborate cleaning in the gin which reduces turnout and may adversely affect the spinning quality of the fiber. Mills often impose additional discounts. Poor weed control, particularly morningglory and other vines, can seriously reduce harvester performance. Vines become entangled with spindle harvested cotton causing it to be left strung out on the plant. In stripping, vines completely choke the machine by becoming entangled with stripper fingers, rolls, and other parts of the system.

The method of weed control, as long as it is effective, should have no effect on mechanical harvesting or quality of the fiber; however, the manner in which mechanical cultivation is done can affect harvesting efficiency. Herbicides have reduced the undesirable practice of "heavy dirting" late in the season which often placed rocks and clods in the drill row. "Heavy dirting" shortens the space between the soil and the lowest fruiting branches. Only a slight elevation in the drill row to prevent the accumulation of defoliated leaves is desired. Where it is necessary to keep a deep irrigation furrow in the row middle, special effort should be made to avoid throwing up a shoulder on each side of the row which forms a depression at the base of the plants to catch leaves. Such shoulders also prevent the harvester from getting low enough to pick low bolls and encourages "plowing" with the picker or stripper head, resulting in rapid wear of lower spindles and stripping rolls. The use of bed shapers and more precision in planting and cultivating will reduce this problem.

Insect and disease control have an indirect bearing on mechanical picking through yield reduction. Tests have shown that picking efficiency increases with yield (Parker et al., 1963). It follows that any insect or disease that reduces yield also reduces picking efficiency. Some of the decrease in efficiency is due to the picker's selectivity in leaving "tight locks" and damaged, shrunken bolls in the field. This result is desirable in most cases, as the tight locks and damaged bolls may detract more in quality value than would be gained from the additional weight. The tight lock problem offers a fertile field for biological research, particularly in the humid areas where tight locks have occasionally represented 30% of the yield.

10-2.5 Plant Characteristics Affecting Harvesting

Cotton breeders have been concerned with the relationship of the characteristics of cultivars to mechanical harvesting since machines became practical. Harvesting engineers have endeavored to describe the plant characteristics desired for mechanical harvesting. Plant characteristics conducive to efficient picking vary drastically from those conducive to stripping. Picker performance tests in a number of cultivars have been conducted in several states. Generally speaking, these tests have shown that a high yielding cultivar of early-maturing, slightly storm-resistant bolls borne well off the ground on a strong central stem, is desired for mechanical picking. Since cotton intended for machine harvest is left in the field until a sufficient amount is open to justify operation of a mechanical picker, storm resistance is a very important characteristic to consider in choosing cultivars for machine harvesting. In areas where tropical disturbances often occur at harvest time, weather loss and overall harvesting efficiency can vary as much as 13% among recommended cultivars (Corley and Stokes, 1964).

A "stripper cultivar" should have enough storm resistance to keep the seed cotton in the boll until harvested. The amount of storm resistance required will vary among areas depending on the normal pattern of fall weather and the length of the growing season. For example, a "storm-proof" boll may be required to withstand normal winds in the High Plains of Texas, whereas a medium storm-resistant boll may satisfactorily withstand normal winds in the mid-South.

Other important plant characteristics for stripping are size and shape. The size of the plants and the maturity of the crop should be uniform, since these conditions promote efficient harvesting and better fiber quality. Large, spreading plants cause more harvesting difficulties than do smaller, compact plants. Most strippers will operate more successfully on plants of moderate height. Plant size and shape is influenced to some extent by cultivar characteristics, but to a greater extent by soil type, fertility, moisture, and plant spacing. Some of these variables can be controlled.

Research has shown that different degrees of storm resistance are needed not only between picker and stripper areas, but also among different areas within so-called picking regions and stripping regions. In the Mississippi Valley area, for example, the most open-boll types with low picking force requirements (peak force of 100 g for seed cotton removal) gave the best overall harvesting efficiency (Barker et al., 1976). In the Southeast Coastal Plains area, the medium storm-resistant types gave the best overall efficiency because of lower pre-harvest losses than open-boll types (Corley, 1970; Fig. 10–1). If research could establish optimum storm resistance or

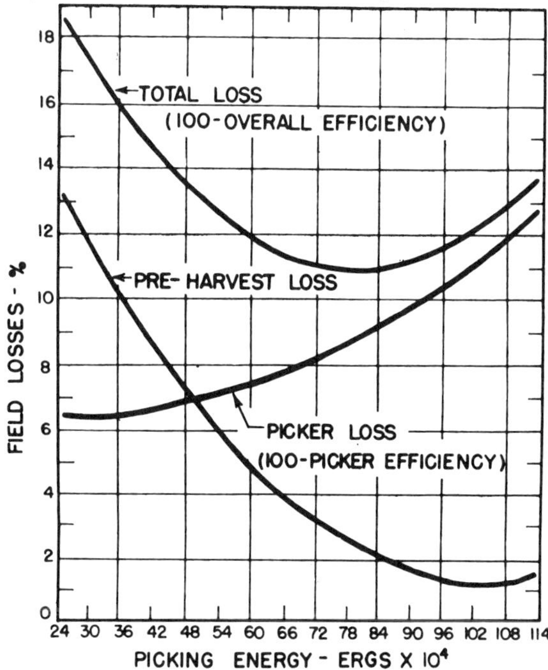

Fig. 10–1. Relation of required lock removal energy with field losses for a once-over harvest of 13 cultivars for 3 years (Corley, 1970).

picking energy requirements in more climatic areas of the U.S. Cotton Belt, they should be of value to breeders. More precise specifications of other plant physical characteristics should be valuable also. Measurements of stalk stiffness in areas where lodging is a problem is a good example.

The development of smooth-leaf cultivars by cotton breeders has helped to minimize grade loss in machine-picked cotton. Cultivar studies under mechanized production and harvesting practices at Stoneville, Mississippi, showed grade advantages for two smooth-leaf cultivars (Dick et al., 1958). In a 3-year study, lint from one smooth-leaf cultivar graded almost a full grade higher than the average of 11 commercial cultivars. Leaf smoothness had no effect on picking efficiency nor on the total amount of wagon trash, but the smooth-leaf trash was easier to remove in the gin.

10–3 PRE-HARVEST PREPARATION AND TIMING OF HARVEST

10–3.1 Plant Preparation

The genetic and cultural aspects of plant configuration are intensively discussed in the chapters on genetics and production practices. When considering harvesting practices and planning harvesting systems, it is necessary to be fully conversant with the entire production system to understand how the performance of the harvesting system can be affected.

Cotton is a perennial and the earliest, unimproved plants used commercially were large, woody, spreading bushes. Genetic selection has lead to the basic plant types in use at present. Some of these are quite small and compact while others can reach heights over 175 cm. Plant size and shape are related to the cultural system as well as to genetics. Restrictions on acceptable plant size and shape are imposed by the need to harvest the crop mechanically. One machine cannot feasibly be made to accomodate all plant types.

Pre-harvest preparation of the crop helps obtain the best performance from mechanical harvesters, in terms of seed and fiber quality and cost effectiveness. Cotton is best harvested when the seed-cotton mass has a moisture content of about 12% or less and so that almost no leaves or other plant parts contaminate the seed cotton. Defoliation by chemicals is an established aid in attaining this objective. Defoliant chemicals hasten formation of an abscission layer in the petioles and thus hasten the leaf-drop phenomenon. When leaves begin to drop due to natural occurrence of abscission, as happens when almost all the bolls have opened, little advantage is gained from defoliation (Stapleton et al., 1967).

The use of harvest-aid chemicals makes harvest management easier because the microclimate in the rows is warmer and dryer in defoliated fields than in undefoliated fields. This permits starting the harvester earlier in the mornings and helps reduce boll rot and fungus growth.

Timing of defoliant application is important. Defoliants should be applied only after 65 to 75% of the bolls are open. Earlier application reduces yield and prevents full maturation of the fiber (Bilbro and Ray, 1974). Later application is not advantageous because natural leaf drop begins to occur. The time required for defoliants to act depends on weather. Formation of the abscission layer is a metabolic process that is slow in cool weather. When plants are killed early by frost, the abscission layer will not form and dead leaves can remain on the plant. When harvest is delayed by rainy weather, especially when temperatures are high, regrowth of defoliated plants occurs, and the new leaves cannot be effectively removed with defoliants. Application of a desiccant is needed to kill regrowth. Regrowth due to weather conditions is sometimes unavoidable. It is therefore prudent not to defoliate an area bigger than can confidently be harvested within 2 weeks. This minimizes the likelihood of regrowth.

When cotton is stripper harvested, vigilance about preventing high moisture in the mass of bur-cotton is essential. Each 1% increase in content of green material causes a 1% increase in moisture content of the harvested mass (Brendel and Miller, 1979). Desiccation by applications of compounds like arsenic acid is an aid to reducing the moisture content of the foreign matter, and hence of the bur-cotton. Bur-cotton that is intended for storage in a rick or module should be desiccated.

Frost also causes desiccation. In the Texas High Plains, application of chemical desiccants is often not needed when a timely, killing frost occurs. An early, killing frost or premature application of desiccants causes yield and quality reduction. Desiccants should be applied only when 95% of the bolls are open and stripping may begin 1 to 2 weeks later when the plants are dry. Where big plants with a dense leaf canopy are to be desiccated, application of a defoliant to remove some leaves before applying the desiccant is recommended.

Should restrictions on the use of desiccants be imposed because of environment-protection considerations, defoliation is a possible alternative to desiccation for stripper harvesting (Brendel and Miller, 1979). Moisture content of defoliated bur-cotton is almost certain to be higher than that of desiccated bur-cotton, so added caution is required when managing a stripper-harvest system without desiccants.

A new type of pre-harvest strategy has been evolving during recent years. It involves the use of growth regulators to affect plant size, to advance maturity, to terminate fruiting, to open bolls, and to cause senescence. Techniques for successful use of this technology are being developed, stimulated by benefits believed to have important implications for yield improvement, insect control, and harvest efficiency.

Control of vegetative growth and improvement of fruit set have been achieved by early-season applications of growth regulators (Walter et al., 1980; Cathey et al., 1982). This procedure not only increased yield and kept the canopy more open, but it also facilitated harvesting, which is always more difficult in a rank-growth crop than in a more compact-growth crop.

Hastening of boll opening and shedding of immature bolls can be achieved by late-season application of growth regulators (Kittock et al., 1979). Early, uniform boll opening results in highest yield at first harvest and aids toward attaining the objective of completing harvest in one picking. This practice brings important cost savings. There is evidence (Luckett and Cathy, 1979) to support the belief that fiber quality might be slightly compromised by this strategy, but the flexibility to trade low harvest cost for a slight decline in market value is a feature that appeals to managers.

Shedding of immature bolls and termination of fruit production have important insect-control implications that are discussed by Kittock et al. (1979). Production of immature bolls was reduced by up to 95%, which significantly removed conditions essential to overwintering of pink bollworms. Termination of fruiting in this manner was found to be better than defoliation or desiccation in controlling pink bollworms (Kittock et al., 1979). Termination is also believed helpful in controlling other insects. Strategic management of irrigation can have the same results as application of growth regulators in terminating a crop, but the result is less certain. Both chemical and cultural termination can have adverse effects on yield (Kittock et al., 1979; Farr and Kittock, 1979). The use of growth regulators and other termination techniques, however, will likely have an important bearing on defoliation, desiccation, and other harvest-preparation strategies. These methods in turn will affect harvester design, harvesting methods, and above all, harvest management.

10–3.2 Timing of Harvest

For any cotton-growing location, profit from the crop depends on a manager's success or failure to carry out harvesting in a timely manner. Deterioration from weathering after bolls open must be avoided. In the Texas High Plains for example, a yield loss of approximately 10% occurred during the first 5 weeks after cotton was deemed ready for harvesting (Ray and Minton, 1973). Data from Arizona, on the other hand, show that weathering did not necessarily have a significant effect on yield of fiber or seed except that ground loss tended to increase with time (Buxton et al., 1973). High temperatures and high humidity after bolls open lead to an increased rate of deterioration in yield and quality. Hence the greater need for timeliness in the rainbelt cotton-growing areas than in the irrigated areas.

Barker et al. (1979) showed that a definite deterioration in fiber quality with respect to time took place in Mississippi. The exact nature and rate of deterioration was not clearly identified, but data trends suggest that most of the deterioration occurred during the first 4 weeks after any given boll opened. Barker's research dealt with the effects of weathering on individual bolls rather than on the crop as a whole. Exposure of about 6 weeks of fall weather in Mississippi caused a drop in classer's grade from middling-plus to low-middling-plus. Barker and his colleagues outlined a procedure and constructed a mathematical model for predicting deterioration of fiber that is exposed to measurable weather patterns.

Burch et al. (1979) observed cotton yield, quality, and market value with respect to delayed harvest for 4 years in South Carolina. The effect of delays is similar to that observed in Mississippi. Yield at first harvest can be reduced by beginning when too few bolls are open. Beginning too soon also causes good, unopened bolls to be shed, and total yield is reduced. Total yield reaches a peak during November in South Carolina, but exact timing depends on weather. Revenue above cost of harvesting, handling, and ginning can drop by 1.0% per day for 15 days after peak revenue is reached and the decline continues for several more weeks at about 0.5% per day. Seed quality also declined as the harvest was delayed.

Climate and weather can also have other effects on the crop. For example, early termination of growth by a killing frost reduces yield and can have very serious effects on quality (Bilbro and Ray, 1974; Burch et al., 1979). Similar effects can be produced by applying defoliants or dessicants too early. When defoliants are used, their action can be drastically modified by weather conditions, even when application of the defoliant has been timely. Cool weather slows the action of the defoliant; good growing weather permits regrowth of leaves. Strategies, including choice of chemicals, that are effective in one location might not be suitable in another.

Stapleton et al. (1967) studied crop development in Arizona during the period approaching harvest time. They observed the results of chemically defoliating the crop, and they based a harvesting-timeliness model on the data collected. The model suggests that for Arizona conditions, defoliation and first harvest when 75 to 90% of bolls are open produced highest total crop value. First harvest done at a later date resulted in less yield and lower grades. When first harvest is deferred, a second harvest may not be an economical operation.

Barker et al. (1976) made a 3-year study of harvesting-ginning systems in the Mississippi Delta. They examined time of harvest and other system parameters. Twice-over harvesting of undefoliated cotton (first harvest at 85% bolls open) gave maximum yields. Delaying first harvest until approximately all the cotton was open reduced fiber strength, length, and reflectance, but it also reduced trash content and seed damage. Even though defoliation did not necessarily produce highest yields or maximum profit, it permited cotton pickers to operate for about one extra hour per day because of a drier microclimate within the crop. The timeliness effect of this procedure (especially if it means that severe weather conditions are avoided) might offset the cost of applying defoliant and could provide extra assurance of avoiding timeliness-related losses.

From the foregoing review, it is evident that harvesting strategies, including choice of chemicals and the timing of their application, can be highly effective in one location or year and might be counterproductive in another location or year. Managers involved with the design of harvesting systems to produce optimum fiber quality and maximum profits should consult the literature and other information services made available by their Cooperative Agricultural Extension Service.

Crop management to synchronize maturity dates and harvesting operations with climate and weather is one aspect of timeliness. The daily management of a harvesting system is another. On a daily basis, the moisture content of the crop and the availability of storage or transportation limit the amount of harvesting that can be done. The constraint from high moisture is common in the rainbelt and in areas with high humidity at night. As a management guide Huitink (1979) suggested that ambient relative humidity needs to be below 70% for moisture content of fiber to be appropriately low for harvesting. In the mid-South, fiber moisture content is seldom low enough for harvesting before 0800 or after 2000 hours. The constraint arising out of early-morning and late-evening humidity is less restricting in defoliated fields than in undefoliated fields. Cotton harvested too damp suffers a disadvantage from which it never fully recovers regardless of the drying effort applied during ginning (Cocke, 1974). Non-lint content and lint color from crops harvested wet are usually inferior to those of the same crops harvested dry. A hand-held moisture meter is a good device for determining when moisture content is low enough for harvesting.

Timing of harvest with respect to moisture content requires special attention when the seed cotton is to be stored for more than a day before ginning. In striving for maximum daily harvesting rate in a system where seed cotton is stored, cotton harvested early and late in the day might be too damp for storage and should be dumped into trailers or made into a module for immediate ginning. Arrangements should be made to gin that cotton before deterioration occurs.

Defoliated crops should be harvested after the defoliant acts and before regrowth of new leaves begins. Temperatures below normal mean that defoliants and dessicants will have a slower action than when temperatures are normal. The defoliating effect of a chemical is usually complete within about 7 days after application. Because defoliants do not kill the plant, regrowth of the leaves can occur and dessication may be necessary to deal with the new leaves. To avoid problems that arise out of regrowth, cotton should not be defoliated more than 2 weeks before harvest. Daily harvesting capacity with respect to time of application of defoliants or dessicants is therefore important.

Chemical desiccants are often used to prepare crops for stripper harvesting. They kill the cotton plants, and their effects are usually complete within about 5 days. More time may be needed for the plants to dry out. Regrowth should not be a problem when the desiccant reaches all the leaves. When harvesting is delayed in a crop that has been desiccated or killed by frost, the plant parts become brittle and the quality of the lint can be reduced by bark and other trash. This condition is a more severe problem when stripping cottons bred for rainfall or irrigated growth than when stripping cottons bred to have small lateral branches that are well suited for stripper harvesting.

It is a management function to ensure that harvesting systems are designed and operated in a way that permits timely efficient movement of the crop to storage and ginning. The number of suitable harvesting hours available during the harvest season varies from year to year at a particular location. Reduction in harvesting hours because of bad weather or inadequacy of harvesting system creates the need for high harvesting capacity per hour. The cost of owning sufficient harvesting capacity should be balanced against the cost of suffering crop losses when insufficient harvesting capacity is available. A method for finding the optimum amount of harvesting machinery in relation to local weather conditions and the area of crop to be harvested has been suggested (Sanders and Lalor, 1972). This approach treats the weather pattern as a stochastic variable that creates a demand for harvesting capacity. Crop area per harvester is optimized with respect to local weather conditions. The theory of this approach fails when deterioration occurs during long periods of bad weather when loss caused by delay cannot be prevented by any amount of harvesting capacity.

McClendon et al. (1981) proposed a different model for managing the harvest system. They used data from Barker et al. (1979) to construct a model that simulated what happened to revenue above harvest cost as the area per two-row harvester was changed. The data apply to the Mississippi Delta. Potential of a two-row harvester to maximize the revenue from a given area is partly dependent on weather. Weather records for 10 years (1960–1969) were available and use of the simulation model with those weather records suggested that a two-row, spindle picker should be allocated to 120 to 160 ha. Allocations at the lower end of the range were usually more profitable, especially when cotton prices are high.

These calculations by McClendon et al. (1981) also show that first harvest at 65 to 70% open bolls gave highest returns when averaging over simulations of 10 years. In the worst-weather year, dollar returns above harvesting cost were independent of harvest initiation date once boll opening had reached 60%.

The available harvesting capacity varies greatly from one producing area to another. In parts of California producers plan to finish their entire first harvest in 30 days, and they own the equipment needed to accomplish this. In parts of Arizona where some gin communities are equipped to pick their seasonal volume in about 14 days, the average area picked by two-row spindle pickers was 96 ha per year (Hathorn et al., 1973). Data collected in the Mississippi Delta show that as much as 65% of the total annual volume can arrive at a gin within a period of 21 days (Lalor, 1976–1977). Seed cotton arrivals at Arizona gins are similarly distributed (Stapleton et al., 1967). Rapid harvesting in South Texas to avoid loss from storms on the Gulf Coast is an important management strategy. Seed-cotton storage is almost essential for the needed harvesting rate to be attained.

A stripper harvester can harvest more area per hour than a spindle picker of similar operating width. Furthermore, stripper harvesters are generally used in areas, such as the Texas High Plains, where relative humidity at harvest time is low. Simulation modeling of harvesting on a Texas High Plains farm showed that optimum harvesting on 365 ha of cotton would be accomplished in 41 days with two, two-row stripper harvesters, four conventional cotton trailers and a ricking system for seed cotton storage (Childers, 1980). This is an area of 183 ha per two-row stripper. Without a ricking system to store seed cotton, the same harvest would take 75 to 80 days. If a palletless module storage system were used in place of the ricks and trailers, harvesting could also be accomplished in 41 days, but at an extra cost of about $11.00/ha (Childers, 1980).

Seed-cotton storage is a method for enhancing the effectiveness and timeliness of a harvesting system. The complement of equipment, however, needed to accomplish maximum profitability is dependent on location and can be most easily identified with the aid of good record-keeping and data-processing capability. The harvesting model developed for Texas (Childers, 1980) has severely limited applicability because of lack of a reliable data base as a starting point. The rapid change in record-keeping and data-processing technology resulting from availability of computers will increase the usefulness of decision models. The aids to managing harvests based on computer models are being modified for adaptation to advances in knowledge about crop development and to new computer technology.

"Time is money" during harvesting. Farmers and ginners are aware of the need for a timely harvest and helping them identify the most profitable way to meet this need continues to be a challenge.

10-4 CARE AND OPERATION OF MACHINES

A large part of field losses is the result of improper adjustment of machines and careless operation in the field. Keeping the harvesting machine in good condition and operating it properly are essential to general harvesting efficiency and the important objective of controlling moisture and trash in the harvested seed cotton. Major machine settings and adjustments should be made by a competent service person in the shop before the harvest season begins. An annual checkup of spindle harvesters to replace worn spindles and other parts and to set the machine to factory specifications is always a good investment as in the thorough training and close supervision of the operator. The machine owner and operator should study the operator's manual to become proficient in detecting and correcting trouble. They should also make use of the material and training available to them from farm machinery dealers and the Extension Service. Extension agricultural engineers in most of the cotton states have published excellent guides for harvesting cotton. Since these guides contain the many details and techniques of machine adjustment and operation, we will cover only some of the highlights of these subjects in a general way as they pertain to efficiency and cotton quality.

With a stripper machine, the plant lifters and stripping rolls or fingers should be adjusted to pick up a minimum of trash from the ground and a minimum of bark and branches from the plants. If there is a boll separating device, it should be adjusted to remove as many unopened bolls as possible.

With a spindle harvester, cleanliness is the watchword. The manufacturer's lubrication guide should be followed, and excess oil and grease should be cleaned from the machine after each lubrication of the picking head. As with strippers, plant lifters and the height of the picker head should be adjusted to prevent picking up soil and trash from the ground. Each time the spindle picker stops to dump the basket, the operator should clean out the picker head doors and the drum area by hand. The picker should also keep the picker basket free of lint streamers and be careful not to let these get into the clean seed cotton. Cleanliness around belt drives will reduce fire danger.

Adjustment of the spindle moistening pads and doffers should be inspected visually every time the machine stops to dump the basket. Poor picking efficiency or dirty spindles may be a sign of improperly adjusted moistening pads and doffers. When this effect becomes obvious from observing the seed cotton left in the field by the picker, the picking efficiency is suffering seriously.

The pressure plates in the picker head should be set with clearance and tension to suit the size of the cotton plants and the condition of bolls. The amount of moisture added to the seed cotton by the picker can be kept at a minimum of only 1 to 2% if care is taken in adjusting the spindle moistening pads and the flow of water to them. Only enough water to keep the spindles clean should be used. The flow should be adjusted throughout the day. Less water will be required early in the day and late in the afternoon when the seed cotton moisture is higher. Wetting agents in the water help keep spindles clean with slightly lower rates of water. In some areas textile oils are used instead of water for moistening spindles. They keep the spindle and picker head clean but may reduce the picking efficiency of the machine; also, the small amount of oil left in the lint after ginning is not desirable in the spinning mills (Colwick and Shaw, 1962).

The skill of the operator naturally plays a large part in harvesting clean, dry seed cotton. Higher wages are justified for a well-trained operator who takes pride in accurate driving and good care and adjustment of machines.

10-5 HANDLING OF SEED COTTON

After the seed cotton has been removed from the plant by the harvester, the harvested material is transported from the field to a gin where the trash and seed are separated from the lint. Cotton that is grown in the USA has been traditionally ginned before it goes into the marketing channels. The gins are basically central processing plants located within cotton producing areas. The volume that is processed through individual gins varies from 3000 to 40 000 bales per year depending upon many factors such as ownership, location, number of gins within the area, and the capacity of the machinery within the gin.

The situation that exists in the production system is that ginning capacities in terms of bales per hour or day are not always equal to the harvesting capacities. In most areas the harvesting capacity is much greater than the ginning capacity, since the cotton should be harvested as rapidly as possible to minimize the loss from field and weather exposure. The gin is independent of the seasonal nature of harvest and weather conditions, and its efficiency is highly dependent upon a steady delivery of seed cotton. The total volume that is processed through a gin must be great enough to keep it operating as an economical unit, and with present fixed costs, it is impossible to obtain a ginning capacity that is equal to harvesting capacity by increasing the gin machinery required to reach this level.

10–5.1 Storage

Storage of the harvested seed cotton is a practical solution to the dilemma involving the two rates of processing that exist between the harvesting and ginning phases. The storage system must encompass receiving, delivering, handling, and maintaining of the quality of harvested seed cotton. The receiving capacity of the system should accept the seed cotton from the harvesters with a minimum delay in discharging the seed cotton accumulated in their baskets. It must be flexible enough to follow the progress of harvesters across a field and among field and farms. The delivery aspects of the system involves transporting the seed cotton to the gin at a steady rate regardless of harvesting operations and weather conditions. Handling is involved from the time the system receives the seed cotton from the harvester until it is conveyed into the gin. Therefore the form in which the seed cotton is contained must permit the loading, unloading, conveyance, and storage with a minimum amount of equipment and energy. Seed cotton is highly vulnerable to moisture from rain or other forms of precipitation, and the system of storage must be economical and free of trouble during the time the seed cotton is in the system.

The storage of seed cotton has been practiced for many years. The form or manner in which it is stored depends upon many factors. Some of these include the relationship of the rates of harvesting and ginning, climatic conditions during the harvest season, method of harvest, availability of labor, and most importantly, the economics of the system. Methods of storage include trailers, baskets, buildings, open fields, ricks, compacted stacks, and modules.

Specially constructed buildings for seed cotton are perhaps the oldest form of storage (Ward et al., 1953). These structures provided maximum protection from weather and were designed to facilitate the handling of the cotton into and out of storage. The seed cotton storage houses were used both on the farms and at the gins. At gins special bins or partitions were used to maintain the identity of the seed cotton of each producer. In some cases a system of floor ducts was used to distribute forced air for aeration and drying. The seed cotton was moved into and out of the houses by hand baskets or forks and in later years with pneumatic systems. The primary disadvantages to this system of storage are the capital investment in the structure, and the amount of labor and cost of loading and unloading operations.

Field storage of seed cotton has been a satisfactory method of adapting ginning capacity to the high rates of stripper type harvesting in the High Plains of Texas and Oklahoma (Smith, 1971). This was the first area of the U.S. Cotton Belt to experience the impact of mechanical harvesting and its effect upon the ginning capacity. In the early stages of field storage the seed cotton was simply dumped in piles in the field. Some of the piles were shaped and rounded over the top by hand for drainage. In most years no problems were encountered with quality losses by rain or snow. Some problems were encountered, however, with sand drifts that accumulated on the sides of the stacks. Various methods have been used to load the seed cotton into trailers for transporting to the gin including hand labor and labor saving devices such as mechanical and pneumatic conveyors. Special hydraulically operated loading devices that were attached to tractors were also used. Storage of seed cotton in the field in this manner is limited to areas of low rainfall and requires the additional cost of reloading into the trailer.

The conventional cotton trailer is the most universal system of providing storage between the harvesting and ginning operations. The number and sizes of trailers within any gin community is generally based upon the amount of seed cotton that can be harvested within an uninterrupted period of 3 to 5 days. The trailers, in most cases, are owned by the producers, but in some cases they may be supplemented with trailers owned by ginners. In most areas the harvesting is interrupted by rains, and the gins are able to "catch-up" with the harvesting rate and thereby make the trailers available again for temporary storage. There are many situations, however, where excellent harvesting conditions exist for long periods, and harvesting operations are slowed or even stopped by the lack of trailers. The investment in additional trailers for the purpose of increasing the storage capacity will result in an exponential rate of cost per bale unless a greater total amount of seed cotton is handled in the system.

Trailers are basically designed as transport units and are not used effectively while waiting loaded at the gin for 3 to 5 days. One approach to reduce the cost of trailers is the basket system (Looney et al., 1965). These four-sided, open-top containers have dimensions similar to those of the larger trailers but are constructed without wheels or running gear. The most common use of this system is at the gin where the seed cotton is transferred from the trailer into the basket. The baskets have also been located in the field to receive the seed cotton directly from the harvester. Special trailers are used in loading, unloading, and for transporting the baskets. The system was not widely accepted because of the cost of transferring seed cotton and the cost of the special purpose trailer chassis.

There are three relatively new systems that have been developed within the past 10 years. These include the mechanical ricking system, the Cotton Caddy, and the high-density modular system. Each system has its inherent advantages and disadvantages, and their use and acceptance by the cotton industry depends upon many factors.

The Cotton Caddy system was developed by McNeal (1966) of the University of Arkansas. The system consists basically of forming a stack of seed cotton on a wooden pallet. The machinery and equipment used in the system is relatively simple and inexpensive. It has been used with varying degrees of success in the Mississippi Delta. The machine is used in the field and receives the seed cotton directly from the harvester. One of the principal disadvantages of the system is that the seed cotton must be compacted manually. This fact has presented a real problem in some areas because laborers are unwilling to do this work. Approximately 5 to 7 bales are normally formed on a 7.3 m long pallet.

The mechanical ricker was developed in a cooperative project between Cotton Incorporated and Texas Tech University (Smith, 1971). The ricker receives the seed cotton directly from the harvester in the field and forms a compacted rick along the turnrow. The principal advantage of the ricking system is the capacity to receive the seed cotton from the harvesters without delaying the harvesting operation. The seed cotton is compacted to a density of about 96 to 112 kg/m³ which is sufficient for the rick to maintain its shape and prevent excessive losses from wind. The amount of seed cotton placed in a single rick will depend upon insurance rates, the yield of seed cotton, and other factors. Normally a rick will contain about 25 bales of cotton. It is recommended that the ricks be covered, particularly in the high rainfall areas. It is also recommended that the rick be completely covered with a plastic material so that no wind can get under the covering and damage or remove it. The ricks should be formed on a high ridge to provide drainage away from the seed cotton because it is stored directly on the ground. The principal disadvantage of the ricking system is that a separate operation is required for loading the seed cotton from the rick into the trailers, however, this system extends the effective use of the conventional trailer system. Both front-end loaders that are mounted on standard farm type tractors as well as commercial loaders adapted and built specifically for this operation have been used for loading the seed cotton into trailers. In most cases the special loader is owned and operated by the gin to reduce the investment that might be required by the producer in handling the seed cotton. Some problems have been encountered with picking up excessive foreign material and dirt when removing the seed cotton from the rick. Hand-labor is also required for gleaning the seed cotton from the ground. This system is used extensively in the low rainfall areas of the U.S. Cotton Belt.

The high density module system was developed in a cooperative project between the Texas Agricultural Experiment Station and Cotton Incorporated (Wilkes et al., 1978). The system is similar to the Cotton Caddy system in that the seed cotton can be formed into a stack on a pallet; however, the stack is formed directly on the ground with the module builder. The principal difference between the two is that the density of the seed cotton in the modules is considerably higher, and the seed cotton is compressed mechanically. The module builder and the transport unit are separate in the module system, whereas the transport unit and the forming unit are one and the same for the Cotton Caddy. The separation of the two units permits the module builder to be used more effectively in the field without interrupting the harvesting operation while transporting the modules to a central storage area or to the gin.

Pallets made of wood or metal can be used to form the floor of the module builder in the areas of high rainfall. The pallets are spaced along the turnrow at intervals dictated by the yield of seed cotton. Modules can also be formed directly on the ground in areas of low rainfall. The module builder moves from one location to the next as the modules are formed. The seed cotton is received directly from the harvester by the module builder. A hydraulically operated compaction device is used to level the seed cotton and compact the modules. The seed cotton is compacted to densities of 192 to 224 kg/m³ for machine stripped seed cotton. The module builders are available in lengths of either 7.3 or 9.7 m. Normally about 9 to 10 bales of machine-picked seed cotton are placed on a 7.3 m pallet. About 8 to 10 bales of machine stripped seed cotton are normally contained in a 9.7 m module.

The uniformity of the modules enhances mechanical handling and feeding at the gin. A mechanical feeder was designed and developed at the Texas Agricultural Experiment Station to capitalize on this feature of the modular system. The seed cotton is mechanically dispersed and conveyed directly into the pneumatic system by the feeder unit. Research with this unit has shown that the labor requirements can be reduced significantly along with reducing the total amount of energy required for conveying the seed cotton from the module into the gin.

A primary factor in selecting a site for storage of the seed cotton is the ability to maintain a steady flow to the gin. This means that the seed cotton should be easily retrieved from the storage area regardless of weather conditions. Easy access should be provided to each unit within the storage area so that the status and condition of the seed cotton can be monitored. The storage sites should be well above flood water levels. The areas around each stack, rick, or module should be graded to provide drainage from normal rainfall away from the storage units.

A primary concern of all producers and ginners is the storability of seed cotton. A tremendous amount of research has been conducted all across the U.S. Cotton Belt on this phase. The results of these studies showed that seed cotton can be stored successfully. The principal limiting factor, regardless of the system used, is the moisture content. The most sensitive component in the seed cotton is the cottonseed. If the moisture content of the seed reaches 12% or more a danger of reduction in germination and increase in free fatty acids exists. The fiber can withstand a higher level of moisture without affecting its quality. The principal source of moisture is the trash in the harvested seed cotton. Therefore, special emphasis should be placed upon good production and harvesting procedures.

Regardless of the system of storage, the seed cotton should be protected from rain. The temperature of the seed cotton should be monitored periodically since an increase in temperature will indicate that the moisture content may be too high for storage. The seed cotton that is harvested during the early morning or late afternoons should not be placed in storage. It should be kept in mind that the gin will be continuing to operate during the harvest season, so therefore, the seed cotton containing excessive trash, as well as the seed cotton harvested during unfavorable conditions, should be ginned as quickly as possible. The excess seed cotton can be accumulated during favorable harvesting conditions and safely stored if it is properly protected.

There is no industry-wise standard for insurance coverage against loss by fire. Most gins insure the seed cotton that is in storage; however, there can be major differences in policies. Some insurance policies require that the seed cotton be covered with a tarp and others require a net-type covering.

Seed cotton should be stored in areas that are not heavily traveled, but are easily accessible for inspections. The storage units should also be spaced far enough apart to allow the removal of a single unit in the event of a fire within the unit. As the modules, rick, and other storage units are formed, the gins should be notified immediately as the seed cotton can be included in the insurance policies furnished by the gin.

10-5.2 Transportation

Until recently the conventional cotton trailer was the most common method of transporting the seed cotton from the field to the gin. It also serves as a temporary storage unit at the gin in the event of a backlog. The trailers are suitable generally for short transport distances because of the limited safe travel speed. Trailers are also used to transport seed cotton stored in ricks from the field to the gin. One of the principal advantages of the module system is the lower transport costs (Smith, 1974). Modules that are formed on pallets are winched onto a specially designed tilt-bed trailer and hauled to the gin or to a storage area. Many areas are suitable for forming and storing the seed cotton in modules directly on the ground. Palletless module trucks and trailers have been developed for handling and transporting these modules. Studies have shown that the cost of handling the palletless modules is considerably less than those with pallets, especially if short haul distances are involved (Lalor et al., 1977; Curley and Kepner, 1977).

10-6 HARVESTING AND QUALITY

Increases in productivity of labor and machines lead to improvements in profits. Introduction of synthetic fibers permitted increases in productivity of labor and equipment in textile mills. Cotton producers and processers were then confronted with the challenge of competing with synthetics in mills where profitability is placed above loyalty to any particular fiber. Regulations limiting the amount of respirable cotton dust in textile mills created a quality factor not known before the 1970s (Lalor, 1982). Quality cotton is, therefore, cotton that permits high productivity in textile processing and manufacture. Quality cotton is also that cotton that can satisfy the need for quality apparel and home furnishings among sophisticated, discriminating consumers.

Losses in both quality and quantity of lint can be reduced by timely harvesting. Timely harvesting often means rapid harvesting, and it creates a need for seed cotton storage.

Performance of the ginning and cleaning subsystems is affected by design and performance of the harvesting system. Excessive drying causes fiber damage during ginning and cleaning. This situation occurs when moisture content from foreign matter in harvested material forces ginners to apply harsh drying and aggressive cleaning. When a gin has to be managed to accommodate a harvesting system that does not contain storage, ginning rates are, of necessity, as high as possible and down time is as low as possible so that harvesting is not delayed. Because of high ginning rates and lack of routine maintenance, quality and quantity of marketable fiber can be reduced.

High moisture content at harvest time is likely the biggest contributor to quality-related problems. Quality of cotton harvested damp never fully matches the quality of similar cotton harvested dry, regardless of the drying and cleaning efforts applied in the gin (Cocke, 1974). The incidence of spindle twist (roping that occurs when spindles are partially doffed) is usually higher under high-moisture harvesting conditions. The existence of spindle twist makes seed cotton difficult to process satisfactorily in the gin and causes reduced ginning rates and deterioration of fiber quality.

Appropriate use of defoliants and desiccants permits seed cotton to be havested dry and with minimum contamination from other plant material such as leaf, bracts, and sticks.

Weed control during the growing season is of great importance in helping produce high quality fiber. Grass control is especially important because if grass is in the field, grass is likely to be in the fiber. The physical properties of fibrillated grass parts are similar to those of cotton fiber, and this fact makes them very difficult to separate. The presence of grass and vines in seed cotton causes ginning difficulties and machine chokages. Some grasses have dark-colored seed parts that cannot be bleached and that have been found as disfigurements in fabric. Effective control of grass and other weeds can only be achieved during the growing season. Little can be done at harvest time to undo the results of neglected weed control.

Setting of harvester components to avoid compromising fiber quality is important. The clearance between doffer plates and picking spindles affects quality. Too much clearance can lead to improper doffing and to spindle twist while lack of adequate clearance leads to abrasion of doffer plates by spindles. This condition results in contamination of the fiber with rubber from doffer-plates. It is often not recognized until the fiber has been constructed into a fabric. The presence of rubber specks in a fabric means that it has to be sold as "second."

Contamination of fiber with oil or grease from harvesters or from gin machines leads to lessening of quality because of subsequent problems in finishing and dyeing. Remnants of sheet plastic such as that used in irrigation or for covering modules, and the remnants of ropes such as those used for tieing tarpaulins are frequently found as disfiguring contaminants in yarn and fabric. Producers should understand that most cotton is destined for a consumer end-product and that the presence of fibrillated foreign material will lead to inferior products.

The need to reduce production costs is causing many producers who now use spindle pickers to include stripper harvesting among their options. Cotton yielding seven bales per hectare is routinely (although not commonly) stripper harvested in California. Spindle picker harvesting has been replaced by stripper harvesting in areas of South Texas. The literature abounds with reports of comparisons between spindle picking and stripper harvesting under equivalent conditions. In some cases, stripper harvesting proves to be inferior from a quality standpoint. In many of these comparisons, however, the crop was not raised in a cultural system appropriate for stripper harvesting, and it is difficult to draw valid conclusions. Scientists in California have been developing cultural practices for stripper harvesting strategies since the early 1970s. This work is still underway and continued improvements are being noted. Results of controlled experiments show little, if any, difference between spindle- and stripper-harvested cotton when each harvesting method is used in the appropriate context. On a practical level, grades from stripper-harvested cotton that was ginned where the appropriate cleaning equipment was available are found to differ little from grades of spindle-harvested cotton from the same producer. As techniques for evaluating cotton quality on the basis of fiber properties of real importance in textile processing become available, textile manufacturers will more easily be able to match the fiber properties of their purchases to the specification that the end-product must meet. This knowledge will permit stripper harvesting to enter into the cotton production system to the extent that market forces permit because its true characteristics will be recognized. If spindle-harvested cotton has a quality advantage in the marketplace, this advantage will be recognized in terms of price. Stripper harvesting is a well-accepted strategy for vast areas of cotton production in the USA.

The increasing consumer demand for quality and performance translates into new demands on cotton as a fiber. Fine yarns are needed for sheer fabrics. Strong yarn is needed for easy-care fabrics because easy-care finishes weaken fabrics. Mature, nep-free fiber is needed to manufacture velours, corduroys, and knitted fabrics that are to be dyed in shades where fabric quality depends on uniform dye uptake. Manufacturing processes used in modern textile mills are designed to be profitable through high productivity. High productivity is easier to attain with high quality cotton. A profitable textile industry is a good customer for cotton farmers, just as profit-making farmers are good sources of feedstock for the textile industry.

REFERENCES

Baker, S. H. 1972. Narrow row cotton—for Georgia? Georgia Agric. Exp. Stn. Res. Rep.

Barker, G. L., R. W. McClendon, R. F. Colwick, and J. W. Jones. 1979. Relationship between cotton lint color and weather exposure. Trans. ASAE 22:470–474.

----, C. S. Shaw, K. E. Luckett, and F. T. Cook, Jr. 1976. Effects of defoliation, plant maturity, and ginning set-up on cotton quality and value in the mid South. ARS-S-120, ARS, USDA, Washington, D.C.

Bilbro, J. D., and L. L. Ray. 1974. Effects of premature crop kill on cotton yields and fiber quality on the Texas High Plains. Texas Agric. Exp. Stn., Prog. Rep. PR3259.

Brendel, T. P., and C. S. Miller. 1979. Defoliation-stripping feasibility tests in Texas. p. 56. Proc. Beltwide Cotton Prod. Res. Conf., Phoenix, Ariz.

Burch, T. A., T. H. Garner, J. B. Davis, S. T. Rayburn, Jr., and W. E. Garner. 1979. Effects of harvest data variations on yield, quality, cost of returns from cotton, South Carolina, 1973–1977. South Carolina Agric. Ext. Stn., Bull. 624.

Buxton, D. R., H. N. Stapleton, Y. Makki, and R. E. Briggs. 1973. Some effects of field weathering of seed cotton in a desert environment. Agron. J. 65:14–7.

----, R. E. Briggs, L. L. Patterson, and S. D. Watkins. Canopy characteristics of narrow-row cotton as influenced by plant density. Agron. J. 69:929–933.

Cathey, G. W., K. E. Luckett, and S. T. Rayburn, Jr. 1982. Accelerated cotton boll dehiscence with growth regulator and desiccant chemicals. p. 113–120. *In* Field crops research. Elsevier Scientific Publishing Co., Amsterdam.

Childers, R. 1980. The use of computer models in extension educational programs on cotton. ASAE No. 80-1051.

Cocke, J. B. 1974. Effect of seed-cotton moisture level at harvest on ginned lint. ARS, USDA, Prod. Res. Rep. 127.

Colwick, R. F., and C. S. Shaw. Spindle moistening agents for mechanical cotton pickers: an evaluation. USDA, ARS 42-68.

----, and E. B. Williamson. 1968. Harvesting to maintain efficiency and to protect quality. p. 433–466. *In* F. C. Elliot, M. Hoover, and W. K. Porter, Jr. (ed.) Advances in production and utilization of quality cotton. The Iowa State University Press, Ames.

----, G. L. Barker, and S. T. Rayburn, Jr. 1979. Row spacing, short season and stripper harvesting of cotton in northeast Mississippi. Mississippi Agric. and For. Exp. Stn. Tech. Bull. 94.

----, and Technical Committee Members S-2 and W-24. 1960. Weed control equipment in mechanized cotton production. South. Coop. Ser. Bull. 71.

----, and ----. 1965. Mechanized harvesting of cotton. South. Coop. Ser. Bull. 100.

Corley, T. E. 1970. Correlation of mechanical harvesting with cotton plant characteristics. Trans. ASAE 13:768–773, 778.

----, and C. M. Stokes. 1964. Mechanical cotton harvester performance as influenced by plant spacing and varietal characteristics. Trans. ASAE 7:281–286, 290.

Curley, R. G., and R. A. Kepner. 1977. Palletless versus pallet systems for handling seed cotton modules. Trans. ASAE 20:828–832.

Dick, J. B., E. B. Williamson, and A. D. Owings. 1958. Three years of variety testing under mechanized cotton production in the Delta. Mississippi Agric. Info. Sheet 573.

Farr, C. R., and D. L. Kottock. 1979. Effect of date of irrigation termination on yield of Upland and Pima cotton. p. 95–97. Proc. Beltwide Cotton Prod. Res. Conf., Phoenix, Ariz.

Glade, E. H., Jr., and R. Cole. 1980. Charges for ginning cotton, costs of selected services incident to marketing and related information, 1979/80 season. USDA, ESCS, Natl. Econ. Div., Washington, D.C.

Hagen, C. R. 1951. Twenty-five years of cotton picker development. Agric. Eng. 32:593–596, 599.

Hathorn, S., Jr., H. N. Stapleton, and F. L. Watson. 1973. Minimizing costs in the cotton harvest-ginning systsem. Arizona Agric. Exp. Stn.

Huitink, G. 1979. Cotton harvesting. 1979. Arkansas Coop. Ext. Ser. EC549.

Kirk, I. W., E. B. Hudspeth, Jr., and D. F. Wanjura. 1964. A broadcast and narrow-row cotton harvester. Texas Agric. Exp. Stn. PR 2311.

----, L. M. Carter, C. K. Bragg, R. G. Curley, and O. D. McCutcheon. 1981. Production and processing performance of narrow-row stripped cotton in the irrigated West. USDA Marketing Res. Rep. 118.

Kittock, D. L., T. J. Henneberry, and L. A. Bariola. 1979. Chemical termination for insect control in cotton: past, present, and future. p. 62–65. Proc. Beltwide Cotton Prod. Res. Conf., Phoenix, Ariz.

Krenz, R. D. 1981. Personal communication of data from the Firm Enterprise Data System, Natl. Econ. Div. EES, USDA, Washington, D.C.

Lalor, W. F. 1976–1977. Unpublished, data collected from gin managers from the Mississippi Delta. Cotton Inc., Raleigh, N.C.

----. 1982. Technology for pre-textile cotton cleaning. p. 11–26. *In* Joseph G. Montalvo, Jr. (ed.) Cotton dust: Controlling an occupational health hazard. ACS Symposium Series No. 189. American Chemical Society, Washington, D.C.

----, J. K. Jones, and G. A. Slater. 1977. Palletless module movers. Agro-Industrial Report 4. Cotton Inc., Raleigh, N.C.

Looney, Z. M., C. A. Wilmot, S. H. Holder, Jr., and C. C. Cable, Jr. 1965. Cost of storing seed cotton. USDA Marketing Res. Rep. 712.

Luckett, K., and G. Cathy. 1979. Effect of harvest-aid chemicals on cotton earliness and yield. p. 114–115. Proc. Beltwide Cotton Prod. Res. Conf., Phoenix, Ariz.

----, J. M. Anderson, and S. T. Rayburn, Jr. 1977. Comparison of a spindle picker and a brush roll stripper for harvesting cotton on mixed clay soils. p. 128–129. Proc. Beltwide Cotton Prod. Res. Conf., Atlanta, Ga.

McArthur, W. C., and F. T. Cooke, Jr. 1980. Cotton production practices and costs in the United States. Georgia Exp. Stn. Res. Rep. 365.

McClendon, R. W., G. L. Barker, J. W. Jones, and R. F. Colwick. 1981. Simulation of cotton harvesting to maximize returns. Trans. ASAE 24:1436–1440.

McNeal, S. 1966. Cotton stacking packing trailer. Arkansas Farm Research, March–April, Univ. of Arkansas, Fayetteville, Ark.

Parker, R. E., J. E. Clayton, and M. M. Lindsey. 1963. Mechanical picker performance related to seed cotton yield. Mississippi Agric. and Exp. Stn. Info. Sheet 822.

Parvin, D. W., J. G. Hamill, and F. T. Cooke, Jr. 1980. Budgets for major crops, Delta of Mississippi, 1980. Mississippi Agric. and For. Exp. Stn. Special Edition, Research Highlights.

----, ----, ----, Ying-Nan Lin, and E. H. Simpson. 1981. Budgets for major crops, Delta of Mississippi, 1981. Mississippi Agric. and For. Exp. Stn. AECM.R No. 115.

Pendleton, A. M., and V. P. Moore. 1968. Ginning for quality preservation. p. 467–486. *In* F. C. Elliot, M. Hoover, and W. K. Porter, Jr. (ed.) Advances in production and utilization of quality cotton. The Iowa State University Press, Ames.

Ray, L. L., and E. B. Minton. 1973. Effects of field weathering on cotton. Texas Agric. Exp. Stn., Miscellaneous Publ. No. 118.

Renoll, E. 1979. Using farm machinery effectively. Alabama Agric. Exp. Stn. Bull. 510.

Sanders, D. W., and W. F. Lalor. 1972. A method for optimizing the machine-size-crop-area relationship. J. Agric. Eng. Res. 17:122–127.

Smith, H. P. 1938. Progress in mechanical harvesting of cotton. Agric. Engin. 19:389–391.

----. 1964. Farm machinery equipment (5th ed.). McGraw-Hill, N.Y.

Smith, M. G. 1971. Open field storage of seed cotton. p. 26–29. Proc. Western Cotton Prod. Conf., Lubbock, Texas.

Smith, M. L. 1974. Economics of systems for farm to gin movement of stored cotton. Report submitted to Cotton, Inc., Raleigh, N.C.

Sprott, M. J., R. D. Lacewell, G. A. Niles, J. K. Walker, and J. R. Gannaway. 1976. Agronomic, economic, energy and environment implications of short season, narrow row cotton production. Texas Agric. Exp. Stn. Misc. Publ. 1250.

Stapleton, H. N., M. D. Cannon, and W. A. LePori. 1967. Cotton-harvest-defoliation scheduling. Trans. ASAE 10:226–229, 232.

Tupper, C. R., J. M. Anderson, and W. S. Spurgeon. 1974. Harvesting cotton with an experimental combine. Mississippi Agric. and For. Exp. Stn. Info. Sheet 1250.

USDA, 1962. Charges for ginning cotton, costs of selected services incident to marketing, and related information, 1961–62 season. Natl. Econ. Div. Econ. Stat. Coop. Serv. Cotton Div., AMS, USDA, Washington, D.C.

----. 1972. Charges for ginning cotton, costs of selected services incident to marketing, and related information. 1971–72 season. Natl. Econ. Div. Econ. Stat. Coop. Serv. Cotton Div., AMS, USDA, Washington, D.C.

----. 1977. U.S. team reports on Soviet cotton production and trade. FAS-M-277.

----. 1981a. Cotton quality crop of 1980. USDA AMS, Cotton Div. 54(7), Memphis, Tenn.

----. 1981b. Cotton price statistics 1980–81. USDA AMS, Cotton Div. 62(13), Memphis, Tenn.

Walhood, V. T., T. J. Henneberry, C. M. Brown, D. L. Kittock, K. M. El-Zik, and B. Taylor. 1979. Short season cultural systems for pest management in cotton. p. 60, Proc. Beltwide Cotton Res. Conf., Phoenix, Ariz.

Walter, H., H. W. Gausman, F. R. Rittig, L. N. Namken, D. E. Escobar, and R. R. Rodruigez. 1980. Effect of mepiquat chloride on cotton plant leaf and canopy structure and dry weights of its components. p. 32–35. Proc. Beltwide Cotton Prod. Res. Conf., St. Louis, Mo.

Ward, J. M., W. E. Paulson, and D. L. Jones. 1953. Storing of seed cotton as an aid to more efficient ginning and marketing. Texas Agric. Exp. Stn. Bull. 765.

Whiteley, E. L., and C. A. Rinn. 1976. Short season, narrow row cotton studies in the Central Blacklands. Texas Agric. Exp. Stn. Prog. Rep. 3384.

----, L. Reyes, and D. L. Longenecker. 1976. Short season, narrow row cotton studies in the Coastal Bend and Brazos River Valley. Texas Agric. Exp. Stn. Prog. Rep. 3381.

Wilkes, L. H., and T. E. Corley. 1968. Planting and cultivation. p. 117–149. *In* F. C. Elliot, M. Hoover, and W. K. Porter, Jr. (ed.) Advances in production and utilization of cotton. The Iowa State Univ. Press, Ames.

----, G. L. Underbrink, and J. K. Jones. 1978. The module system of handling and storage of seed cotton. ASAE Paper No. 78-1553. St. Joseph, Mich.

Williamson, E. B., and F. E. Fulgham. 1956. The influence of spacing on mechanical harvesting of cotton. Miss. Farm Res. 19:6–7.

----. 1963. Unpublished Annual Report. p. 477–481. Regional Project S-2, Stoneville, Miss.

11 Ginning

Roy V. Baker
ARS-USDA
Lubbock, Texas

A. Clyde Griffin, Jr.
ARS-USDA
Stoneville, Mississippi

The cotton gin has as its principal function the conversion of a field crop into a salable commodity. Thus, it is the bridge between cotton production and cotton manufacturing. At one time the sole purpose of a cotton gin was to separate fibers from seed, but today's modern cotton gin is required to do much more. To convert mechanically harvested cotton into a salable product, gins of today have to dry and clean the seed cotton, separate the fibers from the seed, further clean the fibers, and place the fibers into an acceptable package for commerce. The cotton gin actually produces two products with cash value—the fibers and the cottonseed. Cottonseed are usually sold to cotton oil mills for conversion into a number of important and valuable products, but in some cases they may be saved for planting purposes. The fibers are the more valuable product, however, and the design and operation of cotton gins are usually oriented toward fiber production. In essence, the modern cotton gin enhances the value of the cotton by separating the fibers from the seed and by removing objectionable foreign matter, while preserving as nearly as possible the inherent qualities of the fiber.

11-1 HISTORICAL BACKGROUND

Ginning, in its strictest sense, refers to the process of separating cotton fibers from the seeds. Originally, ginning was accomplished entirely by hand; a tedious task that produced less than 1 kg of fiber per worker per day. Pinch-roll devices, such as foot rollers and the ancient churka gin, were

Published in *Cotton,* Agronomy Monograph no. 24, © ASA-CSSA-SSSA, 677 South Segoe Road, Madison, WI 53711.

Fig. 11-1. Model of Eli Whitney's cotton gin.

probably the earliest mechanical devices used for ginning (Bennett, 1959; Brown and Ware, 1958). The churka gin, which was used for centuries in India, employed a pair of small counter-rotating wood or steel rollers to pinch and pull fibers from the seeds. This hand-operated device somewhat resembled a later-day clothes wringer, and variations of this device were used to a limited extent in the USA after the Revolutionary War. Many attempts were made during this era to improve the capacity and performance of pinch-roll devices, but none of the improvements gained wide-scale commercial acceptance. These pinch roller designs were at best only marginally successful, and only then for the low-production ginning of the black-seed Sea Island cultivar.

Mechanization of spinning in England during the late 1700s greatly increased the demand for U.S. cotton (Mirsky and Nevins, 1952). During this period the USA began to export limited quantities of the Sea Island fiber, but increases in production were hampered by the inefficient roller ginning process. Also, the cultivation of Sea Island cotton was restricted to a relatively small area along the South Carolina and Georgia coasts. Although vast areas in the U.S. interior were available for the production of the green-seed Upland cultivars, these fuzzy-seed cottons could not be produced economically at that time because of the difficulty in ginning these short-staple cultivars by hand or with existing roller mechanisms. This situation set the stage for one of the most dramatic developments in the history of the cotton industry: Eli Whitney's invention of the spike-tooth cotton gin in 1794.

11-1.1 The Saw Gin

Whitney's gin (Fig. 11-1) removed the fibers from the seed by means of small spikes driven into a wooden cylinder. The spikes engaged the fibers and pulled them through narrow slots in a metal bar at the back of a seed-cotton roll box. The slots, being too narrow for the seeds to pass, retained the seeds and allowed the fiber to be pulled free. A brush cylinder behind the slotted bar removed the fibers from the spikes. Whitney was granted a U.S. patent, signed by George Washington, on this fundamental principle on 14 March 1794. Whitney's spike-tooth gin successfully ginned the short-staple green-seed cotton in a batch-type operation that was 50 to 100 times faster than hand ginning (Moore, 1977). It was an immediate success. In a letter to Thomas Jefferson, then Secretary of State, Whitney, in 1793, described his machine and its performance (Mirsky and Nevins, 1952):

"The cylinder is only two feet two inches in length, and six inches in diameter. It is turned by hand, and requires the strength of one man to keep it in constant motion. It is the stated task of one negro to clean fifty weight (I mean fifty lbs. after it is separated from the seed) of the green-seed cotton per day. This task he usually completes by 1:00 o'clock in the afternoon. He is paid so much per pound for all he cleans over and above his task, and for ten or fifteen days successively he had cleaned from sixty to eighty weight per day and left work every day before sunset."

Initially, Whitney and his partner, Phineas Miller, elected not to sell or license their gin, but chose instead to establish ginneries throughout the country and to monopolize the ginning industry. Within a couple of years Whitney and Miller established about 30 horse- and water-powered gins in Georgia. Their ginning fee, which was paid in cotton, was high; about 2/5 share of the ginned lint (Mirsky and Nevins, 1952). The cotton planters viewed this fee as an exorbitant tax, and as their resentment mounted it became more and more difficult for Whitney and Miller to maintain their monopoly over ginning. The Whitney gin, because of its simplicity, could be easily duplicated by local mechanics and within a few years copies of Whitney's gin were prevalent throughout Georgia, South Carolina, and the Mississippi Territory. For many years Whitney and his partner were involved in much litigation over patent rights to their gin. In the end, however, they received only a small portion of the monetary rewards they had once envisioned.

Many improvements have been made to Whitney's original gin design over the years. One of the earliest improvements was made by Hodgen Holmes in 1796. Holmes improved Whitney's design by replacing the spikes with circular saws and by using flat-iron ribs instead of the slotted bar. Holmes also opened up the bottom of the roll box for the discharge of ginned seed. This improvement permitted continuous operation, as opposed to Whitney's batch operation (Moore, 1977). These inventions by Whitney and Holmes were the origin of today's modern saw gin stand.

11–1.2   The Roller Gin

Although much of the glamour of early gin development has been focused on the saw gin, important advances have also been made in the development of roller gins. As mentioned previously, early roller gins were of the churka type. The first major improvement in roller gin design occurred in 1840 with the invention of the reciprocating knife principle by Fones McCarthy, of Demapolis, Ala. (Alberson and Stredronsky, 1964). McCarthy roller gins gained wide acceptance in many cotton-producing areas of the world during the late 1800s and early 1900s, and continue to be used extensively in several countries. During the 1960s additional improvements were made to the roller gin. A high-capacity, rotary-knife roller gin is now available and used extensively in the Pima cotton areas of the USA.

Roller gins are used primarily for ginning extra-long-staple cotton, but sometimes they are also used for ginning short-staple, fuzzy-seed cotton. Roller gins typically have lower hourly production rates than do saw gins, but they produce fewer neps (small knots of tangled fibers) during ginning and this is an especially desirable characteristic for some cottons. Roller-ginned extra-long-staple cotton brings a much higher market price than does saw-ginned Upland cotton. In some instances, specialty markets have been developed and are supplied by roller ginning Upland cotton such as the Acalas.

Roller ginning in the USA is limited to the 10 counties in west Texas, New Mexico, and Arizona that produce Pima cotton. However, on a world-wide basis, roller ginning is a widely accepted practice. In some countries, such as India and Egypt, roller ginning is the predominant method of ginning cotton regardless of the cotton's staple length.

11–1.3   Development of Ginning Systems

For many years after Whitney's invention, ginning was simply another farm operation performed at the plantation after harvesting had been completed. These plantation ginneries usually contained only one hand-fed gin stand powered by mules or by water. Bagging and packaging of the lint was generally a separate operation, and all cotton was handled by hand. By the 1880s, mechanical screw presses, gin feeders, and pneumatic cotton-handling systems had eliminated much of the hand labor from ginning, and made it practical to operate several gin stands simultaneously as a single system. As cotton ginning systems became larger and more complex, many planters found that they could no longer afford to buy gins for their individual use. This situation prompted the establishment of custom ginning operations in most local communities by the close of the 19th century, and marked the end of the plantation gin (Bennett, 1962).

During the early days of ginning, cotton was carefully hand-picked by farm workers who had a direct interest in the crop. Since the cotton was very clean, there was little need for cleaning equipment at the gin. As the production of cotton increased, however, it became necessary for farmers to employ additional off-the-farm labor for harvesting. This practice led to faster and rougher harvesting methods that greatly increased the foreign matter content of the cotton (Moore and Merkle, 1953). By the turn of the century these rough hand-harvesting practices had created a need for additional cleaning at the gin, and many types of seed-cotton cleaners and extractors were developed for this purpose during the early 1900s. By the 1930s, gin operators had begun to install seed-cotton drying equipment to improve the performance of their cleaners when handling damp cotton and to prevent rough ginning preparation of lint.

Mechanical harvesting methods that came into extensive use following World War II placed additional pressures on the cotton gin. Mechanical harvesting not only produced "trashier" cotton, it also greatly shortened the ginning season (Moore, 1977). These problems were solved by installing additional seed-cotton extracting and lint cleaning equipment at the gin and by developing higher capacity ginning machinery. Historically, ginning has evolved in response to changing production and harvesting practices, and today's modern cotton gin represents the latest development in that long evolutionary process.

11–2 HANDLING SEED COTTON

11–2.1 Delivery and Handling on the Gin Yard

Harvested seed cotton is normally delivered to the cotton gin in 3- to 6-bale cotton trailers or in 8- to 12-bale modules. The trailer system is the traditional method for transporting seed cotton from the farm to the cotton gin, and continues to be used extensively in many areas despite the growing popularity of module handling systems. Cotton trailers are four- or six-wheel conveyances equipped with wire-mesh sides and open tops. They are usually owned by the cotton producers, but in some areas trailers owned by gins are available to the producers for use during the peak harvesting period. Historically, cotton producers have been responsible for transporting the cotton trailers to the cotton gins.

The cotton module system was initially developed to alleviate harvesting delays caused by shortages of trailers during the peak harvesting periods. In addition to overcoming harvesting delays, the module system has also proven to be an efficient and economical method for handling and storing large volumes of seed cotton. Palletless modules, placed directly on the ground, are the most common type in use today. Palletless modules are loaded in the field and transported from the farm to the gin by specialized trucks or trailers. Module-mover trucks and trailers are typically owned by the gin which often charges a fee for loading and transporting modules. Once the module arrives at the gin it may be fed directly into the ginning system or placed in temporary storage on the gin yard. Special gin-yard module movers are often used to retrieve the module from storage and to move it to the gin's seed-cotton unloading system.

11-2.2 Unloading and Conveying Systems

Pneumatic suction systems have been employed since the late 1800s to feed seed cotton into the ginning process (Bennett, 1962). Over the years, these basic systems have been enlarged and mechanized to increase capacity and to improve efficiency. Currently a seed-cotton unloading system consists of (i) a means of introducing seed cotton into a suction conveying pipe, (ii) vertical and horizontal conveying pipes, (iii) a seed-cotton separator, and (iv) a centrifugal suction fan or fans. Green-boll separators and airline cleaners are also incorporated into many unloading systems, particularly those that handle machine-stripped cotton.

11-2.2.1 Telescopes

Vertical telescoping pipes of 33 to 60 cm in diam are commonly used to feed seed cotton into the suction system. As the suction telescope moves across the upper surface of a load of cotton, the vacuuming action lifts the seed cotton and conveys it through a flexible connection to an overhead conveying pipe. Telescopes are designed (i) to be raised or lowered, (ii) to be repositioned lengthwise along the load of cotton, and (iii) to be swung back and forth transversely across the load. These motions may be accomplished either manually or by various mechanical means. In some designs, the operator's efforts to move the telescope are assisted by counter weights, or by hydraulic or electric motors. The latest innovation in telescope design is a remotely controlled system. With this system the operator controls the motions of the telescope from an air-conditioned cab by means of hand-operated switches and valves.

11-2.2.2 Mechanical Module Feeders

Cotton modules may be fed into the gin with the conventional telescope system, but in recent years more and more gins have installed specialized mechanical feeders for modules (Fig. 11-2). Generally, these feeders utilize several dispersing cylinders to remove cotton at a regulated rate from one end of the module, and a vacuum dropper or conveyor to feed the cotton into an air conveying pipe. These feeders often increase the gin's operating capacity and greatly improve the uniformity of cotton flow into the gin. Module feeders may be arranged to feed cotton into the gin's existing pneumatic suction system or arranged to feed cotton directly into the hot-air line of the first drying system. Also, cylinder cleaners or extractors are sometimes attached to the dispersing units for early removal of soil particles and other foreign matter.

Two different designs of module feeders are available. One design utilizes a stationary dispersing unit to feed seed cotton from modules that are conveyed into the unit by a wire belt or walking-bar type of conveyor. Cotton from the dispersing unit is then fed through a vacuum dropper into an air conveying pipe. The other design utilizes a dispersing unit that travels

Fig. 11–2. Mechanical cotton-module feeder (Continental Conveyor and Equipment Co.).

along the length of a concrete runway. The dispersing unit feeds seed cotton from modules parked on the runway onto a long, belt conveyor that parallels the runway. The belt conveyor then delivers the seed cotton to a suction pickup pipe for movement into the gin. Also, the suction pickup arrangement is usually designed to remove green bolls, clods, and rocks.

11–2.2.3 Green Boll Separators

Early-season, machine-stripped seed cotton often contains green bolls that should be removed from the seed cotton prior to ginning. Seed cotton may also contain other heavy foreign material such as rocks or scrap metal that should be removed early in the ginning process. Gravity separators, located in the pneumatic suction system, are the most common means of separating heavy foreign objects from the cotton (Fig. 11–3).

11–2.2.4 Seed-Cotton Separators

The primary function of a seed-cotton separator is to separate seed cotton from the conveying air of a pneumatic system. Most separators also remove a limited amount of finely divided foreign matter during the separation process. Generally, a separator consists of a stationary screen or a revolving screen drum mounted in an enclosed housing, and a vacuum dropper (Fig. 11–4). The screening device separates the seed cotton from the conveying air and the vacuum dropper feeds the seed cotton out of the housing. Since most separators operate in negative air-pressure systems, the vacuum dropper is necessary to prevent excessive air leakage at the seed-cotton exit.

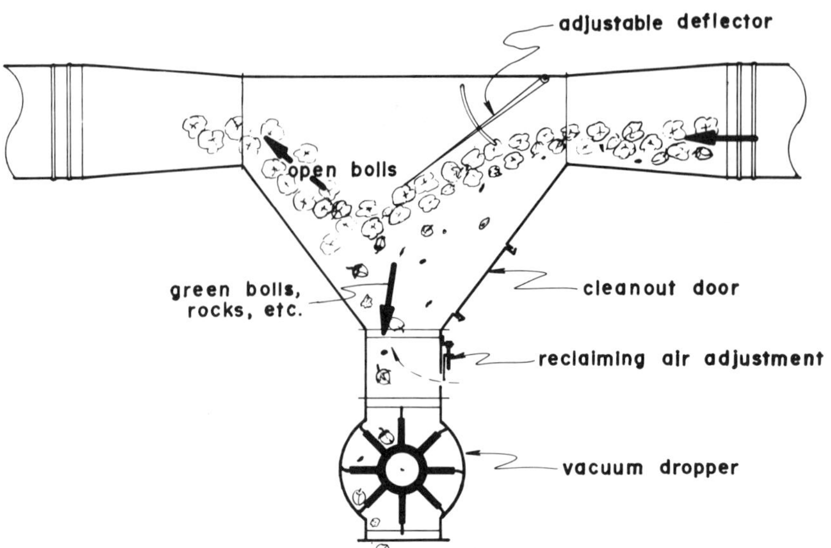

Fig. 11–3. Gravity green boll separator.

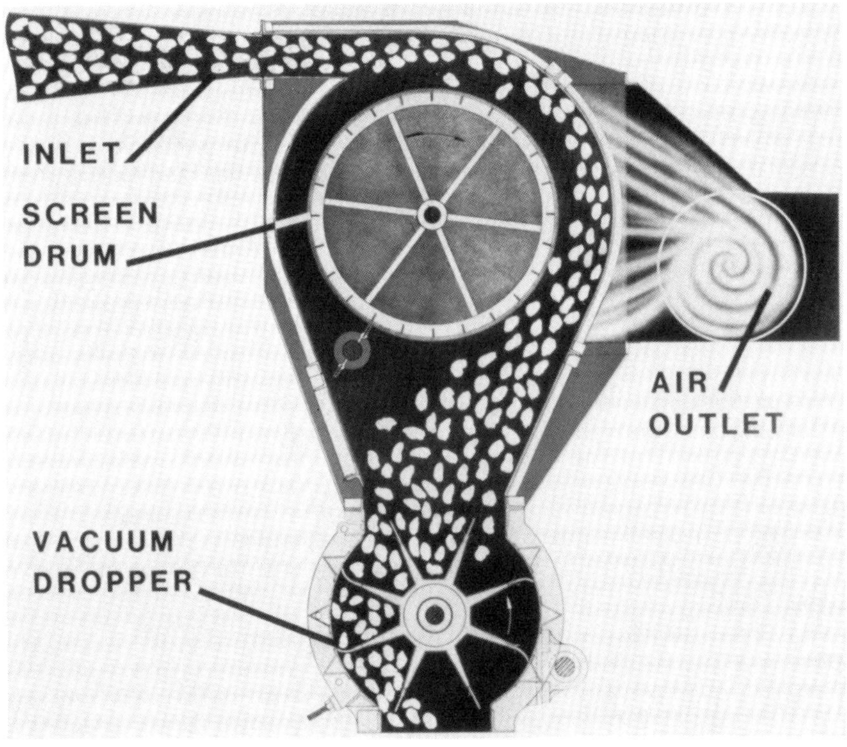

Fig. 11–4. Seed-cotton separator with revolving screen drum (Murray-Carver, Inc.).

11–2.2.5   Feed Control Units

Most cotton gins utilize an automatic feed control unit to regulate the flow of seed cotton from the unloading system into subsequent conditioning and cleaning processes. The feed control unit consists of a surge hopper equipped with variable-speed feed rollers. The unit is usually installed directly under the unloading system's seed-cotton separator. The feed rollers meter the seed cotton into a vacuum dropper which usually feeds the seed cotton into the hot-air line of the first drying system. The feed control unit improves the efficiency of the ginning process by (i) minimizing chokages, (ii) providing an even flow of seed cotton to the conditioning and cleaning equipment, and (iii) reducing the processing time lag between bales of cotton.

11–2.2.6   Conveying System

The modern ginning system consists of a series of individual processes connected by several materials-handling systems. Seed cotton is generally moved from one stage of processing to another by pneumatic conveyors, and distributed to a battery of extractor-feeders over the gin stands by a mechanical screw conveyor. With the pneumatic system, seed cotton is fed into a conveying pipe 30 to 50 cm in diam by a vacuum dropper and conveyed through the piping network by a stream of air traveling at a velocity of 20 to 25 m/sec (4000 to 5000 ft./min). At the point of delivery, the seed cotton is separated from the air stream by a seed-cotton separator or by a cylinder cleaner. The airflow for a pneumatic conveying system is supplied by a straight-bladed centrifugal fan capable of delivering the required volume of air at a static operating pressure that can vary from 3 to 6 kPa (12 to 24 in. of water).

Whenever possible, gin machinery is arranged so that the cotton-handling process between one machine and another is accomplished by gravity. Gravity handling of seed cotton frequently occurs between seed-cotton cleaners and extractors, the conveyor distributor and feeders, and between the feeders and the gin stands.

11–3   MOISTURE CONDITIONING

11–3.1   Fiber/Moisture Relationships

Cotton fibers are naturally hygroscopic. They absorb and desorb moisture depending upon the relative humidity of the surrounding atmosphere. The sorptive processes are driven by vapor pressure differences between the moisture in and on the fiber and the water vapor present in the atmosphere. Changes in atmospheric humidity produce corresponding changes in the moisture content of the fiber. Fibers reach moisture equilibrium when the rate of moisture gain from the atmosphere equals the rate of moisture loss from the fibers. A typical relationship between the cotton fiber moisture content and ambient relative humidity is shown in Fig. 11-5.

Moisture content of seed cotton is one of the most important factors affecting cleaning efficiency, ginning performance, and fiber quality preservation during ginning. Seed cotton that is too wet will not clean and gin properly and will receive low grades due to excessive foreign matter and rough preparation (Leonard et al., 1970). On the other hand, ginning seed cotton that is too dry may adversely affect fiber quality. The apparent strength of cotton fibers is directly proportional to fiber moisture content and is, therefore, greater at higher moisture levels. Consequently, as fiber moisture content is lowered, as by drying, the apparent strength is reduced and the frequency of fiber breakage during ginning is increased (Griffin and Moore, 1965). Also, low fiber moisture contents contribute to the generation of static electricity, causing chokages and decreased operating ef-

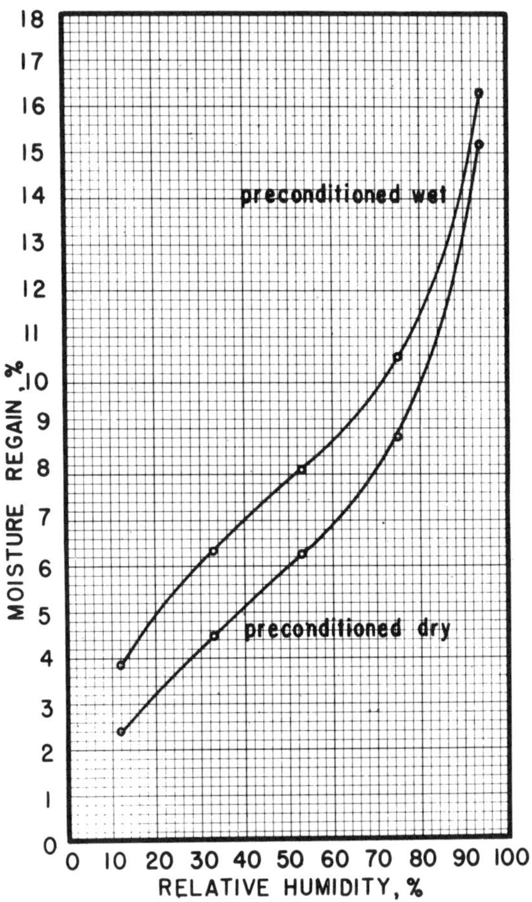

Fig. 11-5. Relationship of cotton fiber moisture content (dry basis) to ambient relative humidity at 21°C.

ficiency. Thus, proper control of fiber moisture content is of utmost importance in the ginning process.

Past research on seed-cotton drying has established an optimum range of fiber moisture content within which satisfactory cleaning and ginning can be achieved while the inherent qualities of the fibers are maintained. The currently accepted optimum fiber moisture content lies within the 6.5 to 8.0% range (Griffin and Moore, 1965). This optimum range is a compromise between effective cleaning and quality preservation on the one hand, and quality preservation and smooth ginning on the other.

11–3.2 Drying Systems

Seed-cotton driers depend for their operation on surrounding the seed cotton with heated air at low relative humidity. The heated air provides the required vapor pressure deficit for drying while at the same time heating the seed cotton for a more rapid release of moisture. During the heating process many of the moisture molecules in the seed cotton reach escape velocity and leave the seed cotton to be swept away by the warm conveying air.

Nearly all systems for drying seed cotton in use today utilize the shelf-type tower drier, Fig. 11–6. This device operates on a parallel-flow principle where the drying air is also the conveying medium. Tower driers are normally 1.2 to 1.8 m wide, about 1.8 m long, and 5 to 6 m high. They usually contain 16 to 24 shelves spaced 21 to 26 cm apart. Conveying air velocities through the passageways between the shelves are generally in the 5 to 10 m/sec (1000 to 2000 ft/min) range.

A typical single-stage drying system is composed of two centrifugal fans, a gas- or oil-fired burner, a tower drier, and connecting air lines, Fig. 11–7. The first fan, called a push fan, takes ambient air from the gin room and discharges it through the burner and conveying lines to the tower drier. Seed cotton is dropped into the heated air stream between the burner and tower drier. The hot conveying air transports the seed cotton through serpentine passageways in the tower drier and delivers it to an air-fed cylinder cleaner or seed-cotton separator while the spent drying air is routed to the second centrifugal fan (pull fan) and discharged back to the atmosphere. The push-pull fan arrangement is necessary to overcome the airflow resistance of the tower drier and air lines.

The seed cotton, as it moves through the drier's serpentine passageways, impacts upon the drier walls as it changes direction between each shelf. This action agitates the seed cotton and lengthens the exposure period, which improves the drying process. The dwell time for seed cotton in a drying system is dependent upon both the velocity of the conveying air and the aerodynamic characteristics of the seed cotton. Well opened seed cotton easily becomes airborne and is transported rapidly through the system; whereas, damp, heavy seed cotton moves through the system more slowly. While the drying period may vary with the seed cotton's moisture content and from one drying system to another, the total exposure is seldom

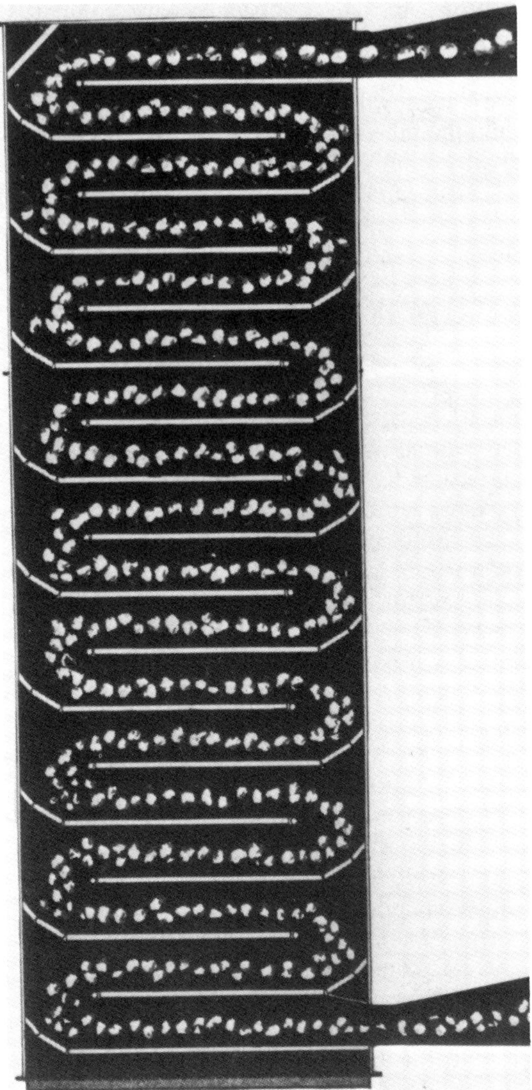

Fig. 11-6. Shelf tower drier for drying seed cotton at cotton gins.

more than 12 sec. Thus, in practice, the moisture content of the seed cotton does not reach equilibrium with the drying air. For extremely wet seed cotton, it is usually necessary to employ two stages of drying for adequate moisture control. From a quality preservation stand point, two stages of drying at low to moderate temperatures is preferable to a single stage at a higher temperature.

Modern drying systems are capable of overdrying and damaging the seed cotton if not used properly. Drying air temperatures in excess of 175°C may adversely affect fiber quality, and temperatures above 230°C will often scorch the fibers or set them afire. Drying temperatures in most systems are

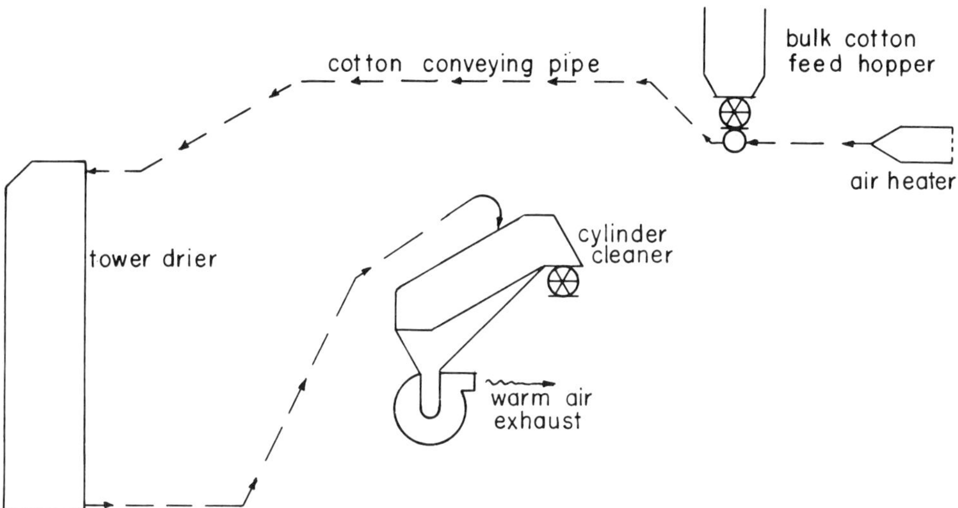

Fig. 11-7. Diagram of a typical seed-cotton drying system.

monitored by one or more strategically located sensors and regulated by an automatic temperature controller. These control systems, when adjusted properly, prevent serious over-drying problems, but often leave much to be desired from the standpoint of precise moisture control. Drying control systems based on the measurement of seed cotton moisture content, made either before or after drying, are also available. These systems provide improved moisture control during ginning and often reduce the energy requirements for drying.

11-3.3 Moisture-Restoration Systems

Cotton harvested under low humidity conditions may be very dry when it is delivered to the cotton gin. Since the fiber breakage rate during ginning is inversely related to the fiber moisture content, the average fiber length of the cotton may be improved by adding moisture before ginning and cleaning lint (Griffin and Moore, 1965). Other benefits resulting from moisture restoration include reducing the static electricity level of the seed cotton and reducing the amount of force required to tramp and press a bale of cotton. Ideally, the fiber moisture content should be increased to the recommended 6.5 to 8.0% level.

One of the most common methods of increasing fiber moisture content is to subject the seed cotton to warm, humid air. This vapor method of moisture restoration may be applied at the extractor-feeders, in hoppers above the feeders, in the lint flue, or at the lint slide before pressing. Another method involves spraying the cotton with a fine mist of water. The spray method is often used at the lint slide as an aid to the pressing operation, but it is not used to any great extent elsewhere in the cotton gin.

For the vapor system, warm, humid air is supplied by a special air humidifier. Ambient air is first heated, and then drawn through a humidifying chamber equipped with water-spray nozzles. The temperature and humidity of the processed air can be varied by adjusting the burner setting and the water-spray pressure. Typically, humidifiers of this type supply about 2 m³/sec (4000 cu. ft./min) of air at temperatures up to 60°C and at relative humidities up to 95%.

11-4 SEED-COTTON CLEANING

11-4.1 Foreign Matter Relationships

The system for cleaning seed cotton in a modern gin serves a dual purpose. First, large foreign matter components such as green bolls, burs, and sticks and stems must be extracted from the seed cotton so that the gin stand will operate at peak efficiency and without excessive downtime. Second, the seed cotton must be cleaned adequately to achieve optimum market values for lint and cottonseed (Baker et al., 1977). The amount of seed-cotton cleaning machinery required to meet these needs vary with the foreign matter content of the seed cotton, which depends in large measure on the method of harvest.

Virtually all cotton produced in the USA today is harvested mechanically, either by pickers or strippers. Hand-picking of cotton, while no longer practiced in the USA, is still the predominate harvesting method in many other countries. Since the foreign matter content of harvested seed cotton can vary from 7 to 635 kg per bale (Table 11–1), a gin's system for cleaning seed cotton must be designed to adequately handle the most severe foreign matter condition that is expected in its trade area.

Several types of cleaning and extracting machines are used to remove foreign matter from seed cotton prior to ginning. The most common of these machines include cylinder cleaners, bur machines, stick machines, combination bur and stick machines, and extractor-feeders. Generally, several of these machines are arranged in series to provide the necessary cleaning and extracting required to ensure satisfactory ginning and maximum bale value.

11-4.2 Cylinder Cleaners

Cylinder cleaners are used for removing leaf, sand, and other finely divided particles and for opening and preparing the seed cotton for more extensive extraction processes (Pendleton and Moore, 1968). The cylinder cleaner consists of a series of spiked cylinders, usually 4 to 7 in number, that agitate and convey the seed cotton across cleaning surfaces containing small openings or slots. The cleaning surfaces may be either concave screen or grid bar sections, or serrated-disk cylinders. Foreign matter that is

Table 11-1. Typical ranges of foreign matter in seed cotton harvested by three methods.

Harvesting method	Range of foreign matter content		
	Low	Normal	High
	kg per bale†		
Hand-picked	7 (15)	14 (30)	68 (150)
Machine-picked	20 (45)	36 (80)	91 (200)
Machine-stripped	208 (450)	318 (700)	635 (1400)

† Based on a standard 218 kg bale of lint, net weight. Numbers in parentheses are pounds.

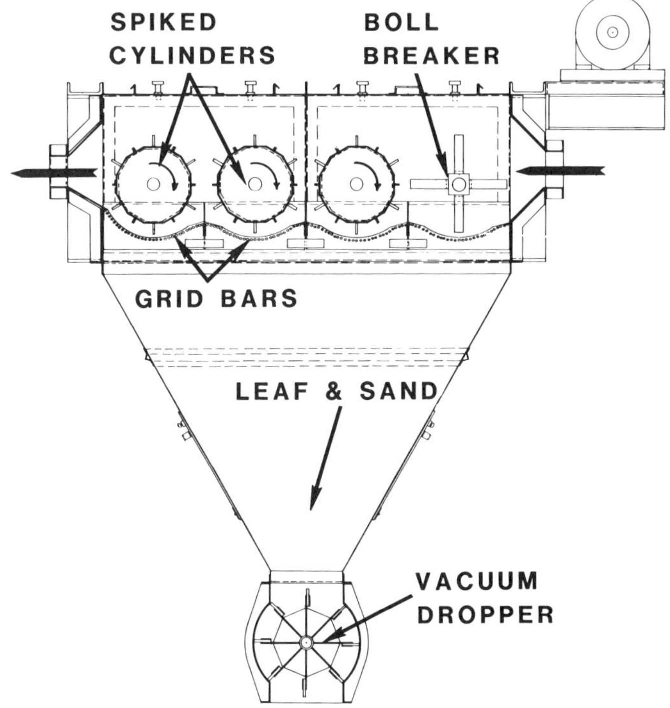

Fig. 11-8. Four-cylinder airline cleaner (Horn and Gladden Lint Cleaner Co.).

dislodged from the seed cotton by the impacting action of the cylinders falls through the screen, grid bar, or disk openings for collection and disposal.

Cylinder cleaners may be designed for installation in either a horizontal or inclined position. Those installed horizontally are referred to as "horizontal cleaners" while those installed at an angle are called "inclined cleaners". Cylinder cleaners are further classified with respect to how they are used in the gin, i.e. airline cleaners, air-fed cleaners, or gravity-fed cleaners (Garner and Baker, 1977).

Airline cleaners are usually mounted in a horizontal position in the unloading-system airline (Fig. 11-8). These installations permit both the air and seed cotton to pass entirely through the cleaner. Airline cleaners have gained wide acceptance in stripper areas as a means for removing soil particles from seed cotton and for opening partially closed bolls and wads of seed cotton for further cleaning (Moore and Merkle, 1953).

Air-fed and gravity-fed cylinder cleaners are usually used immediately ahead of bur and stick extractors, or as finishing cleaners above the distributor (Fig. 11–9). The air-fed machines are generally connected to the drying system, and in these applications they serve as an extension of the drying system and as a means of separating seed cotton from the drying air.

Cylinder cleaners are currently manufactured in widths of 1.8, 2.4, 3.0, and 3.7 m with rated capacities of 5 to 8 bales hour^{-1} m^{-1} (1.5 to 2.5 bales hour^{-1} ft^{-1}) of width. For higher capacities, two cleaners are installed in parallel with each machine cleaning half the seed cotton.

11–4.3 Bur and Stick Extractors

Burs and sticks are extracted from seed cotton by specially designed equipment generally classified as extractors. These machines include the bur machine, stick machine, and combination bur and stick machine.

11–4.3.1 Bur Machine

The bur machine is based on a dislodging or stripping principle (Fig. 11–10). Seed cotton is presented to a large-diameter saw cylinder by a kicker

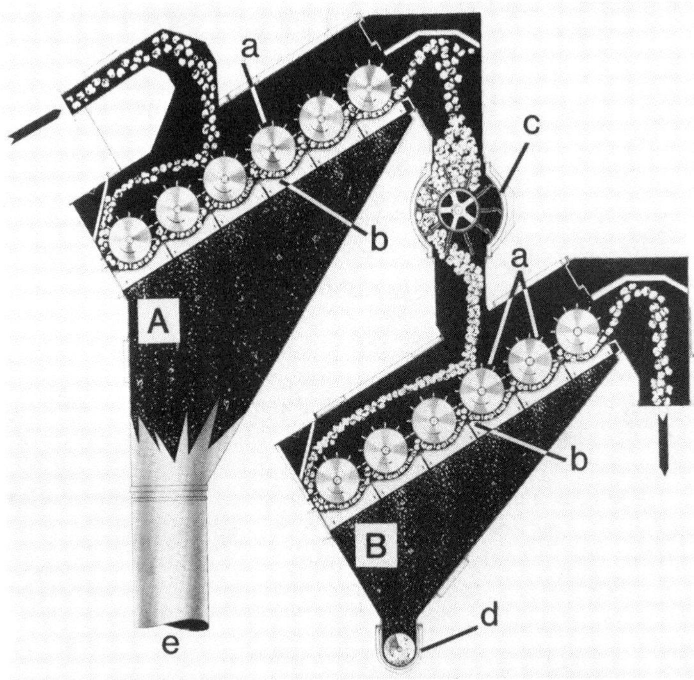

Fig. 11–9. Air-fed (A) and gravity-fed (B) inclined cleaners. (a) Spiked cylinders; (b) grid bars; (c) vacuum dropper; (d) waste auger; (e) air outlet (Lummus Industries, Inc.).

conveyor equipped with special flippers. Seed cotton adheres to the saw cylinder and is carried past a flighted stripper roller which dislodges burs and sticks from the surface of the cylinder. The dislodged material finds its way through the incoming stream of seed cotton to the kicker conveyor which moves the material to one end of the machine. At this point the material falls onto a spiked conveyor and is moved back along the entire length of the saw cylinder so that seed cotton can be reclaimed from the dislodged material. Fine particles and dirt sift through a screened trough to another auger located underneath the spiked conveyor.

Bur machines, which are available in lengths of 3.0, 4.3, and 5.5 m are usually employed as single or multiple parallel units about midway through the seed-cotton cleaning process. They are usually preceded by one stage of drying and one cylinder cleaner. Bur machines generally operate at capacities of 1.5 to 2.5 bales hour^{-1} m^{-1} (0.50 to 0.75 bales hour^{-1} ft^{-1}) of length. In an era of high capacity ginning, this low operating capacity in combination with a relatively low efficiency has made the machine obsolete for most modern ginning applications. However, the bur machine continues to be used in many older, low-capacity cotton gins and is important from a historical viewpoint. The bur machine, more than any other machine, made it possible for a gin to satisfactorily handle hand-snapped and machine-stripped cotton (Pendleton and Moore, 1968).

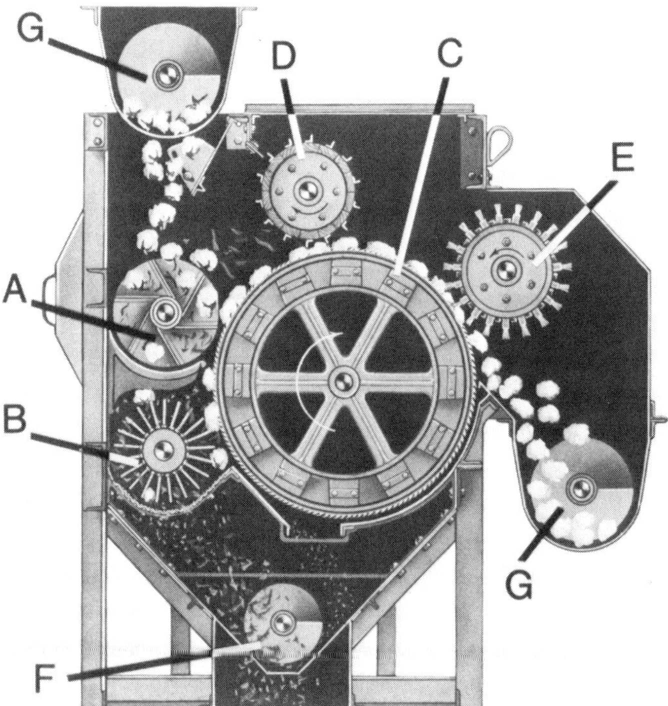

Fig. 11-10. Bur Machine. (A) Kicker conveyor; (B) spiked conveyor; (C) saw cylinder; (D) stripper roller; (E) doffing brush; (F) hull auger; (G) cotton auger (Murray-Carver, Inc.).

11-4.3.2   Stick Machine

Stick machines utilize the sling-off action of high-speed saw cylinders to extract burs and sticks from seed cotton by centrifugal force (Franks and Shaw, 1969). Seed cotton is fed onto the primary sling-off saw cylinder and wiped onto the sawteeth by one or more stationary wire brushes (Fig. 11-11). Foreign matter and some seed cotton is slung off the saw cylinders by centrifugal forces 25 to 50 times that of gravity. Grid bars are stategically located about the periphery of the saw cylinder to help control the loss of seed cotton and to aid in the extraction process. However, some loss of seed cotton is inevitable if satisfactory cleaning efficiencies are obtained. Additional saw cylinders are used to reclaim the seed cotton extracted with the burs and sticks. Reclaimer saw cylinders resemble the primary saw, but usually operate at slower speeds and are equipped with more grid bars.

Commercial stick machines vary widely with respect to number of saw cylinders, grid bar type and spacing, saw speed and size, and location of stationary brushes. However, all models have at least one primary sling-off saw and one reclaimer saw. Generally, stick machines are classified as either two-saw or three-saw machines.

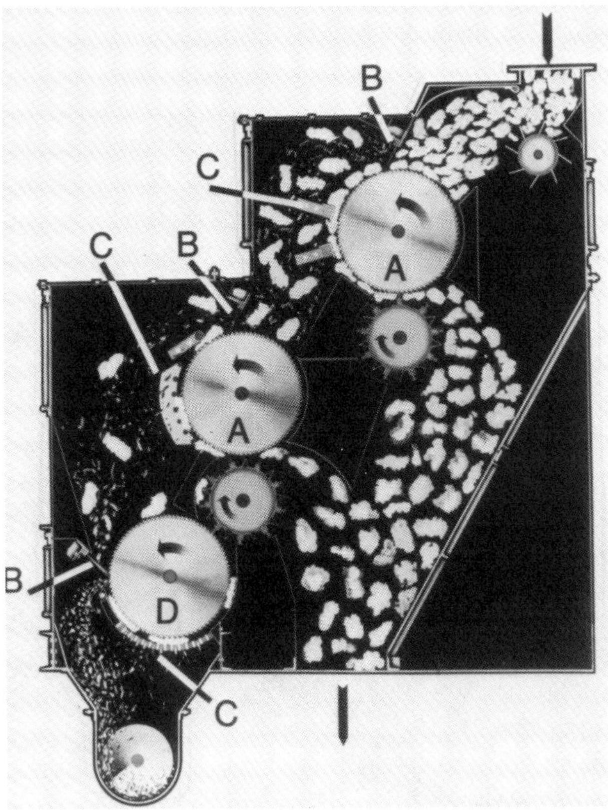

Fig. 11-11. Gravity-fed, three-saw stick machine. (A) Saw cylinders; (B) stationary brush; (C) grid bars; (D) reclaimer saw cylinder (Murray-Carver, Inc.).

Stick machines are available in widths of 2.4, 3.0, and 3.7 m with rated capacities of 5 to 7 bales hour^{-1} m^{-1} of width (1.5 to 2 bales hour^{-1} ft^{-1}). They may be fed by air or by gravity, but the trend in recent years has been to gravity feeding by a separator or cylinder cleaner. Depending on capacity requirements, stick machines may be employed as single or multiple parallel units.

11–4.3.3 Combination Bur and Stick Machine

In recent years a new type of extractor has become available that combines some of the features of both the bur machine and the stick machine. Although these new machines are usually identified by various trade names, generally this class of extractors may be referred to simply as combination bur and stick (CBS) machines.

The upper section of a CBS machine resembles a bur machine in that it is equipped with an auger feed and trash extraction system and a large-diameter saw cylinder (Fig. 11–12). The CBS machine, however, also differs

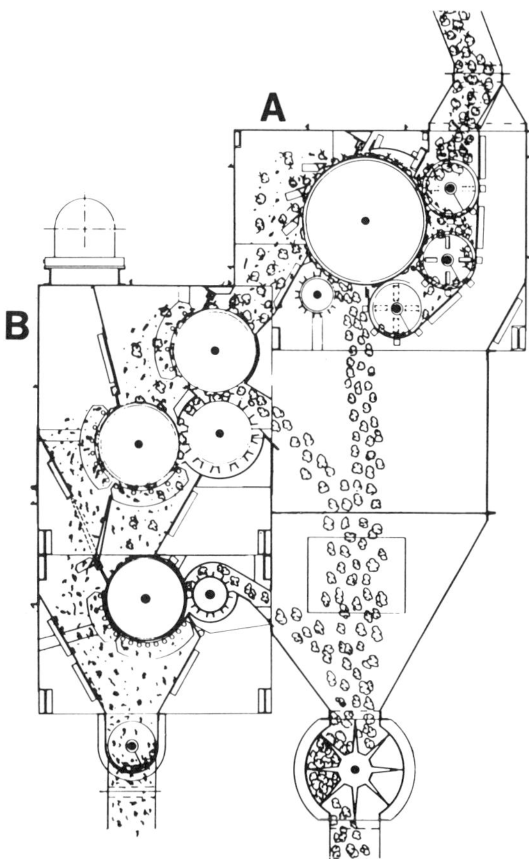

Fig. 11–12. Combination bur and stick machine. (A) Upper feeding and extracting unit; (B) lower stick machine unit (Horn and Gladden Lint Cleaner Co.).

from a bur machine in several important respects. The CBS machine is not as wide as a bur machine although its rated capacity is much higher. Seed cotton is generally fed into a CBS machine across its entire width as opposed to the end-feeding method used for bur machines, and the CBS machine is equipped with a stationary stripper plate rather than the bur machine's steel-flighted stripper roller. Also, the upper section of the CBS machine is designed to provide a foreign matter sling-off feature that is not found in conventional bur machines. Seed cotton and foreign matter that are discharged by sling-off action from the large-diameter saw cylinder enters the lower section of the machine for additional cleaning. The lower section of the CBS machine consists, with minor modifications, of a standard two- or three-saw stick machine. Thus, the upper section of the machine serves as a primary cleaner and feeder for the lower stick-machine unit (Baker et al., 1981).

The CBS machines have gained in popularity in recent years, particularly at gins processing stripper harvested seed cotton. These machines are available in widths of 1.8, 2.4, 3.0, and 3.7 m with operating capacities of 5 to 7 bales hour^{-1} m^{-1} (1.5 to 2 bales hour^{-1} ft^{-1}) of length. They are usually employed as the first stage of extraction in a seed-cotton-cleaning system.

11–4.4 Extractor-Feeders

The primary function of a modern high-capacity extractor-feeder is to feed seed cotton to the gin stand uniformly and at controllable rates, with extracting and cleaning as a secondary function. Gin stand feeders with extracting capabilities have been used since the early 1900s (Bennett, 1962). Early machines were based on the stripping principle of the bur machine, but current models utilize the stick machine's more efficient sling-off principle (Fig. 11–13).

Feed rollers, located at the top of the extractor-feeder and directly under the distributor hopper, control the feed rate of seed cotton to the gin stand. These feed rollers are powered by variable-speed hydraulic or electric motors, controlled manually or automatically by various interlocking systems with the gin stand. The feed rolls may be designed to automatically "start" and "stop" as the gin breast is engaged or disengaged; the system may also be designed to stop feeding seed cotton in cases of gin stand overloads or underloads. Many of the systems are designed to maintain constant seed-roll densities.

11–5 GIN STANDS

A modern cotton gin performs a number of essential services for the farmer, but the basic purpose is to separate fibers from seed. The separation of fibers from seed is accomplished by machines commonly known as 'gin stands', of which there are two basic types. Saw gin stands are used primarily for ginning Upland cotton while Pima and other extra-long-staple cottons are normally ginned by roller gin stands.

11–5.1 Saw Gin Stands

The specific design features of modern saw gin stands vary from one manufacturer to another, but all may be broadly classified as double-rib, huller-front gins. A modern gin stand consists of a huller section, seed roll compartment, gin saws and ribs, gravity and/or overhead moting sections, and a lint doffing mechanism (Fig. 11–14). In addition to these basic components, special features such as seed-roll agitators, seed-extraction tubes, and variations in huller rib and ginning rib construction are common with many current gin stand designs (Fig. 11–15 and 11–16). Gin stands may also be classified with respect to method of doffing. Brush gin stands employ a rotating brush cylinder to remove the fibers from the sawteeth, whereas air-blast gin stands utilize high-velocity airjets for doffing the gin saws. Current models of saw gin stands are equipped with 93 to 158 saws of 30.5 to 45.7 cm in diam. Gin stands with rated capacities as high as 12 bales per hour are now available.

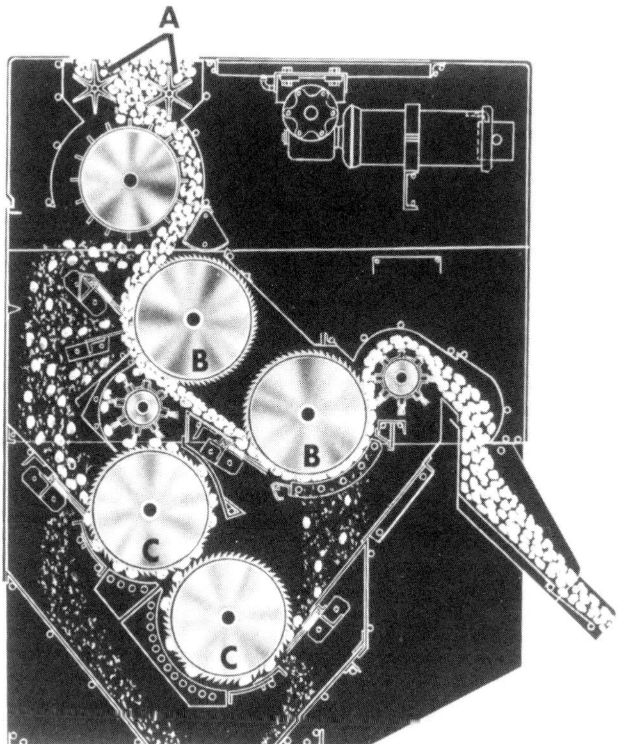

Fig. 11–13. Extractor-feeder for a gin stand. (A) Feed rolls; (B) sling-off saw cylinders; (C) reclaimer saw cylinders (Bush Hog/Continental Gin Co.).

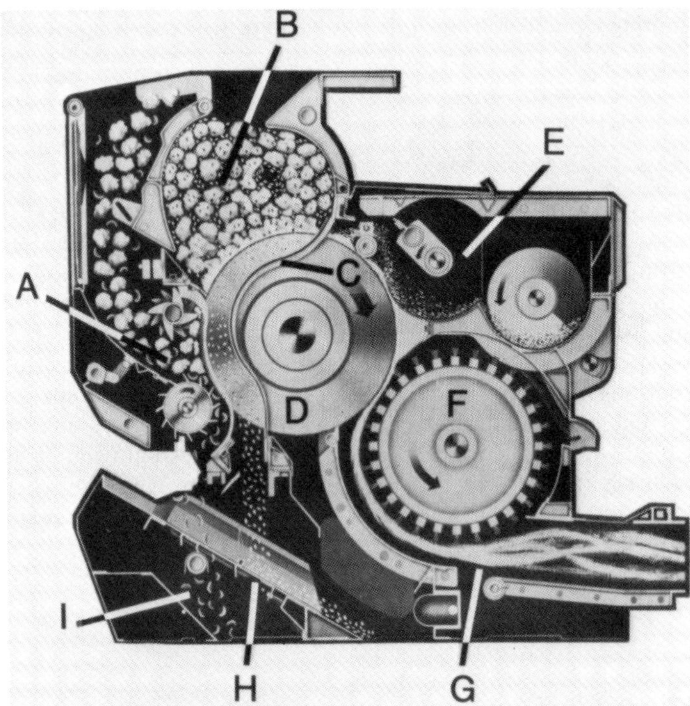

Fig. 11-14. Modern saw gin stand. (A) Huller section; (B) seed roll; (C) ginning ribs; (D) ginning saw; (E) overhead moting section; (F) brush doffing cylinder (Murray-Carver, Inc.).

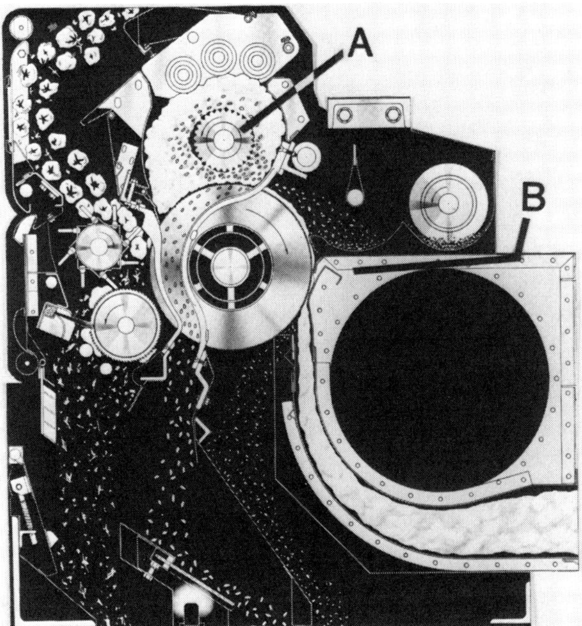

Fig. 11-15. Saw gin stand with seed roll agitator (A) and airblast doffing system (B) (Lummus Industries, Inc.).

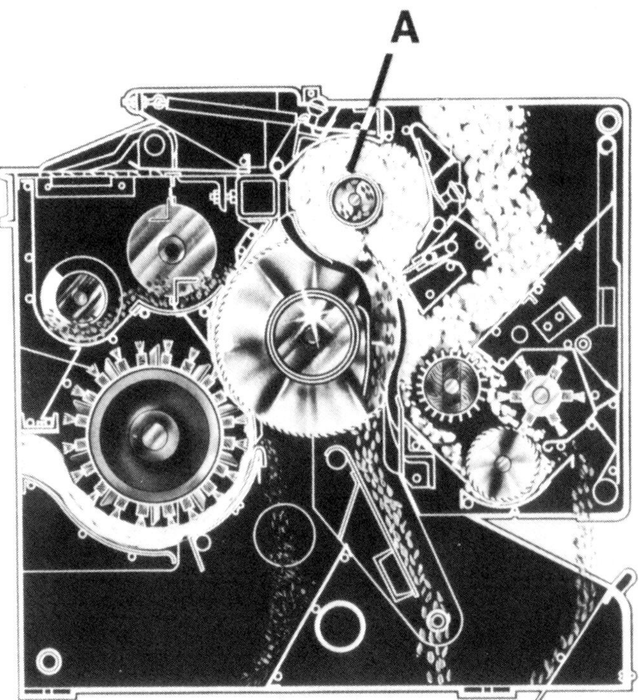

Fig. 11–16. Saw gin stand with seed-extraction tube (A) in seed roll (Bush Hog/Continental Gin Co.).

Although modern saw gins vary substantially in detail, they all operate on basically the same ginning principle. Seed cotton from an extractor-feeder slides down a short feeder apron and falls into the huller front of the gin stand. The huller front performs a limited amount of cleaning, but its primary function is to feed seed cotton into the lower portion of the seed roll compartment. The gin saws grasp the seed cotton and draws it through the widely spaced huller ribs, or in the case of split huller ribs, through a narrow slot near the top of the ribs into the seed roll compartment. The fibers are then drawn through closely spaced ginning ribs near the rear of the seed roll compartment to separate the fibers and seed. After the fibers are removed, the seed exit the seed roll compartment and slide down the face of the ginning ribs and fall out the bottom of the gin stand. The ginned fibers are then whipped across an overhead mote board for removal of motes (small, immature seed with attached short fiber) and foreign matter, and doffed from the sawteeth with a brush cylinder or airblast nozzle. The fibers are then carried from the gin stand by an air stream to a lint cleaner condenser for additional cleaning.

11–5   Roller Gin Stands

Roller ginning has long been the preferred method for ginning extra-long-staple, fine-fibered Sea Island, Egyptian, American-Egyptian, and Pima cottons (Bennett, 1956). While it is possible to gin these types of cotton with a saw gin, the resulting quality is substantially lower than that obtained with roller gins. Saw ginning tends to decrease the fiber length of these types of cotton and to greatly increase their nep content (Chapman and Stedronsky, 1965).

The McCarthy roller gin utilizes a leather or composition roller to draw the fibers between a fixed knife and the roller. The pulling action of the roller on the fibers combined with the pushing action of a moving knife are required to completely remove the fibers from each seed. The seed then falls through a seed grid and the fibers are removed from the roller by a rotating doffer (Fig. 11–17).

One major disadvantage of the McCarthy roller gin is its low ginning capacity. To overcome this capacity problem improved designs of roller gins have been developed. The rotary-knife roller gin stand, which was introduced in the USA in the early 1960s, operates at a capacity that is four to seven times that of the McCarthy gin (Leonard, 1970). The rotary-knife gin stand uses a large-diamter roller and a stationary knife to exert a pulling action on the fibers in a manner very similar to that of the McCarthy gin. However, rather than having a reciprocating knife, the new roller gin stand utilizes a small-diameter flighted roller (rotary knife) to provide the necessary seed-pushing action at the point of ginning (Fig. 11–18). High capacity rotary-knife roller gins have virtually eliminated the use of McCarthy gins in the USA, and are gaining in acceptance throughout the world.

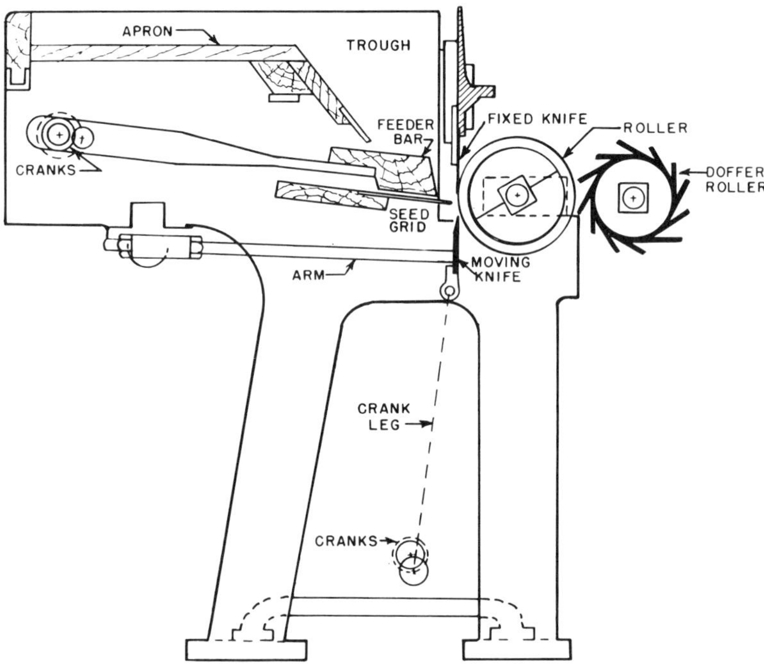

Fig. 11–17. McCarthy roller gin stand.

11–5.3 Cottonseed Handling

Cottonseed from the gin stands is usually delivered to a scale for weighing and then conveyed to a temporary storage facility located outside the gin building. Belt or screw conveyors are normally employed to collect the seed from the gin stands, and screw elevators are frequently used to elevate the seed for loading into the scale. After weighing, the seed are fed into a high-pressure seed blowing system and conveyed by air to the storage facility. The seed conveying line may vary from 12 to 20 cm in diameter, depending on capacity requirements. High-pressure air from a lobe-type, positive-pressure blower is delivered to the system at a nominal conveying velocity of 25 m/sec (5000 ft/min). The blower must be capable of delivering the required volume of air at an operating static pressure within the 7 to 27 kPa (1 to 4 psi) range.

In most regions of the Cotton Belt the seed are stored temporarily in 40- to 160-t overhead storage houses. These bottom-dumping houses facilitate the loading of trucks, which transport the seed to a cottonseed oil mill. In other regions, where the risk of rain damage is minimal, the seed are simply heaped into large piles on concrete slabs. These seed may be loaded onto trucks by front-end loaders or screw-tube loaders. In addition to these bulk seed-storage facilities, most gin are also equipped with small "customer" hoppers for the convenience of individual producers who desire to save small amounts of cottonseed for planting purposes.

Fig. 11–18. Rotary-knife roller gin stand. (A) Diagram of essential components; (B) commercial installation (Hardwick Etter Co.).

11–6   LINT CLEANING

Lint cleaning at the cotton gin developed concurrently with and in response to the widespread adoption of mechanical harvesting in the USA (Brooks, 1947; Moss, 1955; Stedronsky and Shaw, 1950). Most mechanically harvested cotton requires additional cleaning after ginning to achieve maximum returns for the producer and acceptable levels of cleanliness for the spinner. The lint cleaner's ability to remove small particles of leaves, motes, grass, and bark fulfills a need that cannot be obtained with seed-cotton cleaning alone. As a result, lint cleaners have dramatically improved the gin's cleaning capabilities and thereby contributed immeasurably to the success of mechanical harvesting. Lint cleaning has become an accepted and essential part of the modern ginning process.

11–6.1   Saw Lint Cleaners

Saw lint cleaners are used for cleaning saw-ginned Upland cottons (Fig. 11–19). Commercial lint cleaners, although variable in size and design, are all equipped with a lint condenser, a batt feeding mechanism, a 30.5 to 40.6 cm-diameter saw cylinder with grid bars, and a lint doffing mechanism.

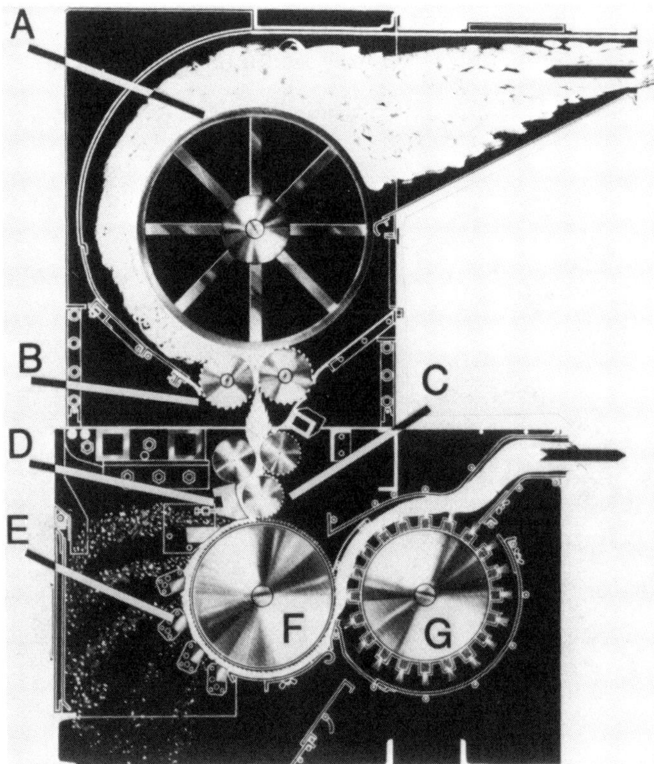

Fig. 11–19.   Controlled-batt, saw-type lint cleaner. (A) Condenser drum; (B) doffing rolls, (C) feed rolls; (D) feed bar; (E) grid bars; (F) saw cylinder; (G) doffing brush (Bush Hog/ Continental Gin Co.).

Lint from a gin stand or another lint cleaner is conveyed by air into the condenser which separates the lint from the air and forms the lint into a batt weighing about 0.2 to 0.4 kg/m² (0.04 to 0.08 lb./ft.²). The batt is then fed through one or more sets of directional rollers to a closely fitted feed roller and matching concave feed bar. The feed bar is positioned directly above the center of the saw cylinder at a distance of 1.5 to 3.0 mm from the sawteeth tips. The surface speed of the saw cylinder is substantially faster than that of the feed roller, creating a combing action as the batt is fed to the sawteeth. The feed roller is spring-loaded to hold the batt firmly against the feed bar to enhance the combing action. Combing ratios of 15:1 to 30:1 are commonly employed.

A high speed saw cylinder (800 to 1300 revolutions/min) whips the fibers across the keen edges of several grid bars, usually four to eight in number, to discharge the foreign matter by a combination of centrifugal force and the stripping action of the grid bars. The foreign matter is collected by an airstream and conveyed from the machine. The cleaned fibers are doffed from the sawteeth by a doffing brush or an air-blast device and conveyed by air to the next stage of lint cleaning or to the press condenser.

Gins handling mechanically harvested cotton are normally equipped with two stages of lint cleaning while those handling hand-picked cotton require only one stage. Saw-type lint cleaners may be employed as single or tandem units directly behind each gin stand, or as battery lint cleaners for several gin stands. A pair of lint cleaners is often arranged in parallel as a means of increasing the capacity of battery installations, or as an alternate split-stream arrangement in lieu of tandem lint cleaners behind a gin stand.

A gin may be equipped with two, or in a few cases, three stages of lint cleaning, but it is not desirable to employ all of this equipment for all cottons. Lint cleaners should be equipped with bypasses and used selectively according to the cotton's cleaning requirements and in concert with prevailing market demand. Excessive lint cleaning can decrease bale weight to an extent not compensated for by improvements in grade, thereby lowering bale value for the producer (Baker et al., 1977). Excessive lint cleaning may also degrade fiber length and lower the cotton's spinning potential (Looney et al., 1963).

Saw lint cleaners are precision machines that operate with close tolerances. Smooth and efficient operation requires careful attention to a number of exacting mechanical adjustments as well as the proper selection or control of several operating parameters. Batt weight, combing ratio, saw speed, feed rate, and fiber moisture content are all factors that influence the performance of a lint cleaner and its effect on fiber quality (Baker, 1978; Griffin et al., 1970; Mangialardi, 1974).

11–6.2 Pneumatic Lint Cleaners

The pneumatic lint cleaner is used to clean both saw-ginned Upland cottons and roller-ginned Pima cotton. For Upland cottons, the pneumatic lint cleaner is installed directly behind a saw gin stand and usually followed by one or more stages of saw lint cleaning. Lint from the gin stand is conveyed by air through a specially shaped duct in the pneumatic lint cleaner. The duct is designed to force the air and cotton to change direction abruptly as they pass a narrow trash-ejection slot (Fig. 11–20). Foreign matter is ejected through the slot by inertial forces. The degree of foreign matter separation can be regulated to some extent by varying the width of the adjustable trash-ejection slot.

For roller-ginned Pima cotton, the pneumatic lint cleaner is usually used in combination with a beater-type opener-cleaner (Alberson and Stedronsky, 1964). Lint from several roller gin stands is conveyed by air to a condenser above the opener-cleaner. The condenser feeds the lint batt to the opener-cleaner which opens and breaks the batt into small tuffs. A lugged cylinder carries the tuffs across a series of closely spaced grids for the removal of motes and foreign matter. From the opener-cleaner the lint then passes through a pneumatic-type lint cleaner. Generally, two stages of this cleaning arrangement are employed for mechanically-harvested Pima cotton (Fig. 11–21).

11–7 HANDLING AND PACKAGING OF LINT

11–7.1 Lint Conveying Systems

Nearly all modern saw gin stands use a revolving brush cylinder to doff lint fibers from the gin saws. The revolving brush also generates a stream of air that transports the lint through a sheet-metal duct to a lint condenser at

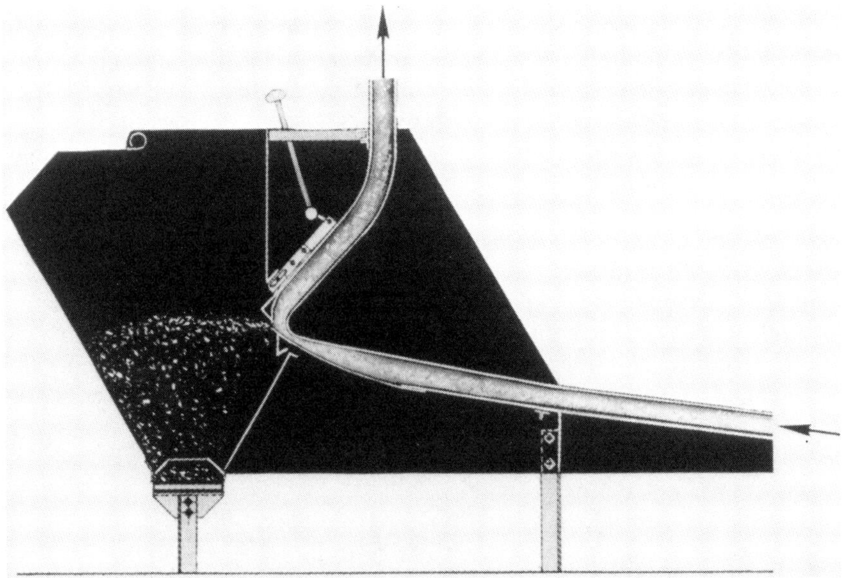

Fig. 11–20. Pneumatic lint cleaner (Lummus Industries, Inc.).

the next stage of processing. An axial-flow suction fan in the air-discharge line of the condenser enhances the conveying process by helping the brush overcome the airflow resistance of the condenser and ductwork. Lint cotton may easily be conveyed with an air velocities of 7 to 10 m/sec (1400 to 2000 ft/min).

Ginned lint is usually delivered first to a lint cleaner condenser. The condenser separates the lint from the conveying air, forms the lint into a batt and delivers the batt to the feed works of a lint cleaner. After cleaning, the lint is doffed from the lint cleaner saws and transported by air to the next stage of lint cleaning, or to a battery condenser at the baling press. Some saw gin stands and lint cleaners are equipped with an air-blast doffing system instead of the brush-doffing system. In either case, the pneumatic conveying of lint from the gin stand to the lint cleaner (or press) is effectively accomplished.

11–7.2   Lint Collection and Baling

The battery condenser at the press forms the lint into a batt and discharges the batt onto an inclined metal slide attached to a lint feeder on the press tramper. The lint feeder, which may be a revolving kicker, a belt conveyor, a hydraulic pusher, or an air-suction device, deposits the lint into the charging box of the press during the upstroke of the tramper. The tramper is a mechanical or hydraulic device which uniformly packs the lint into the press box.

Fig. 11–21.   Opener-cleaner and pneumatic lint cleaner arrangement for extra-long-stable fiber (Lummus Industries, Inc.).

The typical baling press is composed of two press boxes (Fig. 11–22). While cotton is being pressed and packaged in one of the boxes, the other box is being filled and loosely packed by the tramper. After enough lint has accumulated to make approximately a 227 kg bale, the "just-filled" box is swung into position for pressing and the other "now empty" box is rotated into the filling and tramping position. For pressing, a hydraulic ram equipped with a follow block forces the cotton into the typical rectangular bale shape for wrapping and tieing.

Several types of gin presses are available for producing bales of different sizes and densities. However, in recent years the U.S. ginning industry has adopted the universal density bale, which has a nominal density of 448 kg/m³ (28 lb/ft³). The universal density bale may be produced at the cotton gin with a gin universal press, or it may be packaged with a modified flat bale press at the gin and repressed to the required density at a compress. Gin universal density bales are approximately 140 cm long, 53 cm wide, and 66 to 76 cm thick while compress universal density bales are approximately 145 cm long, 64 cm wide, and 56 cm thick.

11–7.3 Packaging Materials

The primary purpose of wrapping cotton at the gin is to protect the bale from foreign matter that may become affixed to the bale before it arrives at the spinning mill. Contaminants in cotton represent a loss of

Fig. 11–22. Modern hydraulic baling press with automatic stapping arrangement (Bush Hog/Continental Gin Co.).

otherwise usable fiber, and if extraneous materials get into the processing line at the mill they can cause serious defects in literally thousands of yards of yarn or fabric. Additional reasons for wrapping the bales are to lessen the risk of fire spread during storage and to protect them from the weather.

For many years cotton bales were packed exclusively with jute bagging weighing 5.4 kg per bale and hot-rolled steel ties weighing 4 kg per bale. Changes in trading practices from gross-weight trading systems (where pricing and sales were made on gross weights of fiber plus tares) to net-weight trading allowed use of improved packaging materials and automated packaging.

Predominant wrapping material in use today is woven polypropylene, however, lightweight jute, burlap, or polyethylene materials are still available and are preferred by many users. The heavy hot-rolled steel bands formerly used for tieing cotton bales have given way to lighter but stronger, high tensile strength steel wires and cold-rolled strapping.

Modified flat bales are commonly fitted with two sheets of wrapping material and bale ties that are applied by hand while the bale is under pressure in the press box. Most universal density bales are tied out naked and subsequently placed in a polypropylene, polyethylene or burlap bag at the gin or warehouse. Some universal density bales, however, are dressed with two sheets of wrapping material in the press box in a fashion similar to that of modified flat bales. Six bale ties are applied to modified flat bales while eight are used for universal density bales. Ties on universal density bales may be applied either mechanically or manually.

11-7.4   Compressing

Modified flat bales produced at the gin are usually repressed to a higher density for efficient and economical shipment to domestic and foreign textile mills. The repressing operation, usually called compressing, is performed at a centralized facility known as a compress warehouse. Universal density bales, which do not need additional compression, are usually also delivered to a compress warehouse for storage prior to final shipment to the textile mill.

In addition to the compressing function, warehouses also perform several other essential services for the cotton industry. Some of these services include bale sampling, insured bale storage, repair of damaged bales, and assembly of bales for final shipment to textile mills. The compress warehouse, which receives bales from many cotton gins, provides the means for concentrating cotton from a large area at a central location. This function greatly simplifies and expedites the orderly marketing and shipment of U.S. cotton.

Modified flat bales are compressed to a final density of about 448 kg/m^3 (28 lb/ft^3) by large steam-powered presses. These presses are much larger and stronger than those used at cotton gins, and they usually operate at higher hourly capacities. In the compressing operation, the thickness of the modified flat bale is reduced to the required dimension without applying side pressure. The absence of side pressure in the operation insures that the layers of lint in the bale remain flat for easy opening and feeding at the textile mill.

11-7.5 Bale Appearance

The American cotton bale has been widely criticized for many years because of its poor appearance when it arrives at domestic mills or at foreign ports. The primary causes of this problem can be traced to the practice of cutting samples out of the bale after it has been packaged and to rough, mechanical handling of the bale during shipment. Progress has been made in recent years in alleviating these problems. The adoption of the universal density bale along with improvements in bale packaging materials has done much to improve the durability and appearance of the U.S. bale. Additional steps are being taken by the industry to reduce problems with bale sampling. Most universal density bales are now sampled prior to packaging. The sample may be mechanically withdrawn during ginning or cut from the naked bale before the bale is placed in a bag. At the compress, these bags may be reopened and the bale resampled without cutting the wrapping material. Some compresses use plastic slip covers in addition to the primary bagging to protect bales from dust and moiture when bales must be temporarily stored outdoors. As these and other innovative bale packaging and handling techniques become more commonplace, the appearance and condition of U.S. bales should improve dramatically.

11-8 WASTE HANDLING AND COLLECTION SYSTEMS

Waste material removed from the cotton by the gin's various processes is transported to outdoor collection facilities by 10 to 20 different pneumatic conveying systems. These systems, which utilize both centrifugal and vane-axial fans, require large volumes of air. The total airflow requirements for handling material at a gin varies from about 18.9 to 82.6 m³/sec (40 000 to 175 000 ft³/min), depending upon the gin's operating capacity and the foreign matter content of the cotton (Wilmot et al., 1974). The amount of waste handled per hour is also a function of ginning capacity and the cotton's foreign matter content. Typical waste handling rates range from 500 to 3000 kg/hour for machine-picked cotton, and from 2000 to 14 000 kg/hour for machine-stripped cotton. A gin's system for handling waste must be designed to (i) remove the waste from the gin building at a rate of commensurate with ginning capacity, (ii) collect the waste from the conveying air streams without causing excessive air pollution problems, and (iii) dispose of the waste in an acceptable manner.

11–8.1 Waste Collection Devices

11–8.1.1 Cyclone Collector

The most widely used device for collecting waste at cotton gins is the cyclone collector. This simple device, which has no moving parts, utilizes centrifugal force to separate waste from the conveying air. The small-diameter, high-efficiency cyclone is the type most commonly employed at cotton gins. Small-diameter cyclones are sized for a 15 m/sec (3000 ft/min) entrance air velocity, and at this velocity the cyclones have a static pressure drop of 0.75 to 1.25 kPa (3 to 5 in of water). Most cyclones used at cotton gins are 0.5 to 1.2 m in diam, and these units are frequently mounted in multiple parallel arrangements for high-volume applications (Fig. 11–23). A properly sized small-diameter cyclone will collect virtually 100% of the waste that is larger than 30 μm in diam, and 99.9% of the total waste (Wesley et al., 1972).

11–8.1.2 Condenser Drum Covering

Short fibers from the exhausts of the lint condensers are one of the cotton gin's most objectionable emissions. These fibers collect on electric lines, rooftops, and surrounding vegetation, creating a nuisance and a fire

Fig. 11–23. A waste collection system featuring small-diameter cyclones and a bur hopper.

hazard. Many cotton ginners control short-fiber emissions by covering their condenser drums with 70 to 100-mesh screen wire or by replacing the standard drum covering with a fine, perforated metal. When properly designed and installed, the condenser drum covering virtually eliminates the short-fiber emission problem with no measurable adverse effect on cotton quality.

11–8.1.3 Unifilter Collection System

The unifilter collection system consists of a settling chamber and a large revolving filter drum (Fig. 11–24). Large waste particles are collected by gravity in the settling chamber while the light material and dust collect on the matt-type filter medium. The filter medium is continuously vacuumed by several suction nozzles to prevent the buildup of excessive accumulations of dust that would restrict airflow. The unifilter system can collect a wide variety of waste products and is suitable for use with centrifugal fan systems as well as the vane-axial fane systems (McCaskill and Wesley, 1976). Unifilter collection systems for airflows up to 82.6 m³/sec (175 000 ft³/min) are commercially available.

11–8.1.4 Miscellaneous Collection Devices

Several other types of waste collection devices are used to a limited extent at cotton gins. Inline air filters, wet scrubbers, dry skimmers, large-volume settling chambers, bag filters, and screen cages are a few of the devices that are occasionally used as a matter of personal preference or as a

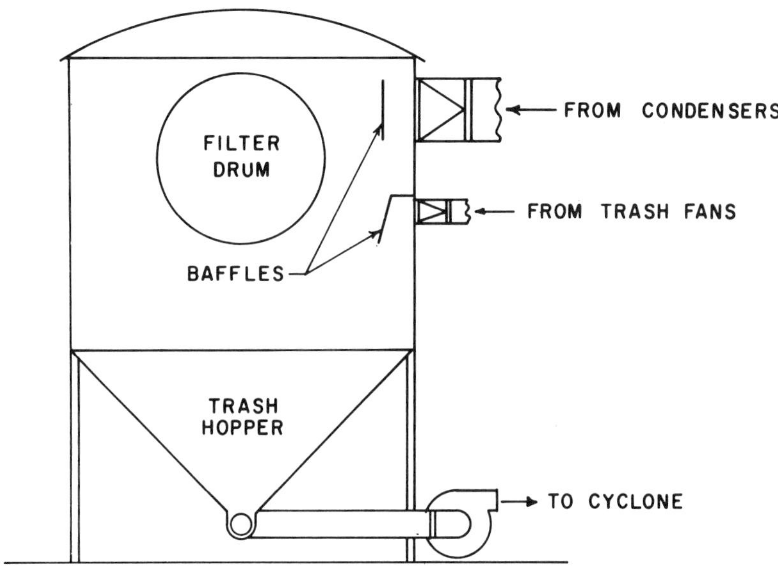

Fig. 11–24. Diagram of a unifilter waste collection system.

means of meeting specialized objectives. Also, various types of industrial dust collection devices have been used on occasion at cotton gins. However, the industrial devices, while very efficient, are usually too expensive for most cotton gin applications. Cotton gins have historically opted for the simplest and least expensive collection device that would meet minimum waste collection and air pollution control requirements.

11-8.2 Waste Disposal

It is usually necessary for gins to store waste temporarily prior to final disposal. Elevated bur hoppers provide a convenient means of accumulating truckloads of waste, and they also expedite the loading operation. Loading the trucks is accomplished by simply opening the bottom of the bur hopper and allowing the accumulated waste to fall into the truck (Fig. 11-23).

Ginning waste may be disposed of by a variety of means. Vegetative waste such as burs, sticks, and leaf material can be composted, applied to the soil as a mulch, fed to livestock, or simply dumped in a landfill. Cotton fibers that are lost during the ginning process are often recovered and sold to mote cleaning or garnetting plants. The fibrous material is used for coarse yars, felts, batting, and other nonwoven products.

For many years incineration was the most common method for disposing of gin waste. This practice was abandoned in the 1960s because the gin-type incinerators of that era failed to meet the new air quality standards resulting from the 1963 U.S. Clean Air Act. Even though incineration of gin waste is now limited to special or unusual circumstances, the practice continues to be of interest from the standpoint of improved energy utilization. Ginning waste is a potential source of vast amounts of energy, and research continues on the development of acceptable processes for converting ginning waste into usable energy for the cotton gin (Griffin, 1976).

11-8.3 Air Pollution and Safety Regulations

States that produce cotton enforce particulate emission regulations for cotton gins in accordance with the provisions of the U.S. Clean Air Act of 1970. Although these regulations vary from state to state, all U.S. cotton gins are now required to control particulate emissions. Some states have specific regulations for cotton gin emissions, while others rely on general regulations covering all agricultural processors. Several methods are used for determining allowable emissions from cotton gins. Generally, allowable emissions are based on the weight of particulates emitted to the atmosphere per (i) unit of time, (ii) unit volume of air discharged, or (iii) weight of material processed by the gin (Wilmot et al., 1974). Some areas also utilize opacity measurements, especially for smoke emissions. One state allows gins to select an "alternate method of control" if the selected method provides a control efficiency that is comparable to the process-weight method.

Strict enforcement of these regulations have produced severe financial hardships for many cotton gins. However, the gins that survived are now reasonably well equipped to control particulate emissions.

United States cotton gins are also required to comply with various state and federal safety regulations as a result of the U.S. Occupational Safety and Health Act of 1970. Generally, gins are expected to provide their employees with a safe working environment that is free of recognized safety hazards. Specifically, gins must comply with applicable sections of the U.S. Occupational Safety and Health Administration's Standards for Agriculture (No. 1928). These standards include requirements for gin machinery guarding and the operation of agricltural tractors. Most existing cotton gin plants have been retrofitted with the required safety equipment, and new gin installations are designed and constructed to meet these requirements.

11-9 MODERN GINNING SYSTEMS

A complete cotton ginning system is composed of a series of machines that condition and clean the seed cotton, separate the fibers from the seed, clean the ginned lint, and package the lint for shipment to the textile mill. The amount of machinery required to satisfactorily accomplish these tasks depends on the initial foreign matter and moisture contents of the harvested seed cotton, which vary with growing and harvesting conditions and with harvesting method. Since harvesting methods has such a large effect on initial foreign matter content, specific gin system designs have been formulated for each of the most common methods of harvesting, i.e., hand picking, machine picking, and machine stripping. These designs are generally flexible enough to accomodate smaller variations in foreign matter and moisture content resulting from vagrancies in the weather during the growing and harvesting season. Since excessive drying and machining of cotton can adversely affect its quality, design of gin systems must also be based on fiber quality considerations. Generally, successful designs are those that maximize bale value for the producer while at the same time preserve the inherent quality of the cotton fiber.

11-9.1 Gin Designs to Handle Specific Types of Cotton

Numerous tests by the USDA Agricultural Research Services' Cotton Ginning Laboratories and by gin machinery manufacturers have formed the basis for recommendations on the sequence and amount of gin machinery required to adequately process cotton harvested by various methods. These machinery recommendations, shown in Fig. 11-25, are general in nature in that they represent typical conditions in most cotton-producing areas. Under these conditions, the recommended machinery arrangements will produce satisfactory lint grades and near-maximum bale values for most cottons. However, some modification of these recommendations may be necessary in some production areas to meet special needs or unusual growing conditions. Cotton containing excessive amounts of foreign matter, particularly weed and grass material, or hard-to-clean cultivars of cotton may require more cleaning than can be obtained with the basic machinery arrangements. On the other hand, very clean cotton can often be satisfactorily cleaned with less machinery. For this reason, seed-cotton cleaners and extractors, and lint cleaners should be provided with bypasses to allow for the selection of less cleaning machinery when processing extra clean, dry cotton. Generally, a ginner should select the minimum amount of machinery required to maximize bale value for the producer. In this way he will minimize ginning costs and at the same time preserve the inherent qualities of the cotton fiber.

11–9.2   Effects of Ginning on Fiber Quality

The term "fiber quality" has different meanings for different interest groups. Producers sometimes view fiber quality strictly in terms of those factors that directly determine market price, i.e., grade, staple length, and micronaire value. To spinners, however, a definition of fiber quality would likely include additional fiber properties that have been found to correlate more precisely than do the market factors with spinning performance and

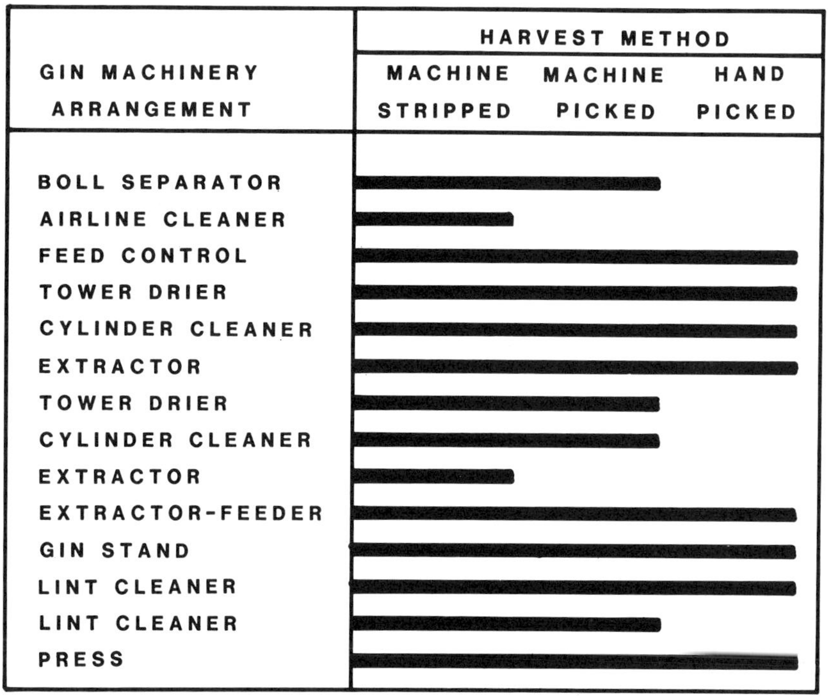

GIN MACHINERY ARRANGEMENT	HARVEST METHOD		
	MACHINE STRIPPED	MACHINE PICKED	HAND PICKED
BOLL SEPARATOR			
AIRLINE CLEANER			
FEED CONTROL			
TOWER DRIER			
CYLINDER CLEANER			
EXTRACTOR			
TOWER DRIER			
CYLINDER CLEANER			
EXTRACTOR			
EXTRACTOR-FEEDER			
GIN STAND			
LINT CLEANER			
LINT CLEANER			
PRESS			

Fig. 11–25. Recommended gin machinery arrangements for hand-picked, machine-picked, and machine-stripped cotton.

yarn quality potential. Depending upon the spinner's particular interests, this latter group might include measurements of fiber length, length uniformity, short-fiber content, strength, maturity, nep content, nonlint content, dust content, and possibly other properties.

Regardless of ones definition of fiber quality, cotton possesses its highest fiber quality when it is still on the stalk in the field. Technically, any mechanical handling of the cotton thereafter has the potential of modifying the natural qualities of the fiber. From a practical standpoint, however, the thermopneumatic and mechanical processes in ginning usually affect only a few fiber properties, principally foreign matter content, the various parameters of fiber length, and the appearance of the lint sample (ginning preparation and nep content). Although ginning affects only a few fiber properties, the ones that are affected have a significant impact on the market price of cotton and on the fibers' ultimate end-use value.

The ginning process influences grade and staple length, but has very little effect on micronaire value. The micronaire value, which is an indirect measure of fiber fineness or maturity, is controlled primarily by genetic factors and cotton growth conditions. Lint grade is determined by color, foreign matter content, and ginning preparation. While the ginning process has little effect on true fiber color, the blending action of lint cleaners can enhance lint-color designation to some extent. Lint cleaning can sometimes change the lint's color category from "spotted" to "light spotted," or from "light spotted" to "white".

Foreign matter removal during ginning has the greatest effect on lint grade. Lint grades increase with increased levels of gin cleaning and drying, but at a declining rate. Levels of cleaning beyond those shown in Fig. 11–25 generally produce only minor improvements in grade. Drying to moisture levels below the recommended 6.5 to 8.0% level may improve lint grade, but at the expense of shortening the staple length and reducing bale weight. Frequently the losses in lint value due to these adverse effects of drying will offset the increases in lint value due to grade improvement. Thus, seed-cotton driers should be adjusted to remove only enough moisture from the cotton to ensure smooth ginning. On the other hand, inadequate drying can be a source of many ginning problems. The handling and cleaning of wet cotton produces a stringy and tangled sample that may be reduced in grade because of rough ginning preparation, and inadequate drying reduces the effectiveness of the gin's cleaning machinery.

Ginning and lint cleaning can adversely affect fiber length and length distribution, especially when the fiber moisture content is very low. Since fiber tensile strength is proportional to fiber moisture content, fiber breakage increases as moisture content decreases. Fiber breakage during ginning decreases fiber length, increases the short-fiber content, and shifts the fiber-length distribution toward the short end of the length spectrum. These length changes adversely affect yarn strength, yarn appearance, and spinning-end breakage. The length characteristics of the fiber can best be preserved by keeping the fiber moisture content within the 6.5 to 8.0% range and by using the minimum amount of machinery needed to produce satisfactory grades and near-maximum bale values.

The presence of an excessive number of neps in lint is a problem for many spinners. Neps are small knots of tangled fibers that adversely affect spinning performance, yarn appearance, and uniformity of dyeing. Since neps do not appear in unginned cotton, their presence in the lint is attributable to the ginning process. The nepping potential of fibers is closely related to fiber fineness and maturity. Fine fibers, especially if they are immature, are more susceptible to nepping than are course or fully matured fibers. While some neps are formed during seed-cotton cleaning, research findings indicate that most neps are created by saw gin stands and saw lint cleaners (Chapman and Stedronsky, 1959). The formation of neps during the saw-ginning process may be minimized by (i) controlling fiber moisture content at the optimum level, (ii) eliminating unnecessary handling and cleaning operations, and (iii) employing moderate processing rates. While these processing remedies are helpful, they do not completely eliminate the problem. Fine-fiber, long staple cultivars of cotton are normally roller ginned to avoid the nepping problem that usually accompanies the saw-type processes.

REFERENCES

Alberson, D. M., and V. L. Stedronsky. 1964. Roller ginning American-Egyptian cotton in the Southwest. USDA Agric. Handbk. No. 257. U.S. Government Printing Office, Washington, D.C.

Baker, R. V. 1978. Performance characteristics of saw-type lint cleaners, Trans. of ASAE 21: 1081–1087, 1091.

----, P. A. Boving, and J. W. Laird. 1981. Compare performances of bur and stick extractors. Cott. Gin. J. Yearb. 49:12–18.

----, E. P. Columbus, and J. W. Laird. 1977. Cleaning machine-stripped cotton for efficient ginning and maximum bale values. USDA Tech. Rep. No. 1540.

Bennett, C. A. 1956. Ginning cotton. USDA Farmers Bull. No. 1748.

----. 1959. Roller cotton ginning developments. Texas Cotton Ginners' Association and Cotton Gin and Oil Mill Press, Dallas.

----. 1962. Cotton ginning systems and auxiliary developments. Texas Cotton Ginners' Association and Cotton Gin and Oil Mill Press, Dallas.

Brooks, E. H. 1947. Lint cotton cleaner. U.S. Pat. 2,418,649. U.S. Patent Office Gaz. 597:252.

Brown, H. B., and J. O. Ware. 1958. Cotton. McGraw-Hill Book Co., Inc., N.Y.

Chapman, W. E., and V. L. Stedronsky. 1959. Ginning Acala cotton in the Southwest. USDA Prod. Res. Rep. No. 27.

----, and ----. 1965. Comparative performance of saw and roller gins on Acala and Pima cottons. USDA Mark. Res. Rep. No. 694.

Franks, G. N., and C. S. Shaw. 1959. Stick remover for cotton gins. USDA Prod. Res. Rep. No. 22.

Garner, W. E., and R. V. Baker. 1977. Cleaning and extracting. p. 18–29. *In* Cotton ginners handbook. USDA Agric. Handbk. No. 503. U.S. Government Printing Office, Washington, D.C.

Griffin, A. C. 1976. Fuel and ash content of ginning waste. Trans. ASAE 19:156–158, 167.

----, P. E. LaFerney, and E. H. Shanklin. 1970. Effects of lint-cleaner operating parameters on cotton quality. USDA Mark. Res. Rep. No 864.

----, and V. P. Moore. 1965. Relation of physical properties of cotton to commerce and ginning research. Trans. ASAE 8:488–490.

Leonard, C. G. 1970. An improved seed cotton feeder for use with a rotary-knife roller gin stand. USDA-ARS Rep. No. 42-174.

----, J. E. Ross, and R. A. Mullikin. 1970. Moisture conditioning of seed cotton in ginning as related to fiber quality and spinning performance. USDA Mark. Res. Rep. No. MRR 859.

Looney, Z. M., L. D. LaPlue, C. A. Wilmot, W. E. Chapman, and F. E. Newton. 1963. Multiple lint cleaning at cotton gins, USDA Mark. Res. Rep. No. 601.

Mangialardi, G. J. 1974. Effects of feed rate and batt density on operation of saw-cylinder lint cleaners. USDA Prod. Res. Rep. No. 156.

McCaskill, O. L., and R. A. Wesley. 1976. Unifilter collecting system for cotton-gin waste materials. USDA, Agric. Res. Serv. Rep. No. ARS-2-144.

Mirsky, Jeannette, and Allen Nevins. 1952. The world of Eli Whitney. The Macmillan Co., N.Y.

Moore, V. P. 1977. Development of the saw gin. p. 1-3. *In* Cotton Ginners' Handbk. USDA Agric. Handbk. No. 503. U.S. Government Printing Office, Washington, D.C.

----, and C. M. Merkel. 1953. Cleaning cotton at gins and methods for improvement. USDA Circ. 922.

Moss, E. E. 1955. Cotton lint cleaners. U.S. Pat. 2,704,862. U.S. Patent Office Gaz. 692:541.

Pendleton, A. M., and V. P. Moore. 1968. Ginning for cotton quality preservation. p. 467-486. *In* F. C. Elliot, M. Hoover, and W. K. Porter (ed.) Advances in production and utilization of quality cotton: Principles and practices. Iowa State University Press, Ames.

Stedronsky, V. L., and C. S. Shaw. 1950. The flow-through lint-cotton cleaner. USDA Circ. No. 858.

Wesley, R. A., W. D. Mayfield, and O. L. McCaskill. 1972. An evaluation of the cyclone collector for cotton gins. USDA Tech. Bull. No. 1439.

Wilmot, C. A., Z. M. Looney, and O. L. McCaskill. 1974. The cost of air pollution control to cotton ginners. USDA, Econ. Res. Ser. Rep. ERS-536.

12 Fiber

Henry H. Perkins, Jr.
ARS-USDA
Clemson, South Carolina

Don E. Ethridge
Texas Tech University
Lubbock, Texas

Charles K. Bragg
ARS-USDA
Clemson, South Carolina

The fibrous mass removed from the seed by the cotton gin is unique as a textile fiber. It has uses in a host of products that sustain and make life more comfortable and aesthetically appealing. It is an agricultural product grown by procedures that have evolved from traditional practices to the high level of efficiency demonstrated by modern agriculture. The high efficiency of the plant in bearing this unique commodity is the culmination of the efforts of countless agricultural scientists in many different disciplines of research. Similarly remarkable is the system for harvesting, ginning, packaging, and distributing the fiber and seed from the plants. Processing of the fiber into products for domestic, personal, and industrial use has always occupied a position of strategic importance. Cotton processing was a key element in the industrial revolution and led the way for the development of modern society. Fiber science grew slowly and largely because of the economic needs of mechanized industry. In more recent years fiber measurement and processing technology have grown into an independent science. The processes involved are difficult to grasp from written descriptions and illustrations. A much clearer understanding of fiber measurement and processing technology can be obtained by visiting a textile mill and observing the operation of instruments and machines (Hamby, 1966).

Published in *Cotton,* Agronomy Monograph no. 24, © ASA-CSSA-SSSA, 677 South Segoe Road, Madison, WI 53711.

12-1   COTTON CLASSIFICATION

Cotton classing has been practiced since the early days of the industry. Classing is done with the hand and eye. The texture and appearance of cotton are used to distinguish the levels of quality. Grade, staple, character, and other descriptive terms now used required extensive time for development to the degree of precision generally associated with them by the cotton industry. Early records indicate that at one time cotton was sold by peddlers and samples were brought to the mill, and the buyer selected those cottons that would perform well based on past experience. The more objective buyers probably kept samples from one year to the next as a basis for comparison; however, standards of quality were not developed for many years. The first real quality distinctions were made on the basis of area of growth or type of cotton. As early as 1775 in England, such terms as Indian, American, and West Indian conveyed specific quality implications. At the beginning of the 19th century, more specific terms began to be used to differentiate among cottons with different qualities because of the rapid increase in cotton production, the widely varying growing conditions in America, and the growth of mechanical processing.

The first distinction in quality made in the sale of cotton grown in America was the separation of the crop into Sea Island and Upland. The terms Sea Island, Santee, and Short Staple were descriptive terms used in Charleston, South Carolina in 1816 as a basis for price quotations. As the use of cotton in fabrics and yarns became more widespread and processing became more mechanized, increased demands on the quality of cotton emphasized the need for quality distinctions, and numerous terms were developed to describe or establish cotton quality. By 1841 the Liverpool Cotton Brokers Association had developed standards for use in classification. In 1853 similar standards were adopted by New York cotton brokers. From 1853 to 1909 standards for cotton classification became widespread among the cotton exchanges. These standards were not uniform, and, in addition to the quality description, it was also necessary to know whose standards were used to fully understand a quality designation. To help solve problems stemming from the use of various standards, USDA established nine standard grades for Upland cotton in 1909. Current standards for classification of cotton were developed from this beginning.

12-1.1   Basis for Cotton Classification

Thousands of farms produce cotton in this country and much variation is found in the quality of this production in any year. Variations occur in the quality of cotton grown on a single farm, and it is not unusual to find considerable variation within a single bale. These variations in quality result from differences in cultivars planted, soils, rainfall, irrigation practices, fertilizers, temperatures, cultural methods, insect damage, length of growing season, exposure of open cotton before harvest, method of harvesting and ginning, and many other variables. Quality is governed to a considerable extent by the cultivar, quality of the seed planted, the weather, and the farming practices followed during the period in which the fibers were developed. The quality factors of color, leaf, and ginning preparation are affected to a great extent by the weather and length of exposure after the bolls open, by plant characteristics, and by harvesting and ginning practices.

Cotton classification, or classing, is the process of describing the quality of cotton in terms of grade, staple length, and micronaire reading according to the Official Cotton Standards of the USA (Cotton Division, AMS, 1980). Cotton may be classed in terms of other standards or by comparison with other types; but in the USA the official standards must be used. For grade, classification is based on appearance and is accomplished chiefly through the sense of sight by integration of color, leaf, and preparation in the sample. Classification for staple length involves sight and touch and is made by pulling out and comparing a typical portion of fibers from the sample with the official staple type. Micronaire reading is determined by an airflow measurement which indicates fiber fineness and maturity in combination.

Classification of grade, staple length, and micronaire reading indicates to a large extent the spinning utility and hence the market value of each bale. In recent years the USDA has classed for farmers about 97% of the cotton produced. Farmers are interested in the classification of each bale to appraise production, harvesting, and ginning practices and to market their cotton advantageously. Classification provides a means for cotton merchants to buy and sell cotton effectively. Uniformity of quality is desired in the processing of cotton. While classification remains essential to the pricing systems for cotton, additional measurements, particularly fiber strength, are often used in commercial transactions and to control processing variations.

12–1.2 Sampling

Cotton is packaged in bales weighing approximately 218 kg. Because of the size and method of packaging, it is impractical for buyers to inspect actual bales of cotton at the time of purchase. Therefore, it is a general practice for market transactions to be based on samples drawn from each bale. Samples are satisfactory for this purpose to the extent that they are representative of the bales from which they are drawn.

A sample should weigh at least 170 g and consist of two parts of 85 g each taken from opposite sides of a bale. In drawing the sample, cuts are made deep enough to reveal a true specimen of the bale and enough to produce a face about 15 cm wide. A sample should not be dressed or trimmed, nor should it be drawn by cutting a second time into an old sample hole. If the sample is too small, or the face too narrow, it is difficult to determine the correct classification.

Bales may be sampled before or after they are wrapped. Originally, all sampling involved cutting the wrapping material. Since many users of cotton objected to losses from the sample holes, and patching the holes is costly, methods were developed for sampling that leave the bale cover intact. Methods which allow the bale to be formed and banded, then cut for sampling before putting on the bale covering, have gained widespread use since their development.

Another way to avoid damaging the protective bale covering is by use of automatic samplers. These machines were developed for sampling bales during the ginning process, and they can be installed in most existing gins. At timed intervals during ginning small portions of the lint are extracted, automatically accumulated, pressed, and packaged in a wrapper bearing the gin bale number. Two or three samples of appropriate size for classing are obtained, and the bale covering can be left intact.

After samples have been properly drawn, they must be handled carefully to prevent loss of leaf, sand, dust, or other material that would change their representativeness. It is also important to assure that no particles of leaf, dust, or sand are dropped upon them, as they will be changed in grade by the accidental accumulation of such trash.

The useful life of a sample depends to a large extent upon the care it is given and the extent to which it is handled. The usefulness of a sample also depends upon the preservation of its identity. A tag or coupon showing bale number, name of gin, compress, warehouse, or other identification should be kept between the sample portions from the two sides of the bale.

If a sample may be needed for future reference after classification, it should be carefully wrapped to preserve its identity. The sample may be preserved by placing it in an individual container, such as a plastic bag, or by storing it with a group of samples carefully placed in a large container. Samples should not be allowed to remain exposed as dust can cause loss of brightness, and light can cause color changes.

Temperature and humidity are important in sample storage. Excessive drying of samples can shorten the apparent staple length. Excessive dampness can accelerate the rate of increase in yellowness of the cotton. An extremely high temperature can magnify the effect of humidity. The validity of any cotton sample is subject to the conditions and length of storage.

12–1.3 Grade

The standards first established by USDA in 1909 were not mandatory and were never used to any great extent in either domestic or export trade. The next Upland standards were promulgated under the U.S. Cotton Futures Act of 1914 (re-enacted, 1916), and their use was made practically compulsory on the cotton futures exchanges in the USA beginning on 18 Feb. 1915. Official standards for Pima cotton were first established in 1918. The U.S. Cotton Standards Act of 1923 gave further impetus to cotton standardization. The grade standards have been expanded in number and revised from time to time since the early days of standardization.

Most of the current standards for grade of Upland cotton were promulgated by USDA in June 1962. The last major revision of grades for Upland cotton occurred at that time, but minor revisions have been made in response to changing production patterns since then. The U.S. Cotton Standards Act provides for a notice of one year before any change or revision of the standards can be made effective.

The Official Cotton Standards of the USA for the grade of Upland cotton are also called Universal Standards. USDA entered into the original Universal Cotton Standards Agreement with nine leading cotton associations in seven major European cotton-consuming countries in 1923 and 1924. The original agreement has been supplemented and revised and now includes cotton associations and exchanges in many foreign countries. The agreement provides for (i) the adoption, use, and observance of the Universal Standards in the classification of U.S. Upland cotton, (ii) the arbitration of settlement of disputes with respect to such classification, and (iii) the preparation, distribution, and protection of key physical copies of the Universal Standards.

Conferences are held every 3 years in the USA to ensure accurate reproduction of the standards and to consider any revisions needed in the standards. United States cotton associations and exchanges participate in the conference and the standardization process. United States and foreign representatives provide vitally important advice to USDA in the development of standards.

Grade is a composite assessment of three factors—color, leaf, and preparation. Although each factor is judged separately, the classer integrates his assessment of the three into a composite grade based on his overall impression of the sample. Color and leaf standards, prepared and maintained by Cotton Division, Agricultural Marketing Service (AMS), provide a basis for maintaining consistency in the assessment of grade. The grade designations of the official standards are shown in Table 12–1. Only a portion of the grades is represented in physical form in grade boxes as shown in Fig. 12–1. The remaining standards for grade are descriptive and no physical form is maintained. The physical forms are used for references

Table 12–1. Official grade designations for Upland cotton.

White	Light gray	Gray	Light spotted	Spotted	Tinged	Stained
GM*	GM	GM	GM	GM	--	--
SM*	SM	SM	SM	SM*	SM	SM
M*	M	M	M	M*	M*	M
SLM*	SLM	SLM	SLM	SLM*	SLM*	--
LM*	--	--	LM	LM*	LM*	--
SGO*	--	--	--	--	--	--
GO*	--	--	--	--	--	--

* Grades for which physical standards are maintained.
  Abbreviations: G = Good, M = Middling, S = Strict, L = Low, and O = Ordinary.
  Example: SLM = Strict Low Middling grade.
  White grades in which the color grade exceeds the leaf grade are designated as "plus."
  Example: Cotton with Middling color and Strict Low Middling leaf is designated SLM plus.

as necessary by classers. The range of grade for each of the physical standards is represented by 12 samples in the official box. Cottons from several areas of production are used in each grade box.

The descriptive standards for grade are based upon the physical standards. The descriptive standards provide descriptions of cottons in which the factors of grade—color, leaf, and preparation—are not illustrated by a physical standard. For example, Middling Gray is Upland cotton which is color is Strict Good Ordinary and which in leaf and preparation is Middling or better, and Middling Plus is Upland cotton which is Middling in leaf and preparation with Strict Middling color.

Pima and Upland grade standards differ. Pima cotton is naturally of a deeper yellow color than Upland cotton. The leaf contents of Pima standards are peculiar to this cotton and do not match that of Upland standards.

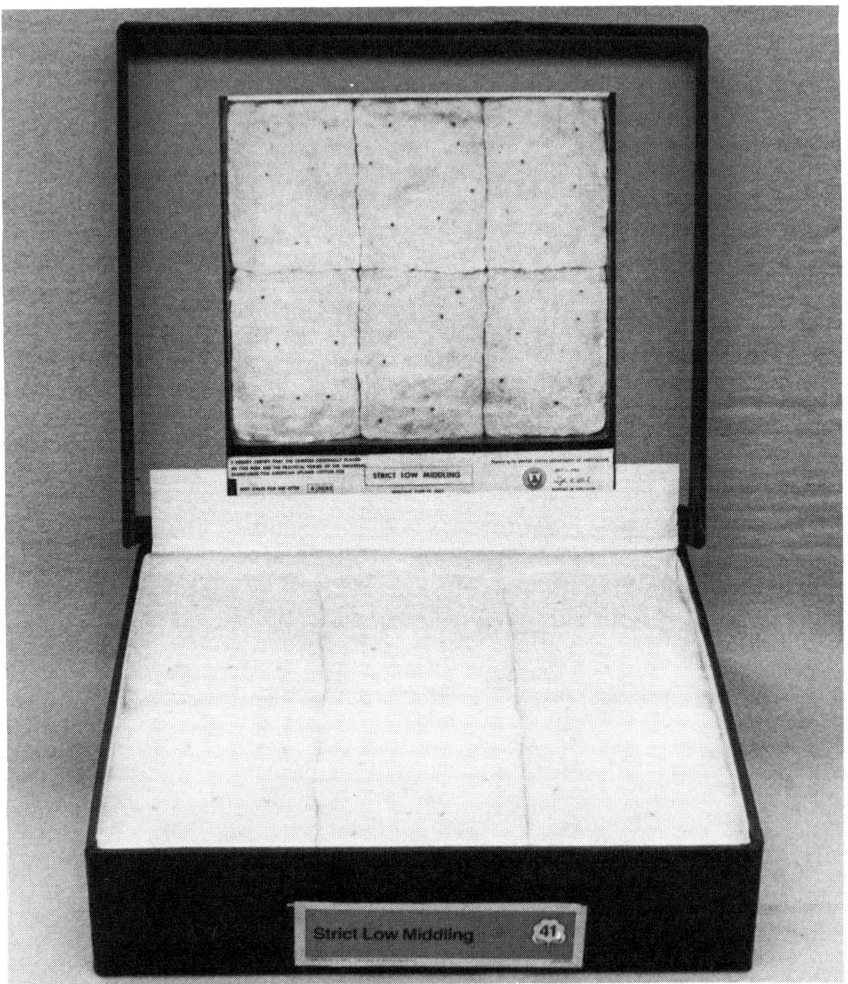

Fig. 12-1. Representative grade boxes of official standards.

The preparation is very different from Upland standards because Pima cotton is ginned on roller gins and is more stringy and lumpy. Pima cotton grades range from the highest grade of 1 to the lowest grade of 10.

12-1.3.1 Color

Upland cotton is naturally white. Continued exposure in the field to weathering and the action of microorganisms can cause this white cotton to lose its brightness and become darker. Under extreme conditions of weather damage, the color can become a very dark, bluish gray.

Upland cotton in which growth is stopped prematurely by a frost, drought, etc., is characterized by increased saturation of yellow color. Cotton can also become discolored or spotted by the action of insects, microorganisms, and soil stains. This cotton will become darker after exposure to adverse conditions in the field.

Discoloration can also be caused by oil or grease from harvesting equipment, or it can come from green leaves or other parts of the cotton plant which have been crushed by the machinery. Any departure from the bright color or normally opened cotton indicates a deterioration in quality.

In the grading of Upland cotton, all of these color differences are recognized, divided into categories, and described. The varying amount of yellow color found in cotton forms the basis for the color groups used in the standards for grading Upland cotton.

These color groups are: White, Light Spotted, Spotted, Tinged, and Yellow Stained. Plus, Light Gray, and Gray designations are used to indicate combinations of color and leaf which are different from those normally found in the White grades.

As the cotton in each of the color groups is exposed to weathering, it becomes progressively darker. The degree of darkness is the principal basis for grade divisions within each color group, the higher grades being brighter in color than the lower grades. However, each grade within a color group also contains differences in reflectance—for example, the difference between the whitest and the creamiest sample within any one of the grade boxes. Strict Middling, Middling, and Strict Low Middling in each color group are examples of grade divisions within color groups.

12-1.3.2 Leaf and Other Trash

Cotton usually becomes contaminated by leaf and other trash in various amounts through exposure in the field and harvesting methods. The amount of foreign matter remaining in the lint after ginning is largely dependent upon the trash content, the condition of the cotton at the time of harvest, and the amount of cleaning and drying machinery used by the gin. Even when cotton is carefully harvested under ideal field conditions, there will be some degree of leaf and trash. Although much of the foreign matter is removed by the cleaning and drying processes during ginning, it is at present impractical to remove all of it.

Leaf includes dried and broken plant foliage of various kinds. Leaf may be divided into two general groups, (i) large leaf and (ii) "pin" or "pepper" leaf. Large leaf is usually less objectionable, as large particles are more easily removed by the manufacturer's cleaning process.

In addition to leaf, stems, burs, bark, whole seeds, parts of seeds, shale (lining of the bur), motes (immature undeveloped seeds), grass—sand, oil, and dust are sometimes found in ginned cotton.

Leaf and other trash in cotton must be removed as waste in the manufacturing process. The small particles of foreign matter that are not removed in manufacturing detract from the quality and appearance of the manufactured yarn and fabric. Cottons with the smallest amount of foreign matter, other properties being equal, have the highest value. Because of this quality relationship, leaf and trash are important in grading cotton.

12–1.3.3 *Preparation*

Preparation describes the relative neppiness of the ginned lint, and the degree of smoothness or roughness of the ginning process.

Various methods of harvesting, handling, and ginning cotton produce differences in roughness or smoothness of preparation that sometimes are very apparent. Laboratory tests do not show that these differences in preparation in the raw cotton produce equally important differences in spinning results. As a general rule, smoothly ginned cotton has less waste, and it produces a slightly smoother and more uniform yarn than roughly ginned cotton. Except for cases in which the roughness is excessive enough to cause the cotton to be reduced in grade materially below that of cotton having normal ginning preparation, yarn tests do not show significantly lower quality. Longer cottons normally will have a rougher appearance after ginning than shorter cottons, but that does not necessarily mean that yarns made from such cottons will be relatively poorer.

Neps are small tangled knots of fiber caused by mechanical processing. They are visible as dots or specks when a thin web of fibers is held to the light or against a dark background. Neps in lint are undesirable because they will appear as defects in the yarn and fabrics. The removal of neps from the lint is difficult, costly, and frequently impossible. The longer, finer cottons tend to have more neps than the shorter, coarser cottons. Lint having a high percentage of thin-walled, immature fibers is especially likely to be neppy. Neps are difficult for the classer to detect or evaluate.

The term "nappy" describes lint that is rough and lumpy, and the term "nap" is applied to large clumps or matted masses of fibers that contribute to the rough appearance of ginned cotton. Their formation is influenced to a large extent by the condition of the seed cotton at the time of ginning. Seed cotton ginned with high moisture content tends to be nappy. Naps are relatively easy for the classer to detect but are not as detrimental to quality as neps.

12-1.3.4 *Special Conditions*

The USDA cotton classing regulations provide that cotton may be reduced in grade because of the presence of extraneous matter or other irregularities that cause the usefulness of the cotton to be below that normally expected for the grade. When such a reduction in grade is made, the original grade and the reduced grade are stated. Notations are normally made on classification memoranda and provide a basis for segregating bales with special grade conditions from bales with normal grades. Conditions which require grade reduction include: rough preparation, gin-cut or reginned cotton, repacked, false packed, and mixed packed, or water packed cotton, and excessive grass, bark, or extraneous matter.

12-1.3.5 *Grade Distribution of U.S. Cottons*

Surveys have shown that differences in cultivars planted and in environment cause grades of cotton grown in the USA to vary widely. The effect of environment may be greater than that of cultivar. Regional differences in temperature, sunlight, rainfall, and other environmental conditions vary from one growing season to the next. These differences cause variations in quality and grade. Table 12-2 is a summary of classification results for three crop years.

Table 12-2. Summary of grade classification for 3 crop years.

| Grade | Bales classified each crop year | | |
	1978–79	1979–80	1980–81
		%	
White			
Strict middling	0.3	1.6	0.8
Middling +	0.1	0.7	0.2
Middling	12.5	21.6	24.3
Strict low middling +	5.4	5.7	7.0
Strict low middling	34.5	26.3	21.5
Low middling +	2.7	2.9	1.8
Low middling	11.5	7.1	8.4
Strict good ordinary +	0.2	0.1	0.2
Strict good ordinary	1.7	1.0	1.3
Good ordinary	0.3	0.2	0.2
Light spotted			
Middling	5.1	11.2	3.5
Strict low middling	16.5	14.1	16.7
Low middling	6.1	2.9	7.0
Spotted			
Middling	0.3	1.6	0.6
Strict low middling	1.1	1.8	2.9
Low middling	0.8	0.5	2.3
Below grade	0.6	0.4	0.9
All other grades	0.3	0.4	0.5
Total bales (1000)	10 459	14 166	10 722

12-1.4   Length of Staple

The length of staple of any cotton is the normal length by measurement, without regard to quality or value, of a typical portion of its fibers at 65% relative humidity and 21°C. The first official standards for length of staple were promulgated by the USDA in 1918. The present official standards provide for various lengths in terms of inches and fractions of an inch from 13/16 to 1 3/4 in., generally in gradations of 1/32 of an inch. For the sake of standardization, each length is generally referred to by the number of 1/32 of an inch that it contains. A staple of 1 inch is designated as "32"; 1 1/16 inches is "34", etc. Cotton that is shorter than 13/16 inch is designated as "Below 26."

Since 1918 the original order has been amended from time to time. Types in physical form are available for Upland cotton of the lengths 13/16, 7/8, 29/32, 15/16, 31/32, 1 1/32, 1 1/16, 1 3/32, 1 1/8, 1 5/32, 1 3/16, 1 7/32, and 1 1/4 in., and for Pima cotton of the lengths 1 5/16, 1 3/8, 1 7/16, and 1 1/2 in.

In classing cotton for length of staple, the classer makes what is known as a "pull." The classer pulls a tuft of fibers from the sample; and by a process of lapping, pulling, and discarding, the fibers are made parallel. A separate pull is made from the sample drawn from each side of the bale. Tufts of cotton for these pulls are taken near the middle of each sample, and thin places are avoided.

If the classer is in doubt about the length of a sample, he or she compares pulls from the sample with a pull from the official staple type that are believed most nearly corresponds with the length of the staple. Length of staple is an important factor of quality because it is related to yield and processing quality. Other factors held constant, cottons with longer staple length provide higher yields and greater returns to farmers. Longer staple

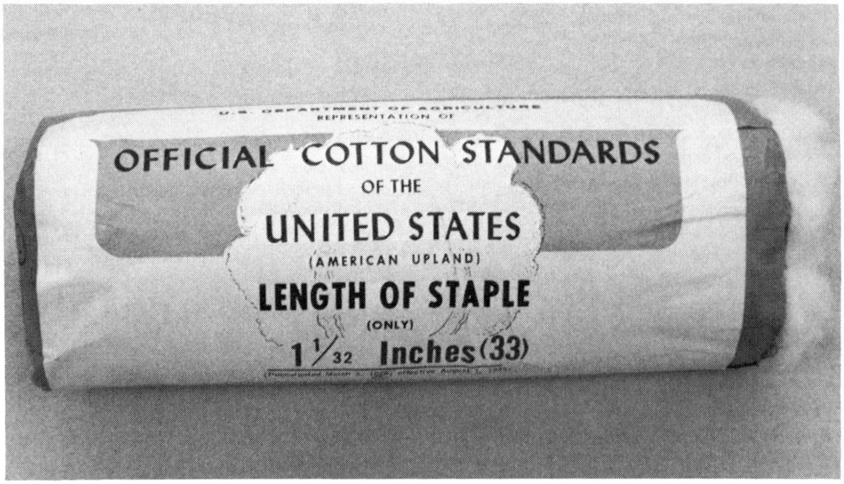

Fig. 12-3. Example of official standards for staple length.

Table 12–3. Staple length classification for 3 crop years.

Staple length	Bales classified each crop year		
	1978–79	1979–80	1980–81
(in.)		%	
7/8 and shorter	0.1	0.2	0.5
29/32	0.8	0.9	1.5
15/16	4.1	4.6	4.2
31/32	8.4	10.3	7.4
1	12.8	12.5	9.7
1 1/32	10.4	8.7	8.4
1 1/16	23.9	10.3	15.6
1 3/32	29.9	30.9	31.0
1 1/8	9.1	20.8	19.1
1 5/16 and longer	0.6	0.8	2.5
Total bales (1000)	10 459	14 166	10 722

cottons process at higher efficiencies in the mill and produce higher quality, finer yarns.

The process of determining staple length is facilitated by the use of practical forms which are also known as staple types. The staple types are particularly useful under variable atmospheric conditions. Cottons selected for use as staple types are carefully screened by a panel of classers and by tests with instruments to ensure consistency. Official types, or standards (as shown in Fig. 12–2) are prepared by and are available from Cotton Division, AMS, USDA, for use in determining staple length. Staple length classification results for 3 crop years are shown in Table 12–3.

Official standards have not been converted to the metric system; therefore, we will use the units of measure in the official standards in discussion of the standards. In other parts of the text, units of measure for fiber properties will be converted to the metric system.

12–1.5 Micronaire Reading

Micronaire reading, as determined by airflow instruments, is a measure of the fineness of cotton fibers, and it is widely used in cotton marketing and manufacturing. Within any cotton cultivar, micronaire reading is also a measurement of the maturity or relative cell wall development. Fineness/maturity measurements are useful because they are directly related to lint yield per acre and to processing and subsequent yarn quality. Micronaire reading is the most widely used measurement by an instrument in domestic and international markets. Mills use this measurement in quality control programs to assure uniform and even-running cotton mixes. Micronaire measurements are reproducible and rapid and have been a part of the USDA Official Classification procedure since 1966.

Micronaire readings for Upland cotton ordinarily range from 2.0 to 6.5. The premium, or no-discount, range for micronaire reading is from 3.5 to 4.9. Optimum micronaire reading depends upon a number of conditions including the relative importance of strength, appearance, and fineness of yarn desired. Low micronaire cottons generally produce more neps in yarn than high micronaire cottons. They also can cause dyeing irregularities and an increase in manufacturing waste. Fine fibers are required for fine, strong yarns. Micronaire readings below 3.5 may indicate immaturity to the degree that an inferior quality yarn will be produced.

Pima is a genetically fine fibered species and, due to its lumpy texture ness on a different scale than Upland cotton and, due to its lumpy texture from roller ginning, must be prepared slightly differently for testing. However, airflow measurements of fineness of Pima cotton are quite useful.

The measurement of cotton fiber fineness and maturity was a very slow and tedious task until airflow instruments were developed in the late 1940s. Airflow values were standardized for micronaire readings through the International Cotton Calibration Standards program in 1957.

A typical airflow instrument for measuring the fineness of cotton fibers is shown in Fig. 12–3. Standards for maintaining proper instrument calibration are available from USDA. Standard values of these calibration cottons are established by an international group of laboratories on a cooperative basis.

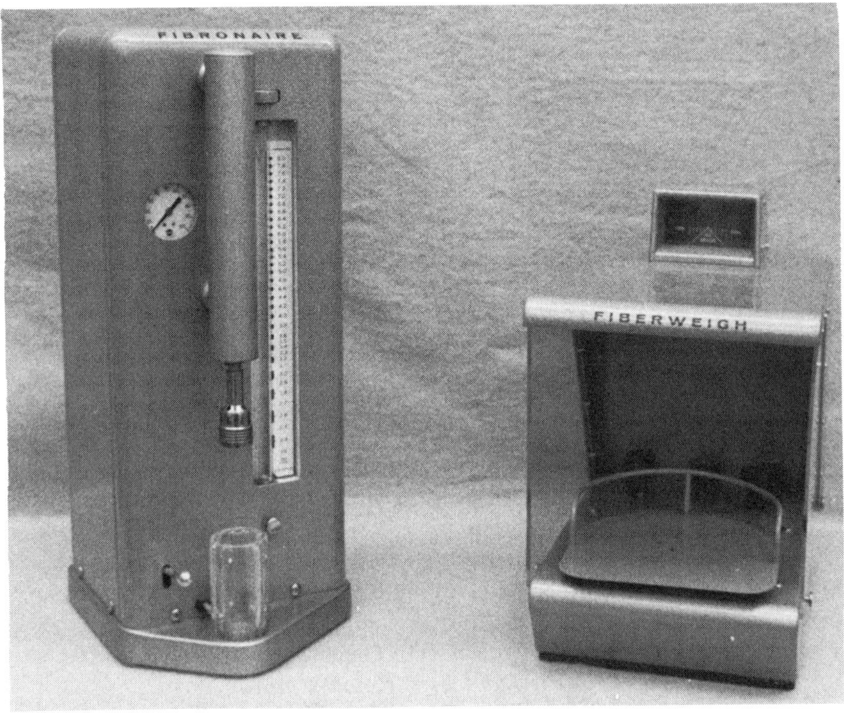

Fig. 12–3. Airflow instrument for measuring fineness.

12-2 PHYSICAL PROPERTIES AND INSTRUMENT
MEASUREMENTS

The need for better communication and more precise definition of the properties of cotton resulted in the development of instruments for measuring physical properties. Agricultural scientists involved in research to improve the yield of cotton soon learned that if objective measurements of physical properties were not available, experiments for plant improvement could not be fully evaluated. Spinners realized that unless measurements were available to monitor the physical properties of cotton, it would be difficult to detect gradual changes in fiber properties before processing and yarn quality reached intolerable levels. When physical properties began to be changed by genetic manipulation of plants, instrument measurements for objective evaluation of results were necessary. The more progressive spinners soon realized that physical property measurements performed under controlled, laboratory conditions could provide information necessary for precision practices for manufacturing yarn.

Fiber properties have been studied since the early 1900s, but only since about 1950 has electronic and other physical science technology been developed sufficiently to provide universal application for physical measurements. Advances in electronics, lasers, and spectral analyses have created opportunities for much more refined measurements of physical properties than have been previously available. These advances also have made using instruments economically feasible and led to measurements of physical properties on a much wider scale than was possible previously.

12-2.1 Preparation of Specimens

The first and most important step in performing instrument measurements of fiber properties is to obtain a representative sample. The most accurate test on a sensitive instrument is useless if the sample tested is not representative of the bale of cotton from which it was taken. Classers' samples, properly drawn, are usually adequate for instrument measurements for marketing and utilization of cotton. Sampling for research investigations might require more elaborate procedures.

After sample validity is assured, sample conditioning must be closely controlled. The sample must be in a condition that permits preparation of the specimen to be presented to the instrument to be consistent and repeatable. Ropy, twisted, lumpy samples often require preblending or mixing to assure that specimen preparation is consistent. Atmospheric conditions must be closely controlled to prevent variations in properties due to moisture and temperature differences. Instrument measurements of fiber properties are usually performed at standard atmospheric conditions of $21.1 \pm 1.1°C$ with a relative humidity of $65 \pm 2\%$.

Only valid, properly conditioned samples should be tested. Specimens prepared from the sample should be selected at random. Specimen selection and preparation are often the key factors in the efficiency of a fiber testing instrument. Instruments used in fiber testing, if properly maintained and calibrated, usually provide very consistent measurements. Instruments for measuring fiber properties will usually provide highly consistent measurements if the same specimen is tested repeatedly. Specimens from the same sample vary. The technique used in the preparation of specimens influences the variability of the measurements.

Some procedures for preparing specimens, such as those used to prepare a fiber beard for length tests or a ribbon of fibers for strength tests, are length-biased. The specimen preparation procedures preferentially favor fibers of a certain length. Other tests, such as micronaire reading, use bulk samples so that length-bias is not apparent. However, in this case the presence or absence of trash contributes to measurement variability. The procedures for the preparation of specimens are critical for all measurements made by instruments. For instruments using beards of fibers the beards should be uniform in density with no gaps. The amount of fiber in the beards should have limits that match the sensitivity of the instrument. For bundle strength tests the fiber bundles should be pre-stressed uniformly before clamping. Faces of specimens for measurements of color should be cleared of particles of dark trash.

Specimens for all instrument measurements require careful preparation to assure consistent results. Technicians performing laboratory tests with instruments are highly trained and develop skill in judging the quality of specimens presented for measurement. However, differences in measurements due to technique are very common, and methods of adjusting for these differences must be developed by individual laboratories. Checking a sample with known physical properties is usually the basis for maintaining consistency among several technicians. Blind checks in which samples with known values are tested serve a useful purpose in guarding against inconsistencies or poor specimen preparation. Such tests also point out changes that may occur because of calibration failures in instruments. Persistence and a good quality control system are required to maintain consistency in preparation. All fiber property measurements are sensitive to the condition of the sample tested.

12–2.2 Length

Length of staple is universally recognized as the premier fiber property because it is so closely associated processing efficiency in manufacturing and the quality of the yarn produced. Length of staple can be measured by several commonly accepted ways depending on the type of information needed. Detailed length information can be obtained using sorting and weighing, as with the Suter-Webb Sorter method. This process is highly labor intensive and is suitable for research investigations where detailed knowledge of the frequency distribution of fiber length is required. This method employs a series of combs on which fibers are separated in length increments of 1.6 mm. Fibers in the length groups are then weighed to determine the weight-length relationship from which estimates of length parameters can be made. Values obtained generally include the mean length of the longest half of the fibers by weight, and mean length, the percentage of the fibers shorter than 12.7 mm and the coefficient of length variation. Several specimens per sample must be measured by skilled technicians to obtain acceptable reliability.

Instrument measurements of fiber length can be made by analysis of the fibrogram. A fibrogram is obtained by forming a beard of fibers and measuring the mass of fibers at various cross sections along the length of the beard. A beard of fibers is formed by grasping a fiber specimen in a clamp and combing or brushing away all of the fibers not held by the clamp. Each fiber is grasped somewhere along its length by the clamp, with one end of the fiber extending from the clamp and contributing to the formation of a beard. Physical measurements are made of the amount of fibers in various cross sections of the beard at different lengths from the clamp to develop the relationship of amount vs. distance from the clamp. One useful measurement resulting from this relationship is the span length. The span length is the length of the fiber beard or the distance from the clamp where the beard has reduced to a certain percentage of the amount at the clamp. Two commonly used parameters are the 50.0 and the 2.5% span lengths, the length of the fiber beard where the cross-sectional amount is reduced to 50.0 and 2.5% of the initial amount.

The 2.5% span length is important because of its close association with staple length as determined by the cotton classer. The 50% span length is importnt because it gives an indication of the uniformity of length of the fibers in the specimen. Other parameters of interest that may be determined from the fibrogram are the mean length of all the fibers and the mean length of the longest half by weight. Two instrument companies, Special Instruments Laboratory, Inc., and Motion Control, Inc., have developed the instruments usually used to measure fiber length. The Spinlab instrument, shown in Fig. 12–4, uses a light beam projected through the fiber beard to assess the amount of fiber in the cross section of the beard. The Motion Control instrument utilizes an airflow principle for measuring beard cross section with the pressure drop being measured across a slotted orifice in which the beard has been inserted. The cross section being measured partially blocks the throat of the orifice causing a pressure drop in proportion to the amount of fiber in the cross section.

The output of the sensing heads of either of these instruments is an electrical signal. The advent of digital computers makes it possible to economically store all measurements recorded from a fibrogram, and the device can be programmed to provide information needed for specific applications. Although the results from the instruments manufactured by the

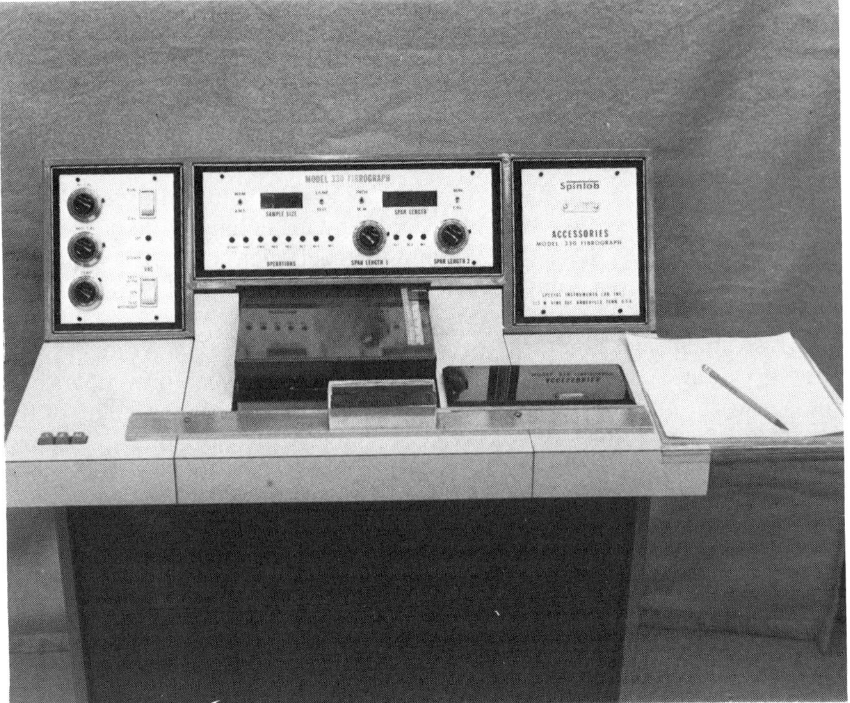

Fig. 12–4. Fiber length tester.

two companies are quite similar, the results from one cannot be directly converted analytically to match the results of the other because of differences in the clamping mechanisms, the beard cross section sensing device, and other basic design parameters. The correlation between test results on the two instruments has been found to be 0.92 or higher.

12–2.3   Strength

Measurements of fiber strength are important because of the significant relationship between fiber strength and yarn strength and because higher strength cottons can endure more vigorous mechanical handling than lower strength cottons. Higher strength cottons can be subjected, without excessive damage, to relatively harsh cleaning treatments to remove foreign matter and to obtain quickly the degree of openness, or fiber-to-fiber separation, necessary for processing into yarn. Fiber strength is related to processing waste and several other processing parameters of importance to manufacturers.

Strength measurements can be performed on single fibers or on bundles of fibers. Strength measurements on single fibers are performed only in research or special projects. These measurements require great care in selecting and mounting specimens and require sensitive measurements to obtain consistent results. Although measurements of single fibers would be useful in controlling quality variations and selecting cottons for specific end-uses, the lack of an automated or semi-automated testing mechanism limits practical use of such measurements. In many situations where single fiber strength measurements are used, a strain gauge apparatus is improvised to make the measurements. The wide variation in single fiber strength requires a large number of observations for statistical reliability of estimates. The practical usefulness of these measurements is limited due to the variability and the tedious and time-consuming nature of the tests.

Fiber strength measurements made on bundles of fibers have found practical application in research, marketing, and quality control. The Pressley tester, shown in Fig. 12–5 is an instrument that was developed to measure the strength of a bundle of cotton fibers. The device consists of a set of jaws and an inclined lever system in which an ever-increasing load is applied to the specimen. In preparing the specimen, tufts of fibers are selected from the sample, combed and formed into a flat ribbon which is clamped between the leather-lined jaws and trimmed. The jaws are then placed in the device, and the load is allowed to roll down the incline until the bundle breaks. An index which indicates the position of the load when the bundle breaks is read from the scale and used, together with the bundle weight, to calculate bundle strength. Strength measurements traditionally have been made at "0" gauge length—that is, no space between the jaws. However, analyses have shown that strength measurements at 3.2 mm gauge length are more highly related to yarn strength and processing param-

Fig. 12–5. Pressley tester.

eters. Other instruments for use in measuring bundle strength are the Stelometer and the Scott-Clemson tester. These instruments use somewhat different loading concepts but require clamped and weighed bundles very much like the Pressley tester. These machines are suitable for laboratory testing and statistical sampling of lots of cotton for use in marketing applications, but are much too labor-intensive and time-consuming for testing cotton rapidly enough for marketing purposes.

Motion Control, Inc., and Special Instruments Laboratory, Inc., have developed testing instruments for rapid strength tests on the same specimen that is prepared and used for length tests. This tapered beard is broken automatically by jaws lined with paper. The force required to break the beard and an estimate of beard mass at a preselected point are used to calculate bundle breaking strength. These measurements are affected by fiber fineness and require correction for micronaire reading, but they can be performed rapidly and are suitable for cotton marketing.

12–2.4 Fineness and Maturity

The fineness of cotton fibers can be expressed in terms of linear density or weight per unit length. Fineness can be determined by sorting fibers into length groups, as with the Suter-Webb method, and weighing a specific number of the fibers on a sensitive balance. This procedure is used in certain research projects. Fineness determined in this fashion has distinct advantages because length-fineness relationships within the sample can be studied. However, it is time-consuming and unsatisfactory for most practical applications.

Micronaire reading was introduced in 1946 as a rapid means for measuring fiber fineness and was adopted readily by the more progressive industry groups. This airflow measurement is performed on a 3.25 g test specimen which is compressed to a specific volume in a porous chamber. Air is forced through the specimen and the resistance to the airflow is proportional to the linear density—which is read from a direct reading scale or a digital readout. Micronaire readings of Upland cotton, because of scale selection, correspond closely with linear density measurements expressed as $\mu g/25.4$ mm of fibers. Fibers with specific gravity or specific surface area different from Upland cotton require a different scale to obtain this relationship to linear density. The Fibronaire, an airflow device for determining micronaire reading, was shown in Fig. 12–3.

Micronaire reading is affected by fiber maturity. Maturity is defined as the relative cell wall thickness of the fibers. Maturity can be determined by viewing microscopic cross sections of cotton fibers that have been treated with a caustic solution. The ratio of cell wall thickness to fiber diameter reflects the degree of maturity. As with other laboratory methods, this microscopic technique is tedious and time-consuming. Other methods with varying degrees of precision have been developed for measuring maturity. These methods include airflow measurements at different densities, caustic swelling in combination with airflow measurements, polarized light measurements, and differential dyeing in which the dye uptake indicates the degree of maturity. None of these methods for measuring maturity, which may be useful in laboratory tests, have been adopted for widespread use by industry in this country.

12–2.5   Color

The colorimeter is an instrument designed to provide objective measurements of the color of cotton samples. In the colorimeter a combination of light sources and filters is used to obtain estimates of the tristimulus values defined color. These estimates are then used to calculate a value (Hunter's + b) indicating the degree of yellowness of the sample, which, when combined with the reflectance (Rd, the degree of lightness or darkness of the samples), can be used for precise cotton color measurements. The readings obtained with the colorimeter are very closely related to the color grade designations assigned by classers. Tests are conducted by exposing the face of a sample to the closely monitored and controlled light sources and are performed very quickly with only moderate technician skill required.

A typical color chart showing how the official grade standards relate to Rd and + b measurements is shown in Fig. 12–6. Color measurements are affected by trash content of the samples because the reflectance values for trash are different from the reflectance values for cotton fibers. This produces a bias that is more pronounced in the lower grade cottons that have higher trash contents. The colorimeter, shown in Fig. 12–7, is a very sensitive instrument and can easily detect subtle differences in the color value of cottons.

Color measurements are important because variations in color cause variations in the color of dyed cotton products. To obtain uniform dyeing, especially for certain color shades and end-products, it is necessary to control the variation in the color of the cotton. The colorimeter is especially useful in this application. Dyeing consistency for pastel shades and certain fabric structures is extremely difficult to maintain unless fiber color variations are controlled.

Color measurements are also important because they are related in a general way to overall cotton quality. Bright, creamy, white cotton is generally associated with high quality. Dull, gray, or yellow tinged cotton is usually associated with weathering and low quality.

12–2.6   Trash Content

The nonlint content of cotton is determined by processing a small amount of cotton on a very thorough cleaning machine such as the Shirley

Analyzer (Fig. 12-8), or the SRRL Non-Lint tester, in which the nonlint material is physically separated from the lint and collected in compartments of the machine. Gravimetric procedures are used to perform a materials balance so that nonlint content can be expressed as a percentage of the weight of the original sample fed to the machine. Unexplained differences that arise from the materials balance are referred to as "invisible waste." Nonlint measurements performed on these machines are the most accurate and useful measurements available for obtaining estimates of the trash

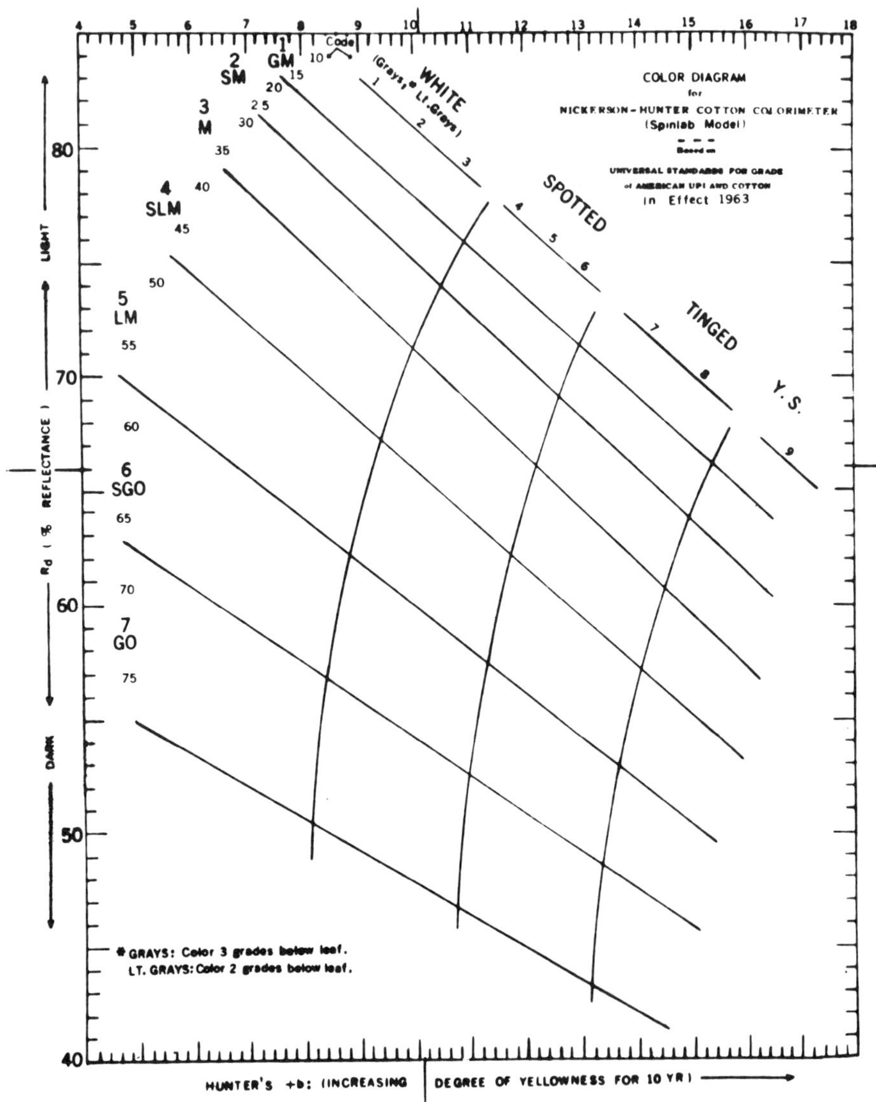

Fig. 12-6. Color chart—relationships of official grade standard to color.

content of cotton and the amount of waste anticipated in processing. There is a very close relationship between nonlint content and the grade of cotton as indicated in Table 12-4 for Upland cotton. A similar relationship exists for Pima cotton. Nonlint measurements on these machines are adequate for laboratory tests where labor and time requirements are less critical. For volume testing of cotton for marketing purposes, no satisfactory test for nonlint content has been developed. Research has shown that it may be feasible to obtain rapid measurements of trash content with video scanning or reflectance measuring devices. However, these methods have not as yet been incorporated into a prototype instrument for evaluation.

12-2.7 Contaminants

Serious and annoying problems caused by sticky cotton occur periodically during the passage of the fiber through the yarn manufacturing

Table 12-4. Relationship between nonlint content and grade.

American Upland Grade		Average Nonlint content
		%
Strict middling	(SM)	1.9
Middling	(M)	2.3
Strict low middling	(SLM)	3.1
Low middling	(LM)	4.4
Strict good ordinary	(SGO)	5.6
Good ordinary	(GO)	7.2

Fig. 12-7. Colorimeter.

processes. The problems are related to buildup of deposits on rolls and other parts of processing machinery. The result is at best a cleaning problem and at worst a serious production and quality problem.

The deposits form on calendar rolls, the card crush rolls, the comber rolls, and the various rolls of the drawing, roving, and spinning frames.

Fig. 12–8. Shirley analyzer.

Laps around rolls may occur in drawing, roving, and particularly, in spinning. One end break that laps in spinning requires three to four times as much effort to repair as a normal end break. Additional trouble may occur at the card in the form of loading of fibers on the card cylinder.

Most of the difficulties are related in some way to the quantity, nature, or relative location of the noncellulosic constituents in the cotton fiber. The normal noncellulosic constituents of cotton are products that are formed by the natural growth processes of the cotton plant. Variations in the relative amounts of these deposits depend upon such factors as cultivar, climate, environment, types of soil, weather, weathering, method and time of harvest, microbiological activity, and other factors.

In addition to these materials which are always present on cotton, other noncellulosic materials appear from time to time as a result of contamination. Contamination may come from natural sources, such as insect or microbiological attack, or from accidental or purposeful introduction of substances to the fiber. Residues from production and harvest-aid chemicals and gin additives are often present. Common and very serious hydrocarbon contamination occurs from misuse of oils and lubricants or carelessness in the operation and maintenance of mechanical harvesters.

Visual examination of bale samples is the first step in examining a cotton for potential sticking problems. If dirty splotches, green smears, or excessive quantities of seedcoat fragments are visible, then problems in processing are likely.

Examination of bale samples under ultraviolet light is sometimes very revealing. Contamination from lubricating oils will show up as brilliant blue-white fluorescence. Sometimes the fluorescence associated with lubricants will be yellow or orange, due to contamination by rust. These hydrocarbons containing iron are often mistaken for microbiological contamination.

The term "cavitoma" is used to describe cotton damaged by microbes. Cavitomic cottons are generally characterized by fiber wall damage and by the high pH and low reducing sugar content of the water extract. A simple test for cavitoma involves a spray application. The spray is an acid-base indicator that changes color with changes in pH. If microbiological damage has occurred, the surface of the cotton fiber usually is alkaline and a characteristic pink/purple color develops when the cotton is sprayed. Cavitoma is often present even though the pH is near normal. These damaged cottons can be detected by the benzidine method; benzidine changes color due to action of water-soluble products on the fiber surface. The benzidine method gives a quantitative estimate of the degree of damage since the intensity of the developed color is proportional to the severity of microbiological damage. The Cavitoma test spray method only indicates whether or not damage is present and does not indicate the degree of damage.

Wax is an important component of the noncellulosics present on cotton. The wax serves as a lubricant and without it cotton would be extremely difficult to process into yarn. When the wax is removed or reduced in quantity prior to processing the cotton, some oil must be added to the fiber before it can be properly processed.

Wax content is often very important in predicting spinning quality of cotton. If the wax content is abnormally high, the chances are good that contamination is present and lapping of cotton on rolls is likely to occur.

The texture of the wax can be an important indicator of the processing potential of a cotton. The normal wax extracted by benzene is relatively hard. If the wax is soft, sticky, or oily, then contamination is surely present and buildup on rolls will probably occur and eventually lead to lapping on rollers. Wax content of cotton fibers can be determined by extraction with benzene or other solvents.

Alcohol extractables are those materials which are removed from the cotton fiber by 95% ethyl alcohol in a 6-hour Soxhlet extraction. Alcohol extractables give a good indication of the general level of noncellulosics present on the cotton. Neither ethyl alcohol nor any other single solvent will remove all of the noncellulosics from cotton, but the alcohol removes sufficient quantities to give meaningful results. Generally, the alcohol extract consists of waxes, organic acids, sugars, and hydrocarbon contaminants. A high level of alcohol extractables indicates that general stickiness and accumulation of deposits on rolls may occur.

The amount of reducing substances on cotton fibers is determined by comparing the reducing ability of the hot-water extract of the cotton to that of a standard reducing substance such as glucose. The method is extremely simple and gives accurate, reproducible results.

Except for laboratory methods requiring special equipment, few specific instruments are available for determining noncellulosic constituents of cotton. Instruments for measuring ultraviolet reflectance and for infrared analysis are available but have not been used widely in industry. This entire area of instrumentation of the surface properties of cotton fibers has been neglected. Significant improvement in the definition of cotton quality will be realized when such instrumentation is adapted for general commercial use.

12–2.8   Other Fiber Properties

Other physical properties of fibers are often of interest in special projects. No standard test method exists for many of these properties. Test methods and procedures must be improvised. Sometimes physical tests for other materials or other fibers can be adapted for testing cotton, but often the process of adaptation takes more time than developing a new procedure. Examples of properties not generally measured specifically are friction, resilience, and elongation. Some testing procedures are standardized and are of interest in special work but are not widely used on a routine basis. Examples are staining tests for fiber damage, pH, and other tests for microbiological activity.

Advances in technology and instrumentation in recent years have provided a technical basis for a wide variety of special purpose measurements in laboratories. Innovative and inquiring minds have little difficulty in identifying methods and techniques for measuring practically any physical property conceivable.

12-2.9 Variation in Physical Properties

Properties of cotton vary to some degree from year to year. While this variation can be significant, it is not as great as could be expected considering the vulnerability of the cotton plant to weather, pests, and weeds. Weather causes most of the annual variation in the physical properties of cotton. The amount of sunlight, daytime and nighttime temperatures, and soil moisture are key factors. Knowledgeable cotton buyers and merchants watch weather patterns closely and keep detailed records on the weather for their major purchasing areas. Experience has shown that information on the weather can be useful in anticipating quality and physical property differences. Weather conditions are particularly important during certain growing periods. For example, low nighttime temperatures during periods of cell wall development can have a pronounced effect on fineness and maturity. The effects of weather on physical properties vary from year to year. The effects of gradual changes in weather patterns are not rapidly discernible; but abrupt, short term weather changes occur more frequently than gradual pattern changes. These rapid weather changes sometimes cause excitement and interest from a scientific as well as economic point of view. For example, the San Joaquin Valley of California is dry and vast systems for irrigation have been installed to make it one of the most productive cotton growing areas in the USA. There was widespread interest in 1965 when it rained in the San Joaquin Valley during the cotton growing season. Considerable attention was given to this area by the entire cotton industry because of this unusual event.

Variations in physical properties are also caused by changes in cultivars of cotton planted. Such changes, in most cases, result in trends rather than transient differences. The change to new cultivars is usually gradual except in areas where one cultivar is predominant. New cultivars are being introduced continually by seed companies and breeders with a resulting gradual improvement in cotton quality. The overall effect of changing cultivars has been a gradual increase in fiber length, strength, and micronaire reading; however, several years are required for a new cultivar to affect crop averages.

Harvesting and ginning practices are sometimes adopted that have a direct and immediate influence on fiber properties. An example is the use of dryers in gins to reduce the moisture content of cotton during ginning. Even though the use of dryers has been thoroughly studied by USDA in Cotton Ginning Research Laboratories and guidelines developed for cotton ginners, the cotton marketing systems, with their heavy emphasis on clean cotton, often provide such incentives that cotton is over-dried to facilitate trash removal to obtain higher grades. Excessive drying and cleaning cause significant fiber damage that shows up quickly in instrument measurements of fiber length and contributes to further variations in quality.

A major contributor to variations in the properties of cotton is contaminants. Excessive sugars, oils, or other substances, such as rubber dust from spindle harvester pads or polypropylene bagging, often appear in cotton very suddenly. Infestations of whiteflies sometimes contaminate cotton from an area for one or two seasons and then disappear.

Contamination from residues may not be apparent until users learn of its presence. While these contaminants do not show up in physical property measurements they sometimes are the cause of variations in processing performance of cotton.

The combined effect of all these sources of variation is generally not cumulative in practice. Rarely does two or more events, such as severe drought or the widespread adoption of a new cultivar, occur during the same year in all cotton growing areas. Even more rare would be widespread contamination in addition to other catastrophies. Actually, when the crop as a whole is considered, year to year variation in physical properties is moderated by the diversification of growing areas. When one growing area experiences a bad season, other areas often seem to have superior seasons, which limits the variation in average properties.

Physical properties and spinning quality of major cotton cultivars in several geographic areas are monitored closely by the USDA on a statistical sampling basis (Cotton Division, AMS, 1982). Typical data for 3 crop years are shown in Table 12–5 for medium staple cottons. More detailed information on a number of cotton fiber properties is available from summary reports prepared by Cotton Division, AMS. These summary reports fill a vital need in that they serve as a continuing record of fiber and spinning quality variations for each crop year.

12–3 HIGH VOLUME INSTRUMENTS

As indicated in section 12–2 instruments have been used with varying degrees of success for a number of years to measure the important properties of cotton. However, use of instruments has been restricted because of high costs and the length of time required for the tests. Even though measurements by instruments are advantageous for cotton marketing, such procedures were impractical until recently. In the mid-1960s USDA and instrument manufacturers began a cooperative effort to modify and develop instruments for high volume use such as would be necessary for cotton classification. To develop these High Volume Instrument (HVI) systems, two basic problems required solutions. First, it was necessary to automate many of the instrument functions and reduce the amount of time required for the instruments to perform the tests on the specimens presented. This was accomplished, although not easily, through use of advanced electronic systems that were mainly a by-product of space-age research. Through repeated prototype construction and laboratory and field evaluation, a system evolved that is capable of measuring the more important fiber properties. Fiber properties measured on HVI systems in use in selected

Table 12–5. Typical values of selected fiber properties for three crop years
for medium staple of Upland cotton.

| | Crop year | | |
Area of growth	1978–79	1979–80	1980–81
2.5% span length (mm)			
Southeast	27.9	27.7	26.9
South Central	27.7	28.4	27.7
Southwest	26.4	26.7	25.9
West	28.2	28.2	27.9
Micronaire reading			
Southeast	4.5	4.4	4.6
South Central	4.6	4.2	4.7
Southwest	4.2	3.9	4.0
West	4.3	4.4	4.3
Fiber strength [Stelometer (mN/tex) = $9.81 \times$ (gf/tex)]			
Southeast	23	23	23
South Central	23	23	23
Southwest	25	25	25
West	25	25	25
Nonlint content (%)			
Southeast	3.5	3.6	3.4
South Central	3.5	3.2	3.3
Southwest	3.6	3.4	3.6
West	2.9	2.2	2.6

Marketing Services Offices by Cotton Division of AMS include length, length uniformity, color, fineness, and strength. Trash content is still determined subjectively by a grader; however, arrangements are being made to incorporate for evaluation video scanning devices for measuring trash content. Except for strength and trash content, very little new technology other than electronics was used in the development of the HVI systems. Principles used in the measurement of length, fineness, and color were well known. Major problems faced in developing the instruments concerned ways to increase testing rate and ways to maintain calibration. These problems were solved by the use of microprocessors and small computers. The instruments are arranged in three testing stations along a conveyor system as shown in Fig. 12–9. Samples are placed on the conveyor and transported past each station. Technicians remove the sample and bale identification card, prepare a specimen for testing, and place the identification card in a reader. The system is controlled by a microprocessor so that the conveyor transporting the samples moves only when a pre-arranged number of specimens are tested from each sample. At this time the conveyor is indexed and the sample is moved forward to the next test station. Test data are compared and analyzed automatically for consistency. Samples with test data outside predetermined tolerances are flagged for retesting. Test data for each sample are printed on a class card which is sent to the owner of the cotton. The test data are also retained in machine readable files to facilitate quality summaries and for use in marketing.

Fig. 12-9. High volume instrument testing system.

In developing these HVI systems two innovative concepts emerged that significantly reduced the amount of time required for the tests. First an automatic specimen preparation mechanism was developed that incorporates a clamping device controlled by air that grabs a tuft of fibers from the bulk sample presented by the technician. The grab sample is held firmly by a clamp and is manually placed on an automatic comber and brusher which straightens and aligns the beard of fibers before it is presented to the instrument for measurement. Length tests are performed on the beard by an airflow method as described previously in section 12-2.2. The clamp is then moved manually to the strength station where a strength test is performed on the same tapered beard. Prior to development of the HVI machines, all strength tests were performed on beards in which the fibers extended the width of the breaker jaws. The second significant innovation involved the use of a tapered beard of fibers for strength tests rather than the traditional beard. This increased the variability of the test data. However, this procedure greatly reduced the amount of time required for specimen preparation so that multiple tests and averaging could be performed to reduce the variability. The net result was a mechanized strength test with high productivity that removed the influence of the technician.

The success of these two automated processes illustrates the potential for further automation and has stimulated interest in combining, automating, and incorporating other tests and instruments into the system. At peak efficiency HVI systems can provide reliable information on six characteris-

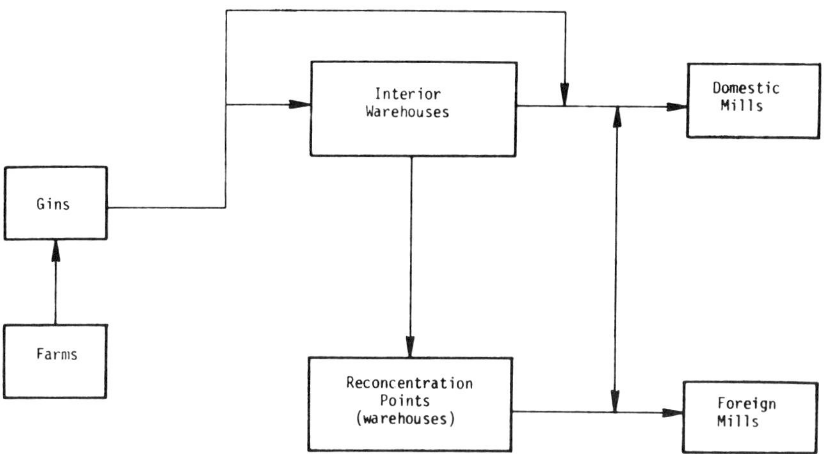

Fig. 12–10. Physical flows of cotton through the U.S. Marketing System.

tics of quality from a cotton sample in approximately 30 seconds that are highly related to the spinning quality and market value of the cotton. Information on every bale of cotton should greatly improve the marketing of cotton and encourage the production of cotton with fiber properties desired by users. HVI systems are now in operation in two USDA Marketing Services Offices and in several marketing cooperative, shipper, and mill groups. The time required to institute use of HVI systems in all Marketing Services Offices will depend on how quickly the systems can be manufactured, how quickly funds required for the purchase of the instruments can be obtained, and how quickly a price quotation system based on fiber properties can be developed. The Secretary of Agriculture has appointed an industry committee to advise USDA on implementation of HVI classification of cotton.

12–4 MARKETING CHANNELS AND PRICING MECHANISMS

As cotton leaves the farm, it moves into the marketing system where numerous marketing functions, including ginning, storage, compression, merchandising, transport, and processing, are performed (Jones et al., 1975). The efficiency of the various functions affects the prices paid by mills and the prices received by producers. To describe the movement of cotton through the system, it is necessary to distinguish between the physical movement of cotton and the transfer of ownership.

Figure 12–10 illustrates the physical flows of cotton. Most cotton moves from a gin to a warehouse, but it may move directly to a domestic mill or a port facility. The ownership of the cotton deviates from the physical flows; the gins and warehouses do not usually own the cotton but rather sell their services to the owners of the cotton. Fig. 12–11 defines alternative marketing channels in terms of ownership. In general, there is a

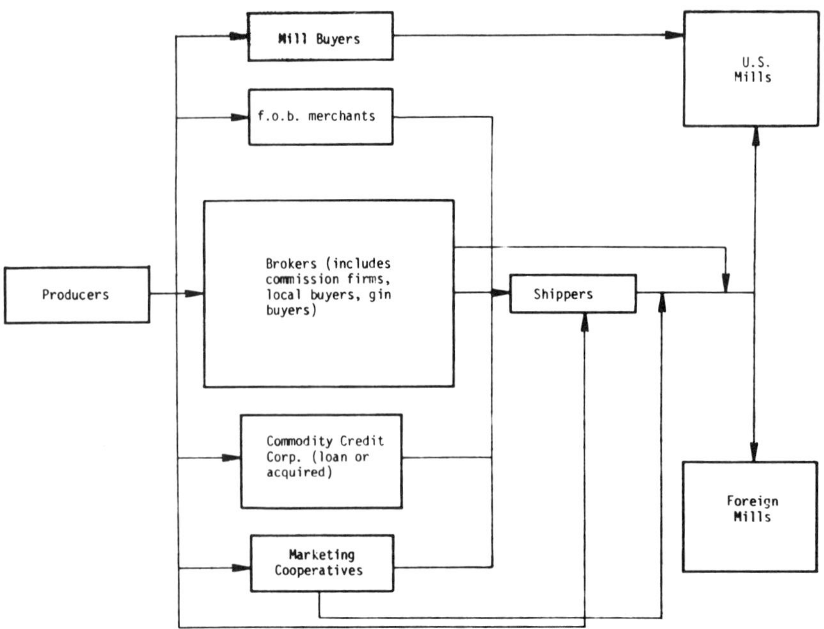

Fig. 12-11. Representative types of marketing channels for U.S. cotton.

pricing point (a price is established) between each pair of traders in Fig. 12-11 except between (i) producers and brokers, (ii) producers and cooperatives, and/or (iii) brokers and the immediately succeeding buyers in the channel. Brokers act as agents for other parties. Cooperatives may take title to the cotton, but establish a price at a point further along the marketing channel.

The marketing system for cotton is a complex system in several respects. The grading system for cotton is more complex than for most agricultural commodities. There are many dimensions of fiber quality, all important insome respect; there are 45 official Smith-Doxey grade designations, 23 official staple designations, and 7 micronaire groups, making over 7000 possible distinct quality combinations. Many private grades exist in the industry. Cotton goes through more, and sometimes more complex, processing steps than other agricultural commodities. The market demand for cotton is also more complicated in some respects than other agricultural commodities; because (i) unlike food products, cotton competes with nonagricultural commodities in the form of synthetic fibers and (ii) in contrast to food products, cotton consumption is responsive to changes in the levels of overall economic activity in the USA and abroad.

Major participants in the cotton marketing system are: (i) cotton producers, of which there are about 175 000 growing cotton on 225 000 farms; (ii) cotton gins numbering about 2250 and declining; (iii) about 500 cotton compresses and warehouses; (iv) merchants and shippers, of which there are some 200 firms which take title for their own or grower accounts; and (v) several thousand textile mills which vary in size from establishments with a few employers to very large textile firms (Sporleder et al., 1978).

12–4.1 Types of Buyers

The middlemen involved in the marketing of cotton may be categorized into four basic types (McArthur et al., 1980).

Merchant-Shippers perform all of the functions involved in moving cotton from the producer to the mill. These firms take title to the cotton from the time it is purchased from farmers, brokers, or cooperatives until it is purchased by a mill or trading firm.

Some merchant-shippers operate in more than one region of the U.S. Cotton Belt, while many shippers specialize their operations in one region. The "f.o.b. merchants" are common in some areas. These firms operate the same as merchant-shippers insofar as financing, purchasing, hedging, or warehousing is concerned, but their sales are made on the basis of the bales being located in a warehouse in the trade territory. Thus, it is the responsibility of the purchasing textile firm or merchant-shipper to arrange for transportation from the production area warehouse to the textile firm's storage facilities. In some areas, trading among f.o.b. merchants and between f.o.b merchants and merchant-shippers is common. This trading involves transfer of warehouse receipts between firms for a stated price. Warehouse receipts are negotiable instruments and ownership of the bale resides with the firm holding the receipt.

Once a sale is made by a shipper, the cotton is purchased and assembled or it is assembled from already existing stocks. Terms of contracts may vary greatly; they may state that quality specifications be based on "green card" classification, or that specifications be based on private types developed by the shipper or purchaser over the years. Merchant-shippers and cooperatives handle the major portion of the cotton produced.

Several foreign trading companies have offices in various locations across the U.S. Cotton Belt. These firms are primarily in the export business, although some make sales to domestic mills as well. These firms tend to buy from growers and cooperatives.

Cooperative Marketing Associations act as shippers, except that they are acting for farmer-members who comprise the association, and any profits are rebated to the grower. Some cooperatives also operate their own compresses and warehouses as part of the marketing process.

Most cooperatives have several sales options available for members' use. Forward contracting is one form in which the cooperative acts as the third party between the grower and purchaser. Another type of sales option is a seasonal pool, which is designed to even out wide fluctuations in prices throughout the year. This evening out is accomplished through the association's blocking cotton from the pool into selected categories and merchandising to firms with narrow quality requirements. A third type of sales option is a call pool where the grower fixes a price in relation to the futures market on a part of his crop. Sales are made on a fixed number of bales. The grower then "calls" the date to fix the futures price. Final price is adjusted according to the contract for quality variations from the base quality specified.

A new system of marketing developed by a Texas-Oklahoma cooperative association allows spot market sales, forward contracting, and other options through the use of an electronic marketing system. This innovation facilitates several marketing options for producers; the system is discussed in a following section on electronic marketing.

Brokers (also designated as Agents, or Commission Buyers, or Sellers) only act as middlemen between growers and merchant-shippers or textile firms or between merchant-shippers and textile firms. They act as intermediaries while receiving a commission for the volume bought or sold. They neither take title to the cotton nor perform any of the associated functions involved in shipping such as financing, hedging, and arranging for transportation. Their function is in the assembling of individual bales or small lots into substantial volumes of cotton for others, in acting as selling agents for shippers or growers, and acting as buying agents for textile mills.

Gin buyers are ginners who have vertically integrated forward into the marketing functions. This type of operation would classify the gin-buyers as a merchant-shipper if he takes title to the cotton. However, most gin buyers today have prearranged outlets for the cotton purchased or are acting as commission buyers.

Direct Mill Buying by domestic mills typically takes the form of forward contracting with growers. Growers with large farms may contract directly with mill buyers to process their crop according to a predetermined set of conditions for a predetermined price to the grower. This practice is not widespread because cotton departments of textile firms do not usually have the personnel to deal with a huge number of growers across the Cotton Belt, and many mills prefer to have a third party to guarantee the performance under any contract.

12–4.2 Export Marketing

Most of the exported U.S. cotton goes to Korea, Japan, Hong Kong, Taiwan, and the PRC. The USA tends to be the residual supplier of cotton to the world market, so exports are subject to considerable year-to-year fluctuations.

Most cotton is exported from the USA by cotton merchant-shippers or by cotton growers' cooperatives. Relatively few cotton exports are handled by agents and brokers. Exporters tend to be located in the main domestic spot cotton markets such as Montgomery, Ala.; Memphis, Tenn.; Greenwood, Miss.; Dallas, Houston, and Lubbock, Tex.; Phoenix, Ariz.; and Fresno and Bakersfield, Calif. They also tend to be more concentrated in the western markets (Dallas, Lubbock, Phoenix, and Fresno) because a higher proportion of western cotton is exported.

The process of exporting cotton involves all of the same steps as shipping cotton to domestic mills plus additional functions. Consequently, more expense is involved in exporting cotton than in domestic sales. Greater distances involve more communication and travel costs and substantially higher transportation costs. Export sales involve establishing relations in importing countries with agents, cotton merchants, or with mills themselves. Travel to importing countries is usually necessary to establish satisfactory contracts and learn about particular terms of trade in those countries, the quality requirements of the mills, and related matters. Exporters also have to supply the foreign contracts with type samples if the universal standards are not used.

12-4.3 Price Discovery Institutions

Buyers and sellers have many ways of establishing prices for individual trades. These methods vary from direct, one-on-one negotiation between buyer and seller to a highly sophisticated computerized electronic market for trading cotton (Henderson, 1976). Three major institutions facilitate cotton trading and pricing.

Futures Markets are organized exchanges which provide for trade in contracts for future delivery of designated commodities. Futures markets trade in a single product quality and quantity. For example, the New York Cotton Exchange trades in cotton futures contracts which specify that each contract is for 22 700 kg of cotton, grade 41, staple 34, and micronaire readings 3.5 to 4.9. Bales with grades below 32, staples below 32, and micronaire readings below 3.5 are not deliverable with the New York contract.

Traders of futures contracts may be hedgers or speculators. Hedgers make simultaneous, opposite transactions in futures and cash markets by buying in one market and selling in the other. Speculators either buy or sell futures contracts without simultaneously completing the opposite transaction in the cash market. In actively traded futures markets, the futures prices and spot prices tend to move together in a generally reliable manner. It is this simultaneous movement of the two prices that makes hedging feasible. Losses in one market generally tend to be offset by gains in the other market. Thus, the hedger uses the futures market to lessen the risks associated with price movements, while speculators assume that risk of price movement. Activity of speculators provides the opportunity and liquidity needed for price risk aversion to hedgers.

Less than 2% of futures market trades involve delivery of the commodity. Most contracts are cancelled by each trader making an offsetting buy or sell action. Consequently, the futures market operates more as a vehicle for price discovery than as a means of trading actual commodities. Merchant-shippers make widespread use of futures markets as an instrument for hedging. Producers make less use of it. However, merchant-shippers often use futures markets to hedge forward crop contract purchases from producers.

Consequently, it is used as an indirect means of allowing growers to hedge prices.

Two futures markets in the USA currently offer cotton contracts. The New York Exchange offers the contract mentioned previously, which has been the standard contract traded for many years. A new contract was established in 1981 by the New Orleans Commodity Exchange. The New Orleans contract is for 22 700 kg, grade 41, staple 32, and micronaire readings 3.5 to 4.9. Unlike the New York contract, grades as low as 42, staples down to 31, and micronaire readings as low as 3.0 are deliverable. Because many of the cottons grown in the Texas-Oklahoma area have these quality characteristics, the New Orleans contract has become known as the "Texas contract."

Forward Crop Contracting involves a written agreement between a producer and a contractor regarding the delivery of a specified product at a future date. Both parties commit themselves to a future course of action by entering into the contract. Forward contracts for cotton are typically signed prior to harvest. Terms contained in a contract may vary widely but usually contain specifications for quality, quantity, and time of delivery. The typical forward contract used in the cotton industry is similar to a spot cotton transaction in the sense that the responsibilities and functions of each party in the transactions are similar. The producer retains title and is responsible for producing and ginning the cotton, and the buyer has no control over the commodity prior to taking title.

In some cotton contracts, the price is set when the contract is signed. The fixed price contract may set one price for all cotton meeting a minimum quality standard, or it may contain provisions for differentials in price based on quality. Not all contracts specify a fixed price; "on-call" and "sellers call" contract and price differentials are based on the government loan rate differentials.

An "on-call" provision ties price to the futures market. Rather than price being fixed at the time of contract signing, a negotiated differential from a specified futures market date is stated in the forward contract; price may be stated as so many points on or off a given month's futures price of cotton. The amount of the negotiated differentials may reflect factors such as an expected quality difference between the cotton to be delivered and the futures contract quality specifications and an allowance for transportation costs to the contractor. The second method of expressing price in a cotton contract in terms of a premium above government loan rates may also reflect quality differentials and transportation costs to a specific location.

In regions where yields are relatively stable from year to year, contracts usually specify quantity and are called "bale" contracts. In regions where yields are less stable, many contracts simply place certain planted acres under contract. These contracts in which the contractor purchases all or a portion of the cotton produced from that given production area are called "acreage" contracts. In regions where yields tend to be most variable, growers avoid bale contracts and buyers avoid acreage contracts because of the risk involved. Data on the extent of forward contracting shown in Table 12–6 bear this out.

Table 12-6. Forward contracting of cotton in the USA by region of production, 1970 through 1981.

Region												

Region	1970	1971	1972	1973	1974	1975	1976	1977	1978	1979	1980	1981
							%					
Southeast†	8	28	23	73	10	5	52	9	15	21	33	10
Mid-South‡	17	59	65	87	31	13	73	19	39	35	63	19
Southwest§	7	37	13	68	6	0	23	19	10	14	20	7
West#	6	23	24	75	48	29	69	34	46	27	41	9
USA avg	11	43	36	75	21	8	49	21	24	21	34	10

Crop forward contracted spans the year columns 1970–1981.

Source: USDA. "Weekly Cotton Market Review", Agricultural Marketing Service, Cotton Division.
† Alabama, Georgia, North Carolina, South Carolina.
‡ Arkansas, Louisiana, Mississippi, Missouri, Tennessee.
§ Oklahoma, Texas.
# Arizona, California, New Mexico.

The primary motivation for producers to enter a forward crop contract is to minimize their price risk within a season. Contracting may also enable producers to obtain production financing more easily. Merchants may also contract to avoid the risk of price change, but they have other motivations for contracting. By contracting merchants have a more secure cotton supply which is important in arranging both foreign and domestic sales. With fixed price contracts in hand, merchants can "forward sell" or sign contracts with buyers. Fixed price contracts may also allow merchants to avoid the expense of futures market transactions if advanced sales are also made to textile mills; when advanced sales are made to mills, merchants may offset the transaction with a crop contract, not subject to margin calls, rather than a futures contract.

Textile mills may contract with selected producers to obtain desired quality from known production areas. This can help establish their production costs and facilitate forward selling of their output; crop contracts allow the mills to know raw material prices in advance, and this procedure allows them to estimate their output cost.

Electronic Marketing: Telcot is a marketing system designed, developed, and operated by Plains Cotton Cooperative Association (PCCA) in Lubbock, Texas (Ethridge, 1978). It consists of an interconnected network of cathode ray terminals and high speed printers attached by telephone lines to a central computer at PCCA. The terminals are located in the office of PCCA, offices of some 52 subscribing cotton merchants located throughout the buying centers of Texas and Oklahoma and in parts of Tennessee, South Carolina, Alabama, and North Carolina, and in 400 gin offices across much of Texas and Oklahoma. Any buyer can examine on his remote terminal recaps of cotton for sale in terms of specific quality characteristics, average characteristics, asking prices, and current selling prices by grade, staple, and micronaire, and/or submit bids to the computer through its terminal at any time during trading hours. Each buyer terminal is separately linked to the central computer to maintain confidentiality. Data printers accompany each merchant's terminal to allow printed market reports if desired. Price, quality, and sale information are stored in the computer and used in updating market price information each day.

Any producer from any of the subscribing gins located in West Texas, South Texas, Eastern New Mexico, and Southwestern Oklahoma can use Telcot to sell cotton. Beginning in 1979, a cost-sharing agreement was reached between PCCA and a private firm, Commodity Exchange Service (CXS), which made Telcot available to producers not served by PCCA. After a producer's cotton has been harvested, ginned, and classed, the class cards for each bale are delivered to PCCA or CXS where the data are coded into the central computer under the producer's identification number(s). The producer using Telcot has essentially three alterative procedures by which to sell the cotton: i) the Telcot "bid" system; ii) the Telcot "firm offer" system; and iii) obtaining bids independent of Telcot and offering it for sale through the bid system to attempt to get a higher bid. Forward crop contracting also can be done over the system.

The bid system operates as a blind auction. If this option is chosen, PCCA offers a producer's lot to all buyers in the system simultaneously. Each buyer may enter his blind bid via a terminal; no buyer knows what other buyers are bidding. Bidding on the lot is open for 15 min; if the high bid is within 25 points (0.25¢/lb.) of the asking price, the computer automatically assigns the cotton to the highest bidder. The asking price is an estimated market price for the lot of cotton at the point in time the cotton is offered for sale; PCCA makes the price estimate using the Telcot computer. If the high bid is more than 0.25¢/lb. *below* the asking price, the bid can still be accepted by the producer providing it is done within 30 min. There is a provision for buyers who do not subscribe to Telcot to also bid on the cotton [alternative (iii) above]. The price results of the blind bid sale are promptly displayed on the monitor tubes of the gin and buyer terminals. If the high bid is a tie, the computer awards the sale to the bid received first. If the lot does not sell, the result shown is a "no sale".

Firm Offer is the other option. The producer may offer a lot to PCCA at any firm price he chooses regardless of how that price relates to the market price estimated by Telcot. The PCCA then offers the cotton at the firm price to all buyer terminals simultaneously. The first buyer to accept the offer is awarded the lot. The firm offer remains in effect until a buyer accepts the offer or until the offer is withdrawn.

The Telcot firm offer option also provides a means for buyers to search all firm offers for specific fiber quality, location, and other characteristics. Buyers can use the system to sort for several criteria. For example, through a remote terminal, a buyer could ask for a display of all lots of cotton for sale with average values of i) grade 42 or better, ii) staple 32 or better, iii) micronaire readings 3.5 to 4.9, iv) lot size of 100 bales or larger, and v) firm offer price of 64¢/lb. or lower. Other criteria such as gin and warehouse location can be used but not more than five criteria can be specified in one inquiry. The computer would then display all lots of cotton in the firm offer system meeting those criteria on the buyer's remote terminal screen. The computer invoices the cotton for both buyer and seller under all three procedures.

Another option added to the system is trading in forward crop contracts. Procedures are similar except merchant-shippers offer the contracts and growers respond. In addition to trading, subscribers (merchants and producers alike) have ready access to very timely price information which facilitates price discovery in the market. Subscribers also receive futures market reports, and several news services over their terminals.

Producers wanting market news, prices, or information about their unsold cotton have this information available from the gin's remote terminal. Gin personnel access the central computer for the desired information through the gin's remote terminal, and the information is displayed on the screen.

12-4.4 Spot Cotton Quotations

The U.S. Cotton Futures Act of 1916 established the USDA system for determining and reporting spot cotton quotations. Spot quotations are reported prices for all specific quality designations, (for example, grade 41, staple 31, 3.3 to 3.4 micronaire reading) in each of the designated spot markets on each trading day. The price quoting procedures are governed by regulations under the U.S. Cotton Futures Act and the understanding between USDA and the cotton exchanges. The purpose was to establish premiums and discounts for settlement of cotton futures contracts. The classification and market news services for cotton were established in 1937 under the Smith-Doxey Act. Over the years the spot quotations have been drafted by the trade as a means for determining spot market prices and for forward contracting. Since the spot cotton price quotations technically are to settle futures contracts, their use as spot market prices may not be valid in all cases.

Data reported in the Daily Spot Cotton Quotations are compiled as follows: Market News Reporters, employed by Cotton Division, AMS, USDA, conduct daily sample surveys of spot sales in nine markets—Montgomery, Memphis, Dallas, Houston, Lubbock, Greenville, Greenwood, Phoenix, and Fresno. Merchants are surveyed to obtain data on prices and quantities traded by quality groups of cotton. These data are compiled and provided to the Quotations Committee in each market. The Market News Reporter is not a member of the Committee, but serves in an advisory role.

The quotations committees are composed of small, rotating groups of spot traders, primarily merchant-shippers. The quotations necessarily depend upon the transactions of the committee members in attendance and what they know about the transactions of others. Cotton delivered to Marketing pools of cooperatives does not enter into the reports of the quotations committees. Spot prices reported by the quotations committees represent the consensus of their members and necessarily depend on their memories, joint judgment, and ability to establish a representative price for each grade, staple, and micronaire combination. Reports of the committees provide no indication into the range of prices for a given quality.

12–5 YARN MANUFACTURING

Yarn manufacturing is a sequence of processes that convert bales of cotton into a product suitable for use in various end-products. A number of processes are required to obtain the clean, strong, uniform yarns required in modern textile markets. Beginning with a dense package of tangled fibers containing varying amounts of foreign matter, trash, motes, etc., continuous operations of opening, cleanng, drafting, and twisting are performed to transform masses of cotton into yarn. The process of yarn manufacturing has evolved over the past 100 years to its current level of technical development. Even though the yarn manufacturing processes are highly developed, competitive pressure spurs industry groups and individuals to seek new, more efficient methods and machines for processing cotton. Automated processes and new approaches to yarn manufacturing are being studied that might supplant conventional systems in future years. However, for the foreseeable future conventional systems of blending, carding, drawing, roving, and spinning as illustrated in Fig. 12–12 will continue to be used. Only the picking process seems clearly destined for elimination in the near future. The objective of yarn manufacturing is to produce yarns for various end-products. This requires that yarns be produced with different diameters, or different weights per unit length. The production of heavier or coarse yarns, which seem to have grown in importance in recent years, possibly might be shifted to the open-end process as opposed to ring spinning. The basic yarn manufacturing process has remained unchanged for a number of years except for increases in processing speeds, control technology, and package sizes (Hunter, 1980).

12–5.1 Blending, Mixing, and Cleaning

Yarn properties and processing efficiency are directly related to the properties of the cotton being processed. As indicated in section 12–2 physical property measurements and classification are necessary to control variations in processing and yarn quality arising from these relationships. Simply performing tests and making measurements of physical properties is not sufficient for control purposes. The test results must be used in selecting bales of cotton for specific "mixes" to obtain the degree of control required. Even though physical property measurements can be obtained for every bale, it is apparent that obtaining a large number of bales with identical physical properties would be impractical because combinations of fiber properties vary from bale to bale. Experience has shown that the only way to maintain uniform processing performance and yarn quality is by blending a number of bales with similar fiber properties into a mix and attempting to keep the average and range of properties of this mix consistent over time. This process helps to assure that the properties of yarn produced at one period of time will be very similar to those of yarn produced at another time. For the purposes outlined above mills routinely blend a number of bales of cotton to obtain a mix. The number and types of bales to be blended depends on several factors. If physical tests are performed and detailed knowledge of the properties of the bales being processed is available, uniform processing performance generally can be maintained with a smaller mix. If little detailed knowledge of physical properties is available, a relatively larger mix is necessary to control variations. Typically, mills will select mixes to maintain a representative sample of the cotton purchased for the crop year. The number of bales used by different mills in each mix ranges from 6 or 12 to over 50 bales. Depending on the type of machinery used, cotton may be fed to the machines from more than one group of bales. For example, a mill might have blending machinery arranged to permit feeding from several 12-bale lots with each 12-bale lot on a staggered re-

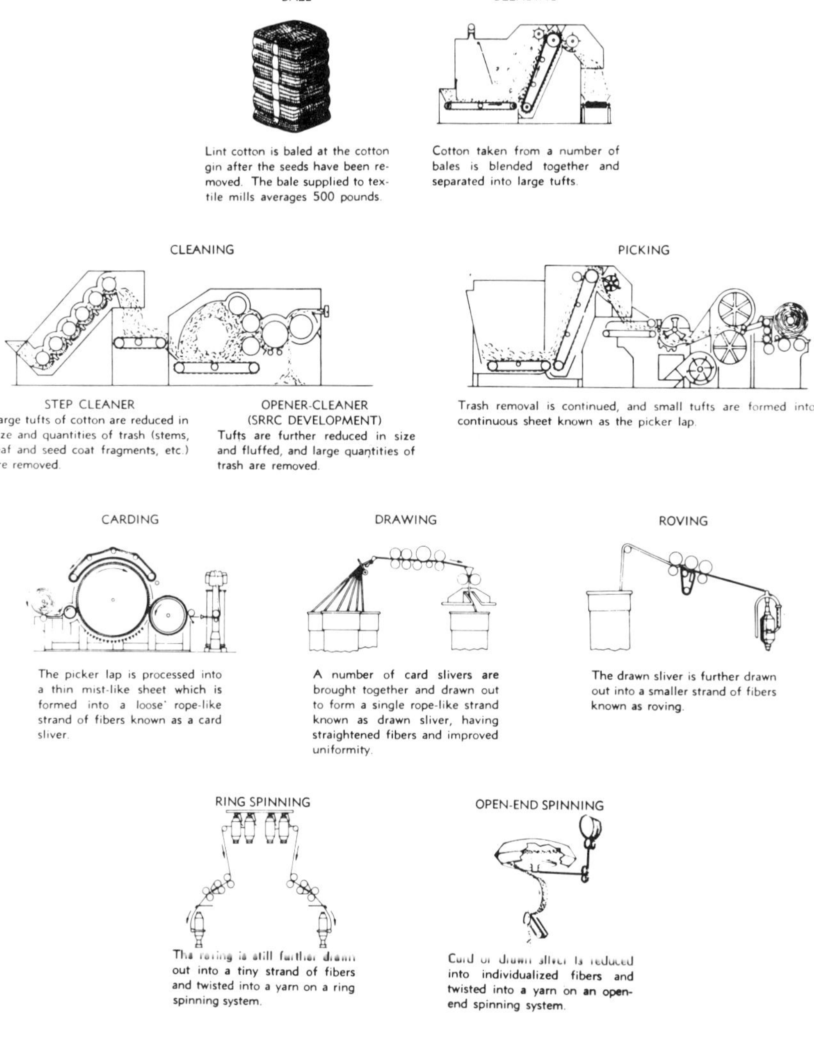

Fig. 12-12. Conventional cotton processing system.

placement cycle. The actual mix then is composed of several 12-bale lots. Maintaining uniform mixes is very important to precision textile manufacturing processes and considerable effort is devoted to blending cottons in textile mills. Since mills run 24 hours a day, cotton is being fed from bales to machines continuously. Once the mix properties and the number of bales to be blended are established, quality control procedures can be used to maintain uniformity. A typical mill opening room is shown in Fig. 12-13.

Processing begins when the bales to be mixed are brought to the opening room. Bagging and ties are removed and either layers of cotton removed from the bales by hand are placed in feeders equipped with lattice conveyors studded with spiked teeth or entire bales are placed on platforms which move the bales back and forth under or over plucking mechanisms. In either method the objective is to begin the sequential process of rendering the compacted layers of cotton into small, light, fluffy tufts to facilitate removal of foreign matter. This process is referred to as opening. Since bales arrive at the mill in various degrees of density, it is helpful to cut bale ties approximately 24 hours before the bales are to be processed to allow them to "bloom." This enhances opening and helps regulate feeding rate. Bales compressed to high densities feed at faster rates than low density bales when processed on pluckers or feeders.

Tufts of cotton emitted from feeders or pluckers are not sufficiently open to permit efficient processing and must be cleaned to remove trash, motes, and foreign matter. For further opening and cleaning the tufts are transported, either by conveyor or aerodynamically, to cleaning machines with intermediate stops in some cases in various containers or machines called blending reserves or auto mixers. The purpose of these devices is to serve as a buffer between the opening and the cleaning machines to compensate for variations in feeding rate of different cotton bales and to accomplish further mixing or blending of the tufts. Mixing is performed by continuously adding layers of tufts to the reserve box and removing the tufts from several layers simultaneously. The cotton is metered at a uniform rate from the reserve boxes or automixers to cleaning machines where the process of opening and cleaning is continued. Cleaning machines are available from several manufactuers, and many different types have been introduced. Most cleaners employ the principle of beating to remove trash and other undesirable material from the cotton. In this method tufts of cotton are fed between two spring loaded or weighted rolls that are grooved to provide powerful griping action. As the feed rolls turn, cotton tufts are presented to a revolving beater. Small tufts of fibers are stripped or beaten from the fringe and carried across slots through which are hurled the loosened trash and foreign matter. The speed of the beater and its closeness to the feed rollers are critical to cleaning efficiency, and they must be precisely set to avoid excessive fiber damage. A number of different types of beaters are available. Their designs range from an array of simple spikes to flat blades or bars to saw-type surfaces wound in the form of stamped wire.

Often more than one cleaning machine is used. The selection of the types of machines and the types of beaters to be used is a complicated process often dominated by personal preference. Technically, the basis for selecting cleaning machinery is the type or grade of cotton to be processed and the quality or type of end-product desired. Low grade cottons with high trash content require more elaborate cleaning than higher grade, cleaner cottons. High quality end-products require more cleaning machinery than lower quality products because cleaning rates are normally slower to preserve the quality of the cotton. Between these extreme conditions a number of combinations of cleaning machines is permissible. Two typical cleaning machines are shown in Fig. 12-14. A coarse opener with a series of spiked

Fig. 12-13. Typical mill opening room.

Fig. 12-14. Two typical cleaning machines.

cylinders receives cotton from the feeders, further reduces the size of tufts, and removes a high percentage of the heavy trash and motes from the cotton. It is followed by a finer opener equipped with a cylinder wound with a saw wire for fine opening and removal of smaller trash particles and foreign matter. At this point the cotton is ready for the subsequent process which may be formation into a batt or for direct feeding into the card.

The opening and cleaning machinery prepares the cotton for processing on the card where a combination of extremely important and related functions are performed. The cleaning machines in the opening and cleaning rooms of mills perform combined functions of opening and first level cleaning. In a similar but more detailed way the card performs combined functions of second level cleaning and transformation of fiber tufts into sliver form to permit the process of fiber parallelization to begin. Historically, cotton has been fed to the card in the form of a picker lap. The picker lap is formed on a picker which is a combination of feed rolls and beaters with a mechanism made up of cylindrical screens on which opened tufts of cotton are collected and rolled into a batt. The batt is removed from the screens in an even, flat sheet and is rolled into a lap. A mechanism for sensing and regulating batt thickness is an essential part of the picker, and controlling picker lap weight variation is a critical quality control process. Pickers are equipped with either one or two beaters and contribute to the opening and cleaning process; however, labor requirements and availability of automated handling systems with the potential for improved quality are contributing to their obsolescence.

The trend in modern mills is to eliminate the picking process. This is accomplished by installing more efficient opening and cleaning equipment to offset the opening and cleaning formerly performed by the picker and by installing chute feed systems on cards. Chute feed systems distribute opened and cleaned tufts of fibers from a reserve box to cards pneumatically through ducts. Narrow chutes at the cards are kept filled to preset levels so that the feed rolls are constantly presented with a supply of cotton in a much more open state than is possible with picker laps. This action contributes to processing consistency, improved quality, and reduction in labor requirements and elimination of lap tails, a source of manufacturing waste.

12–5.2 Carding and Combing

The card is the most important machine in the yarn manufacturing process. It performs second and final level cleaning functions in an overwhelming majority of cotton textile mills. Only a minority of mills, where combed yarn is produced, perform more extensive cleaning than is provided by the card. The card is composed of a system of three wire-covered cylinders and a series of flat, wire-covered bars that successively work small clumps and tufts of fibers into a high degree of separation or openness, remove a very high percentage of trash and other foreign matter, collect the fibers into a rope-like form called a sliver, and deliver this sliver in a container for use in the subsequent process. The first cylinder in the sequence is called the licker-in. It is covered with coarse, saw-toothed wire and works in conjunction with grooved feed rolls similar to cleaning machines. A tufted fringe of cotton is presented to the fast-turning licker-in. Small tufts are combed out by the licker-in, dragged across slotted bars for further trash removal, and transferred to the much larger main cylinder. The main cylinder is covered with much smaller wire teeth and a significant amount of combing and straightening of fibers occurs at the transfer point. The fibers are hooked on the fine cylinder wires and are pulled past a series of wire covered plates, called flats, attached to a chain mechanism that slowly moves, usually in the same direction as the main cylinder. The distance between the wire surfaces of the main cylinder and the flats is approximately 0.25 mm. The carding action that occurs between these two wire surfaces reduces fiber tufts to a very thin web of fibers and removes much of the remaining trash and foreign matter as well as the short fibers whose length does not allow them to remain hooked on the cylinder wire. These short fibers and entrapped fine trash and dust collect on the flat wires and are brushed out after the flats have revolved to a position out of contact with the cylinder. This very thin web, practically in a single fiber layer, is deposited on the third cylinder, called the doffer, which is also covered with fine toothed wire. The web is stripped off the doffer and is drawn through highly polished and pre-cision ground steel rolls that are heavily weighted to crush any large trash particles remaining in the web. These crush rolls perform a vital cleaning function because they break large particles of trash into small pieces that drop out in subsequent processes. After leaving the crush rolls the web passes through a high polished cone called a trumpet and is deposited in coils in a large can in preparation for the drawing process. The carding process in a typical textile mill is shown in Fig. 12–15.

Fig. 12–15. Carding process in a typical textile mill.

At the carding process fiber quality differences begin to make themselves known. Cottons with higher short fiber content or with lower fiber strength usually result in more waste in the form of flat strips and undercard waste. Cottons that are immature or that have low micronaire readings tend to contain more neps. Contaminants such as lubricating oils or excessive sugars cause loading on the cylinder and lapping on the crush rolls.

Mills that produce combed yarn produce the cleanest, most uniform cotton yarn available. The cotton in combed yarns has received the ultimate cleaning possible in cotton yarn manufacturing. The purpose of combing is to remove short fibers and to remove hooks from the longer fibers. In the combing process neps and trash are also removed so that the resulting combed sliver is very clean and lustrous.

The comber is a complicated machine composed of grooved feed rolls and a cylinder, called a half lap, that is partially covered with needles that penetrate the fiber fringe presented by the feed rolls to comb out short fibers. The combed fringe is advanced and gripped by detaching rolls which pull the fibers through another needle arrangement called the top comb. This process is discontinuous and a combed sliver is produced in segments that must be reassembled or pieced together. The fibers in the sliver are nearly parallel and are easily pulled apart when they leave the combing mechanism. For this reason the sliver is supported by a smoothly finished chute until the output from 6 to 8 combing mechanisms can be combined to form one combed sliver. The fibrous material combed out of the sliver by the comber is called noil. Typically, combers remove from 8 to 14% noils depending on the quality of the cotton and the end product desired. Noils have resale value and are used in the production of coarse cotton yarn and in nonwoven and medical products.

12–5.3 Drawing and Roving

Drawing is the first process in the sequence of yarn manufacturing that employs the concept of roller drafting. Very high draft occurs at carding but most of the drafting occurs between the licker-in and the feed rolls. In drawing practically all draft results from the action of rollers. Drafting occurs when a sliver is fed into a system of paired rolls whose surfaces are moving at different speeds. In Fig. 12–12 a typical 4 over 5 drafting system is illustrated for the drawing process. The surface speeds of each pair (top and bottom) of rollers is higher than for the pair of rollers immediately preceding it. The first set of rollers pulls the sliver into the system until it is caught by the faster turning second pair. Fibers are pulled out, or drafted, by the gripping action of the second set of rollers. Further draft occurs as the sliver passes through the remaining rollers of the system. One purpose of drawing is to straighten the fibers in card sliver by drafting to make more of them parallel to the axis of the sliver. Parallelization is necessary to obtain the properties desired when the fibers are subsequently twisted into yarn.

Another purpose of the drawing process is to obtain a sliver that is more uniform in weight per unit length. The averaging effect of combining or uniting laterally several slivers into one reduces the variation in weight per unit length. This process of uniting several slivers into one is called doubling. Typically, six or eight slivers are doubled and fed into the back rolls of the draw frame. These six or eight doublings are typically drawn or attentuated to produce a sliver approximately the same weight per unit length as each of the original ones fed. Typically, mills feed in, or double, eight slivers weighing approximately 4.6 g/m each and draft approximately eight to produce a single sliver weighing approximately 4.6 g/m. For each meter of the doubled mass fed into drawing approximately 8 m are produced.

The final purpose of drawing is to achieve greater blending. It is necessary to use at least two drawing processes to obtain the quality required by most mills. With eight doublings at each process a total of 64 doublings occurs which helps maintain uniformity of raw material in the end product. Fig. 12-16 is an example of the drawing process in a modern mill.

The fibers in the sliver that are produced by the final drawing process, called finisher drawing, are nearly straight and parallel to the axis of the sliver. Weight per unit length of finisher drawing sliver of 4.6 g/m is too high to permit drafting into yarn on conventional ring spinning systems.

Fig. 12-16. Drawing process in modern mill.

Further drafting to weights lower than approximately 3.2 g/m reduces sliver cohesion and increases difficulty of removal of sliver from cans at high rates.

The purpose of the roving process is to reduce the weight per unit length of sliver to a suitable size for spinning into yarn. At this point in the processing sequence it is necessary to twist the fibers together to maintain integrity of the strand. The two functions of drafting to reduce the size of the sliver and inserting twist to maintain the integrity of the drafted strand are performed at the roving process. The term roving is used for the process and also for its product. Weight per unit length of roving is typically one-eighth the weight per unit length of the sliver fed. To produce roving, finisher drawing sliver is fed into the drafting system of the roving frame as shown in Fig. 12-14. The drafting system consists of three pairs of rolls at least one of which is covered with a flat band or belt of material called an apron that helps control the fibers during drafting. The sliver is drafted in a similar manner to that described for drawing, twisted slightly and simultaneously wound onto a bobbin. At this point the material is called roving. It typically has a weight per unit length of 0.6 g/m and a twist of 43 turns/m. The amount of twist in roving is kept to a minimum because it is costly to insert. Only enough twist is required to impart sufficient strength to allow the roving to be pulled from the bobbin without breaking. Longer, finer fibered cottons require less twist in roving and spinning than shorter, coarser cottons for equivalent strand strength, hence a contribution to the price differential commanded by these cottons. The roving processes in a typical textile mill is shown in Fig. 12-17.

Fig. 12-17. Roving process in typical textile mill.

12-5.4 Spinning

Spinning is the most costly single process in converting cotton fibers into yarn. The yarn produced is used in a number of different ways for a number of different products. It can be woven or knitted into apparel or industrial fabrics, or it can be used as sewing thread and cordage. Whatever its end-use the properties of the yarn are a function of the properties of the fibers in it and of the processing conditions to which the fibers were subjected. Because of the high cost of yarn manufacturing, the importance of yarn quality to end-product quality, and the relationship of yarn properties to fiber properties, considerable research and development effort is being devoted to radical, new methods of yarn production. Experts predict that new methods of spinning will be developed soon and that the type of spinning system used in the future will be determined by the end-product desired.

12-5.4.1 Ring Spinning

Over 85% of all yarn produced in the world is produced on ring spinning frames. There are a number of manufacturers of these machines in the world, but they employ the same basic principle of operation as illustrated in Fig. 12-14. Roving is pulled from a bobbin suspended from a rack over a drafting system of three pairs of rollers. Initially the roving is placed by hand between the back rollers whose grip pulls the strand into the system where it is grasped by the apron-covered middle pair of rollers. This back or break draft is usually about 1.2 turns or just enough to overcome the cohesive force imparted to the roving strand by the twist that was inserted at the roving process. The drafted strand is then pulled into the front drafting zone and carried along by the aprons until the tips of the fibers are grasped by the front pair of rollers which are turning with a surface speed about 30 times greater than the middle rolls. The gripping action of the front rolls pulls the fibers from between the aprons whose frictional properties have been specially selected to retard the motion of the fibers and thereby assist in straightening or stretching them out full length. The drafted strand passes through the front rolls through the thread guide, under the hook-like object called the traveller which is free to revolve on a ring and is wound onto a bobbin which is held by friction on a spindle that is revolving approximately 15 to 20 times for each 2.5 cm of drafted strand that is produced by the front rolls.

Yarns produced on the ring spinning system vary from the coarsest for crude products such as mop yarns to the finest for use in specialty fabrics such as printing ribbons or fine apparel. These yarns vary in linear density from 0.600 to 0.006 g/m. The ring spinning system is the most versatile in commercial use and is capable of producing high quality yarns from a wide range of cottons with different fiber properties. Typical properties of commercial ring yarns are shown in Table 12-7. Fig. 12-18 is an example of a ring spinning frame used in a commercial mill.

12-5.4.2 *Open End Spinning*

Although less than 15% of the production of yarn in the world now comes from open end spinning machines, experts predict that as much as one-third of the world spinning capacity will be changed to this system. Open end spinning is attractive to manufacturers for several reasons. It eliminates the necessity for the roving process and in some cases can eliminate one of the drawing processes so that the yarn production sequence is opening-cleaning, carding, drawing, and spinning. The open end spinning

Table 12-7. Properties of typical commercial cotton yarns.

Property	Typical value
Strength, g/Tex	
combed	18
carded	12
open-end	10
Elongation, %	3–8
Variability in mass per unit length, %	
Short term	
combed	15–19
carded	16–20
open-end	12–16
Long term	
combed	2.4
carded	3.0
open-end	2.4
Appearance, grade	C^+–B

Fig. 12-18. Ring spinning frame in commercial mill.

process is one step nearer to automated yarn manufacturing which has long been a goal of progressive textile manufacturers because of low profit margins. Spinning yarn directly from drawing sliver requires several changes in manufacturing practices compared with ring spinning. Very clean cotton is required to maintain spinning efficiency so that it is necessary to purchase high grade cottons and to install more elaborate and thorough cleaning machinery. It is also necessary to produce very uniform drawing sliver with a somewhat lower weight per unit length than is required for ring spinning. Sizes of cans of drawing sliver are somewhat smaller for open-end spinning as compared to ring spinning and require automatic handling to maintain machine production efficiency. However, for yarns with linear density approximately 0.025 g/m and heavier, the cost of production of open end yarn is considerably lower than for ring yarn because the production rate is so high. At speeds of 60 000 revolutions/min the production rate of an open end spinning rotor is three to five times higher than for a ring spindle. To produce the yarn, clean drawing sliver is pulled into the system by feed rolls as shown in Fig. 12–12 where a small opening roller with wire-teeth plucks individual fibers from the fringed mass and hurls them into an airstream which carries the stream of fibers into the rapidly spinning rotor. Air currents and centrifugal force carry the fibers to perimeter of the rotor where they are very evenly distributed in a small groove. Using a starter yarn, the tails of the fibers are twisted together by the spinning action of the rotor and yarn is continuously drawn from the center of the rotor. A major problem with some open end spinning systems is rotor residue. Dust and small trash particles accumulate in the rotor groove and interfere with yarn production. Yarn quality deteriorates as the residue increases to the point where spinning is interupted, and it is necessary for the operator to manually clean the rotor with a brush before production of yarn on the rotor can be resumed. Some open-end spinning machines have cleaning devices that work in conjunction with the opening roller so that rotor residue is minimized. Others have cleaning devices that automatically clean the rotor and restart production but these devices are difficult to justify economically. Yarn from open-end spinning is much more uniform than ring spun yarn and it has fewer neps. However, it is considerably weaker and has a harsher feel. Its properties make it particularly suited for heavier fabrics such as denims, velveteens, and corduroys where its use is growing. Clean cottons with low micronaire reading and high fiber strength are desirable for open end yarns. A commercial open-end spinning machine is shown in Fig. 12–19.

12–5.4.3 Other Spinning Methods

A number of new spinning systems are currently under development that may revolutionize yarn manufacturing and could cause changes in the relative importance of fiber properties as they are now perceived. In addition to high production capacity these new systems promise improved yarn quality (Hunter, 1978).

In general there are four different approaches used in the new system that appear practical for use on cotton. Core spun systems are currently in use to produce a variety of specialty yarns and sewing threads. These core spun systems involve a modification of conventional systems to feed a monofilament core behind the front rolls so that the conventionally prepared and drafted staple fiber is wrapped around the core during twisting. These yarns can be produced at higher production rates than conventional yarns and offer improved yarn quality but have found limited use because of the cost of the monofilament.

Other methods involve conventional drafting of staple fibers and wrapping of the drafted strand with a monofilament to maintain integrity of the yarn until it is converted into fabric. These systems have experienced excessive yarn breakage at spinning and yarn strength that is too low for further processing.

Twistless yarns have been produced commercially on a limited basis by a system that bonds the fibers together with a polyvinyl alcohol or some other bonding agent. Wetted sliver is drafted and passed through a system of rolls where the binder is padded onto the fiber strand. False twist is inserted in a stream vortex mechanism to maintain the integrity of the

Fig. 12–19. Open-end spinning machines.

strand until the adhesive can be solidified on a heated drum. The yarn is then wound onto a package. The twistless yarn system offers potentially high production rates and very uniform yarns. Knit and other apparel fabrics from twistless yarn have excellent appearance.

Air vortex spinning is under study by several machinery manufacturers. In air vortex spinning, drawing sliver is presented to an opening roller, similar to rotor spinning, where individual fibers and tufts are separated and introduced into an air stream which transports the fibers to the vortex chamber where jets of air are directed tangentially along the inner wall of the chamber. The fibers are collected by centrifugal force and condensed to form a ring. The spinning process is started by inserting a piece of starter yarn which unites with the fiber. Twist is inserted by the rotation of the condensed ring and the yarn drawn out of the chamber. Air vortex spinning is capable of very high production speeds but prototype models are particularly sensitive to fiber length variations and foreign matter content such as trash particles.

12-6　FABRIC MANUFACTURING

In almost all situations yarn that is produced at the spinning process must be further processed before it is suitable for manufacturing into fabric. Exceptions to this rule include certain specialty yarns, direct spun filling, and many open-end yarns used in knitting or as filling in woven fabrics. Most yarns are processed on winders after they are spun. Winders are used to form packages of yarn, called cones, that cannot be produced economically at the time of ring spinning. In the winding process yarn is removed by unwinding from the bobbins produced at spinning. Modern winders are costly, automatic, high-production machines. They remove bobbins of yarn from a reserve bin, transport them to winding positions on the machine, install them in a position for unwinding, tie knots in thread ends to join the yarn sequentially from one bobbin to another, and initiate the winding process. Several bobbins of yarn are joined together sequentially in this process to produce a cone of yarn weighing up to 2.25 kg and containing many yards depending on the size of the yarn. If the yarn breaks during the process of unwinding, the machine automatically locates the broken ends, ties them together, and restarts the process. Knots tied are very small and have short tails but still are undesirable because they can interfere with fabric manufacturing. During the winding process yarn is cleared of major defects called slubs, which are thick bunches of fibers, by snick plates or other devices built into the thread guides of the winder. These devices can be adjusted to restrict the passage of thick places so that the yarn is broken and retied. Only slubs, imperfections, and trash particles that are more undesirable than knots are removed.

Depending on end use and properties desired, yarns may be plied after winding. Plying is the process of twisting together two or more yarns. Plying of cotton yarns is usually performed on a ring twister. Yarns from two or more cones or packages are unwound, twisted together in a manner similar to ring spinning and wound onto a bobbin. Plied yarns are more uniform and are stronger than single yarns of equivalent weight. They also have greater abrasion resistance and are used widely in fine apparel and industrial fabrics.

12–6.1 Weaving

Weaving is the process of forming a fabric in which lengthwise (warp) yarns are interlaced with crosswise (filling) yarns. Weaving is performed on a loom. Warp yarn is fed to the loom from a beam which is a cylindrical object shaped like a spool containing thousands of yarns. Warp yarns are threaded through slots in thin metal strips which are attached to bars on frames called harnesses. Each loom uses a number of harnesses depending on the type of fabric being made and the density (threads per inch) desired. The warp yarns then are threaded through the reed which is a rack made of small, very smooth vertical bars and onto a take-up roll. The reed extends the width of the loom. A system of cams controls the harnesses so that some are raised and some are lowered simultaneously during operation of the loom to form a shed, which is an opening between layers of the warp yarns. Filling yarn is inserted by passing a shuttle containing a bobbin of yarn through this shed. Other methods of filling insertion include use of rapiers or jets of air to propel the filling yarn through the open shed. After the filling yarn is inserted the positions of some of the harnesses are reversed so that those formerly raised are now lowered and vice versa. This locks the filling yarn between strands of warp yarn and the reed is moved forward to pack the yarns together. The cycle is repeated continuously to form a fabric. Each cycle is called a pick.

The weaving industry is currently in a state of change. Technology has advanced rapidly in recent years and has made possible significant increases in weaving speeds. Looms typically have been capable of producing fabric at nominal rates of 300 picks/min. Modern, high technology looms are now capable of almost twice this rate. These faster speeds and higher production rates of place added stress on yarn quality, and consequently, fiber property requirements are affected. Yarns used in high speed weaving must be stronger and more uniform than yarns formerly used. These demands for improved strength and uniformity have magnified the need for instrument measurements in the marketing and utilization of cotton.

Preparing yarn for weaving is a critical process in fabric manufacturing because this preparation affects weaving efficiency and fabric quality. Warp yarns are wound onto loom beams as previously described. To produce the loom beams cones of yarn are wound onto an intermediate beam, called a section beam, by a warper. A warper is a machine that simply winds yarns from approximately 600 cones onto a spool-like beam for further processing. The cones are placed in a creel as shown in Fig. 12–20, and the yarns are passed through tension control devices before being wound onto the beam. Yarns must be wound onto the beam in uniform sheets with very little variation in tension.

WARPING

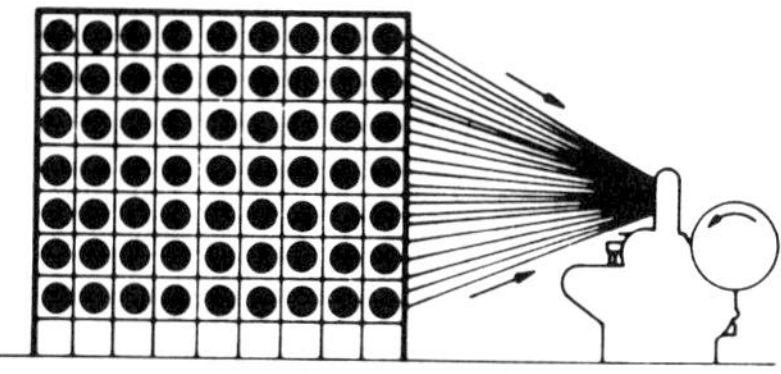

Thousands of yarns are wound on
a giant spool known as a warper
beam.

SLASHING

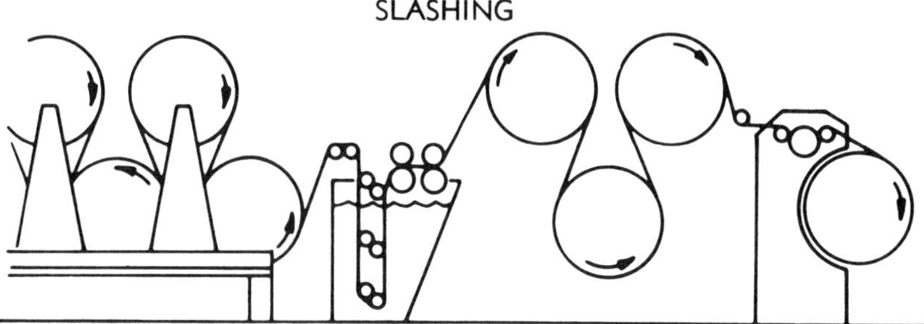

Yarns from a number of warper beams known as warp yarns are
coated with starch, dried, and wound onto a loom beam.

WEAVING

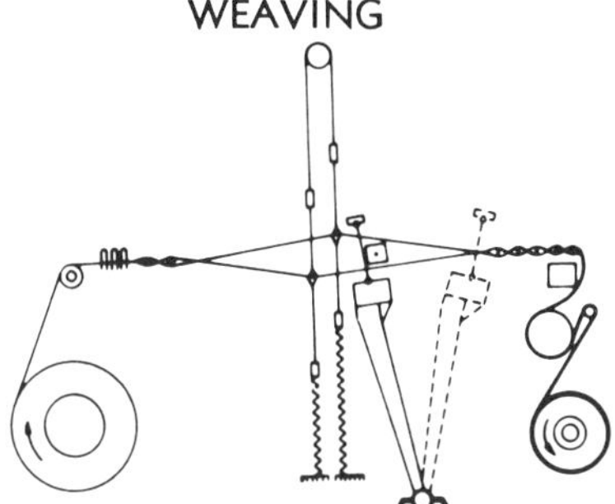

Weaving on a loom is the process
of interlacing warp yarns with
cross or filling yarns to form a
fabric.

Fig. 12–20. Conventional warp preparation and weaving processes.

Section beams are used in slashing to produce loom beams. Sheets of yarn from several section beams are unwound with close control of yarn tension and passed through a box of hot mixture, called size, that is padded onto the yarns by squeeze rolls. As the sheet of sized yarn leaves the size box it is passed over a series of hot, teflon coated cylinders that are steam heated as illustrated in Fig. 12–20. The size mixture is dried and forms a film on the outer surface of the warp yarn to prevent yarn abrasion during shedding in the weaving process. The composition of the mixture used by mills to size warp yarns is sometimes considered proprietary. The mixture may have a number of components such as lubricants, preservatives, adhesives, anti-foaming agents, etc., but the primary component is usually either pearl cornstarch, carboxymethylcellulose (CMC), or polyvinyl alcohol (PVA). The type of sizing material used is determined by a number of complicated factors including cost, type of fabric being produced, and the waste water treatment facility in finishing. Cornstarch requires more extensive treatment before disposal but is lower in raw material cost than CMC. Although PVA is even more expensive, it can be extracted and recycled. Warp preparation and slashing are very important to efficient weaving of cotton fabric. However, variations in cotton fiber properties affect these processes through their contribution to yarn strength and uniformity.

12–6.2 Knitting

Preparation of yarn for knitting is relatively simple compared to that required for weaving. Fabric can be knitted directly from cones of good quality yarn without any preparation other than application of wax or lubricant to help reduce fly and to facilitate movement through thread guides and devices for maintaining uniform tension as the yarn is fed to the machine. Wax or lubricant is sometimes applied to the knitting yarn during the winding process by pulling the yarn acros a felt wick loaded with a liquid wax or lubricant.

Knitting is performed by forming loops with a single, continuous yarn and joining each individual loop to its neighbors to form a fabric. Loops are formed on needles in which an eye is made by a hook and latch mechanism or by a spring device. Hook and latch needles are widely used in the USA. The loops of a knitted fabric form a series of chains, called wales, running lengthwise of the fabric. The loops also form lines at right angles to the wales and are called courses. Wales and courses in knitted fabric are equivalent terms to warp and filling in woven fabrics. Knitted fabrics can be either warp knit or weft knit. In weft knit fabrics the yarns forming the loops generally run crosswise of the fabric. In warp knits the yarns run lengthwise. Knitting machines may be either circular or flat. Flat knitting machines have needles arranged in one plane or in two planes at right angles to each other. Flat knitting machines may produce either flat or tubular fabrics. Circular machines have one or two sets of needles arranged in a circle and produce tubular fabrics. Needle movement in both types of machines is controlled by cams and can be programmed to produce a number of different fabric styles. Cones of yarn are placed on a creel to facilitate unwinding. The yarn is threaded through tension devices and guides into the needle hooks. Either the creel or the needle bed revolves on a circular machine. A jersey knit machine is shown in Fig. 12–21. The creel remains stationary and the needles revolve passing over cams which raise and

Fig. 12–21. Jersey knit machine.

lower them to form loops. Fabric is accumulated on the roll which is revolving underneath with the needles. Knitting loops and stitches are illustrated in Fig. 12–22. Knitted fabrics can generally be produced at higher rates and have properties that make them more suitable for certain end-uses than woven fabrics. Undergarments and hosiery are examples. The hand and drape characteristics of knitted fabrics are quite different from woven fabrics. Knits have lower resistance to abrasion and generally are not as dimensionally stable as woven fabrics. Considerably more cotton is used in woven fabrics than in knitted fabrics.

WOVEN FABRIC

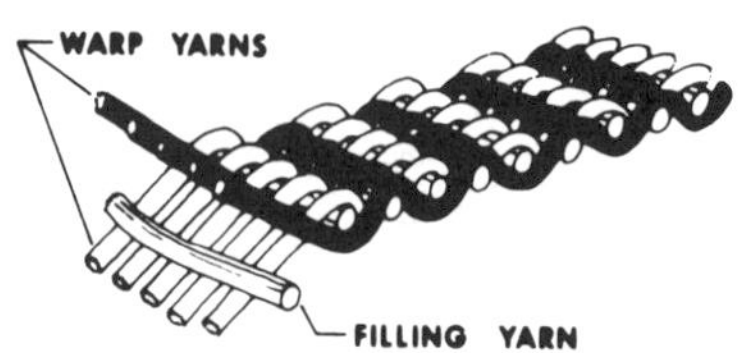

Pictorial view of how yarns are interlaced to form plain fabrics such as broadcloth, printcloth, and sheeting.

WEFT KNITTING

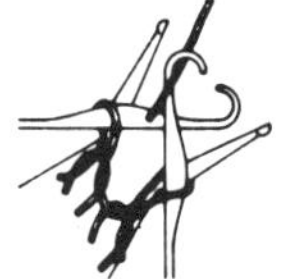

Knitting is a method of producing a fabric by a series of interlocking loops. In weft knitting an individual end of yarn is fed to one or more needles in a crosswise fashion.

WEFT KNIT FABRIC

Pictorial view of how yarns are interlocked by a series of loops running widthwise to form a weft knit fabric.

WARP KNITTING

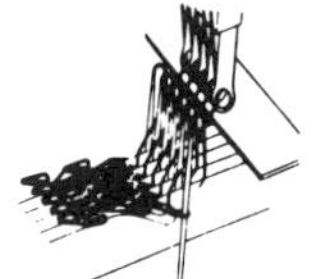

Warp knitting is a system of knitting in which a multiplicity of yarns are fed vertically to a series of needles, one or more yarns to each needle.

WARP KNIT FABRIC

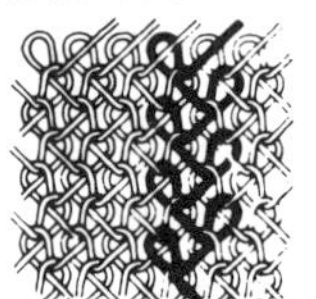

Pictorial view of how yarns are interlocked by a series of loops running lengthwise to form a warp knit fabric.

Fig. 12–22. Knit loops and stitches.

12-6.3 Nonwoven

Felting, a nonwoven process, is probably the oldest method of fabric formation. A nonwoven product might be defined as an assembly of textile fibers held together by the mechanical interlocking or the adhesive bonding of a mat. Animal fibers were the first fibers to be mechanically entangled or felted to form a fabric. The development of the weaving and spinning of staple fibers largely eliminated the use of felts. The woven fabrics were superior in strength and could utilize fibers other than animal.

The modern concept of nonwoven fabric is that of a fibrous web processed on modifications of textile carding equipment and given body through the application of bonding medium or by the fusion of thermoplastic fibers. These nonwovens are distinguished from paper and reinforced plastic film products by the length and amount of the fibers of which they are composed. Nonwovens are generally considered to be greater than 50% fiber in content and formed from textile fibers greater than 12.7 mm (1/2 in.) in length.

The concept of forming a fabric without the many processes involved in spinning and weaving is obviously an attractive one. The webs from modified cards and garnetts are formed into bats by cross layering or by air layering techniques that randomly orient the fibers. An adhesive may be applied by saturation, wet roll, foam, spray, or dry powder techniques. The adhesive solvent is evaporated, a resin is cured, or thermoplastic fibers are fused by heating the bats to complete the nonwoven structure.

Some uses of nonwoven products are apparel interlinings, upholstery, filters, sanitary and medical supplies, and wiping cloths. These products are formed from man-made fibers such as rayon, acetate, nylon, polyester, and acrylics and natural fibers such as cotton and wool. Currently, approximately 360×10^6 kg of textile fibers annually are used in nonwovens. Half of this amount is for the production of durable products and half for disposable. Of the disposable products it is estimated that 23×10^6 kg of cotton fibers are used.

Cotton is desirable for forming nonwovens for several reasons. In many applications cotton that is waste from other textile manufacturing processes can be used to make nonwovens. This processing waste is relatively abundant because of the volume of spun textile processing. The bleached cotton that is used in nonwovens is a very pure form of cellulose and is medically inert. In most nonwoven applications it is cotton's absorbancy that is the key characteristic.

12-7 FINISHING

Fabric finishing comprises all of the processes that involve converting gray state woven or knitted fabrics into fabrics possessing required end-use properties such as whiteness, color, and easy-care characteristics. The first four processes are called preparation processes and include desizing, scouring, bleaching, and mercerizing. These processes systematically and uniformly remove contaminants and produce white fabrics that will easily and uniformly absorb dyes and other chemicals that may be applied later. The final rotproofing, waterproofing, and others. Not every fabric receives every finishing treatment. For example, knitted fabrics do not contain sizes, but do contain knitting oils and lubricants; thus, desizing is not required, but an effective scouring is needed to remove the oils and lubricants. Many fabrics are not mercerized, and fabrics that are to be dyed or printed in dark shades usually are not bleached. Almost all fabrics receive some type of scouring treatment.

Because of the high costs of energy and labor, many changes have occurred in the finishing operations in the past 10 years (Balmforth, 1982). The trend is to automation, continuous processes, energy-efficient systems and techniques, and on-line monitoring and computer control. Use of batch processes are declining except in relatively small operations. For the continuous systems, knitted goods are usually processed in rope form and woven goods are processed either in rope or in open-width form. Energy-efficient systems that use a minimum of water and heat are becoming standdard. Use of innovative foam and solvent systems result in use of less water, thus lowering heating costs and reducing the load on the waste stream. Solvents can be recovered and reused. The finishing processes are critical to consumer acceptance of cotton textile products.

12–7.1 Desizing, Scouring, and Bleaching

Because the trend in these processes is to automation and continuous processing, this summary will be confined to the continuous processes. However, similar chemical treatments are used regardless of whether the processes are batch or continuous. In most systems, desizing, scouring, and bleaching are performed in a single operation without intermediate drying or storage of fabrics. The basic treatments in each of these three individual processes are the same and include: (i) saturation, (ii) digestion or reaction, and (iii) washing. The saturation step simply involves passing the fabric through the proper treating solution and squeezing out excess solution. The digestion or reaction step allows for fabric dwell time in the presence of steam to accelerate the desired reactions. The washing step removes the unused reactants and impurities and prepares the fabric for the next process. A wet processing line would consist of at least three sections with each section containing a saturator, steam chamber or J-box, and washing-rinsing chambers. For rope systems, the digestion chamber is a J-box in which fabric is packed tightly for the steam treatment; whereas, for open-width systems, the digestion chamber is more open and the fabric is loosely plaited for the steam treatment. Because of the packing, dwell times are longer for the J-box systems than for the open-width steam chamber systems (McConnell et al., 1982).

The first step in the finishing process is singeing in which the fabric passes rapidly over an open flame or other source of heat to burn fuzz balls and short fibers from the fabric surfaces. This step gives the fabric a solid, uniform appearance and helps prevent skittery dyeings and pill formation. The hot fabric goes directly from the singer into the desize saturator. If the fabric contains sizing material of starch, the saturator contains an enzyme plus wetting agent to dissolve the starch. If the fabric contains either polyvinyl alcohol or carboxymethyl cellulose, the desize saturator contains a wetting agent, detergent, and alkali to promote solubilization and removal of the sizing material. Dwell times will be 10 to 20 min for J-boxes and less than 1 min for open-width steamers. The fabric passes through the washer-rinser chambers to remove solubilized sizing materials and then goes to the scouring saturator. The scouring solution contains a wetting agent, sodium hydroxide (caustic), a detergent, and probably other alkalies such as sodium carbonate and sodium silicate to facilitate effective removal of cotton waxes and other impurities from the fabric. Dwell times will be about 45 min for J-boxes and about 2 min for open-width steamers. The fabric then passes through the washer-rinser chambers and proceeds to the bleach saturator without intermediate drying. The bleach saturator contains hydrogen peroxide, sodium silicate, sodium hydroxide, sodium carbonate, a wetting agent, and a chelating agent to tie up interferring metals such as iron. Dwell times will be about 60 min for J-boxes and 15 min or less for the open-width steamers. The fabric is then washed, rinsed, and dried unless it is to be mercerized, in which case it goes directly to the mercerization ranges without drying. The fabric is now suitably purified to be dyed or to be used as white goods that would probably receive an easy-care treatment.

12-7.2 Mercerization

Mercerization is the process by which cotton is treated with a strong alkaline solution to alter the surface and crystalline structures of the fibers. Cotton is swelled by the alkali; the fiber becomes more rounded, and the luster of the fiber is greatly increased. Immature fibers are swollen and become more capable of retaining dye molecules, thus decreasing the tendency for skittery or neppy appearance of the fabric after dyeing. The fiber becomes much more receptive to dyes and finishes and the efficiencies and yields of these processes are improved significantly by mercerization. Pastel shades dyed on mercerized fabrics are very bright and lustrous. On continuous dye systems, dark shades are difficult to achieve unless the fabric has been mercerized. If mercerization is carried out while the fabric is under tension, the surface luster is even more pronounced, and this is the normal treatment in the industry today. Mercerized fabrics are also stronger and physically more compact than corresponding unmercerized fabrics.

Until recent years, cotton was mercerized almost exclusively using strong sodium hydroxide solutions (20% to 25%) and this is still practiced widely today. Liquid ammonia processes are now used in some of the larger textile organizations. Liquid ammonia is used on heavier weight goods such as denim and gives the fabric an excellent surface appearance and hand. The liquid ammonia process is also used in combination with certain easy-care finishing treatments to allow use of lower levels of finishing agents to give increased abrasion resistance and wear life to garments. Sodium hydroxide is still the preferred agent for enhancing dyeability of fabrics, particularly heavy, dark shades.

Mercerization with sodium hydroxide is accomplished on one of several types of commercial mercerization ranges by double padding of the sodium hydroxide onto the fabric and then providing a dwell time of 30 sec to 1 min while the fabric is held under tension. The treatment is completed by a series of washings and rinsings, neutralization of the sodium hydroxide, and further rinsing.

Mercerization by liquid ammonia is accomplished on a specially designed unit by saturating the fabric with liquid ammonia ($-35°C$), squeezing out excess ammonia, and allowing a dwell time of less than 10 sec while the fabric is held under tension. The ammonia is evaporated from the fabric by dry cans and is recovered. Any residual ammonia on the fabric is then removed by steaming and washing.

12–7.3 Dyeing and Printing

As in the case for the preparation processes, industry trends in dyeing and printing are toward automated, computer-controlled, energy-efficient processes. Basically, the principles involved in coloration of fabric today are the same as they have been for many years. Only the processes have been upgraded from an engineering and control standpoint. However, a few innovations in dyes and methods are being pursued with some success and may become important commercially in the near future. These include: (i) transfer printing in which a design that has been printed on paper is transferred from the paper to the fabric by contact and application of heat and (ii) use of foam systems for padding dye onto fabric that maximize the transfer of dye from pad bath to fabric with minimum usage of water. The paper transfer systems are in commercial use for cotton/polyester blends and are in the experimental stage for 100% cotton (Neal, 1981). Foam systems are in use commercially on a limited scale.

Color matching is an integral part of any dyeing or printing operation. In the past, color matching was mostly an art form practiced by skilled workers and was a very time consuming process. Today, color matching is done almost exclusively by computers, and even the most difficult shades can be matched with no more than two or three trial formulations. This advance has resulted in cost savings and has given dyehouses versatility to use more complex dye combinations to satisfy fashion requirements.

Several classes of dyes are used on cotton including: directs, fiber reactives, vats, sulfurs, naphthols, and pigments. Direct dyes are applied by exhaust techniques from a dyebath containing high concentrations of a salt such as sodium chloride. These dyes are held in the fiber by intermolecular attraction between the dye molecules and the hydroxyl and other groups in the cotton cellulose. Diffusion time is required and dyeing cycles of 1 hour and longer are common. These dyes are characterized by good light fastness, but poor wash fastness; they are often used to dye upholstery and drapery fabrics that are dry-cleaned rather than washed.

Fiber reactive, vat, and naphthol dyes have good fastness properties and are used on high quality apparel and fashion goods. The fiber reactives and some of the vats give very bright, lustrous dyeings. Vat dyes are somewhat in disfavor because of carcinogenity of intermediates used to make them. Sulfur dyes are low cost, but have a limited range of available colors—the best are greens and browns. They are used in large quantities to dye work clothes. They are alkali solubilized for dye application and then are insolubilized by a neutralization and oxidation step. They have very good washfastness, which is a prime requirement for work clothes.

Dyeing may be accomplished with cotton in raw stock, yarn, or fabric form. Raw stock dyeing is performed on a limited basis in the manufacture of blankets and in the manufacture of denim where a small amount of black dyed raw stock is blended with undyed stock to produce a gray yarn for use in filling in fashion denims.

Yarn can be dyed either on cones or on warp section beams. For cone dyeing, the yarn is wound on perforated tubes that are placed over perforated cylinders in a package dyeing machine that can be closed up and operated under pressure or at atmospheric pressure. The dye is pumped through the cones in alternating cycles from inside-out to outside-in to achieve a uniform dyeing. For beam dyeing, the process is exactly the same as for cones except that the dye chamber is designed to contain the beam. These beams of yarn after dyeing are processed on the slasher for application of size for weaving. Yarn dyeing is practiced widely in the industry today. It is used particularly where fancy checked patterns are required because diverse and intricate patterns can be developed very effectively in weaving by use of colored yarns in both the warp and the filling.

Most cotton is dyed in fabric form by either batch or continuous processes. Short runs of less than 15 000 m usually are dyed as piece goods in batch processes. For longer runs, continuous processes are more economical and easier to control for dyeing quality within the entire lot. Today's fashion requirements are often such that many short runs are necessary. This is a distinct economic disadvantage and results in increased cost for the dyeing process. The common batch methods use becks or kettles and the fabric pieces, with ends sewn together into endless ropes, are reeled in and out of the dye solution usually for several hours until the dyeing is complete. The becks can be enclosed for operation under pressure to accelerate the rate of dyeing. Fabric can be dyed in open width by winding onto a large roll mounted on one side of a U-shaped container called a jig that contains the dye solution. The fabric is threaded under rolls in the bottom of the jig and is rolled up on another roll on the other side of the jig. The cloth is run back and forth through the dye solution until the dyeing is complete.

Another batch process that has been used increasingly in recent years is beam dyeing. In this process fabric is wound onto a perforated beam and the beam is placed in a chamber that can be operated either at atmospheric conditions or under pressure. As in the yarn beam dyeing process, dye is then pumped alternatively inside-out and outside-in until dyeing is complete. In this process, the fabric is held rigidly in place and is not subject to vigorous mechanical treatment as in the case with other batch processes. This handling makes this process very useful for dyeing soft goods and particularly knit goods, which are easily distorted (Turner et al., 1976). Another batch process of recent vintage is jet dyeing. This dyeing is practiced on a machine similar to the becks mentioned earlier. However, the circulation of dye liquor and movement of the rope of fabric is accelerated by use of jet-flow principles and fabric stretch is minimized. The efficiency of dye-fabric contact and rate of dyeing are enhanced; however, fabric distortion may occur, and this process may not be entirely suitable for soft goods.

Continuous dyeing methods are used where possible for long runs. Several of the dye classes are especially well-suited for application by continuous process. Pigments are insoluble colored particles, and they are widely used for dyeing solid shades on cotton. They have no affinity for cotton and must be bonded to the fabric by resins. The continuous process includes simple padding on the pigment-resin solution, intermediate drying, and then heat curing of the resin to fix the pigment. High speeds are possible, because no diffusion or dye reaction steps are required.

Fiber reactive, vat, and naphthol dyes are also uniquely suited for continuous application. Each of these dyes in the unreacted state readily penetrates the cotton fiber. The final form of the dye becomes fixed within the fiber in an insolubilization or reaction step in the process. The fiber reactive dyes actually form chemical bonds with the cotton cellulose, whereas the vats and naphthols form complex insoluble molecules within the physical fiber structure. The steps are essentially the same for each of these dyes: (i) pad on dye, (ii) fix the dye by chemical reaction with the fiber or by insolubilization of dyestuff, and (iii) wash, rinse, and dry.

Warp yarns for denim are dyed in a special process called ball-warp dyeing. Warp yarns are gathered together into a rope or tow containing hundreds of individual yarns. This tow is rolled into a large ball several feet in diameter and is dyed using indigo dye. After dyeing the ball is unrolled and the individual yarns are separated and rewound on warp section beams for slashing. In weaving, the indigo-dyed warp yarns are combined with white or gray filling to give the traditional denim appearance. Large quantities of cotton are dyed by this process.

Printing is carried out to give either intricate, high complexity patterns on high quality goods or to rapidly and easily give shallow surface colored designs to cheap goods. In the past, high quality goods were printed using screen printing techniques that gave excellent pattern definition using several colors. The process was slow and labor intensive. The cheaper goods were printed using roller printing machines that could apply up to eight colors very rapidly. However, pattern definition was often poor and overlapping of colors occurred frequently. Both processes have been largely replaced by rotary screen printing that is in essence an automation of the basic screen printing process. Thus, the entire process of printing, both for high quality and for cheaper goods has been effectively upgraded.

12–7.4 Special Finishes

Cotton receives special finishing to achieve a variety of functional end-use characteristics. By far, the most important of these finishes is easy care. Today's society requires wearing apparel that retains shape during wear, does not wrinkle appreciably, and can be worn with little or no ironing after washing. This easy-care property is variously described by terms such as wash-and-wear, permanent press, no-iron, and others. Easy-care properties are bestowed on cotton by reactive compounds called resins that polymerize within the fiber and physical fabric structure or, more importantly, react with hydroxyl groups on adjacent cellulose chains to cross-link the molecules and ultimately fix the fabric in a desirable geometrical configuration. The fabric has a "memory" for this configuration that is retained during the useful life of the fabric. Most fabrics made for shirts, blouses, and other apparel are treated in flat, unwrinkled state, and the garments from these fabrics tend to retain this state during wear and laundering and require only touch-up ironing. Some fabrics are prepared by a special post-cure process in which the resin is applied and dried onto the fabric, but final fixation of the resin is delayed until the fabric is made into a garment. The garment is then subjected to high temperature in a special oven for fixation and the fabric memory is then in the shape of the garment itself. This process was very widespread and became dominant industry practice for a number of years, but because it is labor intensive, lacks versatility, and is energy inefficient, it has largely disappeared except for making permanent creases in trousers.

Most of the early effective resins released formaldehyde from the finished fabric during normal usage. Because of the toxicity of formaldehyde, procedures have of necessity been implemented to minimize this problem. These include more effective fixation and afterwashing and development of resins that, because of their chemical nature, do not release formaldehyde.

The basic problem with cotton treated with resin to achieve easy-care characteristics was that fabric strength and abrasion resistance were affected adversely and, consequently, wear-life of garments was poor. Improvements have been made by using lower levels of resins on liquid ammonia pretreated fabrics. This combination gives fabrics with excellent easy-care characteristics and good retention of strength and abrasion resistance. Another remedy for the low strength and abrasion resistance is to blend cotton with low levels of polyester. Blends of cotton with polyester, 60% cotton—40% polyester and 85% cotton—15% polyester, are common in shirtings and these blends have excellent easy-care and wear-life properties, yet retain the excellent esthetic and comfort properties associated with 100% cotton.

Another end-use property sometimes given to cotton by special finishing is soil release. For apparel, these finishes are now confined almost exclusively to work clothes that are treated with acrylic compounds. These finishes are dual purpose in that fabrics resist soiling to an extent and the soiling that does occur is easily removed in laundering. Thus, appearance during wearing is improved and wear-life can be extended because less severe laundering conditions can be used. Fluorocarbon finishes that give oil and water repellency and staining resistance are widely used for tablecloths and upholstery fabrics.

Waterproofing and rotproofing treatments are given to heavyweight cotton fabrics prepared for outdoor uses such as tents, awnings, and sand bags. The rotproofing agents are usually copper containing compounds and the waterproofing agents are combinations of aluminum or zirconium salts and waxes or other compounds containing long hydrocarbon chains. For lightweight goods, particularly wearing apparel, silicones, and fluorocarbons are used to give durable water repellency.

Just about all special finishing is performed on fabric. As in the preparation and dyeing processes, the technology is directed toward continuous, automated, energy-efficient processes. Just about all special finishes are applied on continuous ranges by processes that include the following steps: (i) pad, (ii) dry, (iii) cure, and (iv) wash. A typical example is the application of an easy-care resin. The pad bath contains the resin, a catalyst, a softener, and possibly, an optical brightener. The fabric is padded in open width, squeezed to remove excess solution, and is then framed and controlled to the desired width on a tenter frame for intermediate drying at moderate temperature and fixation of resin at high temperature. The fabric is washed, rinsed, and dried and is ready for converting into the final end-product.

12–7.5 Sanforizing

In the ordinary textile processes of spinning, weaving, and finishing, the cotton stock is usually under tension and is subject to stretching. The result is often a fabric containing distortions and tensions that when relaxed by laundering will result in shrinkage. To the consumer, particularly in apparel goods, excessive shrinkage is an overwhelming defect. To counteract this tendency for shrinkage in finished products, cotton fabrics are preshrunk by compressive shrinkage in a process called sanforizing. A special machine uses rubber-covered rolls and felt or rubber blankets to mechanically force steam-treated, damp fabric to shrink in a highly controlled manner both in the warp and in the filling directions. The Sanforized label is a registered trademark of Cluett, Peabody, and Company, Inc. Compressive shrinkage can be applied to fabrics at any stage of finishing. However, in any particular finishing sequence it is always the last step, because any subsequent wet processing and drying probably would counteract the effects of the shrinkage treatment.

Table 12-8. Domestic fiber use.

Fiber	Total consumed	
	1974	1981
	%	
Polyester	27	34
Nylon	19	20
Cotton	33	26
Wool	1	1
Rayon & Acetate	10	6
Olefin	5	8
Acrylic	5	5

12-8 COTTON AND SYNTHETIC FIBER BLENDS

Since their commercialization in the 1950s synthetic fibers have captured 74% of the world-wide textile fibers market that was dominated by cotton. Synthetic fibers are now a component of practically every textile product formerly exclusively made from natural fibers. Also, because of the properties that make them so different from the natural fibers, new textile materials have been developed for special uses that are not possible with natural fibers. In some products, such as carpets, synthetic fibers have practically supplanted the natural fibers formerly used. Chemical companies have worked hard and diligently to develop their fibers and market them in products with consumer appeal. The cotton industry was slow in developing products and an organized marketing program to compete with the chemical companies. End-products, fabric construction, and fiber blends were selected by the chemical companies who also provide technical assistance to mills that processed synthetic fibers. Marketing programs were initiated to increase public awareness of desirable properties of blends of synthetic fibers with cotton or of synthetic fibers alone in fabrics and yarns. Such properties as easy-care apparel fabrics made from all synthetic fibers or blends with cotton; the durability and tensile strength of industrial products; and the soil resistance and cleaning ease of home furnishing products were emphasized in an effective promotional and marketing program. In the course of a few years many standard all-cotton fabrics were replaced by blend fabrics with 65 to 80% synthetic fibers. Before the cotton industry mounted a promotional and marketing program and products to counter these gains cotton's share of the fibers market was seriously eroded. By 1974 cotton's share of the domestic fibers market had dropped to 33% and estimates for 1981 indicated that it would be further reduced to 26%, as indicated in Table 12-8. Cotton's chief competitor is polyester, which is produced in two forms—polyester staple and polyester filament. Polyester staple is processed into yarn on the same type of machines used to process cotton.

Polyester polymer can be extruded directly into yarn and in its texturized form, in which crimp is imparted by heat setting to provide increased bulk and loftiness. It is formidable competition for any staple fiber. Fabrics made from filament yarns tend to pick and fray when snagged by sharp objects. Woven texturized polyester fabrics are not as vulnerable to picks as other filament fabrics, but they still do not have the comfort or aesthetic appeal of fabrics made from cotton. The promotional programs financed by cotton producers through the National Cotton Council of America and Cotton Incorporated in recent years have focused consumer attention on the superior properties of cotton fabrics, such as their comfort in warm and cool weather, resistance to pilling, etc. These promotional efforts together with the development of easy-care, soil-release, and other finishes has done much to maintain market demand for cotton in recent years.

Blends of cotton and staple polyester are obtained in two ways. Draw-frame blends are obtained by processing cotton and polyester separately on different opening and carding machines to obtain card sliver made from 100% cotton and 100% polyester. For 50% cotton/50% polyester equal number of slivers of cotton and polyester are doubled and drafted at the first drawing process. The second drawing process provides additional blending and maintains the desired blend ratio. The fibers in blends produced in this fashion are not randomly mixed. Even after the second drawing process some of the cotton fibers are still bunched together and some of the polyester fibers are still together in streaks. In early applications practically all cotton/polyester blends were draw-frame blends.

More satisfactory, and perhaps more costly, blends of cotton and polyester are obtained by processing cotton through machines to obtain the desired degree of openness and cleanliness and then transporting the cotton back to the opening room to be fed to the initial processing machines in the desired ratio with polyester. Cotton and polyester fibers are processed simultaneously through opening, cleaning, carding, and subsequent machines. Blends obtained using this method are called intimate blends.

Fabrics made from intimate blends dye and finish more uniformly than fabrics made from draw-frame blends.

The easy-care properties of cotton/polyester fabrics have contributed much to the success of polyester in apparel fabrics and in home furnishings such as bed sheets. Many women by necessity are members of the work force. The automatic washer and dryer and easy-care fabrics, primarily cotton/polyester, have made life much easier for working women and men.

12-9 COTTON DUST AND BYSSINOSIS

Cotton dust is airborne particulate matter released into the atmosphere as cotton is handled or processed on machines. Cotton dust is composed of fragments of leaf, stem, bract, burr, shale, fiber, and other plant (not necessarily cotton plant) material as well as inorganic components from soil which contaminate the lint during wind and rain storms and during harvesting. Other components include microbiological material, bacteria and their endotoxins, and fungi. The amount of dust generated during processing is a function of the amount of dust and trash in the cotton, the processing rate, and the processing machinery used. Certain processes in yarn and fabric manufacturing have, in the past, been characterized by relatively high concentrations of cotton dust.

Byssinosis is a respiratory condition induced in some people by continued inhalation of dust produced during processing of cotton and certain other vegetable fibers. Individuals suffering from byssinosis in its early stages experience chest tightness and shortness of breath when returning to work after being away from the job for 2 or more days. Continued exposure can result in respiratory impairment and disability. Apparently, not all individuals are susceptible because many people who work in cotton mills never develop respirtory problems. The particulate matter in cotton dust varies in size from very small to very large particles, but physicians believe that particles in the size range from 0 to 15 μm cause the respiratory problems. An instrument called the vertical elutriator, shown in Fig. 12–23, was designed to measure the concentration of these particles in the air. Dust concentrations are higher in the early stages of yarn manufacturing, opening, carding, drawing, and roving than in the subsequent processes.

The prevalence of byssinosis in textile mills has been estimated by various groups including labor organizations, trade associations, and government agencies. Estimates range from 10 to 40% of the work force depending on the criteria used to classify the symptoms. Diagnosis has been a problem. Until the development and use of spirometry and other equipment for measuring pulmonary function, byssinosis was diagnosed on the basis of stated symptoms. Symptoms are still important to diagnosis but objective pulmonary function tests with spirometers and other equipment are being used to supplement classical diagnostic procedures. In advanced stages of respiratory disorder such as chronic bronchitis or emphysema, it is often impossible to determine if the condition originated as byssinosis.

Byssinosis is an occupational hazard and victims are compensable if respiratory impairment can be associated with exposure to cotton dust. This liability and the health hazard associated with cotton is of great concern to textile manufacturers, and many persons are attempting to reduce their utilization of cotton to avoid the risks involved.

12–9.1 The Cotton Dust Standard

In 1978 the Occupational Safety and Health Administration issued a standard regulating the exposure of workers to cotton dust. The standard specified that the maximum allowable concentration of cotton dust permitted in the workplace was to be gradually reduced over a 7-year period. Permissible concentration of cotton dust, as measured by the vertical elutriator, or any other device proved equivalent, is 200 $\mu g/m^3$ in yarn manufacturing, 750 $\mu g/m^3$ in fabric manufacturing, and 500 $\mu g/m^3$ in all other areas where cotton is handled.

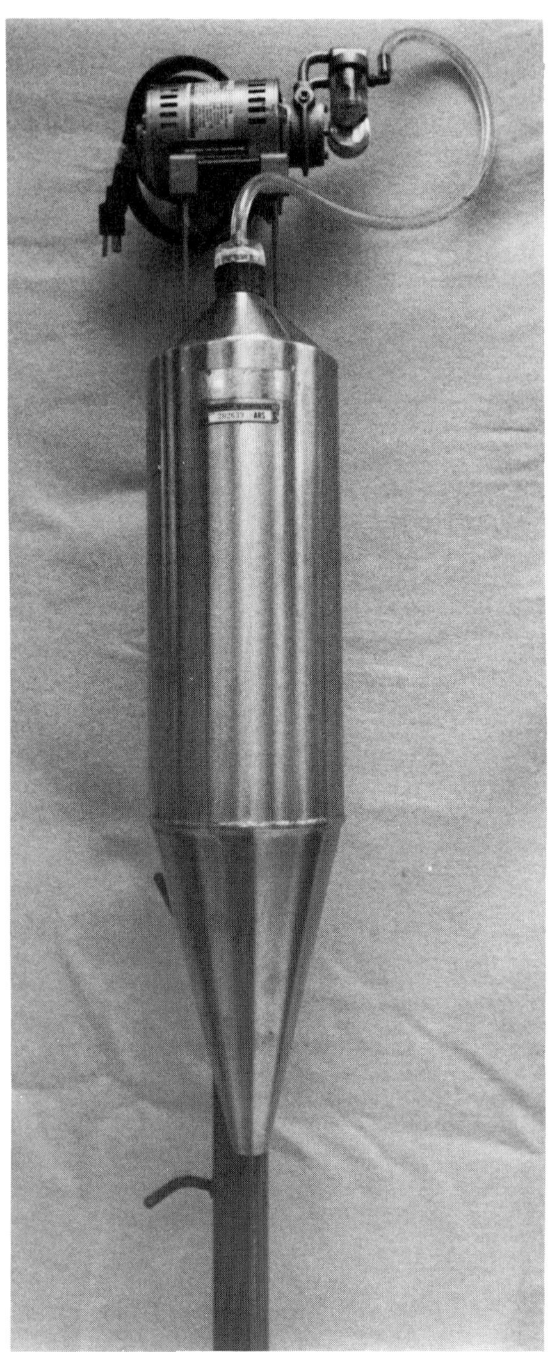

Fig. 12–23. Vertical elutriator.

Systems for ventilation and air filtration were proposed as a method for reducing dust concentrations to the levels required by the standard. Although technology is available that will allow mills to meet these rigorous air quality standards, its cost is significant. Various estimates ranging from $800 million to $2.5 billion have been made for the industry-wide cost of reaching the specified levels. An accurate estimate of the cost is difficult to obtain because many textile mills are continually modernizing processing equipment to increase productivity. Also, costs are much higher in old mills that are modernized than in newly constructed mills that can incorporate better air handling methods. Air handling and dust removal features are included in modern, high production processing machinery so that as mills modernize and automate, dust control will become easier and less costly. Mills with a sufficient capital base to install modern equipment or construct new plants will be able to meet the dust standard. Many small mills with limited capital will not be able to meet the standard and will go out of business, be absorbed by larger companies, or begin using synthetic fibers instead of cotton.

12–9.2 Medical Monitoring

Even though textile mills install equipment and develop methods for maintaining dust concentrations at or below the levels required in the cotton dust standard, there is no guarantee that some employees will not be adversely affected and liability still exists. Epidemiology data indicate that even at the low levels required by the standard, a small percentage of the population would still risk some form of respiratory disorder when exposed to cotton dust. To guard against this risk the standard requires medical monitoring of all employees exposed to cotton dust. Pulmonary function tests (spirometry) are performed on all employees periodically to determine if the gradual loss in pulmonary function that is characteristic of all individuals after they reach adulthood is being accelerated. If significant reductions are found it is assumed to be caused by exposure to cotton dust and the employee must be removed from that environment. Some evidence indicates that smokers are more susceptible to the effects of cotton dust than non-smokers. Questionnaires on symptoms are also completed periodically for all employees. Spirometry tests, symptom questionnaires, and other medical information must be retained in mill files on all employees for 20 years after employment is terminated. The requirement for medical monitoring and maintenance of medical records has resulted in more rigid criteria than previously followed for selecting new employees. Individuals with previous history of asthma, bronchitis, or other respiratory conditions are not hired for positions that will require their continued exposure to cotton dust.

Many in the cotton industry feel that the standard is too severe and that its cost to the industry is not justified. They feel that medical monitor and work practice adjustment will eliminate byssinosis as an occupational disease and that a more moderate standard is warranted. Further review of the standard by U.S. Occupational Safety and Health Administration (OSHA) to consider evidence that would warrant a change in the standard is anticipated.

12-9.3   Potential for Eliminating the Health Hazard

Byssinosis has been recognized as an occupational hazard for a number of years but only recently have significant efforts been made to find the cause of the disease. Several sources have been suspected.

Bract from the base of the boll has been suspected as being associated with the causative agent. Unique compounds have been isolated from bract and synthesized in the laboratory for use in challenge tests, but conclusive proof of relationship to byssinosis has never been obtained. Gram negative bacteria and their endotoxin have been related to changes in pulmonary function, but a causal relationship has never been established. It was the lack of information of knowledge about the relationship of the respiratory disorder and causative agents that prompted OSHA to institute a standard for exposure to cotton dust. Since little or no knowledge on causative agents exists, limiting dust concentration would limit exposure to causative agents.

Recently, research has been directed toward experiments to eliminate or neutralize the causative agents. The USDA, in cooperative with the National Institute for Occupational Safety and Health (NIOSH), has shown that panels of human subjects exposed to dust from cotton harvested in the closed boll stage experience a much smaller decrement in lung function as compared to dust from standard cotton.

Cotton Incorporated, National Cotton Council, USDA, NIOSH, and others have cooperated in studies showing that mild water washing of cotton apparently is effective in removing or neutralizing causative agents. Cotton Incorporated has shown that cottons with low endotoxin levels cause a much smaller decrement in lung function than cottons with high levels. Experimentation involving exposure of panels of human subjects to dusts from experimental cottons is costly and time-consuming, but scientists feel that if such research continues, chances of identifying or discovering a practical method for neutralizing the causative agent are good.

The U.S. National Academy of Sciences has studied the cotton dust/byssinosis problem in detail and recommended a number of approaches to the solution of the problems. These recommendations should be invaluable in concentrating research effort on more promising solutions.

12-10   UTILIZATION

When the number of textile products used by modern society is considered, it is apparent that cotton is suitable for use in practically all of them either as 100% cotton or as blends with synthetic fibers. In some products the properties of other fibers make them more suitable for use than cotton.

Table 12-9. Estimated consumption of cotton and competing materials used in apparel products; 1980.

| | 218 kg bale equivalents, 1000s | |
Items	Total	Cotton
Trousers, slacks, jeans	2016	1156
Shirts, blouses	1657	693
Underwear, diapers	788	425
Dresses, skirts	686	173
Gloves, mittens	130	116
Linings, pocketing	375	103
Coats, jackets	463	84
Overalls, coveralls	127	81
Sleepwear, robes	408	75
Hosiery	369	64
Play garments	129	62
All other	1387	187
Total	8535	3219

However, two major criteria, economics and technical suitability, appear to dominate when decisions are made by users on which textile fibers will be used for specific end-products. In products such as backing material for coated fabrics, cotton apparently is equal in suitability to other fibers. The fiber selection then is likely to be based on economics. The fiber that allows the textile manufacturer to meet the minimum specification of the user at least cost will be selected. Since cotton is an agricultural product whose production is affected by conditions, such as weather, that are often beyond the producer's control, variations in production exist even though the producer's efforts may remain consistent. These variations in supply contribute to variations in prices. Chemical companies have more control over production of fibers than cotton producers. Synthetic fiber producers can control production and can maintain a more consistent price structure than the cotton industry. The assurance of an adequate supply at a stabilized price has a direct effect on fiber selection when decisions are made by users on an economic basis (Malzbender, 1982).

When decisions on fiber utilization are made for technical reasons, the unique combination of desirable properties possessed by cotton makes it the first choice for many textile products. Many textile manufacturers will state that they prefer to use cotton because it generally spins and weaves with fewer difficulties than synthetic fibers. Many consumers prefer cotton fabrics because of their comfort and aesthetic appeal.

The estimated utilization of cotton in various products is shown in Tables 12-9, 12-10, and 12-11. Table 12-9 shows estimated consumption of textile materials in apparel fabrics where fiber selection is made on a technical basis. Approximately 38% of the domestic consumption of textile material for apparel fabrics is cotton. Tables 12-10 and 12-11 show similar data for textile material used in home and industrial products where fiber selection is dominated by factors other than fiber properties. Cotton's share of these markets is approximately 18% (Howell et al., 1981).

Table 12-10. Estimated consumption of cotton and competing materials
used in home furnishing products; 1980.

	218 kg bale equivalents, 1000s	
Item	Total	Cotton
Towels, wash cloths	665	620
Sheets, pillowcases	975	413
Drapery, curtains, upholstery	1 562	343
Piece goods	739	160
Bedpsreads	221	96
Mattresses, comforters, blankets	373	64
Tablecloths, napkins, mats	82	45
Pillows, cushions, pads	90	43
Rugs, carpets	5 167	32
Thread for home use	182	9
Total	10 056	1 825

Table 12-11. Estimated consumption of cotton and competing materials
used in industrial products; 1980.

	218 kg bale equivalents, 1000s	
Item	Total	Cotton
Medical supplies	271	166
Tarpaulins, tents, awnings	205	110
Thread	253	96
Shoes, boots	164	69
Rope, cordage	363	58
Abrasives	60	53
Wiping cloths	46	44
Wall coverings	75	38
Automotive	1485	30
Belts	142	24
Luggage, handbags	107	24
Sleeping bags	36	18
Book bindings	93	17
Filters, tobacco cloth	79	17
Bags	332	14
All other	408	34
Total	4119	812

The elimination of the liability and health hazard associated with
cotton because of byssinosis and the maintenance of an adequate supply of
cotton at a consistent price would do much to increase the percentage of
cotton used in all textile products.

REFERENCES

Balmforth, D. 1982. New Concepts in Finishing Techniques. Text. Chem. Color. 14:118–124.

Cotton Division, AMS. 1980. The classification of cotton. Agric. Handb. no. 566, USDA. U.S. Government Printing Office, Washington, D.C.

----. 1982. Summary of cotton fiber and processing test results—Crop of 1981, USDA. U.S. Government Printing Office, Washington, D.C.

Ethridge, D. E. 1978. A computerized remote-access commodity market: Telcot. South. J. Agric. Econ. 10:177–182.

Hamby, D. S. 1966. The American cotton handbook. Vol. 1 and 2. Interscience Publishers, N.Y.

Henderson, D. R., L. S. Schraeder, and M. S. Turner. 1976. Electronic Commodity Markets. Natl. Public Policy Ed. Comm. Publ. no. 7. Cornell Univ., Ithaca, N.Y.

Howell, J. T., Jr., R. J. Duren, and W. P. Crawford. 1981. Cotton Couns Its Customers. Natl. Cott. Counc. Am., Memphis, Tenn.

Hunter, L. 1978. The production and properties of staple-fibre yarns by recently developed techniques. Text. Prog. 1 and 2:1–163.

----. 1980. Textiles: Some technical information and data V: Cotton. So. Agrican Wool Text. Res. Inst. Counc. Sci. Ind. Res. Port Elizabeth, South Africa.

Jones, A. D. (ed.). 1975. National cotton marketing study committee report. USDA. U.S. Government Printing Office, Washington, D.C.

Malzbender, H. K. 1982. The coming of age of polyester. p. 1–16. *In* Proc. Text. Yarn Assoc. Conf. Text. Yarn Assoc. Am., Inc., Winston-Salem, N.C.

McArthur, W. C., B. Bolton, D. E. Ethridge, A. Heagler, J. L. Ghetti, D. L. Shaw, F. T. Cooke, Jr., and J. Lawlor. 1980. The cotton industry in the United States: Farm to consumer. Texas Tech Coll. Agric. Sci. Pub. no. T-1-186.

McConnell, B. L., P. W. Wolfgang, and B. K. Easton. 1982. Survey of preparation and bleaching equipment. Text. Chem. Color. 14:76–82.

Neal, B. L. 1981. Dyeing problems in poly/cotton blends. Am. Text. Rep. Bull. Ed. AT10: 14–15.

Sporleder, T., J. Haskell, D. E. Ethridge, and R. Firch. 1978. Who will market your cotton? Producer Aternatives. Texas Agric. Ext. Ser. D-1054.

Turner, J. D., S. A. Wolfgang, B. W. Jones, and W. A. Blanton. 1976. Finishing cotton/polyester knits. Text. Chem. Color. 8:135–138.

ACKNOWLEDGMENT

The authors wish to express their gratitude to the following associates and co-workers for their support and assistance in the preparation of this summary:

Mike Stevens, Vice President
E. F. Hutton and Co., Lubbock, Tex.

C. L. Boggs, General Manager
Plains Cotton Cooperative Association

Bert Kyle, Sales Manager
Plains Cotton Cooperative Association

Vern Highley, Administrator
Agricultural Marketing Service, USDA
(formerly with Plains Cotton Cooperative Association)

Jefferson, D. Bargeron, III, Research Textile Technologist
USDA-ARS, Clemson, S.C.

Carolyn L. Simpson, Administrative Technician
USDA-ARS, Clemson, S.C.

Geneva S. Lawless, Secretary
USDA-ARS, Clemson, S.C.

13   Seed

John P. Cherry
ARS-USDA
Philadelphia, Pennsylvania

Harry R. Leffler
ARS-USDA
Stoneville, Mississippi

In the world market, cottonseed production and processing ranks second (17.3% of world oilseed production) among the five major oilseeds, which include soybean seed, cottonseed, sunflower seed, peanut seed, and rape seed (Fig. 13-1; van Waalwijkn van Doorn, 1982; Anon., 1982a). The USA, USSR, People's Republic of China, India, and Pakistan produce approximately 75.3% of the 27.8 Mt of the world cottonseed supply (Table 13-1). Cottonseed is processed into four major products—oil, meal, hulls, and linters. These products and their many uses are summarized in Fig. 13-2 (Anon., 1950, 1978a). Cottonseed yields approximately 16% crude oil, 45% meal, 9% linters, and 26% hulls, with losses of 4% due to handling and processing (Kromer, 1977; Carter et al., 1979). World production of oil and meal (44% protein) for 1981/1982 were predicted to be 3.5 and 8.2 Mt, respectively (Mathews, 1981). In 1981/1982, the USA was expected to process 0.9 and 2.1 Mt of oil and meal, respectively.

Table 13-1. Cottonseed production by major producers.†

	1981/1982 Preliminary production	
	Mt	%
USSR	4 950	17.8
PRC	5 936	21.4
USA	5 803	20.9
India	2 750	9.9
Pakistan	1 470	5.3
Total	20 909	75.3

† From van Waalwijkan van Doorn, 1982; Anon., 1982a.

Published in *Cotton,* Agronomy Monograph no. 24, © ASA-CSSA-SSSA, 677 South Segoe Road, Madison, WI 53711.

13-1 COMPOSITION

13-1.1 Major Constituents

Electron microscope photographs of cottonseed show them to contain a number of distinct structural features, including the following: (a) nuclei; (b) spherosomes, fat storage bodies; (c) protein bodies that contain the storage proteins, globulins; (d) globoids, phytin particles, in the protein bodies; (e) the cellular cytoplasm, in which structures (a) to (d) are embedded with nonstorage proteins, enzymes, and other particulates important to the physiology of the seed; (f) gossypol glands; and (g) cell walls (Fig. 13-3; Yatsu, 1965; Englemen, 1966; Dieckert and Dieckert, 1972, 1976). The fact that storage constituents are in compartments has enabled the development of techniques to separate them during the processing of cottonseed to feed and food ingredients.

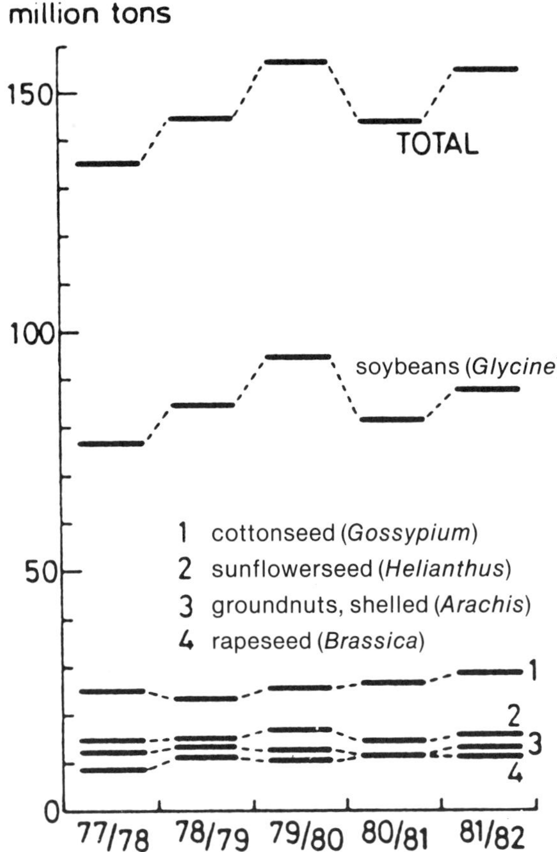

Fig. 13-1. World production of major oilseeds (van Waalwijkn van Doorn, 1982; Anon., 1982a).

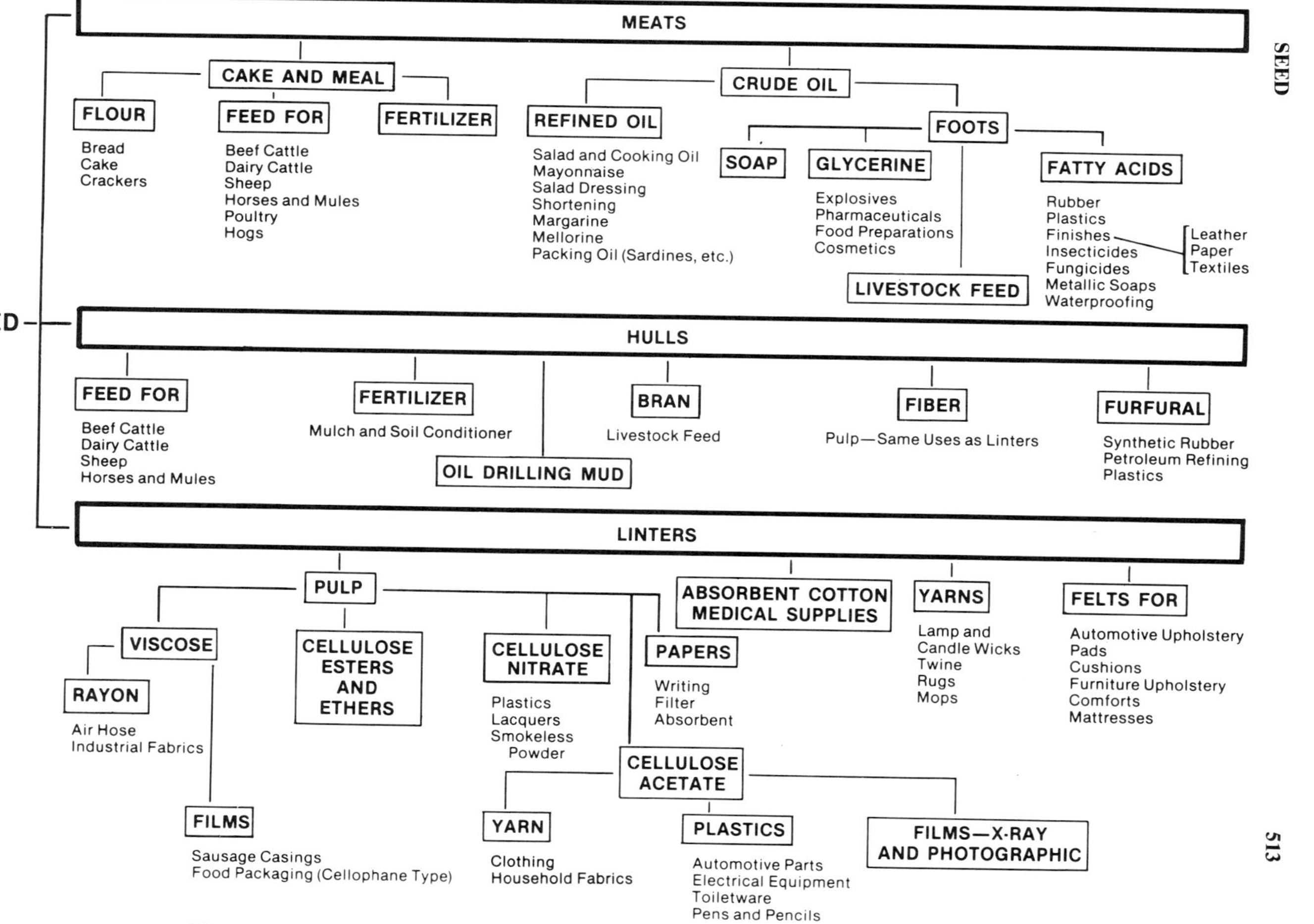

Fig. 13–2. Cottonseed products and their many uses (Anon., 1950, 1978a).

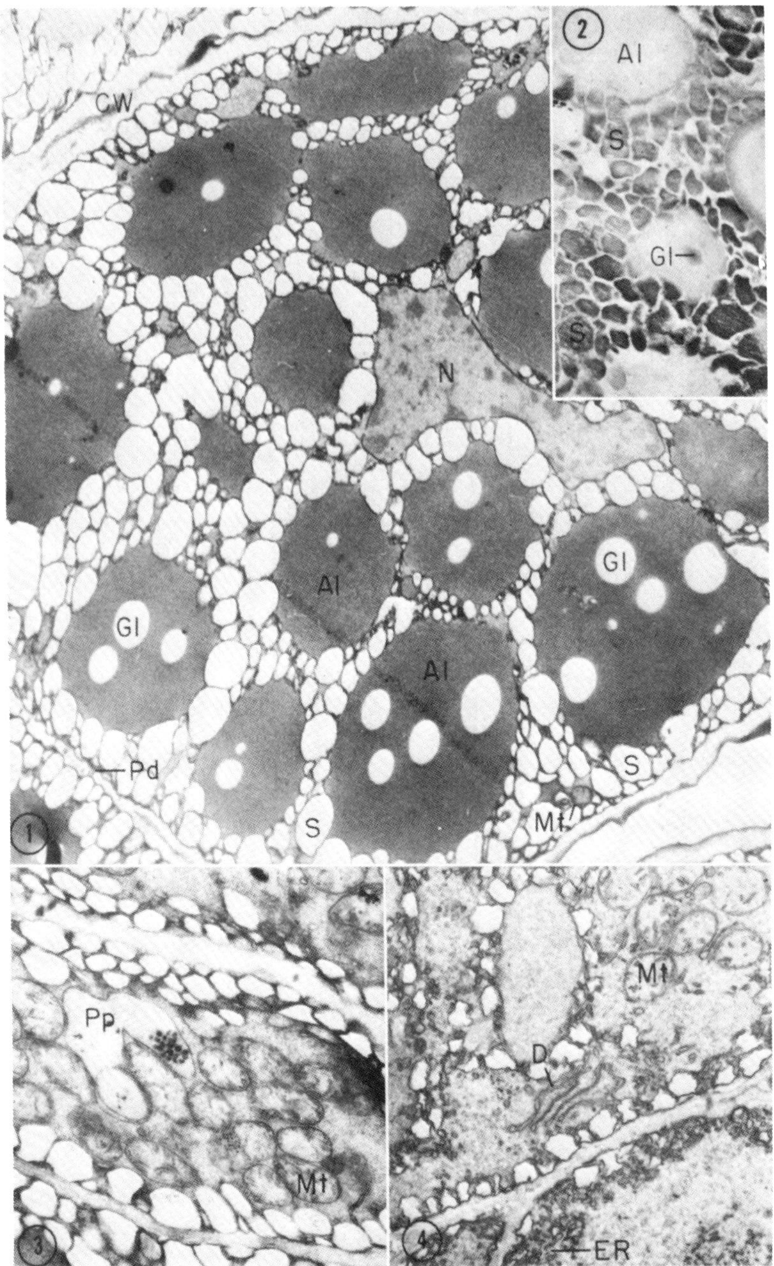

Fig. 13-3. Ultrastructure of cotyledonary tissue (Yatsu, 1965). 1) A typical spongy parenchymal cell from the cotyledon of a dry cottonseed fixed with lithium permanganate, × 7000, and 2) osmium tetroxide fixation, × 7000. 3) Provascular region of cotyledonary tissue fixed with lithium permanganate, × 12 000. 4) Dictyosome profile of (3) fixed with lithium permanganate, × 16 000. Al, aleurone grains; CW, cell wall; D, dictyosome; ER, endoplasmic reticulum; Gl, globoid; Mt, mitochondria; N, nucleus; Pd, plasmodesmata; Pp, proplastid; S, spherosome.

The breeding and production of cotton have traditionally been guided by considerations of fiber yield and quality. Until recently, seed characteristics, except for viability and vigor of planting seed, have generally been ignored. Competition from other seed sources in the oil and feed industries, and the developing prospect of using cottonseed as a food, have increased the awareness of the potential importance of cottonseed to food and feed reserves of the world. As a result, the National Cottonseed Products Association, Inc., Memphis, Tennessee, (Anon., 1980), an association of cottonseed processors, developed nine objectives relating to cottonseed quality improvements (Table 13-2). Research on cottonseed has made it clear that many factors affect their composition. Efforts to improve cottonseed composition must be accomplished without sacrifice to fiber quality and yield or planting performance.

Table 13-2. Identified goals relating to cottonseed quality.†

Goal	Needed change
Increase seed yield per acre	Concern for availability of cottonseed for crushing. Increasing acreage as well as total seed cotton yield per acre are the only two approaches acceptable to the entire cotton industry. The results would be increased yields of fiber, seed, oil, and protein per acre.
Increase oil percentage	Economically important to have seed with high percentages of oil. Variety tests show three percent or more variation between genetic lines which could be used to select for increased oil percentages in seeds of future cultivars.
Eliminate cyclopropenoid fatty acids	Biologically active fatty acids present in minute quantities of cottonseed oil. Although they are no health hazards, it is best to eliminate these fatty acids genetically.
Eliminate or reduce seed gossypol	As well as affecting animal performance when cottonseed meal is used as feed, gossypol pigments discolor oil so that additional processing, refining, and bleaching are necessary, and reduces the nutritional value of the meal when it binds to lysine. Cotton breeders have developed plants producing glandless (essentially gossypol-free) cottonseed.
Increase protein percentage	Increased seed protein percentage would improve the desirability of meal. Variety tests have shown that protein levels can be influenced by genetics.
Increase lysine content	Increased lysine in the protein would improve seed meal nutritional value. It should be possible for plant breeders to select lines capable of providing higher levels of lysine per unit of protein.
Improve mycotoxin resistance	Mycotoxins have cost both producers and crushers heavily. Cultural practices are available to reduce aflatoxin levels in seed cotton. Selection of resistant plants followed by a simple breeding program may be the start of controlling this in insidious production and crushing problems.
Insure production of non-shattering seeds	The ideal seed coat should protect the seed from mechanical damage and the subsequent deterioration of the kernel
Reduce linter percentage	Introduction of varieties with higher seed linter level than are now common is not acceptable to the processes of the cottonseed crushing industry.

† From Anon. (1980).

Research shows that cultivar, location, year, and their interactions are highly significant sources of variation associated with cottonseed composition (Pope and Ware, 1945; Bailey, 1948; Stansbury et al., 1953a, b, 1954, 1956; Cherry et al., 1970; Pandey and Thejappa, 1975; Turner et al., 1976a, b; Lawhon et al., 1976; Cherry et al., 1978a, b, 1979c, 1981a, b; Cherry and Berardi, 1983; Cherry, 1983; Kohel and Cherry, 1983). Breeding and agronomic practices can affect both the physical and chemical properties of cottonseed, and thereby influence the planting seed quality and the efficiency of their processing into oil, feed, and food products without affecting fiber quality.

The composition of cottonseed and the effects of genotype and environment are summarized in Table 13-3 (Cherry et al., 1978a, b, 1981a, b; Cherry, 1983; Cherry and Berardi, 1983). Cottonseed are excellent sources of both oil and protein because of the high composition of these constituents and their fatty acid and amino acid qualities (Tables 13-4 and 13-5; Cherry et al., 1978a, b, 1981a, b; Cherry, 1983; Cherry and Berardi, 1983).

McMichael (1959, 1960) caused considerable excitement in the cottonseed industry when he reported that cottons with seed free of gossypol could be developed by selection for recessive alleles at two loci. Reviews on the genetic development of glandless cottons were presented by Hess (1976, 1977a). Progress has been made in developing glandless cottons that yield competitively with glanded cottons in fiber and seed (Phelps et al., 1979). The protein and oil compositions of seeds of glandless cultivars are at least comparable or exceed those of glanded cultivars (Lawhon et al., 1976; Cherry et al., 1978a, b, 1981a, b; Cherry, 1983; Cherry and Berardi, 1983).

Table 13-3. Summary of mean values of seed quality traits of cultivars grown at various Texas locations in 1974.[†]

Quality trait	Cultivars across locations	Locations across cultivars
Hulls[†]	37.2 –40.0	36.6 –44.8
Kernel[‡]	49.0 –50.9	43.5 –53.4q
Linters[‡]	9.9 –11.6	9.9 –12.4
Seed index[§]	9.3 –10.6	8.6 –11.1
Oil[¶]	23.6 –25.0	23.2 –25.7
Protein[¶]	25.6 –27.5	25.6 –27.6
Total sugars[#]	6.8 – 7.0	6.7 – 7.0
Ash[#]	4.4 – 4.5	4.1 – 4.9
Crude fiber[#]	2.1 – 2.2	2.0 – 2.3
Free gossypol[#]	0.74– 0.87	0.75– 0.86
Total gossypol[#]	0.90– 0.99	0.88– 1.00
N-solubility[††]	96.4 –97.9	96.5 –98.1
Cyclopropenoid fatty acids[‡‡]	0.97– 1.04	0.82– 1.18
ϵ-Free lysine[§§]	3.17– 3.21	3.12– 3.26

[†] From Cherry et al., 1978a, b.
[‡] % of seed.
[§] Number of ginned seed/100 g.
[¶] %, moisture, lint-free seed basis.
[#] %, moisture-free kernel basis.
[††] % of protein soluble in 0.02 N NaOH.
[‡‡] %, oil.
[§§] g/100 g fat-free flour.

Despite many years of interest in improving cottonseed quality, we find little continuity of either interest or research programs. Research on the quality of cottonseed, other than the breakthrough of gossypol-free seed, has not had an impact on the improvement of cotton cultivars; i.e., little has been done to utilize the wealth of information on compositional variability to improve cottonseed quality. For example, Pope and Ware (1945) found oil to range from 16.1 to 26.7%, and protein to range from 20.5 to 26.8% in cottonseed. Thirty years later, Lawhon et al. (1976) found the range of oil to be 16.5 to 25.6% and protein to be 19.6 to 24.0% in cottonseed. Turner et al. (1976a) reported average values of 19.6% for oil, and 24.8% for protein in cottonseed. Similar oil and protein values of 23.2 to 25.7%, and 25.6 to 27.6%, respectively, in cottonseed were reported by Cherry et al. (1978a, b, 1981a, b).

Evidence that cottonseed composition can be improved is shown by the breeding program of the Acala SJ series in California (Cherry et al., 1981a;

Table 13-4. Fatty acid composition of oil from glanded cottonseed kernels of cultivars grown at various Texas locations in 1974.†

Fatty acid	Range	Mean
	%	
Myristic (C14:0)	0.68– 1.16	0.82
Palmitic (C16:0)	21.63–26.18	23.68
Palmitoleic (C16:1)	0.56– 0.82	0.65
Stearic (C18:0)	2.27– 2.88	2.55
Oleic (C18:1)	15.17–19.94	17.41
Linoleic (C18:2)	49.07–57.64	54.54

† From Cherry et al. (1978a, b).

Table 13-5. Amino acid composition of fat-free flour of cottonseed from cultivars grown at various Texas locations.†

Amino acid	Range	Mean
	g/100 g sample	
Alanine	2.00– 2.26	2.12
Valine‡	1.66– 2.61	2.20
Glycine	2.08– 2.35	2.21
Isoleucine‡	1.20– 1.86	1.59
Leucine‡	2.88– 3.84	3.12
Proline	1.83– 2.13	1.96
Threonine‡	1.42– 1.98	1.74
Serine	2.14– 2.59	2.40
Methionine‡	0.59– 0.96	0.74
Phenylalanine‡	2.62– 3.02	2.82
Aspartic acid	4.58– 5.25	4.93
Glutamic acid	9.46–11.15	10.53
Tyrosine	1.29– 1.68	1.51
Lysine‡	2.28– 2.63	2.46
Histidine‡	1.12– 1.62	1.41
Arginine‡	4.39– 6.42	5.43
Half cystine	0.70– 1.08	0.88
Total	44.34–50.82	48.02

† From Cherry et al. (1978a, b).
‡ Essential amino acid.

Table 13-6. Mean values of cottonseed quality traits of Acala cultivars grown at four California locations, 1975–1977.[†]

	Cultivars[‡]	
Quality factors§	SJ-2	SJ-5
Hull	41.60 a‡	36.89 b
Kernel	45.37 a	51.63 b
Lint	19.03 a	21.81 b
Quantity index	97.29 a	109.59 b
Grade	96.73 a	109.29 b
Oil	19.03 a	21.81 b
Protein	22.25 a	23.44 b
Free fatty acids	1.20 a	0.86 b
Free gossypol	1.03 a	0.73 b
Total gossypol	1.09 a	0.80 b
Phosphorus	0.94 a	0.88 b
LCP Flour:		
Free gossypol	0.03 a	0.02 b
Total gossypol	0.06 a	0.04 b

† From Cherry et al. (1981b) and Cherry (1983).

‡ Means among cultivars having the same letter are not significantly different to the Newman-Keuls multiple range test.

§ Hull, kernel, and lint are presented as % of seed; oil and protein are % of linted seed; free fatty acid is % of oil; free and total gossypol, and phosphorus, are % of kernels; and LCP flour protein and free and total gossypol is % of flour. All of these values are presented on an "as is" moisture value which was 9.35 and 9.08 for the 'Acala SJ-2' and 'Acala SJ-5' cottonseed, respectively; values that were not significantly different.

Cherry, 1983). This program increased both oil and protein in Acala cottonseed, and simultaneously reduced gossypol (Table 13-6). Other statistically significant improvements in seed quality include the following: (a) reduced portions of the seed as hull and linters, (b) decreased amounts of cyclopropenoid fatty acids, and (c) higher levels of essential amino acids and select fatty acids (Tables 13-7 and 13-8). These improvements are consistent with those identified as important to the cottonseed crushing industry (Table 13-2).

13-1.2 Gossypol and Other Polyphenolic Pigments

The polyphenolic pigment gossypol, 1,1′,6,6′, 7,7′-hexahydroxy-5,5′-diisopropyl-3,3′-dimethyl (2,2′-binaphthalene)-8,8′-dicarboxaldehyde, is contained within discrete bodies in leaves, stems, and roots, and seeds of cotton plants (Boatner et al., 1947; Adams et al., 1960; Markman and Rzhekhin, 1965; Berardi and Goldblatt, 1980). In this form, the pigment is called free gossypol, because it is readily extracted by 70% aqueous acetone. At least 15 gossypol pigments or derivatives have been identified in extracts of cottonseed or their oils and meals (Berardi and Goldblatt, 1980); only eight have been isolated and characterized.

Gossypol has three tautomeric structures, (a) the hydroxyaldehyde, (b) the lactol, and (c) the cyclic carboxyl forms (Fig. 13-4). These structures are

Table 13-7. Mean values of amino acids of flour from cottonseed of Acala cultivars grown at four California locations, 1975–1977.[†]

| | Cultivars | |
Amino acid	SJ-2	SJ-5
	g/100 g flour	
Alanine	2.00 a[†]	2.03 a
Valine§	2.10 a	2.14 a
Half-cystine	0.79 a	0.82 a
Arginine§	6.47 a	6.62 a
Lysine§	2.24 a	2.26 a
Histidine§	1.50 a	1.59 b
Tyrosine	1.54 a	1.54 a
Aspartic acid	4.84 a	4.84 a
Glutamic acid	10.38 a	10.40 a
Phenylalanine§	2.68 a	2.78 a
Glycine	2.10 a	2.12 a
Isoleucine§	2.55 a	2.57 a
Leucine§	2.96 a	3.01 a
Proline	1.92 a	1.95 a
Threonine§	1.64 a	1.67 a
Serine	1.41 a	1.43 a
Methionine§	0.64 a	0.66 a

[†] From Cherry et al. (1981b).
[‡] Means among cultivars having same letter are not significantly different according to the Newman-Keuls multiple range test.
§ Essential amino acid.

Table 13-8. Mean values of fatty acids of oil from cottonseed of Acala cultivars grown at four California locations, 1975–1977.[†]

| | Cultivars | |
Fatty acid	SJ-2	SJ-5
	%	
Palmitic (C16:0)	23.32 a[‡]	22.69 b
Palmitoleic (C16:1)	0.72 a	0.64 b
Stearic (C18:0)	2.17 a	2.29 b
Oleic (C18:1)	16.63 a	17.26 b
Cyclopropenes	0.90 a	0.84 b

[†] From Cherry et al. (1981b) and Cherry (1983).
[‡] Means among cultivars having the same letter are not significantly different according to the Newman-Keuls multiple range test.

markedly reactive, exhibit strongly acidic properties, and act as phenolic and aldehyde compounds. When dissolved in dilute aqueous alkali, they behave as strong dibasic acids to form neutral salts. In alcoholic solution, gossypol is extremely sensitive to oxidation. The phenolic groups of gossypol react readily to form esters and ethers. The aldehyde groups react with amines to form Schiff bases, and with organic acids to form heat-labile compounds; gossypol reacts with the aromatic amines of aniline to form di anilinogossypol which can be analyzed quantitatively (Pons, 1977; Berardi and Goldblatt, 1980).

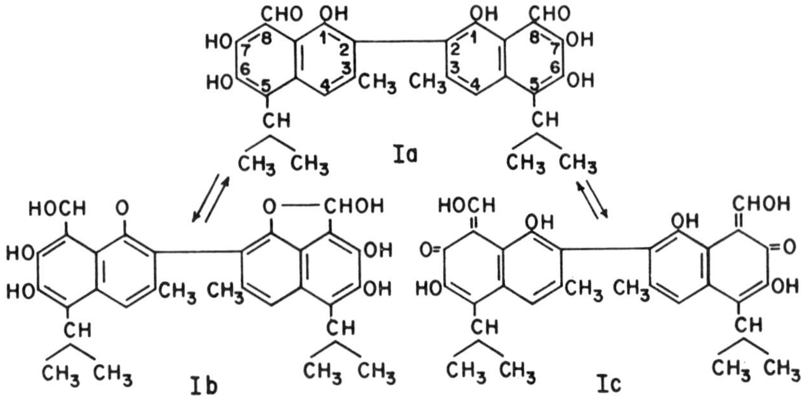

Fig. 13-4. Tautomeric structures of gossypol (Berardi and Goldblatt, 1980).

Since gossypol is highly reactive with other cottonseed storage constituents during processing, it is responsible, in part, for (a) reduction, through binding, of the biological availability of lysine, and thus the nutritive quality of the meal proteins; (b) adverse physiological effects in nonruminants; and (c) production of dark-colored pigments in the oil and meal that are not removed in conventional refining and bleaching operations (Berardi and Frampton, 1957; Pons et al., 1959; Smith, 1972; Jones, 1981; Berardi and Goldblatt, 1980; Bressani et al., 1980; Cherry, 1983).

Adverse physiological effects may be counteracted by making free gossypol a bound form during processing, by the use of heat and certain minerals, especially iron salts (Anon., 1966; Berardi and Goldblatt, 1980). To quantitatively extract bound gossypol from cottonseed, it first must be hydrolyzed to the free form with compounds such as oxalic acid (Pons, 1977). Total gossypol is then determined as the sum of the free and bound gossypol and gossypol-like pigments extracted after oxalic acid hydrolysis.

Thin layer chromatography and gel filtration have been used to fractionate seven major flavonoid pigments in solvent extracts of cottonseed products (Blouin and Cherry, 1980; Blouin et al., 1981a, b, 1982). These major flavonoids have been tentatively identified and are listed in Table 13-9. Diagnostic ultraviolet-visible spectral analysis indicated that flavonoids 1, 2, 5A, 5B, and 6 were 3-o-glycosides of quercetin, while flavonoids 3 and 4 were 3-o-glycosides of kaempferol.

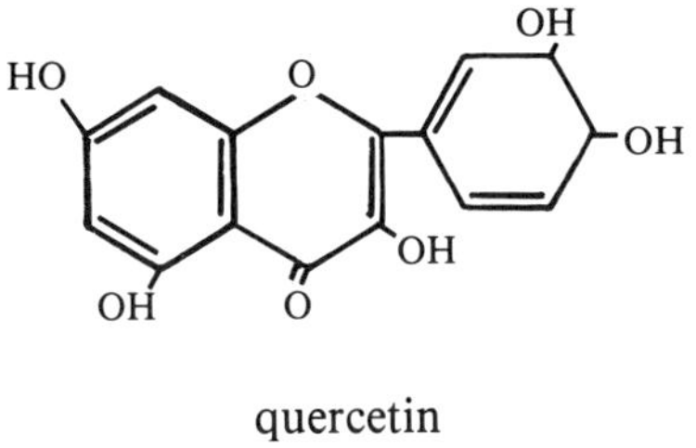

quercetin 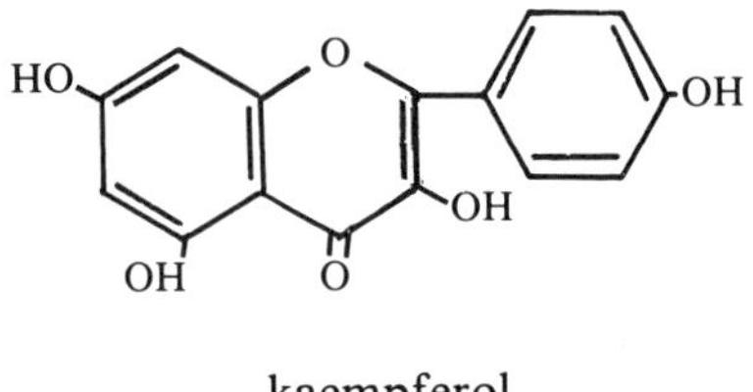   kaempferol

Table 13-9. Cottonseed flavonoids.†

Designation	Tentative identification	Characterization method‡
1	quercetin 3-0-neohesperidoside (2-0-α-L-rhamnosyl-β-D-glucoside)	1,2,4
2	quercetin 3-0-glucoglucoside	1,2
5B	quercetin 3-0-robinoside (6-0-α-L-rhamnosyl-β-D-galactoside)	1,2
5A	quercetin 3-0-rutinoside or rutin (6-0-α-L-rhamnosyl-β-D-glucoside)	1,2,3
6	quercetin 3-0-glucoside or isoquercetrin (β-D-glucoside)	1,2,3
3	kaempferol 3-0-neohesperidoside (2-0-α-L-rhamnosyl-β-D-glucoside)	1,2,4
4	kaempferol 3-0-glucoside	1,2

† From Blouin et al. (1981).
‡ (1)—Ultraviolet—visible diagnostic spectral analysis.
  (2)—Hydrolysis and TLC of aglycones and sugars.
  (3)—Chromatographic mobility with standards.
  (4)—NMR spectra.

When cottonseed products are used in foods, gossypol and flavonoids contribute to a serious discoloration problem. If plant protein products are to be used in food, the color problems must be solved. Yellow colorations are caused by flavonoids, and brown discoloration is mainly bound gossypol. A knowledge of the nature of the bonding between the components causing color and proteins should lead to the development of better methods for prevention or elimination of these problems.

13-1.3 Cyclopropene Fatty Acids

Cottonseed oil contains sterculic and malvalic acids; 18 and 17 carbon cyclopropenoid fatty acids (CPFA), respectively, which contain one double bond at the site of a propene ring, either at the 9, 10, or 8, 9 position (Fig. 13-5). The cyclopropene ring is chemically and physiologically reactive (Phelps et al., 1965; Allan et al., 1967).

The cyclopropene ring reacts with sulfur dissolved in carbon disulfide to produce a chromogen (Halphen reaction) that allows colorimetric quantitation of the CPFA (Deutshoman and Klaus, 1960; Sheehan et al., 1972). Acid titration, in which halogen acids (primarily HBr) are added to the double bond of the cyclopropene ring, is another method for the quantitation of these fatty acids (Mayne et al., 1966; Feuge et al., 1969). Derivatives of CPFA have been quantitated by gas-liquid and high performance liquid chromatography (Fisher and Schuller, 1981; Bianchini et al., 1982; Fisher and Cherry, 1983).

Poultry and animal feeding studies have shown CPFA to change the ratio between stearic and oleic acids; as a result, CPFA cause the hardening of fats in egg yolk and milk (Johnson et al., 1967; Roehm et al., 1970; Bickerstaffe and Johnson, 1972). The fatty acids bind to the thiol groups of

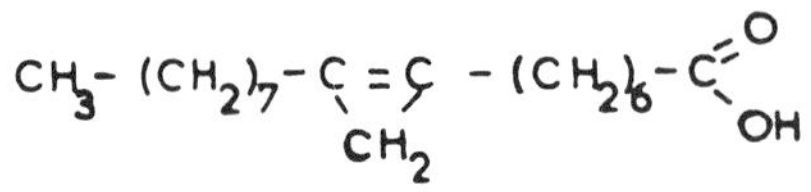

C18:CE malvalic acid (8,9-methylene-8-heptadecenoic acid)

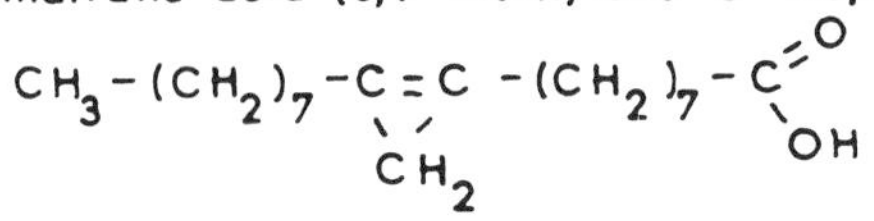

C19:CE sterculic acid (9,10-methylene-9-octadecenoic acid)

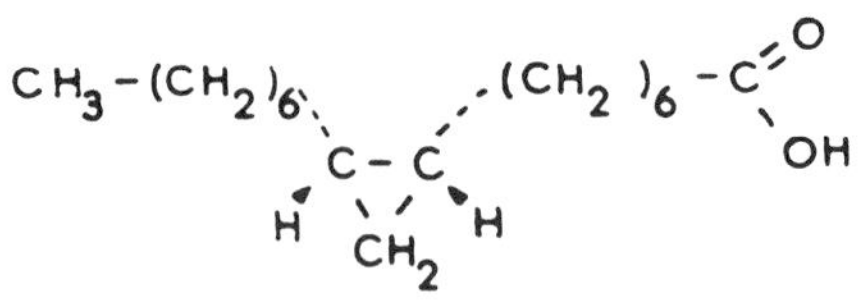

C17:CA *cis*-8,9-methylenehexadecanoic acid

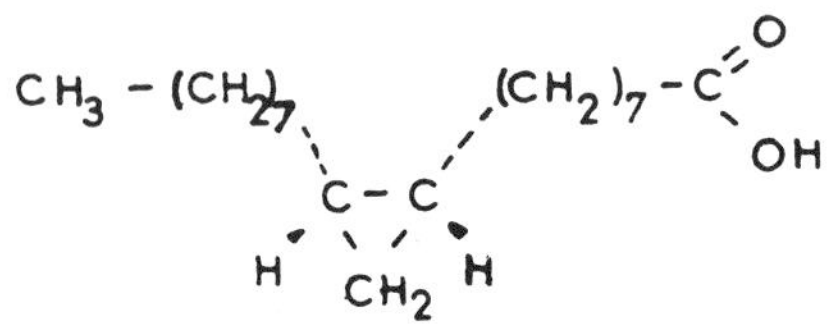

C19:CA dihydrosterculic acid (*cis*-9,10-methyleneoctadecanoic acid)

Fig. 13-5. Cyclopropenoic and cyclopropanoic fatty acids (Bianchini et al., 1982).

acyl desaturase, a complex of enzymes involved in the unsaturation of stearic acid to form oleic acid (Kircher, 1964; Allan et al., 1967). The CPFA also alter permeability of the vitelline membrane of the egg yolk, which leads to diffusional changes shown by altered concentrations of water, fat, and proteins in the yolk. They also affect concentrations of iron and non-protein nitrogen compounds in the white, causing them to become discolored (Phelps et al., 1965). Jones (1981) reviewed other possible effects of CPFA ingested by poultry and animals, that have been reported by a number of investigators. There is evidence that the CPFA act as cocarcinogens with aflatoxins in trout, although feeding studies with rats to determine the cocarcinogenicity of CPFA and aflatoxins have not been conclusive.

In the process of oil production from cottonseed, the deodorization and hydrogenation steps greatly reduce the Halphen response normally detected in crude oil (Phelps et al., 1965). Calculations of human consumption, based on domestic disappearances of cottonseed oils, proportions of vegetable fat in the U.S. diet, and statistics on food consumption, indicate human CPFA intakes of at least one order of magnitude lower than that used in almost all studies with laboratory animals reported in the available literature (Jones, 1981). This finding is consistent with the observation that no deleterious reactions have been noted in the more than a century that cottonseed oil has been available commercially as a vegetable oil for human use, although the reactivity of CPFA is such that it should be given priority in cottonseed improvement efforts.

13-2 SEED DEVELOPMENT

13-2.1 Cultural Conditions

Reproductive development starts in the early stages of plant growth, although vegetative growth predominates during the early stages of development. Gradually, the partitioning of growth is shifted increasingly toward reproductive and away from vegetative development. Finally, vegetative growth is essentially arrested, at least until most bolls are advanced in development. Then, if environmental conditions are favorable, a second cycle of vegetative and reproductive growth is initiated. This regrowth phenomenon is a liability to successful crop production management in most areas of the cotton production zone.

Because of the indeterminant flowering pattern of cotton, fruit forms are initiated over a period of several weeks. Basically, compositional changes during cottonseed development and ripening are like those of other seeds (Altschul, 1948; Grindley, 1950; El-Nockrashy et al., 1976; Elmore and Leffler, 1976; Kajimoto et al., 1979). Even if meteorological conditions are generally considered to be "ideal" for crop growth, many aspects of the environment interact to produce a continuously varying system in which seed development takes place. Seeds in relatively early bolls are nourished through the high photosynthetic capability of a comparatively young plant, while those in later bolls are nourished by the lesser capacity of the more aged canopy. These contrasts between seeds of early and late bolls are made even more striking by shifting weather patterns as summer advances to early fall (Meredith and Bridge, 1973; Benedict et al., 1976; Leffler et al., 1977, 1978). Seasonal changes in seed size and composition are consistent both with the effects of cool night temperatures (Gipson and Joham, 1969; Gipson and Ray, 1970) of late-season development and with the declining photosynthetic activity of an aging leaf structure (Leffler, 1980). For example, Gipson and Ray (1970) showed that percentages of oil tended to respond hyperbolically to temperature, with the optimum being near 20°C. Stansbury et al. (1954) examined eight cultivars at 13 locations for 3 years and found the correlation of percentage of oil to maximum temperature to be r = −0.57 (significant at the 0.01 level). Low temperature changed both the ratio among simple sugars and the time of their maximum accumulation in seed (Conner et al., 1972). The rates of accumulation of minerals and reserves were also altered by low temperatures (Kreig et al., 1973).

Evaluations of the planting quality of cottonseed obtained from stratified harvests (repeated, frequent harvests of seed cotton) at Stoneville, Mississippi, have consistently shown a nearly linear decline with later periods of development and, consequently, times of harvest. As indicated by Leffler et al. (1978), the initial and final stand counts were reduced with seeds formed late, and seedling growth rates and yielding capability of the seedlots were similarly reduced. The very early study of Simpson and Stone (1935) suggested that seed formed late may not germinate as well as those formed early. The contributions of the environment to the development of the quality of cottonseed can be highly significant (Peacock and Hawkins, 1970). Efforts to manipulate the reproductive development of cotton must therefore consider the resultant effects on seed quality.

Within a production environment, the processes of seed development can be adversely affected by meteorological extremes. Leffler (1976) described the temporary cessation of dry matter and protein accumulation by cottonseed during a period of high rainfall and low solar radiation.

Maturing cottonseed responded to elevated nitrogen fertilization by synthesizing additional storage proteins (Sood et al., 1976; Leffler et al., 1977). The degree to which these increases in storage proteins occurred depended on the time during the maturation period that fertilizer was added. If environmental conditions are adequate to influence the patterns of seed composition, it is reasonable to conclude that they might also affect the normal synchrony of development during germination.

Stewart (1980) calculated from data of Leffler et al. (1977) that high nitrogen reversed the decline in oil, but not protein. In further studies, major environmental differences during two seasons at College Station, Texas, influenced intraseasonal variation (Kohel and Cherry, 1983). Weather conditions affected seed development, differentiation, and energy demands within cotton plants as the growing seasons progressed. Seed decreased in weight and in their relative concentrations of oil and free gossypol, but increased in protein with progressively later harvests. Significantly, there appears to be a marked contrast between Texas and Mississippi in the seasonal responses of oil and protein. In numerous evaluations of seed quality, seed of some cultivars consistently rate at or near the top, while seed of other cultivars consistently rate poorly. Since the quality of seed is not usually a characteristic for which direct genetic selections are made, these observed differences among cultivars are probably the natural results of unselected characteristics. There are indications that improvement in seed quality might be made through breeding efforts. El-Zik and Bird (1969) reported that the ability to produce a stand of cotton was a quantitatively inherited trait that could be enhanced through breeding. Similarly, Hess (1977b) described the genetic enhancement of seed density, which is associated with cottonseed quality.

The production of a seed crop is dependent on the leaf canopy. Consequently, anything that influences either the rate or the duration of the canopy development may also affect the quality of the seed that the plants produce. One management factor is the establishment of a population density for the crop. Both vegetative and reproductive developments of the cotton crop are highly influenced by the density of the population of plants. By arranging a fixed population of plants into different competitive patterns, Walhood and Johnson (1976) produced significantly different rates and durations of crop development. The varying competition among plants also influenced the times at which the plants initiated reproductive development. Thus, because of these effects of flowering patterns, earliness, and cutout, the crops from the various spatial patterns matured under different environmental conditions. Both Caldwell (1962) and Maleki (1966) described reductions in the quality of cottonseed produced at above-normal plant population densities. These declines in quality may result from shifts in canopy development related to the density of populations.

13-2.2 Chemical Treatments

Throughout the normal cotton production season, the crop canopy is frequently subjected to numerous chemical treatments. The majority of these applications are made in the effort to control insects, although late-season applications with growth terminators or defoliants are also common. Additionally, there has recently been considerable interest in the early- or mid-season application of plant growth regulators (e.g., 1,1-dimethyl piperidinum chloride, or mepiquat chloride) to manipulate the development of the canopy (Anon., 1978b, 1979; Morgan, 1979).

While it is often acknowledged that canopy development is influenced by the response of plants to insects, it is often overlooked that cotton plants may also respond to the application of chemical insecticides. For example, Brown et al. (1962) reported that accumulation of yield was delayed in cotton treated with methyl parathion. Roark et al. (1963) later attributed this delay to a direct effect of the chemical on the metabolism of the cotton plants. Although the effects of insecticide applications on seed quality have not been evaluated, such an effect could be identified, if for no other reason than the shift in crop maturity.

Growers frequently include harvest-aid and defoliant chemicals among their cotton management practices; these practices have recently been reviewed by Cathey (1979). When properly applied, these chemical treatments can benefit the producer. If, however, applications are made prematurely—when too many bolls have yet to open—these chemicals can lower the quality of cottonseed, particularly those in the second harvest (McMeans et al., 1966; Cathey, 1979).

13-3 POSTMATURATION AND PREHARVEST

13-3.1 Field Environment

As developing bolls complete their filling periods, they progress to the maturation phase. Completion of the maturation phase is identified initially by the dehiscence of the boll and finally by the drying and fluffing-out of the seed cotton. Since seed cotton is usually harvested only once or twice, many open bolls remain in the field for a considerable time before harvest.

During this postmaturation, preharvest interval, the seed cotton is exposed to weathering. This exposure can hurt the quality of the seed for planting, especially in the humid areas of the U.S. Cotton Belt. Weathering deterioration is of little consequence in the arid cotton production zones, because of the absence of the moisture required to support biological activity in the seed.

The first bolls to open are usually low and inside the canopy; they are exposed to prolonged weathering and dry slowly because of the atmospheric humidity "trapped" by the foliage. Further, in those areas where cotton is harvested only once, the duration of the exposure is greater. Simpson and Stone (1935) provided one of the earliest documentations of the association between moisture and the deterioration of seed quality. Cottonseed exposed to weathering deterioration usually contain elevated amounts of free fatty acids, exhibit reduced levels of germinability and seedling vigor, and often have an off-colored (greenish or greenish-brown, rather than creamy white) embryo (Altschul, 1948). Weathered seedlots also often produce a high number of abnormal seedlings, particularly those in which cell division in the root tip does not occur normally (Wiles and Presley, 1960; Hunter and Presley, 1963).

A variety of yellowish, greenish or greenish-brown compounds, including glandular terpenoids (gossypol), flavonoids, and flavonol glucosides are present in cottonseed, which may contribute to autodestruction and discoloring processes during their weathering (Halloin and Bell, 1979; Halloin, 1981, 1982). Nonglandular terpenoid aldehydes have been synthesized in glandless cottonseed containing high moisture.

Continued exposure of seed cotton to high levels of moisture produces conditions that support the resumption of high metabolic activity in the opened bolls, especially early in the fall when temperatures are relatively high. Under these conditions, those seed that do not possess hard seedcoats become hydrated and initiate some phases of normal germination physiology, particularly lipolysis. This metabolism is not usually sustained, both because seeds do not remain hydrated and because residual biochemical dormancy may suppress further development. When open bolls remain unharvested for an extended time, through repeated cycles of wetting and drying, however, the deleterious effects of this exposure to the elements accumulate (Cherry et al., 1978a, b).

13–3.2 Microorganism Infections

In addition to the germinative or pre-germinative deterioration described above, the seed and boll infections by pathogens also contributes to losses of quality of seed in the field. Development of internal infection by microorganisms is also associated with the exposure of cottonseed to moisture and weathering conditions in the field. In the relatively moist eastern areas of the U.S. Cotton Belt, several species of fungi and bacteria have been isolated from field-weathered seed. Fungal infections by *Aspergillus niger*

(Link ex Fries) and species of *Alternaria, Colletotrichum, Diplodia, Fusarium* and *Rhizopus* were indicated by Halloin (1981, 1982) to be common; bacterial infections in this zone include those caused by *Bacillus, Pseudomonus,* and *Xanthomonas.* Conversely, in the more arid western areas of the U.S. Cotton Belt, only the incidence of the aflatoxin-producing fungus *A. flavus* (Link ex Fries) received prominence in the review by Halloin (1981). Not even in arid zones can the influence of moisture be separated from the development of this fungus, however, since Russell et al. (1976) described the effects of water management practices on its incidence.

The symptoms of seed deterioration due to the direct effects of weathering—increases in seed free fatty acids, coupled with decreases in germinability and the suppression of seedling vigor—are also observed in the indirect effects brought about by pathogenic development. Roncadori et al. (1971, 1972) reported significantly positive correlations between the free fatty acid percentage and the fungi in the embryo. Each of these seed quality determinants increases with length of field exposure prior to harvest (Roncadori et al., 1972). Consequently, Halloin (1981) suggested that both lipolysis and other aspects of seed quality deterioration resulted from the combined actions of seed (direct) and microbial (indirect) systems. Whatever may be the main reasons for it, the fact remains that, especially in humid cotton production zones, quality deterioration is proportional to the period of time that elapses between boll opening and seed harvest.

The harvest of cottonseed for planting and processing purposes should obviously be timely to minimize any field deterioration of seed quality. Stratified harvest studies in the Mississippi Delta indicated a progressive reduction in quality of planting seed; even in the absence of weathering deterioration (Leffler et al., 1978). Consequently, seed for use in planting should probably be obtained from only the first two-thirds of the crop to mature. Similarly, because of the rapidity of field deterioration once the seed have been exposed to moisture, any seed that have been exposed to fall rains should (ideally) be removed from the processing chain that provides planting seed, feed, and food products.

13–3.3 Aflatoxins

Throughout the world, aflatoxins are the only contaminants of feeds and foods routinely monitored. The level of aflatoxin contamination cannot exceed 10 μg/kg (Pons and Franz, 1978; Lee and Goldblatt, 1981; Ciegler et al., 1981). Aflatoxins B_1, B_2, G_1, and G_2 are most commonly found in food and feed commodities contaminated by *A. flavus*; M_1 and M_2, metabolic byproducts produced by cows ingesting feeds containing B_1, are found in milk and dairy products. The structures and physicochemical properties of aflatoxins are presented in Fig. 43-6. Spectrophotometric properties enable the precise detection of aflatoxins after their separation by thin layer chromatography (TLC) and high pressure liquid chromatography (Lee and Goldblatt, 1981). In addition to the chemical tests, products suspected of being contaminated by aflatoxin are further checked by a number of biological systems—including chick embryos, ducklings, rats, hamsters, guinea pigs, dogs, trout, catfish cells, brown bullhead, and others (Ciegler et al., 1981).

Swine, cattle, and poultry given feed contaminated with aflatoxin suffer hepatic necroses, fatty infiltration, bile duct proliferation, and

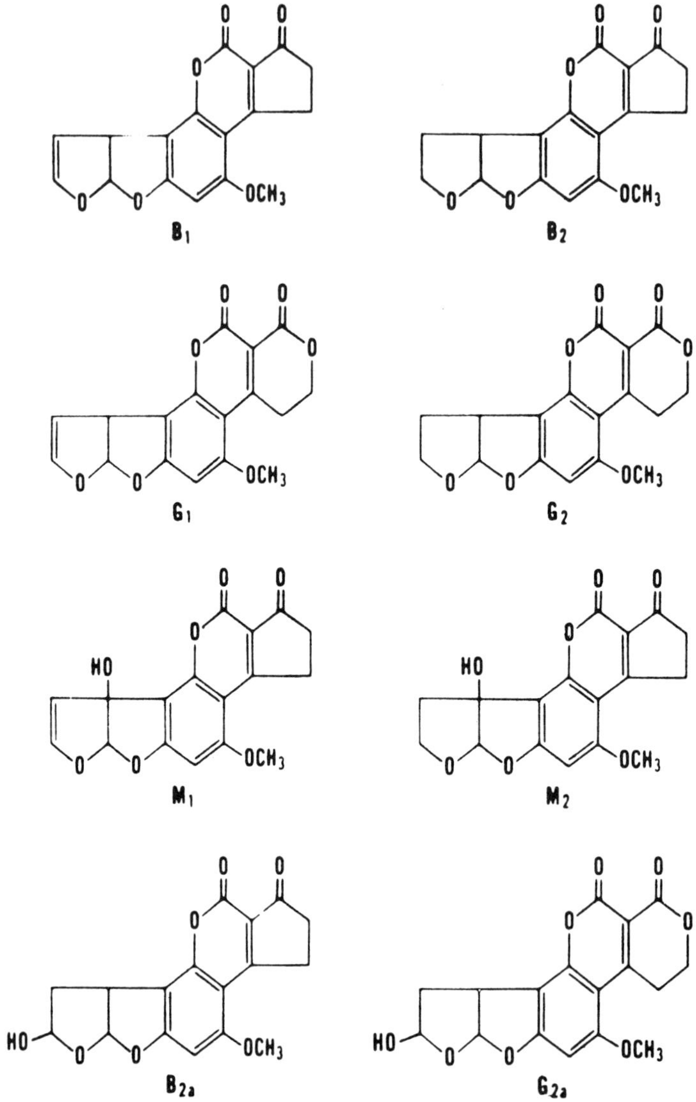

B_1 B_2

G_1 G_2

M_1 M_2

B_{2a} G_{2a}

Fig. 13-6. Structures of aflatoxins (Ciegler et al., 1981).

hepatic failure (Newberne and Rogers, 1981). These acute toxicities of afla-toxins have been experimentally reproduced with laboratory animals. The principal target organ is the liver, although necrosis of spleen, pancreas, heart, and kidney occurs, and fatty infiltration of the latter two organs have been observed. Because of their acute toxicity and proven carcinogenic na-ture, aflatoxins have become a mycotoxin of utmost concern.

When considering factors affecting cottonseed composition, *Aspergillus* spp. (in particular, *A. flavus*) contamination and its production of alfatoxins ranks as the greatest cause of deterioration (Goldblatt, 1969). Be-cause *Aspergillus* species were once considered storage molds, it was believed that aflatoxin contamination of seed arose primarily, if not exclusively, after harvesting. Factors that influence mold growth and toxin production on a commodity such as cottonseed are similar to other factors that affect seed quality: moisture, relative humidity, temperature, seed or substrate composition, competing microorganisms, and fungal strain. Moisture and relative humidity are probably the most important factors involved in fungal growth. It is now known that although *Aspergillus* molds can be a storage problem, they are also a field problem. In cottonseed, the problem is largely confined to a few locations in the far southwestern USA, and there it is primarily a field problem—exactly the reverse of what was thought a few years ago. The presence of aflatoxins in cottonseed and their products prevent the use of the contaminated seed in both feeds and foods.

As a result of these observations, Ramey (1974) presented a system that helped define factors affecting aflatoxin production during the maturation of bolls and seed of cotton. The system identified major research areas that could lead to an understanding of *A. flavus* growth and aflatoxin synthesis in cottonseed as follows: (a) determination of the origin and quantity of *A. flavus* inoculum in the field, (b) study of the environmental conditions leading to *A. flavus* infection of cotton bolls, (c) analysis of the regulatory mechanisms governing the production of secondary metabolites, and (d) de-velopment of additional information on genetic variability in *A. flavus* strains.

Contamination by aflatoxin in freshly harvested bolls of cotton can be detected by long wavelength ultraviolet light (Ashworth and McMeans, 1966); contaminated cotton has a bright greenish-yellow fluorescence (BGYF). This method has received much attention, and is being used to de-tect aflatoxin in the fiber of seed cotton. The highest concentrations of aflatoxin in seed were associated with unfluffed seed cotton. These findings stimulated studies to establish the mode of infection of cottonseed by toxin-causing fungi in the field. Only certain seed have been found to be afla-toxin-contaminated, even though select locks or entire bolls were *A. flavus*-infected. This selectivity might be explained if mold infection occurred during pollination when mold spores carried by insects may have inter-mingled with pollen (Lee and Russell, 1981).

13-4 HARVEST AND POSTHARVEST

13-4.1 Field Handling

The biological quality of cottonseed is maximum when they mature on the plant; handling of the seed during harvesting and processing subjects them to reductions in quality (Delouche, 1981). These quality reductions are caused by two types of factors: biological deterioration during storage and mechanical damage during processing and handling. Mechanical damage can occur at each operation, beginning with harvesting and continuing through the final processing step.

Seed can be damaged in harvesting if the equipment is not maintained, adjusted, and operated properly. Chief problems in the harvesting procedure are associated with excessive speed of operation and impact damage to the seeds (Colwick et al., 1972). Harvesting damage to cottonseed is usually characterized by cracked or broken seedcoats, often leaving portions of the embryo unprotected and exposed (Delouche, 1981).

13-4.2 Storage

Once the seed cotton has been harvested, it may be stored for a while before it is ginned. The conditions under which this storage occurs are major determinants of the desirability of this storage. In many ways the type of deterioration that may occur during storage of seed cotton is similar to that occurring in periods of adverse weather in the field prior to harvest. Consequently, high temperatures and moisture must be minimized during storage of seed cotton. Moisture is particularly important because if moisture in the seed mass is adequate to support elevated respiratory activity, the heat liberated can raise temperatures to extreme levels (Thomson, 1979). In his recent review of seed cotton storage, Baskin (1981a) concluded that field storage of seed cotton with greater than 10% moisture should not exceed 2 or 3 days to ensure the preservation of seed quality.

The module builder, which allows field storage of seed cotton prior to ginning, may transform the cotton industry by breaking the temporal connection between harvesting and ginning, allowing each operation to proceed at its own pace (Roberts et al., 1973; Paxton and Roberts, 1973; Curley et al., 1973; Kepner and Curley, 1976; Eickhoff and Willcutt, 1978; Wilkes, 1978; Cherry et al., 1979b, 1981a). Germinability, mycroflora, and free fatty acid content of high quality cottonseed are not affected significantly during module storage of seed cotton so long as seed cotton moisture remains below 12%. The oil and protein quality deteriorate regardless of seed moisture level, however. The moisture in trash such as leaves, soil, and branches causes localized "hot spots" that damage the seed. Temperatures that exceed 49°C during module storage indicate that seed are deteriorating and that the seed cotton should be ginned immediately. The addition of propionic acid to moist seed cotton stored in modules lessens these changes.

Good management during harvesting and close monitoring of the conditions of the seed cotton during storage are requisite to the use of modules.

13–4.3 Ginning

After the harvesting operation, the seed cotton is conveyed to the gin. Operations at the gin contribute to the mechanical damage observed in seedlots. Watson and Helmer (1964) found that approximately 1% seed damage was attributable to the cleaning, drying, and conveying operations, while an additional 5% damage was associated directly with the ginning operation. They also found that both seed moisture and ginning rate were directly proportional to the incidence of seed damage. The mechanical damage inflicted by gin saws is characterized by cuts or gashes in the seedcoat (Delouche, 1981).

Cottonseed from the gin are held in storage, where efforts must be made to minimize heat accumulation. Aeration of cottonseed stored in bulk should begin as soon as possible after ginning to reduce the heat remaining from the drying operations at the gin. The effects of bulk storage are similar to those described previously for module storage of seed cotton; the combined effects of elevated temperature and moisture can destroy a seedlot within a short period of time (Delouche, 1981; Baskin, 1981b).

13–5 PROCESSING

13–5.1 Planting Seed

The value of cottonseed used for planting is determined initially by their ability to germinate rapidly, support vigorous seedling growth, and produce an acceptable yield. Only insofar as it influences one or another of these principal determinants of seed quality does the seed's chemical composition merit consideration as a factor in planting seed quality.

Because gin-run cottonseed, taken directly from the gin, are still covered by fuzz fibers (linters), they tend to clump together into a mass. This aggregation of the seed effectively prevents controlled handling of single seeds during the planting process. To enhance their handling characteristics, cottonseed are subjected to one of several operations to remove the fuzz. Even though this delinting introduces an additional step to the processing of cotton planting seed, it is essential to the delivery of a product that is suitable for controlled handling. Several methods exist for delinting cottonseed, each falling into one of the three main categories of (a) mechanical delinting, (b) flame delinting, or (c) acid delinting (Delouche, 1981).

Mechanical delinting is a repetition of the ginning operation. This reginning is done with more and finer saws, spaced closely together, which act upon the seeds to remove most of the fuzz. Although mechanically delinted seed are less likely to aggregate than are gin-run seed, they do not separate completely into single seeds because they are not completely free of fuzz. Additionally, mechanical delinting subjects the seed to another cycle of vulnerability to mechanical damage from the saws.

Flame delinting is a second-stage delinting operation, in that seeds which have first been mechanically delinted are then passed through an intense flame to singe most of the remaining fuzz. If conditions are adequately controlled, little deterioration in seed quality can be attributed to the flaming process. However, the tolerances to excessive heat are narrow, and conditions must be carefully controlled. The handling characteristics of flame-delinted seed are superior to those of mechanically-delinted seed, but are not usually as good as those of acid-delinted seed.

The third category of delinting procedures, acid delinting, produces the most completely delinted seed. There are three types of acid delinting, including (a) wet acid, (b) dilute wet acid, and (c) gas acid. In wet acid delinting, concentrated sulfuric acid is used to hydrolyze the fuzz. The dilute acid procedure introduces a dilute solution of sulfuric acid to the seed; this concentration is then effectively increased as the acid-wet seed are dried. Partial delinting is accomplished by lowering the initial concentration of sulfuric acid. The gas acid method employs gaseous anhydrous hydrochloric acid to degrade the fuzz. The degraded fuzz in all three processes are then removed by frictional abrasion in a buffer drum. Because delinting is complete, or nearly so, the handling characteristics of acid-delinted seeds are excellent, so precision metering during planting is possible.

Since acid can adversely affect seed quality, no matter which method of acid delinting is used, residual acidity must be removed, either by washing or by neutralization, or both. Seed quality problems arise when high temperatures occur, reaction times are too long, or the neutralization process is incomplete. Since acid can enter small cracks and holes in the seedcoats, complete removal of acid is particularly important, but difficult, in seedlots that have a high level of mechanical damage.

Once the cottonseed have been delinted, they are ready for the conditioning treatments that generally precede their use as planting seed. This conditioning process actually refers to the upgrading of a seedlot's performance through the removal of contaminants and some of the poorly performing seed from that seedlot. Two basic types of conditioning treatments are used in the processing of cotton planting seed. The first of these treatments is sizing by screening seed from debris on the basis of their length and width. Passage of seed through length/width separators enables the removal of contaminants that are outside the normal size range for cottonseed. Removal of immature or low-density seeds from the seedlot is also accomplished with the gravity table. This procedure, conducted on acid delinted seed, can significantly improve the germination performance of a seedlot by removing the smaller, weaker, lower density seeds.

Before cottonseed are used for planting purposes they often are coated with one or more seed treatments; these usually include a fungicide, and sometimes an insecticide as well. While these chemicals may enhance the field performance of a seedlot, especially under adverse conditions, they can reduce the longevity of the seed. The chemicals have a degree of phyto-toxicity which may be magnified in seedlots containing high levels of mechanical damage; consequently, storage of treated seed should be mini-mized except where seed are of highest quality (Thomson, 1979; Davis et al., 1981).

13-5.2 Oil Extraction

Extraction of oil from cottonseed involves the following steps: (a) elimination of leaves, twigs, pieces of bolls, and sand; (b) removal of fuzz fibers, (c) dehulling; (d) screw press extraction, solvent extraction or pre-press-solvent extraction; (e) solvent removal; (f) toasting; and (g) grinding or making into pellets (Harrison, 1977). The preparation and separation processes necessary to achieve maximum extraction of oils from various seed including cottonseed, are summarized in Fig. 13-7 (Cherry and Berardi, 1983; Cherry, 1983). The crude oil is warmed and treated with sodium hydroxide to enable removal of the soapstock. This refining process also removes the darker coloring materials, such as gossypol, leaving a clear yellow oil. Bleaching clay is used to remove any remaining colored sub-stances. A winterization step at 3° to 4°C removes stearine, a material that turns the oil cloudy at 4° to 10°C. The compositions of feed-grade meals prepared by screw press, pre-press solvent, and direct solvent extraction methods are presented in Table 13-10 (Anon., 1982b); percentages of fatty acids in the oils extracted by these three processes were similar to those pre-sented in Tables 13-4 and 13-8. Reviews of research programs having the objective of improving the competitiveness of cottonseed as a viable source

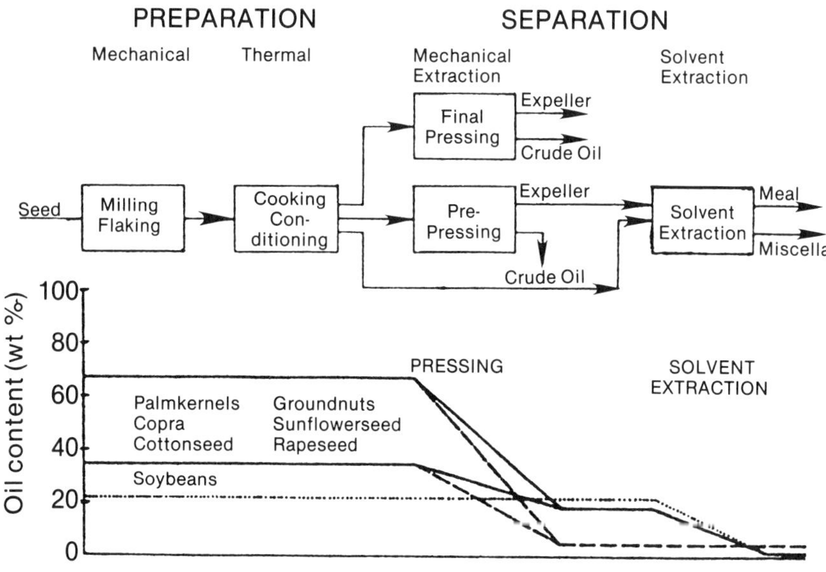

Fig. 13-7. Oil recovery processes (Cherry, 1983).

of edible oil products were presented by Cherry et al. (1981a, b), Cherry and Berardi (1983), and Cherry (1983).

Glandless cottonseed can be processed much more efficiently than glanded seed to high quality kernels, oil, and meal products (Fig. 13–8;

Table 13–10. Adjusted mean analytical composition of cottonseed meal by oil extraction process.†

Components‡	Prepress solvent	Screw press	Direct solvent
Dry matter	89.9	91.4	90.4
Ash	6.4	6.2	6.4
Crude fiber	13.6	13.5	12.4
Ether extract	0.58	3.72	1.51
Crude protein	41.4	41.0	41.4
Gossypol-free	0.05	0.04	0.30
Gossypol-total	1.13	1.02	1.04
N-solubility§	54.4	36.8	69.4
Calcium	0.15	0.16	0.15
Iron	0.011	0.010	0.009
Magnesium	0.40	0.42	0.40
Potassium	1.22	1.20	1.16
Sodium	0.04	0.04	0.04
Phosphorus	0.97	0.93	0.98
Lysine	1.71	1.59	1.76
Histidine	1.10	1.07	1.10
Arginine	4.59	4.33	4.66
Aspartic acid	3.72	3.65	3.68
Threonine	1.32	1.30	1.34
Serine	1.74	1.68	1.78
Glutamic acid	8.30	8.55	8.08
Proline	1.54	1.42	1.45
Glycine	1.70	1.69	1.69
Alanine	1.62	1.58	1.62
Valine	1.88	1.84	1.82
Methionine	0.52	0.55	0.51
Isoleucine	1.33	1.31	1.33
Leucine	2.43	2.23	2.41
Tyrosine	1.13	1.09	1.14
Phenylalanine	2.22	2.20	2.23
Cystine	0.64	0.59	0.62
Tryptophan	0.47	0.50	0.52
Copper	17.8	16.7	16.3
Manganese	20.0	21.6	20.7
Zinc	62.3	57.4	57.4
Cobalt	1.3	1.5	1.5
Biotin	0.55	0.53	0.55
Choline	2932.6	2807.2	2706.0
Folic acid	2.66	2.73	2.79
Niacin	40.3	37.8	39.2
Pantothenic acid	7.0	7.7	9.9
Pyridoxine	4.0	4.8	4.8
Riboflavin	4.0	4.2	4.4
Thiamine	3.3	9.7	7.7

† From Anon. (1982b). Adjusted means calculated by determining mean less one-half standard deviation. Exceptions: ash, fiber, and gossypol. In these cases a one-half standard deviation is added to mean value. Values for vitamins and energy are unadjusted.

‡ Components dry matter through trytophan are expressed as %, and copper through thiamine are expressed as mg/kg.

§ Nitrogen soluble in 0.02 N NaOH.

Lusas et al., 1977; Rosenblum, 1980). Glandless kernels can be prepared for direct consumption or for use in food as nut replacements because of the absence of gossypol. Kernels and meal fines can be flaked and extracted with hexane to produce flour. To improve oil removal, kernels are conditioned to a moisture content of 8 to 10%, heated to 71° to 82°C, and rolled into flakes. Percolation of solvent through the flakes yields high quality oil and defatted meal. With glandless cottonseed, extensive heat treatments are not needed during processing to bind gossypol in the meal or to reduce color reversion in crude oil. Binding of gossypol with proteins is not a problem with glandless products. Further, a refined oil from glandless cottonseed requires less purification processing, and as a result, less loss of neutral oil.

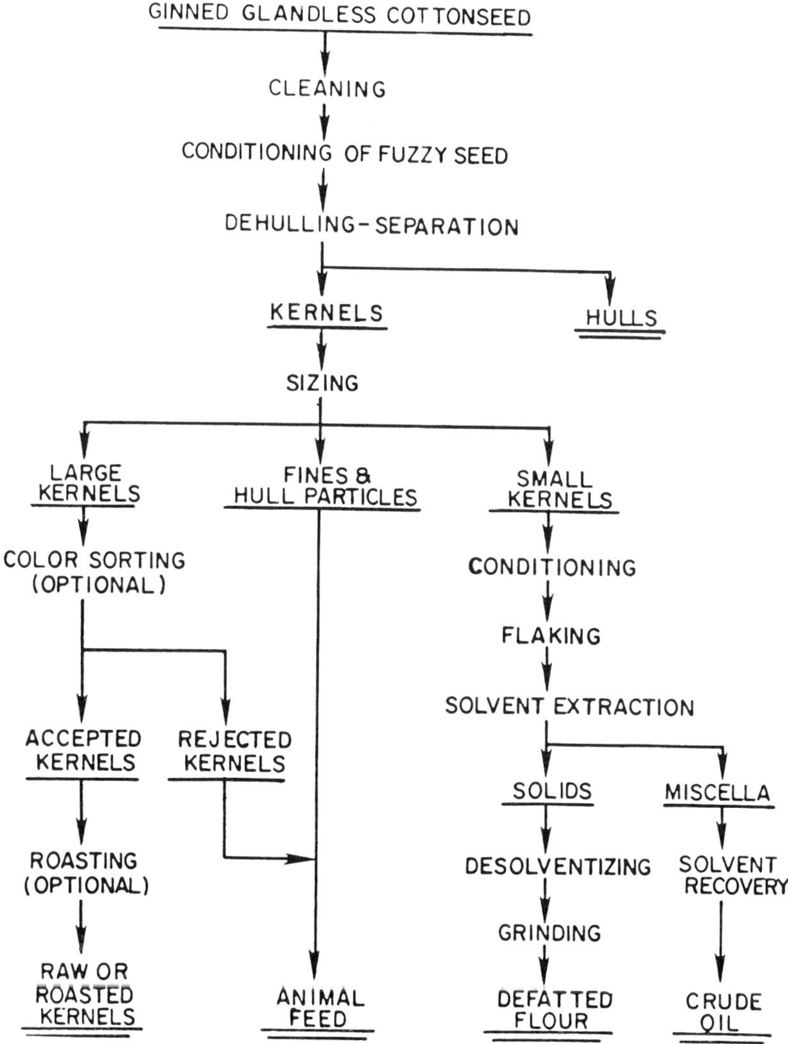

Fig. 13-8. Flow chart for production of glandless cottonseed kernels and flour (Lusas et al., 1977).

13-5.3 Phospholipid Extraction

Industrial lecithin can be fractionated as phospholipids and glycolipids after neutral lipids and protein-containing contaminants are removed (Cherry et al., 1981a; Cherry, 1983). Their purity and special properties can be improved by a number of methods, including solvent fractionation, hydrogenation, sulfonation, and ethoxylation. Lecithin, which contains both polar and nonpolar groups, has high surface activity and is reactive with both oil and protein, making it an excellent emulsifying agent in food systems.

Among common oilseeds, except for soybean, cottonseed has the highest content of phospholipids (Cherry et al., 1981a). Cottonseed phospholipids are superior to those of other oilseeds since most of the fatty acids present do not contain more than two double bonds, making them more stable to oxidation and rancidity processes (Cherry et al., 1981a; Cherry, 1983). Cottonseed phospholipids have been marketed to a small extent, however. The heat and moisture produced from the hydraulic press method of oil extraction causes gossypol to bind to phospholipids and other constituents of the meal. Solvent processing also includes heating steps that bind gossypol to the phospholipids. The availability of gossypol-free cottonseed provides an opportunity to produce food-grade cottonseed phospholipids as a byproduct of the production of edible oil.

13-5.4 Protein Extraction

Cottonseed proteins are widely recognized as potential sources of nutrients for human consumption. Because their potential as food supplements has been known for many years, cottonseed proteins have been the subject of numerous investigations. Osborne and Voorhees (1894), Jones and Csonka (1925), Fontaine et al. (1945, 1946), and Arthur and Karon (1948) showed that proteins in cottonseed meal could be extracted as albumins and globulins with water, salt, and alkaline solutions. Rossi-Fanelli et al. (1964) devised nonchromatographic methods to isolate from cottonseed a homogeneous monodispersed globulin, acalin A, with a sedimentation coefficient of 9.2 and a molecular weight of approximately 180 000. Karavaeva (1972) isolated a similar protein, globulin A, that had a molecular weight of 170 000; globulin A dissociated into two subunits of approximately 80 000 molecular weight with sodium dodecyl sulfate polyacrylamide gel electrophoresis (SDS-PAGE). Reduction and alkylation of the subunits split them further into smaller components of 40 000 molecular weights.

Based on the fact that a group of proteins in cottonseed is in compartments in aleurone grains (Fig. 13-3), a two-step extraction procedure was developed (Berardi et al., 1969; Martinez et al., 1970; Martinez, 1979). Alkaline-soluble (storage) globulins of high molecular weight precipitable at pH 7, and low molecular weight water-soluble (nonstorage) proteins precipitable at pH 4 are isolated. Alkali or salt is needed to rupture the membrane structure of cottonseed protein bodies and dissolve the storage globulins. The water-soluble proteins are predominantly the functional proteins of the seed cytoplasm. The water-insoluble proteins are essentially the storage globulins of the protein bodies that contribute polypeptide fragments, amino acids, and nitrogen to the germinating seedling.

The classical procedure for isolating proteins was developed to extract the nonstorage and storage proteins together (Martinez et al., 1970). The proteins are extracted with dilute alkali at pH 10, then acidified to the isoelectric pH 5 to precipitate and collect them as an isolate. These proteins can also be selectively precipitated from the pH 10 extract at pHs 7 and 4 to prepare storage and nonstorage protein isolates, respectively.

Podgornov et al. (1974) showed that a 10% NaCl extract of defatted glandless cottonseed flour could be fractionated by gel filtration chromatography into five components; two major components had molecular weights of approximately 130 000 and 300 000. Dieckert and Dieckert (1976, 1978) and Dieckert et al. (1981) similarly isolated two major storage globulins from cottonseed and labeled them acalin A and acalin B; these proteins are also known as the 7S and 12S (or 11S) globulins, respectively. Reduction of acalin A with 2-mercaptoethanol (2-Me) produced polypeptide subunits having molecular weights of approximately 49 000, 31 000, and 28 000 as determined by SDS-PAGE. The predominant subunit of acalin A has a molecular weight of 53 500 and does not exist in a disulfide-bridged subunit arrangement. A homogeneous subunit with a sedimentation coefficient of 2S and a molecular weight of 22 000 was isolated from the 7S globulin in the presence of 8 M urea (Ibragimov et al., 1974). Acalin B appears to have a native molecular weight of 240 000 to 250 000. Based on SDS-PAGE with 2-Me, acalin B contains polypeptides having molecular weights of 23 000, 20 000, and 15 000.

Other studies have shown that the acalin globulins consist of a large number of repetitive acidic and basic subunits bound together by disulfide bonds (Dieckert et al., 1981). The 7S globulin was shown to consist of eight polypeptides of two types which differ on the basis of carbohydrate content and partial amino acid sequences (Dieckert et al., 1981; Kuchenkova et al., 1977; Ovchinnikova et al., 1977; Redina et al., 1977). Kuchenkova et al. (1977) developed the amino acid sequence of two 7S globulin subunits (I and II; Fig. 13–9). Pronase hydrolysis of the 7S globulin allowed isolation of a glycopeptide which contained aspartic acid, glucosamine, and mannose in a ratio of 1:2:5 (Fig. 13–9; Kuchenkova et al., 1981). The carbohydrate chain has a branched structure with an N-glycoside type of oligosaccharide-protein bond.

In further studies, acalin B (11S-globulin; molecular weight of 260 000) was shown to consist of three subunits, A, B, and C, of molecular weights of 28 000, 24 000, and 17 000, respectively (Asatov et al., 1977). Since the globulin did not dissociate in 8 M urea and SDS, these researchers suggested that its quaternary structure may be formed by noncovalent bonds such as ionic, hydrogen, and hydrophobic interactions. The ϵ-amino group of lysine residues was shown to play an important role in the stabilization of the 11S-globulin's quaternary structure (Shadrina et al., 1979). Asatov et al. (1978a, b) sequenced the amino acids of subunit C of the 11S globulin.

By using gel filtration chromatography, Zarins and Cherry (1981) separated the globulins contained in a 10% NaCl-extract of defatted glandless cottonseed flour into six fractions (I to VI; Fig. 13–10A and B). Column and thin-layer gel filtration methods demonstrated that these fractions had molecular weights of $> 600\,000$, $280\,000$, $127\,000$, $63\,500$, $10\,700$, and < 2000, respectively. Fractions II and III are probably equivalent to the 12S (alcalin B) and 7S (acalin A) globulins of studies discussed above. At pH 10.5, fraction I decreased while fraction VI increased, which indicated the release of a small constituent from the large protein (Fig. 13–10A). Fractions I and VI were yellow at neutral pH; at alkaline pH, they became green-gold, which suggested that flavonol and gossypol or gossypol-like pigments were bound to these proteins (Blouin et al., 1982). Because of its large size ($> 600\,000$), and low protein content (16.6%), fraction I was thought to be the globulins that contain fragments of aleurone grain membrane.

```
                           10
Arg-Gln-Gln-Lys-Ser-Ala-Pro-Gln-Gly-Phe-Gln-Leu-Asn-Arg-Val-Pro-

         20                                    30
Val-Ala-Gly-Phe-Thr-His-Gln-Asn-Lys-Val-Ser-Gln-His-Pro-CmCys-Leu-

                   40
Ala-Arg-Phe-His-Asn-Gly-Gln-Arg-Phe-Gly-Ile-Asn-Phe-Glx  -Arg-Leu-

     50                                  60
Gly-Tyr-Gly-Ser-Gln-Arg-His-Ser-Asn-Gln-Trp-Gln-Arg-Ser-Gly-Gln-

              70                                     80
Tyr-Phe-Ala-Pro-Gln-Asn-Leu-Val-Met-Asn-Asn-His-Gln-Ile-Arg-Leu

                          90
Ala-Asn-Glx  -Asn-Lys-CmCys-Phe-Tyr-Gln-Ser-Gly-Asn-Arg-Val-Arg

           100                              110
Leu-Ala-Gln-Thr-Arg-Gly-Ala-Val-Lys-Leu-Glx  -Ala-Phe-Val-Gly-Ser-

                   120
Arg-Gln-Asn-Phe-Leu-Val-Gly-Gln-Asn-Ser-Ala-Lys-Phe-Glx  -Gln-Asx  -

              130
Val-Asn-Ile-Thr-Ser-Ala-Leu-Val


           [ (GlcNAc)       (Man)     ]
                    3-4         4-5
```

---

[a] x denotes an amide in subunit I and an acid in subunit II.

Fig. 13–9. Amino acid sequence of two 7S globulin subunits (Kuchenkova et al., 1977).

Gradient polyacrylamide gel electrophoresis under non-denaturing conditions showed that proteins in each fraction (Fig. 13–10B) were represented in the pattern of the storage protein isolate (Fig. 13–11).

The proteins in each fraction separated by gel filtration (Fig. 13–10B) were treated with SDS plus 2-Me to dissociate them to polypeptide subunits (Marshall et al., 1984). The subunits were then separated by 0.1% SDS-PAGE. Fractions I and II have similar electrophoretic patterns, but are different from those of the other four fractions (Fig. 13–12). Fractions III and IV are similar especially in the 24 000 to 63 000 molecular weight range. Seven and eight polypeptides ranging in average molecular weight from 11 750 to 63 000 were noted in the gel patterns of fractions I and II (Table 13–11). Fraction III contains eight polypeptides with average molecular weights from 15 800 to 80 200; fraction IV, which is closely associated with fraction III, has four polypeptides with average molecular weights from 26 000 to 63 000. These gels have similarities in polypeptide mobilities indicating incomplete separation, or some type association-dissociation, of fractions III and IV during gel filtration chromatography. Fractions V and VI have two and one very small polypeptides, respectively.

The amino acid compositions of the proteins in the defatted glandless cottonseed flour, nonstorage protein isolate, storage protein isolate, and fractions I, II, III, V, and VI are presented in Table 13–12 (Zarins and Cherry, 1981). The nonstorage protein isolate had a much higher amount of

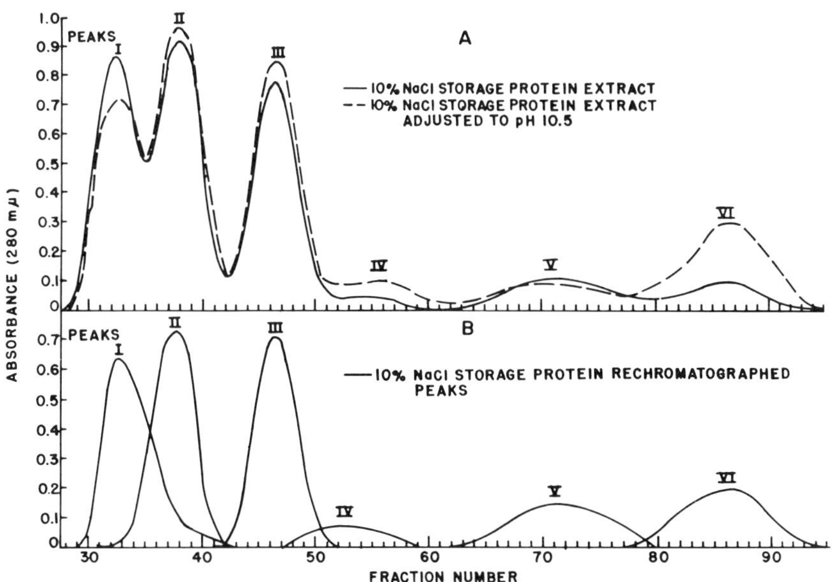

Fig. 13–10. Sephadex G-200 gel filtration chromatography of salt-soluble cottonseed protein isolate: A) protein separation of native (solid line) and alkali-treated (dashed line) salt-soluble extracts; B) composite chromatogram of the six isolated peaks (Zarins and Cherry, 1981).

Table 13-11. Molecular weights of polypeptides present in the 10% NaCl isolate and in the individual fractions I through VI (Fig. 13-10) as estimated by SDS-electrophoresis.†

Peak‡	Fraction	Molecular weight
1	Isolate	82 300
2	III	80 200
3	III	79 000
4	Isolate	75 000
5	Isolate	70 000
6	Isolate, I, II, III, IV	63 000
7	Isolate, I, II, III, IV	56 000
8	Isolate, I, II, III	41 000
9	Isolate, I, II, III, IV	35 000
10	Isolate, I, II, III, IV	26 000
11	Isolate, I, II, IV	21 200
12	Isolate, I, II, III, V	15 800
13	Isolate, II, V	11 700
14	Isolate, VI	10 000

† From Marshall et al., 1984.
‡ Corresponding to number in Fig. 13-12.

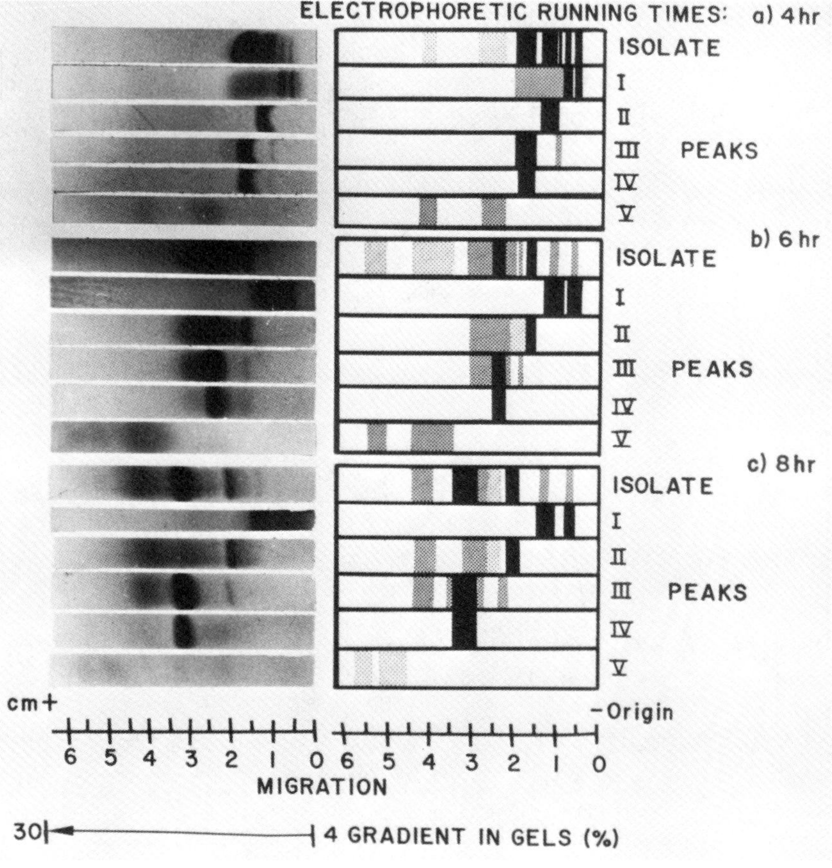

Fig. 13-11. Gradient polyacrylamide gel electrophoresis (for 4-, 6-, and 8-hr periods) of proteins in the salt-soluble isolate, and fractions obtained by gel filtration column chromatography shown in Fig. 13-10 (Zarins and Cherry, 1981).

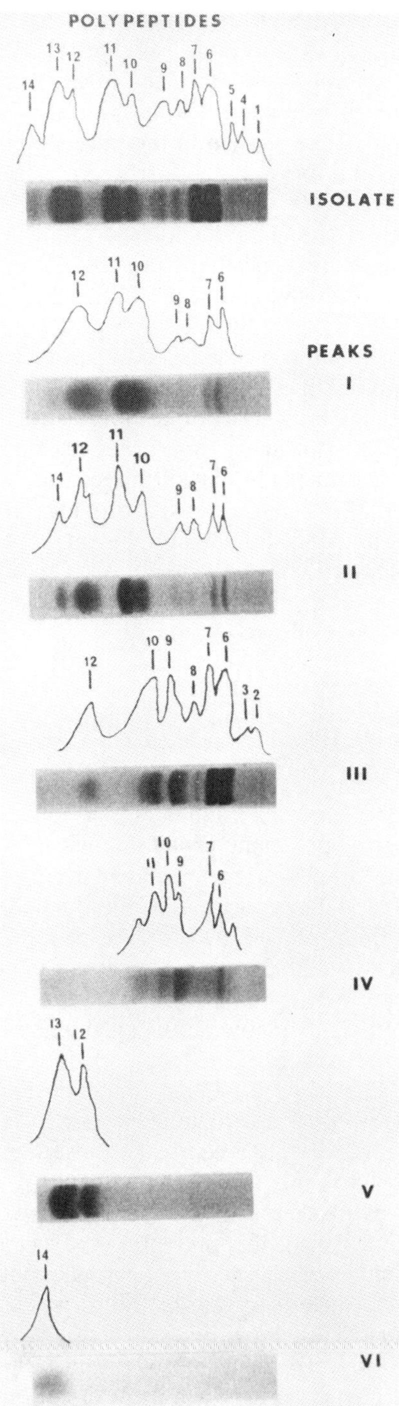

Fig. 13-12. SDS, 2 Me-gel electrophoresis of the purified fractions of 10% NaCl extract of cottonseed storage protein shown in Fig. 13-10 (Marshall et al., 1983).

lysine than the flour and the storage protein isolate. Fractions II and III have a similar amino acid composition, except that III contains very low amounts of sulfur-containing amino acids (methionine and half-cystine) and tryptophan, while II is low in tyrosine and phenylalanine. Fraction III is approximately 60% of the protein in the salt-soluble fraction of cottonseed; its amino acid composition most closely resembles those of the storage protein isolate and flour. Except for the quantities of threonine, proline, and tyrosine, fractions I and II have similar amino acid compositions, adding further support to the findings that these proteins may be alike. Fraction VI contains an extremely high amount of lysine, which explain the presence of bound pigments.

Understanding the basic physicochemical properties of cottonseed is essential to the improvement of cottonseed utilization. In particular, knowledge of protein and polypeptide compositions, structures, and contributions to functional and nutritional properties is essential to the development of technological advancements in the cottonseed industry and the maintenance of farm value. This research will provide information both for geneticists to use in breeding for quality proteins that provide high value raw agricultural commodities, and also for seed technologists to develop new and improved harvesting, handling, and processing techniques for production of quality commodities to the consumers.

13–5.5 Protein Products

Cottonseed proteins are sources of low-cost, food-grade products that can provide functional and nutritional properties in food formulations. These derivatives can supplement protein-deficient diets in areas of the world suffering from food shortages, particularly in developing countries where cotton is extensively cultivated. It is estimated that approximately one-quarter of the flour potentially available from world production of cottonseed could alleviate the protein shortages of underdeveloped nations (Gillham, 1969).

Excellent reviews are available on the production technology of edible protein products from cottonseed (Meinke, 1952; Martinez et al., 1970; Lusas et al., 1977; Cherry et al., 1978c; Experience, Inc. Report, 1978a; Berardi and Cherry, 1979a; Spadaro and Gardner, 1979; Rosenblum, 1980; Cherry and Berardi, 1983). Until recently, cottonseed protein products were used mainly as fertilizer and feed for ruminants. Proflo, a food-grade cottonseed flour, was marketed in the USA as an additive to condition bread dough (Altschul, 1958); it is presently used as a source of protein in culture media for microorganisms. Incaparina, containing mixtures of vegetable proteins that include cottonseed flour, was used as food in South America (Bressani and Elias, 1969; Bressani et al., 1961, 1969, 1980; Scrimshaw, 1980; Wise, 1980). Excellent improvement in the growth of children fed Incaparina that contained as much as 38% cottonseed flour has been reported. The flour used in these studies was processed at a mill that ordinarily produced cottonseed meal as feed and subsequently upgraded for food according to the guidelines established by the Protein Advisory Group of WHO/FAO/UNICEF (Milner, 1965). During normal processing, heat was used to bind gossypol in the meal, making it physiologically inactive and producing a light-colored oil. Gossypol is thought to bind to the free ϵ-amino group of lysine, and possibly also to arginine and cysteine of proteins, during heating (Damaty and Hudson, 1975). Thus processing denatures proteins and causes interactions among various constituents, and lowers the nutritive value of the meal.

A primary technical objective in the processing of edible cottonseed flour, as well as products used for feeding monogastric animals, such as swine and poultry, is the reduction of the toxic impact of gossypol. Several solvents and azeotropes effectively removed gossypol, but problems with flavor, color, and adapting the processes to commercial practice discouraged their use. Pons and Eaves (1971) patented a process using aqueous acetone to lower free gossypol levels to approximately 0.01 to 0.02%. Damaty and Hudson (1975) prepared cottonseed flour with a free gossypol level of 0.01 to 0.03% and a total level of 0.20 to 0.36% by sequential extraction first with aqueous, and then dry acetone. However, during acetone extraction the sulfur-containing amino acids decompose to produce hydrogen sulfide which reacts with an acetone condensation product, mesityl oxide, to produce a "catty odor" (Alyevand et al., 1967). Canella and Sodini (1966) removed gossypol from cottonseed meal with a butanol-hydrochloride acid solution, but this process reduced the free gossypol level in the meal only to 0.070%, which was higher than the 0.045% level permitted in cottonseed products for food use.

A process described as differential settling was developed to fractionate oil, pigment glands, and hull fragments from cottonseed flour (Fig. 13–13; Vix et al., 1947, 1949, 1969, 1971; Gardner et al., 1976). The method uses the principle of mixed solvent (hexane) flotation to separate a low weight, high protein fraction from oil, gossypol, and other coarse materials. It produces edible grade cottonseed flour with free gossypol levels of less than 0.045%. The underflow fraction contains most of the gossypol and hull materials. The method has been engineered to pilot plant scale as the liquid cyclone process (LCP) (Gardner et al., 1976).

Commercial plants with stringent sanitary conditions can be developed for LCP. The edible-grade LCP flour produced at pilot plant scale at the Southern Regional Research Center meets FDA approval in the USA and the guidelines established by the Protein Advisory Group of WHO/FAO/UNICEF for food use (Experience, Inc., 1978a). A representative LCP flour contains 65 to 68% protein with solubility greater than 95% in 0.2 N NaOH solution, total gossypol less than 0.060%, free gossypol less than 0.045%, and lipid less than 1.000%; it also contains 2 to 3% crude fiber, 7% ash, and 3.9 g available lysine/16 g nitrogen. The product is bland in flavor, light in color, and production costs are reasonable (Experience, Inc., 1978b). On a dry weight basis, the flour qualifies as a concentrate containing approximately 70% protein. Considerable interest in the potential of LCP cottonseed products is developing throughout the world.

Methylene chloride reduces the amounts of free and total gossypol in meal extracted with hexane and the LCP underflow protein fraction of glanded cottonseed from 2.600 to 3.400% to 0.013 and 0.150%, respectively (Cherry and Gray, 1981); the free gossypol percentage is well below the 0.045% value. The cottonseed meals were pretreated in one of three ways to rupture the gossypol glands: (a) equilibrated with additional water, (b) suspended in various water–propylene glycol mixtures, or (c) mixed with an

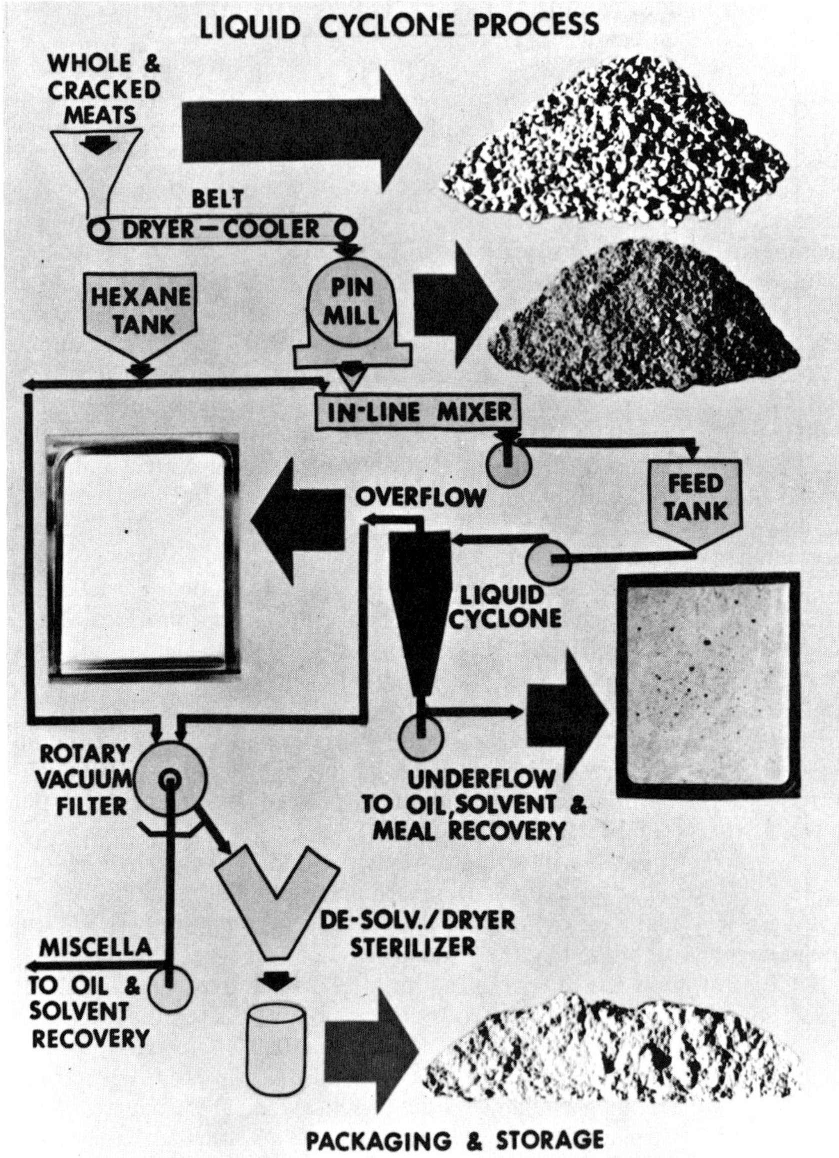

Fig. 13–13. Pilot plant LCP flowsheet (Gardner et al., 1976).

acetic acid–water–propylene glycol solution. The gossypol was then readily extracted from the meals with methylene chloride. Low levels of water and acetic acid in propylene glycol aided methylene chloride in the removal of free and total gossypol; they did not alter proximate composition, solubility, gel electrophoretic properties of proteins, amino acid content, or chemical scores of treated meals. Success with this process should improve the potential of using the underflow of the LCP as feed or food.

Edible cottonseed protein products containing as much as 60.4% protein, and as low as 0.05% free and 0.11% total gossypol can be obtained from glanded, defatted cottonseed flour by air classification (Kadan et al., 1979, 1980; Freeman et al., 1979). Optimal results are obtained if there is minimal pigment gland damage during the lipid extraction and various milling steps. The defatted material must be ground as finely as possible, without disrupting the structure of the pigment glands. Other important factors include maximum removal of hulls, a high velocity through-flow type dryer to dry flakes to 2% moisture before oil extraction, and solvent extraction of flakes to about 2% residual lipids. Defatted flour prepared this way can be fractionated into edible products containing high protein and low gossypol by multiple air classification steps.

Laboratory and pilot plant studies show that protein products from glandless cottonseed are of high quality because the difficulties attributed to gossypol are either minimized or avoided (Fig. 13–8; Lusas et al., 1977; Rosenblum, 1980). Vegetable protein products from the processes described above can be used as such in several food applications, ground into flour, or used in the production of protein concentrates and isolates (Cherry and Berardi, 1983).

Bitter-tasting, odorous, off-flavored compounds, trypsin inhibitors, aflatoxins, and flatulence sugars are usually lowered or removed in the preparation of protein concentrations from cottonseed products (Cherry and Berardi, 1983). Process variations, including water leaching at the isoelectric point, aqueous alcohol leaching, dilute salt leaching, liquid cyclone fractionation, air classification, ultrafiltration, reverse osmosis, and aqueous extraction have been developed at the experimental level to produce concentrates with 60 to 70% protein content.

All processes, except the LCP and aqueous extraction process, usually begin with defatted flours or flakes free from gossypol. The aqueous extraction process involves simultaneous separation of a pH 4.0 aqueous extract of seed into oil, solid, and aqueous phases. Acid soluble proteins, carbohydrates, oil, and other constituents are extracted, leaving the fiber and acid insoluble substances in the protein concentrate. Low molecular weight nonprecipitable proteins, sugars, and salts may be recovered from the whey by ultrafiltration and reverse osmosis (Lawhon et al., 1977a, b; 1978). Approximately 96% of the oil in seeds can be recovered by this process.

Production of cottonseed isolates with 90% or more protein involves extracting the water-insoluble polysaccharides, water-soluble sugars, and other minor constituents. These components usually remain in concentrates

(Cherry and Berardi, 1983). Several methods have been developed for producing cottonseed protein isolates (Berardi et al., 1969; Martinez et al., 1970; Lusas et al., 1977; Cherry and Berardi, 1983). These methods have been discussed earlier in this chapter. Briefly, they include the classical method, in which nonstorage and storage proteins are extracted together from flour with 0.034 N NaOH. The alkaline extract recovered after centrifugation is then acidified to pH 5.0 to provide a protein curd containing both storage and nonstorage proteins. The curd is collected by centrifugation and lyophilized or spray dried to form the classical protein isolate. Secondly, the selective precipitation method, the flour is mixed with 0.034 N NaOH, centrifuged, and the soluble fraction adjusted to pH 7.0 to selectively precipitate storage proteins. After centrifugation or filtration, the soluble portion is further adjusted to pH 4.0 to isolate the nonstorage proteins. The third method, the selective extraction, flour is suspended in water to dissolve the nonstorage proteins. After centrifugation, these proteins are precipitated and collected by centrifugation or filtration after adjusting the soluble extract to pH 4.0. The spent flour collected after the initial centrifugation step is mixed with 0.015 N NaOH, centrifuged, and the storage proteins are precipitated and collected at pH 7.0.

Protein, lipid, fiber, and ash are 93.4, 1.1, 0.5, and 3.4%, respectively, in a representative classical isolate; 86.2, 2.8, 0.2, and 5.9% in nonstorage protein isolates; and 98.0, 0.8, 0.1, and 1.4% in storage protein isolates. The nonstorage protein isolates are the low molecular weight components that are high in essential amino acids (Cherry and Berardi, 1983). The storage globulins with high molecular weight and low essential amino acid content are present mainly in the storage protein isolate (Table 13–12).

An improved process employing a combination of isoelectric precipitation and ultrafiltration to isolate glandless cottonseed protein was developed (Lawhon et al., 1980). Potassium hydroxide and sodium hydroxide each dissolved approximately 88% of the flour nitrogen including storage protein (SP) at pH 10, 87% at pH 9.5, and 85% at pH 9.0. Nonstorage protein (NSP), extracted prior to SP extraction with water, was separated by acid precipitation into curd and whey. The lysine-rich NSP whey was then added to SP extract and the mixture ultrafiltered to obtain a higher yield of SP isolate having a lighter color and increased lysine content.

Coprecipitated protein isolates have been prepared from blends of glandless or LCP cottonseed flour, other oilseed flours, and animal proteins (Berardi and Cherry, 1979b, 1981). More specifically, coisolates were made by coprecipitating the cottonseed proteins with soybean and peanut proteins. The coisolate method involves protein extraction with 0.034 N NaOH, acidification of the protein extract to pH 2.5, and adjustment of the resulting mixture to pH 5.0 to precipitate the protein curd. The recovered curd is resuspended in water, neutralized, and lyophilized. Disc-gel electrophoresis showed that some of the proteins in the extract dissociated into subunits at pH 2.5, then reassociated into their original or new protein forms as the pH was adjusted to 7.0. The coisolates contain 95% protein and accounted for more than 67% of the total nitrogen in the flour blends.

Table 13-12. Amino acid composition of glandless cottonseed flour storage proteins.†

Amino acids	Flour	Nonstorage protein isolate	Storage protein isolate	Gel filtration peaks of storage protein isolate§				
				I	II	III	V	VI
				g/100 g dry weight protein				
Aspartic acid	9.97	10.51	9.46	9.58	9.69	9.18	8.75	7.85
Glutamic acid	21.87	19.35	21.29	18.06	20.45	20.04	19.25	11.91
Alanine	4.40	6.03	3.73	4.78	5.26	3.57	4.20	4.80
Isoleucine‡	3.16	4.47	3.67	3.58	3.75	3.26	2.75	3.02
Phenylalanine‡	5.55	4.69	8.35	6.05	5.46	10.54	5.94	3.85
Threonine‡	3.73	4.65	3.02	4.00	2.63	3.04	3.38	4.19
Proline	1.19	4.78	1.08	3.51	1.26	1.48	3.60	0.80
Valine‡	4.67	2.99	4.69	4.00	4.86	5.49	3.76	4.12
Leucine‡	6.03	4.33	6.34	6.80	6.72	6.36	4.99	5.45
Histidine‡	3.12	2.48	3.10	2.63	2.97	3.71	3.29	2.49
Arginine‡	12.78	8.53	12.67	12.13	15.99	12.46	9.67	23.06
Serine	4.88	4.14	5.00	6.21	4.33	5.37	8.31	7.31
Glycine	4.61	4.52	4.19	5.68	5.38	3.85	5.91	6.66
Methionine‡	0.67	1.59	1.45	1.14	1.22	0.21	0.63	1.46
Tyrosine	2.09	3.15	3.88	4.27	0.66	4.06	4.11	2.56
Lysine‡	4.86	8.09	3.13	2.10	2.27	3.12	2.65	8.43
Half-cystine	2.22	3.25	1.53	--	2.03	0.57	--	--
Tryptophan	2.05	--	1.16	--	2.49	1.31	--	0.07
Ammonia	2.14	2.48	2.27	5.29	3.43	2.18	6.76	2.01

† From Zarins and Cherry (1981).
‡ Essential amino acids.
§ Corresponding to numbers in Fig. 13–10.

13-6 UTILIZATION

13-6.1 Planting Seed Properties

The processes that result in germination are initiated by the uptake of water by the seed. When plotted against time, under laboratory conditions, the water content of delinted cottonseed increases sigmoidally during the first few hours (Dewez, 1964; Krieg and Bartee, 1975; Leffler and Williams, 1983). After an initial rapid increase in the rate of water uptake, further absorption occurs, but at a reduced rate (Dewez, 1964; Krieg and Bartee, 1975). This rate change indicates not only that water uptake is sustained, but also that it is probably regulated, either osmotically or metabolically. With many seeds, a hard-seed character can severely restrict the imbibitional hydration of cottonseed. In seed with this characteristic, the chalazal end is highly compacted and is effectively plugged by substances that are impermeable to water (Christiansen and Moore, 1959). This hard-seededness can be broken by subjecting the seed to a hot (85°C) water treatment (Walhood, 1956). The vast majority of cottonseed are not hard-seeded, however, so this character normally presents little problem in seedlots of cultivars destined for use as planting seed.

Upon hydration of the embryonic tissues, an environment is created within the embryo in which rapid metabolic activity can resume. The principal storage reserves in the cotyledons of the cotton embryo are proteins and oils; the initial phases of germinative metabolism are dominated by the mobilization of the oil reserves. Lipase action upon the glycerides of the embryo produces free fatty acids, low levels of which are detected during the early stages of germination (Christiansen and Moore, 1961; St. Angelo and Altschul, 1964). Only low levels of free fatty acids accumulate because of the rapid development of high amounts of glyoxylate cycle activity, through which carbohydrates are formed. These carbohydrates are utilized as substrates for the respiratory production of energy; they are also translocated, along with amino acids produced by proteolysis, to the developing axis.

Once germination has been initiated, sustained vigorous growth by the developing axis becomes the next critical period in the establishment of a cotton plant. Axis growth rates are influenced by the initial quality of the seed, the metabolism of the cotyledonary storage reserves, and the moisture and temperature regime of the seedbed. Selection of the seed within a seedlot that have an optimum density results in the maximum expression of axis growth (Leffler and Williams, 1983). Under favorable temperature conditions, radicle growth between 3 and 12 days was significantly correlated with the depletion of both lipid and non-lipid cotyledonary reserves (Krieg and Carroll, 1978). When temperatures were less than optimum, however, lipid use was correlated with radicle growth only between 3 and 6 days, while non-lipid utilization assumed primary importance between 6 and 12 days. Except under extreme moisture and temperature conditions, the axis growth rate is generally proportional to the seedbed temperature. Guinn (1965) reported that low temperatures suppressed root growth more than hypocotyl growth. Wanjura et al. (1967) identified a consistent relationship between the initial emergence of cotton seedlings and the cumulative number of hours, since planting, that soil temperature exceeded 18°C. Their data suggested the presence of a low temperature threshold for axis growth.

Although it is not the final step in the germination sequence, the emergence of the seedling is a decisive factor in the process of stand establishment. While the initiation of germination and rate of axis growth are fundamental processes associated with emergence, they are not entirely responsible for the success or failure of emergence. Among the other factors that often negatively affect the emergence of seedlings is the formation of a crust on the soil. The characteristics of the soil primarily related to formation of crust are the texture and organic matter content (Russell, 1957; Wilkes and Corley, 1968). The after-effects of a rain between planting and emergence can be detrimental to the establishment of a stand (Wanjura and Minton, 1981). Crusting poses a more serious problem in soils that are light-textured and low in organic matter (Wilkes and Corley, 1968). Soil crusting may be avoided by the selection of planting dates to minimize the likelihood of adverse weather and by the adjustment of the planting depth. Currently, there are two main ways to overcome soil-crusting problems: (a) the use of high-quality seed that emerge rapidly, before a crust fully forms, and (b) the use of implements, such as the rotary hoe, to destroy the crusted layer so that seed that germinate slowly can survive and produce seedlings that will emerge. Although seeding at excessive rates is practiced in some areas to overcome soil crusting problems, this practice introduces the potential problem of excessive stands when growing conditions are optimum.

Because of the importance of the germination capacity of a seedlot as a production resource, many attempts have been made to develop assays to estimate this quality. While each procedure undoubtedly measures certain aspects of seed quality, each also possesses deficiencies in the accurate prediction of the actual field performance of the seedlot (Delouche, 1981). The most often observed contrast between these laboratory procedures and field performance is that the former overestimates the latter; this contrast becomes more pronounced as field conditions become increasingly severe.

A commonly accepted reason for this divergence of results is that the standard germination test (AOSA, 1978) is conducted under highly favorable conditions that are not realized in field plantings. The temperature regimes accepted for the standard test are either (a) alternating cycles of 20° and 30°C, or (b) a constant 30°C. Because field plantings are often made before soil temperatures remain consistently above 20°C, even the alternating-temperature regime provides an unusually favorable environment for germination. This differential performance in these environments is indicative of the fact that the individual seed within a seedlot do not fall into discrete classes. Conversely, the field expression of germination potential by a given seedlot forms a continuous distribution that ranges from zero to a relatively high value. Whereas germination per se may, in fact, fall into discrete classes, field peformance is dictated by the success or failure of seedling emergence and survival. Survival is predicated upon an expression of vigorous axis growth under adverse environments.

Two vigor assays, the tetrazolium test and the Texas cool test, have been used in the evaluations of planting cottonseed quality (Delouche, 1981). The tetrazolium vigor test consists of a subjective interpretation of the tetrazolium viability test (Baskin, 1981c). With this test, seeds are classified as either non-germinable or low-, medium-, or high-vigor on the basis of the intensity and distribution of red stain and tissue turgidity after they have been immersed in a solution of 2,3,5-triphenyl tetrazolium chloride. Because of the subjective nature of this evaluation, however, interpretation and scoring of results are both time-consuming and difficult to make. Therefore, this quality test will probably not gain widespread adoption (Baskin, 1979). The Texas cool test, on the other hand, utilizes the same procedure as does the standard germination test, but it is conducted at 18°C and covers a period of either 6 days, for acid-delinted seeds, or 7 days, for either machine-delinted or gin-run seeds. Because of the similarities between the cool test and the standard test, the cool test will likely gain wider acceptance (Baskin, 1979).

Wanjura et al. (1969) measured the productivity of cotton plants that had emerged at different times; they found that the vast majority of the yield came from the first plants to emerge. In comparisons of fully and partially developed cottonseed, Ferguson and Turner (1971) found that the degree of development was positively associated with both emergence and early seedling vigor. A favorable association between the rate of initiation of germination and the rate of early seedling growth has been attributed to seed density selection (Tupper et al., 1970, 1971; Krieg and Bartee, 1975; Krieg and Carroll, 1978; King and Lamkin, 1979). More recently, Leffler and Williams (1983) found that the growth advantage developed by the earliest germinating seedlings increased throughout the first month of seedling growth. Similarly, both Turner and Ferguson (1972) and Minton and Supak (1980) reported that seed quality influenced yield in ways that extended beyond the direct influence on stand. Turner and Ferguson (1972) detected significant differences in seedling populations due to the degree of seed development; they then thinned plots to a uniform population and measured the yield from these plots. Those plots that had initially contained the highest population subsequently produced the highest yield, even after the plots had been adjusted to a common seedling population. Similarly, Leffler et al. (1978) identified significant effects of seed quality on both stand and yield; seed quality differences were associated with the portion of the season in which the seeds developed. Using seeds of selected densities, Minton and Supak (1980) identified influences of the seedlots on stand, disease susceptibility, and yield.

The growth of a cotton crop is continuously influenced by all the events that have affected the preceding phases of crop development. Plant populations are affected by seed quality and seedling growth rates. In addition, plant growth is influenced by plant density per unit area. Young seedlings growing in thick stands tend to grow taller than those in thinner stands. Light penetration into the canopy in thinner stands may cause the plants to effectively initiate reproductive development sooner than those in thicker stands. By altering both inter-row and intra-row spacings to maintain a constant total population density, Walhood and Johnson (1976) created significantly different competitive environments for seedling growth. Uniform spacing of plants ultimately resulted in yields 40% greater than those produced by plants grown in the "standard" planting pattern.

13–6.2 Food Functional Properties

Research has shown that cottonseed protein derivatives compete well with other vegetable protein products in food formulations (Meinke, 1952; Martinez et al., 1970; Lawhon and Cater, 1971; Lusas et al., 1977; El-Sayed et al., 1978a, b; Experience, Inc., 1978a; Spadaro and Gardner, 1979; Berardi and Cherry, 1979a; Terrell et al., 1979; Rosenblum, 1980; Manak et al., 1980; Cherry, 1983; Cherry and Berardi, 1983). Numerous potential

Table 13-13. Potential uses of cottonseed protein products in food and industrial products.†

Product	Function	Effect
Meat		
Ground beef; sausages (frankfurters and wieners); frozen meat patties; canned meats with sauces and casseroles; simulated meats	Extender and substitute as flours, concentrates, isolates, and texturized proteins; emulsifier; binder; fat and water absorption; thickening agents	Reduces stickiness, increases juiciness; improves texture, firmness, and snappiness; enhances oil and water absorptivity
Baked goods		
White bread; health and dietetic breads; donuts; biscuits; cookies; cakes; waffles, pancake mixes, crackers; milk replacers	Protein fortifier as flour concentrates, isolates and extrudates; color; flow control; dough conditioner; synergistic properties of composite proteins	Nutritional supplements; specialty products for dietetics; improves yellow color; maintains size uniformity; enhances texture
Pasta		
Macaroni, spaghetti and noodles	Protein fortifier as flours, concentrates and isolates; meat substitutes; color	Nutritional supplements; improves yellow color
Tortillas and tacos	Protein fortifier as flours, concentrates, and isolates	Nutritional supplements
Snack foods (potato chips, pretzels, etc.)	Protein fortifier as flours, concentrates and isolates; texture	Nutritional supplements; improves textural properties
Dairy		
Caseinate, albumin, and nonfat milk solids replacers; cheese substitutes; soft frozen desserts, puddings, ice cream and yogurts	Flours, concentrates, and isolates as milk replacers; rennin activity; substitutes and extenders; texture, body, color	Nutritional supplements; improves texture, body and color
Breakfast foods; diet and health foods	Protein fortifiers	Nutritional supplements
Beverages	Protein fortifiers; acid pH solubility	Nutritional supplements
Confections	Substitutes and extenders	Cocoa and chocolate substitutes
Industrial uses—calf milk replacers; pet foods; plywood glues	Substitutes; extrudates; protein fortifiers; extenders	Nutritional supplements; viscosity

† From Experience Incorporated (1978a).

uses were developed for cottonseed proteins in foods (Table 13-13). Functional properties, flavor, and color of cottonseed products meet consumer acceptance.

Cottonseed flour in water can be emulsified with oil to viscosities that simulate salad dressing and mayonnaise products (Cherry et al., 1978c, 1979a; Cherry and McWatters, 1981; McWatters and Cherry, 1981). Suspensions that are low in soluble protein (especially at the isoelectric point) do not form emulsions (Fig. 13-14). Mayonnaise-like products are formed with suspensions adjusted to low acidic or alkaline pH, where most storage proteins are soluble. Excellent foam volumes with good stabilities are obtained from protein suspensions adjusted to pHs between 3.0 and 3.5 (Fig. 13-15; Cherry et al., 1979a; Cherry and McWatters, 1981). The volumes of these foams are greater than those of egg white solids at comparable protein concentration and pH. Storage proteins produce greater foam volumes than the nonstorage fraction. The excellent whippability, and emulsifiability of cottonseed isolates at low pH may be partially attributed to many of the proteins being acid-dissociated. Between pH 4.5 and 8.0, the egg white standard performs better than the cottonseed proteins. Isolates from cottonseed flours produce cream-colored foams and emulsions at acid pH, while they are tan at alkaline pH. The capacity of cottonseed products to perform functionally in these ways suggests that they may be used to improve the protein content of food formulations such as frozen desserts, whipped toppings, and mayonnaise or salad dressing-type products.

Water and oil absorptivity of cottonseed flours are not appreciably affected by pH changes. Absorption values of 2.5 g of water or 1.5 g of oil per g of flour are noted (Berardi and Cherry, 1979a). Nonstorage and storage protein isolates respond differently in water and oil absorption tests when their pH vlaues are between 3 and 8. The storage protein isolate absorbs

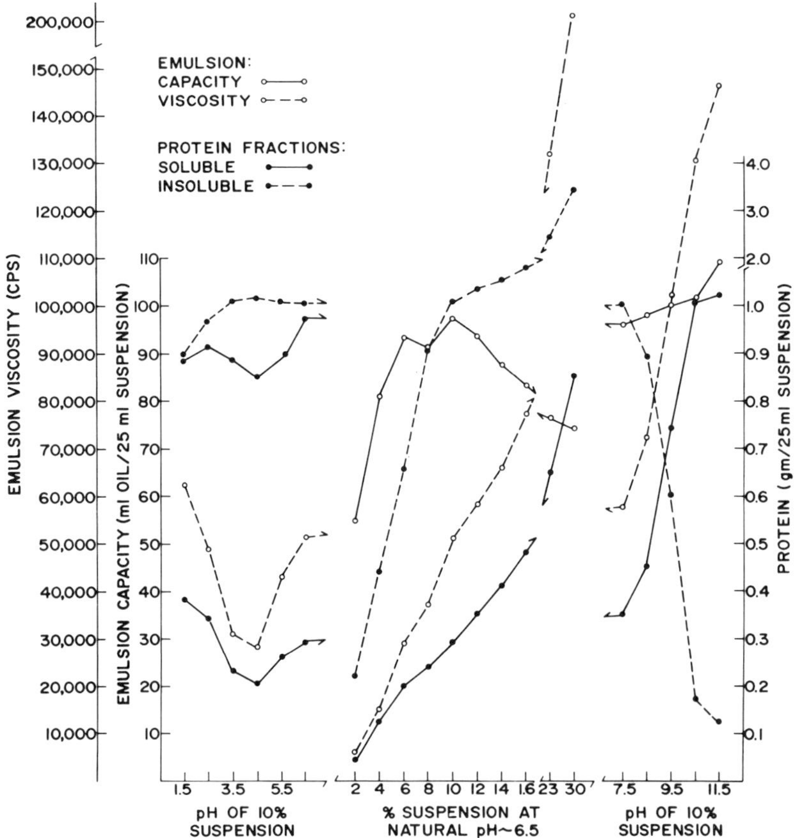

Fig. 13-14. Emulsion properties and content of soluble constituents related to pH and percentage of glandless cottonseed flour in aqueous suspensions (Cherry et al., 1978c).

more oil than water between pHs 3 and 5, but the reverse is true between pHs 5 and 8. The nonstorage proteins absorb more water than oil at all pHs. An example of performance differences among isolates is shown at neutral pH: 1 g of storage protein absorbs 1 g of water and 0.95 g of oil, and 1 g of nonstorage protein absorbs 3 g of water and 2 g of oil at about 25°C. Water binding capacity reflects the ability of flours and isolates to be incorporated into aqueous food formulations. Oil binding capacity determines the performance of cottonseed protein derivatives as meat analogs or extenders. Differences in water and oil absorptivity of cottonseed protein products are probably due to compositional variations (molecular weight, configuration, amino acid content, and solubility properties).

Minor structural changes of cottonseed proteins, through acylation with various acid anhydrides including acetic, dimethylglutaric, maleic, and succinic anhydrides, can improve extractability of proteins and certain

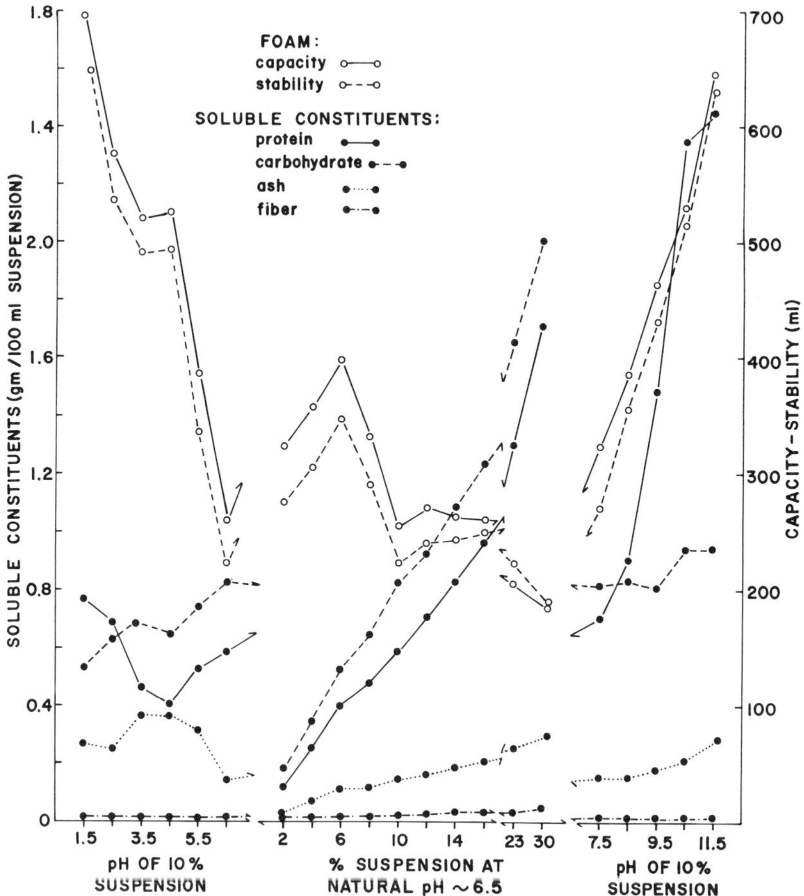

Fig. 13-15. Foam properties and content of soluble constituents related to pH and percentage of glandless cottonseed flour in aqueous suspensions (Cherry and McWatters, 1981).

functional properties of cottonseed products required for food application (Cherry and McWatters, 1981; Choi et al., 1981, 1982a). In the production of isolates, bulk density was decreased, which resulted in fluffy products. Succinylated glandless cottonseed protein isolates showed markedly improved characteristics (bulk density, whitening capacity, cream separation, and oil retention capacity) as coffee whiteners when compared to nonsuccinylated products (Choi et al., 1982b). Replacement of approximately 50% of the sodium caseinate with succinylated cottonseed protein isolate did not affect the quality of whiteners relative to that of 100% sodium caseinate-based whitener.

Spinning and thermoplastic extrusion processes further expand the potential utilization of cottonseed protein products as meat extenders (Taranto et al., 1975, 1978a, b; Cegla et al., 1978; Rhee et al., 1981). Berardi and Cherry (1980) and Cherry and Berardi (1982) developed a simple heat-stir procedure to produce texturized products and self-sustaining gels from cottonseed protein isolates. Storage protein isolates produce self-sustaining gels in 6% or higher suspensions adjusted to pHs between 3.5 and 4.0 or between 9.5 and 10.0 while being heat-stirred to 90°C. Texturized products form when the suspension is adjusted to pHs between 4.5 and 9.0. Classical isolates, which contain both nonstorage and storage proteins. Texturized proteins have a sponge-like texture; taste panelists have suspension prior to heating. The storage protein isolate absorbs up to 2.5 times its weight in water during heating, while producing gels or texturized proteins. Texturized proteins have a sponge-like texture; tase panelists have judged them to have properties like cooked hamburger in tests for chewiness and mouth feel. The texturized proteins can be dehydrated to provide a stable product, then rehydrated to their original forms when needed for food use.

Compared to all-meat controls, frankfurters made with 5% cottonseed proteins (glandless cottonseed flour and storage protein isolate, and LCP flour) generally had comparable pH values, cured color, skin firmness, texture, and desirability as judged by sensory panels (Terrell et al., 1981). At 10 to 15% levels of replacement, however, panels judged these properties to be less desirable. At all percentages of replacements, storage protein isolates were judged to be the most desirable. Further studies comparing 10% flour, 10% concentrate, and 10% isolate-extended beef patties with all-beef patties showed that the seed protein ingredients retarded oxidative rancidity development, improved cooked yields, retained less fat, and were not significantly different in sensory quality (Ziprin et al., 1981).

Physical tests (mix viscosity, overrun, penetrability, and Hunter color values) and sensory evaluations (texture, color, flavor, odor, and overall acceptance) showed that up to 40% of nonfat milk solids could be replaced with glandless cottonseed flour and storage protein isolate and LCP flour with satisfactory results in frozen desserts (Simmons et al., 1980). Of all ingredients tested, except glandless cottonseed storage protein isolate, acceptability and functionality in the desserts tended to fall at or above 40%

substitution. The glandless storage protein isolate might successfully replace up to 80% of the nonfat milk solids in frozen desserts. Protein isolates produced from glandless cottonseed flour by ultrafiltration with an industrial membrane system had functional properties equal or superior to those of a commercial soybean isolate (Manak et al., 1980). Incorporation of whey proteins into the membrane-separated isolates markedly enhanced their functionality. Cottonseed protein products have reasonably bland flavors. Flours and isolates of glandless cottonseed are lighter colored than those of corresponding LCP products. The color of all cottonseed protein derivatives is pH dependent; it is white in the acid pH range, but yellow to green at alkaline pH (Blouin et al., 1982).

Vegetable protein products of glandless and LCP cottonseed have the potential of becoming useful food ingredients. These products are (a) free from foreign matter, microorganisms, and toxic substances; (b) bland in flavor and light colored; (c) stable under most processing conditions; (d) processed at competitive costs; (e) nutritious; (f) functionally desirable; (g) available in various forms suitable for different food uses; (h) useful as additives in traditional as well as new food formulations; and (i) with unique marker characteristics that are distinguishable and readily detectable in food blends.

13–6.3 Other Uses

The seed fuzz that are not removed during the ginning operation are known as linters (Van Wyck, 1948). During preparation of cottonseed for processing into edible oil and flour, and animal feeds, the linters are removed from the seeds. Three major types of linters are recovered: first-cuts, 0.15 to 0.38 kg/10 kg of seeds; second cuts, 0.90 to 1.05 kg/10 kg of seeds; and mill-runs, 0.35 to 1.00 kg/10 kg of seeds. First-cut linters are used in surgical, paper, and packing products; second-cut linters, in chemical cellulose for preparation of regenerated fibers, films, lacquers, explosives, plastics, and paper; and mill-run linters in chemical cellulose and padding products (Dunning, 1948; Glade and Ghetti, 1979).

Cottonseed hulls are used primarily as roughage in livestock feed products, and have been used as fuel for oil mills, insulation material, soil conditioner, thickener for oil well drilling muds, filler for phenolic plastics to improve impact strength, cellulose for regenerated fiber production, and a source of xylose and furfural (Dunning, 1948). Cottonseed raffinose has been used in certain culture media (Dunning, 1948).

In addition to its use as a source of feed and vegetable protein products for food, cottonseed meal is a source of technical proteins (Bailey, 1948), and it has been used as an adhesive (Arthur, 1950; Arthur and Hogan, 1953a, b; Hoffman and Arthur, 1952; Hogan and Arthur, 1951a, b, 1952). Its proteins have been extruded to form fibrous material for use by the textile industry (Arthur et al., 1949).

13-7 NUTRITION

Guidelines for edible, high nutritional value cottonseed protein flours and related products were recently updated (Milner, 1980). Cottonseed, to be suitable for preparing edible, high nutritional value products, should have less than 1.0% foreign matter and 10.0% moisture, not more than 1.8% free fatty acids in the oil, and fewer than 5.0% discolored kernels. The total gossypol should not be subjected to conditions before processing that cause gossypol to bind to lysine. During storage, cottonseed should contain less than 12% moisture and stored below 50°C. The kernels should be free from insect damage and bacterial and mold contamination. Aflatoxin-free, high-quality cottonseed should be cleaned, delinted, and dehulled to produce meats. The meats should be suitably prepared for oil recovery employing either continuous screw-pressing, pre-press solvent extraction, or direct solvent extraction. Minimization of heat, moisture, and processing time is essential to maintain protein quality and to obtain a free gossypol content of not more than 0.045% in the meal. Processing methods must be engineered to provide flours with high nitrogen solubility, low lipid content, not more than 60 ppm hexane residue, less than 30 to 50 ppb total aflatoxins, and less than 0.2 ppm arsenic. In the USA, new installations or satellite plants operating under stringent sanitary conditions for food production appear to be the only avenue for preparing edible-grade cottonseed flours.

The nutritional value for cottonseed oil primarily provides food energy, essential fatty acids, particularly linoleic acid, and vitamin E (Cherry and Berardi, 1983). Cottonseed meal is used principally as a feed that provides proteins to ruminant livestock (Smith, 1970). Table 13–10 presents the compositions of cottonseed meals prepared by different processes. Extensive research, supported by practical experience, suggests that cottonseed meal, if heated to inactivate physiologically active free gossypol, can be a major oilseed supplemental protein source in nonruminant rations (Anon., 1966). Increasing amounts of currently produced commercial meals are being utilized for nonruminant rations, supplemented with iron salt (ferrous sulfate) to inactivate free gossypol.

When treated with ferrous sulfate, glanded cottonseed kernels containing 0.75% free gossypol can be safely used to replace 50% and 17% of the protein in rat and pig diets, respectively (Clawson et al., 1975a, b). Glandless cottonseed kernels fed as raw, cooked, or roasted ground flours at 20% of the diet caused no significant detrimental effects to the number of litters born, litter size, or weights of the young of sexually mature rats or their offspring (Reber and Pyhe, 1980; Reber, 1981). Growth and food consumption were similar among rat offspring in all treatments.

Food-grade cottonseed protein products were successfully used to rehabilitate malnourished infants in Central and South America (Bressani et al., 1961; Behar and Bressani, 1966; Bressani and Elias, 1969; Milner, 1969; Graham et al., 1969, 1970; Bressani et al., 1980; Schrimshaw, 1980; and Wise, 1980). In India, Srikantia and Sahgal (1968) found successful use of cottonseed flour in reducing protein caloric malnutrition when the free gossypol content was low. For example, Behar and Bressani (1966) and Bressani et al. (1980) showed that the nutritional value of a product identified as Incaperina Mixture No. 9 was of high quality and supported growth of children and infants.

Alford et al. (1977) and Thomas et al. (1979) demonstrated that cottonseed proteins maintained nitrogen balance in women when fed at the higher levels of protein intake consumed in the USA. Alford and Onley (1978) found that the minimum cottonseed protein required to maintain nitrogen balance in women was 0.106 g N per kg of body weight. Using the factor for estimating protein from nitrogen results in 0.66 g of protein per kg of body weight. This factor is greater than the FAO recommendation of 0.52 g for a safe level of protein intake required for adult women. Glandless cottonseed protein and LCP cottonseed values for the minimum required intakes agreed closely.

Careful processing of cottonseed products is important for the maintenance of protein quality (Lawhon and Cater, 1971; Harden and Yang, 1975; Olsen, 1973). Martinez and Hopkins (1975) reported on the calculated nutritive quality and protein efficiency ratio (PER) of protein isolates prepared by classical, selective precipitation, or selective extraction processes from glandless cottonseed and LCP flours. Isoleucine is the first limiting essential amino acid; threonine, lysine, and methionine are present in amounts that can be limiting in certain products.

Rats fed a diet containing casein had greater serum and high-density lipoprotein (HDL)-cholesterol concentrations as well as increased lecithin:cholesterol acyltransferosa activities than those fed a diet containing glandless cottonseed flour (Park and Liepa, 1982). Animals fed an arginine-supplemented casein diet to make its arginine-to-lysine ratio similar to that found in cottonseed protein showed a decrease in both serum and HDL-cholesterol when compared to the casein control group. The addition of lysine to the cottonseed protein diet to duplicate the arginine-to-lysine ratio of casein caused an increase in the same two cholesterol fractions. The hypothesis that a plant protein is hypocholesterolemic because of its ratio of arginine-to-lysine is supported by the results obtained with the feeding of cottonseed protein diets.

13–8 CONCLUSIONS

Cottonseed quality is the cumulative product of the plant's genetics, the physiology of the plant upon which development occurs, the environment to which the seed is subjected between boll opening and harvest, and the effects of the harvest and post-harvest practices. Although the post-maturation and pre-harvest environment is not usually subject to management influences, several factors can be manipulated to affect the physiological competence of the parental canopy and the quality–preservation efficiency of the harvesting and processing procedures. Since the quality and the quantity of the seed crop often appears to be inversely related, production of a cottonseed crop might reasonably be expected to incorporate different management practices than those for the production of fiber.

Interest in the potential of cottonseeds as a source of edible vegetable protein products has been stimulated by an increased understanding of constituents; e.g., protein, oil, etc., physicochemical, functional, and nutritional characteristics of food products. Further expansion in the processing and utilization of cottonseed products may be constrained by economic conditions rather than by limitations in functionality, nutritional quality, or consumer acceptability. Should changes occur to improve the competitive position of cottonseed, the potential contributions of their products may be realized.

REFERENCES

Adams, R., T. A. Geissman, and J. D. Edwards. 1960. Gossypol, a pigment of cottonseed. Chem. Rev. 60:555–574.

Alford, B. B., and K. Onley. 1978. The minimum cottonseed protein required for nitrogen balance in women. J. Nutr. 108:506–513.

----, S. Kim, and K. Onley. 1977. Nitrogen balance of women consuming cottonseed protein. J. Am. Oil Chem. Soc. 54:71–74.

Allan, E., A. R. Johnson, A. C. Fogerty, J. A. Pearson, and F. S. Shenstone. 1967. Inhibition by cyclopropene fatty acids of the desaturation of stearic acid in hen liver. Lipids 2:419–423.

Altschul, A. M. 1948. Biological processes of the cottonseed. p. 157–212. *In* A. E. Bailey (ed.) Cottonseed and cottonseed products. Interscience Publishers, Inc., N.Y.

----. 1958. Processed plant protein feedstuffs. Academic Press, N.Y.

Alyevand, F., G. Coleman, and D. R. Haisman. 1967. Catty odours in food: the reaction between mesityl oxide and sulfur compounds in food stuffs. Chim. Ind. (Paris) 37:1563–1564.

Anon. 1950. Cottonseed and its products. p. 16–17. 8th ed. Natl. Cottonseed Prod. Assoc. Inc., Memphis, Tenn.

----. 1966. Proceedings of a Conference on inactivation of gossypol with mineral salts. New Orleans, La. Natl. Cottonseed Prod. Assoc. Inc., Memphis, Tenn.

----. 1978a. Cottonseed and its products. p. 12–13. 8th ed. Natl. Cottonseed Prod. Assoc. Inc., Memphis, Tenn.

----. 1978b. Growth regulators for cotton. Agric. Res. 27:10–12.

----. 1979. Pix. BASF Aktiengesellschaft, D-6700 Ludwigshafen, Federal Republic of Germany.

----. 1980. Cottonseed quality goals. Natl. Cottonseed Prod. Assoc. Inc., Memphis, Tenn.

----. 1982a. World forecast: record oilseed production. J. Am. Oil Chem. Soc. 59:752A–766A.

----. 1982b. Cottonseed meal. Natl. Cottonseed Prod. Assoc. Inc., Memphis, Tenn.

Arthur, J. C. 1950. Stabilization of viscous dispersions of cottonseed proteins. U.S. Patent 2,531,383.

----, and J. T. Hogan. 1953a. Cottonseed meal glues of low alkalinity. U.S. Patent 2,646,408.

----, and ----. 1953b. Cottonseed meal glue. U.S. Patent 2,662,023.

----, and M. K. Karon. 1948. Preparation and properties of cottonseed protein dispersions. J. Am. Oil Chem. Soc. 25:99–102.

----, ----, A. F. Pomes, and A. M. Altschul. 1949. U.S. Patent 2,462,933.

Asatov, S. I., E. G. Yadgarov, T. S. Yunusov, and P. Kh. Yuldashev. 1978a. The globulins of cottonseeds. XVII. Primary structure of the carbohydrate containing submit C of the 11 S globulin. Khim. Prir. Soedin 4:541–542.

----, T. S. Yunusov, and P. Kh. Yuldashev. 1978b. The globulins of cottonseeds. XV. Primary structure of a glycopeptide of submit C 11 of the 11 S globulin. Khim. Prir. Soedin. 4:539–540.

----, ----, and ----. 1977. A study of the structure of the 11 S globulin from cottonseeds. XIV. Tryptic peptides of submit C. Khim. Prir. Soedin. 2:272–273.

Ashworth, L. J., and J. L. McMeans. 1966. Association of *Aspergillus flavus* and aflatoxins with a greenish yellow fluorescence of cottonseed. Phytopathology 56:1104–1105.

Association of Official Seed Analysts. 1978. Rules for testing seed. J. Seed Technol. 3:1–126.

Bailey, A. E. (ed.). 1948. Cottonseed and cottonseed products. Wiley, N.Y.

Baskin, C. C. 1979. Germination test is not enough. p. 53–54. Proc. Beltwide Cott. Prod. Mech. Conf., Phoenix, Ariz.

----. 1981a. Storage of seed cotton. p. 310–311. *In* J. M. Brown (ed.) Proc. Beltwide Cott. Prod. Mech. Conf., New Orleans, La.

----. 1981b. Storage of bulk cottonseed. p. 308–310. *In* J. M. Brown (ed.) Proc. Beltwide Cott. Prod. Res. Conf., New Orleans, La., Natl. Cotton Council, Memphis, Tenn.

----. 1981c. Tetrazolium evaluation of cottonseed. p. 312. *In* J. M. Brown (ed.) Proc. Beltwide Cott. Prod. Res. Conf., New Orleans, La., Natl. Cotton Council, Memphis, Tenn.

Behar, M., and R. Bressani. 1966. Experience in development of incaparina for the pre-school child. p. 213–219. *In* Proc. 1964 Conf. on Malnutrition in the Pre-school Child. Natl. Res. Coun.-Natl. Acad. Sci., Washington, D.C.

Benedict, C. R., R. J. Kohel, and M. A. Schubert. 1976. Transport of ^{14}C-assimilates to cottonseed: integrity of funiculus during seed filling stage. Crop Sci. 16:23–27.

Berardi, L. C., and J. P. Cherry. 1979a. Cottonseed protein derivatives and their potential food uses. Cott. Gin Oil Mill Press 89:14–16.

----, and ----. 1979b. Preparation and composition of co-precipitated protein isolates from cottonseed, soybean, and peanut flours. Cereal Chem. 56:95–100.

----, and ----. 1980. Textural properties of cottonseed proteins. J. Food Sci. 45:377–380, 387.

----, and ----. 1981. Functional properties of co-precipitated protein isolates from cottonseed, soybean, and peanut flours. Can. Inst. Food Sci. Technol. J. 14:283–288.

----, and V. L. Frampton. 1957. Note on gossypol and its relation to color fixation in cottonseed oil. J. Am. Oil Chem. Soc. 34:399–401.

----, and L. A. Goldblatt. 1980. Gossypol. p. 183–237. *In* I. Liener (ed.) Toxic constituents of plant foodstuffs. Academic Press, Inc., N.Y.

----, W. H. Martinez, and C. J. Fernandez. 1969. Cottonseed protein isolates: two-step extraction procedures. Food Technol. 23:75–82.

Bianchini, J. P., A. Ralaimanarivo, and E. M. Gaydow. 1982. Reversed-phase high-performance and liquid chromatography of fatty acid methyl esters with particular reference to cyclopropenoic and cyclopropanoic acids. J. High Resol. Chromat. Chromat. Comm. 5:199–204.

Bickerstaffe, R., and A. R. Johnson. 1972. The effect of intravenous infusions of sterculic acid on milk fat synthesis. Br. J. Nutr. 27:561–570.

Blouin, F. A., and J. P. Cherry. 1980. Identification of color-causing pigments in biscuits containing cottonseed flour. J. Food Sci. 45:953–957, 961.

----, Z. M. Zarins, and J. P. Cherry. 1981a. Role of flavanoids in the production of color in biscuits prepared with wheat and cottonseed flours. J. Food Sci. 46:266–271.

----, ----, and ----. 1981b. Color. p. 21–39. *In* J. P. Cherry (ed.) Protein functionality in foods. ACS Symposium Series 147. Am. Chem. Soc., Washington, D.C.

----, ----, and ----. 1982. Discoloration of proteins by binding with phenolic compounds. p. 67–91. *In* J. P. Cherry (ed.) Food protein deterioration: mechanisms and functionality. ACS Symposium Series 206. Am. Chem. Soc., Washington, D.C.

Boatner, C. H., C. M. Hall, M. L. Rollins, and L. E. Castillon. 1947. Pigment of glands of cottonseed. II. Nature and properties of gland walls. Bot. Gaz. 108:484–494.

Bressani, R., J. E. Braham, and L. G. Elias. 1980. Human nutrition and gossypol. Food Nutr. Bull. 2:24–32.

----, and L. G. Elias. 1969. Effect of pH on the free and total gossypol and nutritive value of cottonseed and protein concentrate. Arch. Latinomer. Nutr. 19:367–379.

----, ----, A. Aquirre, and N. S. Scrimshaw. 1961. All-vegetable protein mixtures for human feeding. III. The development of INCAP vegetable mixture nine. J. Nutr. 74:201–208.

----, ----, J. E. Braham, and M. Erales. 1969. Long-term rat feeding studies with vegetable mixtures containing cottonseed flour produced by different methods. J. Agric. Food Chem. 17:1135–1138.

Brown, C. L., G. W. Cathey, and C. Lincoln. 1962. Growth and development of cotton as affected by toxaphene-DDT, methyl parathion, and calcium arsenate. J. Econ. Entom. 55:298–301.

Caldwell, W. P. 1962. Relationship of pre-harvest environmental factors to seed deterioration in cotton. Ph.D. Dissertation. Mississippi State Univ., Mississippi State, Miss.

Canella, M., and G. Sodini. 1966. Extraction of gossypol and oligosaccharides from oilseed meals. J. Food Sci. 42:1218–1219.

Carter, M. E., J. P. Cherry, and P. A. Miller. 1979. Science and Education Administration's program on maintaining and improving the quality of cottonseed for processing. Oil Mill Gaz. 83:22–27.

Cathey, G. W. 1979. Harvest-aid chemicals and practices for cotton. Outlook Agric. 10:191–197.

Cegla, G. F., M. V. Taranto, K. R. Bell, and K. C. Rhee. 1978. Microscopic structure of textured cottonseed flour blends. J. Food Sci. 43:775–779.

Cherry, J. P. 1983. Cottonseed oil. J. Am. Oil Chem. Soc. 60:360–367.

----, and L. C. Berardi. 1982. Heat-stir denaturation of cottonseed proteins: texturization and gelation. p. 163–200. *In* J. P. Cherry (ed.) Food protein deterioration: mechanisms and functionality. ACS Symposium Series 206. Am. Chem. Soc., Washington, D.C.

----, and ----. 1983. Cottonseed. p. 187–256. *In* I. A. Wolff (ed.) CRC handbook of processing and utilization in agriculture, Vol. II: Part 2, Plant products. CRC Press, Inc., Boca Raton, Fla.

----, and M. S. Gray. 1981. Methylene chloride extraction of gossypol from cottonseed products. J. Food Sci. 46:1726–1733.

----, ----, and L. A. Jones. 1981a. A review of lecithin chemistry and glandless cottonseed as a potential commerical source. J. Am. Oil Chem. Soc. 58:903–913.

----, R. J. Kohel, L. A. Jones, and W. H. Powell. 1981b. Cottonseed quality: factors affecting feed and food uses. p. 266–283. *In* J. M. Brown (ed.) Proc. Beltwide Cott. Prod. Res. Conf., New Orleans, La. Natl. Cotton Council, Memphis, Tenn.

----, F. R. H. Katterman, and J. E. Endrizzi. 1970. Comparative studies of seed proteins of species of *Gossypium* by gel electrophoresis. Evolution 24:431–447.

----, and K. H. McWatters. 1981. Whippability and aeration. p. 149–176. *In* J. P. Cherry (ed.) Protein functionality in foods. ACS Symposium Series, No. 147. Am. Chem. Soc., Washington, D.C.

----, ----, and L. R. Beuchat. 1979a. Oilseed protein properties related to functionality in emulsions and foams. p. 1–26. *In* A. Pour-El (ed.) Functionality and protein structure. ACS Symposium Series No. 92. Am. Chem. Soc., Washington, D.C.

----, J. G. Simmons, A. H. Hyer, R. H. Garber, L. M. Carter, and H. B. Cooper. 1979b. Quality of module-stored cottonseed in California. p. 36–40. Proc. Beltwide Cott. Prod. Res. Conf., Phoenix Ariz. Natl. Cotton Council, Memphis, Tenn.

----, ----, R. J. Kohel, H. B. Cooper, M. Lehman, J. Dobbs, K. E. Fry, D. L. Kittock, and T. J. Henneberry. 1979c. Relationship of cottonseed quality to genetic and agronomic practices. Cott. Gin Oil Mill Press 80:18–21.

----, ----, and R. J. Kohel. 1978a. Potential for improving cottonseed quality by genetic and agronomic practice. p. 343–364. *In* M. Friedman (ed.) Nutritional improvement of food and feed proteins. Plenum Press, N.Y.

----, ----, and ----. 1978b. Cottonseed composition of national variety test cultivars grown at different Texas locations. p. 47–50. *In* J. M. Brown (ed.) Proc. Beltwide Cott. Prod. Res. Conf., Dallas, Texas. Natl. Cotton Council, Memphis, Tenn.

----, ----, Z. M. Zarins, J. I. Wadsworth, and C. H. Vinnett. 1978c. Cottonseed protein derivatives as nutritional and functional supplements in food formulations. p. 767–796. *In* M. Friedman (ed.) Nutritional improvement of food and feed proteins. Plenum Press, N.Y.

Choi, Y. R., E. W. Lusas, and K. C. Rhee. 1981. Succinylation of cottonseed flour: effect on the functional properties of protein isolates prepared from modified flour. J. Food Sci. 46:954–955.

----, ----, and ----. 1982a. Effects of acylation of defatted cottonseed flour with various acid anhydrides on protein extractability and functional properties of resulting protein isolates. J. Food Sci. 47:1713–1716.

----, ----, and ----. 1982b. Formulation of nondairy coffee whiteners with cottonseed protein isolates. J. Am. Oil Chem. Soc. 59:564–567.

Christiansen, M. N., and R. P. Moore. 1959. Seed coat structural differences that influence water uptake and seed quality in hard seed cotton. Agron. J. 51:582–584.

----, and ----. 1961. Temperature influence on the in vivo hydrolysis of cottonseed oil. Crop Sci. 1:385–386.

Ciegler, A., H. R. Burnmeister, R. F. Vesonder, and C. W. Hesseltine. 1981. Mycotoxins: occurrence in the environment. p. 1–50. *In* R. C. Shank (ed.) Mycotoxins and n-nitroso compounds: Environmental risks. CRC Press, Inc., Boca Raton, Fla.

Clawson, A. J., J. H. Maner, G. Gomez, Z. Flores, and J. Buitrago. 1975a. Unextracted cottonseed in diets for monogastric animals. II. The effect of boiling and oven vs. sun drying following pretreatment with a ferrous sulfate solution. J. Anim. Sci. 40:648–654.

----, ----, ----, O. Mejia, Z. Flores, and J. Buitrago. 1975b. Unextracted cottonseed in diets for monogastric animals. I. The effect of ferrous sulfate and calcium hydroxide in reducing gossypol toxicity. J. Anim. Sci. 40:640–647.

Colwick, R. F., T. H. Garner, G. D. Christenbury, G. B. Welch, R. L. Clark, J. C. Delouche, C. C. Baskin, J. W. Sorenson, L. H. Wilkes, N. K. Person, and H. W. Schroeder. 1972. Factors affecting cottonseed damage in harvesting and handling. ARS, USDA, Prod. Res. Rep. No. 135.

Conner, J. W., D. R. Krieg, and J. R. Gipson. 1972. Accumulation of simple sugars in cotton bolls as influenced by night temperatures. Crop Sci. 12:752–754.

Curley, R. G., R. A. Kepner, M. Hoover, U. U. McCutcheon, L. K. Stronberg, and E. A. Yeary. 1973. Seed cotton storage, an aid to both growers and ginners. Calif. Agric. 27:7–9.

Damaty, S., and B. J. F. Hudson. 1975. Preparation of low-gossypol cottonseed flour. J. Sci. Fd. Agric. 26:109–115.

Davis, R. G., L. S. Bird, A. Y. Chambers, R. H. Garber, C. R. Howell, E. B. Minton, R. Sterne, and L. F. Johnson. 1981. Control of seedling diseases. p. 17–20. *In* G. M. Watkins (ed.) Compendium of cotton diseases. Am. Phytopathol. Soc., St. Paul, Minn.

Delouche, J. C. 1981. Harvest and post-harvest factors affecting the quality of cotton planting seed and seed quality evaluation. p. 289–305. *In* J. M. Brown (ed.) Proc. Beltwide Cott. Prod. Res. Conf., New Orleans, La. Natl. Cotton Council, Memphis, Tenn.

Deutshoman, A. J., and L. S. Klaus. 1960. Spectrophotometric determination of sterculic acid. Anal. Chem. 32:1809–1810.

Dewez, J. 1964. Water uptake and heat evolution by germinating cotton seed. Plant Physiol. 39:240–244.

Dieckert, J. W. and M. C. Dieckert. 1972. The deposition of vacuolar proteins in oilseeds. p. 52–85. *In* G. E. Inglett (ed.) Seed proteins. AVI Publishing Co., Inc., Westport, Conn.

----, and ----. 1976. The chemistry and cell biology of the vacuolar proteins of seeds. J. Food Sci. 41:475–482.

----, and ----. 1978. The comparative anatomy of the principal reserve proteins of seeds. *In* Abl. Akad. Wiss. DDR., Abt. Math., Naturwiss., Tech. p. 73–86.

----, R. W. Wallace, and M. C. Dieckert. 1981. Chemistry and biology of the cottonseed globulins. p. 351–355. *In* J. M. Brown (ed.) Proc. Beltwide Cott. Prod. Res. Conf., New Orleans, La. Natl. Cotton Council, Memphis, Tenn.

Dunning, J. W. 1948. Cotton hulls. p. 873–893. *In* A. E. Bailey (ed.) Cottonseed and cottonseed products: their chemistry and chemical technology. Interscience, N.Y.

Eickhoff, W. D., and M. H. Willcutt. 1978. Guidelines—seed cotton modules. p. 58–63. *In* Proc. West Cotton Prod. Res. Conf., Phoenix, Ariz.

Elmore, C. D., and H. R. Leffler. 1976. Development of cotton fruit. III. Amino acid accumulation in protein and nonprotein nitrogen fractions of cottonseed. Crop Sci. 16: 867–871.

El-Nockrashy, A. S., H. M. Mostafa, M. M. El-Fouly, and V. El-Shattory. 1976. Biochemical changes in cottonseed during development and maturity. Nahrung 20:125–132.

El-Sayed, K., A. E. Salem, and A. A. Abdel-Bary. 1978a. Utilization of cotton flour in human food. I. Effect of cotton flour on physical properties of wheat flour dough. Alex. J. Agric. Res. 26:327–332.

----, ----, and ----. 1978b. Utilization of cotton flour in human food. II. Effect of cotton protein fractions on the physical properties of wheat flour dough. Alex. J. Agric. Res. 26:609–617.

El-Zik, K. M., and L. S. Bird. 1969. Inheritance of final seedling stand ability and related components in cotton. p. 119–120. *In* J. M. Brown (ed.) Proc. Beltwide Cott. Prod. Res. Conf., New Orleans, La. Natl. Cotton Council, Memphis, Tenn.

Engleman, E. M. 1966. Ontogeny of aleurone grains in cotton embryo. Am. J. Bot. 53:231–237.

Experience Incorporated. 1978a. Feasibility study to determine specific market opportunities for cottonseed flour, fractions, and derivatives in domestic markets. Contract No. 12-14-1001-1235, Phase I. Experience, Inc., Minneapolis, Minn.

----. 1978b. The economic feasibility of modifying and operating the Lubbock cottonseed flour plant to supply potential domestic markets. Contract No. 12-14-1004-1285, Phase II. Experience, Inc., Minneapolis, Minn.

Ferguson, D., and J. H. Turner. 1971. Influence of unfilled cotton seed upon emergence and vigor. Crop Sci. 11:713–715.

Feuge, R. O., Z. M. Zarins, J. L. White, and R. L. Holmes. 1969. Titration of cyclopropene esters with hydrogen bromide. J. Am. Oil Chem. Soc. 46:185–188.

Fisher, G. S., and J. P. Cherry. 1983. Variation of cyclopropenoid fatty acids in cottonseed lipids. Lipids (in press).

----, and W. H. Schuller. 1981. Gas chromatographic analysis of cyclopropenoid acids in cottonseed oils. J. Am. Oil Chem. Soc. 58:943–946.

Fontaine, T. D., S. B. Detwiler, and G. W. Irving, Jr. 1945. Imrpovement in the color of peanut and cottonseed proteins. Ind. Eng. Chem. 37:1232–1236.

----, G. W. Irving, Jr., and K. S. Markley. 1946. Peptization of peanut and cottonseed proteins. Effects of dialysis and various acids. Ind. Eng. Chem. 38:658–662.

Freeman, D. W., R. S. Kadan, G. M. Ziegler, and J. J. Spadaro. 1979. Processing factors affecting air classification of defatted cottonseed flour for production of edible protein products. Cereal Chem. 56:452–454.

Gardner, H. K., R. J. Hron, and H. L. E. Vix. 1976. Removal of pigment glands (gossypol) from cottonseed. Cereal Chem. 53:549–560.

Gillham, F. E. M. 1969. Cotton in a hungry world. S. Afr. J. Sci. 65:173–179.

Gipson, J. R., and H. E. Joham. 1969. Influence of night temperature on growth and development of cotton (*Gossypium hirsutum* L.) IV. Seed quality. Agron. J. 61:365–367.

----, and L. L. Ray. 1970. Temperature-variety interrelationships in cotton. 1. Boll and fiber development. 2. Seed development and composition. Cott. Grow. Rev. 47:247–271.

Glade, E. H., and J. L. Ghetti. Feb. 1979. Collecting gin waste for sale. p. 24–44. *In* Cotton and wool situation. USDA-ESCS Rep. CWS-18. Washington, D.C.

Goldblatt, L. A. 1969. Aflatoxin: scientific background, control, and implications. Academic Press, N.Y.

Graham, G. G., E. Morales, G. Acevedo, J. M. Baertl, and A. Cordano. 1969. Dietary protein quality in infants and children. II. Metabolic studies with cottonseed flour. Am. J. Clin. Nutr. 22:577–587.

————, ————, ————, ————, and ————. 1970. Dietary protein quality in infants and children. III. Prolonged feeding of cottonseed flour. Am. J. Chem. Nutr. 23:165–169.

Grindley, D. N. 1950. Changes in composition of cottonseed during development and ripening. J. Sci. Food Agric. 1:147–151.

Guinn, G. 1965. Root development as affected by seed quality and environment. p. 9–11. Proc. Beltwide Cott. Prod. Mech. Conf., Atlanta, Ga.

Halloin, J. M. 1981. Weathering: changes in planting seed quality between ripening and harvest. p. 286–289. *In* J. M. Brown (ed.) Proc. Beltwide Cotton Prod. Res. Conf., New Orleans, La. Natl. Cotton Council, Memphis, Tenn.

————. 1982. Localization and changes in catechin and tannins during development and ripening of cottonseed. New Phytol. 90:651–657.

————, and A. A. Bell. 1979. Production of nonglandular terpenoid aldehydes within diseased seeds and cotyledons of *Gossypium hirsutum* L. J. Agric. Food Chem. 27:1407–1409.

Harden, M. L., and S. P. Yang. 1975. Protein quality and supplementary value of cottonseed flour. J. Food Sci. 40:75–77.

Harrison, J. R. 1977. Review of extraction process: emphasis cottonseed. Oil Mill Gaz. 82:16–24.

Hess, D. C. 1976. Prospects for glandular cottonseed. Oil Mill Gaz. 82:22–26.

————. 1977a. Genetic improvement of gossypol-free cotton varieties. Cereal Foods World 22:98–105.

————. 1977b. Selecting for increased seed density in cotton. p. 84–86. *In* J. M. Brown (ed.) Proc. Beltwide Cott. Prod. Res. Conf., Atlanta, Ga. Natl. Cotton Council, Memphis, Tenn.

Hoffman, W. H., and J. C. Arthur. 1952. Cottonseed meal as a tire cord-to-rubber adhesive. Rubber Agric. 71:354–356.

Hogan, J. T., and J. C. Arthur. 1951a. Preparation and utilization of cottonseed meal glue for plywood. J. Am. Oil Chem. Soc. 28:20–23.

————, and ————. 1951b. Cottonseed and peanut meal glues: permanence of plywood glue joints as determined by interior and exterior accelerated cyclic service tests. J. Am. Oil Chem. Soc. 28:272–274.

————, and ————. 1952. Cottonseed and peanut meal glues. Resistance of plywood bonds to chemical reagents. J. Am. Oil Chem. Soc. 29:16–18.

Hunter, R. E., and J. T. Presley. 1963. Morphology and histology of pinched root tips of *Gossypium hirsutum* L. seedlings grown from deteriorated seeds. Can. J. Plant Sci. 43:146–150.

Ibragimov, A. P., P. Tursunbaev, A. V. Tuichiev, and Sh. Yunuskhanov. 1974. Study of 7S globulin subunits from cotton seeds. Biokhimiya 39:225–229.

Johnson, A. R., J. A. Pearson, F. S. Shenstone, and A. C. Fogrty. 1967. Inhibition of stearic to oleic acid by cyclopropene fatty acids. Nature 214:1244–1245.

Jones, D. B., and F. A. Csonka. 1925. Proteins of the cottonseed. J. Biol. Chem. 64:673–683.

Jones, L. A. 1981. Natural antinutrients of cottonseed protein products. p. 77–98. *In* R. L. Ory (ed.) Antinutrients and natural toxicants in foods. Food and Nutrition Press, Inc., Westport, Conn.

Kadan, R. S., D. W. Freeman, G. M. Ziegler, and J. J. Spadaro. 1979. Air classification of defatted, glanded cottonseed flours to edible protein product. J. Food Sci. 44:1522–1524.

————, ————, ————, and ————. 1980. Protein displacement during classification of glanded cottonseed. J. Food Sci. 45:1566–1569, 1572.

Kajimoto, G., H. Yoshida, A. Shibahara, and S. Yamashoji. 1979. Changes in the composition of lipids and fatty acids in cottonseeds during maturation. Nippon Nageikagaku Kaishi 53:317–320.

Karavaeva, N. N. 1972. Study of structure of cottonseed globulin A. Biochemistry [Translation] Biokhimiya 37:603–608.

Kepner, R. A., and R. G. Curley. 1976. Handling seed cotton modules without pellets. Costs of hauling from field to storage. Calif. Agric. 30:6–8.

King, E. E., and G. E. Lamkin. 1979. Uniform quality cottonseed for laboratory and field use. p. 32. *In* J. M. Brown (ed.) Proc. Beltwide Cott. Prod. Res. Conf., Phoenix, Ariz. Natl. Cotton Council, Memphis, Tenn.

Kircher, H. W. 1964. The addition of mercaptans to methyl sterculate and sterculene: a hypothesis concerning the nature of the biological activity exhibited by cyclopropene derivatives. Am. Oil Chem. Soc. 41:4–8.

Kohel, R. J., and J. P. Cherry. 1983. Variation of cottonseed quality with stratified harvests. Crop Sci. 23:1119–1124.

Krieg, D. R., and S. N. Bartee. 1975. Cottonseed density: associated germination and seedling emergence properties. Agron. J. 67:343–347.

————, and J. D. Carroll. 1978. Cotton seedling metabolism as influenced by germination temperature, cultivar, and seed physical properties. Agron. J. 70:21–25.

————, J. R. Gipson, and L. W. Barnes. 1973. The chemical composition of cotton bolls and seeds developed under controlled night temperatures. p. 42–45. *In* J. M. Brown (ed.) Proc. Beltwide Cott. Prod. Res. Conf., Phoenix, Ariz. Natl. Cotton Council, Memphis, Tenn.

Kromer, G. W. 1977. Current status and future market potential for cottonseed. p. 1–9. *In* Glandless cotton: its significance, status, and prospects. Proc. Conf. ARS, USDA, Dallas, Texas.

Kuchenkova, M. A., E. F. Redina, N. L. Ovchimnikova, and P. Kh. Yuldashev. 1977. The globulins of cottonseeds. XII: The structure of the 7S globulin. Chem. Nat. Compd. (USSR) 13:570–572.

————, ————, and P. Kh. Yuldashev. 1981. A study of the globulin of cottonseeds. Isolation and characterization of a glycopeptide of the 7S globulin. Chem. Nat. Compd. (USSR) 17:171–173.

Lawhon, J. T., and C. M. Cater. 1971. Effect of processing method on the yields and functional properties of protein isolates from glandless cottonseed. J. Food Sci. 36:372–377.

————, ————, and K. F. Mattil. 1976. Evaluation of the food use potential of sixteen varieties of cottonseed. J. Am. Oil Chem. Soc. 54:75–80.

————, D. W. Hensley, D. Mulsow, and K. F. Mattil. 1978. Optimization of protein isolate production from soy flour using industrial membrane systems. J. Food Sci. 43:361–364, 369.

————, S. H. C. Lin, D. W. Hensley, C. M. Cater, and K. F. Mattil. 1977b. Processing whey-type byproduct liquids from cottonseed protein isolation with ultrafiltration and reverse osmosis membranes. J. Food Process. Eng. 1:15–35.

————, L. J. Manak, and E. W. Lusas. 1980. An improved process for isolation of glandless cottonseed protein using industrial membrane systems. J. Food Sci. 45:197–199, 203.

————, D. Mulsow, C. M. Cater, and K. F. Mattil. 1977a. Production of protein isolates and concentrates from oilseed flour extracts using industrial ultrafiltration and reverse osmosis. J. Food Sci. 42:389–394.

Lee, L. S., and L. A. Goldblatt. 1981. Contributions of Walter A. Pons, Jr., to development of methodology for mycotoxins. 1981. J. Am. Oil Chem. Soc. 58:928A–930A.

————, and T. E. Russell. 1981. Distribution of aflatoxin-containing cottonseed within intact locks. J. Am. Oil Chem. Soc. 58:27–29.

Leffler, H. R. 1976. Development of cotton fruit. I. Accumulation and distribution of dry matter. Agron. J. 68:855–857.

————. 1980. Leaf growth and senescence. p. 133–142. *In* J. D. Hesketh and J. W. Jones (ed.) Predicting photosynthesis for economic models. Vol. II. CRC Press, Inc., Boca Raton, Fla.

————, and R. D. Williams. 1983. Seed density classification influences germination and seedling growth of cotton. Crop Sci. 23:161–165.

————, C. D. Elmore, and J. D. Hesketh. 1977. Seasonal and fertility-related changes in cottonseed protein quantity and quality. Crop Sci. 17:953–956.

----, W. R. Meredith, Jr., and J. M. Chandler. 1978. Canopy physiology and seasonal patterns affect the composition and yield capacity of cottonseed. Agron. Abst. 70:79-80.

Lusas, E. W., J. T. Lawhon, S. P. Clark, S. W. Matlock, W. W. Meinke, D. W. Mulsow, K. C. Rhee, and P. J. Wan. 1977. Potential for edible protein products from glandless cottonseed. p. 31-43. *In* Glandless cotton: its significance, status, and prospects. Proc. Conf., ARS, USDA, Dallas, Texas.

Maleki, P. 1966. Microenvironmental influence on cottonseed deterioration in the field. M.S. Thesis, Mississippi State Univ., Mississippi State, Miss.

Manak, L. J., J. T. Lawhon, and E. W. Lusas. 1980. Functioning potential on soy, cottonseed, and peanut protein isolates produced by industrial membrane systems. J. Food Sci. 45:236-238, 245.

Markman, A. L., and V. P. Rzhekhin. 1965. Gossypol and its derivatives. IPST Press Binding: Wiener Bindery Ltd., Jerusalem, Israel.

Marshall, H. F., M. A. Shirer, and J. P. Cherry. 1984. Characterization of glandless cottonseed storage proteins by SDS-polyacrylamide gel electrophoresis. Cereal Chem. (in press).

Martinez, W. H. 1979. Functionality of vegetable proteins other than soy. J. Am. Oil Chem. Soc. 56:280-284.

----, L. C. Berardi, and L. A. Goldblatt. 1970. Potential of cottonseed: products, composition, and use. p. 248-261. Third Internat. Congr. Food Sci. and Technol. SOS/70 Proceed. Instit. of Food Technol., Chicago, Ill.

----, and D. T. Hopkins. 1975. Cottonseed protein products: variation in protein quality with product and process. p. 355-374. *In* M. Friedman (ed.) Nutrition quality of foods and feed. Part II. Quality factors plant breeding, composition, processing, and anti-nutrients. Marcel Dekker, N.Y.

Mathews, J. L. 1981. World and U.S. oilseeds and products outlook. p. 183. *In* Agric. Outlook, Commit. on Agric., Nutr., and Forestry. U.S. Government Printing Office, Washington, D.C.

Mayne, F. C., J. A. Harris, R. A. Pittman, and E. L. Skau. 1966. Methods for the determination of cyclopropenoid fatty acids. VII. The dilution HBr-titration technique as a general method. J. Am. Oil Chem. Soc. 3:319-524.

McMeans, J. L., V. T. Walhood, and L. M. Carter. 1966. Effects of greenpick, defoliation, and desiccation practices on quality and yield of cotton, *Gossypium hirsutum.* Agron. J. 58:91-94.

McMichael, S. C. 1959. Hopi cotton, a source of cottonseed free of gossypol pigments. Agron. J. 51:630.

----. 1960. Combined effects of glandless genes gl_2 and gl_3 on pigment glands in the cotton plant. Agron. J. 52:385-386.

McWatters, K. H., and J. P. Cherry. 1981. Emulsification: vegetable proteins. p. 217-242. *In* J. P. Cherry (ed.) Protein functionality in foods. ACS Symposium Series No. 147. Am. Chem. Soc., Washington, D.C.

Meinke, W. W. 1952. Food uses for cottonseed. Texas A&M College System Res. Rep. No. 34. Texas A&M Univ., College Station.

Meredith, W. R., Jr., and R. R. Bridge. 1973. Yield, yield component and fiber property variation of cotton (*Gossypium hirsutum* L.) within and among environments. Crop Sci. 13:307-312.

Milner, M. 1965. Problems and opportunities in utilizing cottonseed proteins to meet world protein deficits. p. 8-13. *In* Proc. 7th Conf. Cottonseed Protein Conc. USDA ARS Rep. 72-38.

----. 1969. Status of development and use of some unconventional proteins. p. 97-104. *In* M. Milner (ed.) Protein-enriched cereal foods for world needs. Assoc. Cereal Chem., St. Paul, Minn.

----. 1980. Guideline for edible cottonseed protein flours and related products. Food Nutr. Bull. 2:51-54.

Minton, E. B., and J. R. Supak. 1980. Effects of seed density on stand, *Verticillium* wilt, and seed and fiber characteristics. Crop Sci. 20:345-347.

Morgan, P. W. 1979. Plant growth regulators. Texas Agric. Progr. 25:22–23.

Newberne, P. M., and A. E. Rogers. 1981. Animal toxicity of major environmental myco-toxins. p. 51–106. *In* R. C. Shank (ed.) Mycotoxins and n-nitroso compounds: environ-mental risks. CRC Press, Inc., Boca Raton, Fla.

Olsen, R. L. 1973. Evaluation of LCP cottonseed and flour. Oil Mill Gaz. 77:7–8.

Osborne, J. B., and C. W. Voorhees. 1894. The proteins of cottonseed. J. Am. Chem. Soc. 14:778–785.

Ovchinnikova, N. L., M. A. Kuchenkova, T. D. Kasymova, and P. Kh. Yuldashev. 1977. The globulins of cottonseed. XI. The structure of the chymotryptic peptides of the 7S globulin. Chem. Nat. Compd (USSR) 13:566–569.

Pandey, S. N., and N. Thejappa. 1975. Study on the relationship between oil, protein, and gossypol in cottonseed kernels. J. Am. Oil Chem. Soc. 52:312–315.

Park, M.-S., and G. U. Liepa. 1982. Effects of dietary protein and amino acids on the meta-bolism of cholesterol-carrying lipoproteins in fats. J. Nutr. 112:1892–1898.

Paxton, K. W., and D. L. Roberts. 1973. Economic feasibility of storing seed cotton in modules under Louisiana conditions. Field storage processing costs. Dep. Agric. Econ. Agribus., Los Angeles State Univ., Agric. Mech. Coll. Agric. Exp. Stn. Rep. No. 456.

Peacock, H. A., and B. S. Hawkins. 1970. Effect of seed source on seedling vigor, yield, and lint characteristics of upland cotton, *Gossypium hirsutum* L. Crop Sci. 10:667–670.

Phelps, R. A., J. P. Cherry, A. H. Hyer, G. C. Cavanaugh, G. A. Harper, J. E. Jernigan, D. C. Hess, and D. G. Lorance. 1979. Advancements in cottonseed quality—industry's viewpoint. Cott. Gin Oil Mill Press 80:14–15.

————, F. S. Shenstone, A. R. Kemmerer, and R. J. Evans. 1965. A review of cyclopropenoid compounds: biological effects of some derivatives. Poult. Sci. 44:358–394.

Podgornov, G. M., M. A. Kuchenkova, and P. Kh. Yuldashev. 1974. Cottonseed globulins. 2. N-terminal amino acids. Chem. Nat. Compd. (USSR) 10:855.

Pons, W. A. 1977. Gossypol analysis; past and present. J. Assoc. Off. Anal. Chem. 60:252–259.

————, L. C. Berardi, and V. L. Frampton. 1959. Kinetic study of gossypol fixation in cotton-seed oil. J. Am. Oil Chem. Soc. 36:337–339.

————, and P. H. Eaves. 1971. Aqueous acetone extraction of cottonseed. U.S. Patent 3,557,168.

————, and A. O. Franz. 1978. High pressure liquid chromatographic determination of afla-toxins in peanut products. J. Assoc. Off. Anal. Chem. 61:793–800.

Pope, O. A., and J. O. Ware. 1945. Effect of variety, location, and season on oil, protein, and fuzz of cottonseed and on fiber problems of lint. USDA Tech. Bull. 903.

Ramey, H. H. 1974. A systems analysis approach to establishing research objectives for con-trol of aflatoxin production in cottonseed. Phytopathology 64:1451–1454.

Reber, E. F. 1981. Safety of low gossypol cottonseed kernels. J. Food Sci. 46:593–596.

————, and R. E. Pyke. 1980. Sub-acute toxicity studies of glandless cottonseed kernels fed to rats. J. Food Safety 2:87–95.

Redina, E. F., M. A. Kuchenkova, L. G. Petrosyan, and P. Kh. Yuldashev. 1977. The globulins of cottonseeds. X. Structures of tryptic peptides of the 7S globulin. Chem. Nat. Compd. (USSR) 13:563–565.

Rhee, K. C., C. K. Kuo, and E. W. Lusas. 1981. Texturization. p. 51–88. *In* J. P. Cherry (ed.) Protein functionality in foods. ACS Symposium Series No. 147. Am. Chem. Soc., Wash-ington, D.C.

Roark, B., T. R. Pfrimmer, and M. E. Merkl. 1963. Effects of some formulations of methyl parathion, toxaphene, and DDT on the cotton plant. Crop Sci. 3:338–341.

Roberts, D. L., K. W. Paston, and R. L. Gill. 1973. The module system: a new approach in storing seed cotton. Louisiana Agric. 16:4–5.

Roehm, J. N., D. J. Lee, J. H. Wales, S. D. Polityka, and P. D. Sinnhuber. 1970. The effect of the dietary sterculic acid on the hepatic lipids of rainbow trout. Lipids 5:80–84.

Roncadori, R. W., O. L. Brooks, and C. E. Perry. 1972. Effect of field exposure on fungal invasion and deterioration of cotton seed. Phytopathology 62:1137–1139.

----, S. M. McCarter, and J. L. Crawford. 1971. Influence of fungi on cotton seed deterioration prior to harvest. Phytopathology 61:1326-1328.

Rosenblum, D. 1980. Glandless cottonseed opens new route for developing protein-fortified foods. Food Prod. Develop. 14:34, 38, 40, 72.

Rossi-Fanelli, A., E. Antonini, M. Brunori, M. R. Bruzzesi, A. Caputo, and F. Satriana. 1964. Isolation of a monodisperse protein fraction from cottonseeds. Biochem. Biophys. Res. Commun. 15:110-115.

Russell, M. B. 1957. Physical properties. p. 31-38. *In* A. Stefferud (ed.) Soils. USDA Yearbk Agric., Washington, D.C.

Russell, T. E., T. F. Watson, and G. F. Ryan. 1976. Field accumulation of aflatoxin in cottonseed as influenced by irrigation termination dates and pink bollworm infestation. Appl. Environ. Microbiol. 31:711-713.

St. Angelo, A. J., and A. M. Altschul. 1964. Lipolysis and the free fatty acid pool in seedlings. Plant Physiol. 39:880-883.

Scrimshaw, N. S. 1980. A look at the Incaparina experiment in Guatemala. The background and history of Incaparina. Food Nutr. Bull. 2:1-2.

Shadrina, T. Y., T. S. Yunusov, and P. Kh Yuldashev. 1979. A study of the globulins of cottonseed. XX. Stability of the quaternary structure of the 11S globulin. Khim. Prir. Soedin. 4:554-557.

Sheehan, E. T., D. L. Schneider, and M. G. Vavich. 1972. Improved halphen method for measuring cyclopropenoid fatty acids. J. Agric. Food Chem. 22:119-121.

Simmons, R. G., J. R. Green, C. A. Payne, P. J. Wan, and E. W. Lusas. 1980. Cottonseed and soy protein ingredients in self-serve frozen desserts. J. Food Sci. 45:1505-1508.

Simpson, D. M., and B. M. Stone. 1935. Viability of cottonseed as affected by field conditions. J. Agric. Res. 50:435-477.

Smith, F. H. 1972. Effect of gossypol bound to cottonseed protein on growth of weanling rats. J. Agric. Food Chem. 20:803-804.

Smith, K. J. 1970. Practical significance of gossypol in food formulation. J. Am. Oil Chem. Soc. 47:448-450.

Sood, D. R., V. Kumar, and K. S. Dhindsa. 1976. Composition of cottonseed as affected by N, P, and K application. Agrochimica 20:77-81.

Spadaro, J. J., and H. K. Gardner. 1979. Food uses for cottonseed protein. J. Am. Oil Chem. Soc. 56:422-424.

Srikantia, S. G., and S. Sahgal. 1968. Use of cottonseed protein in protein-calorie malnutrition. Am. J. Clin. Nutr. 21:212-216.

Stansbury, M. F., C. L. Hoffpauir, and T. H. Hopper. 1953a. Influence of variety and environment on the iodine value of cottonseed oil. J. Am. Oil Chem. Soc. 30:120-123.

----, A. F. Cucullu, and G. T. Den Hartog. 1954. Cottonseed contents variation: influence of variety and environment on oil content of cottonseed kernels. Agric. Food Chem. 2: 692-696.

Stansbury, M. F., W. A. Pons, and G. T. Den Hartog. 1956. Relations between oil, nitrogen, and gossypol in cottonseed kernels. J. Am. Oil Chem. Soc. 33:282-286.

----, ----, and C. L. Hoffpauir. 1953b. Phosphorus compounds in cottonseed kernels. Influence of variety of cottonseed and environment. Agric. Food Chem. 1:75-78.

Stewart, J. Mc D. 1980. Integrated developmental events and their response to environment in cotton bolls. p. 322-341. *In* J. M. Brown (ed.) Proc. Beltwide Cott. Prod. Res. Conf., St. Louis, Mo. Natl. Cotton Council, Memphis, Tenn.

Taranto, M. V., G. F. Cegla, and K. C. Rhee. 1978a. Morphological, ultrastructural and rheological evlauation of soy and cottonseed flours texturized by extrusion and nonextrusion processing. J. Food Sci. 43:973-979, 984.

----, ----, K. R. Bell, and K. C. Rhee. 1978b. Textured cottonseed and soy flours micro scopic analysis. J. Food Sci. 43:767-771.

----, W. W. Meinke, C. M. Cater, and K. F. Mattil. 1975. Parameters affecting production and character of extrusion texturized defatted glandless cottonseed meal. J. Food Sci. 40: 1264-1269.

Terrell, R. N., J. A. Brown, Z. L. Carpenter, K. F. Mattil, and C. W. Monagle. 1979. Effects of oilseed proteins, at two replacement levels on chemical, sensory, and physical properties of frankfurters. J. Food Sci. 44:865–868.

----, R. L. Swasdee, P. J. Wan, and E. W. Lusas. 1981. Cottonseed proteins in frankfurters: effect on pH, cured color, and sensory and physical properties. J. Food Sci. 46:845–849.

Thomas, M. R., J. Ashby, S. M. Sneed, and L. M. O'Rear. 1979. Minimum nitrogen requirement from glandless cottonseed protein for nitrogen balance in college women. J. Nutr. 109:397–409.

Thomson, J. R. 1979. An introduction to seed technology. Wiley, N.Y.

Tupper, G. R., L. E. Clark, and O. R. Kunze. 1970. The measurement of certain physical characteristics related to rapid germination and seedling vigor in cotton seed. Proc. Assoc. Off. Seed Anal. 60:138–148.

----, O. R. Kunze, and L. H. Wilkes. 1971. Physical characteristics of cottonseed related to seedling vigor and design parameters for seed selection. Trans. Am. Soc. Agric. Eng. 14:890–893.

Turner, J. H., and D. Ferguson. 1972. Field performance of cotton grown from filled and partially filled seeds. Crop Sci. 12:868–871.

----, H. H. Ramey, and S. Worley. 1976a. Influence of environment on seed quality of four cotton cultivars. Crop Sci. 16:407–409.

----, ----, and ----. 1976b. Relationship of yield, seed quality, and fiber properties in upland cotton. Crop Sci. 16:578–580.

Van Waalwijkn van Doorn, J. J. L. 1982. U.S. sun oil exports expected to raise. J. Am. Oil Chem. Soc. 59:588A–597A.

Van Wyck, P. 1948. Cotton linters. p. 894–905. *In* A. E. Bailey (ed.) Cottonseed and cottonseed products: their chemistry and chemical technology. Interscience, N.Y.

Vix, H. L. E., H. P. Dupuy, and M. G. Lambou. 1969. Critical evaluation of the use of acetone in solvent extraction processes. p. 62–69. *In* Conference on protein-rich food products from oil seeds. New Orleans, La. USDA ARS Rep. 72-71.

----, P. H. Eaves, H. K. Gardner, and M. G. Lambou. 1971. Degossypolized cottonseed flour —the liquid cyclone process. J. Am. Oil Chem. Soc. 48:611–615.

----, J. J. Spadaro, C. H. Murphy, R. M. Persell, E. F. Pollard, and E. A. Gastrock. 1949. Pilot-plant fractionation of cottonseed. II. Differential settling. J. Am. Oil Chem. Soc. 26:526–530.

----, ----, R. D. Westbrook, A. J. Crovetto, E. F. Pollard, and E. A. Gastrock. 1947. Pre-pilot plant mixed-solvent flotation process for separating pigment glands from cottonseed meats. J. Am. Oil Chem. Soc. 24:228–236.

Walhood, V. T. 1956. A method of reducing the hard seed problem in cotton. Agron. J. 141–142.

Walhood, V. T., and R. E. Johnson. 1976. Inception of solar radiation by a constant population of cotton plants. p. 70–71. *In* J. M. Brown (ed.) Proc. Beltwide Cott. Prod. Res. Conf., Las Vegas, Nev. Natl. Cotton Council, Memphis, Tenn.

Wanjura, D. F., E. B. Hudspeth, Jr., and J. D. Bilbro, Jr. 1967. Temperature-emergence relations of cottonseed under natural diurnal fluctuation. Agron. J. 59:217–219.

Wanjura, D. F., E. G. Hudspeth, Jr., and J. D. Bilbro, Jr. 1969. Emergence time, seed quality, and planting depth effects on yield and survival of cotton (*Gossypium hirsutum* L.). Agron. J. 61:63–65.

----, and E. B. Minton. 1981. Delayed emergence and temperature influences on cotton seedling vigor. Agron. J. 73:594–597.

Watson, H., and J. D. Helmer. 1964. Cottonseed quality as affected by the ginning process—a progress report. ARS USDA Publ. 42-107.

Wiles, A. B., and J. T. Presley. 1960. Seed deterioration as a factor in nub-root production in cotton. Plant Dis. Rep. 44:472–473.

Wilkes, L. H. 1978. Seed cotton storage: effects on seed quality. p. 215–217. *In* J. M. Brown (ed.) Proc. Beltwide Cott. Prod. Res. Conf., Dallas, Texas. Natl. Cotton Council, Memphis, Tenn.

Wilkes, L. H., and T. E. Corley. 1968. Planting and cultivation. p. 117–149. *In* F. C. Elliot, M. Hoover, and W. K. Porter (ed.) Advances in production and utilization of quality cotton: principles and practices. Iowa State Univ. Press, Ames.

Wise, R. P. 1980. The case of Incaparina in Guatemala. Food Nutr. Bull. 2:3–8.

Yatsu, L. Y. 1965. The ultrastructure of cotyledonary tissue from *Gossypium hirsutum* L. seeds. J. Cell Biol. 25:193–199.

Zarins, Z. M., and J. P. Cherry. 1981. Storage proteins of glandless cottonseed flour. J. Food Sci. 46:1855–1859, 1867.

Ziprin, Y. A., K. S. Rhee, Z. L. Carpenter, R. L. Hostetler, R. N. Terrell, and K. C. Rhee. 1981. Glandless cottonseed, peanut and soy protein ingredients in ground beef patties: effect on rancidity and other quality factors. J. Food Sci. 46:58–61.

14 Marketing and Economics

Arlie L. Bowling
National Cotton Council
Memphis, Tennessee

Historically cotton has played a greater role in the economy and politics of the world than any other commodity. Its relative importance today is reduced by increased trade in commodities such as grain, and competition from synthetic fibers. Still, cotton is an agricultural and industrial commodity of worldwide importance. Growing world consumption of cotton insures that it will continue to be a significant commodity in future world trade.

Marketing of cotton is complex compared to that of some other commodities. Cotton is supplied from agriculture, where it must compete with other crops for land. The demand from consumers comes through the textile mills. Cotton competes in the fiber markets primarily with synthetic fibers derived from the petrochemical industry.

Figure 14-1 illustrates the materials flow for cotton and helps to identify some of the channels of distribution. Though cotton production involves only a few tens of thousands of growers, the handling of cotton after it leaves the farm—marketing, transportating, spinning, weaving, manufacturing of products, and distributing cotton goods—requires one of the largest segments of labor in our society.

14-1 WORLD TRADE IN COTTON

The U.S. cotton industry functions as a segment of the world market. Therefore, what happens to markets and production in the rest of the world can create a significant change in U.S. cotton markets. A review of world trends in consumption, production, and trade provides the framework for

Published in *Cotton,* Agronomy Monograph no. 24, © ASA-CSSA-SSSA, 677 South Segoe Road, Madison, WI 53711.

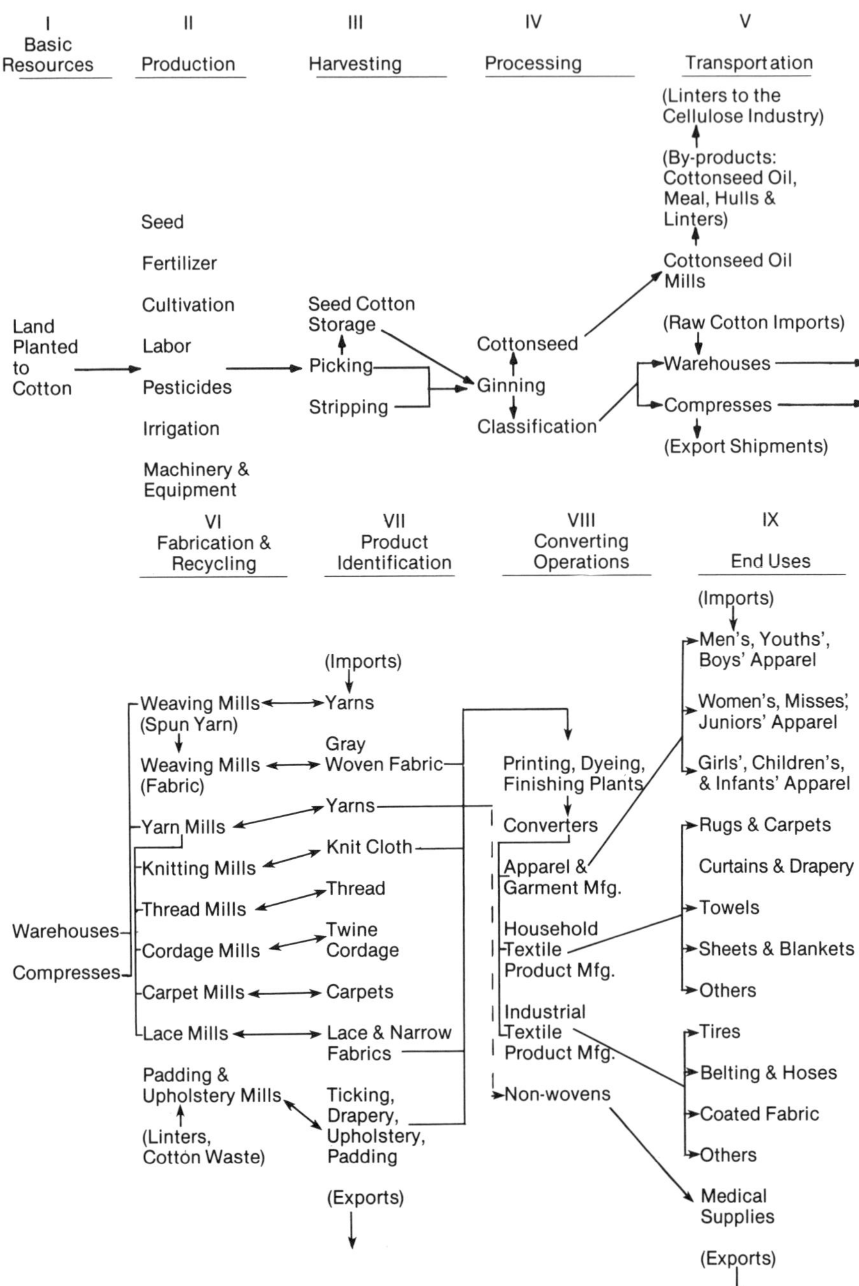

Fig. 14-1. Materials flow for renewable fiber resources: cotton (National Academy of Sciences, 1976, p. 132).

the performance of U.S. cotton (International Cotton Advisory Committee, 1950–1982).

World cotton consumption has increased at over a million bale average annual rate since 1947 (Fig. 14-2). The growth rate in cotton consumption during the 1950s was at the rapid rate of 1.1 million bales/year. This very large demand for textiles helped to stimulate the growth of synthetic fiber production. During the 1960s, the increase in world consumption of cotton continued at a 1.0 million bale annual rate. Synthetic fibers were able to fill a significant portion of the rapidly increasing world demand for fibers during the 1960s and the 1970s (Fig. 14-3). During the 1970s, world demand for cotton continued to increase at an average annual rate just over 1 million bales/year.

Each decade since World War II, the volume of cotton in world trade has increased. However, the proportion of cotton in world trade has remained at approximately one-third of world cotton consumption. The increase in world export trade increases the potential markets for U.S. cotton (Table 14-1).

By the 1950s cotton had recovered from the serious trade constraints of World War II, and the average annual world exports during the 1950s was 13.63 million bales. During the 1960s, a decade of strong economic growth, world exports averaged 17.05 million bales, a 17% increase. During the decade of the 1970s, world exports averaged 19.33 million bales, a 13% increase over the prior decade.

Changes in yield could significantly influence future world trade in cotton. As the developing countries manufacture more of their raw cotton into textiles, less cotton is available for export unless either their yield or production area increases. World cotton yield has increased rather consistently from 234 kg/ha in 1950 to an average of 424 kg/ha for the period 1976 through 1980. The U.S. producers will need a yield increase to properly supply the continually increasing world market for cotton.

The increase in world trade during the 1960s paralleled the increase in world mill consumption. Both world trade and world mill consumption increased in the 1960s by 24.0% over the 1950s. However, during the 1970s, world average mill consumption increased 19.5% above the 1960s average, but average world cotton exports increased only 13.0%. The slower rate of growth in world trade may be attributed to two factors:

1. The level of world stocks of cotton was reduced in proportion to consumption; and

2. Many of the developing cotton-producing countries expanded textile manufacturing operations such that they export textile products rather than raw cotton. This change improves their value added and trade balance situation, but it reduces the amount of raw cotton in world trade.

World cotton exports have remained near one-third of world cotton consumption during the past 30 years (Table 14-1).

Table 14-1. World raw cotton consumption and trade averaged by decade.

Source	1950s	1960s	1970s
		218 kg bales $\times$ 10^6	
World exports	13.63	17.05	19.33
World mill consumption	40.65	50.51	60.38
World exports/world mill consumption	33%	34%	32%

14-2 WORLD COTTON CONSUMPTION

To this point we have been examining cotton's market position in terms of absolute volume which showed that demand for the natural fiber, cotton, continues to increase. However, the world market for total fibers has grown much more rapidly, and cotton has lost market share since the 1950s. Cotton's share of the world market was 72.7% in 1950 and 48.0% in 1980 (Table 14-2).

Total world fiber consumption increased 51% during the 1960s when compared to the 1950s average (Fig. 14-2). The 1970s decade percentage growth rate is only 35%, largely because of the larger base on which the growth is computed.

Cotton has provided 25% of the fibers for the world market growth while synthetic and other fibers have provided the remaining 75% of the rapid growth in world fiber consumption.

The world outlook for cotton is for continued volume growth, but synthetic fibers seem likely to continue to capture a large proportion of fiber market increases. However, cotton's market share of world fiber consumption has stabilized in recent years at about 48% (Table 14-3).

14-3 WORLD COTTON PRODUCTION

How has world cotton production grown to meet this increasing demand? The agricultural production area has remained nearly constant. It has fluctuated to meet market requirements but has stayed close to 32.4 million ha during each decade since the 1930s except for a sizable reduction during World War II (Fig. 14-3). However, yield has improved such that this 32.4 million ha now produces twice as much cotton as it did at the end of World War II.

During the 1970s, world production area fluctuated as much as 10% in response to market changes. For example, the production area was 33.6 million ha in 1973 and 33.4 in 1974. Then it dropped to 30.1 million ha in 1975 because the recession reduced the demand for fibers.

The land devoted to cotton has remained constant for several decades even though world population has increased rapidly, and the demand for food crops has increased greatly.

Table 14–2. Cotton's world market share.†

Decade	Beginning of decade	End of decade
	%	
1950s	72.7	69.1
1960s	69.1	56.0
1970s	56.0	48.0

† Source: International Cotton Advisory Committee, 1950–1980.

Table 14–3. World fiber consumption.

Calendar year	Total	Cotton	Wool	Synthetic	Cotton
	bales (218 kg) × 1000				%
1970	99 944	55 359	7 216	37 369	55.0
1971	105 580	56 756	7 193	41 631	53.0
1972	112 203	59 319	7 248	45 636	52.0
1973	119 529	61 000	6 628	51 901	51.0
1974	117 107	60 701	5 796	50 610	51.0
1975	112 344	58 744	6 237	47 363	52.0
1976	112 711	61 629	6 834	54 248	50.0
1977	123 951	60 205	6 655	57 091	48.0
1978	129 496	61 560	6 619	61 317	47.0
1979	135 392	64 274	6 885	64 233	47.0
1980	136 221	66 360	6 800	63 061	48.0
1981	137 850	67 000	6 850	64 000	48.0

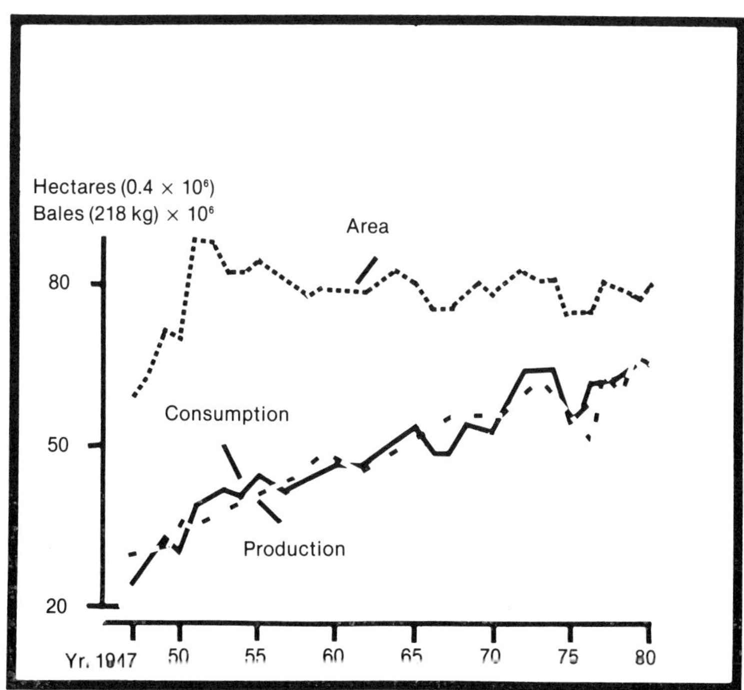

Fig. 14–2. Area planted, consumption, and production in the world.

The increased yield per area may be attributed to improved technology and production management as well as a shift of much of the cotton area to land more suitable for cotton production. However, much of the increase in yield was closely related to increased energy consumption in the forms of fertilizer, fuel, and pesticides. Availability of energy and its price will influence future production in the USA and other countries.

The approximate 32.4 million ha of land historically available for the production of cotton has continued to be available during recent years. Future world cotton production increases will depend primarily upon three things:

1. Improved technology and management to obtain higher yields from the land available for planting cotton;

2. Cotton's competitive strength in the fiber market—the demand must be strong enough for the price to be at a level that cotton can compete effectively with other crops for land; and

3. Governmental decisions on resource allocation, particularly energy, capital, land, and water.

World cotton has been able to compete for land on a stable basis for the last three decades through critical social changes as well as significant fiber market changes. This fact indicates that the world market demand is reasonably favorable and that cotton has good expectations of being able to compete in the years ahead.

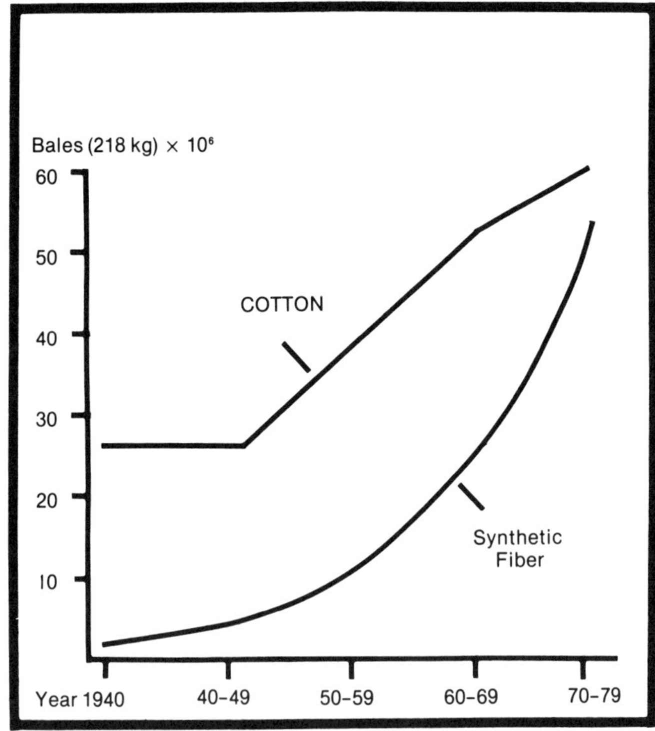

Fig. 14–3. World fiber consumption, annual average.

14-4 THE U.S. RAW COTTON EXPORT MARKET

Cotton exports fluctuate greatly from year to year as shown in Fig. 14-4, but when averaged by decades such that the short run impact of economic changes and weather are reduced, U.S. cotton exports have been reasonably constant and have increased in recent years. Average exports by decades were as follow: 1950s, 4.7 million bales; 1960s, 4.3 million bales; and 1970s, 5.7 million bales.

However, there is some indication that the foreign production response to the market has changed during the 1970s. For the decade of the 1970s, the annual average rate of increase in foreign production was 1.4%, which is somewhat below the annual average rate of increase in foreign consumption of 1.9%. This moderation in foreign production increase to 1.4% is significantly different from the annual average foreign production increases of 9.0% of the 1950s and 4.0% in the 1960s.

Though the annual average rate of increase in foreign production for the 1970s of 1.4% is only slightly below the foreign consumption of 1.9%, the extent to which foreign consumption exceeds production offers potential export markets for U.S. cotton. If this trend continues, the future export markets for U.S. cotton appear good.

Cotton production requires more inputs than most competing crops thus requiring more capital and management. As labor in some developing countries becomes less available for hand harvesting and crop care, the decision whether or not to invest capital in machinery must be made. Capital

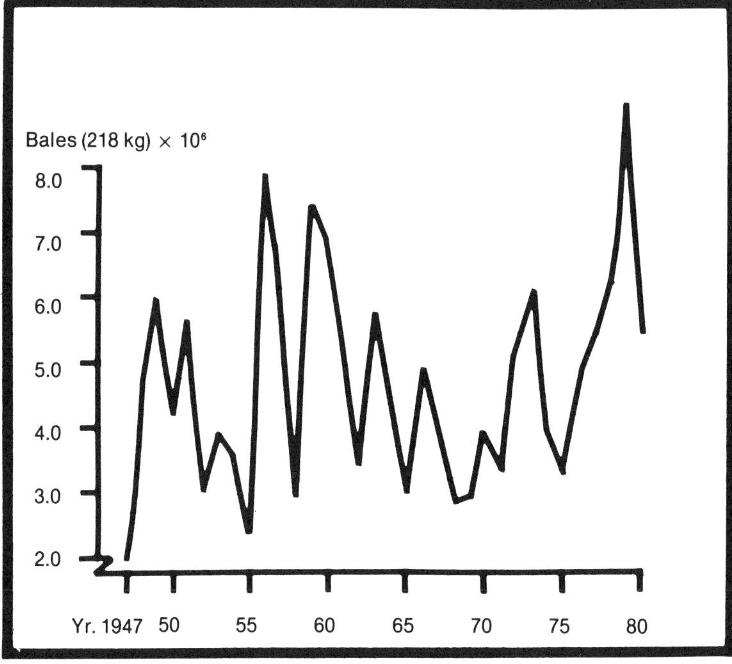

Fig. 14-4. U.S. Cotton exports.

is more readily available in the USA than in many other cotton-producing countries. Further, if fields are extremely small, machine operations are difficult. This fact has provided some opportunity for U.S. export expansion.

But foreign trade is complex and depends on many factors. Many countries find that exports of cotton are essential to their balance of trade. The USSR is an example of this principle, and encourages cotton production on that land best suited for cotton. Mexico, like many other countries, allocates a portion of its resources according to the nation's perceived economic needs. Mexico diverted an increased amount of water to food production in 1975, but as cotton trade opportunities improved in 1976, more irrigation was allocated to cotton.

Exports also fluctuate widely from year to year, and individual years could be much higher or lower than the yearly averages shown above.

14-5 THE U.S. DOMESTIC MARKET

The world cotton market provides the framework for the performance of the U.S. domestic market.

14-5.1 Domestic Mill Consumption

While total world consumption of cotton has continued to grow at a significant rate, in the USA domestic mill consumption of cotton has been declining (Fig. 14-5). Mill consumption of cotton had been at 8 to 9 million bales annually from World War II through 1971. Cotton has lost about 2 million bales of its U.S. domestic market to synthetic fibers from the post World War II levels. But, the nature of the fiber market changed during the 1970s, and these market changes will affect domestic mill consumption of cotton in the future.

14-5.2 The Changing Nature of Fiber Markets

During the decade of the 1960s, cotton lost a share of the U.S. market to synthetic fibers at a rapid, continuous rate. Cotton's market share was 67.0% in 1960 and 39.5% in 1970. However, cotton's share of the fiber market has stabilized in recent years at near 25.0% as shown in Table 14-4.

Significantly the changes in fiber markets account for cotton's improved market performance. Synthetic fibers went through the marketing introductory stage prior to the 1960s. During the decade of the 1960s, people were fascinated with the newly developed synthetic products, which offered change, easy-care characteristics, and considerable appeal to consumers. Synthetic fibers were in the growth stage. They gained market share each year almost without regard to:

Table 14–4. Ten-year U.S. textile mill fiber consumption (Economic Research Service, 1982).

Calendar year	Total bales × 10⁶†	Cotton		Synthetic‡		Silk & wool	
		bales	%	bales	%	bales	%
1977	25.4	6.6	26.0	18.5	72.9	0.3	2.7
1978	25.9	6.4	24.5	19.2	74.4	0.3	3.1
1979	26.4	6.4	24.0	19.7	74.9	0.3	1.1
1980	24.9	6.3	25.3	18.3	73.5	0.3	1.2

† Due to rounding, individual numbers may not add to total shown.
‡ Includes acetate staple and synthetic fiber waste.

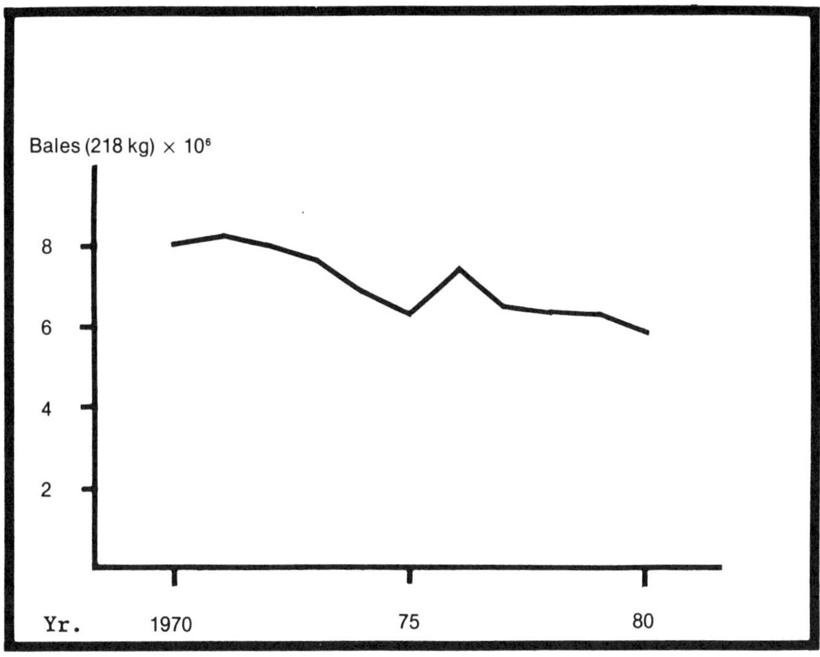

Fig. 14–5. U.S. Mill consumption.

1. Changes in prices to consumers;
2. Changes in relative price with natural fibers; or
3. Changes in cost of production.

The consumer's preference for easy care and strength was a dominant market characteristic that diminished the importance of cost of production and price. The synthetic fibers industry profited well and allocated much of its capital investment to plant capacity expansion during the growth stage. During the growth stage of synthetic fibers, expansion of capacity enhanced profits.

Since 1973, the markets for fibers have changed. After the 1973 commodity boom, the markets for all fibers leveled off partially because of the serious recession of 1974 to 1975, but due also because of changes in consumer desires and expectations.

For example, some products made of synthetic fibers, such as double knits, lost their aura of newness and fascination. This change resulted in lower market demand. The lower level of demand for such products, when combined with the expanded fiber plant capacity, resulted in very low plant operating rates, which were at an unprofitable level. This change had severe impact on fiber producers.

Synthetic fibers have moved from a period of rapid growth into the product maturity stage, and they now compete in that stage with cotton and wool. In this mature stage of market competition, success normally requires very effective cost control to permit a competitive price. Because of the lower gross profit margins customarily experienced in the market maturity stage, a low cost of production is essential to having a competitive price, and it becomes more important in the battle for market share. Consequently, synthetic fiber producers have diverted much of their capital investment in recent years toward cost reduction rather than plant expansion. In contrast to this mature marketing stage, during the synthetic fibers growth stage of the 1960s, neither high prices for synthetic fibers nor low prices for cotton substantially influenced cotton's market share losses.

Furthermore, since synthetic fibers have reached market maturity, cotton has held market share better during the past 4 years at prices considerably above those of polyester staple. Price differences and production cost differences between fibers will have more impact on today's markets and the markets in the years ahead. This change can occur even though the price level at which cotton can compete with polyester is now relatively higher because of the market shift in consumer tastes and preferences toward natural fibers.

Though the mature market experience is new for some producers of synthetic fibers, it is not new for producers of cotton. Considerable cotton industry effort has been made in recent years to provide the technological change for innovation. This change is essential to provide both strong promotion and product differentiation. Sanfor-set and Natural Blend are examples of such technological success. However, in a mature market, the need for technological change to reduce cost, improve the product, and provide product differentiation is a continuous requirement for a healthy industry.

14-5.3   Domestic Cotton Market Competition

A major factor in favor of cotton and other natural fibers is that they possess very desirable characteristics of feel and comfort which synthetic fibers cannot duplicate. This automatically gives cotton some advantage in product differentiation and, therefore, some demand advantage for end-use markets such as apparel.

Price combined with technological improvements give synthetic fibers advantages in other markets such as in industrial goods. However, recent market contacts indicate that even in the industrial markets, where buyers

are concerned with minute price changes in fibers, cotton becomes somewhat more desirable because of its special characteristics when the price of
polyester increases in relation to the price of cotton.

The changing mature market situation gives strong recognition to
cotton's quality characteristics. The demand for U.S. cotton has the
opportunity to increase during the next few years if cotton can provide
technological fiber improvement along with adequate promotion to maintain or improve its level of acceptance to mills and consumers.

Many other factors influence domestic mill consumption. The synthetic fibers industry is confronted with escalating costs and price increases.
Because of the energy intensity of synthetic fibers, cotton has an energy advantage (National Academy of Sciences, 1976, p. 26). It also derives from
14 to 20% of its crop revenue from cottonseed for feed and food products.
These factors along with the very high capital and plant costs for future
expansion of synthetic fiber plants may cause a relatively greater price increase for synthetic fibers than for cotton during the next decade.

14-5.4   Textile Imports

Domestic consumption of cotton (domestic mill consumption + imports of cotton textile goods − exports of textile goods) indicates that consumers desire to purchase cotton in quantities in excess of that available
from domestic mills (Fig. 14-6).

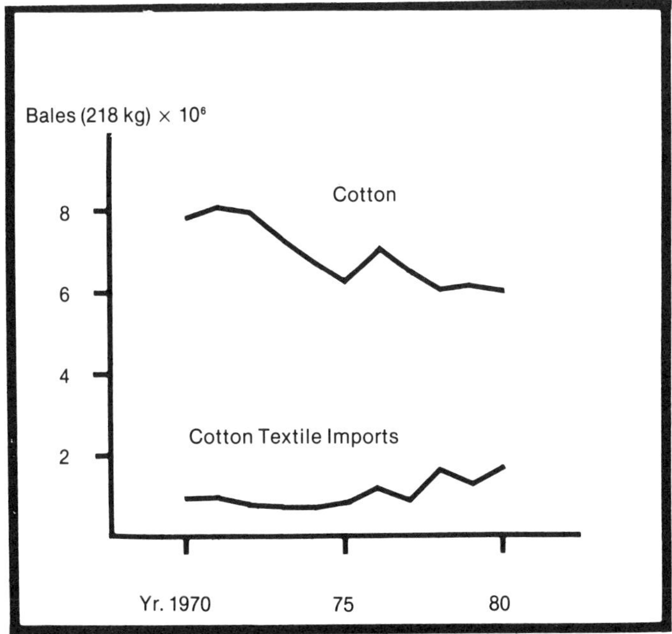

Fig. 14-6. Domestic textile consumption.

Since 1975, cotton textile imports have increased rapidly (Fig. 14-7). Current expectations are that imports will continue to increase at a rate faster than our growth in Gross National Product (GNP). This poses a serious problem to the market opportunities of domestic textile mills. The U.S. mill consumption will be adversely affected until trade negotiations provide for imports at a level more consistent with market growth.

14-5.5   Government Regulations

Government regulations pose serious potential impact on future mill consumption of cotton as well as on other fibers. Areas of primary concern for cotton are cotton dust and flammability standards along with those regulations which affect the availability and cost of cotton crop inputs such as pesticides, energy, water, and others. The impact of anticipated regulations cannot be projected, so it must be assumed that the industry will adequately cope with the problem such that cotton's competitive position with other fibers for markets and other crops for production will not be adversely affected. However, the current level of regulation indicates that this may be a serious oversimplification.

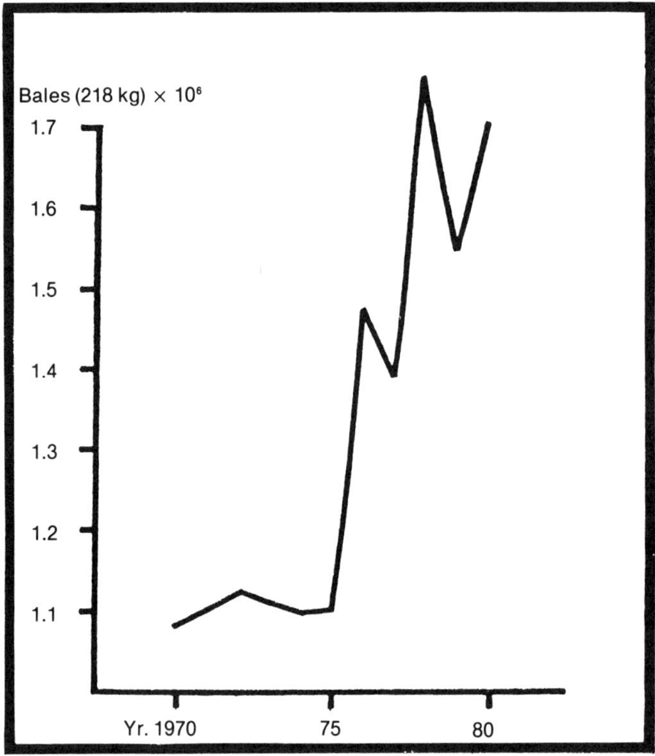

Fig. 14-7. Cotton textile imports.

The developing nature of the fiber market indicates that the demand for cotton remains strong. Cotton is perceived by consumers and mills as a high-quality fiber with desirable characteristics. This perception is supported by the stable share of cotton in the world fiber market since 1976. Cotton will likely become more price-competitive in the next few years as energy costs and plant expansion costs for synthetic fibers continue to increase.

14-6 DOMESTIC COTTON PRODUCTION

The Cotton Belt spans the southern half of the USA stretching from the Carolinas to California. Cotton is grown in 17 states, and is a major crop in 14 (Fig. 14-8). Its growing season of about 150 days is the longest of any annually planted crop in the country. Because of much variation in climate and soil, production practices differ from region to region. In the western half of the Belt, for example, much of the crop is irrigated. The demand for cotton, both domestic and export, determines the level of cotton production. The strength of the demand provides the market signals and production incentive price. The cotton needed by the mills will normally be produced by the cotton farmers, barring unusual weather or other crop disaster.

If world cotton demand continues to increase at a 1 or 2 million bale annual rate, the profit incentive for production would seem to continue to be available for the following reasons:

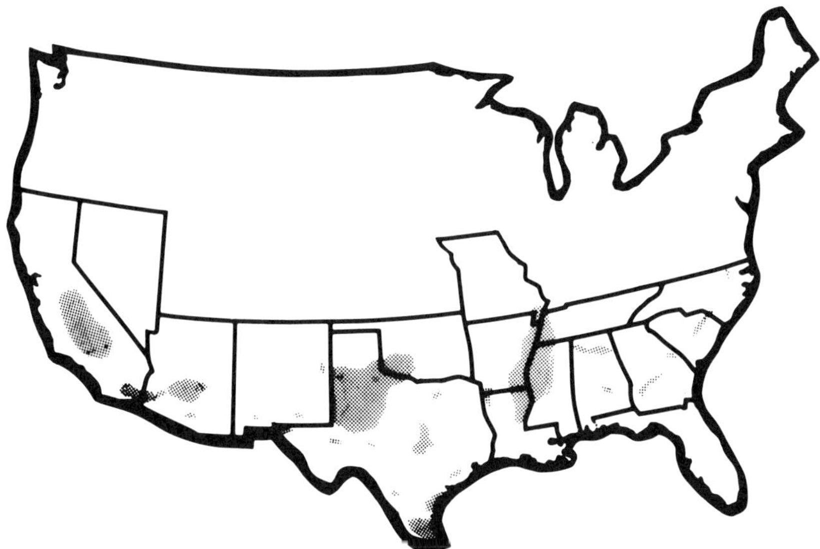

Fig. 14-8. Cotton Belt stretches from coast to coast, with production density varying from region to region.

1. Land used for cotton production has responded to the supply needs of the cotton industry during the past few years (Fig. 14-9). When the market has indicated the need for more cotton, it has been produced. For example, 5.67 million ha were planted in 1972, 5.55 million ha were planted in 1974, and 5.79 million ha were planted in 1981. Likewise, when excess stocks were on hand in 1967 and 1975, planting dropped to 3.81 million and 3.85 million ha, respectively.

2. Sufficient land will be available for the production of both fiber and food crops in the years ahead. A study by the USDA Economic Research Service (1975) shows that additional land can be brought into crop production when it is needed in the USA, and that it will be adequate for many years. Land availability is essential to avoid significant increases in production costs for cotton, and cotton would remain competitive with other fibers.

3. The market demand for cotton seems adequate to keep cotton's price competitive with other crops for land and still be competitive with synthetic fibers.

Technological changes affecting cost of production, and processing costs for both cotton and competing fibers can influence considerably the profit incentive for producing cotton. Since the synthetic fiber industry has in recent years changed its investment course from plant expansion to product improvement and cost reduction, the cotton industry must likewise improve its product and its production efficiency.

The cost of producing cotton is affected most by yield. Technological breakthroughs to improve yield and reduce costs will be essential to keep cotton competitive with synthetic fibers.

Automation, the use of more fertilizers and chemicals, improved technology, improved cultivars, and improved farm management resulted in a high rate of yield increase from 1952 to 1967 (Fig. 14-10). During this period, the yield for cotton increased more rapidly than did the yield for other crops with which cotton competes for land. However, the yield trend line for cotton shows no yield improvement for 1967 to 1981. Additionally, improvement in U.S. cotton yield will be necessary in the years ahead if U.S. cotton adequately supplies the potential world market available to it.

Weather has always been a significant factor in yield. Developments such as short season cotton and growth regulators may reduce weather risks and aid in cost reduction in some cotton-producing areas.

The scarcity of resources such as water, energy, and capital influences cotton production and agriculture in general.

The scarcity of resources is closely related to governmental decisions that will influence future levels of world cotton production. Energy allocation is currently being monitored or regulated by nearly all countries. In the USA, regulations on pesticides, energy, water, cotton dust, and flammability result in higher costs for production of cotton goods. When cost of production increases, competition for land from other crops becomes more severe. Government regulations are also used to enhance a country's economy by increasing value added by encouraging processing and manufactur-

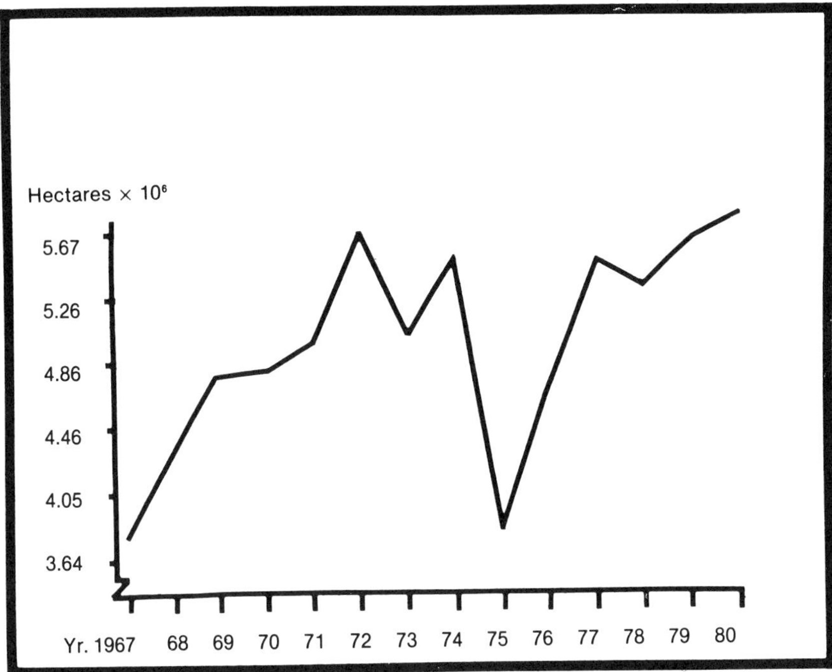

Fig. 14-9. U.S. area planted to cotton.

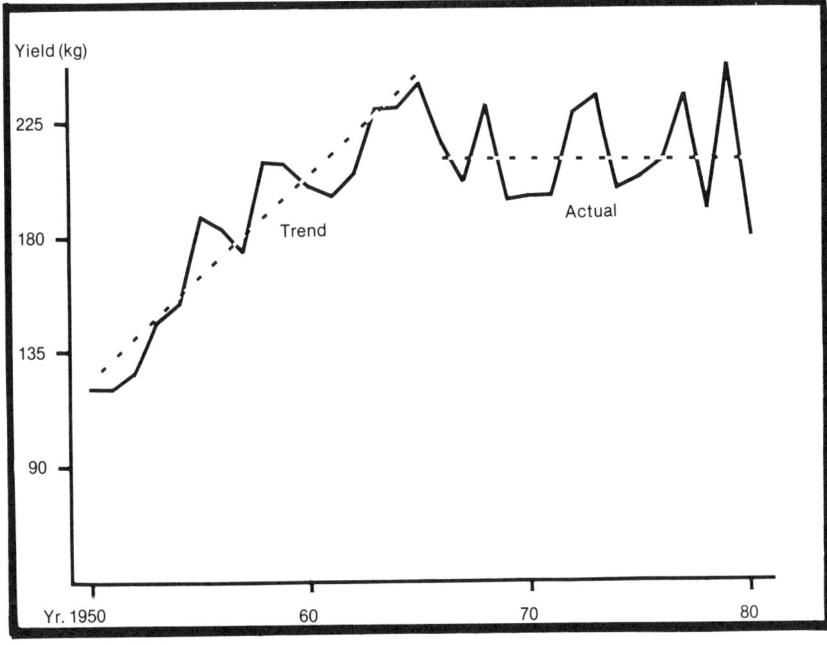

Fig. 14-10. Cotton yield per harvested hectare.

ing such as spinning and weaving. To several countries this effort is essential to their balance of trade.

In our present mature fiber market, where cost reduction is a dominant element of a strong competitive position, research to lower production cost is essential. Both the market demand and recent history of land use indicate that cotton can be produced to supply the market's needs.

14-7 GOVERNMENT POLICIES ON COTTON

During the past two decades, the philosophy regarding farm programs has changed considerably. The farm programs prior to 1970 set artificially high price supports that made cotton non-competitive with foreign growth and synthetic fibers. As a result, surpluses overhung the market, and the cotton growers had to depend on the U.S. government to subsidize exports and reduce cotton production at heavy cost.

A change began with the U.S. Agricultural Act of 1970, and another step was made in the Agriculture and Consumer Protection Act of 1973, which substituted the target price concept for guaranteed payments. Another came in 1977 when a unique mechanism for moderating prices worked out by the cotton industry was adopted.

Simply put, this mechanism for moderating prices provides an extended 8-month loan period for producers to withhold cotton from the market when prices are low. If prices rise above a trigger point determined by legislative formula, the option closes and a special raw cotton import quota is then opened for foreign cotton. The concept has worked and seems to be a far more desirable remedy for price and supply problems than a reserve or an embargo.

The cotton leadership feels that the government should not price cotton out of the market. This position explains why the cotton title sets the loan rate so precisely at a percentage of the market price.

Under legislation adopted by the U.S. Congress in the past decade, government cotton programs in fiscal 1980 cost only one-tenth what they did a decade ago. If inflation is taken into account, the cost is less than one-twentieth. Cotton program expenditures in 1970 accounted for 22% of total government spending to stabilize farm income. In 1981 cotton program expenditures accounted for only 3%.

Support by Congress and USDA for agriculture research, including cotton research, is essential to the future health of the cotton industry. This research has paid big dividends in greater productivity at costs much lower than they would have been without it. It is essential to the future health of the cotton industry.

REFERENCES

International Cotton Advisory Committee. 1950–1982. Cotton-world statistics. Quarterly Bulletins of the International Cotton Advisory Committee (Special Base Book Issues), International Cotton Advisory Committee, Washington, D.C.

National Academy of Sciences. 1976. Renewable resources for industrial materials. Printing and Publishing Office, National Academy of Sciences, Washington, D.C.

Economic Research Service. 1975. Farmland, will there be enough? USDA Rep. ERS-584.

Economic Research Service. 1982. Statistics on cotton and related data 1920–81. USDA-ERS Rep. Stat. Bull. 535.

Subject Index